The Aqueous Chemistry of Oxides

The Aqueous Chemistry of Oxides

BRUCE C. BUNKER

AND

WILLIAM H. CASEY

OXFORD
UNIVERSITY PRESS

OXFORD
UNIVERSITY PRESS

Oxford University Press is a department of the University of Oxford. It furthers the University's objective of excellence in research, scholarship, and education by publishing worldwide. Oxford is a registered trade mark of Oxford University Press in the UK and certain other countries.

Published in the United States of America by Oxford University Press
198 Madison Avenue, New York, NY 10016, United States of America.

First Edition published in 2016

Cataloging-in-Publication data is on file at the Library of Congress

9780199384259

9 8 7 6 5 4 3 2

Printed by Sheridan Books, USA

Bruce Bunker dedicates this book to his wife, Cathy, whose numerous suggestions and infinite patience were critical to the completion of the project. He also thanks those scientists who mentored him throughout his career, including Professor Frank DeHaan (Research Advisor, Occidental College), Professor Russell Drago (Advisor, University of Illinois), Professor Galen Stucky (now at University of California at Santa Barbara), Dr. Robert Eagan (Sandia National Laboratories), Dr. Gary McVay (Pacific Northwest National Laboratory), and Professor Delbert Day (University of Missouri-Rolla).

From Bill Casey: Do mo bhean chéile álainn, Sorcha, a bhfuil mé pósta léi le 33 bliain agus do mo mhac iontach, Séamus Óg–tiomnaím an leabhar seo daoibh, an bheirt a chuireann gliondar ar mo chroí.

CONTENTS

COVER ART FIGURE CAPTIONS

Oxide–water reactions define many of Earth's environments and provide us with critical materials and technologies. Front cover: On the coast of the island of Hawaii, breaking waves encounter hot lava, creating a plume of wind-swept steam. (Image provided by CJ Kale at Lavalight.com, with permission). The resulting oxide–water reactions contribute to the formation of Hawaii's famous black sand beaches. Similar reactions occur on a massive scale along the entire length of spreading centers such as the Mid-Atlantic Ridge, exerting a strong influence on the elemental composition of our oceans. Back cover: The complex chemical behaviors exhibited by a claylike layered cobalt hydroxide illustrate several fundamental water–oxide reactions. Some of these reactions—including ion exchange, and electrochemical and photochemical processes—are critical to many energy and environmental technologies. Reactions discussed in this book include electrochemical energy storage in batteries, light harvesting for solar energy, optical communication systems, catalytic conversions, water purification, and environmental remediation.

The Aqueous Chemistry of Oxides

PART ONE

An Introduction to Oxides

1

The Importance of Oxides and Oxide–Water Reactions

1.1 GEOCHEMICAL IMPORTANCE

Earth is composed primarily of oxides [1, 2] (Fig. 1.1, Plate 1, Table 1.1). The interior is a iron-nickel core that represents 35% of the planetary mass. The core is encapsulated within a plastic zone called the *mantle.* Floating on top of the mantle, like the slag on top of a blast furnace, is a thin skin called the *crust.* Both the mantle and the crust (65% of the planetary mass) are primarily oxides. Although the crust is thin (8–13 km thick under the ocean basins and up to 70 km thick under the continents) [2], it still represents a massive amount of material, corresponding to 7 billion km^3, with a weight of $2.5 \cdot 10^{19}$ metric tons. Of this mass, 46% is tied up in the relatively light element oxygen, predominantly in oxides containing silicon, aluminum, and iron. Almost every material that we humans encounter or use is derived from the oxide building blocks that comprise Earth's crust.

Water is by far the most abundant liquid on the planet [3]. In fact, it can be difficult to comprehend just how much water is contained in Earth. Most of our liquid water is in the oceans, which cover a total surface area of 361 million km^2 (70% of Earth's surface) at an average depth of almost 4 km (Plate 1). The volume of water in the oceans is 1.4 billion km^3, which is one-fifth the volume of the entire crust. The total mass of the hydrosphere is $1.7 \cdot 10^{18}$ MT. However, the total

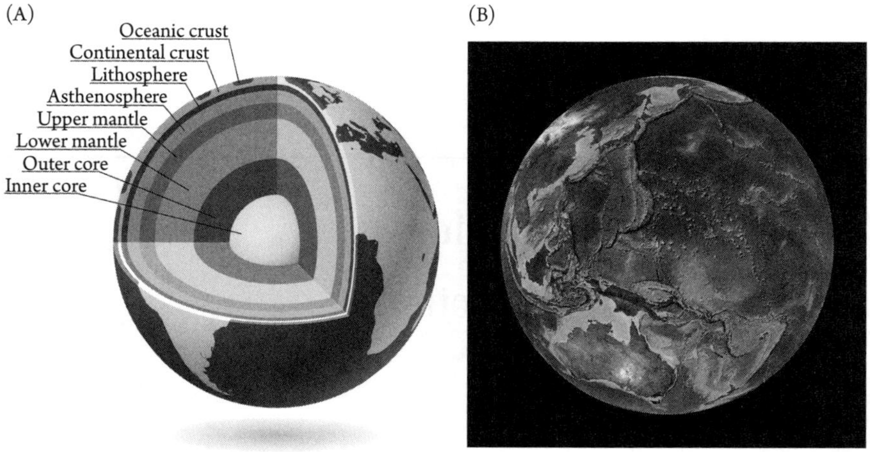

Figure 1.1 The geochemical importance of oxide–water reactions. (A) A depiction of Earth's interior, showing the different planetary layers. The center of Earth is a solid iron-nickel inner core ($r = 1{,}300$ km) that is surrounded by an outer core comprised of liquid iron-nickel (thickness, 950 km). All successive layers (total thickness, 4,150 km), including mantle layers, the asthenosphere, and the lithosphere, are composed primarily of oxides, representing 95% of Earth's total volume (SOURCE: Alhovic/Shutterstock.com). (B) A topographical map of the floor of the western Pacific Ocean. Light shading represents shallow water and dark indicates deep regions, such as the Marianas Trench, which has a maximum depth of 11.5 km (SOURCE: Martin Jacobsson/Science Source.com with permission). In the oceans alone, the contact area between the crust and water is more than 361 million km^2 (70% of Earth's surface), indicating the enormous scope over which oxide–water reactions modify our planet. (See Plate 1.)

water content of the planet is estimated to be 100 times greater than the liquid we see; most of the water on Earth ($2 \cdot 10^{20}$ MT) is tied up in hydrous oxide minerals in Earth's interior. We all know that life on Earth could not exist without water. However, to put things in perspective, the total mass of all living things on Earth is $3.6 \cdot 10^{11}$ MT [4], which is 5 million times less than the mass of liquid water on our planet.

Reactions between water and oxide minerals are the most prevalent reactions on the surface of Earth. Water drives the weathering processes that render mighty mountains into boulders, pebbles, sand, soil, and, finally, to dust that blows away in the wind. However, water is also the primary liquid involved in the formation of new oxide minerals via dissolution, precipitation, and deposition of dissolved minerals. For example, the early oceans of Earth were green instead of blue because they contained large concentrations of dissolved Fe(II) species. When oxygenic bacteria appeared and produced Earth's oxygen-rich atmosphere, these soluble Fe(II) species were oxidized into insoluble oxide and hydroxide minerals

Table 1-1. MAJOR ELEMENTS IN THE EARTH'S CRUST

Element	Weight Abundance (ppm)	Atomic Abundance (%)
O	472,000	60.4
Si	28,800	20.6
Al	79,600	6.25
Fe	43,200	1.86
Ca	38,500	1.90
Na	23,600	2.56
K	21,400	1.38
Mg	20,900	1.79
Ti	4,010	0.19
C	1,900	2.90
P	757	0.08
Mn	716	0.04
S	697	0.03
Ba	584	0.06
Cl	472	0.02

Adapted with permission from Wedepohl, K. H. The composition of the continental crust. *Geochim. Cosmochim. Acta* **1995**, *59*, 1217–1232.

containing Fe(III), which represents most of the massive iron ore deposits we mine today [5]. To help put things in perspective, over geologic time, the volume of material that water decomposes continuously and (with tectonic subduction) helps recycle is on the scale of entire continents. Yet, many of these reactions occur at such slow rates and on such a massive scale that humans scarcely notice their enormous impact.

1.2 THE TECHNOLOGICAL IMPORTANCE OF OXIDES

Throughout history, people have exploited reactions between water and oxides to survive and enhance their quality of life. The first *Homo sapiens* were using water to help crack, chip, grind, and polish rocks to make their tools and weapons as early as 2.6 million years ago [6] (Fig. 1.2, Plate 2). People in the Stone Age learned how to use water to consolidate colloidal particles such as soils and clay minerals into adobe homes and "stone-like" materials such as ceramic pots as early as 12,000 years ago and 7,000 years ago, respectively [7] (Fig. 1.3, Plate 3). As the ability to control fire and heat improved, humankind developed the technologies for melting oxides into glass by at least 3,500 BC [8]. As glass fabrication improved, beautiful works of art such as stained glass were possible, as well as the production of inexpensive windows for homes (Fig. 1.4, Plate 4) and parts

(A)

(B)

Figure 1.2 (A) A collection of arrowheads fashioned in North America out of obsidian and chert before the arrival of Europeans (SOURCE: I. Pilon/Shutterstock.com). Humans have used water to assist in the fabrication of stone tools such as these as early as 2.6 million years ago (see Chapter 16 for mechanisms). (B) A view of the prehistoric Stonehenge monument in Wiltshire, England, under falling rain (SOURCE: Pitcha Torranin/Science Source.com with permissions). The oldest slabs in Stonehenge have been exposed to the elements (water in the form of rain, snow, and ice) since they were erected in 3300 BC. (See Plate 2.)

for our machines. However, in all of these applications, whether in cooking or windows, the corrosion and weathering of glass by water was a concern. The characteristic iridescence of many ancient glass artifacts is a consequence of such weathering over many centuries (Plate 4).

Applications for oxide materials have expanded rapidly and have become more complex until they affect almost every aspect of our daily lives. Oxides are found in electrical energy systems ranging from high-voltage insulators to energy storage devices such as batteries, capacitors, ultracapacitors, and photoelectrochemical cells (Fig. 1.5, Plate 5). Oxides have a wide range of applications in electronic and microelectronic systems (Fig. 1.6, Plate 6). More recently, oxides have been developed for advanced optics, including fiber-optic

Figure 1.3 (A) A view of the Long House at Mesa Verde in Colorado. Such cliff dwellings are invariably made of oxide mineral aggregates such as consolidated clay minerals or volcanic tuff (SOURCE: kravka/Shutterstock.com). (B) Clay amphorae uncovered in the excavation of the city of Akrotiri on the Santorini Island in Greece (SOURCE: Will & Demi McIntyre/ Science Source.com with permission). The site was buried by a volcanic eruption around 2000 BC, and thus is quite well preserved. Both the formation and stability of adobe buildings and clay pots are influenced by the colloidal interactions described in Chapter 8. (See Plate 3.)

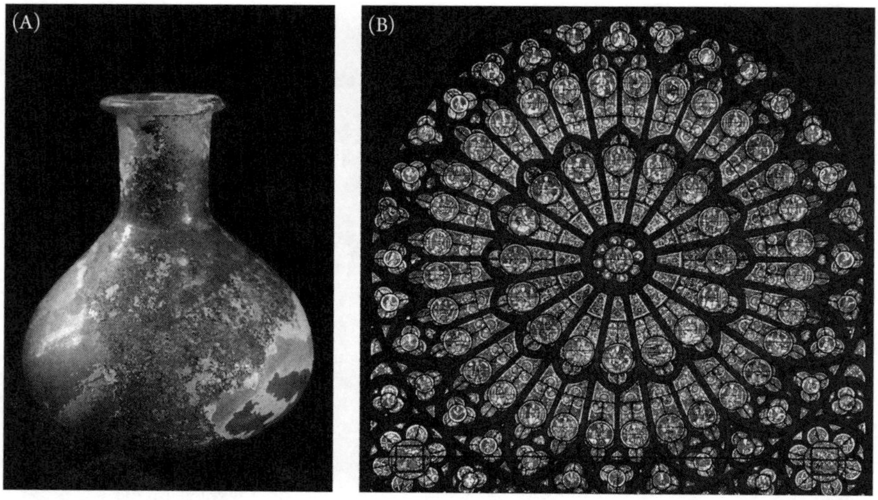

Figure 1.4 (A) A Roman perfume bottle found in Pompeii, Italy, which was destroyed by a volcanic eruption in 79 AD (SOURCE: Patrick Landmann/Science Source.com with permission). Exposure to water for 2,000 years created an iridescent layer of corrosion products on this glass artifact (see Chapter 15 for corrosion mechanisms). (B) Stained glass in the Rose Window of the Notre Dame Cathedral in Paris, France (SOURCE: Vlad G/Shutterstock.com), illustrating the importance of glass in human culture. (See Plate 4.)

transmission lines and optoelectronic switches (Fig. 1.7, Plate 7). Oxides are also used widely in environmental technologies, including water-treatment modules, and even in materials that we use to dispose of our most hazardous wastes. In all these applications, interactions between oxides and water play important roles in the fabrication, use, and decomposition of these critical glass and ceramic materials.

1.3 THE PHYSICAL PROPERTIES OF OXIDES

Although this book deals primarily with the chemistry of oxides, we provide a brief examination of the truly diverse physical properties of oxides to help illustrate why oxides are of technological interest, and how physical and chemical properties are coupled.

1.3.1 Electrical Properties

The electrical conductivities of oxides span an enormous range [9]. At one extreme are some of the best insulators known to humankind [10], including silica glass (up to 10^{18} ohm-cm) and Mg_2SiO_4 (forsterite) (10^{17} ohm-cm). Many of the most widely used semiconductors, including TiO_2, are oxides. Several oxides,

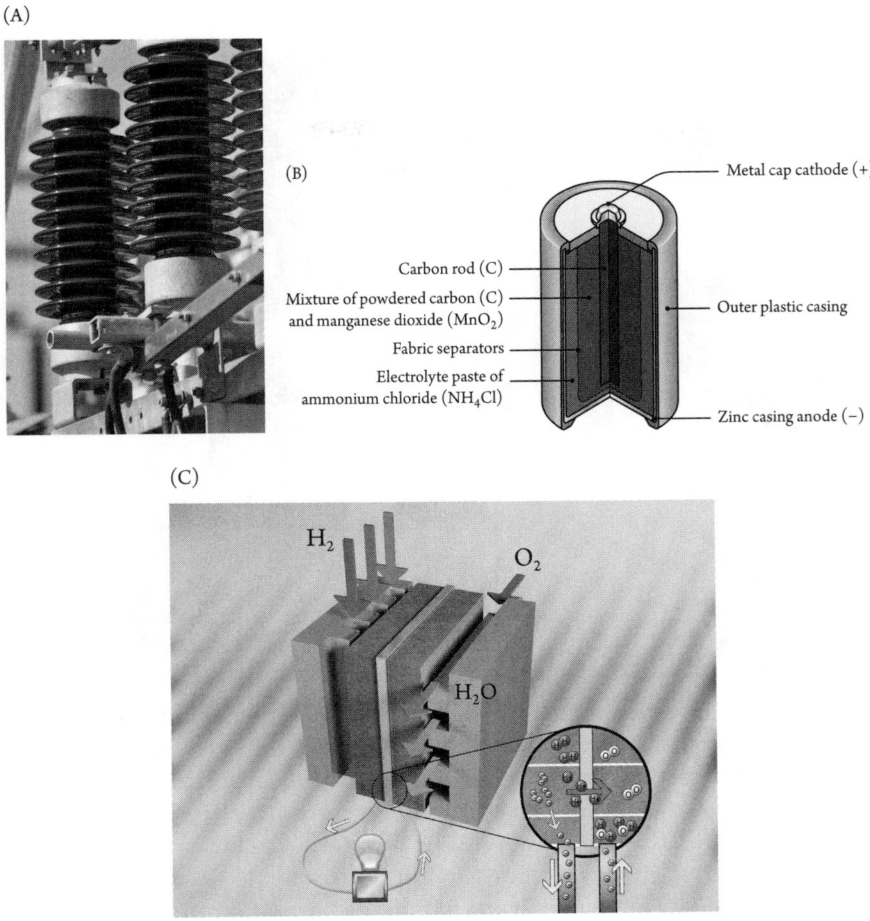

Figure 1.5 Oxide materials used for the generation, storage, and use of electrical energy. (B) A cross section of a manganese-dioxide dry-cell battery (SOURCE: Charles D. Winters/Science Source.com with permission). (A) A stack of ceramic insulator bodies incorporated into a high voltage power line (SOURCE: D. Russell 78/ Shutterstock.com). (C) A schematic representation of a hydrogen-oxygen fuel cell that extracts electrical energy by reacting hydrogen and oxygen fuels with each other to form water. Oxides are often used as catalysts and separators in such devices. (SOURCE: Equinox Graphics/Science Source.com with permission,) (see Chapter 10). (See Plate 5.)

including RuO_2, exhibit a conductivity that is defined as metallic ($\sigma > 1$ S/cm) [10, 11]. In fact, materials exhibiting superconductive behavior (i.e., no resistance whatsoever) at the highest known transition temperatures are compounds, such as the Y–Ba–Cu oxides [12]. Dielectric constants for the oxides range from around 4.5 for Al_2O_3 to more than 12,000 for some of the titanates [13]. Oxides are used as insulators, capacitors, varistors, ferroelectrics, and sensor materials [14].

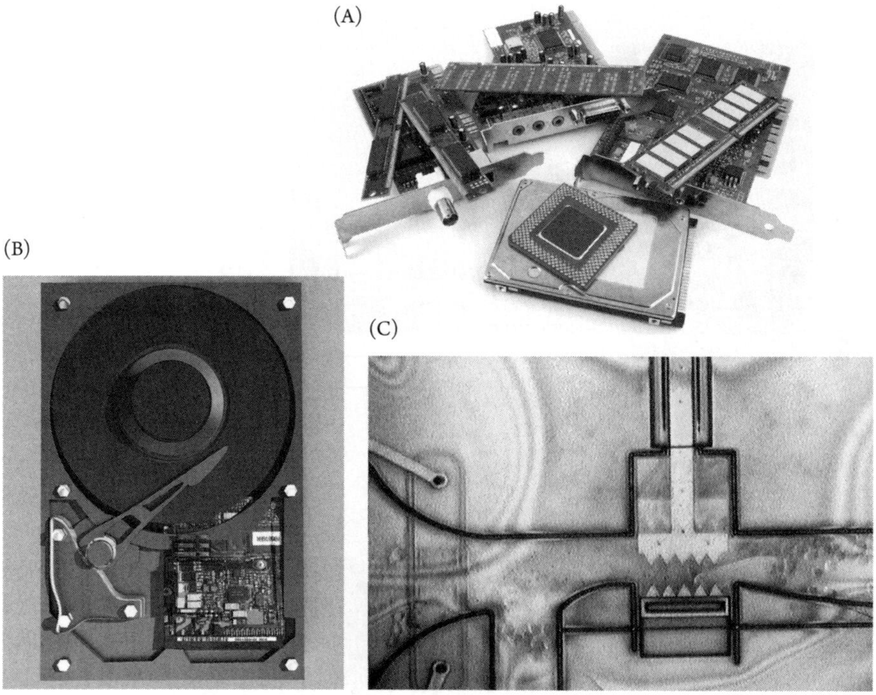

Figure 1.6 Oxides used in microelectronics applications. (A) A range of computer components (SOURCE: Deserg/Shutterstock.com). Oxide ceramics are used in the packaging of many of these components, as well as in insulating materials. (B) A hard disk memory device (SOURCE: Christian Darkin/Science Source.com with permission). The magnetic materials used in both the storage disk and in the head used to read and write information into the disk are often ferromagnetic ceramic materials such as ferrites. (C) A microjaw device for lysing cells on a microfluidic chip developed at Sandia National Laboratories (SOURCE: DOE/SNL/Science Source.com with permission). Aqueous solutions containing cells enter the region between the microjaws, where an oscillating piston drives the jaws together to smash the cells and release their contents for subsequent on-chip analyses. Moving parts in microfluidic systems can be actuated by piezoelectric ceramics in which applied electric fields induce volume changes and motion. Microfluidic systems based on silicon chips (such as the one shown) rely on the integrity of protective thermal silicon dioxide coatings to protect both static and moving components from attack by water (see Chapters 15 and 16 for mechanisms regarding both corrosion and stress corrosion of silicates in water). (See Plate 6.)

1.3.2 Optical Properties

The transparency of oxides to light probably spans the widest range of any known class of materials. At one extreme is silica, which is completely transparent to visible light and is used in applications ranging from windows to optical fibers. At the other extreme are compounds such as Fe_3O_4 that are completely opaque to all

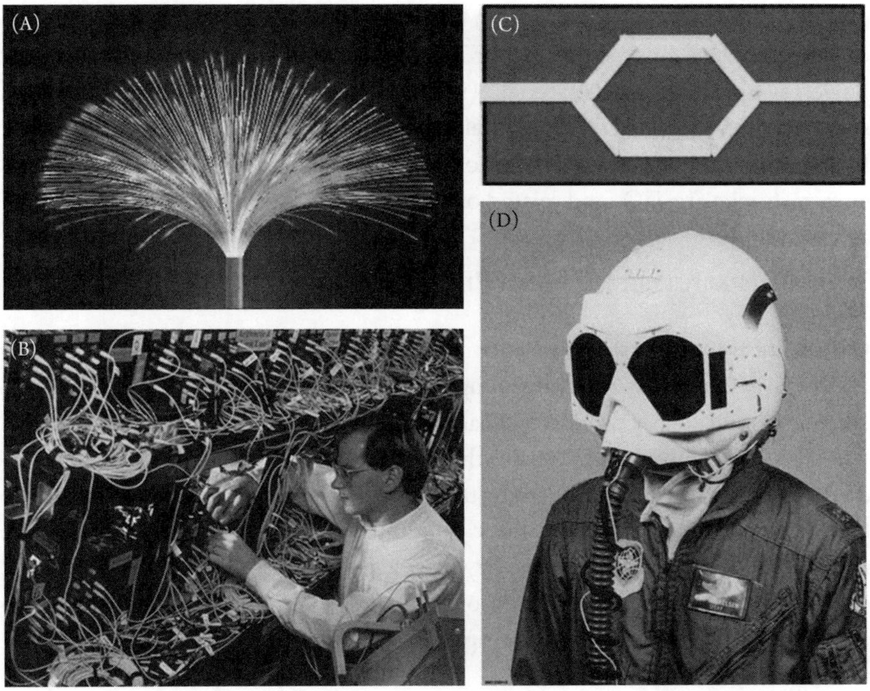

Figure 1.7 Applications of oxides in optical systems. (A) Passive optical fibers based on silica glass transmit most optical wavelengths with minimal optical losses (SOURCE: D Russell 78/Shutterstock.com). The primary failure mode for such fibers in optical communication systems involves water-assisted stress corrosion cracking (Chapter 16). (B) A technician working on the first optical computer that stores and manipulates data as pulses of light (SOURCE: David Parker/Science Source with permission). (C) The optical switches that guide the light and perform data processing functions are often based on electro-optic ceramics such as lithium niobate ($LiNbO_3$). Shown is a simple schematic of one such optical switch. Light enters the waveguide from the left and is split into two branches. In the absence of an applied voltage, the left and right branches are in phase, so when they recombine on the right, the light is unchanged. However, application of a voltage changes the refractive index and speed of light in the upper branch. With the proper voltage, the two branches can be made to be out of phase, so they cancel each other out on recombination, and no light leaves the device. The on and off states can be used to define 1 and 0 logic states in an optical circuit. (D) Flash-blindness goggles developed for the U.S. Air Force using the electro-optic ceramic PLZT (lead–lanthanum–zirconate–titanate) (Photo courtesy of Sandia National Laboratories). The optical–birefringence properties of PLZT can be switched with an applied voltage. When placed between two polarizer plates, this switching can reversibly change the optical transmission of the goggles from 20% to less than 0.0015% (the state shown) in less than 150 μs. (See Plate 7.)

optical wavelengths, with extinction coefficients exceeding 10^6. Refractive indices for the oxides range from 2.56 for TiO_2 down to 1.45 for borax [15]. The most powerful optoelectronic materials (i.e., materials that can be used as optical switches via voltage mediation of refractive index) are oxides such as $LiNbO_3$. Finally, the most widely studied inorganic materials for photoelectrochemical energy storage are TiO_2 and related oxides.

1.3.3 Mechanical and Thermal Properties

Oxides are among the strongest and most durable materials known, as attested by the survival of ancient structures ranging from pyramids to Stonehenge. For example, the silica glass used in optical fibers has a tensile strength of more than 9,000 MN/m^2 (more than 10 times the strength of most metal alloys) [16]. Paradoxically, the same materials are brittle and can be quite fragile, as anyone who has dropped a glass on a hard floor can confirm. In terms of thermal stability, oxides are among the most refractory materials known. For example [17], ThO_2 exhibits a melting temperature of more than 3,300°C. What is not always appreciated is that other oxides can be temperature sensitive. For example [17], Mn_2O_7 melts at only 6°C.

1.4 THE CHEMICAL PROPERTIES OF OXIDES

The previous examples represent just a few of the interesting physical properties of oxides that can be exploited in materials applications. Perhaps even more impressive is the range of chemical properties exhibited by oxides. The primary concern of this book involves oxide reactions that occur in water. Although each of the following topics has at least one chapter devoted to it, it is instructive to highlight how sensitive many of the chemical properties of oxides, oxyhydroxides, and hydroxides really are to local bonding configurations.

1.4.1 Acid–Base Chemistry

Surface sites in oxide materials exhibit both Brønsted and Lewis acidity. What is amazing is the range of acidities that can be generated by the O^{2-} anion (see Chapter 4). In terms of Brønsted acidities (involving the capture or release of protons), species containing O^{2-} anions can be more acidic than the most corrosive acids and more basic than the most caustic bases that can be added to water. Acid–base equilibrium constants for species containing O^{2-} anions can be varied by more than 15 orders of magnitude (a factor of 1,000 trillion). The local pH that can be generated by O^{2-}sites exceeds the normal pH range available in water of pH 0 to 14.

1.4.2 Ion Exchange and Transport

Oxide materials exhibit some of the most exceptional properties known for ion exchange materials. In terms of ionic conductivity, ceramics in the NASICON (Sodium superionic conductor = NaSICon or NASICON) family exhibit sodium ion conductivities exceeding 0.1 S/cm at 300°C and 10^{-3} S/cm/s at room temperature (see Chapter 10). Oxides can be tailored to be highly selective for the adsorption of specific ions and are, in fact, among the most selective exchangers known. For example, silicotitanates prefer Cs^+ to Na^+ by factors exceeding 10^5 (see Chapter 10). As far as ion-exchange capacity is concerned, oxides can contain one ion-exchange site for almost every network-forming cation, approaching theoretical limits for exchange density.

1.4.3 Electrochemical Properties

Depending on what cations are present, oxide materials can be strong oxidizing agents, strong reducing agents, and anything in between. Reduction potentials for converting one oxide and/or hydroxide to another range from +1.93 V for NiO_2 to −0.4 V for $Mn(OH)_3$ (see Chapter 11). Some oxides can even decompose water via both oxidation and reduction processes. In electrochemical systems such as batteries, oxides function as anodes, cathodes, and ionic conductors that are exploited in every viable commercial battery system that uses supporting electrolytes based on water.

1.4.4 Chemical Stability

The rates at which oxide materials are attacked by water can vary by more than 10 orders of magnitude, and more than 20 orders of magnitude if we include the rates at which oxides in water exchange ligands. At one extreme are substances such as phosphate glasses, which are transformed into unstable gels when exposed to ambient humid air (see Chapter 16). At the other extreme are oxides such as zircon ($ZrSiO_4$) that exist unweathered in the oldest rocks on Earth, or ions such as $Ir^{3+}(aq)$ that exchanges waters of hydration with aqueous solutions only once every 200 years at room temperature (see Chapter 4).

Many of the chemical properties presented here are coupled to one another. Selecting an oxide that is optimized for a specific application can be tricky; each chemical property of interest can vary over many orders of magnitude. Compositional changes that improve one reaction parameter often degrade other parameters by orders of magnitude. For example, use of the layered hydrated cobalt–oxide material shown in Figure 1.8 and Plate 8 as a battery electrode [18] requires that (1) the cations exhibit the proper redox potentials, oxidation states,

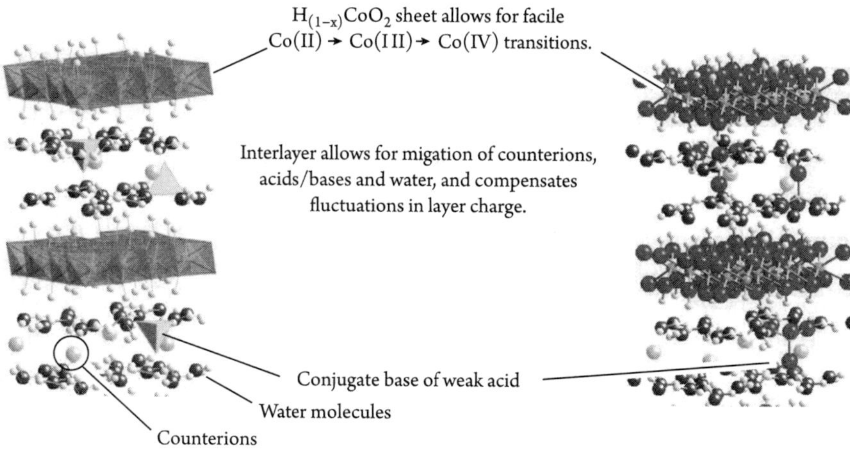

Figure 1.8 A schematic drawing of the critical elements of cobalt oxyhydroxide CoOOH materials under investigation as electrocatalysts. The cobalt ions within the sheets have access to three oxidation states within the stability window of water—Co(II), Co(III), and Co(IV)—allowing the sheets to be oxidized and reduced under an applied voltage. Thin layers of aqueous solutions exist between the sheets of this clay-like material. When the cobalt ions within the oxyhydroxide layers are oxidized or reduced, changes in layer charge are compensated by the intercalation of counterions that migrate between the sheets. (See Plate 8.)

and concentrations; (2) the layers exhibit a high electrical conductivity to facilitate charge extraction; (3) the space between the layers allows for the rapid transport of counterions and protons to compensate for changes in layer charge that accompany redox reactions; and (4) the material must be stable chemically and mechanically under standard battery operating conditions. Simultaneous optimization of all these properties requires understanding many fundamental reactions that occur between water and oxides.

1.5 THE AQUEOUS CHEMISTRY OF OXIDES: A ROAD MAP

When examining the interactions between water and oxides in this book, we explore reactions between two media that appear, at first blush, to be completely distinct: a highly dynamic fluid represented by water, and a group of rigid and brittle materials called *oxides*. Because these materials are so different, we tend to lose sight of the fact that water itself is, in fact, an oxide. Compositionally, the primary difference between water and a rock involves the cations that charge-compensate the negative charges on the oxygen anions. If the oxygen anions are charge-compensated by protons, we normally get fluid structures that are held together with hydrogen bonds, although solid phases such as ice and clathrate hydrates can also form (see Chapter 3). If any other cation predominates, we get the rigid, solid phases described in Chapter 2. Many of the critical reactions that we encounter in this book involve transitions in which one set of cations is exchanged for another as we move from water to hydrated cations to soluble hydrolysis products and, finally, to extended hydroxide and oxide solids. All these reactions can go in either direction. However, *regardless of where we are in the sequence of phases, the fundamental mechanisms that control the critical chemical reactions are essentially the same.*

In Part Two of this book (Fig. 1.9, Plate 9), we describe the basic phenomena that control the key water–oxide reactions, starting with water (Chapter 3) and

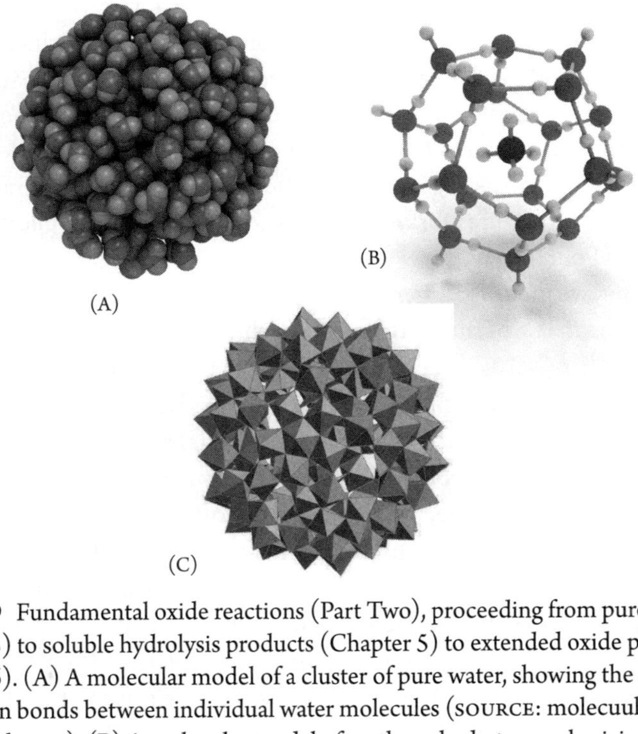

(A) (B) (C)

Figure 1.9 Fundamental oxide reactions (Part Two), proceeding from pure water (Chapter 3) to soluble hydrolysis products (Chapter 5) to extended oxide phases (Chapter 6). (A) A molecular model of a cluster of pure water, showing the orientations of hydrogen bonds between individual water molecules (SOURCE: molecuul.be/Shutterstock.com). (B) A molecular model of methane hydrate, emphasizing the distribution of ring structures between water molecules that encapsulate the methane solute in a zeolite-like cage (SOURCE: Mikkel Juul Jensen/Science Source.com with permission). This model illustrates how the extended structure of water adapts to the presence of ions and other foreign species (Chapter 4). (C) A representation of a water-soluble polyoxometalate cluster that contains 72 molybdenum and 30 iron cations (SOURCE: Reproduced with permission from Balogh, E.; Todea, A. M.; Muller, A.; Casey, W. H. Rates of Ligand Exchange Between $>Fe^{III}\text{-}OH_2$ Functional Groups on a Nanometer-Size Aqueous Cluster and Bulk Solution *Inorg. Chem.* **2007**, *46*, 7087–7092; p. 7088, Fig. 1). This cluster contains many of the oxygen and cation bonding motifs found in extended crystalline oxides. (See Plate 9.)

progressing toward aquated metal ions (Chapter 4), more complex soluble species called *hydrolysis products* (Chapter 5), and last to the chemistry of extended oxide surfaces (Chapter 6). In this sequence of chapters, we introduce the fundamentals of acid–base, ligand exchange, and hydrolysis and condensation reactions that are used throughout the rest of the book.

In Part Three, we explore several dimensions of the synthesis and processing of oxide materials in aqueous solutions (Fig. 1.10, Plate 10). Chapter 7 describes how oxide species make the transition from an aqueous solution to precipitates via nucleation and growth phenomena. Chapter 8 explains how surface chemistry principles introduced in Chapter 6 affect interactions between the resulting colloidal particles. Such interactions are critical in the formation of macroscopic ceramic materials from oxide particles as well as use in applications ranging from paint to nuclear-waste disposal. Chapter 9 applies concepts introduced in Chapters 6 through 8 (and in Chapter 10 on ion exchange) to explain how nucleation, growth phenomena, and surface chemistry control biomineralization processes that form hard tissues such as teeth and bones. Chapter 9 also provides an introduction of how scientists are exploiting the concepts of biomineralization to create oxide nanostructures for the widespread technological applications highlighted in Part Four.

In Part Four, we move from generic oxide reactions to more specialized chemical phenomena that are critical for specific technological applications (Fig. 1.11, Plate 11). In Chapter 10, we apply the principles outlined in Chapter 6 to reactions that control the ion-exchange properties of oxides. Such reactions are critical for a wide range of separations and environmental remediation technologies. Chapter 11 introduces electrochemical reactions by which oxides are oxidized or reduced in water. Oxide electrochemistry is critically important to energy-storage technologies, as will be discussed in Part Four. Metal oxidation leads to the highly destructive and expensive corrosion phenomena described in Chapter 12. This chapter highlights how defect chemistry and hydration reactions couple with electrochemistry to dictate corrosion kinetics. The photochemical reactions of oxides (Chapter 13) are important for their use to capture sunlight as a source of renewable energy and for their role in chemical conversions involving pollutants in the atmosphere and hydrosphere.

Part Five is devoted to the chemistry of oxides such as the silicates that contain tetrahedral cations (Fig. 1.12, Plate 12), rather than the octahedral cations that dominate Parts Two through Four. Topics covered include the formation of silicates and aluminosilicates from aqueous media (Chapter 14), the destruction of glass and other silicate materials by water (Chapter 15), and how water promotes the fracture and mechanical degradation of glass and other silicates (Chapter 16). Such degradation reactions affect phenomena such as the stability of nuclear

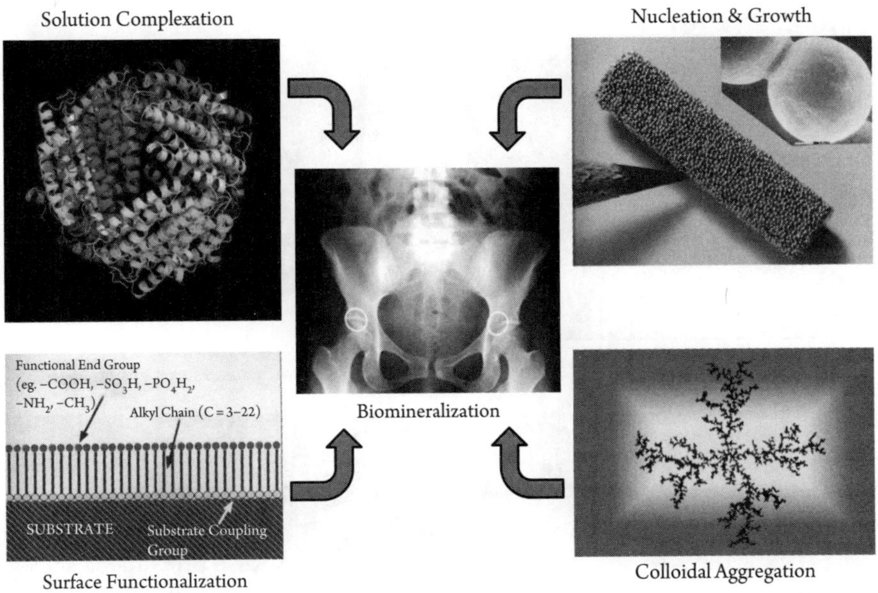

Figure 1.10 The aqueous synthesis of oxides (Part Three) is illustrated by key attributes of biomineralization processes (Chapter 9), such as those involved in bone formation. (Center) An X-ray image of a broken pelvis (SOURCE: fotografos/Shutterstock.com). (Top left) The aqueous synthesis of oxides typically involves stimulating reactions between soluble precursor complexes by inducing a condition of supersaturation (Chapter 7). Shown is the biological complex ferritin (SOURCE: Laguna Design/Science Source.com with permission). Here, a protein shell stimulates the nucleation of small iron-oxide clusters within a protein cage. (Bottom left) Whether in a protein or on an extended surface, the placement of functional groups can lower the barrier to the nucleation and growth processes leading to oxide formation (SOURCE: Reproduced with permission from Bunker, B. C.; Rieke, P. C.; Tarasevich, B. J.; Campbell, A. A.; Fryxell, G. E.; Graff, G. L.; Song, L.; Liu, J.; Virden, J. W.; McVay, G. L et al. *Science* **1994**, *264*, 48–55; p. 50, Fig. 3). (Top right) The stimulation of nucleation and growth is illustrated by the formation of the bone mineral apatite within a porous titanium bone implant material (Reproduced with permission from Bunker, B. C.; Rieke, P. C.; Tarasevich, B. J.; Campbell, A. A.; Fryxell, G. E.; Graff, G. L.; Song, L.; Liu, J.; Virden, J. W.; McVay, G. L. *Science* **1994**, *264*, 48–55; p. 53, Fig. 12). (Bottom right) Aqueous synthesis can also involve the consolidation of colloidal oxide suspensions via controlled aggregation processes. Shown is the result of a computer simulation of diffusion-limited aggregation (SOURCE: Scott Camazine/Science Source.com with permission). Surface functionalization plays a key role in such aggregation processes. (See Plate 10.)

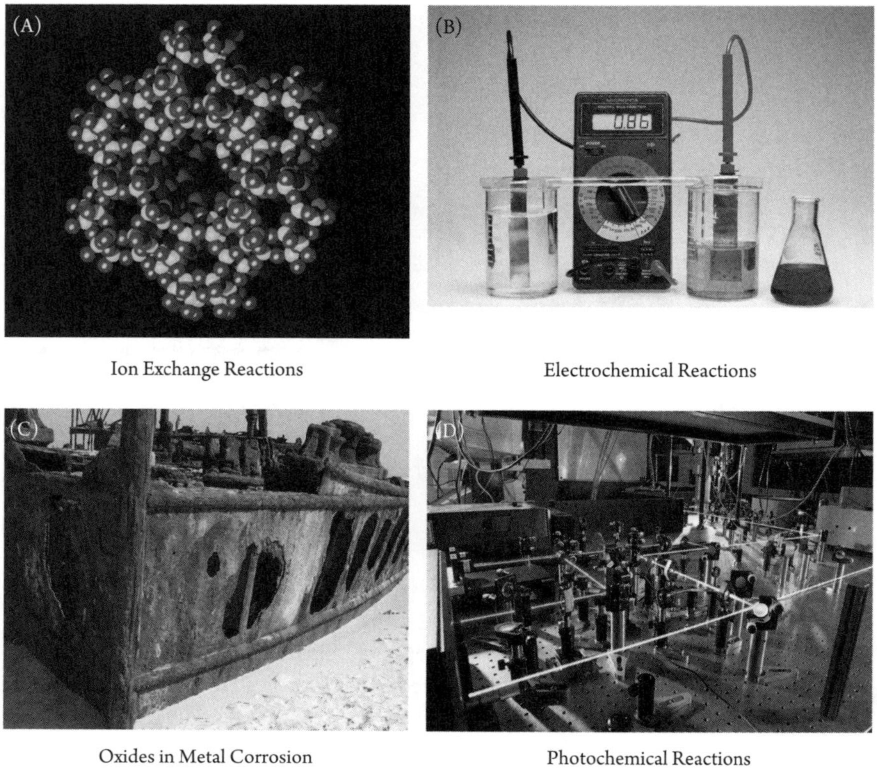

Figure 1.11 Technologically important oxide reactions (Part Four). (A) A molecular model of the oxide exchange material zeolite Y (SOURCE: Clive Freeman/Science Source with permission). Such structures exemplify the charge distribution and steric attributes that make oxides such as zeolites among the most important materials used in ion exchange and water treatment technologies (Chapter 10). (B) Electrodes immersed in aqueous solutions play a critical role in electrochemical energy storage systems such as voltaic electrochemical cells (SOURCE: Charles D. Winters/Science Source.com). Oxides can be both oxidized and reduced. The coupling of such reactions forms the basis for energy storage systems such as most commercial batteries involving aqueous electrolytes (Chapter 11). (C) The electrochemical conversion of metals into oxides can also be highly destructive, as evidenced by the corrosion seen on the rusted *SS Maheno* shipwreck in Australia (SOURCE: Walter Quirtmire/Shutterstock.com) (Chapter 12). (D) Oxide materials are active in photoelectrochemical reactions used for the harvesting of solar energy (Chapter 13). Shown is flash photolysis equipment (SOURCE: Colin Cuthbert/Science Source.com with permission) used to study such reactions. Here, lasers are used to stimulate optical transitions within the material, and the products of the resulting reactions are identified using techniques such as optical absorption spectroscopy. (See Plate 11.)

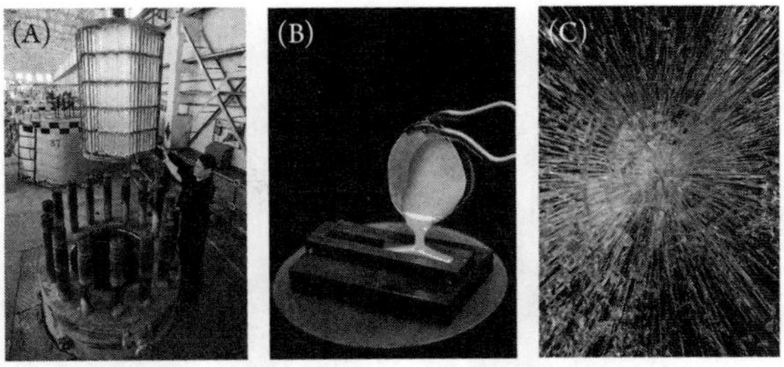

Hydrothermal Synthesis Nuclear Waste Glass Glass Fracture (Stress Corrosion)

Figure 1.12 Reactions involving tetrahedral oxides (Part Five). (A) Autoclaves are key to the aqueous synthesis of tetrahedral oxides at high temperatures and pressures (Chapter 14). Shown is a hydrothermal system used in the synthesis of large quartz crystals (SOURCE: RIA NOVOSTI/Science Source.com with permission). (B) Suppressing the corrosion and dissolution of glass (Chapter 15) is of paramount importance in nuclear waste disposal. Shown is a sample of molten nuclear waste glass being poured into a mold (SOURCE: U.S. Department of Energy/Science Source.com with permission). The resulting glass ingot is used for testing the chemical durability of the waste in a range of hostile aqueous environments to help optimize compositions used in waste repositories. (C) The fracture of glass (SOURCE: bookpc/Shutterstock.com) is often a process that is mediated by water. The role of water is most pronounced in slow crack growth, or stress corrosion cracking, when stress-induced chemical reactions at the crack tip determine failure rates for components such as optical fibers (Chapter 16). (See Plate 12.)

wastes, the erosion of buildings, the development of cracks in car windshields, and failures in optical communication systems.

Last, most of the reactions described in this book are relevant to environmental geochemistry. Part Six (Fig. 1.13, Plate 13) summarizes how the basic chemical principles described in detail in previous chapters affect Earth's major environments. Chapter 17 covers how reactions associated with the weathering of oxides create many of the rock and soil minerals that cover the surface of Earth. Chapter 18 then highlights how oxides mediate key chemical reactions of importance in many of our planet's major environments. By studying oxide–water reactions in nature, geochemists are unearthing many of the basic principles used to develop remediation technologies for ensuring we preserve clean air, water, and soil for future generations.

The content of this book is colored by the fact that both of us have spent much of our careers working with oxides. Although it is impossible to be comprehensive regarding all topics in oxide chemistry, we provide a framework and references to guide you to more extensive fundamental knowledge, and key information to meet a wide range of technological challenges and to appreciate how oxide reactions affect the natural world.

Figure 1.13 The environmental geochemistry of oxides (Part Six). Most of the minerals and other materials that cover Earth's surface are produced as a result of weathering reactions involving oxides and water (Chapter 17). When produced, weathered oxides play important roles in the chemistry of all of Earth's major environments (Chapter 18). On land, weathering products such as clay minerals play a major role in mediating environmental contaminants in soil and groundwater. A view of the Chernobyl power plant in Pripyat, Ukraine, 25 years after the nuclear disaster (June 27, 2010) (A) illustrates such environmental recovery (SOURCE: Helen Sergeyeva/ Shutterstock.com). Clays and other soil minerals are carried into the atmosphere by the wind, (B) where they play key roles in both water condensation in clouds and photochemical reactions. Shown is a tornado in the process of sucking soil into the air from a Kansas farm field (SOURCE: Jim Reed/Science Source.com with permission). Hot lava and water exhibit extensive reactions on Earth's surface (SOURCE: Claudio Rossol/ Shutterstock.com) (C). The reverse reactions occur as weathered and hydrated minerals are pulled beneath the surface in subduction zones that regenerate igneous rocks and release entrapped water in violent volcanic eruptions. All of these processes affect the chemistry of the oceans (D) (SOURCE: Jim Agronick/Shutterstock.com). (See Plate 13.)

References

1. Rudnick, R. L. Composition of the Continental Crust. *Treatise in Geochemistry* **2003**, 3, 1–64.
2. Wedepohl, K. H. The Composition of the Continental Crust. *Geochim. Cosmochim. Acta* **1995**, 59, 1217–1232.
3. Weast, R. C., Ed. *CRC Handbook of Chemistry and Physics, 58th ed.*; CRC Press: Cleveland, OH, 1977, Table B, p F-196.

4. Weast, R. C., Ed. *CRC Handbook of Chemistry and Physics, 58th ed.*; CRC Press: Cleveland, OH, 1977, Table B, p F-197.
5. Benton, M. Foundations: Life in the Oceans. In *The Book of Life*; Gould, S. J., Ed.; W. W. Norton: New York, 1993; pp 37–65.
6. Fletcher, F. What Is Humankind? In *The First Humans: The Illustrated History of Humankind*; Burenhult, G., Ed.; Weldon Owen Pty Ltd: McMahons Point, NSW, Australia, 1993; Vol. 1, pp 17–29.
7. Palmovist, L. The Great Transition. In *The People of the Stone Age: The Illustrated History of Humankind*; Burenhult, G., Ed.; Weldon Owen Pty Ltd: McMahons Point, NSW, Australia, 1993; pp 17–37.
8. Tite, M. S.; Monti, P.; Shortland, A. J. A Technological Study of Ancient Faience from Egypt. *J. Archeological Sci.* **2007**, *34*, 1568–1583.
9. Samsonov, G. V., Ed. *The Oxide Handbook*; IFI/Plenum: New York, 1973; pp 263–280.
10. Buchanan, R. C. Properties of Ceramic Insulators. In *Ceramic Materials for Electronics*; Buchanan, R. C., Ed.; Marcel Dekker: New York, 1986; pp 1–79.
11. Goodenough, J. B. Metallic Oxides. *Prog. Solid State Chem.* **1971**, *5*, 145–399.
12. Wu, M. K.; Ashburn, J. R.; Torng, C. J.; Hor, P. H.; Meng, R. L.; Gao, L.; Huang, Z. J.; Chu, C. W. Superconductivity at 93 K in a New Mixed-Phase Y-Ba-Cu-O Compound at Ambient Pressure. *Phys. Rev. Lett.* **1987**, *58*, 908–910.
13. von Hippel, A.; Breckenridge, R. G.; Chesley, F. G.; Tisza, L. High Dielectric Constant Ceramics. *Ind. Eng. Chem.* **1946**, *38*, 1097–1109.
14. Buchanan, R. C., Ed. *Ceramic Materials for Electronics*; Marcel Dekker: New York, 1986.
15. Samsonov, G. V., Ed. *The Oxide Handbook*; IFI/Plenum: New York, 1973; pp 333–335.
16. Michalske, T. A.; Bunker, B. C. The Fracturing of Glass. *Sci. Am.* **1987**, *257*, 122–129.
17. Samsonov, G. V., Ed. *The Oxide Handbook*; IFI/Plenum: New York, 1973; pp 105–115.
18. Gerken, J.; McAlpin, J. G.; Chen, J.; Rigsby, M.; Casey, W. H.; Britt, R. D.; Stahl, S. Electrochemical Water Oxidation with Cobalt-Based Electrocatalysts from pH 0–14: The Thermodynamic Basis for Catalyst Structure, Stability, and Activity. *J. Am. Chem. Soc.* **2011**, *133*, 14431–14442.

2

An Overview of Oxide Structures and Compositions

2.1 LOCAL OXIDE STRUCTURES: THE OXIDE (O^{2-})ANION

This entire book is devoted to exploring the chemistry of compounds that contain one simple anion: the O^{2-} ion. Except under high-vacuum conditions [1] (see Chapter 6), the species in oxides that interact with water and other environmental chemicals are O^{2-} ions, because the charge-compensating cations are invariably buried beneath an oxide surface layer. One might wonder how we can fill an entire volume discussing the chemical interactions between this single anion and a single chemical: the water molecule. *The single most important concept that must be appreciated to understand the contents of this book is that the chemistry and properties of O^{2-} anions are critically dependent on all the cations to which the O^{2-} ions are bound.* Each bound cation modifies the electron distributions around the O^{2-} site, changing its local charge, local bonding configurations, acid–base chemistry, ion exchange chemistry, electrochemical properties, chemical stability, and electrical and optical properties. None of these changes are subtle, and in fact most oxide properties are staggering in their diversity. Before considering the chemistry of oxides, it is important to gain an appreciation of just how diverse the structures of oxide materials really are. As this introduction makes clear, there is no such thing as a single, simple O^{2-} ion. There are a myriad of different O^{2-} sites found in the oxides that we encounter often in our daily lives, each of which exhibits its own unique properties. The purpose of this book is to provide a framework that can be used to predict, rationalize, and exploit the rich chemistry associated with those sites.

The number of different structures and compositions that can be generated for oxides is almost limitless. The O^{2-} ion forms compounds with more than 90 elements in the Periodic Table that are capable of losing electrons to form cations [2]. The oxide anion combines with cations with charges that range from +1 to +7. Many elements exhibit more than one stable oxidation state, pushing the total number of chemically distinct cations with which O^{2-} can interact to well more than 120. The number of cations to which O^{2-} can bond (called the *coordination number*) ranges from one to eight. In most known crystal structures, cations representing many different elements can substitute into the same lattice, producing an enormous range of different oxide, oxyhydroxide, and hydroxide compositions. In this chapter, we highlight just some of the major structures that oxides can form, including close-packed crystalline arrays, less-dense three-dimensional (3D) lattices, layered two-dimensional structures, and engineered nanoarchitectures. Because most oxides contain network-forming cations that are in *octahedral* or *tetrahedral* coordination (bonded to six or four O^{2-} anions, respectively), the structures to be discussed focus on oxides containing these two coordination geometries. However, before exploring extended oxide structures, we first introduce the formal-charge models used to rationalize local bonding configurations and their effect on oxide chemistry.

2.2 PREDICTING OXYGEN-BONDING CONFIGURATIONS: AN INTRODUCTION TO THE FORMAL-CHARGE MODEL

The local bonding configurations around O^{2-} anions exert the strongest influence on most of the chemical and physical properties to be discussed in this book. Here, we introduce how the local charge distributions around each anion dictate the nature and stability of extended oxide structures. A limited number of researchers with access to powerful computers can calculate exactly how electron distributions should change based on techniques, such as density functional theory [3, 4], that probe local bonding using quantum mechanics. In this book, we use models for predicting local charge distributions that are at the complete opposite end of the computational spectrum. Rather than requiring a supercomputer, these models require only a piece of paper and the ability to add and subtract. This simple approach is derived from the electrostatic valence principle of Linus Pauling [5, 6]. Pauling's Second Rule for bonding in ionic crystals leads to a definition of *formal charges*. This rule was used to predict the strength of oxyacids and led to the *bond–valence principle* that is an essential tool in structural chemistry [7]. For example, this principle allows one to distinguish between a structural water molecule and a structural hydroxide ion based on X-ray data alone.

The beauty of formal charge models is that they are intuitive and lead to immediate insights about reactivity that transfer across both chemical and physical properties. Given how simple the models are, it is amazing how accurate and quantitative they can be. In the remainder of this book, we use formal charge models repeatedly to illustrate concepts for water–oxide reactions, taking care to point out where these models break down. Next, we introduce the simplest and most generic version of the formal charge model that we apply to everything we discuss about oxide–water systems. Those of you who are interested are urged to extend this treatment with more detailed investigations of the development of bond–valence sums [7–9] and applications to acid–base chemistry [10, 11].

Formal charge models are based on the simple assumption that stable structures are dictated by bonding configurations that come closest to achieving charge neutrality around each oxygen in an oxide, hydroxide, or hydrolysis product structure. Neutrality is achieved if the sum of positive charges donated to each oxygen anion by adjacent cations is equal to the formal charge on the oxygen anion of −2 (Fig. 2.1). The charge donated by different cations is taken as the

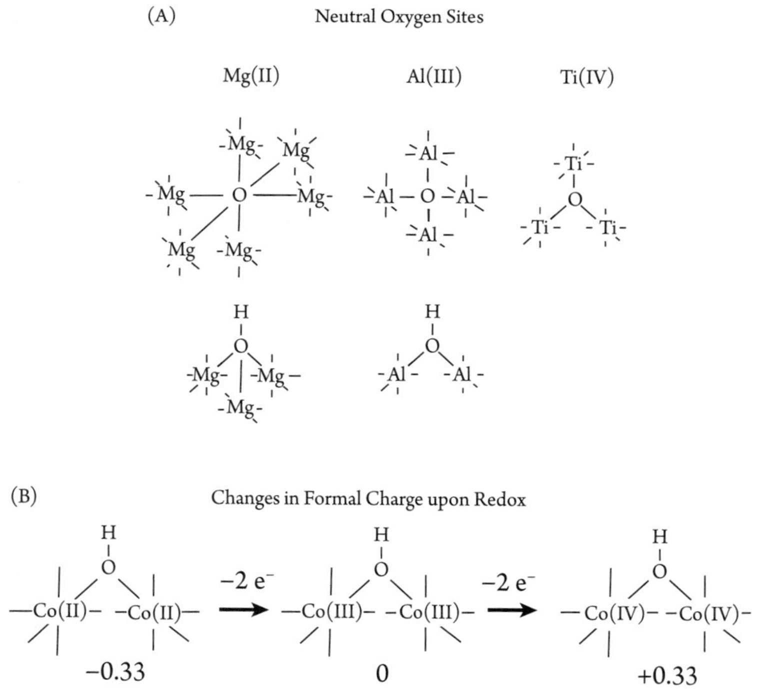

Figure 2.1 Examples of formal charge on an oxygen in different coordination geometries. (A) A range of neutral oxygen anion sites in both oxides and hydroxides. (B) The net charge on bridging hydroxyl groups in $Co(OH)_3$ as a function of the valence charge on the metal. Oxidation of the metal makes bonded oxygens stronger acids.

Table 2-1. FORMAL CHARGES ON OXYGEN IONS CALCULATED BY ADDING THE NATIVE OXYGEN ANION FORMAL CHARGE OF −2 TO THE CHARGES DONATED BY ALL CATIONS BONDED TO THAT OXYGEN

Cation charge donation					
Cation charge	+2	+3	+4	+5	+6
Octahedral cation charge donation	+0.33	+0.50	+0.67	+0.83	+1
Tetrahedral cation charge donation		+0.75	+1		
Oxygen formal charges (involving identical cations having the charge shown)					
Cation charge	+2	+3	+4	+5	+6
Nonbridging oxygen (6)	−1.67	−1.5	−1.33	−1.17	−1
Nonbridging oxygen (4)		−1.25	−1		
μ_2 Oxygen bridge (6 + 6)	−1.33	−1	−0.67	−0.33	0
μ_2 Oxygen bridge (4 + 4)		−0.5	0		
μ_2 Oxygen bridge (6 + 4)		−0.75	−0.33		
μ_3 Oxygen bridge (6 + 6 + 6)	−1	−0.5	0		
Hydroxyl group formal charges (when the proton donates a charge of +1)					
Cation charge	+2	+3	+4	+5	+6
Nonbridging hydroxyl (6)	−0.67	−0.5	−0.33	−0.17	
Nonbridging hydroxyl (4)		−0.25	0		
μ_2 Hydroxyl bridge (6 + 6)	−0.33	0	+0.33		
μ_3 Hydroxyl bridge (6 + 6 + 6)	0				
Formal charges on representative mixed-cation species					
Cation charges	+2 and +3	+3 and +4	+4 and +5		
μ_2 Oxygen bridge (6 + 6)	−1.17	−0.83	−0.5		
μ_2 Oxygen bridge (4 + 4) (e.g., Al-O-Si)		−0.25			
μ_2 Hydroxyl bridge (6 + 6)	−0.17	+0.17			
Formal charges on additional oxygen sites in clays					
μ_3 Oxygen Bridge (Al(III)(6) + Al(III)(6) + Si(IV)(4)) = 0					
μ_4 Oxygen Bridge (Mg(II)(6) + Mg(II)(6) + Mg(II)(6) + Si(IV)(4)) = 0					

Cation charge donation is simply the cation charge divided by its coordination number. In the oxygen formal charge listing, the coordination number of each cation bonded to the given oxygen is shown in parentheses (e.g., octahedral cation = 6, tetrahedral cation = 4). Only common species are listed. Unstable cation coordination geometries are not listed, nor are unstable oxygen sites having a formal charge of +0.5 or greater. Bridging oxygens are given the notation μ_n, where *n* corresponds to the number of metals bridged).

cation charge divided by the cation coordination number, as in the electrostatic valence rule [5] and Brown and Shannon's treatment of bond strengths [12] (Table 2.1). For example, an octahedrally coordinated Ti^{4+} donates a net charge of $+4/6 = +0.67$ to each oxygen in its inner coordination sphere.

General rules used to evaluate the stability of different oxygen bonding configurations include the following:

1. Oxygen sites with a net positive charge greater than around +0.7 are unstable. For example, oxygen anions bonded to three tetrahedral Si^{4+} ions (each donating a cation charge of +1) would have a net charge of $+3 - 2 = +1$, and should be much less stable than the charge-neutral Si—O—Si bonds that are observed in almost all silicates.
2. Oxygen sites can carry a net negative charge, but only when this charge is compensated by ion pairing with low-valence cations such as protons, alkali–metal cations (e.g., Na^+), or alkaline–earth cations (e.g., Ca^{2+}). For example, in silicate structures, nonbridging oxygens (with a formal charge of −1) can exist, provided that each is paired with cations such as Na^+ or a proton. Divalent cations such as Ca^{2+} can charge-compensate two nonbridging oxygens, forming cation bridges between the anions.
3. Oxides, hydroxides, and hydrolysis products contain mixtures of structural units that minimize the anionic character on the maximum number of oxygens. For example, in aluminosilicate structures, the model predicts that substitution of Al^{3+} cations into tetrahedral sites to make four Si—O—Al bonds (when each oxygen has a formal charge of −0.25) is preferred to the formation of a single Si—O nonbridging oxygen (with the same net negative charge of −1). This spreading of charge over multiple structural units is favorable energetically.
4. The number of bonds formed to all species must be consistent with the formula unit and both cation and anion coordination geometries. For example, for the formula $Na_2Ti_4O_9$, if the Ti is bonded to six oxygens in an octahedral coordination geometry, the mix of possible oxygens in the structure must account for a total of $(6\ \text{bonds/Ti}) \cdot (4\ \text{Ti}) = 24$ Ti—O bonds. All nine oxygens in the formula unit cannot be neutral oxygens coordinated to three Ti cations as in TiO_2, because this would result in a net of 27 Ti—O bonds. In addition, the nine oxygens must have a net charge of −2 to compensate for the two Na^+ ions in the formula unit. The actual structure for $Na_2Ti_4O_9$ contains one Ti—O—Ti bond (with a formal charge of −0.67), one nonbridging Ti—O bond (with a formal charge of −1.33), and seven neutral tribridging oxygens (see Chapter 10).

(In this book, we use the standard notation in which non-bridging oxygens are denoted as η species, while bridging oxygens are given the notation μ_n, where n corresponds to the number of metals bridged).

2.3 CLOSE-PACKED CRYSTALLINE LATTICES

In addition to local charge distributions, steric effects based on ion size dictate how the cations and anions pack to form extended structures. The simplest oxide structures to consider are those consisting of close-packed lattices of O^{2-} ions [13] with "spherical radii" of 0.14 nm. Considering the O^{2-} ion as a sphere, the most-dense structures that can be produced via the stacking of perfect two-dimensional sheets of anions packed like marbles are hexagonal and cubic close-packed arrays (Fig. 2.2). In these arrays, each ion has 12 nearest neighbors (i.e., six in the plane: three above and three below). In hexagonal arrays, the third layer lies directly below the first in an ABA stacking arrangement, whereas in cubic packing, the third layer is offset in an ABC stacking sequence. Both hexagonal and cubic arrays contain both octahedral and tetrahedral interstices between the spheres, amounting to one octahedral and two tetrahedral voids per oxygen anion. Cations are inserted into these interstices to compensate for the anionic lattice charge.

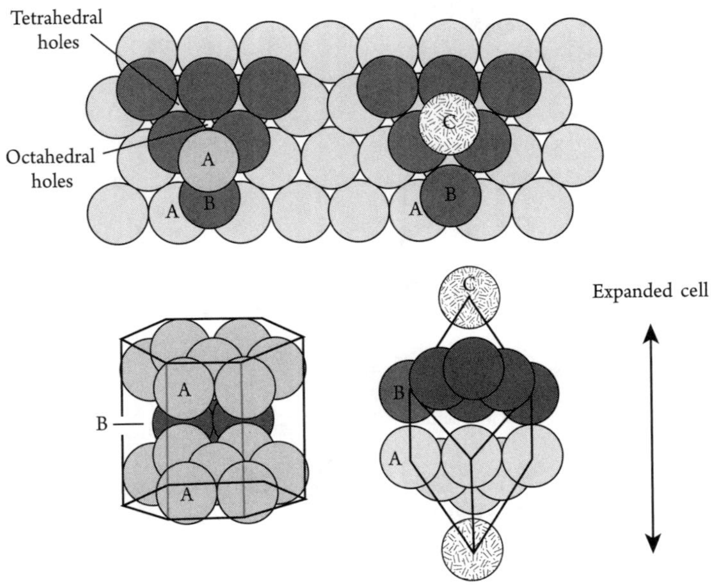

Figure 2.2 Oxide structures based on close packing of oxygen anions. Placement of the first two rows are identical, but there are two options for the placement of the third row. (top) This row covers the tetrahedral voids to make a hexagonal close-packed structure. (bottom) The octahedral voids are covered to make a cubic close-packed structure.

The most prevalent oxide crystal structures are based on hexagonal close-packed arrays, with different structural families being defined by how the cations are distributed. Examples include (1) the corundum structure [13], exemplified by α-Al_2O_3 and α-Fe_2O_3, in which two-thirds of the octahedral interstices in a hexagonal close-packed array are filled with cations; (2) the NaCl structure found for oxides, such as MgO, in which all octahedral interstices are filled with cations; (3) spinels such as $MgAl_2O_4$, in which one-eighth of the tetrahedral holes are filled by +2 cations, and half of the octahedral holes are filled by +3 cations; and (4) perovskites such as $CaTiO_3$, in which both the large cation and O^{2-} occupy sites in a dense but interpenetrating cubic close-packed array in which the smaller cation occupies octahedral holes defined by O^{2-} anions (e.g., the rocksalt structure shown in Fig. 2.3).

Cations are not necessarily optimized to fit within the octahedral and tetrahedral holes of close-packed oxygen anion arrays. Other extended lattices with somewhat lower net O^{2-} densities can be created by packing geometries with different void geometries and sizes as determined by the number of oxygen anions that can be packed around a given cation (Fig. 2.3). For example, the

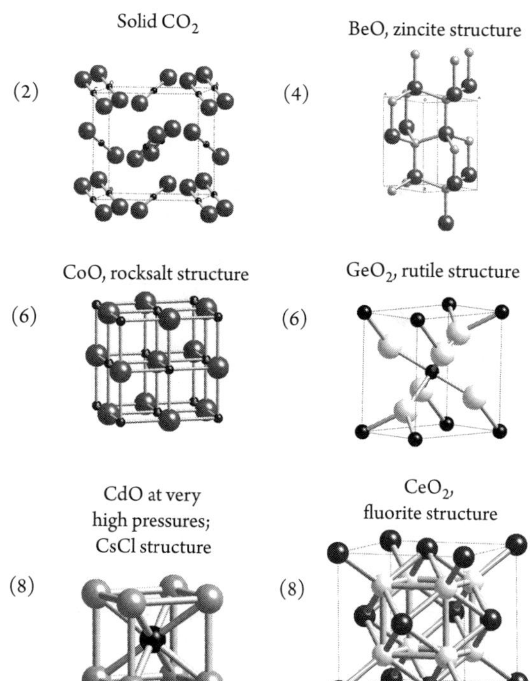

Figure 2.3 Additional oxide packing geometries as dictated by cation size relative to oxygen and the local charge distributions. Larger ions allow for a larger number of coordinated oxygens, and very high pressures can cause higher packing numbers to reduce volume. The coordination number of the metal is shown in parentheses next to the structure.

zincblende and wurtzite structures contain regular arrays in which all cations assume a tetrahedral coordination geometry. The factor that determines the most efficient packing and preferred crystal structure is the ratio of the cation radius to the O^{2-} radius (r^+/r^-). Simple geometric arguments [5, 6] show that cations with r^+/r^- less than 0.22 can fit within trigonal voids. This suitable void size explains why small cations such as B^{3+} ($r^+ = 0.02$ nm and $r^+/r^- = 0.14$) are often found coordinated to three O^{2-} ions. Preferred cation coordination numbers increase to 4, 6, 8, and even 12 for $r^+/r^- = 0.22-0.41$, $0.41-0.73$, $0.73-1$, and more than 1, respectively. Examples of cations exhibiting these coordination numbers include Si^{4+} ($r^+ = 0.041$ nm), Fe^{3+} ($r^+ = 0.076$ nm), Sr^{2+} ($r^+ = 0.11$ nm), and Cs^+ ($r^+ = 0.17$ nm), respectively. For the series of +4 cations from Si^{4+} ($r = 0.041$ nm) to Ge^{4+} ($r = 0.053$ nm) to Ti^{4+} ($r = 0.068$ nm), the pure oxides exhibit structures in which the cation is predominantly tetrahedral (in SiO_2 or quartz), octahedral (in TiO_2 or rutile), or both (GeO_2 exhibits phases with both the quartz and rutile structures.) Cation radii, r+/r− values, and packing numbers predicted to be the most stable are listed in Table 2.2.

Table 2-2. Selected Ionic Radii for Cationic Species

Ionic Species	**Ionic Radius (Å)**	**r^+/r^-**	**Coordination No.**
Li(I)	0.76	0.54	6
Na(I)	1.02	0.71	6
K(I)	1.38	0.98	8
Be(II)	0.35	0.25	4
Mg(II)	0.72	0.51	6
Ca(II)	1.00	0.71	6
Ba(II)	1.35	0.96	8
B(III)	0.23	0.16	3
Al(III)*	0.51	0.36	4
Ga(III)	0.62	0.44	6
Si(IV)	0.4	0.28	4
Sn(IV)	0.71	0.51	6
Ti(IV)	0.68	0.49	6
Fe(II)	0.76	0.54	6
Fe(III)	0.60	0.42	6

Shown are the radii (in angstroms), the ratio of each radius to that of the O^{2-} ion (r^+/r^-), and the coordination number predicted to be stable based on that radius ratio. (The ionic radius of the O^{2-} ion is taken to be 1.4 Å.) (See Table 10.1 for more alkali metal and alkaline earth cation data.)

*Actual coordination numbers in oxides may differ from the predictions. For example, Al(III) is commonly found in both octahedral and tetrahedral geometries.

Although specific cations usually assume the same local coordination geometry as dictated by cation size, the oxygen anions in both oxide and hydroxide phases can adopt a wide range of coordination geometries. Within a close-packed O^{2-} lattice, anions can have different coordination numbers depending on how charge-compensating cations are distributed among the tetrahedral and octahedral voids. Considering all crystal structures, oxygen anions can assume coordination numbers ranging from 1 (called a *nonbridging oxygen*) up to 8, although 2, 3, and 4 are most common. As is shown later, the behavior of a given oxygen anion is critically dependent on both the coordination number and the charges on each cation in the anion's immediate vicinity. *Local bonding configurations are what distinguish one oxygen anion from another.* An understanding of local charge distributions as mediated by bonds to local cations is the most important parameter that controls the aqueous chemistry of oxides.

2.4 OXIDES CONTAINING OCTAHEDRAL CATIONS

These geometries are among the most common in the Earth for oxide solids containing trivalent metals. The oxides of trivalent metals are important because these metals have an interesting acid-base chemistry in the pH region of natural waters. Many transition-metal oxides form a lattice in octahedral coordination to oxygens and can undergo changes in valence, making them very important to the cycling of mass in soil.

2.4.1 Three-Dimensional Networks

In this section, we provide examples of oxide structures containing octahedral cations, moving from the densest and purest single-component oxides to increasingly lower density structures containing foreign cations. Most oxides that are stable in water contain cations with oxidation states of +2, +3, or +4 (see Chapter 5). (The rarer +5 and +6 oxidation states can also exhibit low solubility.) The +3 cations prefer to occupy octahedral sites in hexagonal and cubic close-packed arrays, such as the corundum structure. The maximum packing density calculated for spherical oxygen anions in such arrays is $6.5 \cdot 10^{22}$ O/cm^3, or around 0.1 mole/cm^3. Packing densities for oxides containing +3 cations cluster around this value, with α-Al_2O_3 having the highest packing density of any oxide ($7.1 \cdot 10^{22}$ O/cm^3). The Al^{3+} cations are distributed such that each oxygen anion is bonded to four Al^{3+} cations. Because each octahedral Al^{3+} has a formal charge of +0.5, all oxygens in the structure are neutral and highly stable, consistent with formal charge predictions. Cations with a valence of +4 are found most often in the slightly less dense rutile structure (with an average density of $6.3 \cdot 10^{22}$ O/cm^3) represented by TiO_2 (Fig. 2.3). Here, all oxygen anions are bonded to three Ti^{4+} cations, each of which donates a formal charge of +2/3 to neutralize exactly the charge on the O^{2-} anion. This trigonal bonding arrangement creates a slight distortion in the close-packed O^{2-} lattice, resulting in slightly lower densities. The +2 cations tend to form phases having the close-packed rocksalt structure that is less dense still (with an average of $5.1 \cdot 10^{22}$ O-atoms/cm^3). In the rocksalt structure, each oxygen anion is bonded to six +2 cations (each with a formal charge of +1/3), again yielding charge-neutral oxygen sites. This lower density arises because the +2 cations tend to be slightly larger than the octahedral voids between O^{2-} anions, which spread the anions apart.

2.4.2 Layered Two-Dimensional Structures

Close-packed O^{2-} lattices can be distorted to create layers as opposed to the uniform 3D distribution seen in the NaCl structure. For example, cations with a valence of +5 or +6 can interact so strongly with O^{2-} that multiple bonds form between the cation and one or more of the oxygens in the inner coordination sphere. In the case of V_2O_5, the net result is that sheets are formed of edge-shared VO_6 octahedra in which the apical oxygens have shorter high-order bonds [14] (Fig. 2.4). These multiple bonded oxygens are a departure from the ionic bonding typical of many oxides as they form from orbital hybridization. In vanadates, this hybridization involves the metal d orbitals and the p orbitals on the apical oxygens. Because so much charge is removed from the apical oxygens, they bond weakly to cations in adjacent layers, resulting in an oxide structure consisting of stacked sheets. The unreactivity of the basal sheets can be used by scientists to make nanosheets of useful vanadates.

Similar stacked-sheet structures are also quite common for niobates and tantalates. Even +4 cations sometimes form layered structures. For example, although the mineral pyrolusite assumes the rutile structure, most other MnO_2 phases have a layered structure. Birnessite, which is a layered mineral ubiquitous in soils, contains a mixture of Mn(III) and Mn(IV). There, alkali metal cations insert between the layers to compensate for the lowering of network charge associated with the reduction of Mn(IV) to Mn(III) [15]. "Electrolytic" manganese dioxide formed by the reduction of aqueous $KMnO_4$ solutions assumes a similar nonstoichiometric composition with a layered crystal structure

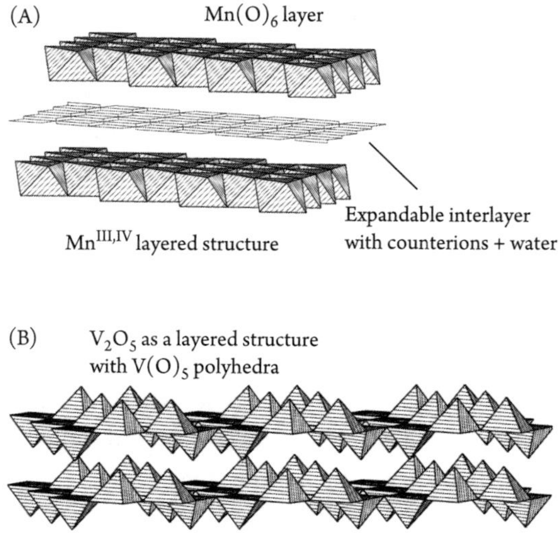

Figure 2.4 Layered oxide networks. (A) The structure of γ-MnO_2 showing sheets of manganese oxide octahedral separated by layers containing water and charge-compensating ions. (B) The structure of V_2O_5 showing the impact of double-bonded oxygens, resulting in a five-coordinate geometry around each cation and a layered structural motif.

(Fig. 2.4) [16]. Both V_2O_5 and this nonstoichiometric MnO_2 are used as electrode materials for batteries precisely because the zones between the layers allow for easy intercalation of protons, Li^+, and other mobile species essential to electrode function (see Chapter 11).

Counterions that can insert into or disrupt the bonding within a dense oxide network are often called *modifier cations*. Common modifier cations have a valence of +1 and +2, and include the alkali metal and alkaline–earth cations. The most important modifier cation is the proton. Bonding to protons is extremely localized. Protons disrupt the oxygen anion lattice by replacing bonds to network-forming cations with nonbridging hydroxyl groups (OH^-). For example, simple formal charge arguments show that bonding to one proton is equivalent to bonding to three octahedral Mg^{2+} cations (Fig. 2.1). At low hydroxyl group concentrations, the parent oxide phase is preserved, although some critical properties such as ionic conductivity are altered dramatically. At sufficiently high hydroxide concentrations, dense oxides are transformed into lower density oxyhydroxide phases such as boehmite [17, 18] (γ-AlOOH), diaspore [19] (α-AlOOH), and goethite [19, 20] (α-FeOOH) (Fig. 2.5).

At high hydroxyl group concentrations, the covalent interactions between adjacent layers in a close-packed oxide structure can be disrupted completely to form less-dense structures based on layers having the thickness of only two

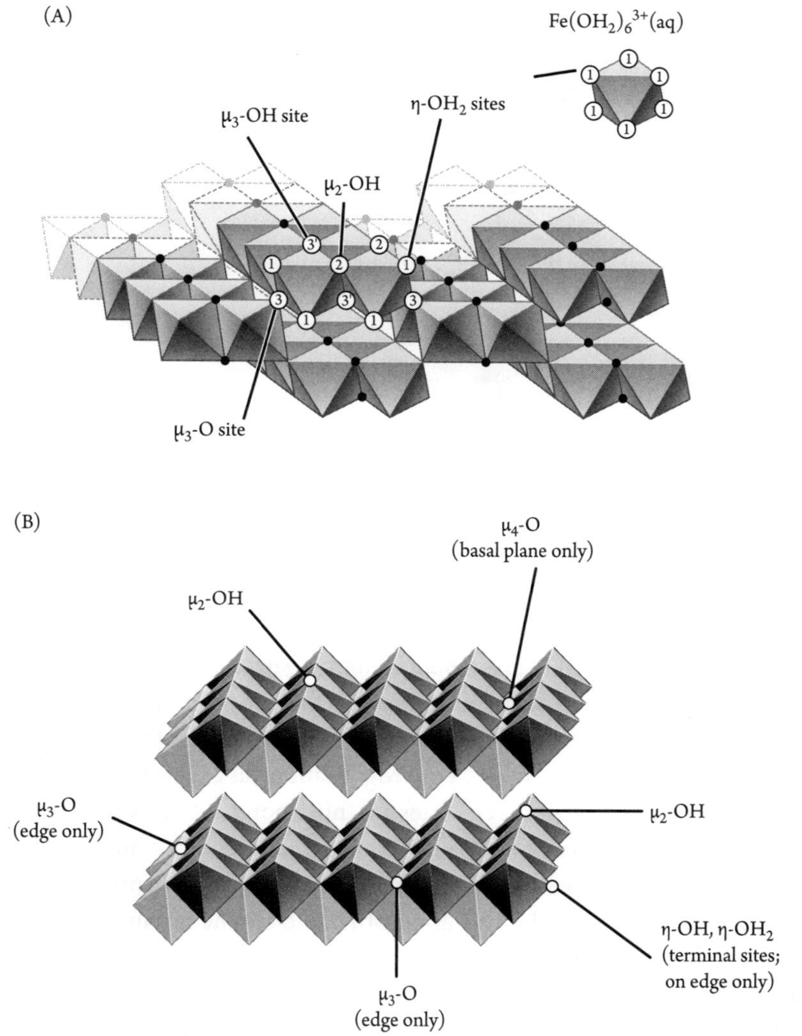

Figure 2.5 The oxyhydroxide phases of +3 cations such as Al(III) and Fe(III) can form similar structures. (A) The mineral diaspore as AlOOH(s) and goethite as FeOOH(s). Numbers indicate the coordination number of the oxygens and black dots identify structural protons on μ_3-OH. (B) The boehmite structure as AlOOH(s) and lepidocrocite as FeOOH(s), again with the coordination number of the oxygens identified.

anions, including the hydroxides (Fig. 2.6) and the layered double-hydroxides [21, 22]. Oxygen packing densities in the series α-Al_2O_3 to γ-AlOOH to $Al(OH)_3$ drop from 7.1 to 6.0 to $5.5 \cdot 10^{22}$ O/cm^3. The most common and important layer structures in hydroxides are *trioctahedral* and *dioctahedral* sheets. The trioctahedral sheet consists of two adjacent layers of O^{2-} anions in which every octahedral

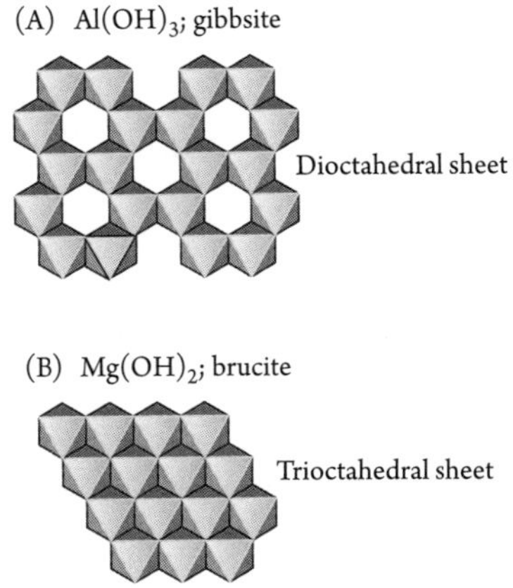

Figure 2.6 (A, B) Simple metal-hydroxide precipitates commonly form simple sheet hydroxide structures. For divalent metal, these form dioctahedral sheets in $Al(OH)_3$ (A) and trioctahedral sheets in $Mg(OH)_2$ (B), which are the building blocks of clays.

vacancy is filled by a +2 cation, and every oxygen anion is capped with a proton, as exemplified by $Mg(OH)_2$. In a dioctahedral sheet, only two-thirds of the octahedral vacancies are filled with +3 cations, as in the mineral gibbsite (γ-$Al(OH)_3$). A formal charge analysis reveals why each of these two distinct sheet structures is stable (see Fig. 2.1). In $Mg(OH)_2$, each Mg^{2+} cation has a formal charge of +1/3, so the most stable and neutral O^{2-} sites are those bonded to three Mg^{2+} and one proton. In $Al(OH)_3$, charge neutrality requires that each O^{2-} should be bonded only to two Al^{3+} (each with a formal charge of +0.5) and one proton. The local bonding requirements of the O^{2-} anions then dictate the number of octahedral sites that can be occupied within dioctahedral and trioctahedral sheets to achieve charge neutrality (Fig. 2.6).

All oxides can contain protons. Many, including TiO_2 and MnO_2, and other oxides containing +4 cations, can be transformed into layered oxides (see Chapter 7) at high proton concentrations. These protonated oxides are often described as *hydrated* or *hydrous* oxides, particularly if they are amorphous rather than crystalline. Proton concentrations in hydrated oxides can exceed concentrations required to form metal hydroxides. The O^{2-} sites can be doubly protonated to form *waters of hydration* (or bound water molecules). For example, precious metals such as rhodium can be precipitated from aqueous solutions as "active hydroxides" having the stoichiometry $M(OH)_3(H_2O)_3$ in an expanded bayerite-like ($Al(OH)_3$) layered structure [23–28]. The beauty of the active hydroxides is that they allow coordinated metals to be precipitated from contaminated solutions, washed, and then quickly redissolved in acids that disrupt the hydrogen bonds between layers to release the purified cations. Active hydroxides are useful for purifying metals with hydroxides that are relatively inert to ligand exchange reactions (see Chapter 4), as dissolution and precipitation processes only involve the disruption and formation of hydrogen bonds rather than hydrolysis and condensation.

In layered systems, oxide chemistry is dominated by the spaces between the layers. Water, ions, and other molecules present in adjacent environments can sometimes penetrate between these layers, resulting in some of the rich chemistry described later in this book. Here, we briefly introduce layered oxides that contain foreign cations and anions other than protons. Modifier cations (most of which have a charge of +1 or +2) can enter an oxide lattice by disrupting metal–oxygen bonds in the same way that protons do. Incorporation of these cations is accompanied by the creation of anionic sites within the lattice that have their charge compensated by the modifier cations. However, because these cations are much larger than protons, they create larger spaces between adjacent oxide layers. A classic example of such a layered oxide is $Na_2Ti_4O_9$ (Fig. 2.7) [29]. Cations from solution can sometimes replace the cations within the oxide, making the oxide a *cation exchanger* (see Chapter 10).

Layered oxides can also contain positively charged sites that make them *anion exchangers*. The classic oxide anion exchangers are the layered double hydroxide minerals [21, 22] (Fig. 2.7). The fundamental structure for these materials is a trioctahedral sheet, as exemplified by $Co(OH)_2$. Under certain circumstances, it is possible to replace a fraction of the +2 cations in the sheet by +3 cations without inducing a phase transformation. For example, some of the Co^{2+} in the sheet can be oxidized to Co^{3+} or even Co^{4+}. When this happens, the sheet assumes an excess positive charge, which can occur without restructuring the sheet. This charge is neutralized by the intercalation of anionic species between the sheets. Pyroaurite [30], a classic layered double-hydroxide mineral, has the

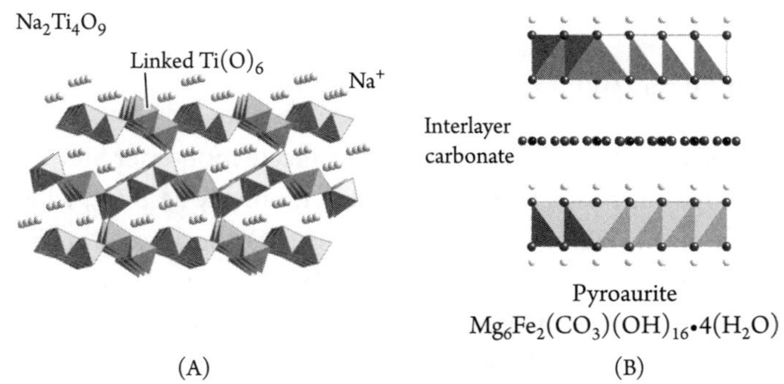

Figure 2.7 Layered cation and anion exchangers. (A) $Na_2Ti_4O_9$. (B) Pyroaurite. The species between the layers in the sodium titanates are Na^+ ions whereas the species shown between the layers in pyroaurite are anionic carbonate groups.

composition $Mg(II)_6Fe(III)_2(CO_3)(OH)_{16}(H_2O)_4$. The Mg(II) and Fe(III) are distributed randomly among the trioctahedral sheets and the excess positive charge is compensated by carbonate ions in the interlayer, which can be exchanged for other anions in solution. Layered oxides that function as cation or anion exchangers are of critical importance for the water treatment (see Chapter 10), environmental remediation, and electrochemical energy-storage technologies (Chapter 11).

2.5 OXIDES CONTAINING TETRAHEDRAL CATIONS

Organic and biological chemistry are both dominated by the element carbon. Carbon most often assumes a tetrahedral coordination geometry. Silicon, the element immediately below carbon in the Periodic Table, also prefers a tetrahedral bonding geometry. However, instead of bonding to hydrogen to form hydrocarbons, silicon prefers to bond to oxygen to form oxides. The oxides of silicon and other +4 tetrahedral cations form a wide range of structures that are analogous to (and almost as diverse as) the hydrocarbons. The major differences between Si^{4+}-based tetrahedral oxides and hydrocarbons are as follows. First, carbon–carbon bonds that interconnect cations are replaced by bridging oxygens such as the neutral Si—O—Si linkage. Second, the neutral hydrogen atoms in hydrocarbons are replaced with nonbridging oxygens with a charge of −1. The charge on these oxygen anions must be charge-compensated by other cations, which are typically protons, alkali metal cations, and alkaline earth cations. In all silicate polymers the ratio of bridging to nonbridging oxygens can be inferred directly based on the charge on the silicate unit. For example, $Si_3O_9^{6-}$ contains three bridging and six nonbridging oxygens consistent with a ring structure containing three tetrahedra. Next, we describe the tetrahedral oxides using the progression found in most organic chemistry textbooks in moving from monomers to dimers, chains, rings, sheets, and extended 3D networks.

2.5.1 Monomers, Dimers, and Longer Chains

The silicate analogue to methane is the orthosilicate monomer SiO_4^{4-}. *Orthosilicate* minerals are ionic salts that have the stoichiometry M_2SiO_4, where M is an alkaline earth or any other +2 cation [31–33]. Orthosilicates such as olivine ($(Mg, Fe)_2SiO_4$) form at high temperatures in lava flows, in deep rocks at the base of Earth's crust, and in the mantle. In *sorosilicate* minerals [32], silicate tetrahedra are linked together by a single bridging oxygen to form $Si_2O_7^{6-}$ dimers in direct analogy to ethane. Once again, the dimers are commonly charge-compensated by alkaline earth cations. Adding more monomers can create silicate chains called *pyroxenes* [32, 34] that are analogous to alkanes [35] (Fig. 2.8).

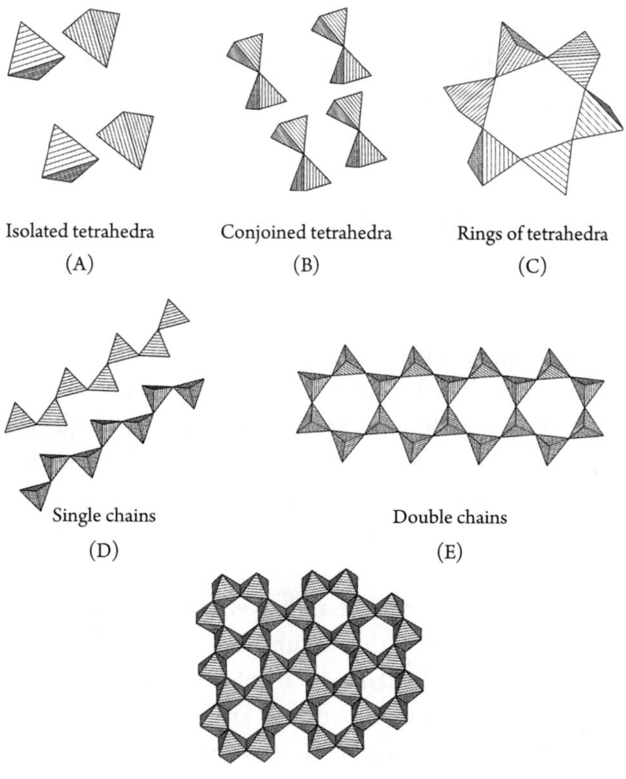

Figure 2.8 (A–F) Polymerization motifs for silicate tetrahedra, including monomers (A), dimers (B), isolated rings (C), linear chains (D), double chains (E), and extended two-dimensional sheets (F).

2.5.2 Rings, Double Chains, and Sheets

The two ends of a silicate chain can react with each other to form silicate rings in the *cyclosilicate* minerals [33]. The most common rings contain three, four, and six silicate tetrahedra with the stoichiometries $(Si_3O_9)^{6-}$, $(Si_4O_{12})^{8-}$, and $(Si_6O_{18})^{12-}$, respectively (Fig. 2.8). The latter ring is analogous to cyclohexane. The most common ring-silicate mineral is the gem mineral beryl [32] $(Be_3Al_6Si_6O_{18})$. In addition to isolated rings, structures containing interconnected rings are common. For example, adjacent pyroxene chains can react with each other to form double chain structures called *amphiboles* [32, 34, 36] (Fig. 2.8). The amphiboles contain infinite chains comprised of interconnected six-member rings. The six-member rings can be linked together into two-dimensional sheets called *tetrahedral sheets* (Fig. 2.8) and have a structure that is the silicate equivalent of graphite. These sheets are found in layered minerals called *phyllosilicates,* which contain the semi-infinite anion $(Si_4O_{10}{}^{4-})_n$. These layers are a major component in clays and micas [37].

2.5.3 Three-Dimensional Networks: Tectosilicates, Tectoaluminosilicates, and Zeolites

Extending polymerization from sheets into the third dimension yields silicates that contain fewer than one nonbridging oxygen per silicon. These silicates are called *tectosilicates*. Pure silica (SiO_2) exhibits a wide range of crystalline tectosilicate phases [32, 38], including the normal phases quartz, tridymite, cristobalite; the high-pressure phases keatite, coesite, and stishovite; and a few unusual phases including melanophlogite [39] and Silica W [40]. A few of these phases are depicted in Figure 2.9. The structure of quartz consists of interconnected six-member rings that form a network similar to that found in ice (see Chapter 3). Although dense for a silicate, quartz exhibits an oxygen packing density ($2.65 \cdot 10^{22}$ O/cm^3) that is substantially lower than a close-packed oxygen anion lattice ($6.5 \cdot 10^{22}$ O/cm^3). This lower density reflects the fact that the voids in the center of six-member rings are larger than the octahedral and tetrahedral voids in a close-packed lattice. Melanophlogite contains a mixture of five- and six-member rings arranged in a complex structure that is even less dense ($2 \cdot 10^{22}$ O-atoms/cm^3). Stishovite is an unusual high-pressure polymorph that is the only known silicate phase containing octahedral rather than tetrahedral silicon centers. As such, it has the highest density of the silica polymorphs ($4.2 \cdot 10^{22}$ O-atoms/cm^3). The least-dense silica phase ($1.97 \cdot 10^{22}$ O-atoms/cm^3) is a water-sensitive material called *Silica W* that contains chains

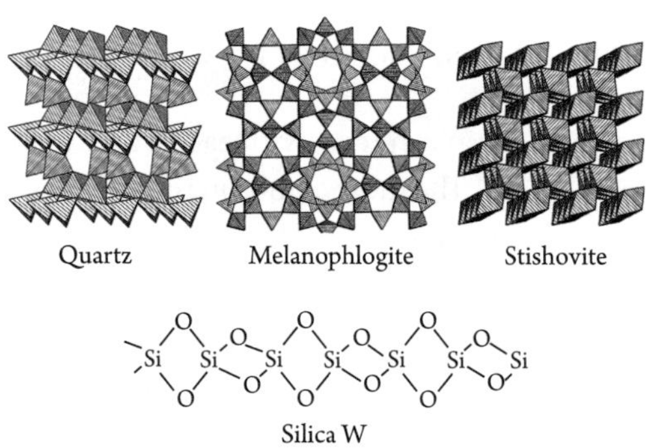

Figure 2.9 The structures of selected polymorphs of silica (SiO_2). Polymorphs such as quartz that contain tetrahedral Si^{4+} in larger unstrained rings are by far the most common and stable structures.

of edge-shared silicate tetrahedra. This phase is highly strained and has been used to study stress-activated chemical reactions (see Chapter 16).

Some of the most common minerals on Earth are *tectoaluminosilicates* in which Al(III) substitutes for Si(IV) in the tetrahedral tectosilicate framework [41, 42]. The feldspars found in granitic rocks, including albite ($NaAlSi_3O_8$) and anorthite ($Ca_2Al_2Si_2O_8$), fall into this mineral class. The first thing to notice regarding these minerals is that they contain modifier cations. However, these cations are not present to neutralize nonbridging oxygen sites, which are absent in the densest tectoaluminosilicates. Instead, the modifiers neutralize the net negative charge on bridging oxygens resulting from the replacement of Si(IV) with Al(III). Because tetrahedral Al(III) has a formal charge of $+0.75$ (instead of $+1.0$ for Si(IV)), each Al$-$O$-$Si bond has a formal charge of -0.25 (instead of 0.0 for Si$-$O$-$Si bonds). The presence of an anionic network charge means the tectoaluminosilicates can function as cation exchangers.

Cation exchange is just one of many useful properties associated with the presence of Al$-$O$-$Si bonds. As discussed in later chapters, such sites also exhibit unique acid–base reactivity and catalytic properties. The subset of the tectoaluminosilicates within which these properties have been exploited most heavily are the *zeolites* (Fig. 2.10) [43]. Zeolites represent rigid frameworks with interconnected cavities and channels defined by unique distributions of silicate ring topologies and architectures. Ring sizes define the openings that are available for the passage of ions, water, other solvents, and a wide range of chemical reactants.

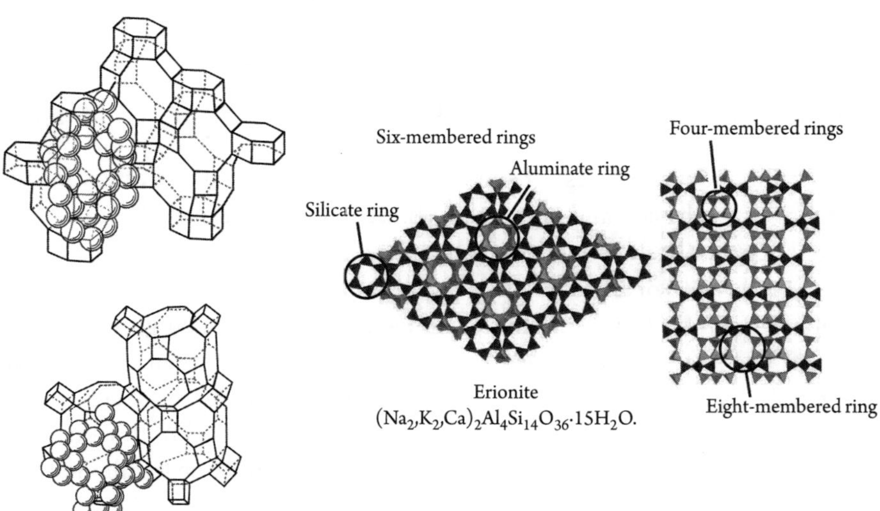

Figure 2.10 Selected zeolite structures showing interconnected ring architectures. In some images, the ionic radii of the structural oxygens are shown as spheres to emphasize the pores.

Common rings contain 4, 6, 8, 10, and 12 tetrahedra, with corresponding openings within these "ideal" planar rings of 0.12 nm, 0.28 nm, 0.45 nm, 0.63 nm, and 0.80 nm, respectively [43]. The sizes, geometries, and pores defined by these rings can dictate specific reaction pathways on the basis of both molecular size and reactant orientations within zeolite cages (see Chapters 10 and 14). When the 3D networks are placed in water, water can penetrate many of their nooks and crannies. As in the two-dimensional layered structures, this means that essentially all atoms within the 3D network become "surface" atoms. However, in contrast to the layered oxides, the 3D networks do not have the ability to swell. The lack of swelling means two things. One, steric effects that influence ion transport and exchange phenomena become important. As is shown in Chapter 10, such steric effects often involve the sizes of solvated rather than bare metal cations. And two, to a large extent, electrolytes are excluded from the zeolite interior (see Chapter 10). As a result of these properties, zeolites and other open oxide networks are among the most important materials used for water treatment, catalysis, and solid electrolytes.

2.6 OXIDES CONTAINING BOTH OCTAHEDRAL AND TETRAHEDRAL CATIONS

Many of the oxide minerals in the Earth's crust and mantle have at least two metals in the structure. These minerals often contain metals that are present in both octahedral and tetrahedral coordination geometries. Of these, the aluminum cation is relatively unique in that it is just the right size to adopt both coordination geometries within the same mineral structure.

2.6.1 Layered Structures: Clays and Micas

The largest and richest class of silicate materials is the phyllosilicate minerals that include both clays and micas [44]. These minerals are made by linking infinite sheets of silicate tetrahedra (called *tetrahedral sheets*) to either trioctahedral (as in brucite or $Mg(OH)_2$) or dioctahedral (as in gibbsite or $Al(OH)_3$) sheets (Fig. 2.11). In making these connections, the hydroxyl groups on the surfaces of the octahedral sheets are replaced by the nonbridging oxygens in the tetrahedral sheets to create oxygens that bridge between the sheets. In the trioctahedral case, the bridging oxygens are typically bonded to three octahedral M^{2+} cations (each with a formal charge of $+1/3$) and one tetrahedral Si^{4+} (with a formal charge of $+1$), resulting in neutral oxygen sites. Similarly, the linkages involving dioctahedral sheets are also neutral; each bridging oxygen is bonded to two M^{3+} cations (each with a formal charge of $+0.5$) and one tetrahedral Si^{4+}.

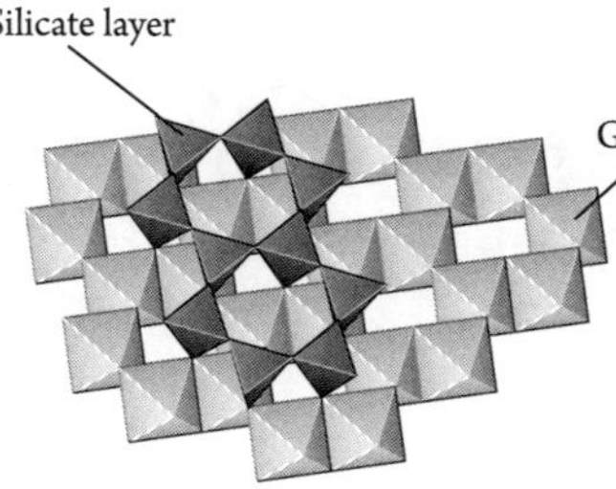

Figure 2.11 A top view of the alignment of the tetrahedral and octahedral sheets in clay minerals, as exemplified by kaolinite ($Al_2Si_2(OH)_5$).

It is important to note that it is rare for either the octahedral or tetrahedral sheets to contain only one type of cation. The substitution of foreign cations into each sheet is common in nature and often introduces charges into the layers, as in layered double hydroxides. The substitution of tetrahedral Al(III) for Si(IV) in the tetrahedral sheet makes this layer anionic, and requires the introduction of modifier cations to achieve charge neutrality. Similarly, the substitution of octahedral Mg(II) for Al(III) in a dioctahedral sheet creates anionic sheets that require the presence of charge-compensating cations. As in the layered double hydroxides, substitution of octahedral Al(III) for Mg(II) in a trioctahedral sheet requires the presence of anions in the interlayer regions to compensate the charge and achieve electrical neutrality. For clays, water can enter the interlamellar regions, often as waters of solvation for the charge-compensating cations. Because individual layers are not attached covalently to one another, water intrusion can also cause the clays to swell reversibly or even exfoliate. Such swelling can generate tremendous pressures of importance in the weathering of rocks. Swelling also creates aqueous environments within the clays that are substantially different from those found within the rigid frameworks of zeolites. Even when clays and zeolites contain similar reactive sites, their patterns of reactivity can be substantially different as a result of steric effects (see Chapter 10).

The complexity of clays is not only associated with their ability to accommodate foreign network-forming cations, modifier cations, and anions, but also by how the layers within a given clay are stacked (Figs. 2.12 and 2.13). Stacking sequences are dependent on the clay composition [44]. The simplest clays, called *T-O clays* or *tetrahedral–octahedral clays*, are based on repeat layers containing a single tetrahedral and octahedral sheet. The simplest T-O clays are kaolinite ($Al_2Si_2O_5(OH)_4$) and serpentine ($Mg_3Si_2O_5(OH)_4$) (Fig. 2.12). If both faces of the octahedral sheet are capped with tetrahedral sheets, the result is the T-O-T clays, such as pyrophyllite ($Al_2Si_4O_{10}(OH)_2$) (Fig. 2.13). Illite ($KAl_2(AlSi_{13}O_{10}(OH)_2)$ is a common T-O-T clay containing Al(III) doping within the tetrahedral sheets. The resulting anionic sheets are charge-compensated by K^+ incorporation between the sheets.

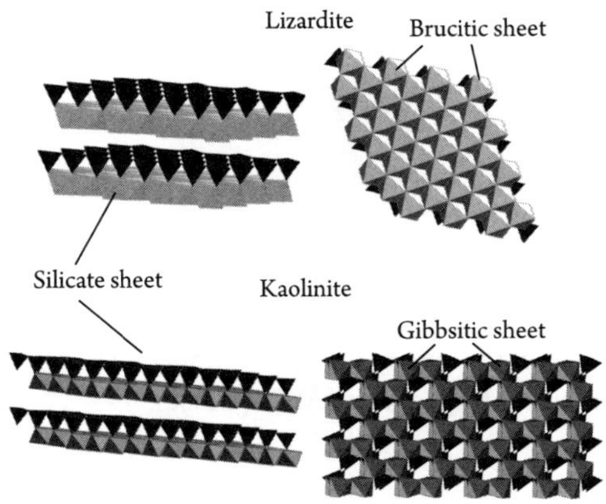

Figure 2.12 Side views of layer stacking in the simplest 1:1 clay structures, as exemplified by lizardite (containing trioctahedral $Mg(OH)_2$ sheets) and kaolinite (containing dioctahedral $Al(OH)_3$ sheets). Compare these with the building blocks shown in Figure 2.6. These structures are sometimes referred to as T-O clays because they have a single tetrahedral and octahedral layer in the basic structure.

As a final example, chlorite contains T-O-T sheets that are separated by additional $Mg(OH)_2$ octahedral sheets (Fig. 2.13).

2.6.2 Three-Dimensional Networks

Both tetrahedral and octahedral cations can be present in 3D networks, some of which resemble zeolites. However, in contrast to standard zeolites, in which the predominant active sites are Al—O—Si bonds, the diversity of oxygen types, local charge distributions, ring topologies, and chemical properties found in octahedrally substituted zeolites is almost staggering. Mixed zeolitic networks such as those found in the family of oxide materials called *NASICON* [45] (with compositions such as $Na_{1+x}Zr_2Si_xP_{3-x}O_{12}$) (Fig. 2.14) exhibit some of the highest ionic conductivities known within a solid phase (see Chapter 10). In part, these high conductivities are associated with the fact that the anionic charge in these materials is highly diffuse. All oxygen anions in *NASICON* bridge between one octahedral and one tetrahedral cation, including Zr—O—Si and Zr—O—P bonds with formal charges of $-1/3$ and only $-1/12$, respectively. Silicotitanates (Fig. 2.14) represent another important family of oxides in which titanium substitutes for silicon within a zeolitic framework [46]. Here, Ti(IV) can assume coordination numbers of 4, 5, and/or 6, again leading to unique local charge and ring distributions. These materials can exhibit exceptional ion-exchange properties (see Chapter 10).

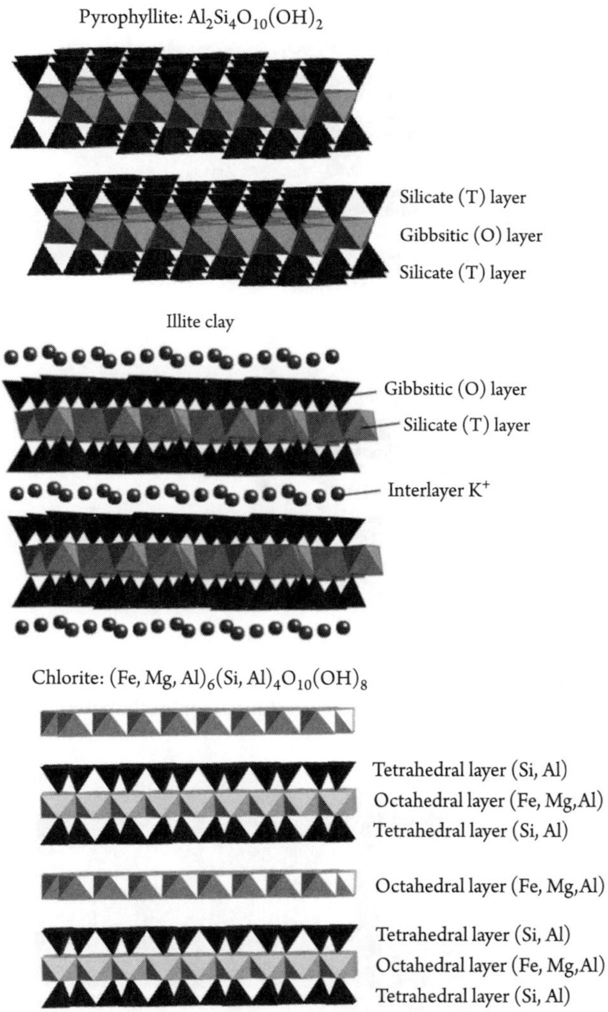

Figure 2.13 Side views of layer stacking in representative 2:1 clays, including neutral pyrophyllite sheets, charged sheets in illite that are charge-compensated by interlayer cations, and the charged layers in chlorite, which are actually charge-compensated by entire octahedral sheets. These are sometimes called T-O-T clays because they have the octahedral sheet sandwiched between tetrahedral silicate layers.

2.7 ENGINEERED OXIDE NANOSTRUCTURES

One of the strategies used to create zeolites and other open networks involves the hydrothermal synthesis or solution-based polymerization of the oxide around molecular templates (see Chapter 14). Recently, researchers have been exploiting templates that are larger than single small molecules to create larger and more complex structures. These strategies are based on an understanding of both

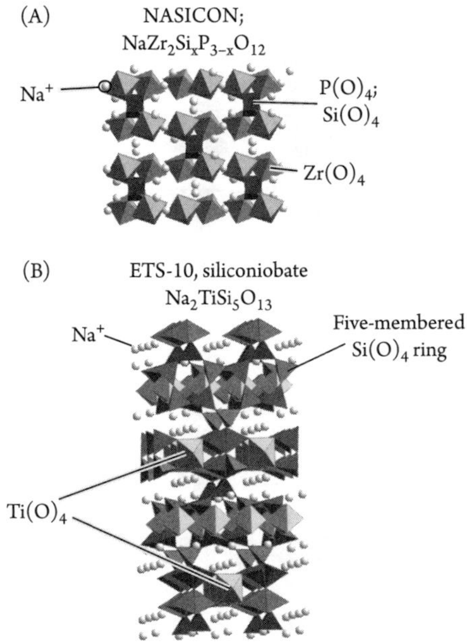

Figure 2.14 (A, B) Three-dimensional oxide networks containing both tetrahedral and octahedral cations, as exemplified by the exceptional *Na+ superionic conductor* (*NASICON*) (A) and a highly selective silicotitanate ion exchanger (B).

Figure 2.15 (A–I) Scanning electron micrographs of a range of complex hierarchical structures that can be made to form from mesoporous silica particles. They are not formed by biota, yet such assembly replicates many of the attributes of nucleation and growth processes important in biomineralization.

fossilization and biomineralization processes during which oxides replace and replicate complex biological structures [47–49]. It is now possible to control oxide structures at length scales ranging from atomic dimensions up to tens of microns (Fig. 2.15). In fact, robust oxide materials can be created that are hierarchical in nature, and structure and porosity can be engineered across multiple length scales. Using these techniques, materials scientists can now combine different oxides at nanometer-length scales at which each oxide component performs its own unique function. These technologies represent the inorganic equivalent of tissue engineering (see Chapter 9), when structures resembling exquisite biological architectures are made using abiotic methods to control the nucleation, growth, and ultimate structure of the oxide.

References

1. Henrich, V. E.; Cox, P. A. *The Surface Science of Metal Oxides*; Cambridge University Press, Cambridge, UK, 1994.
2. Cotton, F. A.; Wilkinson, G. *Advanced Inorganic Chemistry, 6th ed.*; Wiley: New York, 1999.
3. Dronskoski, R. Theoretical Increments and Indices for Reactivity, Acidity, and Basicity with Solid-State Materials. *J. Am. Chem. Soc.* **1992**, *114*, 7230–7244.
4. Cramer, C. J. *Essentials of Computational Chemistry: Theories and Models, 2nd ed.*; Wiley: Chichester, UK, 2004.
5. Pauling, L. The Principles Determining the Structure of Complex Ionic Crystals. *J. Am. Chem. Soc.* **1928**, *51*, 1010–1026.
6. Pauling, L. *The Nature of the Chemical Bond, 3rd ed.*; Cornell University Press: Cornell, NY, 1960.
7. Brown, I. D. The Bond–Valence Method: An Empirical Approach to Chemical Structure and Bonding. In *Structure and Bonding in Crystals*; O'Keeffe, M. O.; Navrotsky, A., Eds.; Academic Press: New York, 1981; pp 1–29.
8. Brese, N. E.; O'Keeffe, M. O. Bond–Valence Parameters for Solids. *Acta Cryst.* **1991**, *B47*, 192–197.
9. Shannon, R. D.; Chevanas, J.; Joubert, J. C. Bond Strength Considerations Applied to Cation Coordination in Normal and High-Pressure Oxides. *J. Solid State Chem.* **1975**, *12*, 16–30.
10. Bickmore, B. R.; Tadanier, C. J.; Rosso, K. M.; Monn, W. D.; and Eggett, D. L. Bond–Valence Methods for pK_a Prediction: Critical Reanalysis and a New Approach. *Geochim. Cosmochim. Acta* **2004**, *68*, 2025–2042.
11. Bickmore, B. R.; Rosso, K. M.; Tadanier, C. J.; Bylaska, E. J.; and Doud, D. Bond–Valence Methods for pK_a Prediction: II. Bond–Valence Electrostatic Molecular Geometry and Solvation Effects. *Geochim. Cosmochim. Acta* **2006**, *70*, 4057–4071.
12. Brown, I. D.; Shannon, R. D. Empirical Bond-Strength–Bond-Length Curves for Oxides. *Acta Cryst.* **1973**, *A29*, 266–282.
13. Cotton, F. A.; Wilkinson, G. *Advanced Inorganic Chemistry, 5th ed.*; Wiley: New York, 1988; pp 3–10.

14. Enjalbert, R.; Galy, J. A Refinement of the Structure of V_2O_5. *Acta Cryst.* **1986**, *C42*, 1467–1469.
15. Post, J. E. Manganese Oxide Minerals: Crystal Structures and Economic and Environmental Significance. *Proc. Natl. Acad. Sci. India* **1999**, *96*, 3447–3454.
16. Linden, D.; Reddy, T. B. *Handbook of Batteries, 3rd ed.*; McGraw-Hill: New York, 2002; pp 8.12–8.13.
17. Hill, R. J. Hydrogen Atoms in Boehmite: A Single Crystal X-ray Diffraction and Molecular Orbital Study. *Clays Clay Miner.* **1981**, *29*, 435–445.
18. Walter, T. H.; Oldfield, E. Magic Angle Spinning Oxygen-17 NMR of Aluminum Oxides and Hydroxides. *J. Phys. Chem.* **1989**, *93*, 6744–6751.
19. Hoppe, W. Uber die Kristallstruktur von α-AlOOH (diaspore) und α-FeOOH (nadeleisenerz). *Zeit. Kristallogr.* **1940**, *103*, 73–89.
20. Cornell, R. M.; Schwertmann, U. U. *The Iron Oxides*; Wiley-VCH: Weinheim, Germany, 2003; p. 19–27.
21. Wang, Q.; O'Hare, D. Recent Advances in the Synthesis and Application of Layered Double Hydroxide (LDH) Nanosheets. *Chem. Rev.* **2012**, *112*, 4124–4155.
22. Rives, V., Ed. *Layered Double Hydroxides: Present and Future*; Nova: New York, 2001.
23. Giovanoli, R.; Stadelmann, W.; Feitknecht, W. Crystalline Chromium(III) Hydroxide. *Helv. Chem. Acta*, **1973**, *56*, 839–847.
24. Spiccia, L.; Marty, W. The Fate of Active Chromium Hydroxide, $Cr(OH)_3{\cdot}3H_2O$, in Aqueous Suspension: Study of the Chemical Changes Involved in Its Aging. *Inorg. Chem.* **1986**, *25*, 266–271.
25. Spiccia, L.; Marty, W.; Giovanoli, R. Morphological and Thermal Investigations of the "Active" Dimer Hydroxide of Chromium(III). *Helv. Chem. Acta* **1987**, *70*, 1737–1744.
26. Spiccia, L.; Stoeckli-Evans, H.; Marty, W.; and Giovanoli, R. A New Active Chromium(III) Hydroxide: $Cr_2(\mu\text{-}OH)_2(OH)_4{\cdot}2H_2O$: Characterization and Use in the Preparation of Salts of the $(H_2O)_4Cr(\mu\text{-}OH)_2Cr(OH_2)_4{}^{4+}$ Ion: Crystal Structure of $[(H_2O)Cr(\mu\text{-}OH)_2Cr(OH_2)_4][(H_3C)_3C_6H_2SO_3]_4{\cdot}4H_2O$. *Inorg. Chem.* **1987**, *26*, 474–482.
27. Spiccia, L. Solubility of Chromium(III) Hydroxides. *Inorg. Chem.* **1988**, *27*, 432–444.
28. Rustad, J. R.; Casey, W. H. A Molecular Dynamics Investigation of Hydrolytic Polymerization in a Metal-Hydroxide Gel. *J. Phys. Chem. B* **2006**, *110*, 7107–7112.
29. Tournoux, M.; Marchand, R.; Brohan, L. Layered $K_2Ti_4O_9$ and the Open Metastable TiO_2(B) Structure. *Prog. Solid State Chem.* **1986**, *17*, 33–52.
30. Ingram, L.; Taylor, H. F. W. The Crystal Structures of Sjoegrenite and Pyroaurite. *Mineral. Mag.* **1967**, *36*, 465–469.
31. Deer, W. A.; Howie, R. A.; Zussman, J. Orthosilicates, 2nd ed.; In *Rock Forming Minerals*; Deer, W. A.; Howie, R. A.; Zussman, J., Eds.; Longmans: London, 1982, Vol. 1A. 919 pages.
32. Mason, B.; Berry, L. G. *Elements of Mineralogy*; W. H. Freemen: San Francisco, CA, 1968; Chapter 15, p. 396–509.
33. Deer, W. A.; Howie, R. A.; Zussman, J. Ortho- and Ring Silicates, 1st ed.; In *Rock Forming Minerals*; Deer, W. A.; Howie, R. A.; Zussman, J., Eds.; Longmans: London, 1962, Vol. 1, 333 pages.

34. Deer, W. A.; Howie, R. A.; Zussman, J. Chain Silicates. In *Rock Forming Minerals*; Deer, W. A.; Howie, R. A.; Zussman, J.; Eds.; Longmans: London, 1965, Vol. 2, 379 pages.
35. Ray, N. H. *Inorganic Polymers*; Academic Press: London, 1978; pp 110–113.
36. Deer, W. A.; Howie, R. A.; Zussman, J. Double-Chain Silicates. In *Rock Forming Minerals*; Deer, W. A.; Howie, R. A.; Zussman, J., Eds.; Longmans: London, 1978, Vol. 2B, 794 pages.
37. Fleet, M. E. Sheet Silicates: Micas, 2nd ed. In *Rock Forming Minerals*; Deer, W. A.; Howie, R. A.; Zussman, J., Eds.; Longmans: London, 2003, Vol. 3A, 780 pages.
38. Iler, R. K. *The Chemistry of Silica*; Wiley: New York, 1979; 3–115.
39. Gies, H. Studies on Clathrasils: III. Crystal Structure of Melanophlogite, a Natural Clathrate Compound of Silica. *Zeit. Kristallogr.* **1983**, *164*, 247–257.
40. Greenwood, N. N.; Earnshaw, A. Chemistry of the Elements; Pergamon Press: Oxford, 1984, pp 393–399.
41. Zoltai, T.; Stout, J. H. *Mineralogy: Concepts and Principles*; Macmillan: New York, 1984.
42. Deer, W. A.; Howie, R. A.; Zussman, J. Framework Silicates: Silica Minerals, Feldspathoids, and the Zeolites, 2nd ed. In *Rock Forming Minerals*; Deer, W. A.; Howie, R. A.; Zussman, J., Eds.; Longmans: London, 2004, Vol. 4B, 982 pages
43. Barrer, R. M. *Hydrothermal Chemistry of Zeolites*; Academic Press: London, 1982; p. 1–42.
44. Bailey, S. W. Classification and Structures of the Micas. In *Micas*; Bailey, S. W., Ed. In *Reviews in Mineralogy*; Ribbe, P. H., Ed.; Mineralogical Society of America: Washington, DC, 1984, Vol.13, p. 1–12.
45. Goodenough, J. B.; Hong, H. Y.- P.; Kafalas, J. A. Fast Na^{+}-Ion Conduction in Skeleton Structures. *Mater. Res. Bull.* **1976**, *11*, 202–220.
46. Anderson, M. W.; Terasaki, O.; Ohsuna, T.; Pilippou, A.; MacKay, S. P.; Ferreira, A.; Rocha, J.; and Lidin, S. Structure of the Microporous Silicotitanate ETS-10. *Nature* **1994**, *367*, 347–351.
47. Mann, S. Biomineralization: Principles and Concepts in Bioinorganic Materials Chemistry. In *Oxford Chemistry Masters*; Compton, R. G.; Davies, S. G.; Evans, J., Eds.; Oxford University Press: New York, 2001 p. 1–188.
48. Mann, S.; Webb J.; Williams, R. J. P., Eds. *Biomineralization: Chemical and Biological Perspectives*; VCH: New York, 1989.
49. Lowenstam, H. A.; Weiner, S. *On Biomineralization*; Oxford University Press: New York, 1989.

PART TWO

Fundamental Oxide Reactions in Aqueous Solutions

3

The Structure and Properties of Water

Water is one of the most complex fluids on Earth. Even after intense study, there are many aspects regarding the structure, properties, and chemistry of water that are not well understood. In this chapter, we highlight the attributes of water [1, 2] that dictate many of the reactions that take place between water and oxides. We start with a single water molecule and progress to water clusters, then finally to extended liquid and solid phases. This chapter provides a baseline for evaluating what happens when water encounters simple ions, soluble oxide complexes called *hydrolysis products,* and extended oxide phases. The primary phenomenon highlighted in this chapter is *hydrogen bonding*. Hydrogen bonding dominates the structure and properties of water and influences many water–oxide interactions.

A single water molecule has eight valence electrons around a central oxygen anion [3]. These electrons are contained in four sp^3-hybridized molecular orbitals arranged as lobes that extend from the oxygen in a tetrahedral geometry. Each orbital is occupied by two electrons (Fig. 3.1 [4]). Two of the lobes are bonded to protons; the other two lobes are referred to as *lone pairs of electrons.* The H–O–H bond angle of 104.5° is close to the tetrahedral angle of 109.5°. The O–H bond length in a single water molecule is 0.96 Å. It is important to recognize that this bond length is really a measure of the electron density associated with the oxygen lone pair bonded to the proton. This is because a proton is so incredibly small (with an ionic radius of only $1.3 \cdot 10^{-5}$ Å) that it makes no

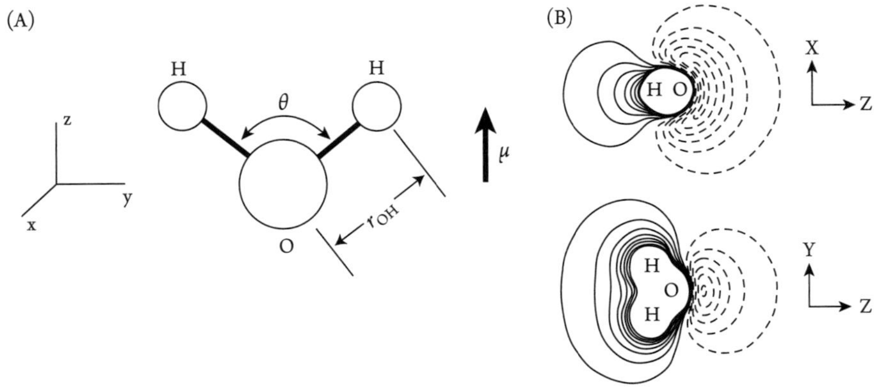

Figure 3.1 (A) A geometric depiction of the water molecule showing a bond length of 0.96 Å and a bond angle of 104.5°. (B) Contours of electrostatic potentials around the water molecule, where dashed lines indicate negative potential and solid lines indicate positive potential [4].
SOURCE: Thiel, P. A.; Madey, T. E. The Interaction of Water with Solid Surfaces: Fundamental Aspects. *Surf. Sci. Rep.* **1987**, 7, 211–385; 217, Table 1; 221, Fig. 4. Reprinted with permission.

contribution to the net bond length. The entire water molecule has a hard sphere diameter of 2.9 Å, which is fairly typical for an oxygen anion. This means the unoccupied lone pairs are distended relative to the protonated lone pairs, extending out to roughly 1.9 Å. The unequal distribution of charges introduces a dipole within the water molecule that facilitates electrostatic interactions with other molecules. Water in a vacuum has a large dipole moment [5] of $6.17 \cdot 10^{-30}$C/m (1.85 Debyes), indicating there is a large negative potential in the region of the lone pairs, and a positive potential corresponding to the positions of the protons. Such a large dipole moment and a small size means water molecules orient in the presence of charges, such as other water molecules, ions, or surfaces. The presence of both lone pairs of electrons and protons on the water molecule means this molecule can (1) exhibit significant activity as both an acid and a base, and (2) participate in extensive hydrogen-bonding interactions that control the structure, transport, and mechanical properties of water.

The water molecule is unusual in that it can function as both an acid and a base [6]. It can donate and accept both protons and electrons. Thus, its acid–base properties are described by both Brønsted and Lewis concepts, respectively. The lone pairs of electrons on the oxygen are Lewis bases that are capable of donating electrons to acidic metal cations. These lone pairs can displace other metal–ligand bonds via the nucleophilic attack mechanisms described in Chapter 4 (i.e., attack by a basic electron donor). If a lone pair of electrons in water coordinates to a proton, the result is the formation of the hydronium ion H_3O^+, whereas if water donates a proton, the result is the formation of the hydroxide ion OH^-. Water molecules can undergo proton transfer reactions with each other. This autodissociation reaction is described by the equilibrium expression [7]:

$$2H_2O \rightleftarrows H_3O^+ + OH^-, K_w = [H_3O^+][OH^-] = 10^{-14} \quad \textbf{(3.1)}$$

In pure water, $[H_3O^+] = [OH^-] = 10^{-7}$M (where M represents the molar concentration in moles per liter), resulting in a solution pH = –log$[H_3O^+] = 7$. In other words, the water molecule is both a weak acid and a weak base. Such molecules are called *ampholytes*. Although pure water has a pH of 7, protons and hydroxide ions can be added to water to create solutions that are highly acidic (with a pH less than 0), highly basic (with a pH greater than 14), or anywhere in between. The most common pH values range from 0 to 14, or 1 M H_3O^+ to 1 M OH^-. In later chapters, we see the acid–base properties of individual water molecules can be tuned through the entire pH range available in aqueous environments via interactions with other ions or materials. This tuning is what gives rise to the rich and complex acid–base properties of the oxides.

As indicated by Eq. 3.1, the number of protons and hydroxide ions created by proton transfer reactions in pure water is limited to one out of every 550 million water molecules (10^{-7}M relative to the 55 M present in liquid water). However, essentially all the protons and lone pairs of electrons in water are interacting constantly with acid–base sites on other water molecules via weak interactions called *hydrogen bonds*. In a hydrogen bond, one proton interacts with lone pairs of electrons on two distinct oxygen anions, thus forming a linear bridge between the two water molecules. The proton in the hydrogen bond is not equidistant between the two oxygen anions. Neutron diffraction studies of ice [8] show that the proton is within 1.01 Å of the parent or donor oxygen (close to the value for the O–H bond length in a single water molecule), whereas the O–H distance in the hydrogen bond to the acceptor oxygen is 1.75 Å (which is close to the length of an unoccupied lone pair). The strength of the hydrogen bond in ice is around 5 kcal/mole (21 kJ/mole), with values measured in the liquid and in other crystalline phases ranging from 15 to 25 kJ/mole [9]. These bonds are relatively weak compared with thermal energy at room temperature (2.5 kJ/mole) and are thus readily disrupted. However, they are sufficiently strong that they can create rings, clusters, and extended solid and liquid phases. Next, we discuss the nature of hydrogen-bonded architectures for small clusters, followed by what can be inferred for the structure of liquid water based on known crystalline compounds.

Theoretical studies [10,11] and limited experiments have been conducted to determine what happens when a single water molecule encounters other water

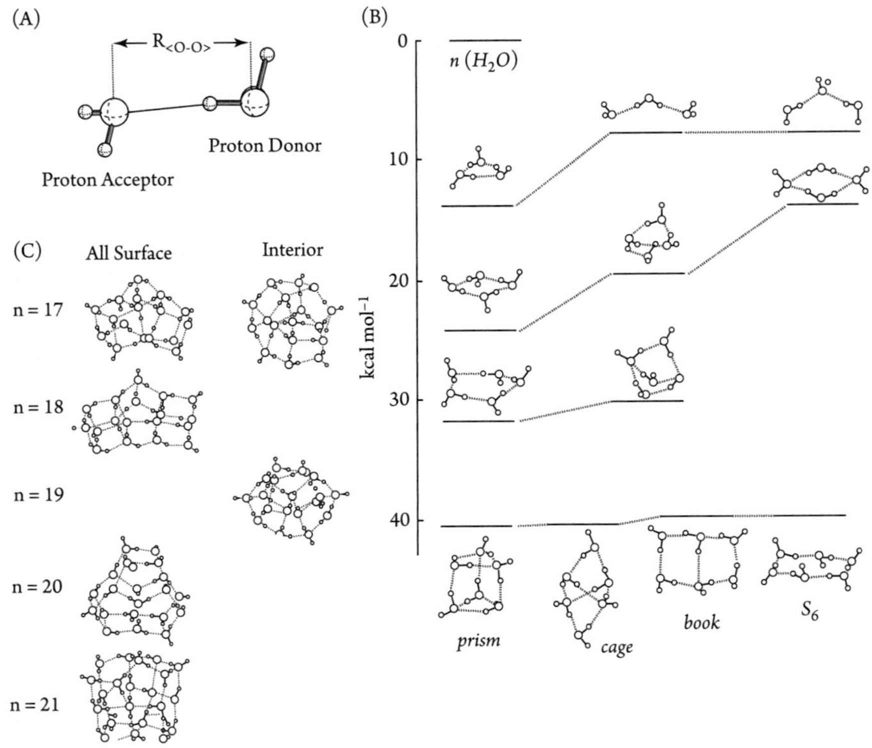

Figure 3.2 (A) Geometry calculated for the water dimer [10] (SOURCE: Xantheas, S. S.; Dunning, T. H., Jr. Ab Initio Studies of Cyclic Water Clusters $(H_2O)n$, n = 1– 6: I. Optimal Structures and Vibrational Spectra. *J. Chem. Phys.* **1993**, *99*, 8774–8792; p. 8777, Fig. 3. Reprinted with permission.). (B) Calculated binding energies for small water clusters [12] (SOURCE: Xantheas, S. S. Cooperativity and Hydrogen Bonding Network in Water Clusters. *Chem. Phys.* **2000**, *258*, 225–231; p. 228, Fig. 2. Reprinted with permission.). (C) Minimum energy geometries calculated for water clusters containing 17 to 21 water molecules [11] (SOURCE: Lagutchenkov, A.; Fanourgakis, G. S.; Niedner-Schattenburg, G.; Xantheas, S. S. The Spectroscopic Signature of the "All-Surface" to "Internally-Solvated" Structural Transition in Water Clusters in the n = 17 – 21 Size Regime. *J. Chem. Phys.* **2005**, *122*, 194310–194319; 194310–3, Fig. 1. Reprinted with permission.). All molecules are either at the cluster surface (left) or the cluster contains a central water molecule that is totally solvated (right).

molecules in the gas phase to create clusters containing two, three, four, or more molecules (Fig. 3.2 [10–12]). A water dimer (n = 2) exhibits a geometry in which a proton on one molecule forms a linear hydrogen bond that interacts with an unoccupied lone-pair orbital on the other molecule. For n = 3, multiple geometries are possible, including a linear chain and two closed trimeric rings. In the most stable ring, all water molecules are equivalent, with each donating and accepting a hydrogen bond within the ring, leaving each with an additional nonbridging proton. The other trimer has one molecule that donates two hydrogen bonds to the ring (designated donor–donor or dd), one molecule with one donor and one acceptor (da), and one molecule with two acceptors (aa).

As more molecules are added, the number of possible structures for water clusters increases exponentially, and the energy differences between the possibilities decrease. However, general trends in water clusters include the following:

1. Small clusters always contain closed hydrogen-bonded rings of three, four, five, or six water molecules.
2. Larger clusters establish networks based on the interconnection of these rings. The interconnected rings resemble the oxide lattice structures found in crystalline clathrate hydrates (described later in this chapter), in oxides such as zeolites (see Chapter 14), in carbon-based fullerenes [13], and in many other systems built from atoms or molecules with a tetrahedral geometry.
3. Ring distributions are determined by arrangements that maximize the number of hydrogen bonds within ring structures. For example, the most stable $n = 6$ cluster is predicted to consist of a closed prism with four-membered rings on the sides and three-membered rings on the end caps rather than a single six-membered ring [12]. This is because the former ring only has three protons that do not participate in hydrogen bonds whereas the latter contains six. However, even for clusters that are this small ($n = 6$), the energy difference between the prism and the hexagonal ring is predicted to be less than 2 kcal/mole.

As more and more water molecules are added, the number of possible hydrogen bonding configurations and cluster topologies becomes almost astronomical. It is truly amazing how a molecule that is as small and simple as a water molecule can generate such complex structures. However, before considering the structure of liquid water, it is useful to examine ordered crystalline structures to learn more about the hydrogen bonds in water that dictate the structure and properties of this complex fluid.

Hydrogen bonds are the glue that holds together the 11 known crystalline phases of ice [14] as well as the clathrate hydrates [15] (Fig. 3.3 [15, 16]). Most of the ice phases consist of planar arrays of interconnected puckered rings containing six water molecules. These rings resemble the rings formed by interconnected silicate tetrahedra in minerals such as cristobalite (see Chapter 2) as well as the silicate glasses. Although hexagonal rings dominate the structure of ice, clathrate hydrates illustrate that crystallized water can also incorporate all the other ring topologies and networks observed in water clusters in the vapor phase. The clathrate hydrates also highlight how solutes modify the hydrogen-bonding networks in water, as well as how water networks adapt to fit the size and shape of almost any foreign solute. In classic clathrate hydrates, hydrophobic molecules such as methane are encapsulated inside hydrogen-bonded cages and rings. These clathrates are found in low-temperature/high-pressure environments such as the ocean depths and are thought to represent one of the largest reserves of hydrocarbons on Earth [17]. The catastrophic melting of these clathrates has been invoked to explain both huge temperature changes and methane releases that appear in the geologic record [18]. The clusters in clathrates, which exhibit the same interconnected ring architectures found in zeolites (see Chapter 2), are thought to represent the types of transitory clusters found in liquid water (Fig. 3.3). As a final note on water rings, as anyone who has examined a soccer ball knows, networks containing five- and six-membered rings exhibit curvature (as in the clathrate hydrates), in contrast to the planar sheets of six-membered rings found in ice. As discussed later in this book, the surfaces of oxides or any other material are expected to have a bias toward ring distributions for geometric reasons, in addition to satisfying local hydrogen-bonding requirements.

Water clusters, ice, and the clathrate hydrates all contain interconnected ring structures based on the tetrahedral arrangement of electron pairs around the

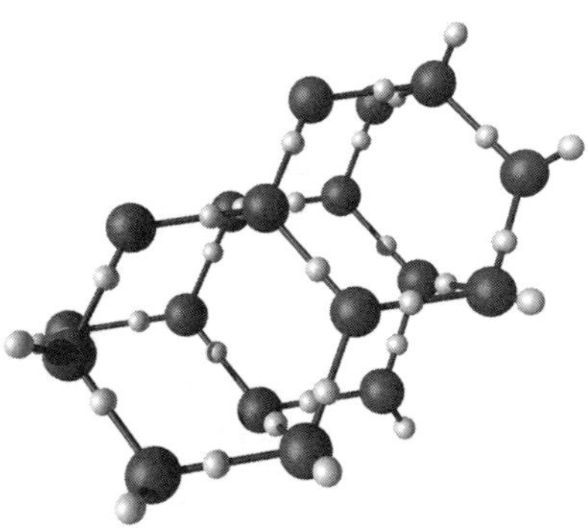

(A)

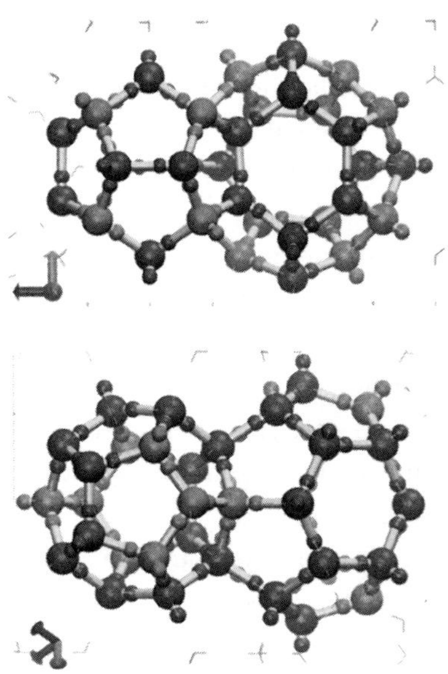

(B)

Figure 3.3 (A, B) Crystalline phases of water, including ice [16] (SOURCE: 3divan/Shutterstock.com) (A) and two CO_2 clathrate hydrate phases (sI and sII) (SOURCE: Siuda, P.; Sadlej, J. *Chem. Phys.*, **2014**, *433*, 31–41; Fig. 1; and Fig. 2. p. 32, Reprinted with permissions.) (B). A wide range of clathrate hydrates are known [15], in addition to the methane hydrate shown in Figure 1.9.

central oxygen. As indicated earlier, these ring structures can be exceedingly complex. As if that complexity was not enough, the fact that the hydrogen-bonding interactions are weak means that, in liquid water, all cluster and network structures are transitory. Because this book deals with reaction pathways and kinetics in water, it is important to have an appreciation of how quickly water molecules move and respond in their liquid environment. Next, we examine the kinetics and mechanisms for the motions of water, protons, and hydroxide ions. These mechanisms highlight the fact that the movements of each species involve the cooperative interactions and mobility of all other species.

Depicting what goes on in liquid water can be conceptualized by considering water molecules as 18th-century court dancers on a checkerboard dance floor. Each dancer can rotate on his or her individual square, exchange places with the dancer on an adjacent square, interact with adjacent dancers using their hands, or pass objects from dancer to dancer. For proficient dancers, each of these moves can be done several times per second. Now consider a 1-L beaker of water. The rotations, translations, and hand-off motions in the dance correspond to the rotations, translations, and proton-hopping phenomena that occur in liquid water. However, in our beaker, we are dealing with more than $3 \cdot 10^{25}$ dancers who all move in three dimensions rather than just two. Most important, we are dealing with processes that occur at such rapid rates they are hard to comprehend. Moves occur at speeds of up to 100 billion times a second. With dynamics like this, it is easy to see why researchers are still fighting over what the structure of water really means.

Although water is extremely dynamic, it has been possible to measure the rates at which vibrations, hydrogen bond formation, rotations, translations, and proton transfers occur in water to help evaluate the mechanisms by which dynamic phenomena in water occur (Fig. 3.4 [19]). The fastest motions in water are associated with the stretching and bending vibrations of O–H bonds [20]. Of these, the O–H stretching vibration is the most rapid, with the 3,400 cm^{-1} mode detected in the infrared spectrum of liquid water corresponding to $1 \cdot 10^{14}$ vibrations per second.

The next most rapid process involves the rate at which hydrogen bonds form reversibly and break at the contact point between two water molecules. The rate of this process has been determined via pulsed vibrational-spectroscopy experiments [21]. Hydrogen bonds in water have an average lifetime of $8 \cdot 10^{-13}$ seconds, which means these bonds are breaking and reforming more than 1 trillion times every second. To put this number in perspective, a hydrogen bond breaks and reforms once for roughly every 100 times that it vibrates.

Rotation rates have been determined via dielectric relaxation experiments [22]. There, the dipoles on each water molecule orient themselves in response to an alternating electric field until the frequency of the alternating current field becomes so high the water molecules can no longer keep up. By examining

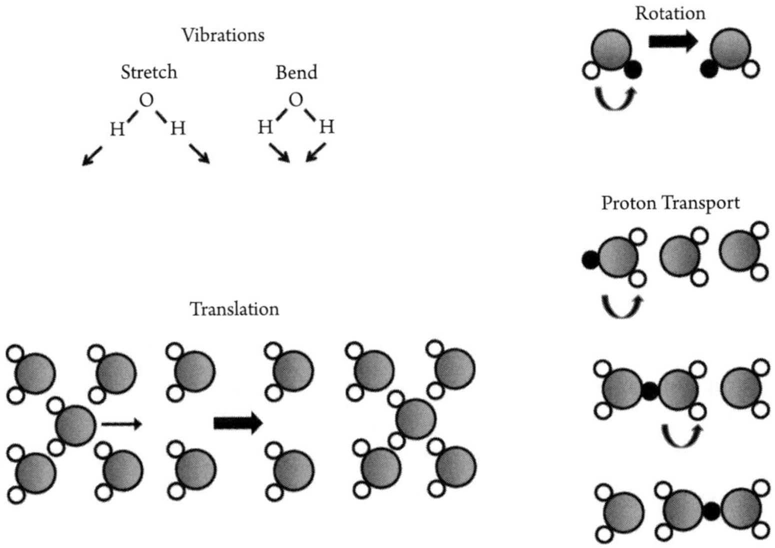

Figure 3.4 Depictions of motions in water, including stretching and bending vibrations, rotations, translations, and proton conduction via the Grotthuss mechanism [19]. For rotations, one of the protons on the water molecule is indicated in black to highlight its motion. Similarly, one of the protons in H_3O^+ is shaded in black to show how water rotations and reversible hydrogen bond formation result in proton transport as a function of time (from top to bottom).

the frequency dependence of the dielectric constant, it is possible to extract a dielectric relaxation time corresponding to the rotational frequency of the water molecule. Because multiple hydrogen bonds must be broken to accommodate this rotation, they occur at a rate of $1.2\cdot10^{11}$/s, which is roughly 10 times slower than the rupture rate for a single hydrogen bond.

Slower still is the rate at which water molecules translate through water [23]. Translation rates have been determined by measuring the self-diffusion coefficient for water. Based on the diffusion coefficient $D = 2\cdot10^{-5}$ cm^2/s at 25°C, the time required for a water molecule to hop to an adjacent site is $4\cdot10^{-11}$ seconds, which corresponds to an attempt frequency of $2.5\cdot10^{10}$/s. This means that each water molecule exchanges places roughly once for every five times a single water molecule rotates in place. Water diffusion is relatively slow (although still very fast) because translation of one water molecule invariably requires the cooperative motions of several water molecules (Fig. 3.4).

Last, it has been observed that protons and hydroxide ions exhibit anomalously high transport rates, as measured via solution conductivity measurements [24]. These rates are fast relative to the self-diffusion of water, which indicates that transport does not involve the motions of intact H_3O^+

or OH^- molecules (both of which would require oxygen anion transport). However, the rates are slower than expected for transport of free protons (or proton vacancies, in the case of OH^-). In fact, measured proton hopping times of $1.5 \cdot 10^{-12}$ seconds obtained via ^{1}H and ^{17}O nuclear magnetic resonance (NMR) are comparable with water rotation rates [25]. A wide range of models, such as the Grotthuss mechanism [19], explains this agreement by postulating that all protons must always be associated with hydrogen bonds, and that transport involves breaking the hydrogen bond on one side, rotating the water molecule, and reestablishing a new hydrogen bond on the other side (Fig. 3.4). Variable temperature studies of rotations, translations, and proton hopping indicate that all the kinetic processes described here have an activation energy consistent with mechanisms involving the rupture of hydrogen bonds. In bulk water, the motions of all H- and O-containing species are cooperative and coupled, and are dictated by the kinetics of dynamic hydrogen bond formation. In Chapter 4, we examine how this behavior is modified by the presence of foreign ions.

References

1. Horne, R. A., Ed. *Water and Aqueous Solutions*; Wiley-Interscience: New York, 1972.
2. Chaplin, M. F. A Proposal for the Structuring of Water. *Biophys. Chem.* **1999**, *83*, 211–221.
3. Pauling, L. *The Nature of the Chemical Bond*; Cornell University Press: Ithaca, NY, 1993; pp 464–469.
4. Thiel, P. A.; Madey, T. E. The Interaction of Water with Solid Surfaces: Fundamental Aspects. *Surf. Sci. Rep.* **1987**, *7*, 211–385.
5. Eizenberg, D.; Kauzmann, W. *The Structure and Properties of Water*; Oxford University Press: New York, 1969.
6. Mahan, B. H. *University Chemistry*; Addison-Wesley Publishing: Reading, MA, 1975; pp 212–217.
7. Mahan, B. H. *University Chemistry*; Addison-Wesley Publishing: Reading, MA, 1975; p 218.
8. Peterson, S. W.; Levy, H. A. A Single-Crystal Neutron Diffraction Study of Heavy Ice. *Acta Cryst.* **1957**, *10*, 70.
9. Pauling, L. *The Nature of the Chemical Bond*; Cornell University Press: Ithaca, NY, 1993; p. 449–504.
10. Xantheas, S. S.; Dunning, T. H., Jr. Ab Initio Studies of Cyclic Water Clusters $(H_2O)n$, $n = 1 - 6$: I. Optimal Structures and Vibrational Spectra. *J. Chem. Phys.* **1993**, *99*, 8774–8792.
11. Lagutschenkov, A.; Fanourgakis, G. S.; Niedner-Schatteburg G.; Xantheas, S. S. The Spectroscopic Signature of the "All-Surface" to "Internally-Solvated" Structural Transition in Water Clusters in the $n = 17 - 21$ Size Regime. *J. Chem. Phys.* **2005**, *122*, 194310–194319.

12. Xantheas, S. S. Cooperativity and Hydrogen Bonding Network in Water Clusters. *Chem. Phys.* **2000**, *258*, 225–231.
13. Kroto, H. W.; Heath, J. R.; O'Brien, S. C.; Curl, R. F.; Smalley, R. E. C60: Buckminster Fullerine. *Nature*, **318 1985**, 162–163.
14. Kamb, B. Structure of the Ices. In *Water and Aqueous Solutions*; Horne, R. A., Ed.; Wiley-Interscience: New York, 1972; p. 9–24
15. Davidson, D. W. Clathrate Hydrates. In *Water: A Comprehensive Treatise*; Franks, F., Ed.; Plenum Press: New York, 1973; Vol. 2, p. 115–234
16. Pauling, L. *The Nature of the Chemical Bond*; Cornell University Press: Ithaca, NY, 1993; p 465.
17. Englezos, P. Clathrate Hydrates. *Ind. Eng. Chem. Res.* **1993**, *32*, 1251–1274.
18. Monastersky, R. The Mother Lode of Natural Gas. *Science News* **1996**, *150*, 298–299.
19. Agmon, N. The Grotthuss Mechanism. *Chem. Phys. Lett.* **1995**, *244*, 456–462.
20. Scherer, J. R. The Vibrational Spectroscopy of Water. *Advances in Infrared and Raman Spectroscopy*, Clark, R. J. H. and Hester, R. E., Eds; Heyden, Philadelphia, 1978, p. 149–216.
21. Woutersen, S.; Bakker, H. J. Resonant Intermolecular Transfer of Vibrational Energy in Liquid Water. *Nature* **1999**, *402*, 507–509.
22. von Hippel, A. The Dielectric Relaxation Spectra of Water, Ice, and Aqueous Solutions and Their Interpretation. *IEEE Trans. Elect. Insul.* **1988**, *23*, 801–816.
23. Jones, J. R.; Rowlands, D. L. G.; Monk, C. B. Diffusion Coefficient of Water in Water and in Some Alkaline Earth Chloride Solutions at 25°C. *Trans. Farad. Soc.* **1965**, *61*, 1384–1388.
24. Alberti, G.; Casciola, M. Solid State Protonic Conductors: Present Main Applications and Future Prospects. *Solid State Ionics* **2001**, *145*, 3–16.
25. Meiboom, S. Nuclear Magnetic Resonance Study of the Proton Transfer in Water. *J. Chem. Phys.* **1961**, *34*, 375–388.

4

Solvated Ions in Water

4.1 INTRODUCTION

In most undergraduate chemistry classes, students are taught to consider reactions in which cations and anions dissolved in water are depicted as isolated ions. For example, the magnesium ion is depicted as Mg^{2+}, or at best $Mg^{2+}(aq)$. For anions, these descriptions may be adequate (if not accurate). However, for cations, these abbreviations almost always fail to describe the critical chemical attributes of the dissolved species. A much more meaningful description of Mg^{2+} dissolved in water is $[Mg(H_2O)_6]^{2+}$, because Mg^{2+} in water does not behave like a bare Mg^{2+} ion, nor do the waters coordinated to the Mg^{2+} behave anything like water molecules in the bulk fluid. In many respects, the $[Mg(H_2O)_6]^{2+}$ ion acts like a dissolved molecular species. In this chapter, we discuss the simple solvation of anions and cations as a prelude to exploring more complex reactions of soluble oxide precursors called *hydrolysis products*. The two key classes of water–oxide reactions introduced here are *acid–base* and *ligand exchange*.

4.2 SOLVATED ANIONS

First, consider how simple anions modify the structure and properties of water. As discussed in Chapter 3, water is a dynamic and highly fluxional "oxide" containing transient rings and clusters based on tetrahedral oxygen anions held together by linear hydrogen bonds. Simple halide ions can insert into this structure by occupying sites that would normally be occupied by other water molecules because they have radii (ranging from 0.13 to 0.22 nm in the series from F^- to I^-) that are comparable to that of the O^{2-} ion (0.14 nm). Such substitution is clearly seen in the structures of ionic clathrate hydrates [1], where the anion can replace one and sometimes even two water molecules. Larger anions can also replace water molecules within clathrate hydrate cages. For example, carboxylate hydrate structures [2] incorporate the carboxylate group within the water framework whereas the hydrophobic hydrocarbon "tails" occupy a cavity within the water framework, as in methane hydrate (see Chapter 3).

Water molecules form hydrogen bonds to dissolved halide ions just as they can to other water molecules, as designated by $OH{-}Y^-$. Neutron diffraction studies [3] show that these hydrogen bonds orient the contacting water molecules such that the protons point directly at the anions. In liquid water, neutron diffraction results indicate individual halide ions tend to be surrounded by six water molecules. These water molecules are called the *water of hydration* of the halide ion. Next-nearest water molecules do not exhibit preferential orientations and they exhibit the structure of bulk water.

The halide ion differs from O^{2-} in terms of its hydrogen bonding requirements. Because halide ions have a net charge of -1 rather than -2, they can only participate in a net number of two hydrogen bonds rather than the four hydrogen bonds that surround each molecule in bulk water. In addition, because the halide anions represent spherical closed shell structures rather than exhibiting the tetrahedral lobes seen in O^{2-}, hydrogen bonds can form anywhere on the halide ion surface. The fact that neutron diffraction results, which represent the water structure averaged over a long period of time, show six hydrogen-bonded water molecules surrounding the central anion indicates that each of these hydrogen bonds is transient and highly dynamic.

The kinetics of reversible hydrogen bond formation around halide ions have been detected using the same pulsed laser Fourier-Transformed Infrared (FTIR) experiments discussed in Chapter 3. These experiments show that the hydrogen bonds to halide anions are stronger than those to other water molecules. Lifetimes of oscillating O-H$-$Y hydrogen bonds ($Y = F^-$, Cl^-, Br^-, or I^-) can be up to 50 times longer ($12{-}25 \cdot 10^{-12}$ seconds for F^-) compared with those observed for OH$-$H bonds in bulk water ($5 \cdot 10^{-13}$ seconds) [4]. However, the facts that the structures of solvated anions are fairly similar to that of bulk water, and that exchange rates and fluxional behavior are so fast (even for hydrogen-bonded halide–water molecule pairs) mean that notations such as Cl^- or $Cl^-(aq)$ are probably sufficient to describe the chemistry of simple anions most of the time.

4.3 SOLVATED CATIONS

Cation solvation is completely different from the solvation of anions. Cation solvation is not dictated by weak hydrogen bonds. Here, the anionic oxygen lone pairs on the O^{2-} anion in water are attracted to cations via Lewis acid–base interactions. The lone pairs of electrons represent *Lewis bases* (or electron donors) whereas the cation represents a *Lewis acid* (or electron acceptor). In the Lewis acid–base interaction, the cation captures electron density from the water molecule to create a metal–oxygen bond. Water molecules in direct contact with metal cations are said to reside in their *inner-coordination sphere.* Shifting electron density into the metal–oxygen bond shifts electron density out of the O−H bonds for these coordinated water molecules. The net result of these shifts in electron density and bond formation is that the structure, acid–base chemistry, and dynamic behavior of water molecules in the inner coordination sphere can be completely different from what is experienced in bulk water.

4.4 THE STRUCTURE OF SOLVATED CATIONS

The proton is so small ($1.3 \cdot 10^{-5}$ Å) that it can only sit at the junction between two anions (see Chapter 3). In contrast, all metal cations have a finite size, which has profound consequences in terms of how cations are solvated. As many water molecules as possible will attempt to crowd around the cation until they contact both the cation and/or each other. The rules that determine how many water molecules fit within the first solvation shell around a given cation are exactly the same as the rules that determine how cations fit into close-packed arrays of oxygen anions in a solid oxide (see Chapter 2). Pauling's First Rule of coordination chemistry states that the cation will fill the smallest void space between water molecules that can accommodate its diameter. This means that the number of water molecules in the first solvation shell can normally be predicted on the basis of cationic radius (see Figs. 2.2 and 2.3).

The ionic radii [5] of representative metal cations are indicated in the Periodic Table presented in Figure 4.1 compared with the sizes of the octahedral and tetrahedral voids in a tightly packed array of water molecules. General trends in cation size include (1) sizes increase as one moves down any column in the Periodic Table and (2) sizes decrease as the charge on the cation increases. For example, Cu^{+} has an ionic radius of 0.77 Å compared with 0.57 Å for Cu^{2+}. An extreme example is the contraction that affects I^{-} (2.2 Å) as it is oxidized to reside within the periodate ion, $[IO_4]^{-}$ (0.53 Å). Ionic radii also exhibit a slight dependence on coordination number. For example, the ionic radius of the potassium ion changes from 1.37 Å to 1.38 Å as its coordination

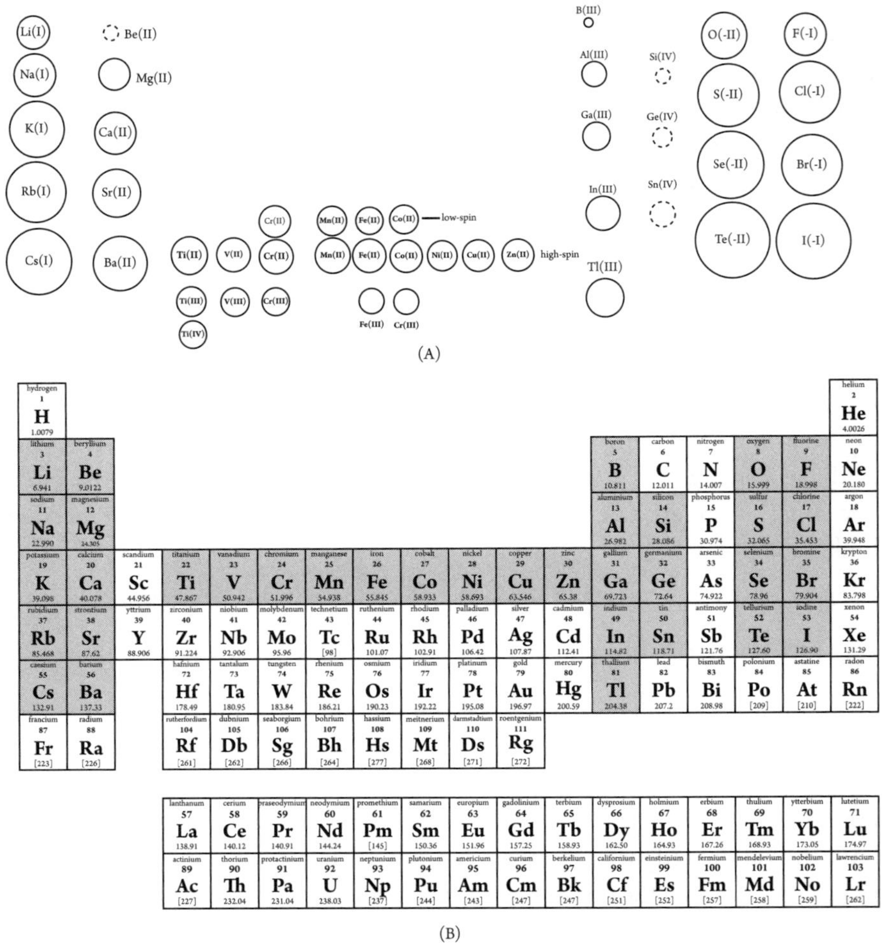

Figure 4.1 The ionic radii of various cations and anions based on their positions in the Periodic Table [5]. The elements shaded in gray in the Periodic Table (B) are shown as circles (A) with their relative ionic radii.

number changes in an oxide crystal from four to six [6]. Although octahedral complexes dominate much of the Periodic Table, the general trends in coordination number for the first solvation shell of aqueous ions are identical to those seen in oxide crystal structures (see Chapter 2) and are predictable based on the ratio of cation radius to O^{2-} (or water molecule) radius. For the example of the alkaline–earth cations, the smallest member Be^{2+} fits within tetrahedral voids to form $[Be(H_2O)_4]^{2+}$. The larger Mg^{2+} fits in octahedral voids to form $[Mg(H_2O)_6]^{2+}$, whereas the largest cation Ba^{2+} is too big to fit within an octahedral vacancy. This large cation can either bond to six waters in a "loose" geometry in which there are gaps between the water molecules, or can actually bond to eight molecules depending on factors such as ionic strength and the activity of water in solution [7].

In closing this section on structure, we have taken our first step toward creating an oxide in water. From the perspective of a cation, the local structure looks like that of an oxide that is only one solvation shell thick. Conversely, from the bulk water looking in, the local environment looks pretty much like water. Water molecules in the first solvation shell may not have exactly the same charge distributions and molecular orientations as bulk water, but adjacent water molecules readily create new hydrogen-bonded ring structures to adapt to this situation (see Chapter 3). The water in contact with the first hydration shell is called the *second hydration shell.* This shell can sometimes be bound to the point where some of the molecules in that shell are counted among the waters of hydration of the particular cation. However, secondary hydration shells are almost as transitory as bulk water structures and normally do not have much of an effect on oxide chemistry. Most of the action involves the water molecules that are in the first hydration shell. As you will see, these waters can be very strange indeed.

4.5 THE ACID–BASE PROPERTIES OF THE FIRST SOLVATION SHELL

The O^{2-} ion in the water molecule and in oxides is supplied with a fixed number of valence electrons. How those electrons are distributed ultimately controls much of the structure and chemistry of all species we consider in this book. The first chemical property we examine to illustrate this point is the acidity of protons on water molecules that are bound to metal cations in the inner solvation shell (Fig. 4.2). By acting as a *Lewis base* and interacting with a *Lewis acid* (the metal cation), the O^{2-} anion donates some of its electron density to the electron-poor cation to create an M—O bond. This means that less electron density is available to the two O—H bonds in the bound water molecule. Weakening of the O—H bond means that bound waters can donate their protons more readily to other species, which means the bound waters are stronger *Brønsted acids* (or proton donors) than the molecules in bulk water. In other words, the Lewis and Brønsted acidities of water molecules are strongly coupled to one another.

The proton transfer reaction between the first solvation shell (Brønsted acid) and bulk water (Brønsted base) is often written without making the bound waters explicit:

$$Mg^{2+} + 2H_2O \rightleftarrows MgOH^+ + H_3O^+ \quad \textbf{(4.1)}$$

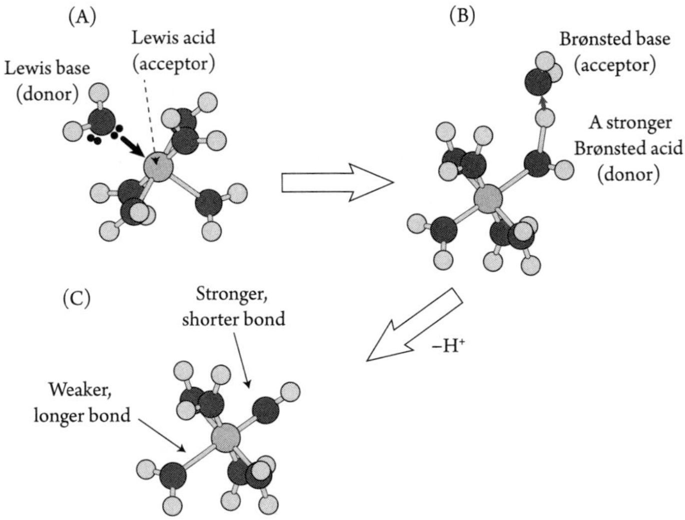

Figure 4.2 (A) When an electron lone pair on a water molecule (a Lewis base) interacts with an open coordination site on a hydrated metal cation (a Lewis acid), a chemical bond is formed. (B) Withdrawal of electron density into the metal–oxygen bond results in a weakening of the O−H bonds in the coordinated water molecule, which increases its Brønsted acidity. (C) If a coordinated water molecule loses a proton to form a bound hydroxyl group, electron density is shifted into the M−OH bond, which weakens the Lewis acid–base interactions to other coordinated water molecules and facilitates ligand exchange.

However, what really happens is

$$\mathrm{Mg}\left(\mathrm{H_2O}\right)_6^{2+} + \mathrm{H_2O} \rightleftarrows \mathrm{Mg}\left(\mathrm{H_2O}\right)_5 \mathrm{OH}^+ + \mathrm{H_3O}^+ \qquad \textbf{(4.2)}$$

Notice that Eq. 4.2 shows exactly the same reaction as the autodissociation of water introduced in Eq. 3.1 except the active water molecule is now associated with a metal cation. These acid–base reactions are treated using conventional equilibrium constants:

$$\mathrm{K_a} = [\mathrm{Mg(H_2O)_5OH^+}][\mathrm{H_3O^+}]/[\mathrm{Mg(H_2O)_6^{2+}}], \mathrm{pK_a} = -\log \mathrm{K_a} \qquad \textbf{(4.3)}$$

The pK_a value corresponds to the pH at which $[\mathrm{Mg(H_2O)_5OH^+}] = [\mathrm{Mg(H_2O)_6^{2+}}]$. In solutions with a pH less than the pK_a, bound water molecules are predominant in the first solvation shell, whereas with a pH greater than the pK_a, each Mg^{2+} is associated with a bound hydroxide ion. The stronger the metal–oxygen bonds in the inner solvation shell are, the more acidic the bound water molecules are, and the lower the pK_a values will be.

4.6 USING FORMAL CHARGE MODELS TO PREDICT AND RATIONALIZE pK_a VALUES

The simplest models that can be used to predict and rationalize the pK_a values for hydrated cations are the formal charge models introduced in Chapter 2. We begin our foray into formal charge predictions of Brønsted acidities by considering species that might be present in deionized water, including H_4O^{2+}, H_3O^+, H_2O, OH^-, and O^{2-}. The formal charges for each of these species is given by

$$\text{Formal charge} = -2 + n(+1)$$

where n is the number of protons bonded to the O^{2-}. Here, the formal charge and species charge are identical. Because no counterions are present, the formal charge model predicts that the neutral water molecule should be the most stable of the species listed by far. The H_3O^+ ion is overcharged and should be eager to donate its proton to another species as a strong Brønsted acid whereas the OH^- ion is undercharged and should be eager to accept a proton as a strong Brønsted base. The autodissociation of water to convert two neutral water molecules into a strong acid and a strong base should not occur to any great extent, as confirmed by the measured autodissociation constant for water of 10^{-14} (see Eq. 3.1). The ions H_4O^{2+} and O^{2-} are so grossly over- and undercharged, respectively, that they should not exist in water at any measurable concentrations.

For water molecules coordinated to cations, the formal-charge model defines trends that should be observed based on both cation charge and coordination number. First, the model predicts that the higher the cation charge, the lower the pK_a value should be. Considering octahedral coordination geometries, the formal oxygen charges associated with water coordinated to Na^+, Mg^{2+}, Al^{3+}, and Zr^{4+} are $+1/6$, $+1/3$, $+1/2$, and $+2/3$, respectively. The expectation is that, although water molecules coordinated to Na^+ should have a pK_a value close to that of molecules in bulk water, pK_a values should decrease as the cation charge increases, until by the time we reach Zr^{4+}, we are dealing with a strong acid. In terms of coordination number, a tetrahedral Be^{2+}, having a formal charge of $+1/2$, should have a lower pK_a value than the octahedral Mg^{2+} ion with its formal charge of $+1/3$. In fact, tetrahedral Be^{2+} should exhibit a pK_a value similar to that of octahedral Al^{3+}. All these features are in agreement with measured pK_a values [8,9]. Amazingly, the measured pK_a values for simple aquated cations span the entire range of values that can be observed in a normal aqueous solution (from pH 0–14). In fact, the binding of a water molecule to a dissolved cation can change its acid–base equilibrium constant by more than a factor of 10^{14}, ultimately making the water of solvation 100 trillion times more acidic than a molecule in bulk water.

Formal-charge models have been refined to provide quantitative predictions of the pK_a values of water molecules bound to metal cations over the enormous range of pK_a values mentioned earlier. For simple cations, pK_a values have been measured directly using pH titration techniques, and the pK_a values have been correlated with different formal-charge models. One such correlation [6] (Fig. 4.3) shows how the pK_a varies with an effective charge parameter given by

$$\text{Charge} = -2 + \Sigma[z/(n \cdot d)] + H/2 \tag{4.4}$$

where -2 is the formal charge on oxygen; z/n is the cation charge divided by the coordination number, as in the simple model; d is the cation–oxygen distance in angstroms; and H is the number of protons on the oxygen site after deprotonation. As indicated in Figure 4.3, this simple formal charge correlation can predict measured pK_a values to within approximately 2 pH units. (*Note*: Throughout this book we use the Shannon ionic radii, recognizing that alternative scales exist and have their proponents. The essential results described here are not affected by the choice of different ionic radii scales.)

The $[Mg(H_2O)_6]^{2+}$ cation is surrounded by six water molecules. Do all these waters of hydration lose their protons at the same pH? And, if so, do all these protons come off simultaneously? Is it possible to remove protons from bound hydroxyl groups? If so, how do the acid–base properties of bound hydroxyl

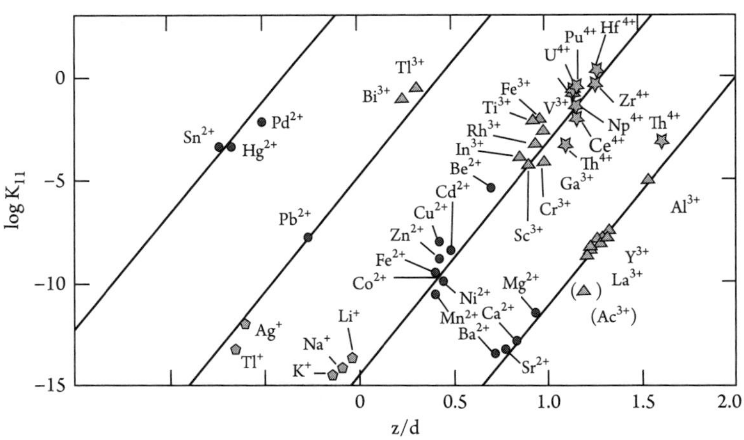

Figure 4.3 Plots of the log of the first acid dissociation constant of bound water molecules (K_{11}) as a function of formal charge (z) and cation–oxygen bond distance (d) [8]. Note that K_{11} for K^+ is essentially identical to the autodissociation constant of pure water (10^{-14}), whereas K_{11} for a highly charged ion such as Zr^{4+} indicates an increase in the Brønsted acidity of the bound water molecules by 14 orders of magnitude.
SOURCE: Adapted from Baes, C. F. J.; Mesmer, R. E. *The Hydrolysis of Cations*; Krieger: Malabar, FL, 1986; p 408, Fig. 18.4.

groups compare with those of bound water molecules? Answers to these questions are known and can be rationalized using formal-charge models. First, waters of hydration are deprotonated in a *stepwise* rather than a *simultaneous* fashion, with each subsequent water molecule exhibiting progressively higher pK_a values. This is a result of the fact that if a bound water molecule is deprotonated, more electron density is localized in the M–OH bond than in the remaining $M-OH_2$ bonds. In the context of the formal charge model, the presence of a bound hydroxyl group effectively lowers the net charge on the solvated metal complex by 1 unit. In simplistic terms, the remaining water molecules sense they are bound to Mg^+ rather than Mg^{2+}. For octahedral cations, a decrease in cation charge by 1 unit corresponds to a shift in the formal charge on oxygen of $-1/6$. According to Figure 4.3, this means the removal of a proton from each successive water molecule should result in an increase in the pK_a value for the remaining molecules of around 2 units. This trend is observed qualitatively for most metal cations [9] (Fig. 4.4).

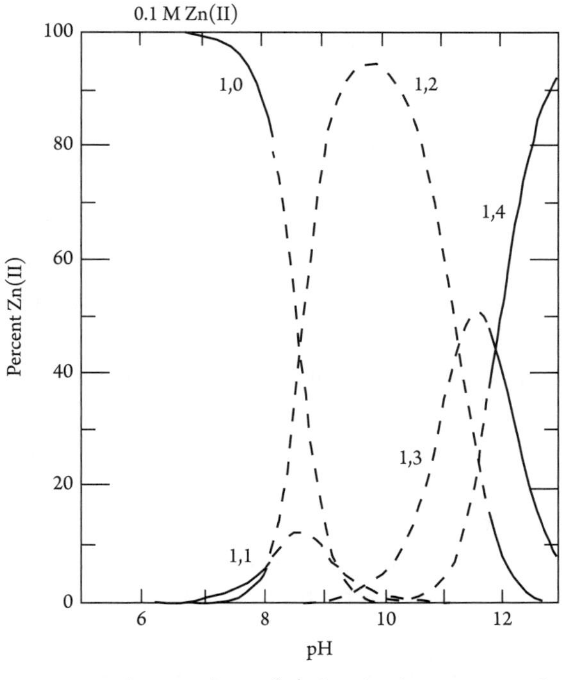

Figure 4.4 The speciation diagram for Zn(II) dissolved in water as a function of solution pH [6]. The species (1,0) represents $[Zn(H_2O)_6]^{2+}$, whereas (1,1), (1,2), (1,3), and (1,4) represent the same species from which one, two, three, and four protons have been removed respectively to leave bound hydroxide ions.
SOURCE: Adapted from Baes, C. F. J.; Mesmer, R. E. *The Hydrolysis of Cations*; Krieger: Malabar, FL, 1986; p 294, Fig. 13.2a.

A second prediction of formal charge models is that it should be easier to deprotonate coordinated water molecules than bound hydroxyl groups. For example, the formal charge on a water molecule bound to Mg^{2+} is $+1/3$, whereas that on a bound hydroxyl group is $-2/3$. In other words, the loss of a proton drops the formal charge on the oxygen anion by 1 unit, which should increase the pK_a value by 10 to 12 units (see Fig. 4.3).

The next question is: How many coordinated water molecules around a hydrated cation can be deprotonated ? To deprotonate all six water molecules in an octahedral complex, the previous discussion suggests that the first dissociation would have to be pH 2 or less for all six deprotonation steps to fall within the normal pH range of water. This is because if each of the six water molecules is to be deprotonated, the range of pK_a values for deprotonating all of them should be 2·6 or 12 units. The requirement for pK_1 to be 2 or less dictates that the central cation have a charge of +3 or greater (see Fig. 4.3). However, even for +3 cations, the maximum number of hydroxyl groups ever found in the inner coordination sphere is four as a result of geometric transitions that are triggered by deprotonation. Hydroxide ions are bound more tightly to the central cation than are water molecules, and they exhibit shorter M—O bonds. Eventually, these bonds become short enough that it is no longer possible to crowd six oxygens around the central cation. For example, as protons are removed from the $[Al(H_2O)_6]^{3+}$ ion, it assumes a five-coordinate geometry [10] as $[AlOH(H_2O)_4]^{2+}$. Eventually, there is a second geometric transition leading to the formation of the tetrahedral $[Al(OH)_4]^-$ ion, which has no waters of hydration in its inner coordination sphere.

As the charge on the metal cation increases from +3 to +4, more water molecules are transformed into hydroxyl groups at increasingly lower pH values. Increases in cation charge and corresponding decreases in cation size trigger a lowering of cation coordination numbers, as seen earlier for the deprotonation of hydrated +3 cations. For example, the neutral species $[Si(OH)_4]^\circ$ is highly stable, with each of its oxygens exhibiting a formal charge of zero. The Si—O bonds in $[Si(OH)_4]^\circ$ are sufficiently covalent and strong that it is now possible to deprotonate the bound hydroxyl groups in the complex to form both $[SiO(OH)_3]^-$ ($pK_a = 9.46$) and $[SiO_2(OH)_2]^{2-}$ ($pK_a = 12.6$) in water [11]. The +5 cation P^{5+} pushes this trend even further by capturing the electron density of two lone pairs into a hybridized P = O bond to create phosphoric acid, or H_3PO_4. The three remaining hydroxyl groups bound in H_2PO_4 can all be deprotonated successively in water, with $pK_{a1} = 2.1$, $pK_{a2} = 7.1$, and $pK_{a3} = 12.1$. Waters bound to +6 and +7 metals are such strong acids that several of the oxygens bound in their inner coordination spheres are completely deprotonated to form oxyanions such as permanganate (MnO_4^-) and sulfate (SO_4^{-2}). These oxyanions are examined in more detail in Chapter 5.

4.7 AN INTRODUCTION TO LIGAND-EXCHANGE REACTIONS

In bulk water (Chapter 3), the transport and/or exchange of water molecules is a fairly facile process involving the reversible rupture of relatively weak hydrogen bonds. Anion transport and exchange within bulk water is equally facile, and is again controlled by hydrogen bonding interactions. One might expect water transport and exchange within the first solvation shell of a cation to follow similar ground rules. Nothing could be further from the truth. In the presence of cations, the dipoles in adjacent water molecules align themselves such that the anionic lone pairs of electrons are facing the metal rather than the bound protons. This means that, for the first time, we are dealing with a dissolved species with interactions dominated by Lewis acid–base interactions rather than hydrogen bonds. Lewis acid–base interactions are more complex than hydrogen bonds, spanning the range from weak electrostatic interactions to strong covalent metal–oxygen bonds [12–14]. In all these interactions, the water molecule is now referred to as a *ligand* that is bound to the metal cation. (The word *ligand* comes from the Latin word *ligare*, meaning "to bind," as in *ligature* or *ligament*.)

We have now arrived at one of the most critical reactions in water–oxide systems: *the ligand exchange reaction*. In a ligand-exchange reaction, one species in the inner coordination sphere of a metal cation is replaced by another species. The incoming and outgoing species can be water molecules, hydroxide ions, oxygen anions in an oxide lattice, or any other *nucleophilic* species that can function as a Lewis base. For example, the dissolution of an oxide in water can be considered to occur via a series of ligand-exchange reactions in which the bridging oxygens that bind the metal cation to the dissolving structure are replaced successively by nonbridging oxygens represented by bound hydroxide ions or water molecules until the cation is released into solution. Conversely, in oxide formation, nonbridging hydroxyl groups and water molecules are replaced by bridging oxygens in a cluster or extended oxide structure.

Regardless of whether ligand-exchange reactions lead to dissolution or precipitation, the coordination number of the metal cation is usually conserved. For example, the Mg^{2+} ion in MgO is coordinated to six oxygens whereas the species released into solution is $[Mg(H_2O)_6]^{2+}$. However, it is important to note that, although cation coordination numbers are conserved, the metal coordination numbers around each oxygen anion can vary dramatically. For example, although the oxygen in water is bound only to one cation in $[Mg(H_2O)_6]^{2+}$, each oxygen is coordinated to six Mg^{2+} in the cubic oxide periclase (MgO). The factors dictating oxygen coordination numbers are discussed in Chapter 5.

In water, most ligand exchange reactions, whether in solution or on an oxide surface, start with the formation of an outer sphere complex. This weak complex forms largely by solvation and electrostatic forces such that the incoming, soon-to-be-exchanged ligand hovers around the inner hydration sphere but cannot yet displace an inner sphere water molecule. Transport and exchange within this outer hydration shell is quite rapid, with rates [12] that compare with transport rates within bulk water (see Chapter 3). Outer sphere complexes are typically in equilibrium with the solution and can be characterized by an outer sphere equilibrium constant (K_{OS}) calculable from simple statistical–mechanical models based on Coulomb's Law.

The real action that controls ligand exchange reactions occurs after the outer-sphere complex forms. Each complex follows a reaction trajectory that starts from an initial ground state consisting of the inner-sphere complex and a ligand in the outer sphere. The ground state moves to a higher energy unstable state called an *activated complex* or transition state that involves the departure of a host ligand and/or the insertion of a foreign ligand from the outer sphere. The activated complex then decays to a new ground state in which the foreign ligand has replaced the host ligand. In some cases, multiple steps are involved to form metastable intermediates, each will have an activated complex between steps.

Although both the departing and entering ligands invariably play a role in the formation of the activated complex, extensive studies of aquated metal complexes [15–19] indicate that ligand-exchange processes can be conceptualized by considering two different extreme reaction mechanisms (Fig. 4.5). As discussed below, the actual steps for ligand exchange around most of these ions are more complicated than we show here and fall somewhere between the two extremes. In a *dissociative (or D) mechanism*, the rate-limiting step involves the departure of one of the host ligands to form a transition-state complex in which the metal cation is bonded to *one fewer* ligands than the parent complex. For the example of water exchange around octahedral Mg(II), the formation of such a five-coordinate transition state can be represented by

$$\left[Mg\left(H_2O\right)_6^{2+}\right]\cdot\left[H_2O\right]_n \rightleftarrows \left[Mg\left(H_2O\right)_5^{2+}\right]^{\ddagger}\cdot\left[H_2O\right]_{n+1} \qquad (4.5)$$

where n indicates the number of water molecules in the outer hydration shell. The vacancy left behind is then filled by an incoming ligand hovering in the outer coordination sphere to reestablish the octahedral geometry of the parent complex:

$$\left[Mg\left(H_2O\right)_5^{2+}\right]^{\ddagger}\cdot\left[H_2O\right]_{n+1} \rightarrow \left[Mg\left(H_2O\right)_6^{2+}\right]\cdot\left[H_2O\right]_n \qquad (4.6)$$

In this mechanism, the rate of ligand exchange is controlled primarily by the basicity of the departing ligand. The basicity of the incoming ligand is less important because any available ligand can fill the vacant coordination site.

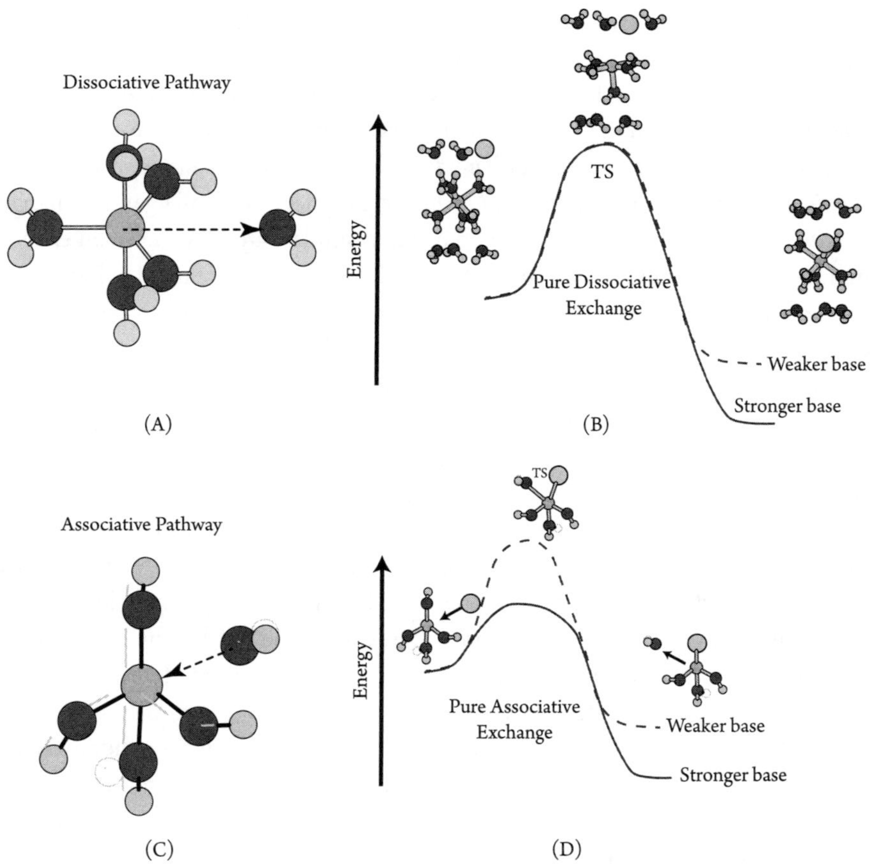

Figure 4.5 Depictions of ligand-exchange mechanisms. (A) In dissociative mechanisms, an octahedral hydrated cation loses one of its waters of hydration to form a five-coordinated intermediate. This activated complex often readjusts to achieve a trigonal bipyramidal geometry, which is here not shown. (B) A new ligand from the outer coordination sphere (shown as a large circle) then enters the open coordination site to regenerate an octahedral complex. The character of the incoming ligand has little effect on a dissociative ligand-exchange reaction except via the stability of the product because loss of an existing ligand is the rate-controlling step. (C) In associative mechanisms, the first step involves the attack of a ligand (shown as a hydroxide ion), which enters the inner coordination sphere and increases the cation coordination number. If the initial complex is tetrahedral, the five-coordinated activated complex can achieve a trigonal bipyramidal geometry as in the previous dissociative case. The higher coordinate intermediate then ejects one of its ligands to reestablish its initial coordination geometry. (D) In associative processes, the energy of the transition state depends critically on the nature of the entering ligand, shown here as a sphere, with more basic ligands affecting both the transition-state and final-state energies.

Dissociative processes are most important for octahedral complexes because the oxygen ligands tend to be so tightly packed around the central cation that it is difficult for the entering ligand to gain access to the metal unless one of the host ligands leaves. For aquated cations, an important consequence of the dissociative process is that *the inherent loss of bound water molecules from the parent complex should dictate the overall rate of ligand exchange.* In other words, all ligand exchange reactions on any given aquated should occur at similar rates. However, an examination of experimentally determined second-order rate constants for ligand exchange (k_2) spans several orders of magnitude and depends on the incoming and outgoing ligands [20, 21] (Table 4.1). The discrepancy between predicted and observed rates is often resolved by the classical Eigen–Wilkins mechanism [22]. Here, it was recognized that the differences in apparent rates are actually related to ligand availability as reflected in the outer-sphere equilibrium constant (K_{OS}). A pseudo first-order rate constant for ligand exchange involving the inner coordination sphere can be obtained by dividing k_2 by K_{OS}. For most divalent cations, including $[Ni(H_2O)_6]^{2+}$ and $[Mg(H_2O)_6]^{2+}$, the resulting k_1 values are almost identical to each other and to the rates of self-exchange of water molecules between inner and outer coordination spheres as predicted by the dissociative mechanism. Actual exchange rates are similar regardless of the identity or Lewis basicity of the incoming ligand (Table 4.1).

Associative (or A) mechanisms for ligand exchange represent the other extreme for ligand exchange processes. Here, the rate limiting step involves the addition of the foreign ligand to the existing metal complex to form an activated complex in which the metal cation is bonded to one extra ligand (Fig. 4.5). A complex illustrating this mechanism for ligand exchange is tetrahedral $[Be(H_2O)_4]^{2+}$:

$$\left[Be\left(H_2O\right)_4^{2+}\right]\cdot\left[H_2O\right]_n \rightleftarrows \left[Be\left(H_2O\right)_5^{2+}\right]^{\ddagger}\cdot\left[H_2O\right]_{n-1} \quad \textbf{(4.7)}$$

Note that here the coordination number around Be(II) increases from four to five. This unstable transition state then decays via the ejection of one of the ligands in the inner-coordination sphere to reestablish the initial coordination number of four:

$$\left[Be\left(H_2O\right)_5^{2+}\right]^{\ddagger}\cdot\left[H_2O\right]_{n-1} \rightarrow \left[Be\left(H_2O\right)_4^{2+}\right]\cdot\left[H_2O\right]_n \quad \textbf{(4.8)}$$

The key point is that the kinetics of associative processes are dominated by the basicity of the incoming ligand, which must force itself into the inner coordination sphere to form the activated complex. In contrast to dissociative processes, ligand-exchange kinetics are now strongly ligand dependent. Although any ligand can be ejected from the activated complex, ejection of the weakest base is typically preferred.

Table 4-1. KINETIC DATA ILLUSTRATING A TWO-STEP REACTION FOR LIGAND EXCHANGE

Ligand	**k_2 (M/s)**	**K_{os} (M)**	**k_1/s**
	Ni(II)		
N-methylimidazole$^+$	$2.3\cdot10^2$	0.02	$1.2\cdot10^4$
Imidazole H^+	$3\cdot10^2$	0.02	$1.5\cdot10^4$
Ammonia	$5\cdot10^3$	0.15	$3.3\cdot10^4$
Hydrogen fluoride	$3\cdot10^3$	0.15	$2.0\cdot10^4$
Imidazole	$2.8-6.4\cdot10^3$	0.15	$1.9-4.3\cdot10^4$
1,10-Phenanthroline	$4.1\cdot10^3$	0.15	$2.6\cdot10^4$
Diglycine	$2.1\cdot10^4$	0.17	$1.2\cdot10^4$
Fluoride$^-$	$8\cdot10^3$	1	$0.8\cdot10^4$
Acetate$^-$	$1\cdot10^5$	3	$3.0\cdot10^4$
Glycinate$^-$	$2.0\cdot10^4$	2	$1.0\cdot10^4$
Oxalate H^-	$5\cdot10^3$	2	$0.3\cdot10^4$
Oxalate^{2-}	$7.5\cdot10^4$	13	$0.6\cdot10^4$
Malonate^{2-}	$4.5\cdot10^5$	95	$0.5\cdot10^4$
Methylphosphate^{2-}	$2.9\cdot10^5$	40	$0.7\cdot10^4$
Pyrophosphate^{3-}	$2.1\cdot10^6$	88	$2.4\cdot10^4$
Tripolyphosphate^{4-}	$6.8\cdot10^6$	570	$1.2\cdot10^4$
Water exchange			$3.2\cdot10^4$
	Mg(II)		
Oxine	$1.3\cdot10^4$	0.2	$0.7\cdot10^5$
Oxinate$^-$	$6.0\cdot10^5$	2.1	$2.9\cdot10^5$
Fluoride$^-$	$5.5\cdot10^4$	1.6	$0.4\cdot10^5$
5-Nitrosalicylate$^-$	$7.1\cdot10^5$	2	$3.6\cdot10^5$
Bicarbonate$^-$	$5.0\cdot10^5$	0.9	$5.6\cdot10^5$
Carbonate^{2-}	$1.5\cdot10^4$	3.5	$0.4\cdot10^5$
Pyrophosphate	$5.4\cdot10^5$	13	$0.4\cdot10^5$
$HADP^{2-}$	$1.0\cdot10^6$	9	$1.1\cdot10^5$
$HATP^{3-}$	$3.0\cdot10^6$	30	$1.0\cdot10^5$
ATP^{4-}	$1.3\cdot10^7$	$1.2\cdot10^2$	$1.1\cdot10^5$
Water exchange			$6.7\cdot10^5$

Second-order rate coefficients (k_2) for binding of a given ligand to typical bivalent hydrated metal ions can be reduced to a pseudo first-order rate coefficient (k_1) through division by the estimated equilibrium constants for formation of the outer-sphere complexes (K_{os}). The resulting value of k_2/K_{os} is approximately equal to the rate coefficient for loss of a water molecule from the $[Ni(H_2O)_6]^{2+}$ and $[Mg(H_2O)_6]^{2+}$ complexes, respectively. ATP = Adenosine triphosphate. Data from [20, 21].

Table 4-2. RATE CONSTANTS, ACTIVATION PARAMETERS, AND INFERRED MECHANISMS OF EXCHANGE OF BOUND WATERS FOR SOME TRANSITION METALS

Molecule	k_{ex}^{298}/s	$\Delta H^{\ddagger}$ (kJ/mol)	$\Delta S^{\ddagger}$ (kJ/mol/K)	$\Delta V^{\ddagger}$ (cm^3/mol)	Mechanism
$[Be(H_2O)_4]^{2+}$	$7.3 \cdot 10^2$	59.3	8.4	−13.6	A, I_a
$[Mg(H_2O)_6]^{2+}$	$6.7 \cdot 10$	49.1	31.1	6.7	D, I_d
$[V(H_2O)_6]^{2+}$	$8.7 \cdot 10^1$	1.8	−0.4	−4.1	I_a
$[Mn(H_2O)_6]^{2+}$	$2.1 \cdot 10^7$	32.9	5.7	−5.4	I_a
$[Fe(H_2O)_6]^{2+}$	$4.4 \cdot 10^6$	41.4	21.2	3.8	I_d
$[Co(H_2O)_6]^{2+}$	$3.2 \cdot 10^6$	46.9	37.2	6.1	I_d
$[Ni(H_2O)_6]^2$	$3.2 \cdot 10^4$	56.9	32.0	7.2	I_d
$[Cu(H_2O)_6]^{2+}$	$4.4 \cdot 10^9$	11.5	−21.8	2.0	I_d
$[Ru(H_2O)_6]^{2+}$	$1.9 \cdot 10^{-2}$	87.8	16.1	−0.4	I_d
$[Pd(H_2O)_4]^{2+}$	$5.6 \cdot 10^2$	49.5	−26	−2.2	I_a
$[Pt(H_2O)_4]^{2+}$	$3.9 \cdot 10^{-4}$	89.7	−9	−4.6	I_a
$[Al(H_2O)_6]^{3+}$	1.3	84.7	41.6	5.7	I_d
$[Ti(H_2O)_6]^{3+}$	$1.8 \cdot 10^5$	43.4	1.2	−12.1	A, I_a
$[V(H_2O)_6]^{3+}$	$5.0 \cdot 10^2$	49.0	−27.8	−8.9	I_a
$[Cr(H_2O)_6]^{3+}$	$2.4 \cdot 10^{-6}$	108.6	11.5	−9.6	I_a
$[Fe(H_2O)_6]^{3+}$	$1.6 \cdot 10^2$	64.0	12.1	−5.4	I_a
$[Ga(H_2O)_6]^{3+}$	$4.0 \cdot 10^2$	67.1	30.1	5.0	I_d
$[Ru(H_2O)_6]^{3+}$	$3.5 \cdot 10^{-6}$	89.8	−48.3	−8.3	I_a
$[Rh(H_2O)_6]^{3+}$	$2.2 \cdot 10^{-9}$	131.2	29.0	−4.2	I_a
$[Ir(H_2O)_6]^{3+}$	$1.1 \cdot 10^{-10}$	130.5	2.1	−5.7	I_a
$[GaAl_{12}]$	$2.3 \cdot 10^2$	63.0	13.0	3.0	I_d

All structures are octahedra, or distorted octahedra, save for $[Cu(H_2O)_6]^{2+}$, which is a bipyramidal distorted octahedron resulting from Jahn–Teller effects, the square–planar $[Pd(H_2O)_4]^{2+}$ and $[Pt(H_2O)_4]^{2+}$ complexes, and the $[Be(H_2O)_4]^{2+}$ ion, which is tetrahedral. $GaAl_{12} = [GaO_4Al_{12}(OH)_{24}(H_2O)_{12}]^{7+}$, a nanometer-size ion with waters bound to octahedral metals.

The associative mechanism tends to be more important for small cations that start out with a low coordination number but that are large enough to accept another ligand, such as the tetrahedral species $[Si(OH)_4]^{\circ}$. For $[Si(OH)_4]^{\circ}$, it is much easier to add one hydroxide to the complex than it is to remove one hydroxide to make a planar three-coordinate intermediate. (In contrast, borate ion is tetrahedral and forms boric acid by losing a hydroxide ion to form a trigonal molecule.) Cations such as Pt(II) that form square planar complexes have two open axial coordination sites that can participate in the associative mechanism [23]. Removal of one of the four ligands in the parent complex to make a three-coordinate intermediate via the dissociative mechanism does not occur.

Dissociative and associative mechanisms represent extremes in ligand-exchange behavior. In practice, many exchange reactions follow *interchange reaction (or I)* pathways involving the simultaneous movement of outgoing and incoming ligands [24, 25] (Table 4.2). Interchange pathways with a considerable dissociative character (I_d) tend to be more important for octahedral complexes because it is difficult to crowd seven oxygen anions around most cations. Interchange pathways with an associative character (I_a) are also found, particularly for larger cations that can accommodate an extra incoming ligand more readily. The I_a mechanism is exhibited by some environmentally important transition metal complexes [23], including $[Fe(H_2O)_6]^{3+}$ and $[Cr(H_2O)_6]^{3+}$. Here, exchange rates *are* sensitive to the identity of the incoming ligand; bond formation to the incoming ligand and the breaking the of metal–OH_2 bond occur simultaneously.

4.8 THE THERMODYNAMICS OF LIGAND EXCHANGE

The transition-state complexes described earlier for both dissociated and associative ligand exchange reactions form as a result of elementary reactions involving the concerted motions of a few atoms to a saddle point on a potential energy surface (Fig. 4.6). The rates of these elementary reactions are often treated using absolute rate theory [26]. In absolute reaction rate theory, the rate (k_{ex}) of the elementary reaction is described by an expression that includes the Gibbs energy for activation ($\Delta G^{\ddagger}$) for forming the unstable transition state from stable reactants:

$$k_{ex} = \frac{kT}{h} e^{-\left[\Delta G^{\ddagger}/RT\right]} \tag{4.9}$$

Here, k is Boltzmann's constant, h is Planck's constant, and R is the ideal gas constant. The energy to reach the top of the transition-state barrier relates to other thermodynamic parameters:

$$\Delta G^{\ddagger} = \Delta H^{\ddagger} - T\Delta S^{\ddagger} + P\Delta V^{\ddagger} \tag{4.10}$$

where $\Delta H^{\ddagger}$ is the activation enthalpy, P is the applied pressure, $\Delta V^{\ddagger}$ is the activation volume, and $\Delta S^{\ddagger}$ is the activation entropy. All these parameters can be obtained by studying the kinetics of ligand exchange as a function of environmental factors such as temperature, pressure, and species concentrations. Eq. 4.10 is discussed in more detail in Chapter 16, which deals with pressure, or stress-activated reactions.

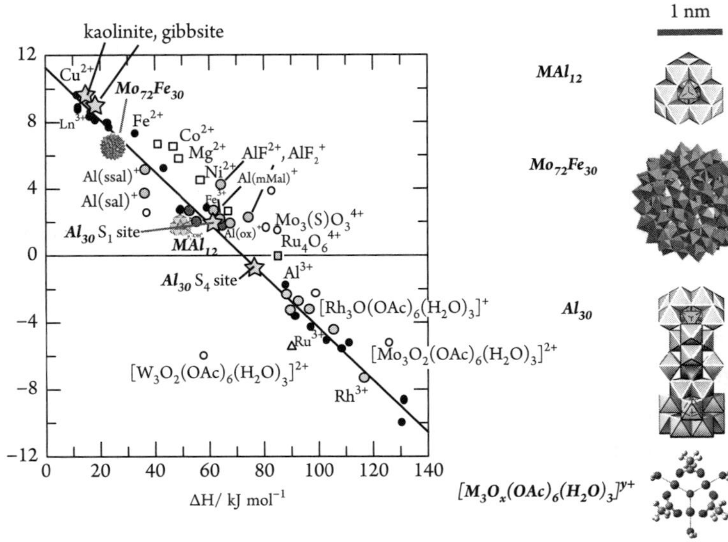

Figure 4.6 The experimental correlation between the rate of water exchange and the enthalpy of the M–O bond in a wide range of solvated cations, hydrolysis products, and even solid oxide phases.
SOURCE: Ohlin, C. A.; Villa, E. M.; Rustad, J. R.; Casey, W. H. Dissolution of insulating oxide materials at the molecular scale *Nat. Mater.* **2009**, *9*, 11. Reproduced with permission.

Values of $\Delta H^\ddagger$, $\Delta S^\ddagger$, and $\Delta V^\ddagger$, can be used to predict ligand-exchange rates as a function of parameters such as temperature and pressure. These values can also provide insight with regard to reaction mechanisms. For starters, it is not surprising there is a strong correlation (Fig. 4.7) between activation enthalpy, $\Delta H^\ddagger$, and the kinetics of water exchange because this term is the largest factor in the exponent:

$$k_{ex} = \frac{kT}{h} e^{-(\Delta H^\ddagger - T\Delta S^\ddagger + P\Delta V^\ddagger)/RT} \qquad (4.11)$$

The activation entropy and volumes terms are typically much smaller than the activation enthalpy for aqueous reactions, in which ordering is relatively small and the bulk solvent is nearly incompressible. The expectation that stronger bonds should yield lower exchange rates is clearly met.

In principle, $\Delta S^\ddagger$, which reflects changes in molecular ordering between the ground state and the activated complex, can indicate whether a given ligand exchange reaction is dissociative or associative in nature. The degree of order should decrease as a result of water loss in the dissociative mechanism and increase for the associative mechanism, in which the activated complex contains more bound water

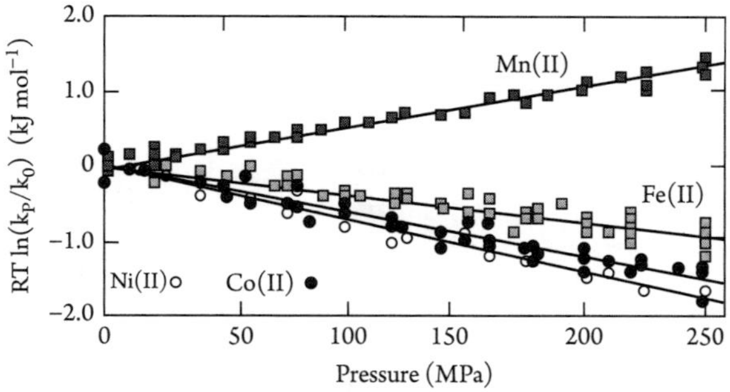

Figure 4.7 Evidence for both ligand-exchange processes based on activation volume values obtained from studies of the pressure dependence of water-exchange reaction rates (see text).

SOURCE: Ducommun, Y.; Newman, K. E.; Merbach, A. E. High-pressure oxygen-17 NMR evidence for a gradual mechanistic changeover from Ia to Id for water exchange on divalent octahedral metal ions going from manganese(II) to nickel(II) *Inorg. Chem.* **1980**, *19*, 3696–3703. Adapted with permission.

molecules than the parent complex. Unfortunately, $\Delta S^\ddagger$ values have not proved to be a reliable guide to mechanisms because $\Delta S^\ddagger$ values are difficult to gauge experimentally. This is a result of the inaccuracies in determining the 0°Kelvin intercept of data curves associated with extrapolating exchange rates over large temperature ranges.

A more reliable means of determining the D, A, or I character of a given exchange reaction is through the use of $\Delta V^\ddagger$. Activation volumes are obtained by measuring the pressure dependence of the reaction rate constant for water exchanges (k) at constant temperature:

$$\left[\frac{d\ln(k)}{dP}\right]_T = \frac{-\Delta V^\ddagger}{R \cdot T} \quad \textbf{(4.12)}$$

Volumes differ in magnitude and sign with mechanism because waters pack more tightly in inner hydration spheres than they do in bulk water. A *dissociative* reaction has a *positive* change in activation volume whereas an *associative* reaction has a *negative* change. In the dissociative mechanism, the release of a water molecule from an ordered solvation sphere into bulk water creates volume in the transition state (i.e., $\Delta V^\ddagger$ is positive). For example, $\Delta V^\ddagger = +6.7\ \text{cm}^3/\text{mol}$ for the exchange of water around $[Mg(H_2O)_6]^{2+}$ in Table 4.2. Conversely, in the associative mechanism, the abstraction of a water molecule into the inner solvation sphere eliminates volume in the transition state (e.g., $\Delta V^\ddagger = -13.6\ \text{cm}^3/\text{mol}$ for water exchange around $[Be(H_2O)_4]^{2+}$ in Table 4.2).

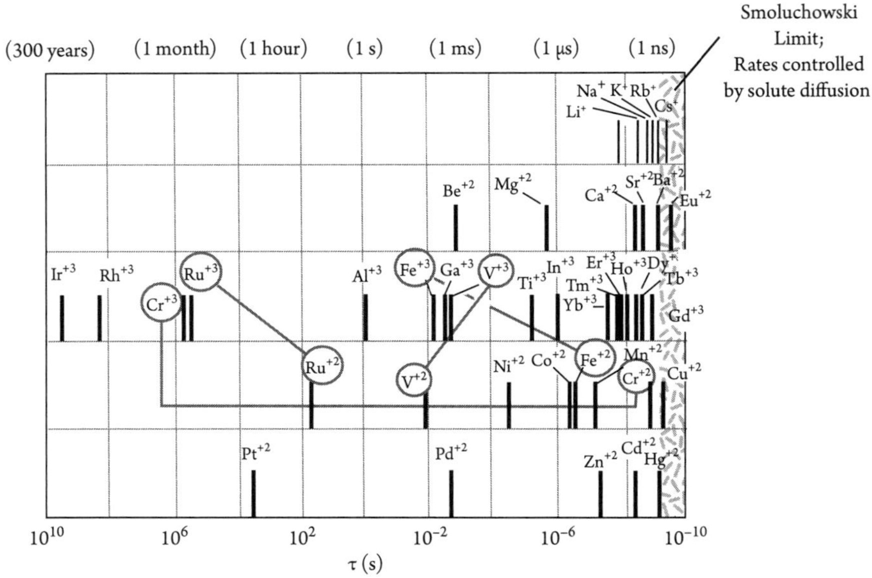

Figure 4.8 Water exchange rates around aquocomplexes of common metal cations, in exchange events per second. Adapted from Helm, L.; Merbach, A. E. Inorganic and bioinorganic solvent exchange mechanisms *Chem. Rev.* **2005**, *105*, 1923–1959.

Classic examples of the utility of $\Delta V^{\ddagger}$ in probing reaction mechanisms is provided by measurements that have been performed on the rate of water exchange around Mn^{2+}, Fe^{2+}, Co^{2+}, and Ni^{2+} as a function of applied pressure [27] (Fig. 4.7). The slopes of the curves shown in Figure 4.7 are negative for Ni^{2+}, Co^{2+}, and Fe^{2+}, indicating a positive $\Delta V^{\ddagger}$ that is consistent with an interchange mechanism with a considerable dissociative character typical of octahedral complexes. In contrast, the slope for Mn^{2+} is positive, indicating an interchange mechanism with considerable associative character.

4.9 THE KINETICS OF LIGAND-EXCHANGE REACTIONS

We have already seen how rapidly water exchange and transport phenomena occur in bulk water (Chapter 3) and in the vicinity of anions as dictated reversible hydrogen bond formation. How do these rates compare with the rates of ligand exchange reactions as dictated by Lewis acid–base bond formation? First, consider the rate at which one water molecule changes places with another in the inner solvation sphere of an octahedrally coordinated metal cation. Amazingly, *ligand exchange rates for octahedral cations with a charge of* $+1$ *to* $+3$ *can vary by more than 20 orders of magnitude* (10^{20}) [16, 28] (Fig. 4.8). As in the case of acid–base properties, many exchange rates can be rationalized on the basis of

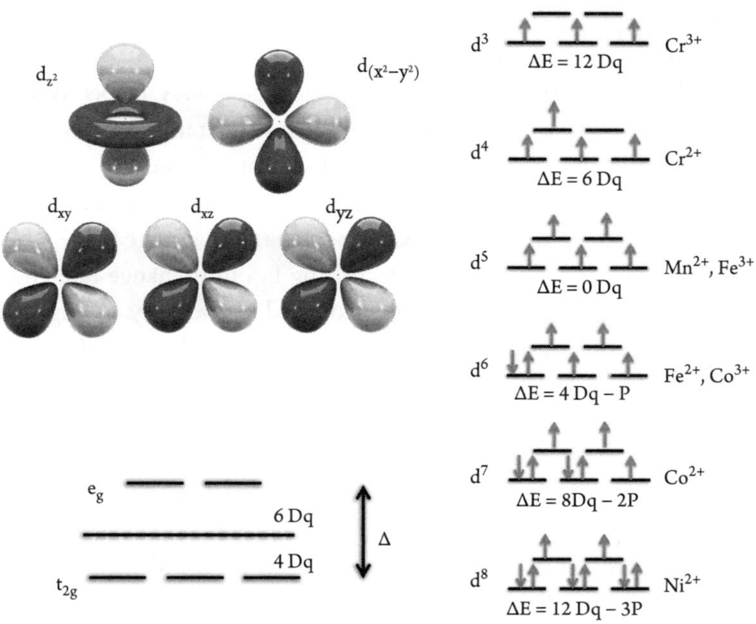

Figure 4.9 (left) Generic energy levels for octahedral transition metal complexes, with images of the corresponding orbitals (SOURCE: general.fmv/Shutterstock.com) shown above. (right) Depictions of electron occupancy in the d-orbital manifolds of transition metal ions for high-spin complexes involving weakly basic ligands such as water. Total stabilization energies for each configuration are shown below each example based on the orbital splitting energies shown on the top left (in units of Dq) and the energy (P) required for pairing electrons within a given orbital. Higher stabilization energies translate into slower ligand-exchange rates.

simple formal charge arguments. At one end of the spectrum, the metal–oxygen bonds to large monovalent cations such as Cs^+ and K^+ are relatively weak. Correspondingly, the rates of exchange of water molecules in their inner coordination shells are so rapid $(3 \cdot 10^9/s)$ that water exchange is only 10 times slower than water exchange in the bulk solution $(2.5 \cdot 10^{10}/s$, see Chapter 3). In other words, the alkali ions and most other +1 cations exert a relatively minor perturbation on the structure and chemistry of water, having effects that are comparable with those exerted by dissolved anions. This is why such ions are often referred to as *charge-compensating ions* in dealing with many water–oxide reactions. The only reactions for which the descriptions Na^+ or $Na(aq)^+$ are inadequate for explaining behavior are ion exchange processes that involve changes in ion hydration numbers (see Chapter 10). Moving to the +2 cations, exchange rates slow appreciably. For the alkaline earth cations, observed rates drop from around $10^9/s$ to $10^8/s$ in moving from Ba^{2+} to Ca^{2+} (decreasing cation size).

Continuing to aquated trivalent ions, the group IIIA cations (Al^{3+}, Ga^{3+}, and In^{3+}) exhibit even slower exchange rates (1/s, 10^3/s, and $10^5/s^1$, respectively, Table 4.1) that scale with cation size (0.68, 0.76, and 0.94 A, respectively). Note that the exchange rate for Be^{2+} falls within the normal range for +3 cations. As discussed earlier, this is consistent with the fact that aquated Be^{2+} is very small, assumes a tetrahedral rather than an octahedral geometry, and thus has oxygen formal charges that are similar to those expected for an octahedral +3 cation.

The simple correlation between a decreasing ligand exchange rate and an increasing charge-to-ion radius is not always obeyed, particularly among transition metal ions. Although it is not surprising that water molecules in $[Cr(H_2O)_6]^{3+}$ exchange more slowly (~10^9/s than those around $[Cr(H_2O)_6]^{2+}$ (~10^{-6}/s), it is amazing just how slow the latter exchange rate really is. *The +3 cation reacts approximately 10^{15} times slower than the +2 cation of the same metal.* To put the slow exchange rate for Cr(III) in perspective, although Cr^{3+} and Ga^{3+} have identical ionic radii, Cr^{3+} exchanges water molecules 1 billion times slower than Ga^{3+}. Among the Group IX trivalent ions, Ir^{3+} is larger but less reactive than Rh^{3+}, which in turn is larger and less reactive than Co^{3+} (Fig. 4.8). In fact, *a water molecule bound to Ir^{3+} exchanges with bulk water only once every 200 years at room temperature.*

Excursions from expected exchange rates for the transition metals are rationalized on the basis of the symmetries of the d orbitals, the number of d electrons, and ligand–field stabilization energies [17, 29, 30]. The valence electrons in transition metals are contained in d orbitals that have lobes of specific symmetry. On an octahedral grid consisting of x-, y-, and z-axes, the d_{xy}, d_{xz}, and d_{yz} orbitals have their lobes oriented between the axes whereas the remaining d orbitals ($d_{x^2-y^2}$ and d_{z^2}) have their lobes aligned on the axes. For a transition metal ion in a perfect vacuum, the environment around the ion is symmetric spherically, and all orbitals have exactly the same energy. However, in an octahedral complex such as $[Cr(H_2O)_6]^{2+}$, the basic lone pairs of each water molecule point along the axes toward the central cation. Electrostatic repulsion between electrons in these lone pairs and electrons in the d orbitals result in a destabilization of $d_{x^2-y^2}$ and d_{z^2} relative to d_{xy}, d_{xz}, and d_{yz}, so that the latter set (referred to as t_{2g}) is lower in energy than the former set (referred to as e_g) by an amount called Δ (or in spectroscopic notation, 10 Dq). Relative to the initial spherical case (or, for that matter, a nontransition metal cation such as Ca^{2+}), the t_{2g} orbitals are stabilized by an energy of 4 Dq whereas the e_g orbitals are destabilized by 6 Dq.

The net degree to which a given transition metal complex is stabilized or destabilized depends on how many electrons occupy the d-orbital manifold, and whether those electrons are forced to pair up within single orbitals (resulting in a destabilizing pairing energy P, see Fig. 4.9). It is important to note that each cation and oxidation state exhibits its own unique values for Δ and P. In addition, values for Δ depend on the basicity of the ligands in the inner coordination sphere, with stronger bases yielding larger Δ values. Last, for octahedral coordination geometries, cations with more than three d electrons can arrange those electrons in multiple configurations. The configuration of choice is that having the lowest energy. If Δ is less than P, electrons fill the higher energy e_g orbitals before pairing up (called a *high-spin configuration*), whereas if Δ is greater than P, electrons pair up and fill the t_{2g} set before occupying the e_g orbitals (called the *low-spin configuration*).

Hexaquo complexes are almost always high-spin complexes because water is a relatively weak base. Even so, stabilization energies can be substantial and can dictate the rate of water exchange. For $[Cr(H_2O)_6]^{3+}$, the value of Dq is 20.8 kJ/mole [24]. Because the stabilization energy for this d^3 complex is 12 Dq, the net degree of stabilization is 250 kJ/mole, resulting in Cr—O bonds that are more than 40 kJ/mole stronger than those of a "normal" M—O bond. The number, distribution, and net stabilization energy of the d electrons varies systematically for the oxidation states of the first-row transition metals. The higher the stabilization energies, the slower the exchange rates. However, it is important to note that what really determines the rate of ligand exchange is the energy difference between the ground state and the activated complex. For the dissociative mechanism, the activated complex is a five-coordinate intermediate. The d orbitals associated with various five-coordinate geometries all exhibit different splitting patterns than the $e_g + t_{2g}$ manifold exhibited by octahedral complexes. However, if the geometry and ligand field strength are known, actual stabilization energies can be estimated. For example, the ligand-field stabilization energy associated with the transformation of $[Cr(H_2O)_6]^{3+}$ into $[Cr(H_2O)_5]^{3+}$ corresponds to around 2 Dq, or 42 kJ/mole. Increasing the bond dissociation energy by 5.7 kJ/mole results in an order of magnitude decrease in exchange rates at room temperature (see Eq. 4.8). Thus, water exchange on $3d^3[Cr(H_2O)_6]^{3+}$ is much slower than that exhibited by $3d^5[Fe(H_2O)_6]^{3+}$, which has no net ligand-field stabilization energy.

We have seen that the strength of the Lewis acid–base interactions influences the Brønsted acidity of water molecules coordinated to metal cations. The inverse relationship is also true: protonation and deprotonation of coordinated water molecules can mediate Lewis acid–base interactions within the first solvation shell. For example, the water exchange rate for $[Fe(H_2O)_5OH]^{2+}$ is several hundred times faster than the rate measured for $[Fe(H_2O)_6]^{3+}$ (Fig. 4.9). For other cations, increases in the exchange rates of between 10^2 and 10^3 are common upon hydrolysis and reduction of net charge. This point is important geochemically because the deprotonation of many important hydrated cations occurs under natural pH conditions $(4 < pH < 10)$. Again, using simple formal charge arguments, deprotonating one water molecule leads to an increased electron density in the M–OH bond, and a corresponding decrease in electron density to the remaining $M-OH_2$ bonds. Stated differently, deprotonation of one water molecule in the solvation shell of Fe^{3+} creates a complex in which the cation looks more like Fe^{2+} than Fe^{3+} to the remaining water molecules, which results in weaker $M-OH_2$ bonds and more rapid exchange rates.

In a few instances, changes in solution pH can induce dramatic changes in exchange rates by changing the local coordination geometry around the metal cation. For example, $[Al(H_2O)_6]^{3+}$ is an octahedral complex that exchanges water via the I_d mechanism at a very slow rate (around 1/s, see Tables 4.1 and 4.2). However, at pH 8 and greater, the dominant complex is the tetrahedral ion: $[Al(OH)_4]^-$. This tetrahedral complex exchanges hydroxyl groups at a rate that is too fast to measure using ^{17}O-NMR methods (i.e., a rate greater than 10^6/s). Because Al(III) can exist in four-, five-, or six-coordination to oxygens, the pathway for this rapid exchange is presumed to be associative, although no experimental data exist to confirm this assignment. Dramatic changes in exchange rates involving local bonding and/or pH effects for solid oxides such as $Al(OH)_3$, gibbsite, are discussed further in Chapters 6 and 14.

Last, the relative kinetics of proton exchange and ligand exchange rates can influence reaction pathways. In many instances, protons exchange faster than the bound water molecules around hydrated cations [31]. For example, proton hopping between the bound waters and the hydroxyl group in $[Rh(H_2O)_5OH]^{2+}$ occurs on microsecond to millisecond timescales whereas the oxygens do not undergo exchange for many hours (Fig. 4.9 [16]). Although significantly slower than proton transport in bulk water, this proton hopping is still sufficiently fast that the metal does not always perceive a hydroxide ion at a fixed position, but six bonds that each represent a population-weighted average of the character of both bound water and bound hydroxide. In other words, one can view each metal–oxygen bond in the $[Rh(H_2O)_5OH]^{2+}$ ion as having five-sixths the character of a bound water molecule and one-sixth the character of a bound hydroxyl group.

References

1. McMullan, R. K.; Bonamico, M.; Jeffrey, G. A. Polyhedral Clathrate Hydrates: V. Structure of the Tetra-n-butyl Ammonium Fluoride Hydrate. *J. Chem. Phys.* **1963**, *39*, 3295–3310.
2. Jeffrey, G. A. Hydrate Inclusion Compounds. In *Comprehensive Supramolecular Chemistry*; Atwood, J. L.; Davies, J. E.; MacNicol, D. D.; Vogtle, F., Eds.; Pergamon: Oxford, UK, 1996; Vol. 8, pp 757–788.
3. Soper, A. K.; Weckstrom, K. Ion Solvation and Water Structure in Potassium Halide Aqueous Solutions. *Biophys. Chem.* **2006**, *124*, 180–191.
4. Kropman, M. F.; Bakker, H. J. Dynamics of Water Molecules in Aqueous Solvation Shells. *Science* **2001**, *291*, 2118–2120.
5. Shannon, R. Revised Effective Ionic Radii and Systematic Studies of Interatomic Distances in Halides and Chalcogenides. *Acta Crystallogr. A* **1976**, *32*, 751–767.

6. Lang, P. F.; Smith, B. C. Ionic Radii for Group 1 and Group 2 Halide, Hydride, Fluoride, Oxide, Sulfide, Selenide and Telluride Crystals. *Dalton Trans.* **2010**, *39*, 7786–7791.
7. Hewish, N. A.; Neilson, G. W.; Enderby, J. E. Environment of Ca^{2+} Ions in Aqueous Solvent. *Nature* **1982**, *297*, 138–139.
8. Kragten, J. *Atlas of Metal-Ligand Equilibria in Aqueous Solution*; Wiley: New York, 1978.
9. Baes, C. F. J.; Mesmer, R. E. *The Hydrolysis of Cations*. Krieger: Malabar, FL, 1986.
10. Swaddle, T. W.; Rosenqvist, J.; Yu, P.; Bylaska, E.; Phillips, B. L.; Casey, W. H. Kinetic Evidence for Five-Coordination in $AlOH(aq)^{2+}$ Ion. *Science* **2005**, *308*, 1450–1453.
11. Bilinski, H.; Ingri, N. A Determination of the Formation Constant of $SiO(OH)_3^-$. *Acta Chem. Scand.* **1967**, *21*, 2503–2510.
12. Jensen, W. B. *The Lewis Acid–Base Concepts: An Overview*; Wiley: New York, 1980.
13. Drago, R. S. Quantitative Evaluation and Prediction of Donor–Acceptor Interactions. *Struct. Bond.* **1973**, *15*, 73–139.
14. Pearson, R. G. Hard and Soft Acids and Bases. *J. Am. Chem. Soc.* **1963**, *85*, 3533–3539.
15. Ma, H.; Wan, C.; Zewail, A. H. Dynamics of Ligand Substitution in Labile Cobalt Complexes Resolved by Ultrafast T-Jump. *Proc. Natl. Acad. Sci. USA* **2008**, *105*, 12754–12757.
16. Richens, D. T. *The Chemistry of Aqua Ions*; Wiley: New York, 1997.
17. Wilkins, R. G. *The Study of Kinetics and Mechanism of Reactions of Transition Metal Complexes, 2nd ed.*; VCH: London, 1991.
18. Helm, L.; Nicolle, G. M.; Merbach, A. E. Water and Proton Exchange Processes on Metal Ions. *Adv. Inorg. Chem.* **2005**, *57*, 327–379.
19. Langford, C. H.; Gray, H. B. *Ligand Substitution Processes, 2nd ed.*; Benjamin Press: New York, 1974.
20. Burgess, J. *Ions in Solution: Basic Principles of Chemical Interactions*; Ellis-Horwood: Chichester, UK, 1988.
21. Margerum, D. W. Cayley, G. R.; Weatherburn, D. C.; Pagenkopf, G. K. Kinetics and mechanisms of complex formation and ligand exchange. *ACS Monogr.* **1978**, *174*, 1–220.
22. Burgess, J. *Metal Ions in Solution*. 1978. Ellis-Horwood, Chichester, UK.
23. Wilkins, R. G. *The Study of Kinetics and Mechanism of Reactions of Transition Metal Complexes*; Allyn & Bacon: Boston, MA, 1974.
24. Swaddle, T. W. Ligand Substitution Dynamics in Metal Complexes. In *Physical Inorganic Chemistry: Reactions, Processes and Applications*; Bakac, A., Ed.; Wiley: New York, 2010; pp 339–393.
25. Casey, W. H.; Swaddle, T. W. Why Small? The Use of Small Inorganic Clusters to Understand Mineral Surface and Dissolution Reactions in Geochemistry. *Rev. Geophys.* **2003**, *41*, 4/1–4/20.
26. Glasstone, S.; Laidler, K. J.; Eying, H. *The Theory of Rate Processes*; Princeton University Press: New York, 1941.
27. Ducommun, Y.; Newman, K. E.; Merbach, A. E. High-Pressure Oxygen-17 NMR Evidence for a Gradual Mechanistic Changeover from Ia to Id for Water Exchange

on Divalent Octahedral Metal Ions Going from Manganese(II) to Nickel(II). *Inorg. Chem.* **1980**, *19*, 3696–3703.
28. Helm, L.; Merbach, A. E. Inorganic and Bioinorganic Solvent Exchange Mechanisms. *Chem. Rev.* **2005**, *105*, 1923–1959.
29. Cotton, F. A.; Wilkinson, G. *Advanced Inorganic Chemistry*; Wiley: New York, 1988.
30. Basolo, F.; Pearson, R. G. *Mechanisms of Inorganic Reactions, 2nd ed.*; Wiley: New York, 1967.
31. Houston, J. R.; Phillips, B. L.; Casey, W. H. Residence Times for Protons Bound to Three Oxygen Sites in the $AlO_4Al_{12}(OH)_{24}(H_2O)_{12}^{7+}$ Polyoxocation. *Geochim. Cosmochim. Acta* **2005**, *70*, 1636–1643.

5

The Hydrolysis Products: Soluble Multi-cation Clusters

5.1 INTRODUCTION

The term *hydrolysis* describes the acid–base reactions that remove protons from hydrated metal cations [1]. This word derives from two Greek words: *hydro,* meaning "water," and *lysis,* meaning "to loosen." Hydrolysis provides two mechanisms for converting one soluble metal complex into another. First, hydrolysis converts coordinated water molecules into coordinated hydroxo- and oxo-anions. For isolated monomers, the extent of hydrolysis depends on cation charge, cation coordination number, and pH (see Chapter 4). For example, as the cation charge increases, complexes can be generated that go from water rich to hydroxide rich to oxide rich, as exemplified by the sequence of species $[Na(H_2O)_6]^+$, $[Mg(H_2O)_5(OH)]^+$, $[Al(H_2O)_4(OH)_2]^+$, $[Si(OH)_4]^\circ$, and finally $[MnO_4]^-$. Second, hydrolysis activates hydrated cations to participate in ligand-exchange reactions.

When the entering ligand is not another isolated water molecule, but involves a hydroxo- or oxo-ligand that is bound to another dissolved metal complex, ligand-exchange reactions are called *condensation reactions* (or sometimes *olation* reactions) because they condense small oxide species into larger ones. In condensation reactions, hydroxo- or oxo-ligands form bridges between two or more cations. Species created via condensation range from dimers, trimers, and tetramers to larger species such as $[Al_{13}O_4(OH)_{24}(H_2O)_{12}]^{7+}$, enormous clusters such as the 2-nm-diameter $Mo_{72}Fe_{30}O_x(OH)_y$ species, and eventually to colloidal particles and extended oxide and hydroxide phases.

The complete set of soluble metal-ion complexes containing only water molecules, hydroxide ions, and oxo-anions are called the *hydrolysis products*. The hydrolysis products represent the minimum basis set of complexes that must be taken into account to rationalize the aqueous chemistry of any oxide or hydroxide phase. The acid–base and ligand-exchange chemistry of monomeric hydrolysis products is described in Chapter 4. In this chapter, we emphasize the polymerization and depolymerization reactions that lead to the formation and disappearance of larger multi-cation clusters. An entire chapter is devoted to these clusters because *reactions involving the formation and disappearance of oligomeric hydrolysis products are the same as those reactions that lead to the precipitation and dissolution of oxides in water.*

The acid–base, ligand-exchange, and electrochemical properties of hydrolysis products and extended oxides are also strikingly similar. These similarities are not exotic, but are expected, because both the clusters and surfaces often exhibit similar local structures and charge distributions. However, although it can be exceedingly difficult to characterize reactions on solids interfacing with a solution, the structures of soluble clusters are well defined. Their reactions can be followed on a molecular scale with clarity. In this chapter, we first provide a survey of the structures and occurrence of all known hydrolysis products. We then focus on the chemistry of just those species containing more than one cation, explaining the impact that multiple cations have on oxygen acidities, ligand-exchange reactions, and species stabilities.

5.2 STRUCTURES OF THE HYDROLYSIS PRODUCTS

Representative structures for most common (and a few uncommon) hydrolysis products are depicted in Figure 5.1. These beautiful species represent the transition between aqueous solutes and solid oxides, with exquisite symmetries and rich chemistries that rival the complexities of organic molecules. In terms of general trends, cation-coordination geometries in multi-cation clusters are almost always identical to the coordination geometry established for the monomeric complexes for the same cation. For example, all hydrolysis products for Mg^{2+} contain octahedral cations, whereas those of Be^{2+} contain tetrahedral cations. Exceptions to this rule are rare but do exist. Aluminum changes coordination from six to five to four with pH increase [2], whereas boron changes coordination from three to four. The cluster $[Al_{13}O_4(OH)_{24}(H_2O)_{12}]^{7+}$ contains 12 octahedral Al^{3+} surrounding a single, central tetrahedral Al^{3+} cation.

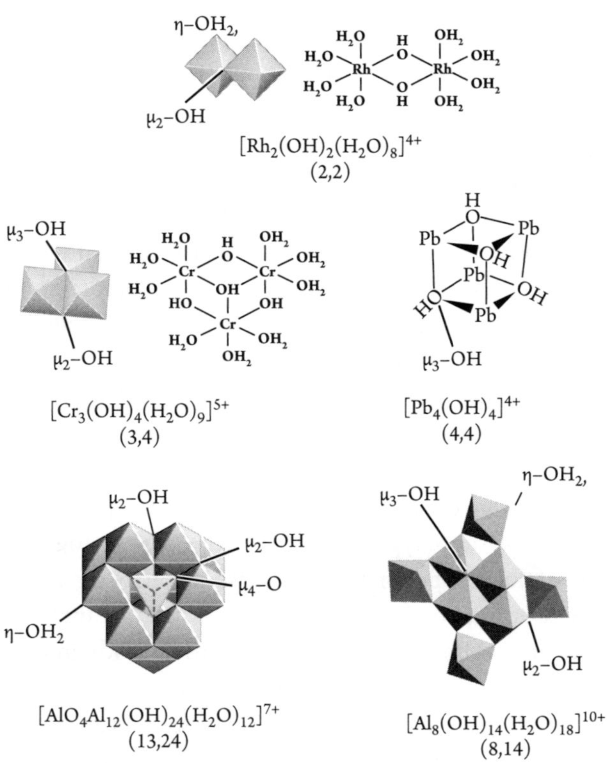

Figure 5.1 Depictions of representative soluble hydrolysis products for some +2, and +3 cations, including the doubly bridged, μ_2-OH, Rh(III) dimer of (2,2), a μ_2-OH and μ_3-OH trimer of Cr(III), (3,4), the μ_3-OH bridged cubane (4,4) molecule of Pb(II), a (13,24) cluster of Al(III) in the ε-Keggin structure, and an octamer of Al(III). Bound waters are labeled as η-OH_2.

In contrast to the narrow range found for cation coordination numbers, oxygen coordination geometries in hydrolysis products can vary considerably within a single cluster. Although clusters can contain water molecules, hydroxide ions, and oxo-anions that are bound to only one cation (called *nonbridging oxygens*), as in monomers, hydroxide ions can readily bridge between two or three metal centers whereas the oxo-anions can interact with up to six cations, as illustrated by MgO(s). The chemistry of the oxygen anions is key to understanding the aqueous chemistry of both soluble hydrolysis products and solid oxides in water.

Given the complexity of the multi-cation hydrolysis products, it is important to have notations that describe the structures of individual species without having to resort to cartoons such as those shown in Figure 5.1. At the highest level, we use standard (x,y) notation [1] to describe the overall composition of each hydrolysis product, such as (3,4) to describe clusters such as

$[Pb_3(OH)_4]^{2+}$. Here, the first and second numbers refer to the number of cations (e.g., three Pb(II) ions) and hydroxide ions (four) found in the species, respectively. For many hydrolysis products, which contain only water molecules and hydroxide ions, this shorthand notation can be used to determine the net charge on a given species by multiplying the cation valence by the first number and subtracting the second number. For example, for divalent cations, (3,4) clusters such as $[Pb_3(OH)_4]^{2+}$ have a net charge of $(+2)(3)-4=+2$. However, cluster charge cannot be derived from the simple notation for clusters containing oxo-bridges. For example, the aluminum cluster denoted by (13,24) would have a calculated cluster charge of $(+3)(13)-24=+15$, whereas the actual cluster charge for $[Al_{13}O_4(OH)_{24}(H_2O)_{12}]^{7+}$ is $+7$. This is because the four O^{2-} bridges that do not appear in the notation contribute an anionic charge of -8.

Another shortcoming of (x,y) notation is that it does not indicate the bonding configurations of the oxygens that hold the cluster together. One must always be aware of the number of metal cations bonded to a given oxygen center, because cation coordination is what determines the distinctive chemistry of each oxygen site. In this book, we use the standard notation in which nonbridging oxygens are denoted as η species, sometimes with a superscript to indicate the number of bonding atoms, whereas bridging oxygens are given the notation μ_n, where n corresponds to the number of metals bridged [3]. For example, the (2,1) complex $[(H_2O)_5Zn\text{-}\mu_2OH\text{-}Zn(H_2O)_5]^{3+}$ contains a single μ_2 hydroxyl bridge.

5.3 THE OCCURRENCE OF HYDROLYSIS PRODUCTS: READING HYDROLYSIS DIAGRAMS

Each element has its own unique set of hydrolysis products with relative stabilities that vary uniquely with pH and metal ion concentration. If a metal cation exhibits more than one stable oxidation state, each oxidation state has its own unique set of hydrolysis products. Hydrolysable metals typically exhibit a mix of cationic, neutral, and anionic complexes. Generally speaking, positively charged species contain an insufficient number of bound hydroxyl groups to neutralize the charge on all cations, whereas the anionic species contain an excess of anionic charge on hydroxyl groups. As described earlier, the net charge on each species can often be calculated from the species notation by multiplying the cation valence by the first number and subtracting the second number. For example, for a divalent cation, the (3,4) cluster has a net charge of $+2$, whereas the species (1,3) has a net charge of -1.

The distribution of all hydrolysis products for a given cation and oxidation state is typically captured in a plot called a *hydrolysis diagram* [1]. The hydrolysis diagram depicts the base 10 logarithm of the concentration of each species as a function of solution pH, where the dissolved species are in equilibrium with a specified solid oxide or hydroxide phase. The sum of the concentrations of all hydrolysis products defines the *solubility* of that specified phase. A typical hydrolysis diagram is shown in Figure 5.2. Notice that in hydrolysis diagrams, the concentration of each species is defined with a straight line, but that most lines have

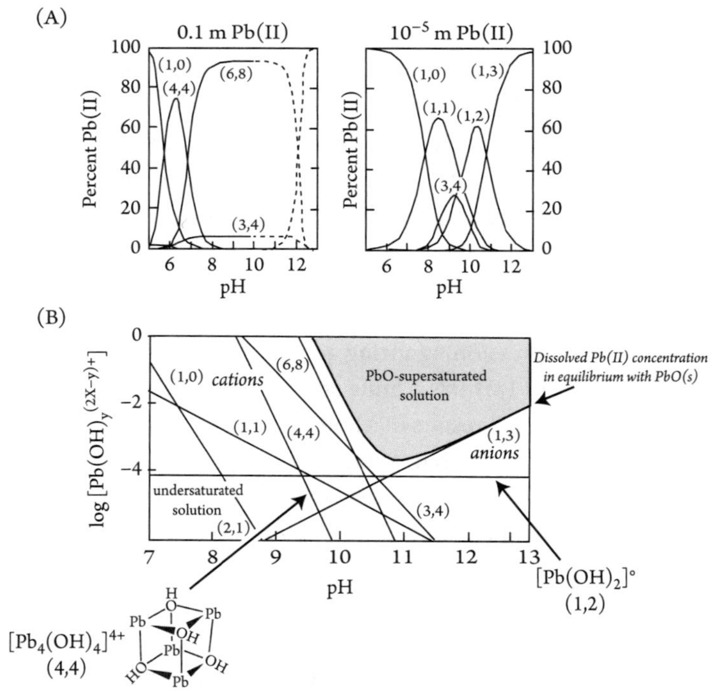

Figure 5.2 A hydrolysis diagram for Pb(II) dissolved in water [1]. (A) Species distributions as percentages of dissolved cation for total lead concentrations of 0.1 M (left) and 10^{-5} M(right). The dashed lines indicate solutions in which a solid is supersaturated. (B) A hydrolysis diagram for Pb(II) equilibrated with PbO(s). Arrows indicate the structures of two of the species present. Here, the logarithm in base 10 of the concentration of each soluble species is represented as a function of pH. The bold line corresponds to the sum of the species concentrations and is the total dissolved Pb(II) concentration in equilibrium with solid PbO. The minimum in this solubility is often also the point of zero charge for solid PbO. As solution pH is lowered relative to that line, cationic species predominate, surfaces become cationic, and solubility increases. As the solution pH is raised to more basic values, anionic species predominate in solution and at the PbO(s) surface, and the solubility of PbO also increases. The chemistry is amphoteric—meaning, that Pb(II) can act as either an acid or a base. The lightly shaded region indicates a supersaturated solution. Adapted from Baes, C. F.; Mesmer, R. E. *The Hydrolysis of Cations*; Krieger Publishing: Malabar, FL, 1986.

different slopes. This is because every line corresponds to a different equilibrium constant expression, and the slopes reflect the number of protons involved in each unique reaction.

Each hydrolysis diagram contains one line with a slope of zero that corresponds to the dissolution of neutral species from the solid reference phase. For example,

$$PbO(s)+H_2O(1)\rightleftarrows Pb(OH)_2^{\circ}(aq) \tag{5.1}$$

Here, $[Pb(OH)_2]^{\circ}$ is shorthand for the octahedral metal complex $[Pb(H_2O)_4(OH)_2]^{\circ}(aq)$. The activities of the pure solid PbO(s) and the pure solvent, $H_2O(1)$, are set to unity on the mole-fraction scale (in Raoult's Law). This means that the activity of the solid and solvent can be dropped from the equilibrium constant expression, which is given by

$$K_{1,2}=\left[Pb(OH)_2^{\circ}\right]=10^{-4}\,M \tag{5.2}$$

This equilibrium expression, ignoring activity coefficients, indicates that the species $[Pb(OH)_2]^{\circ}$ or (1,2) has a finite solubility of $\sim 10^{-4}$ M that is independent of solution pH. Eq. 5.2 results in a line having zero slope and an intercept at the y-axis in the hydrolysis diagram at $\log[M]=-4$.

In contrast to the neutral species described by Eqs. 5.1 and 5.2, the equilibrium concentrations of all charged and hydrolytic species exhibit clear pH dependencies. The behavior of cationic and anionic species is exemplified by the following reactions:

$$PbO(s)+H_3O^+(aq)\rightleftarrows Pb(OH)^+(aq)+H_2O(l) \tag{5.3}$$

$$K_{1,1}=\frac{\left[Pb(OH)^+\right]}{\left[H_3O^+\right]}=1.5\cdot 10^{-16} \tag{5.4}$$

$$3PbO(s)+2H_3O^+(aq)\rightleftarrows Pb_3(OH)_4^{2+}(aq)+H_2O(l) \tag{5.5}$$

$$K_{3,4}=\frac{\left[Pb_3(OH)_4^{2+}\right]}{\left[H_3O^+\right]^2}=10^{-23} \tag{5.6}$$

$$PbO(s)+3H_2O(l)\rightleftarrows Pb(OH)_3^-(aq)+H_3O^+(aq) \tag{5.7}$$

$$K_{1,3}=\left[Pb(OH)^+\right]\cdot\left[H_3O^+\right]=10^{-29} \tag{5.8}$$

As illustrated by Eqs. 5.3 through 5.8, dissolution reactions leading to the formation of cationic species result in a net consumption of protons, whereas those reactions leading to the formation of anionic species result in a net release of protons (or consumption of hydroxide ions). For this reason, as the solution pH increases, the concentrations of cationic species decrease and concentrations of anionic species increase.

The logarithms of concentrations plotted versus pH (i.e., the $-\log$ of $[H^+]$) in hydrolysis diagrams yield straight lines with slopes that indicate how many protons are involved in the reaction to make a given species from the neutral solid phase. For example, in Eq. 5.6, the concentration of (3,4) is proportional to $1/[H_3O^+]^2$. If we take the logarithm of this relation, the slope of the line appearing in the hydrolysis diagram is -2:

$$\log\left(K_{3,4}\right)=\log\left[Pb_3(OH)_4^{2+}\right]-2\cdot\log\left[H_3O^+\right]=-23 \qquad \textbf{(5.9)}$$

$$\log\left[Pb_3(OH)_4^{2+}\right]=-23-2\ \mathrm{pH} \qquad \textbf{(5.10)}$$

The sign of the slope is determined by the position of the hydronium ions in the equilibrium expression. Negative slopes indicate that protons are consumed in the reaction whereas positive slopes indicate the reverse. In most cases, the slope is equal to $-z$, where z is the net charge on the complex. The y-intercepts of each line correspond to the species concentration extrapolated to a condition where $[H_3O^+]=1$ M (i.e., at pH 0). Finally, the intersection point between any two lines represents the pH at which the concentrations of the two crossing species are equal. For simple monomeric hydrolysis products, the intersection point is equal to the pK_a for the acid–base equilibrium between the two species. For $[Pb(H_2O)_6]^{2+}$, pK_1, pK_2, and pK_3 are 7.9, 9.6, and 10.8, respectively, corresponding to the sequential deprotonation reactions that eliminate protons from bound waters to form $[Pb(H_2O)_5OH]^+$, $[Pb(H_2O)_4(OH)_2]^\circ$, and $[Pb(H_2O)_3(OH)_3]^-$, respectively (Fig. 5.2) [1].

5.4 TRENDS IN CATION HYDROLYSIS

Before exploring what hydrolysis diagrams reveal about the aqueous chemistry of oxides, we first describe trends in the occurrence of various hydrolysis products as a function of two important parameters: the charge and size of the metal cation. Most aquated monovalent ions (e.g., $Na^+(aq)$ and $Ag^+(aq)$) do not deprotonate to produce hydrolysis products except under extremely basic conditions. In other words, $[M(H_2O)_6]^+$ (or (1,0)) can be converted into monomeric hydroxide species such as (1,1) and (1,2) only at very high pH. The divalent cations do not undergo hydrolysis except in basic solutions, but at lower pH values than those required to hydrolyze +1 cations. At low solution concentrations and elevated pH, divalent cations tend to be dominated by monomeric hydroxide species. The species (1,1), (1,2), and (1,3) are commonly observed. The species (1,4) is seen at high pH for the smallest cations (such as Be(II) and the first series of transition metals), but not for large cations such as Pb(II).

For divalent cations, multi-cation clusters form at high concentrations, with the most common clusters being the (2,1) dimer and (4,4) tetramer depicted in Figure 5.1. Other clusters found for +2 cations include (2,2), (2,3), (2,6), (3,3), (3,4), and (6,8). With the exception of (2,6), all these clusters are cationic. The +3 and +4 cations exhibit more complex hydrolysis diagrams that tend to favor cationic clusters at low pH and monomeric hydroxide species at high pH. They have a much stronger tendency to form multimeric clusters than metals with a lower charge. Oxides containing cations having a charge of +5 or higher tend to form stable anions such as sulfate $[SO_4]^{2-}$ and permanganate $[MnO_4]^-$. These oxo-anions form because the metals are so small and have such high charges that the bonded oxygens are extremely strong acids. Waters or hydroxide ions bonded to these metals deprotonate immediately, leaving a large unprotonated oxyanion in solution. Metals such as Nb(V) and Ta(V) have no aqueous monomers, but are only present as anionic oligomers such as $[H_xNb_{10}O_{28}]^{(6-x)-}$ and $[Ta_6O_{19}]^{8-}$ (see Section 5.11).

5.5 HYDROLYSIS DIAGRAMS AS PREDICTORS OF OXIDE CHEMISTRY

Hydrolysis diagrams not only provide information regarding the concentrations of specific hydrolysis products, but also supply a wealth of information regarding the chemistry of extended oxide and hydroxide phases. First, the sum of the concentrations of all the soluble hydrolysis products defines the aqueous *solubility* of the selected oxide and hydroxide phase (e.g., AlOOH (boehmite) or $Al(OH)_3$ (gibbsite)). In general, the pH at which the net charge on all species is zero is the pH where the hydroxides and/or oxides in equilibrium with solution are the least soluble. In moving away from the neutral species, if the predominant soluble species have a charge of +1 or −1, oxides and hydroxides become 10 times (10^1) more soluble for every unit pH change away from the point of minimum solubility. Note that the power of 10 corresponds to the charge on the complex. In pH regimes in which the predominant species have a charge of +2 or −2, the oxides and hydroxides become 100 times more soluble per unit pH change. In the extreme case where $[Zr_4(OH)_8(H_2O)_{16}]^{8+}$ is predominant, the solubility of ZrO_2 increases by a factor of 100 million in response to dropping the pH by only 1 unit.

The solubility of oxides is normally defined by trends seen in hydrolysis diagrams (Fig. 5.3). As with hydrolysis, solubility reflects the acid–base chemistry of oxygen sites, which in turn is controlled by the size and charge of associated metal cations. Oxides of the +1 cations, such as Na_2O and Ag_2O, tend to exhibit high solubilities at all pH values. The oxides of +2 cations, such as MgO and ZnO, are soluble in acidic and neutral solutions, but have moderately low solubility in basic solutions. Solids of the +3 cations (e.g., $Al(OH)_3$, gibbsite) are *ampholytes* that exhibit low solubility in neutral solutions, but can be dissolved in both strong acids and bases. The +4 cations (e.g., Ti(IV), Si(IV), and Zr(IV)) tend to form oxo-bridges, and exhibit some of the lowest known solubilities for oxide and hydroxide materials. Oxides containing metal cations with a valence of +5 or greater form such strong metal–oxygen bonds they either form highly soluble anions such as $[MnO_4]^-$ and $[SO_4]^{2-}$ or polyoxometalate (POM) clusters. POM anions of Group 5 and 6 elements tend to exhibit their minimum solubility in acidic solutions but are highly soluble in neutral or basic solutions.

Hydrolysis products are not the only oxide species that are charged. Colloidal oxides and even extended oxide surfaces immersed in water almost always exhibit a positive or negative charge that is pH dependent. *The surface charge on oxide surfaces arises from the same local charge distributions and acid–base properties exhibited by oxygen sites in the hydrolysis products.* The net surface charge is controlled by the adsorption and desorption of protons (or hydroxide ions), which are called the *potential-determining ions* for oxides [4]. The solution pH at which the oxide surface is electrically neutral is called the *point of zero charge* (PZC). At pH values below the PZC, surfaces are positively charged, whereas above this pH, surfaces are anionic. The origin and manipulation of surface charge is covered in detail in Chapter 8 on colloid chemistry. However, in the context of this chapter on hydrolysis, it is important to note that the PZC almost always coincides with the pH at which the net charge on the hydrolysis products is zero and oxide or hydroxide solubility is at a minimum. The similarity between surface charge and hydrolysis product charge is not a coincidence (see Chapters 6 and 8); it is a reflection of the fact that the acid–base properties of species dissolved in solution resemble the acid–base properties of species found at the oxide–water interface.

5.6 SPECIATION DIAGRAMS INVOLVING FOREIGN LIGANDS

It is important to remember that hydrolysis diagrams represent the *minimum* basis set, or smallest set of complexes, that define the solubility and surface chemistry of an oxide. Many other ligands can be present in aqueous solutions that can compete with hydroxide ions and water molecules to form metal complexes via reactions, such as these [5]:

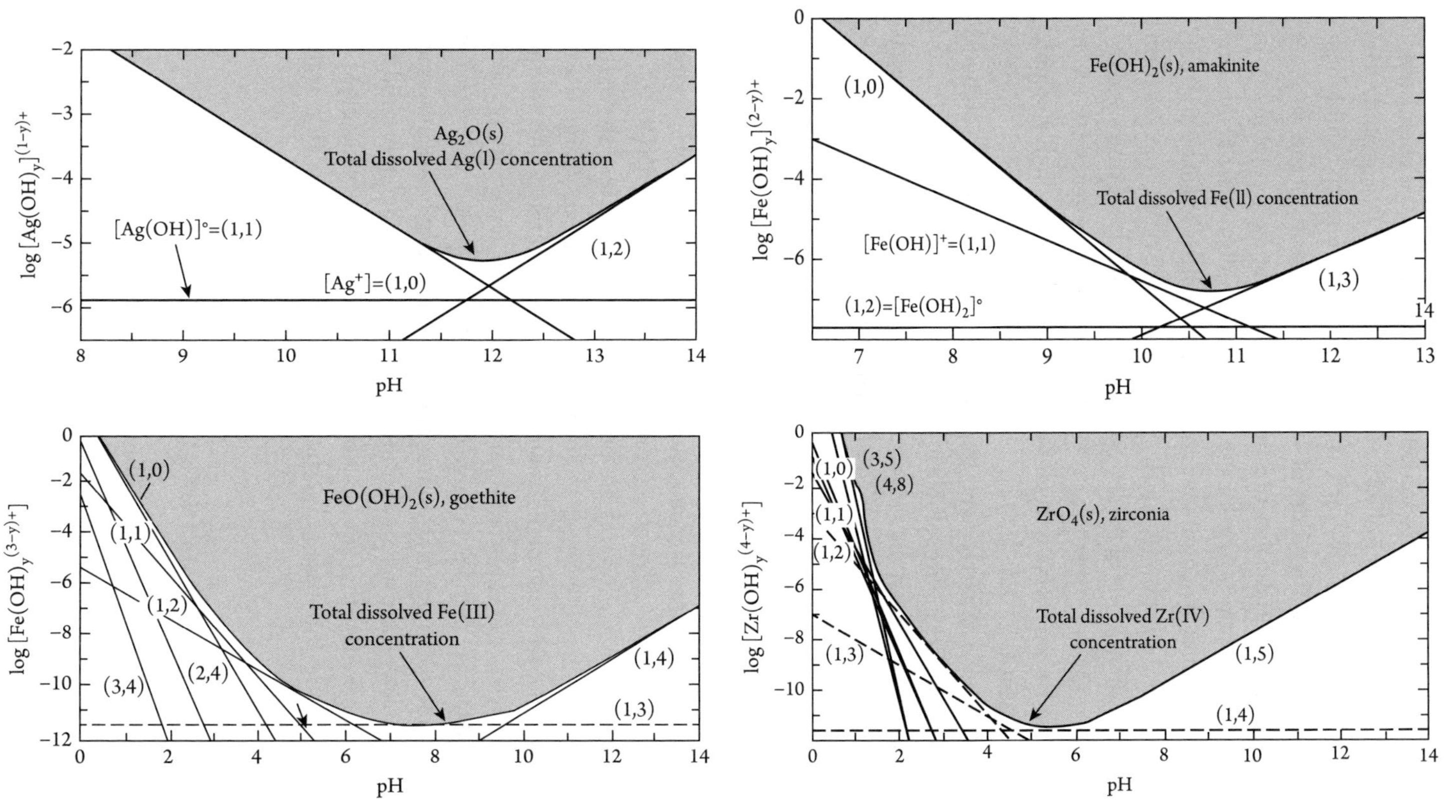

Figure 5.3 Representative hydrolysis diagrams for +1, +2, +3 and +4 cations showing trends in species distributions and oxide solubility as a function of solution pH. Note that the minimum in solubility of each solid decreases toward more acidic pH values with an increase in the metal charge. The lightly shaded regions indicate a solution supersaturated with the indicated solid phase. Adapted from Baes, C. F.; Mesmer, R. E. *The Hydrolysis of Cations*; Krieger Publishing: Malabar, FL, 1986.

$$\mathrm{Fe(H_2O)_6^{3+} + F^- \rightleftarrows Fe(H_2O)_5F^{2+} + H_2O} \qquad \textbf{(5.11)}$$

$$\mathrm{K = \frac{\left[Fe(H_2O)_5F^{2+}\right]}{\left[Fe(H_2O)_6^{2+}\right]\left[F^-\right]} = 1.5 \cdot 10^5} \qquad \textbf{(5.12)}$$

The impact of these foreign or spectator ligands can be substantial. For example, the solubility of iron oxides can be enhanced dramatically by exposing them to the cyanide ligand (CN^-) to form highly soluble species such as $[Fe(CN)_6]^{3-}$. The resulting complexes may not contain *any* oxygen anions in their inner coordination sphere. Foreign ligands change the charge and electron distributions for both the dissolved metal complexes discussed here and the surfaces of insoluble oxide phases (see Chapters 6 and 8). Here, we introduce several features associated with complexes containing foreign ligands, including acid–base chemistry, ligand exchange rates, and the impact these phenomena have on speciation diagrams and oxide solubility. The bottom line is that all ligands follow the same general rules that have been outlined earlier to describe the specific behavior of water and hydroxyl groups as ligands.

First, we need to consider the stability of a given ligand complex. One critical factor that influences this stability is the strength of the interaction between the central cation (the Lewis acid) and the ligand (the Lewis base). Lewis acid–base interactions are complex, and involve both electrostatic and covalent components. A ligand that is a stronger base toward one cation may actually be a weaker base with respect to another cation. The interaction strength is often determined by how well the molecular orbitals on the acid and base match up. Basic ligands can donate electron density into unoccupied metal orbitals on the metal cation. Conversely, metal cations (particularly transition metals) can donate electron density from occupied orbitals into unoccupied π-orbitals on ligands such as CO and CN^- (called *π back-bonding*). In Chapter 4, we saw that for the transition metals, ligand binding can remove the degeneracy within the d-orbital manifold, resulting in ligand-field stabilization that is ligand dependent. Parameterized models developed to quantify the net results of all these interactions include "hard and soft" acid–base correlations [6] and donor–acceptor models [7, 8].

A second key factor that controls ligand stability is the *chelate effect*. The term *chelate* comes from the Greek word for *claw*, evoking an image of a lobster grabbing a morsel of food. The chelate effect says that the more points of attachment a ligand has with a metal cation, the harder it is to displace that ligand. Monodentate ligands interact only with a cation at one binding site, so one ligand exchange reaction is sufficient to displace that ligand. However, for a bidentate ligand with two points of attachment, detachment at one site does not totally displace the ligand. In fact, the detached site can often reattach before the second site is displaced.

Table 5-1. EQUILIBRIUM CONSTANTS (K_1, β_n) FOR THE COMPLEXATION OF NI(II) BY MONODENTATE AMMONIA AND THEN BY POLYAMINES WITH THE SAME NUMBER OF BONDING NITROGENS

Reaction	**Equilibrium Constant**	**Centricity**
$Ni^{2+} + 2\,NH_3 = [Ni(NH_3)_2]^{2+}$	$\log \beta_2 = 5.08$	Monodentate
$Ni^{2+} + en = [Ni(en)]^{2+}$	$\log K_1 = 7.35$	Bidentate
$Ni^{2+} + 4\,NH_3 = [Ni(NH_3)_4]^{2+}$	$\log \beta_4 = 8.12$	
$Ni^{2+} + 2\,en_3 = [Ni(en)_2]^{2+}$	$\log \beta_2 = 13.54$	*Bis*-bidentate
$Ni^{2+} + trien = [Ni(trien)]^{2+}$	$\log K_1 = 14.4$	Tetradentate
$Ni^{2+} + 6\,NH_3 = [Ni(NH_3)_6]^{2+}$	$\log \beta_6 = 9.08$	
$Ni^{2+} + 3\,en = [Ni(en)_3]^{2+}$	$\log \beta_3 = 17.71$*	*Tris*-bidentate
$Ni^{2+} + penten = [Ni(trien)]^{2+}$	$\log K_1 = 19.1$	Hexadentate

*en, ethylenediamine or ethane-1,2-diamine ($H_2NCH_2CH_2NH_2$); †dien, diethylenetriamine or 1,2-ethanediamine,N-(2-aminoethyl) or $HN(CH_2CH_2NH_2)_2$; ‡trien = *N,N′*-bis(2-aminoethyl)ethane-1,2-diamine, $[CH_2NHCH_2CH_2NH_2]_2$; §trien = *N,N′*-bis(2-aminoethyl) ethane-1,2-diamine, $[CH_2NHCH_2CH_2NH_2]_2$; ¶penten = $(NH_2CH_2CH_2)_2NCH_2CH_2N(NH_2CH_2CH_2)_2$.

Data From: Martell, A. E.; Hancock, R. D. Metal Complexes in Aqueous Solution. In *Modern Inorganic Chemistry*; Fackler, J. P. J., Ed.; Plenum Press: New York, 1996. p.63.

The impact of the chelate effect is illustrated by simple metal complexes involving ligands that occupy more than one site in the inner coordination sphere [5]. All ligands in Table 5.1 contain amine groups that can bind to Ni(II). The differences between these ligands involve how many amine groups are present within a single coordinating ligand. For each series shown, the total number of amine groups bound to Ni(II) is identical. However, dramatic increases are observed in the equilibrium constant for ligand binding as one moves from mono- to di- to tri- and finally to hexadentate ligands. In fact, the nitrogen donors in the hexadentate chelating ligand *penten* bind to Ni(II) *10 billion times* more effectively than the simple monodentate ligand ammonia. This chelate effect not only applies to foreign ligands, but to the stability of bonding motifs within both multi-cation hydrolysis products and extended oxide surfaces.

Although parameterized acid–base models and the chelate effect can be used to rationalize ligand binding, nothing beats actual experimental data. Numerous compendia of equilibrium constants are available for a wide range of water-soluble metal–ligand complexes [5–9]. Several important trends emerge from an examination of equilibrium constants involving foreign ligands [9]. First, just as in the case of the OH^- ligand (Chapter 4), ligand exchange proceeds in a stepwise rather than a simultaneous fashion. For example, the cyanide ligand substitutes into $[Cd(H_2O)_6]^{2+}$ one ligand at a time to form $[Cd(H_2O)_5(CN)]^+$,

$[Cd(H_2O)_4(CN)_2]^\circ$, $[Cd(H_2O)_3(CN)_3]^-$, and finally $[Cd(H_2O)_2(CN)_4]^{2-}$. The logarithms of the equilibrium constants for each step in this sequence are 5.48, 5.12, 4.63, and 3.65, respectively, indicating that subsequent exchanges are progressively harder to carry out.

As in the case of water, deprotonated (and therefore more anionic) ligands are more basic and are thus more capable of displacing water molecules in the inner coordination sphere around most metal cations. This means that complexation by foreign ligands tends to be low for pH values below the pK_a of the ligand. From the ligand's perspective, protons outcompete most metals for occupying basic sites below the ligand's pK_a. Another effect associated with the binding of anionic ligands is that ligand binding often reduces the net charge on the aqueous complex (as shown clearly by the Cd(II)/CN example presented earlier). For the specific example of the OH^- ligand, we have already seen how the net complex charge has a dramatic impact on the Brønsted acidity of the remaining water molecules in the inner coordination sphere. In Chapters 8 and 10, we see how such charge manipulation impacts both the colloid and ion exchange chemistry of insoluble metal oxide surfaces.

Foreign ligands have a dramatic effect on the solubility (as well as the precipitation and dissolution) of all oxides. For example, consider solubility diagrams for Al(III) in the presence of different ligands [9] (Fig. 5.4). In acidic solutions, the fluoride ion can readily displace waters of hydration around Al(III) to make $[AlF_x(OH_2)_{6-x}]^{3-x}$ complexes, in analogy to Eq. 5.11, because fluoride is a stronger nucleophile than the water molecule. In contrast, fluoride complexes are not dominant in basic solutions because (1) the hydroxide ion is a stronger nucleophile than water and can compete with fluoride ions for binding sites, and (2) in basic solutions, hydroxide ions are present at high concentrations. For these reasons, complexation by fluoride ions results in dramatic increases in solubility in acidic solutions, whereas the solubility in basic solutions is largely unchanged. Similar trends are observed for most other ligands. However, for many ligands, binding is negligible below the pK_a of the ligand because protons outcompete the dissolved metal cations for active ligand binding sites. In other words, protons compete with metals for ligands, and ligands compete with one another for metals.

Foreign ligands affect rates of ligand exchange reactions via many of the same factors that control acid–base and complexation equilibria involving water molecules, including (1) ligand basicity, (2) complex charge, (3) the chelate effect, (4) hydrogen bonding, and (5) electron-spin pairing in transition metal complexes (see Chapter 4). In terms of ligand basicity, the replacement of one water molecule by a more basic molecule such as ammonia creates a stronger $M-NH_3$ bond while weakening all the remaining $M-OH_2$ bonds. Many geochemically relevant ligands, including $[OH]^-$, halides, $[SO_4]^{2-}$, $[CO_3]^{2-}$, and $NH_3(aq)$,

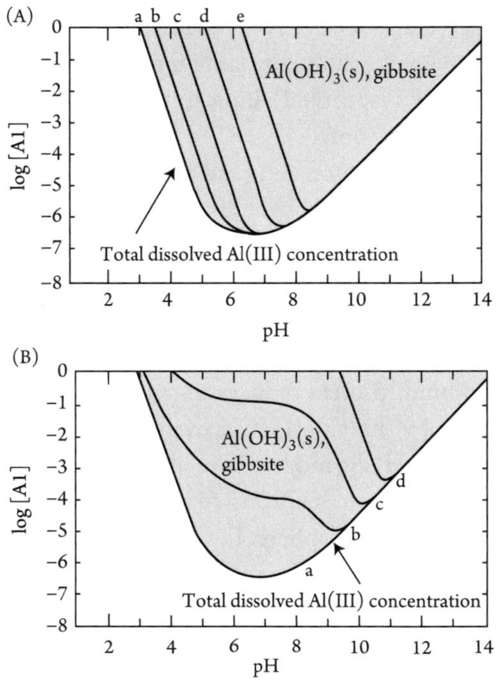

Figure 5.4 Solubility diagrams for $Al(OH)_3$ (gibbsite) in the presence of foreign ligands. (A) Solubility in the presence of various F^-, concentrations ($[F^-] = 0, 10^{-5}, 10^{-4}, 10^{-3}$, and 10^{-2} M for curves a, b, c, d, and e, respectively). (B) Solubility in the presence of the acetylacetonate ligand at concentrations of 0, 10^{-3}, 10^{-2}, and 0.1 M for curves a, b, c, and d, respectively. Adapted from Kragden, J. *Atlas of Metal Ligand Equilibria in Aqueous Solution*; Wiley: New York, 1978.

and biodecay products such as carboxylates exhibit such behavior. For example, water-exchange rates increase with each stepwise replacement of a bound water molecule by NH_3 in the series of complexes $[Ni(NH_3)_x(H_2O)_{6-x}]^{2+}$ (Table 5.2).

The rates of exchange of a bound water in a series of $[AlF_x(OH_2)_{6-x}]^{3-x}$ complexes increases by about a factor of 100 for every added fluoride ion or unit reduction in complex charge (Table 5.2). In terms of complex charge, the addition of an anionic ligand such F^- or Cl^- has roughly the same impact as replacing a bound water molecule by a bound hydroxyl group. As seen when replacing water molecules by hydroxyl groups (Chapter 4), reducing the net charge on the complex is equivalent to reducing the charge on the central metal cation, which weakens bonds and increases the exchange rates. Therefore, increased rates of exchange are observed in moving from $[Cr(H_2O)_6]^{3+}$ to $[Cr(H_2O)_5Cl]^{2+}$ to

Table 5-2. RATE COEFFICIENTS FOR EXCHANGE OF A SINGLE WATER MOLECULE FROM AROUND Ni(II) AMMONIA COMPLEXES WITH VARYING LIGAND-TO-METAL RATIOS AND Al(III) FLUORIDE COMPLEXES WITH VARYING LIGAND-TO-METAL RATIO AND CHARGES

Complexes	**k_{298}/s**	**Complexes**	**k_{298}/s**
$[Ni(H_2O)_6]^{2+}$	$0.32 \cdot 10^5$	$[Al(H_2O)_6]^{3+}$	1.2
$[Ni(NH_3)(H_2O)_5]^{2+}$	$2.5 \cdot 10^5$	$[AlF(H_2O)_5]^{2+}$	240.
$[Ni(NH_3)_2(H_2O)_4]^{2+}$	$6.1 \cdot 10^5$	$[AlF(H_2O)_4]^{+}$	16,500
$[Ni(NH_3)_3(H_2O)_3]^{2+}$	$25 \cdot 10^5$		

SOURCE: Data from Burgess, J. *Ions in Solution: Basic Principles of Chemical Interactions;* Ellis-Horwood: Chichester, 1988; Margerum, D. W.; and Rosen, H. M.; The Effect of Coordinated Ligands on the Rate of Replacement of Bound Water by Ammonia in Nickel(II) Complexes. *J. Am. Chem. Soc.* **1967**, *89*, 1088–1092; Yu, P.; Phillips, B. L.; Casey, W. H. Water Exchange in Fluoroaluminate Complexes in Aqueous Solution: a Variable Temperature Multinuclear NMR Study *Inorg. Chem.* **2001**, *40*, 4750–4754.

$[Cr(H_2O)_4Cl_2]^+$. Aquation of the *trans*-isomer of $[Cr(H_2O)_4Cl_2]^+$ to form $[Cr(H_2O)_5Cl]^{2+}$ in acidic solutions proceeds 300 times faster than the subsequent aquation of $[Cr(H_2O)_5Cl]^{2+}$ to $[Cr(H_2O)_6]^{3+}$.

For transition metals, ligands can induce dramatic changes in reactivity on the basis of the ligand-field effects described in Chapter 4. This is particularly true if the incoming ligand is highly basic and induces a large ligand-field splitting in the d-orbital manifold. Such splitting can transform a high-spin aquated complex into a low-spin complex in which more d electrons are paired and strong metal–ligand bonds make the complex less reactive. For example, the $3d^6$ Fe(II) cation in $[Fe(H_2O)_6]^{2+}$ has four unpaired electrons (see Fig. 4.10), is paramagnetic, and is highly reactive. Water exchange in this complex occurs in less than a microsecond at room temperature. In contrast, in the diamagnetic $[Fe(CN)_6]^{4-}$ complex, all six electrons are paired within the low-energy e_g orbitals. This complex is as kinetically inert as the $[Co(H_2O)_6]^{3+}$ ion [10] described in Chapter 4 (see Fig. 4.9), which reacts in years rather than in microseconds.

5.7 THE ACID–BASE CHEMISTRY OF MULTI-CATION HYDROLYSIS PRODUCTS

Both metal–oxide clusters and solid oxide surfaces look like close-packed arrays of oxygen anions when viewed from the aqueous solution. However, although each anion looks the same, they certainly do not act the same. The acid–base chemistry of each oxygen anion can vary dramatically depending on what metal cations are below the surface and how many cations are bonded to a given oxygen. We saw, for simple monomeric solvated cations, that as the cation charge increases from +2 up to above +5, the K_a values (i.e., Brønsted acidity) for bound water molecules can be lowered by more than 14 orders of magnitude to span the entire pH range available in water. Here, we examine the range of Brønsted acidities that can be generated within multi-cation clusters.

The starting point for rationalizing pK_a values for oxygen sites in clusters involves calculating the formal charges for oxygens that bridge between two or more cations. When the formal charge is known, correlations of pK_a with formal charge can be established [1]. In a few instances, pK_a values for specific oxygen types have been measured, which can be used to determine how accurate formal charge models are for predicting the acidities of a broad range of oxygen types. The formal charges for possible aquo-, hydroxo- and oxo-species are provided in Table 5.3 based on the charge of the metal and the number of protons bonded to the oxygen. In Figure 5.5, qualitative predictions of pK_a values are indicated for each acid–base reaction that is expected to occur within the "normal" pH range for water of $pH\ 0 < pH < 14$.

The pK_a values for multi-cation hydrolysis products are dictated by both cation valences and oxygen coordination numbers. Starting with the role of cation valence, we have already seen that for octahedral complexes, an increase in the cation valence by 1 unit increases the formal charge donated by the cation by +1/6. For coordinated water molecules and nonbridging hydroxyl groups, this

Table 5-3. Formal Charges on the Oxygens Calculated for Aquo, Hydroxo, and Oxo Species of Generic Metals That Differ in Charge*

Charge	**$^1\eta$-OH_2**	**pK_a**	**$^1\eta$-OH**	**pK_a**
+1	+1/6	~13	−5/6	>>14
+2	+1/3	~9	−2/3	>>14
+3	+1/2	~5	−1/2	>>14
+4	+2/3	~1	−1/3	>14
	μ_2-OH_2	**pK_a**	**μ_2-OH**	**pK_a**
+2	+2/3	~1	−1/3	>14
+3	+1	<0	0	~14
+4	+4/3	<<0	+1/3	~ 9
	μ_3-OH	**pK_a**		
+2	0	~14		
+3	+1/2	~5		
+4	+1	< 0		

*The formal charge on the oxygen in $[Ni(H_2O)_6]^{2+}$, for example, is +1/3. The estimates of pK_a values from formal charges are all qualitative only. The pK_a value in any column corresponds to the oxygen site to the immediate left. Any entry << or >> notation is expected to fall outside the normal pH range of water.

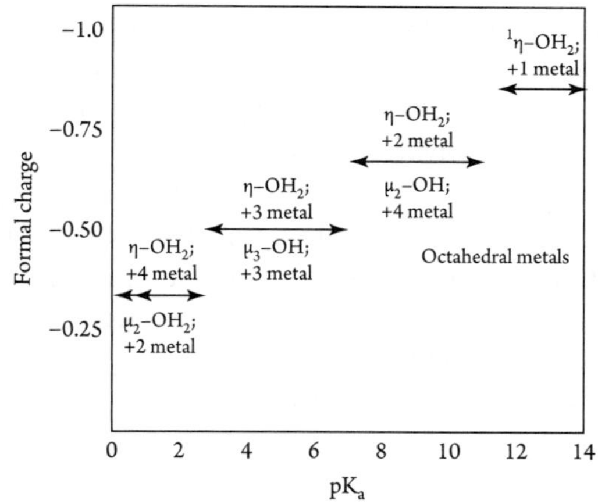

Figure 5.5 A depiction of the range of acidities (or pK_a values) found for both nonbridging (η-OH_2) and bridging oxygen types as a function of the formal charge on the deprotonated oxygen species in the given acid–base reaction. Note that acid–base chemistry reflects the charge on the metal as well as the oxygen-coordination geometry.

formal charge increase lowers the pK_a by roughly 4 units. As the cation charge increases from +1 to +2, +3, and +4 in the series Na^+, Mg^{2+}, Al^{3+}, and Ti^{4+}, the pK_a drops from around 13 to 9, 5, and 1, respectively (i.e., from strongly basic to mildly basic to mildly acidic to strongly acidic). For μ_2-bridging oxygens, *two* cations contribute to the oxygen formal charge, whereas for μ_3 bridges, *three* cations contribute to the charge, amplifying the impact of cation valence on the acid–base chemistry. For these respective bridges, an increase in cation valence by 1 unit increases the formal charge on oxygen by $+1/3$ and $+1/2$, which means that a unit increase in cation charge should lower the pK_a for the μ_2-and μ_3-hydroxyl bridges by 8 units and 12 units, respectively. The reactivity trend makes sense. As more cations bind to an oxygen anion, less electron density is available for binding to protons. This eventually lowers the pK_a to the point where protonation is no longer possible.

Let us now take a survey of both the occurrence and acidity of different oxygen types as a function of cation charge, starting with nonbridging oxygens and progressing to μ_2 and μ_3 bridges. For cations having a charge of +1, nonbridging oxygens in the form of coordinated water molecules are the predominant species regardless of pH. Increasing the cation charge to +2 makes it easier to deprotonate these water molecules to form bound hydroxide ions (see Chapter 4). The pK_a values for bound water molecules decrease as the cation charge increases from +2 to +3 to +4. For octahedral cations, it is difficult to deprotonate bound hydroxyl groups even at pH 14 until the cation charge is +5 or greater. However, such deprotonation is possible for tetrahedral species such as $[Si(OH)_4]^\circ$.

Moving on to μ_2-bridges, no such bridges have been observed in aqueous solution for +1 cations. Here, M−O bonds are so weak, and exchange rates are so rapid, that if such species form, they are highly transient. The μ_2-bridges that could conceivably form involving +2 cations include bridging water molecules (at low pH), bridging hydroxyl groups, and bridging oxo-groups (at extremely high pH values). For +2 cations, doubly protonated water bridges have not been observed in solution (although they can be found in molecular crystals). This trend makes sense from a formal charge perspective because the oxygen sites in such bridges would have a highly unstable formal charge of +2/3. However, loss of a proton to form a bridging hydroxyl group results in a lowering of the formal charge on the bridge to −1/3, which constitutes a stable configuration. Concentrated solutions of +2 cations often contain significant populations of species having μ_2-OH bridges at intermediate pH values. However, the μ_2-OH bridges for +2 cations are near neutral and are difficult to deprotonate to form μ_2-oxo bridges (having a formal charge of −4/3) regardless of solution pH. For octahedral +3 cations, neutral μ_2-OH bridges are extremely stable and are the dominant bridge configuration over all but the most extreme pH conditions. These bridges are difficult either to protonate or deprotonate. For octahedral +4 cations, the μ_2-OH group (formal charge =+1/3) can be deprotonated under mildly basic conditions to make oxo-bridged structures.

Finally, formal-charge distributions can be used to evaluate the relative occurrence and pK_a values of μ_3-bridge configurations. There are no stable μ_3–water bridges for any metal cation in water, because these bridges would be greatly overbonded and overcharged. For +2 cations, neutral μ_3-OH bridges such as those found in the $[Pb_4(OH)_4]^{4-}$ ion (Figs. 5.1, 5.2) are highly stable and are difficult to protonate or deprotonate regardless of pH. The μ_3-OH bridge for +3 cations is slightly less stable (with a formal charge of +1/2), and exhibit pK_a values in the neutral pH regime. These bridges deprotonate in basic solutions to form μ_3−oxo-bridges. For +4 cations, μ_3 hydroxy-bridges are so acidic (with a formal charge of +1) they lose their protons to form neutral oxo-bridges regardless of pH.

At this point, it is important to note that the acidity of both μ_2 and μ_3 bridges can be substantially different if the cation coordination changes or if one is dealing with mixed bridges. The classic example of this behavior is shown by the Al(III)−μ_2-O−Si(IV) bridges found in most aluminosilicates, such as zeolites. Here, both Al(III) and Si(IV) are in tetrahedral coordination, which imparts a formal charge of −0.5 to the oxo-bridge. Referring to Figure 5.5, this means that the bridge has a pK_a of around 5. This intermediate pK_a value is critically important for the ion exchange, catalysis, and dissolution behaviors of aluminosilicates (see Chapter 10 and Part Five).

Because the pK_a ranges for nonbridging, dibridging, and tribridging oxygens overlap, it is possible to have distinctly different oxygen sites exhibiting similar acid–base properties in a large molecule or on a surface, as illustrated in Table 5.3 and Figure 5.5. For example, for +3 cations, formal-charge analyses suggest that a terminal water molecule and a μ_3-OH group (both with a formal charge of +0.5) should exhibit similar pK_a values. It is difficult to assign specific acid–base reactions and pK_a values to specific oxygens in an olimeric cluster. In rare instances, it is possible to make assignments based on the consequences of site protonation and deprotonation, such as the activation of site-specific condensation or dissolution reactions.

As difficult as it is to assign pK_a values to oxygens in a multimeric cluster ion, it is even more difficult to assign pK_a values to oxygens at the aqueous interface of a metal–oxide solid, which contains a distribution of oxygen types and proton affinities. An additional complication arises at interfaces where the surface charge itself affects the acid–base equilibria. Such interactions are separated from intrinsic proton affinities at surfaces but not for ions (see Chapter 6). Addressing these complications is a major goal of computational chemistry and quantum mechanical methods of simulation.

It is instructive to compare the predictions of formal-charge models with experimental results that have been obtained for specific hydrolysis products. Some metals (such as Cr(III) and Rh(III)) are so unreactive that complexes formed by hydrolysis can be separated using ion-exchange columns and then titrated as distinct species [11–13]. The pK_a values for these species can be determined without worrying about competition from other species, or whether acid–base chemistry leads to rapid and irreversible condensation reactions.

Consider the double-bridged Cr(III) (2,2) dimer $[(H_2O)_4Cr\text{-}(\mu_2\text{-}OH)_2\text{-}Cr(H_2O)_4]^{4+}$. This dimer has three symmetrically distinct oxygens: two μ_2-OH bridges, four water molecules that are *cis* (or adjacent to) to the bridges, and four water molecules that are *trans* (on opposite sides from the bridges). Acid–base titrations of the dimer indicate that it exhibits two distinct pK_a values of 3.7 and 6.0 within the cluster's stability window. Which of the three oxygen types are most likely to account for the two pK_a values? Formal charge arguments indicate that it should be difficult either to protonate or deprotonate the neutral μ_2-OH bridges to create sites with formal charges of +1 and −1, respectively. Experimentally, water-bridged and oxo-bridged dimers are almost never observed unless they are supported by other organic ligands. (One notable exception is the oxo-bridged dimer of Fe(III), which can only form during hydrolysis.) This means that the two observed pK_a values discussed earlier are undoubtedly associated with the deprotonation of bound water molecules and not the bridge.

Comparisons among the pK_a values for the dimers with those of the corresponding Cr(III) monomer (which exhibits two pK_a values at 4.3 and 6.1, corresponding to the sequence $(1,0) \rightleftarrows (1,1) \rightleftarrows (1,2)$ suggest that the dimeric cluster undergoes sequential deprotonation of its water molecules just like the monomer. However, the first pK_a is substantially lower than that of the monomer. This pK_a shift is a reflection of the so-called "*trans* effect" that is observed in some transition-metal complexes, including those of Co(III) [10]. For our specific (2,2) dimer, withdrawal of electron density into a metal–hydroxyl bond has the effect of weakening the M—O bond that is directly opposite. Weakening the M—O bond strengthens the O—H bonds in the *trans* water molecules, resulting in an increased pK_a. Conversely, the *cis* water molecules experience a slight increase in their Lewis acid–base interaction with the cation, which should make their O—H bonds more acidic, giving rise to the observed pK_a of 3.7. This is yet another example of the coupling between Lewis and Brønsted acidities within oxide and hydroxide systems.

The trimeric (3,4) cluster exhibits three pK_a values [12]. Two of the values (6.0 and 4.3) are similar to those seen for the dimer and are probably associated with the waters of hydration, as in the case of the dimer. The new pK_a value of 5.6 is believed to be associated with the presence of a new oxygen type in the trimer, which is a single μ_3-OH group. As discussed earlier, this site has exactly the same formal charge as a nonbridging water molecule, and it is expected to have a similar pK_a value (5.6 compared with 6.0). However, as the hydrolysis product examples illustrate, spectroscopic methods are required to determine exactly which site exhibits which pK_a value. One difficulty among many is that changes in pH can stimulate the ligand-exchange reactions described later, including condensation reactions, precipitation, and proton-enhanced dissociation.

5.8 MECHANISMS FOR REVERSIBLE CONDENSATION REACTIONS

Multi-cation hydrolysis products form via *condensation reactions* in which an oxygen-containing ligand in the inner coordination sphere of one species exchanges for a water molecule or other host ligand within a second species to form an oxide bridge. What are the mechanisms that lead to oxide polymerization via bridge formation? What factors determine cluster stability and dictate both the cation content and the coordination numbers of the structural bridging oxygens? To answer these questions, we address the fundamental chemistry of condensation reactions starting with the formation of a single bridge, by reactions leading to multiple bridges, and finally considering factors that allow for the growth of dimers into trimers, tetramers, and larger oligomers. Once again, these condensation reactions illustrate the strong coupling between the Brønsted acidity, which controls oxygen protonation, and the Lewis acidity that controls the ligand exchange reactions.

First, consider what happens when two monomeric hydroxide species encounter each other:

$$2\,Zn(H_2O)_5(OH)^+ \rightleftarrows \left[(H_2O)_5\,Zn\text{-}\mu_2(OH)\text{-}Zn(H_2O)_5\right]^{3+} + OH^- \quad \textbf{(5.13)}$$

In shorthand notation, this reaction is equivalent to $2(1,1) \rightleftarrows (2,1) + OH^-$, which corresponds to the intersection point between (1,1) and (2,1) in the hydrolysis diagram for Zn^{2+}. For this reaction, the dimer becomes more predominant than the monomer as the zinc concentration increases or the pH decreases relative to this intersection. Formation of the μ_2-OH bridge in (2,1) represents a ligand exchange reaction in which a water molecule coordinated to one Zn^{2+} is replaced with a hydroxyl group that is already attached to another Zn^{2+}. This reaction is followed by the protonation of the remaining hydroxyl group (which is equivalent to the release of OH^- into solution).

The first important observation regarding bridge formation is that condensation reactions between dissolved metal cations are catalyzed by their hydrolysis (i.e., conversion of bound water molecules into bound hydroxides). For example, rates of dimer formation versus the extent of hydrolysis have been measured for $[Cr(H_2O)_6]^{3+}$ (Table 5.4) [14]. For the series shown in Table 5.4, the rate of condensation can ultimately be increased by a factor of more than 1 million by increasing the extent of hydrolysis on one or both of the metal cations.

Several other factors contribute to enhanced condensation rates. First, the rate-limiting step for ligand exchange on octahedral complexes often involves the loss of one water molecule to form an open coordination site that can be filled by

Table 5-4. Numerical Values for the Rate Coefficient, $k_{p,q}$, to Form a Cr(III) Dimer, $[(H_2O)_4Cr\text{-}(\mu_2\text{-}OH)_2\text{-}Cr(H_2O)_4]^{4+}$, at 298 K and $I = 1.0$ M ($NaClO_4$)

Reaction	$\mathbf{k_{p,q}}$ **($M\,s^{-1}$)**
$[Cr(H_2O)_6]^{3+} + [Cr(H_2O)_6]^{3+} \rightarrow$ dimer	Too slow to measure
$[Cr(H_2O)_5OH]^{2+} + [Cr(H_2O)_6]^{3+} \rightarrow$ dimer	$k_{1,0} = 6 \cdot 10^{-6}$
$2[Cr(H_2O)_5OH]^{2+} \rightarrow$ dimer	$k_{1,1} = 2(\pm 0.4) \cdot 10^{-4}$
$[Cr(H_2O)_4(OH]_2]^{+} + [Cr(H_2O)_5OH]^{2+} \rightarrow$ dimer	$k_{1,2} = 3.8(\pm 1.0) \cdot 10^{-2}$
$2[Cr(H_2O)_4(OH]_2]^{+} \rightarrow$ dimer	$k_{2,2} = 1.8(\pm 0.2)$

Source: Rotzinger, F. P.; Stünzi, H.; Marty, W. Early Stages of the Hydrolysis of Chromium(III) in Aqueous Solution. 3. Kinetics of Dimerization of the Deprotonated Aqua Ion *Inorg. Chem.* **1986**, 25, 489–495.

another ligand. For octahedral aquo complexes, the kinetics of both water and ligand exchange increase by a factor of around 10^2 as each bound water molecule is converted into a bound hydroxyl group. These rate enhancements arise as a result of Lewis acid–base reactions that shift electron density into the M−OH bonds and out of the M−OH_2 bonds, which in turn weakens the M−OH_2 bonds (see Chapter 4). Second, a bound hydroxyl group is a stronger nucleophile than a bound water molecule [15], and can compete more effectively with neutral bulk water molecules for the open coordination site. For the example of Zn^{2+}, a bound water molecule has a formal charge of +1/3 compared with the charge on the bound hydroxide ion of −2/3. Third, electrostatic repulsion between the two dissolved metal cations is lowered as complex charges are reduced via deprotonation, facilitating approach and contact between the reacting species. This trend is readily apparent in the data for Cr(III) complexes forming the doubly bridged dimer (Table 5.3) [14].

Although deprotonation of a coordinated water molecule activates bridge formation, what activates the reverse reaction, which is bridge disruption? Again, detailed studies of inert-metal oligomers identify the key steps [16], as adapted to describe the (2,1) Zn(II) dimer shown in Eq. 5.13. Disruption of the dimer requires a lowering of the solution pH, which in turn implies that depolymerization occurs when the dimer is protonated. However, we know from formal charge arguments that all the bound water molecules have a charge of +1/3 and are difficult to protonate. The bridging hydroxyl group, on the other hand, has a charge of −1/3 and should have a much higher pK_a value than that of the coordinated water molecules. Therefore, the depolymerization reaction can be written as

$$\begin{aligned}&\left[(H_2O)_5Zn-\mu_2OH-Zn(H_2O)_5\right]^{3+}+H_3O^+\rightleftarrows\\&\left[(H_2O)_5Zn-\mu_2OH_2-Zn(H_2O)_5\right]^{4+}+H_2O\end{aligned}\tag{5.14}$$

$$\left[(H_2O)_5Zn-\mu_2OH_2-Zn(H_2O)_5\right]^{4+}+H_2O\rightleftarrows 2Zn(H_2O)_6^{2+}\tag{5.15}$$

Here, it is clear that protonation of μ_2-OH to create μ_2-OH_2 results in the creation of an overbonded oxygen, weakening the Zn−O bonds in the bridge and facilitating a ligand exchange in which the bridge dissociates as a water molecule enters from solution.

Equations (5.14) and (5.15) illustrate of one of the key premises of reaction kinetics involving metal oxides. It is not sufficient to know that protons are involved in a reaction; it is also critical to know where the protons are. In the case of bridge formation, *removal* of a proton from a terminal water molecule accelerates the reaction, but in the case of bridge dissociation, *association* of a proton to the bridging oxygen is the accelerating factor [11].

5.9 HYDROLYSIS PRODUCT STABILITY: DIMER REACTION PATHWAYS

If polymerization and depolymerization reactions involving hydrated cations occurred in a random fashion, the number of hydrolysis products that could potentially form for a given cation would be almost endless. Indeed, in a gel in which transport and the rapid establishment of chemical equilibria are inhibited, it has been shown that such random distributions of polymeric oxides can be generated [17]. Yet, one of the amazing features of the hydrolysis products is how few species are typically present in water. How does an element "decide" how many cations will be contained in a cluster and how those cations are to be connected to one another? There must be something about the chemistry of hydrolysis products interacting in water that leads to a small set of well-defined molecules. Cluster structures in solution represent a balance between local charge distributions and the drive to maximize the number of bridging bonds. These two factors tend to produce a relatively small number of structures, as illustrated next by the pathways that are available to simple dimers. These pathways have been elucidated by elegant studies of reaction dynamics in inert-metal oligomers [18–25].

Formation of a single bridge is often just the first step in a series of reactions leading to the most stable configuration for a dimer (Fig. 5.6). After a reaction such as $2(1,1) \rightleftarrows (2,2)$ takes place, the initial species formed contains one

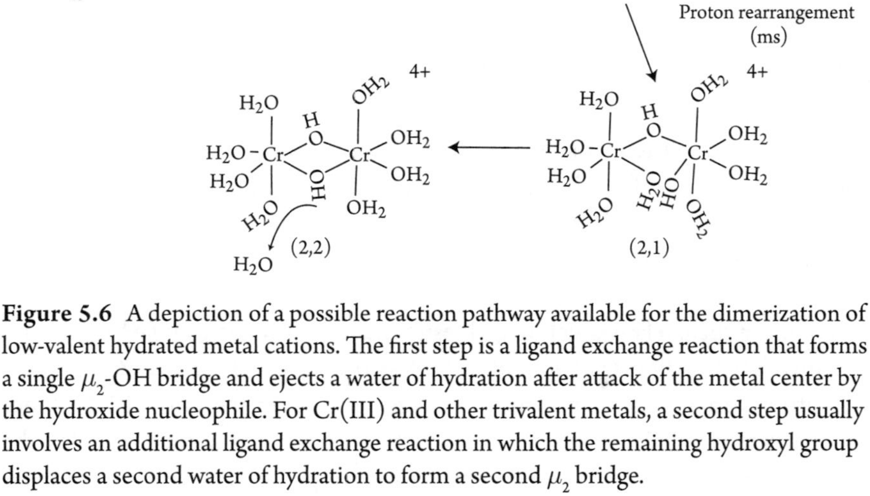

Figure 5.6 A depiction of a possible reaction pathway available for the dimerization of low-valent hydrated metal cations. The first step is a ligand exchange reaction that forms a single μ_2-OH bridge and ejects a water of hydration after attack of the metal center by the hydroxide nucleophile. For Cr(III) and other trivalent metals, a second step usually involves an additional ligand exchange reaction in which the remaining hydroxyl group displaces a second water of hydration to form a second μ_2 bridge.

μ_2-OH bridge and one nonbridging hydroxyl group. One option available to this (2,2) species involves additional *hydrolysis* reactions that are triggered by dimer formation, such as the protonation of the nonbridging hydroxyl group (see the discussion of Eq. 5.13 presented earlier). The net reaction for dimer formation then becomes $2(1,1) \rightleftarrows (2,1) + OH^-$. Eq. 5.13 is known to occur for divalent cations such as Zn(II).

A second option is that the dimer can undergo an *internal condensation* reaction to generate a new and more stable bonding configuration. Here, the nonbridging hydroxyl group that remains after dimer formation can swing around, hydrogen-bond to a water molecule on the adjacent metal center, and form a second μ_2-OH bridge (Fig. 5.6) accompanied by the loss of a water molecule. This hydrolysis product is also a (2,2) species. However, for the case of Cr(III), both simulations and experiments show that the closed-ring structure in the doubly bridged $[(H_2O)_4Cr\text{-}(\mu_2OH)_2\text{-}Cr(H_2O)_4]^{4+}$ species forms quickly and is much more stable than its singly bridged $[(H_2O)_5Cr\text{-}\mu_2OH\text{-}Cr(H_2O)_4OH]^{4+}$ parent [11, 21]. For many trivalent metals, equilibrium constants have even been measured for such ring closures [18, 19, 21]. Two primary factors dictate the pathways a species such as $[(H_2O)_5Cr\text{-}\mu_2OH\text{-}Cr(H_2O)_4OH]^{4+}$ takes to reach the most stable cluster configuration. The first factor involves formal charge distributions. Ensembles of oligomers will gravitate toward those configurations that minimize the formal charge on the maximum number of oxygens. For the Cr(III) example shown in Figure 5.6, the μ_2-OH bridge is electrically neutral and is more stable than either a coordinated water molecule or a nonbridging hydroxyl group (with formal charges of $+0.5$ and -0.5, respectively). Therefore, on the basis of formal charge, $[(H_2O)_4Cr\text{-}(\mu_2OH)_2\text{-}Cr(H_2O)_4]^{4+}$, in which 20% (2 of 10) of the oxygen sites are neutral, is expected to be preferred to $[(H_2O)_5Cr\text{-}\mu_2OH\text{-}Cr(H_2O)_4OH]^{4+}$, in which only 9% of the oxygen sites (1 of 11) are neutral.

The second factor favoring the closed-ring structure of $[(H_2O)_4Cr\text{-}(\mu_2OH)_2\text{-}Cr(H_2O)_4]^{4+}$ is a manifestation of the chelate effect (see Section 5.6). In an analogy to a bidentate ligand, the $[(H_2O)_4Cr\text{-}(\mu_2OH)_2\text{-}Cr(H_2O)_4]^{4+}$ dimer is stabilized because each Cr site is holding on to the other site via two bridges as opposed to just one. If one bridge is disrupted, it is often the case that the broken bridge can be reestablished before the other remaining bridge is disrupted. In contrast, if the single bridge in $[(H_2O)_5Cr\text{-}\mu_2OH\text{-}Cr(H_2O)_4OH]^{4+}$ is broken, the dimer is destroyed.

In light of the chelate effect, why doesn't the formation of a single bridge tend to evolve into the dibridged dimer for the case of +2 cations such as Zn^{2+}? Starting with a formal charge analysis, the oxygen species in the initial (2,2) dimer have charges of $+1/3$, $-1/3$, and $-2/3$ for the coordinated water molecules, the μ_2-OH bridge, and the nonbridging hydroxyl group, respectively. None of these oxygens are satisfied, and bridge formation does not result in the formation of a neutral group as was seen for +3 cations. Based on charge alone, the nonbridging hydroxyl group is least stable. Protonation of this hydroxyl group immediately improves the local charge imbalance. The nonbridging hydroxyl group in (2,2) is expected to have a more basic pK_a than that found in the parent (1,1) complex, because (2,2) has an extra hydroxyl group in its inner coordination sphere. At the pH where (1,1) forms, the free hydroxyl group in (2,2) is protonated immediately, as indicated in Eq. 5.13. In other words, the nonbridging hydroxyl group is transformed into a coordinated water molecule before it has a chance to react with the other Zn^{2+} to form a second bridge. However, at higher Zn^{2+} concentrations, the relatively unstable (2,2) species can condense with other species to form larger and more stable clusters, such as (4,4), that contain neutral μ_3-OH bridges.

5.10 THE STABILITY OF LARGER HYDROLYSIS PRODUCTS

Now consider what can happen to the (2,2) species in the presence of other dissolved monomers. As seen for dimers, experimental evidence indicates that larger species form sequentially via the creation of a single bridging bond between the monomer and the cluster, followed by an evolution toward more compact structures and/or more neutral local charge distributions. As for dimers, the stability of larger clusters involves the balance between local charge distributions and the chelate effect. Charge distributions can be estimated using simple formal charge models. The chelate effect can be evaluated by determining the fraction of bonds present in a given cluster that are bridging bonds.

As an example, consider three of the possible isomers for simple tetrameric (4,6) clusters containing octahedral cations (Fig. 5.7). The total number of bonds in an octahedral tetramer is $4\times6=24$ bonds. For each (4,6) isomer, the total number of bridging bonds is equal to twice the number of μ_2 bridges plus three times the number of μ_3 oxygen bridges. Isomer A in Figure 5.7 contains a total of 12 bridging bonds (involving six μ_2 bridges), for a bridging bond fraction of $12/24=50\%$; isomer B contains 13 bridging bonds $\left(5\mu_2+1\mu_3\right)$, for a bridging bond fraction of $13/24=54\%$; and isomer C contains 14 bridging bonds $\left(4\mu_2+2\mu_3\right)$ for a bridging bond fraction of 58%. Based on the chelate effect alone, isomer C is predicted to be more stable than the other two isomers, which is in agreement with experimental observations. However, charge distributions also have to be taken into consideration. If the (4,6) cluster contains +3 cations such as Cr(III), then the μ_2 bridges are electrically neutral, whereas the μ_3 bridges are slightly overbonded with a formal charge of +1/2. On the basis of formal charge distributions, isomer A is predicted to be the most stable, with eight bonds to neutral oxygen sites (33%), followed by isomer B (29%) and isomer C (25%).

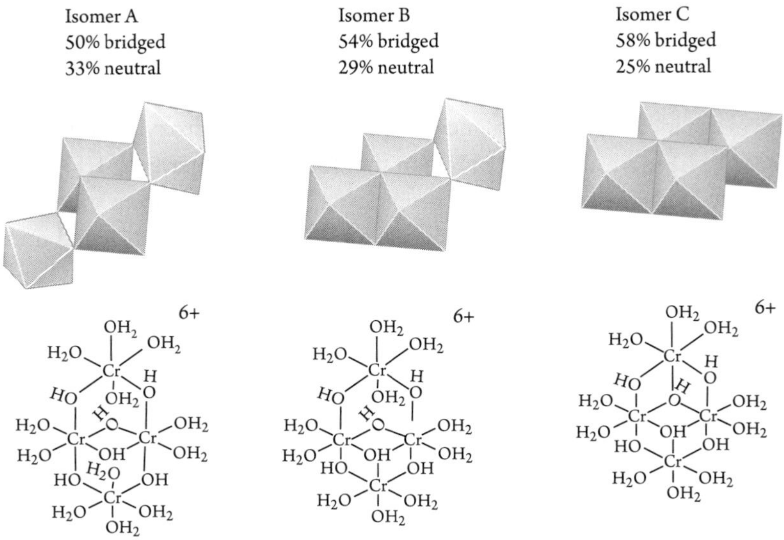

Figure 5.7 Depictions of possible isomeric structures for the (4,6) hydrolysis products of a +3 metal (shown as octahedral cations in the middle and as bonded atoms in the bottom) [12]. The percentages of M—O bonds that are bridging and of oxygen sites that have a formal charge of zero are indicated.

This example illustrates that the ultimate driving forces influencing cluster stability are sometimes opposed to one another. The consequences of this creative tension are explored in specific experimental examples presented later.

Another factor biasing cluster formation and stability has to do with the relative acidities and reactivities of the water molecules around a given cluster, as illustrated by the addition of a (1,1) Cr(III) monomer ($[Cr(H_2O)_5OH]^{2+}$) to the (2,2) Cr(III) dimer ($[(H_2O)_4Cr\text{-}(\mu_2\text{-}OH)_2\text{-}Cr(H_2O)_4]^{4+}$) (Fig. 5.8). As discussed earlier, dimerization creates a condition in which water molecules that are *cis* to the μ_2-OH bridges are more acidic than those that are *trans*. This means that at the pH for trimer formation, the *cis* sites are preferentially deprotonated to make nonbridging hydroxyl groups that participate in condensation reactions with other monomers. Condensation on *trans* sites, which is required to form linear chains, does not appear to happen, either here or in the formation of tetramers (discussed later). Note that two adjacent *cis* sites can participate in condensation, as occurs during the formation of the actual (3,4) trimer.

Three experimental examples illustrate the interplay of the chelate effect, local charge distributions, and site acidities in determining the stability of hydrolysis product structures. The first example involves the internal evolution that occurs within a newly formed trimeric hydrolysis product of Rh(III) corresponding to the (3,4) species described earlier (Fig. 5.9). Rh(III) exhibits

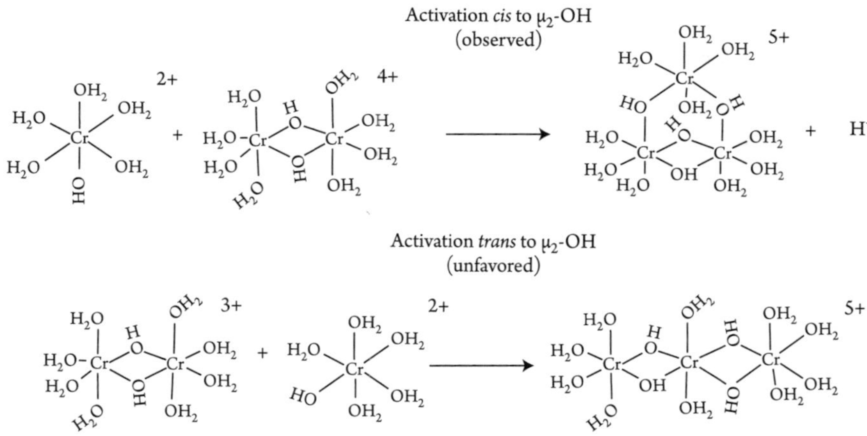

Figure 5.8 Two possible reaction pathways for the addition of a Cr(III) (1,1) monomer to the (2,2) dimer. The top reaction depicts the structure that forms if attack occurs on sites that are *cis* to the μ_2-OH in the dimer, whereas the bottom structure indicates the linear chain that would be produced via attack at *trans* sites. Experimentally, *cis* attack occurs almost exclusively, resulting in more compact hydrolysis products.

such slow hydrolysis kinetics that reactive intermediates can be separated from each other using ion exchange resins. Here, reaction pathways can be determined by examining species distributions as a function of time [25]. The first trimeric species to appear in Rh(III) solutions is the (3,3) species. In terms of charge distributions, (3,3) contains three neutral μ_2-OH bridges plus 12 coordinated water molecules (each with a formal charge of $+0.5$), yielding a cluster with a charge of $+6$. The crosslink density within (3,3) is $6/18 = 33\%$. Experiments show that (3,3) evolves into (3,4) via a series of both internal condensation reactions (replacing nonbridging water molecules with bridging bonds) and proton transfers [25]. The net result is the formation of an isomer of (3,4) in which three water molecules have been replaced by a central μ_3-OH bridge. As discussed earlier, this (3,4) isomer contains a total of nine bridging bonds, yielding a substantial increase in crosslink density (from 33%–50%). The (3,4) species contains a total of 10 overbonded oxygen sites ($9H_2O + 1\mu_3$-OH), each with a formal charge of $+0.5$. The slight decrease in overbonded oxygen content in evolving from (3,3) (80%) to (3,4) (77%) may further bias stabilization in favor of (3,4).

A second experimental example (Fig. 5.10) illustrates the reaction pathways that monomers take when adding to existing clusters. Here, the relatively inert model complexes are those of Cr(III) rather than Rh(III). Starting with solutions that contain both (3,4) trimers and (1,2) monomers, the first step in the

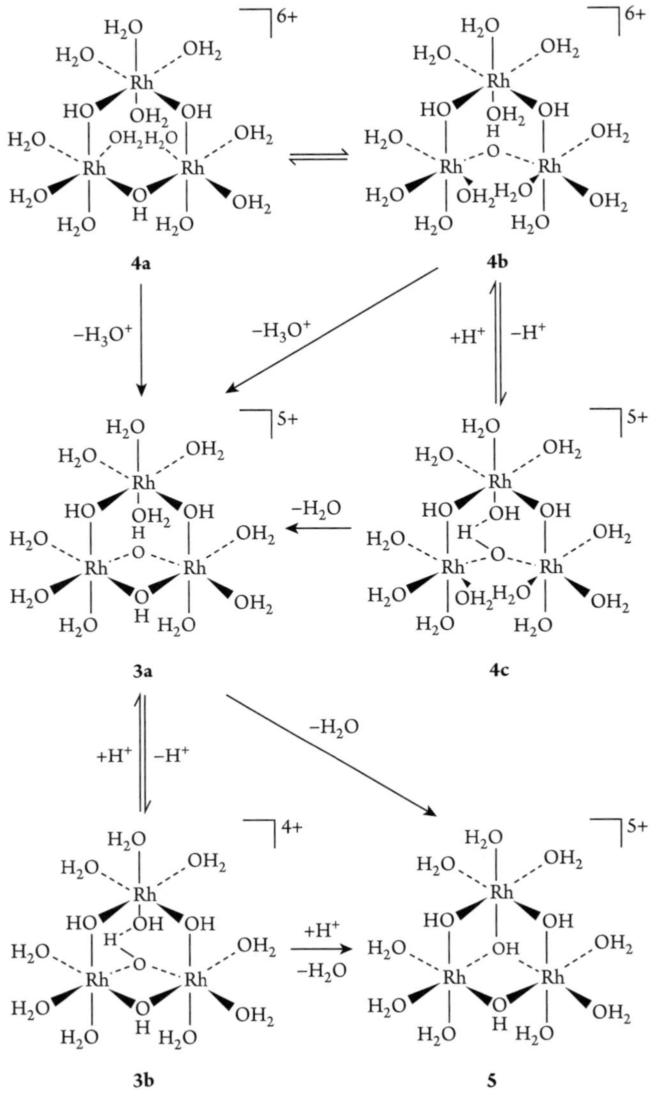

Figure 5.9 Steps in the condensation of a (3,4) Rh(III) trimer from the μ_2-OH-linked ring structure. The central μ_2-OH core forms via a series of acid–base reactions that are reversible and that lead to a stable condensed molecule with stoichiometry $[Rh_3(OH)_4(H_2O)_9]^{5+}$. Each of these intermediates was separated by ion exchange chromatography and characterized because the Rh(III) metal reacts so slowly. Adapted from Spiccia, L.; Aramini, J.M.; Crimp, S.J.; Drljaca, A.; Lawrenz, E.T.; Tedesco, V.; and Vogel, H.J. Hydrolytic Polymerization of Rhodium(III). Characterization of Various Forms of a Trinuclear Aqua Ion *J. Chem. Soc. Dalton Trans.* **1997**, *1997*, 4603–4610.

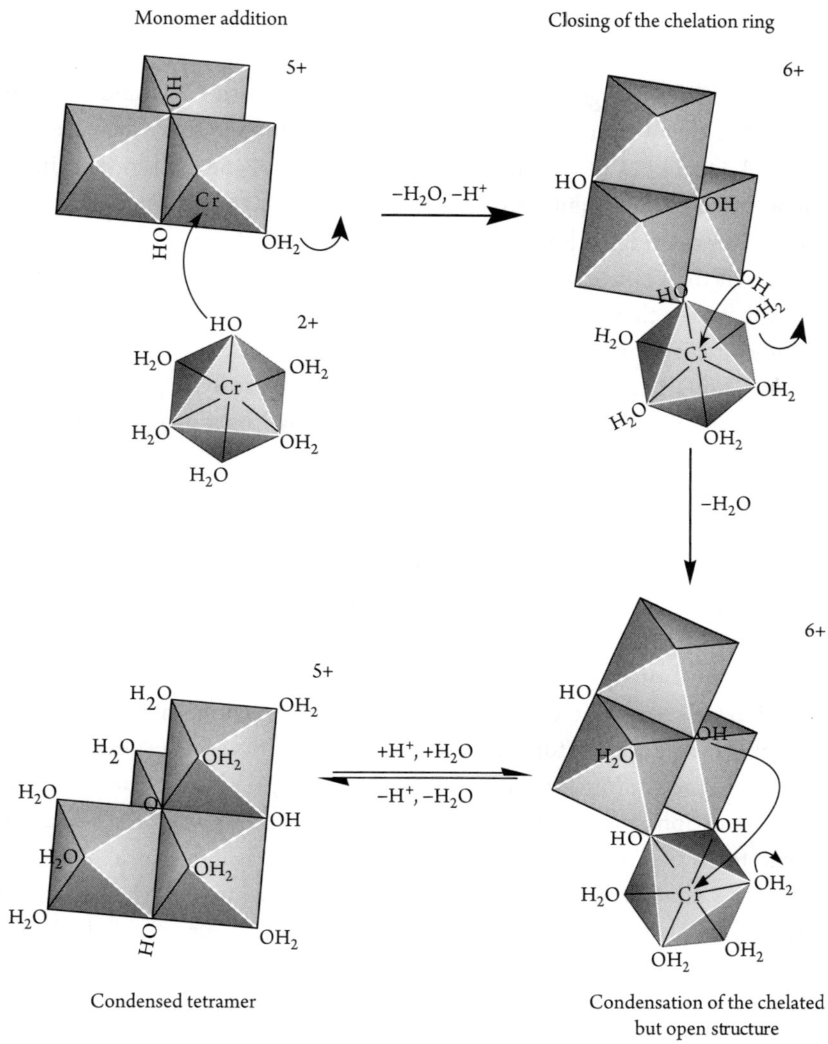

Figure 5.10 A depiction of the reaction pathway by which the Cr(III) trimer reacts with another monomer eventually to form a tetramer [12]. Once again, the first step involves the exchange of a coordinated hydroxyl group on the monomer for a terminal water molecule on the trimer, followed by subsequent internal ligand exchange reactions to form more compact structures.

reaction involves a condensation reaction between a deprotonated *cis* site and the monomer to form a single μ_2-OH within a metastable (4,6) tetramer structure. Second, the newly attached monomer folds over and condenses with a second *cis* site in the cluster (with the loss of a second water molecule) to form a second μ_2-OH. Finally, the monomer can fold over once again, lose a third water molecule, and bond to the μ_3-OH group to form a highly condensed tetramer. This final reaction involves (1) the loss of a proton by the μ_3-OH; (2) the formation of a neutral μ_4-oxo, such as those found in bulk Cr_2O_3 eskolaite; and (3) the reduction in cluster charge from +6 to +5. Starting with (3,4) and (1,2), the fraction of bridging bonds present in this entire reaction sequence increases from 9/24 (38%) to 11/24 (48%) to 13/24 (54%) to 14/24 (58%). Once again, this sequence reflects the evolution within hydrolyzed clusters to form denser structures with more bridging bonds, in keeping with the chelate effect.

Although the chelate effect is clearly important, our final experimental example illustrates that local charge distributions can sometimes override the tendency to maximize the number of bridging bonds. Hydrolysis diagrams show that, for +2 cations, the common tetrameric hydrolysis product is (4,4). X-ray crystallography clearly shows that all four hydroxyl groups in (4,4) are μ_3-OH, forming a densely packed cube (Figs. 5.1 and 5.2). However, the (4,4) isomer is not indicated as a primary species in the hydrolysis diagram for any +3 cation [1]. Why is this? The (4,4) isomer is certainly highly condensed, with a bridging bond density of 50%. For +2 cations, μ_3-OH bridges have a formal charge of zero, and are highly stable. In contrast, for +3 cations, μ_3-OH bridges have a formal charge of +0.5 and are much less stable. Not only that, in (4,4), all the cationic μ_3-OH sites would be in close proximity and repel each other. Here, electrostatic considerations win out over the chelate effect. The most stable configuration for bridges involving +3 cations is the neutral μ_2-OH bridge, which is predominant in the more stable (3,4) cluster. The conclusion is that +2 cations prefer structures that contain μ_3-OH whereas +3 cations prefer to form μ_2-OH because both bridges are electrically neutral. As we saw in Chapter 2, the initial precipitates of these metals, as insoluble hydroxide and oxyhydroxide phases, exhibit the same preferences as their soluble hydrolysis products.

We close this section by reviewing general reaction trends among the larger cationic hydrolysis products. An *increase* in solution pH causes (1) deprotonation of coordinated water molecules and/or hydroxyl groups, (2) a lowering of cationic species charge, and (3) the loss of coordinated water molecules. These factors all promote condensation reactions that create bridging bonds, larger clusters, and denser oxide structures. Conversely, a *decrease* in solution pH leads to protonation of oxygen sites, an increase in cationic charge, proton-induced dissociation of bridging bonds, and the hydration of metal cations, all of which eventually lead to oxide dissolution.

The small clusters shown in Figures 5.8 and 5.9 can be thought of as fragments of extended hydroxide and oxide phases. The reactions shown for both the formation disruption of M—O bonds in these figures are the same as the reactions that lead to both precipitation and dissolution of oxide materials. This point is illustrated in Figure 5.11, in which a small section of the goethite

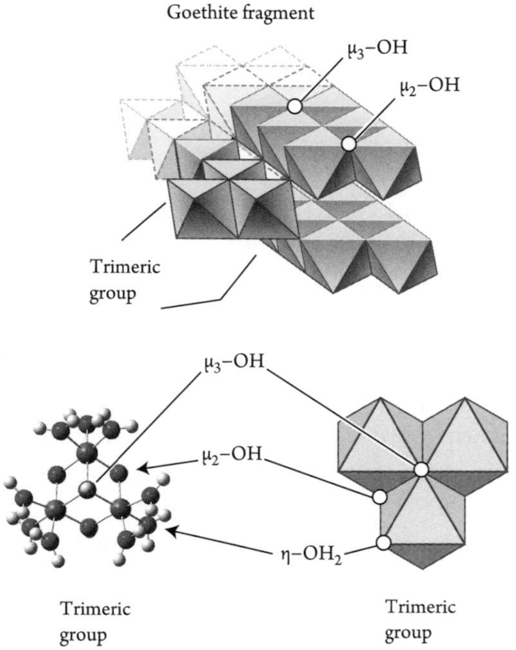

Figure 5.11 A comparison between the structures of a fragment of an extended surface of FeOOH (goethite) and the (3,4) trimer of Fe(III). Although relative site populations differ, the local structures and chemistry of specific oxygen sites (η-OH_2, μ_2-OH, and μ_3-OH) in both extended oxide phases and hydrolysis products are essentially identical.

(FeOOH) mineral structure is compared with the hydrolytic trimers discussed earlier. There is a profound similarity in the chemistry of the hydrolysis products, colloidal precipitates, and extended oxide phases because their localized structures, charge distributions, and acid–base chemistry are similar regardless of length scale. These similarities are highlighted in both Chapter 6 and Chapter 14.

5.11 THE POLYOXOMETALATES: COVALENT OXIDE CLUSTERS

5.11.1 An Introduction to the Polyoxometalates

Highly charged metal cations, such as Nb(V) and W(VI), bond so strongly to oxygens that they can form discrete classes of large, predominantly anionic clusters containing multiple cations imbedded in an oxo network. These *polyoxometalates* (or POMs) represent the most diverse and technologically important class of the hydrolysis products. POMs are typically formed via +5 or +6 cations, although some contain +3 cations as well. Examples involving basic molybdate solutions include [26]

$$7MoO_4^{2-} + 8H_3O^+ \rightleftarrows Mo_7O_{24}^{6-} + 12H_2O \quad \textbf{(5.16)}$$

$$HPO_4^{2-} + 12MoO_4^{2-} + 23H_3O^+ \rightleftarrows \left[PMo_{12}O_{40}\right]^{3-} + 35H_2O \quad \textbf{(5.17)}$$

The dominant elements forming POM structures are the transition metals Mo, W, U, V, Nb, and Ta. If all cations in the structure are identical (Eq. 5.16), the clusters are called *isopolyanions* or *isopolyoxometalates* [27]. However, more than half the elements in the periodic table are known to substitute into such clusters (Eq. 5.17) to form *heteropolyanions*. Given the possible range of substitutions and the fact that clusters can contain anywhere from six to hundreds of metal cations, the number of possible POMs is almost endless.

From a chemical perspective, POMs provide an enormous range of properties [28], including acidities that span the entire pH range available in water and the widest range of redox potentials of any class of oxides (see Chapter 11). The physical properties of POMs are also diverse and highly tunable, including optical (particularly for POMs such as the mixed-valence molybdate blues [29]) (Chapter 13) and magnetic behavior [30,31]. The technological impact of POMs spans an enormous range of applications [32–35], including oxidation catalysts, separations media, analytical reagents, antiviral agents, and even nuclear-waste disposal.

Next, we provide a brief survey of the basic structures of several POM families. We then focus on several examples that illustrate the unusual acid–base and ligand exchange properties of POMs. As you will see, the chemistry of the POMs is clearly distinct from that exhibited by most other hydrolysis products. This distinction arises because the M—O bonds in POMs can be so strong and covalent that simple formal-charge models are inadequate for describing their physical and chemical attributes. Accurate predictions of POM properties require the use of quantum mechanics and molecular-orbital theory [36]. For more details regarding the structures, chemistry, and applications of POMs, there is a series of excellent reviews [28, 29, 37–40].

5.11.2 A Survey of Polyoxometalate Structures

The simplest isopolyoxometalates are the Lindqvist ions [39] represented by the clusters $[Nb_6O_{19}]^{8-}$ and $[W_6O_{19}]^{2-}$ for +5 and +6 cations, respectively (Fig. 5.12). These structures, described in more detail later, contain six distorted octahedral cations that are interconnected by μ_2-oxo-bridges around a central μ_6-oxo-anion. Additional octahedra can be added to form larger clusters containing

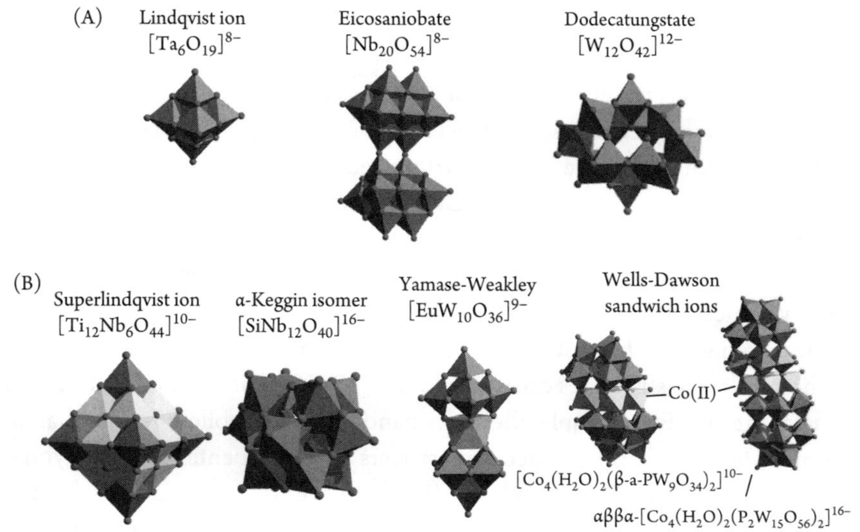

Figure 5.12 Representative structures of simple polyoxometalates. (A) Isopolymetalates are presented, including the Lindqvist ion (left), eicosaniobate (center), and the dodecatungstate ion (right). (B) Heteropolymetalates are shown, including the Superlindqvist ion (left), an isomer of the Keggin ion (center left), the Weakley-Yamase ion (center right) that traps a lanthanide in a chelating array, and two different Wells-Dawson ions (right) that sandwich a layer of Co(II)-oxide. Images provided by Prof. C. Andre Ohlin, with permission.

more than six cations. Clusters having 7, 8, 10, 12, and 19 metal centers are relatively common. In some instances, simple clusters can bridge to each other to form larger clusters, as exemplified by the 24-cation isopolyniobate species [41]. The addition of octahedra to the hexametallate species is usually accompanied by the formation of μ_3-bridges on the cluster surface. Such bridges are overbonded and are relatively inert with respect to acid–base activity.

Heteropolyoxometalates can be formed via substitution of foreign octahedral cations into isopolyoxometalate structures. However, heteropoly species can also contain cations in nonoctahedral geometries. One common class of such heteroclusters is the Baker-Figgis-Keggin series [42] (Eq. 5. 17) (Figs. 5.12 and 1.9). These clusters contain 13 metal ions in which a tetrahedral cation such as B(III), Al(III), Si(IV), P(V), or As(V) is completely encapsulated by four sets of oxy-bridged octahedral trimers. The aluminum hydrolysis product $[AlO_4Al_{12}(OH)_{24}(H_2O)_{12}]^{7+}$ is an unusual Keggin ion formed by a +3 cation that contains hydroxy- rather than oxy-bridged octahedral cations. Here, either Ga(III) or Ge(IV) can substitute for the central tetrahedral atom. Aluminum Keggin ions are important in water treatment plants, consumer products, and natural stream and soil chemistry.

One can also have POM molecules with missing atoms. These *lacunary* structures (from the word *lacuna*, which is the Latin for "cavity") function as ligands that form the basis for even more complex heteropolyanions [43]. In Weakley-Yamase ions, one apical cation is replaced in two Lindqvist ions that cap a central eight-coordinate lanthanide ion. Two truncated Keggin ions containing tetrahedral cations can dimerize to form the Dawson ions [44] (Fig. 5.12) in which each half of the cluster contains a tetrahedral cation. Lacunary heteropolyanions can also form structures in which layered metal-hydroxide sheets are capped on both sides by truncated Keggin ions [44–46].

The extreme sizes that soluble heteropolyoxometalates can attain are illustrated by a broad class of structures called *Keplerates* [47–50] that can contain hundreds of metal cations. For example, there is a nanometer-size hollow sphere that can accommodate 30 transition metals as spacers between pentahedral molybdate tiles (Fig. 1.9). In this $Mo_{72}Fe_{30}$ cluster (i.e., $[Mo_{72}Fe_{30}O_{252}(CH_3COO)_{12}[Mo_2O_7(H_2O)_2[H_2Mo_2O_8(H_2O)](H_2O)_{91}]\text{-}150\,H_2O)]$, Fe(III)-$OH_2$ sites dominate the weak acidity exhibited by the cluster surface. Solutions containing these enormous molybdate-blue ions can age to form even larger hydrogen-bonded arrays called *blackberry structures* [49, 50]. Blackberries can contain tens or even hundreds of individual spheres arranged in a myriad of architectures, including toroidal shapes and disks. Because these enormous blackberries are porous, they can serve as high-capacity soluble ion exchangers. These porous, spherical oxide shells have been compared with enclosed cell membranes [48–50].

We close this survey of POM structures with an example that illustrates the wonders of soluble oxide clusters containing foreign ligands. For years, it was known empirically that the extraction of uranium ore was facilitated by the addition of hydrogen peroxide to the extraction solution. Pioneering work in the past decade revealed that the reason for enhanced extraction involves the generation of an unusual class of peroxide-containing POMs that resemble buckminsterfullerenes [38, 40,51]. Cluster structure types are wide ranging and unusual (Fig. 5.13) as a result of the fact that the U(VI)–peroxide bond is bent. As with most soluble oxide species, cluster types are highly dependent on pH, uranium concentrations, and ligand availability. Pure uranyl-peroxide clusters form at high pH and peroxide concentrations. Ironically, these conditions are those found during the weathering of uranium-rich fuel, with important implications for the safe storage of nuclear wastes. For fuel rods stored in aqueous cooling basins, the radiolysis of water creates peroxide, whereas normal hydrolytic conditions cause the pH to increase. These clusters probably represent one of the few examples in which cations actually create their own solubilizing ligands. Although far from life, such clusters illustrate how complex POM chemistry can be. In fact, POMs and other oxide clusters are found in most living systems (see Chapter 9).

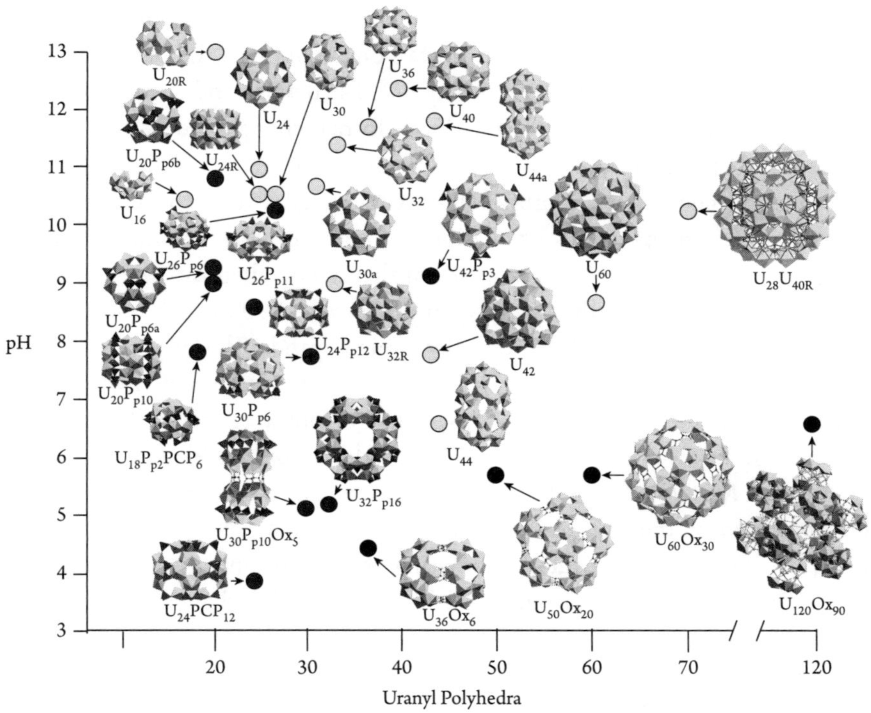

Figure 5.13 A rich array of U(VI)-peroxide-oxide polyoxometalate clusters as a function of formation pH. Light-gray dots represent peroxide-containing clusters; black dots involve clusters that also contain other anions, such as oxalate, diphosphate, and/or methyldiphosphonate polyhedral.
SOURCE: Nyman, M. D. and Burns, P. C. A Comprehensive Comparison of Transition-Metal and Actinyl Polyoxometalates *Chem. Soc. Rev.*, **2012**, *41*, 7354–7367 with permission.

5.11.3 Covalent Bonding in Polyoxometalates: The Lindqvist Ions

Let us now examine the structures of POMs in more detail using the Lindqvist ions as simple examples [52, 53]. These clusters contain six octahedral cations associated with six nonbridging oxygens, 12 μ_2-O bridges, and a central oxygen that is loosely bonded to all six cations. X-ray crystal structures reveal that bonding within these clusters is unusual relative to the oxides and hydrolysis products we have been discussing up to now. First, the nonbridging oxygens exhibit bond lengths that are substantially shorter (1.75–1.8 Å) than the bonds to bridging oxygens (around 2 Å). The short terminal bond lengths are a reflection of the fact that +5 and +6 cations can interact so strongly with oxygen anions that they can form double and even triple bonds to nonbridging oxygens (called "yl" oxygens).

Multiple bond formation reflects the fact that there is extensive hybridization between d orbitals on the metals and p orbitals on the oxygens. This orbital hybridization and bond covalency explains why POM structures and chemistry can no longer be described adequately using simple formal-charge models.

A second unusual but related structural attribute of POMs is that the oxygen coordination geometries around high-valence cations tend to be highly distorted. In fact, the oxygen coordination geometry around +5 and +6 cations can more closely resemble a pentahedron than an octahedron. The O−M−O bond angles adjacent to the "yl" oxygen often exceed 90°, which opens up the octahedral sites to nucleophilic attack by foreign ligands. The "yl" bonds to nonbridging oxygens also tend to remove electron density preferentially from sites that are *trans* to the multiple bond. This is why the central oxygen in the Lindqvist cluster is so far from the metal that it resembles a nonbonded O^{2-} anion. For W(VI) and Mo(VI) POMs, bond hybridization can cause periodicity in bridging bonds within the large structural rings or cages that define the cluster exterior [36]. For large molecules, alternating short and long μ_2-oxo-bonds also create openings for foreign ligands and facilitate ligand-exchange reactions (discussed next).

5.11.4 The Acid–Base Properties of Polyoxometalates

It is easier to rationalize the acid–base chemistry of the POMs on the basis of *electron distributions* than on formal charges. As a starting point, consider phosphoric acid (H_3PO_4), which is a simple monomeric acid containing a +5 cation. This tetrahedral species contains one short P=O bond (1.38 Å) and longer (1.62 Å) single bonds to each of the three hydroxyl groups. This acid can lose all three of its protons sequentially, with pK_a values of 1.8, 6.7, and 11.7. As each proton is removed, the remaining hydroxyl groups maintain a P−O single-bond character. However, the bond order on the remaining oxygens changes from 2 to 1.5 to 1.33 to 1.25 in the series $[H_3PO_4]^{\circ}$ to $[H_2PO_4]^-$ to $[HPO_4]^{2-}$ to $[PO_4]^{3-}$. In other words, the excess electron density that started out in the nonprotonated double bond ends up being shared equally by all nonprotonated oxygens until finally (for $[PO_4]^{3-}$) all four oxygens are identical. In larger and more complex POMs, the movement of electron density from multiple metal–oxygen bonds can involve all the oxygens in the cluster, which impacts both the acid–base and ligand-exchange properties of these hydrolysis products.

Let us now consider how bond hybridization impacts the acid–base chemistry of the POMs. First, the consumption of electron density that is required to create multiple bonds means that less electron density is available for other bonds that can potentially form to both the oxygen and the metal. For the case of the "yl" oxygen, some of this electron density is provided at the expense of electrons that would normally reside in its basic, nonbonded lone-pair orbitals. What this means is that in contrast to hydrolysis of divalent and trivalent metals, terminal oxygens in POMs are almost always less basic than bridging oxygens. Protons tend to interact most strongly with μ_2 bridging oxygens. The lack of acid–base activity for the "yl" oxygens mirrors that seen for the P = O bond in phosphoric acid.

The pK_a values associated with bridging oxygens in POMs span the entire pH range of water. What determines the pK_a values of these bridging oxygens? As in the case of $[H_3PO_4]^{\circ}$, the electron density associated with "yl" oxygens responds to acid–base reactions in such a way that its effects are spread throughout the cluster, rather than being highly localized on a single site. For the clusters in Table 5.5 [52, 54], pK_a values are estimated for the removal of the last two or three protons from each cluster, where the highest pK_a value corresponds to the formation of the completely deprotonated species (such as $[Nb_6O_{19}]^{8-}$). The clusters being compared are this niobate Lindqvist ion, the $[Mo_7O_{24}]^{6-}$, and three isostructural clusters having the stoichiometries (in their fully deprotonated forms) of $[Nb_{10}O_{28}]^{6-}$, $[TiNb_9O_{28}]^{7-}$, and $[Ti_2Nb_8O_{28}]^{8-}$. In the latter three clusters, Ti(IV) was substituted for Nb(V) to investigate links between cluster charge and cluster chemistry. Here, an increase in anionic charge on the cluster suppresses deprotonation (i.e., increases the pK_a and proton affinity of oxygen sites) (Fig. 5.14). This is regardless of whether the increase is a result of substitution of Ti(IV) for Nb(V) or proton removal from the cluster (i.e., from $[H_3Nb_{10}O_{28}]^{3-}$ to $[H_2Nb_{10}O_{28}]^{4-}$ to $[HNb_{10}O_{28}]^{5-}$ to $[Nb_6O_{19}]^{8-}$). In other words, the acid–base chemistry reflects the overall molecular charge.

Table 5-5. Estimated Equilibrium Constants for the Deprotonation of Several Polyoxometalate Anions

Complexes	pK_{a1}	pK_{a2}	pK_{a3}
$[H_xNb_6O_{19}]^{(8-x)-}$		10.9	13.8
$[H_xNb_{10}O_{28}]^{(6-x)-}$	1.6	3.74	6.0
$[H_xTiNb_9O_{28}]^{(7-x)-}$		6.88	8.28
$[H_xTi_2Nb_8O_{28}]^{(8-x)-}$	6.73	7.72	9.5
$[H_xMo_7O_{24}]^{(6-x)-}$	2.5	3.6	4.5

SOURCE: Data from Villa, E. M.; Ohlin, C. A.; Casey, W. H. Isotope-Exchange Dynamics in Isostructural Decametalates With Profound Differences in Reactivity *J. Am. Chem. Soc.* **2010**, *132*, 5264–5272; and Black, J. R.; Nyman, M.; Casey, W. H. Rates of Oxygen Exchange Between the $[H_xNb_6O_{19}]^{(8-x)-}$ Lindquist Ion and Aqueous Solutions. *J. Am. Chem. Soc.* **2006**, *128*, 14712–14720.

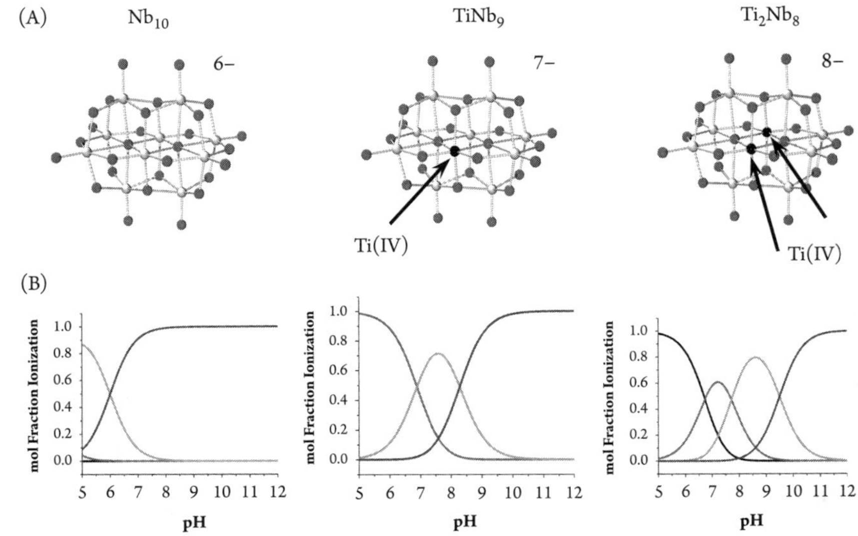

Figure 5.14 (A) Structures and charges associated with the decametalate ions $[H_xNb_{10}O_{28}]^{(6-x)-}$, $[H_xTiNb_9O_{28}]^{(7-x)-}$, and $[H_xTi_2Nb_8O_{28}]^{(8-x)-}$. (B) Acid–base speciation diagrams for the ions shown directly above them. For all three plots, the far right trace corresponds to the unprotonated species, with other traces (from right to left) corresponding to the singly, doubly, and triply protonated species, respectively. With substitution of Ti(IV) for Nb(V), the overall charge of the ions increases, with predicable changes in the pK_a values. Adapted from Villa, P. C.; Ohlin, C. A.; Rustad, J. R.; and Casey, W. H. Isotope-Exchange Dynamics in Isostructural Decametalates with Profound Differences in Reactivity *J. Am. Chem. Soc.* **2010**, *132*, 5264–5274.

5.11.5 Ligand-Exchange Reactions Involving Polyoxometalates

Ligand-exchange reactions involving POMs and water are unusual when compared with those of more conventional hydrolysis products containing lower valence metals. Rates for such oxygen anion exchange have been quantified for a few POMs using ^{17}O NMR experiments [54, 55]. Given the high charge on the metal cations, it is not unexpected that POMs can exhibit negligible rates of oxygen-isotope exchange involving either water molecules or hydroxide ions. However, some POMs, including the polyoxoniobates, are surprisingly labile to such exchange. As in more conventional hydrolysis products, the rates of these exchange reactions vary considerably with pH. In strongly acidic solutions, exchange rates are rapid, but are difficult to measure. In part, these difficulties arise because extensive protonation results in cluster neutralization and the precipitation of oxide gels. However, experiments have been conducted at intermediate pH before precipitation is reached. Here, it is seen that cluster protonation accelerates oxygen isotopic exchange. For the singly protonated Lindqvist cluster

$[HNb_6O_{19}]^{7-}$ that is present at pH 12.5, exchange rates on bridging and nonbridging sites are $1.8 \times 10^{-4}/s$ and $1.2 \times 10^{-5}/s$, respectively. At pH 9.5 (below pK_1, where $[H_2Nb_6O_{19}]^{6-}$ is predominant), exchange rates on both sites are too fast to measure $(>10^{-3}/s)$.

Similar ^{17}O NMR experiments have been performed for the Nb_{10}, $TiNb_9$, and Ti_2Nb_8 clusters with pK_a values that appear in Table 5.5 [54]. Isotope-exchange rates have been measured for all seven (or more for the Ti(IV)-containing clusters) chemically distinct oxygen sites, including two central μ_6-oxo, four μ_3-oxo, three distinct μ_2-oxo, and two distinct nonbridging oxygens (four each) (Fig. 5.13). Once again, the results indicate that ligand-exchange rates are enhanced by protonation as pH decreases. Exchange is slowest for the μ_6-oxo in the center of the molecule, which only occurs when the cluster dissociates completely and then reforms. Exchange is slow for the μ_3-oxo-bridges, as one might expect based on both the chelate effect and the fact that such sites are difficult to protonate. Faster rates of ligand exchange are seen for the μ_2-oxo-bridges, with the fastest exchange rates being observed for the oxo-bridges at the ends of the molecule and all nonbridging oxygen sites.

Rates also increase in basic solutions when the clusters are completely deprotonated. Here, hydroxide ions are involved in a direct nucleophilic attack on metal centers that is facilitated by the opening up of O—M—O bond angles around +5 and +6 cations. To summarize, rates of isotopic exchange at all the reactive oxygens increase as pH is either lowered or raised on each side of a minimum in reactivity where rates are nearly independent of pH. This amphoteric reactivity trend is seen for many of the lower-valence hydrolysis products and oxides discussed previously in this chapter. In other words, POMs containing as few as 40 atoms already exhibit the chemical attributes of an extended solid oxide.

5.11.6 Ligand-Exchange Mechanisms in Polyoxometalates

Three different reaction pathways can be envisioned to help rationalize ligand exchange processes in the POMs: (1) completely localized exchange, (2) bond hydrolysis followed by localized rearrangements, and (3) scrambling of many sites via concerted motions and isomerization of the entire cluster (Fig. 5.15). The first mechanism has been observed for low-valence cations that occupy sites within a heteropolyoxometalate structure. For example, consider the rate of water exchange around the Co(II) sites bound within the Wells-Dawson sandwich structure $[\alpha\beta\beta\alpha\text{-}Co_4(OH_2)_2(P_2W_5O_{56})_2]^{16-}$ shown in Figure 5.12. Here, the electronic structure of the Co(II) is so different from the two lacunary tungstate ligands that it is chemically decoupled from the surrounding POM environment. This means that water exchange on the Co(II) site proceeds via the

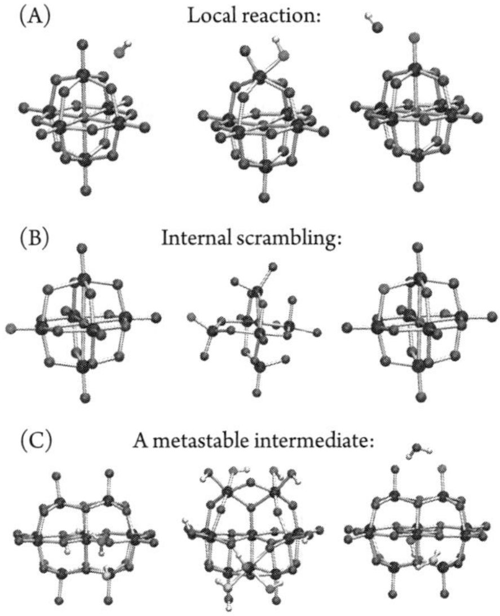

Figure 5.15 Three conceptual pathways leading to oxygen isotope exchanges involving polyoxometalate clusters and water. (A) Reactions can be limited to a single functional group. Here, the nucleophilic attack by the hydroxide ion (left) forms a higher coordinate intermediate (center) that subsequently ejects what was the nonbridging "yl" oxygen on a localized site to return to the parent configuration (right). (B) Scrambling between bridging and nonbridging oxygens can occur as a result of a reversible hydrolysis reaction in which a μ_2-bridge is broken to create two active sites containing two nonbridging oxygens each (center). Reformation of the bridge and the cluster results in scrambling if the unlabeled nonbridging oxygen formed via bond rupture is the oxygen that reattaches to reform the bridge and the original cluster structure. (C) Extensive scrambling of many oxygen sites in a given cluster can occur if bond hydrolysis creates a metastable intermediate (middle) within which oxygens hop from site to site until a stable bridging bond is established to regenerate the rigid parent structure. Reproduced with permission from Ohlin, C. A.; Villa, E. M.; Rustad, J. R.; Casey, W. H. The Dissolution of Insulating Oxides at the Molecular Scale *Nature Materials*, **2010**, 9 11–19.

classic dissociative-interchange I_d exchange mechanism seen for isolated Co(II) hydrolysis products. The extended lacunary tungstate ligands have no effect on the reaction kinetics at these Co(II) sites, which exchange water molecules at about the same rate as the $[Co(H_2O)_6]^{2+}$ (aq) ion [45].

A completely localized oxygen-exchange mechanism involving the exchange of the nonbridging "yl" oxygens adjacent to +5 and +6 cations is illustrated in Figure 5.15A for the simple Lindqvist ion. Because of the open bond angles adjacent to the "yl" oxygen, nucleophiles such as the hydroxyl ion can attack the site to form a higher coordinate intermediate. The bonding electrons required to form this intermediate are supplied in part by the "yl" oxygen. The M—O bond in the "yl" oxygen then resembles the bond to the adjacent M—OH group. Next, the intermediate ejects a ligand, which sometimes involves the rupture of the remaining bond to the original "yl" oxygen. Electron density then flows back into the remaining nonbridging hydroxyl group to form a new "yl" oxygen to complete the exchange process.

The second ligand-exchange mechanism resulting in local oxygen scrambling involves the local disruption of μ_2-oxo-bridges. These bridges can be opened by either dissociative or associative ligand exchange reactions similar to those seen in standard (i.e., lower valence) hydrolysis products. However, the dominant and unique exchange mechanism for POMs is thought to involve a proton-activated associative exchange. When the solution pH drops below the highest pK_a for the cluster, a μ_2 bridge can abstract a proton from an adjacent water molecule. Consequences of this protonation are (1) creation of a μ_2-OH bridge having weaker M—O bonds to the cations in the cluster and (2) the generation of a nucleophilic hydroxide ion that is perfectly situated to attack the metal cation center via an associative mechanism. Both factors accelerate ligand exchange and oxygen scrambling consistent with the ^{17}O NMR results described earlier.

Hydrolysis and rupture of a μ_2-oxo-bridge creates a nonbridging oxygen or hydroxyl group that is adjacent to the "yl" oxygen. As in the completely localized exchange process, electron density associated with the covalent "yl" bond rebalances to make both nonbridging oxygens (i.e., from the old "yl" oxygen and from the newly broken bridge) equivalent. However, in contrast to the completely localized exchange on isolated nonbridging oxygens, either of these nonbridging oxygens can reattach the adjacent cation to reform the bridge (Fig. 5.15B). This results in an isotopic scrambling between bridging and nonbridging oxygens within the cluster. Such scrambling may be why one reason why experimental exchange rates on the terminal oxygens and μ_2-oxo-bridges can be so similar.

Finally, exchange processes associated with the hydrolysis of μ_2-oxo need not be local to those atoms in the immediate vicinity of the break. Bond rupture processes can trigger reaction pathways involving the formation of metastable clusters having highly fluxional structures (Fig. 5.15 C). These metastable intermediates can exchange oxygen anions rapidly both with other sites in the cluster and with solution species, including water molecules, hydronium ions, and hydroxyl ions. The loosely bonded intermediate eventually collapses back into the rigid parent structure, thus trapping the new isotopic distribution within the cluster into place.

Formation of such metastable intermediates is influenced by species such as counterions that lie within the outer coordination sphere of the cluster, as separation of charges is required. Evidence for the "metastable intermediate" mechanism is provided by studies that show that one isomer of a POM cluster can be converted into another by changing the counterions present in solution [56]. Here, oxygen-isotope exchange rates at different structural sites are dramatically affected by otherwise inert background electrolyte ions. In addition, counterions clearly stabilize one metastable intermediate relative to another [57]. A cluster may have several metastable forms that are accessible, depending on its structure, composition, and symmetry. The metastable structures also suggest pathways for forming lacunary forms of the POMs.

In closing this chapter, it is important to note that isotopic scrambling between bridging and nonbridging oxygens resulting from reversible bridge hydrolysis is not unique to the POMs. For example, such exchange has been documented in surface science studies on α -Fe_2O_3 in which water adsorption and reversible hydrolysis triggers mixing between terminal and μ_3-oxo surface sites [58]. Exchange rates are also dramatically affected by the injection or removal of electrons involving redox-active oxides [59] (see Chapter 11). In fact, *the hydrolysis-induced scrambling between distinct oxygen anion sites almost certainly occurs on all extended oxide surfaces immersed in water*, as detailed in Chapters 6, 8, and 15. Such scrambling is just one of many of the chemical attributes that can be predicted by understanding the chemistry of soluble hydrolysis products. Therefore, this chapter represents an important resource for clarifying the chemistry presented in many of the remaining chapters in this book.

References

1. Baes, C. F. J.; Mesmer, R. E. *The Hydrolysis of Cations*; Krieger Publishing: Malabar, FL, 1986.
2. Swaddle, T. W.; Rosenqvist, J; Yu, P; Bylaska, E; Phillips, B. L.; Casey, W. H. Kinetic Evidence for Five-Coordination in $AlOH(aq)^{2+}$ Ion. *Science* **2005**, *308*, 1450–1453.
3. Cotton, F. A.; Wilkinson, G. *Advanced Inorganic Chemistry, 6th ed.* ; Wiley: New York, 1999.
4. Sposito, G. *The Surface Chemistry of Natural Particles*; Oxford University Press: Oxford, 2004.
5. Martell, A. E.; Hancock, R. D., Metal Complexes in Aqueous Solution. In *Modern Inorganic Chemistry*; Fackler, J. P. J., Ed; Plenum Press: New York, 1996:1–253.
6. Pearson, R. G. Hard and Soft Acids and Bases. *J. Am. Chem. Soc.* **1963**, *85*, 3533–3539.
7. Drago, R. S. Quantitative Evaluation and Prediction of Donor–Acceptor Interactions. *Struct. Bond.* **1973**, *15*, 73–139.
8. Weinhold, F.; Landis, C. R. *Valence and Bonding*; Cambridge University Press: Cambridge, 2005.
9. Kragden, J. *Atlas of Metal Ligand Equilibria in Aqueous Solution*; Wiley: New York, 1978.
10. Wilkins, R. G. *The Study of Kinetics and Mechanisms of Reactions of Transition Metal Complexes*; Allyn & Bacon: Boston, 1974.

11. Springborg, J. Hydroxo-Bridged Complexes of Chromium (III), Cobalt (III), Rhodium (III), and Iridium (III). *Adv. Inorg. Chem.* **1988**, *32*, 55–169.
12. Stuenzi, H.; Marty, W. Early Stages of the Hydrolysis of Chromium (III) in Aqueous Solution: 1. Characterization of a Tetrameric Species. *Inorg. Chem.* **1983**, *22*, 2145–2150.
13. Stünzi, H.; Rotzinger, F. P.; Marty, W. Early Stages of the Hydrolysis of Chromium (III) in Aqueous Solution: 2. Kinetics and Mechanism of the Interconversion between Two Tetrameric Species. *Inorg. Chem.* **1984**, *23*, 2160–2164.
14. Rotzinger, F. P.; Stünzi, H.; Marty, W. Early Stages of the Hydrolysis of Chromium (III) in Aqueous Solution: 3. Kinetics of Dimerization of the Deprotonated Aqua Ion. *Inorg. Chem.* **1986**, *25*, 489–495.
15. Crimp, S. J.; Spiccia, L.; Krouse, H. R.; and Swaddle, T. W. Early Stages of the Hydrolysis of Chromium (III) in Aqueous Solution: 9. Kinetics of Water Exchange on the Hydrolytic Dimer. *Inorg. Chem.* **1994**, *33*, 465–470.
16. Rotzinger, F. P.; Marty, W. A Unified Interpretation of Kinetic Data on the Acid-Induced Cleavage and of Product-Analysis Data on Spontaneous Cleavage of the Mono-ol Cation μ-Hydroxo-bis[pentaamminecobalt(III)]($[(NH_3)_5CoOHCo(NH_3)_5]^{5+}$). *Helv. Chim. Acta* **1985**, *68*, 1914–1922.
17. Rustad, J. R.; Casey, W. H. A Molecular Dynamics Investigation of Hydrolytic Polymerization in a Metal–Hydroxide Gel. *J. Phys. Chem. B* **2006**, *110*, 7107–7112.
18. Merakis, T.; Spiccia, L. Early Stages of the Hydrolysis of Chromium (III) in Aqueous Solution: V. Measurement of the Equilibrium between Singly and Doubly Bridged Dimer. *Aust. J. Chem.* **1989**, *42*, 1579–1589.
19. Stuenzi H.; Spiccia, L.; Rotzinger, F. P.; and Marty, W; Early Stages of the Hydrolysis of Chromium (III) in Aqueous Solution: 4. The Stability Constants of the Hydrolytic Dimer, Trimer, and Tetramer at 25°C and I = 1.0 M. *Inorg. Chem.* **1989**, *28*, 66–71.
20. Spiccia, L. Early Stages of the Hydrolysis of Chromium (III) in Aqueous Solution: 7. Kinetics of Cleavage of the Hydrolytic Dimer in Acidic Solution. *Polyhedron* **1991**, *10*, 1865–1872.
21. Spiccia, L.; Marty, W. Early Stages of the Hydrolysis of Chromium (III) in Aqueous Solution: VI. Kinetics of Intramolecular Interconversion between Singly- and Doubly-Bridged Hydrolytic Dimers. *Polyhedron* **1991**, *10*, 619–628.
22. Drljaca, A.; Spiccia, L. Early Stages of the Hydrolysis of Chromium (III) in Aqueous Solution: Kinetics of Formation of Trimer from Monomer and Dimer. *Polyhedron* **1995**, *14*, 1653–1660.
23. Drljaca, A.; Spiccia, L. Early Stages of the Hydrolysis of Chromium (III) in Aqueous Solution: XI. Kinetics of Formation of Hexamer from Trimer and Tetramer from Monomer and Trimer. *Polyhedron* **1996**, *15*, 2875–2886.
24. Drljaca, A.; Spiccia, L. Early Stages of the Hydrolysis of Chromium (III) in Aqueous Solution: XII. Kinetics of Cleavage of the Trimer and Tetramer in Acidic Solution. *Polyhedron* **1996**, *15*, 4373–4385.
25. Spiccia, L.; Aramini, J. M.; Crimp, S. J.; Drljaca, A.; Lawrenz, E. T.; Tedesco, V.; and Vogel, H. J. Hydrolytic Polymerization of Rhodium (III): Characterization of Various Forms of a Trinuclear Aqua Ion. *J. Chem. Soc. Dalton Trans. Inorg. Chem.* **1997**, *1997*, 4603–4610.

26. Pope, M. T. Polyoxo Anions: Synthesis and Structure. *Comprehens. Coord. Chem. II*, **2004**, *4*, 635–678.
27. Pope, M. T. *Heteropoly and Isopoly Oxometalates*; Springer Verlag: Berlin, 1983.
28. Hill, C. L. Polyoxometalates: Reactivity. *Comprehens. Coord. Chem. II* **2004**, *4*, 679–759.
29. Müller, A.; Meyer, J.; Krickemeyer, E.; and Diemann, E. Molybdenum Blue: A 200 Year Old Mystery Unveiled. *Angew. Chem. Int. Ed.* **1996**, *35*, 1206–1208.
30. Kortz, U.; Müller, A.; van Slageren, J.; Schnack, J.; Dalal, N. S.; and Dressel, M. Polyoxometalates: Fascinating Structures, Unique Magnetic Properties. *Coord. Chem. Rev.* **2009**, *253*, 2315–2327.
31. Botar, B.; Koegerler, P.; Müller, A, Garcia-Serres, R.; and Hill, C. L. Ferrimagnetically Ordered Nanosized Polyoxomolybdate-Based Cluster Spheres. *Chem. Comm.* **2005**, *45*, 5621–5623.
32. Pope, M. T.; Müller, A., Eds. Polyoxometalates: From Platonic Solids to Anti-Retroviral Activity. Top. Mol. Organ. Eng. 1994, 10.
33. Pope, M. T.; Müller, A., Eds. *Polyoxometalate Chemistry from Topology via Self-Assembly to Applications*; Springer: New York, 2001.
34. Ruhle, J. T.; Hill, C. L.; Judd, D. A.; and Schinazi, R. F. Polyoxometalates in Medicine. *Chem. Rev.* **1998**, *98*, 327–357.
35. Yamase, T. Photo- and Electrochromism of Polyoxometalates and Related Materials. *Chem. Rev.* **1998**, *98*, 307–326.
36. Yan, L.; López, X.; Carbó, J. J.; Sniatynsky, R.D.; Duncan, D. C.; Poblet, J. M. On the Origin of Alternating Bond Distortions and the Emergence of Chirality in Polyoxometalate Anions. *J. Am. Chem. Soc.* **2008**, *130*, 8223–8233.
37. Hill, C. L. Progress and Challenges in Polyoxometalate-Based Catalysis and Catalytic Materials Chemistry. *J. Molec. Catal. A Chemical* **2007**, *262*, 2–6.
38. Nyman, M.; Burns, P. C. A Comprehensive Comparison of Transition-Metal and Actinyl Polyoxometalates. *Chem. Soc. Rev.* **2012**, *41*, 7343–7367.
39. Nyman, M. Polyoxometalate Chemistry in the 21st Century. *Dalton Trans.* **2011**, *40*, 8049–8058.
40. Burns, P. C.; Kubatko, K.-A.; Sigmon, G.; Fryer, B. J.; Gagnon, J. E.; Antonio, M. R.; and Soderholm, L. Actinyl Peroxide Nanospheres. *Angew. Chem. Int. Ed.* **2005**, *44*, 2135–2139.
41. Bontchev, R. P.; Nyman, M. Evolution of Polyoxometalate Cluster Anions. *Angew. Chem. Int. Ed.* **2006**, *45*, 6670–6672.
42. Keggin, J. F. Structure of the Molecule of 12-Phosphotungstic Acid. *Nature* **1933**, *131*, 908–909.
43. Constant, R.; Thouvenot, R. Dawson-Type Heteropolyanions: 2. Synthesis of Lacunary Polyoxotungstoarsenates Derived from Octadecatungstodiarsenate: Tungsten-183 NMR Structural Study of Related Octadeca(molybdotungstovanado)diarsenates. *Can. J. Chem.* **1991**, *69*, 1498–1506.
44. Mbomekalle, I. M.; Keita, B.; Nadjo, L.; Neiwert, W. A.; Zhang, L.; Hardcastle, K. I.; Hill, C. L.; and Anderson, T. M. Lacunary Wells–Dawson Sandwich Complexes: Synthesis, Characterization, and Stability Studies of Multi-Iron Species. *Eur. J. Inorg. Chem.* **2003**, *2003*, 3924–3928.
45. Ohlin, C. A.; Harley, S.J.; McAlpin, J. G.; Hocking, R. K.; Mercado, B. Q.; Johnson, R. L.; Villa, E. M.; Fidler, M. K.; Olmstead, M. M.; Spiccia, L.; Britt, R. D.; and

Casey, W. H. Rates of Water Exchange for Two Cobalt (II) Heteropolyoxotungstate Compounds in Aqueous Solution. *Chem. Eur. J.* **2011**, *17*, 4408–4417.
46. Anderson, T. M.; Hardcastle, K. I.; Okun, N.; and Hill, C. L. Asymmetric Sandwich-Type Polyoxoanions: Synthesis, Characterization, and X-ray Crystal Structures of Diferric Complexes $[TM^{II}Fe^{III}{}_2(P_2W_{15}O_{56})(P_2TM^{II}_2W_{13}O_{52})]^{16-}$, TM = Cu or Co. *Inorg. Chem.* **2001**, *40*, 6418–6425.
47. Müller, A.; Henry, M. Nanocapsule Water-Based Chemistry. *Compt. Rend. Chim.* **2003**, *6*, 1201–1208.
48. Müller, A.; Sarkar, S.; Shah, S.Q.N.; Bogge, H.; Schmidtmann, M.; Sarkar, S.; Kogerler, P.; Hauptfleisch, B.; Trautwein, A. X.; and Schunemann, V. Archimedean Synthesis and Magic Numbers: Sizing Giant Molybdenum-Oxide-Based Molecular Spheres of the Keplerate Type. *Angew. Chem. Int. Ed.* **1999**, *38*, 3238–3241.
49. Liu, T.; Langston, M.L.K.; Li, D.; Pigga, J.M.; Pichon, C.; Todea, A.M.; and Müller, A Self-Recognition among Different Polyprotic Macroions during Assembly Processes in Dilute Solution. *Science* **2011**, *331*, 1590–1592.
50. Müller, A.; Diemann, E.; Shah, S. Q. N.; Kuhlmann, C.; and Letzel, M. C. Soccer-Playing Metal Oxide Giant Spheres: A First Step Towards Patterning Structurally Well Defined Nano-Object Collectives. *Chem. Commun.* **2002**, *2002*, 440–441.
51. Forbes, T. Z.; McAlpin, J. G.; Murphy, R.; and Burns, P. C. Metal–Oxygen Isopolyhedra Assembled into Fullerene Topologies. *Angew. Chem. Int. Ed.* **2008**, *47*, 2710–2711.
52. Nyman, M.; Alam, T. M.; Bonhomme, F.; Rodriguez, M. A.; Frazer, C. S.; and Welk, M. E.. Solid-State Structures and Solution Behavior of Alkali Salts of the $[Nb_6O_{19}]$ Lindqvist Ion. *Cluster Sci.* **2006**, *17*, 197–204.
53. Klemperer, W. G. Introduction to Early Transition Metal Polyoxoanions. *Inorg. Synth.* **1990**, *27*, 71–74.
54. Villa, E. M.; Ohlin, A. C.; Rustad, J. R.; and Casey, W. H. Isotope-Exchange Dynamics in Isostructural Decametalates with Profound Differences in Reactivity. *J. Am. Chem. Soc.* **2010**, *132*, 5264–5274.
55. Alam, T. MNyman, M. D.; Cherry, B. R.; Segall, J. M.; and Lybarger, L. E. Multinuclear NMR Investigations of the Oxygen, Water, and Hydroxyl Environments in Sodium Hexaniobate. *J. Am Chem. Soc.* **2004**, *126*, 5610–5620.
56. Klemperer, W. G.; Shum, W. Synthesis and Interconversion of the Isomeric α– and $\beta\text{-}Mo_8O_{26}^{4-}$ Ions. *J. Am. Chem. Soc.* **1976**, *98*, 8291–8293.
57. Villa, E. M.; Ohlin, C. A.; Casey, W. H. Borate Accelerates Oxygen-Isotope Exchange for Polyoxoniobate Ions in Water. *Chemistry* **2010**, *16*, 8631–8634.
58. Henderson, M. A.; Joyce, S. A.; Rustad, J. R. Interaction of Water with the (1x1) and (2x1) Surfaces of $\alpha\text{-}Fe_2O_3$ (012). *Surf. Sci.* **1998**, *417*, 66–81.
59. Yanina, S. V.; Rosso, K. M. Linked Reactivity at Mineral–Water Interfaces through Bulk Crystal Conduction. *Science* **2008**, *320*, 218–222.

6

The Chemistry of Extended Oxide Surfaces

6.1 INTRODUCTION

In Chapters 4 and 5, we demonstrated that local structures and charge distributions have an enormous impact on the equilibrium constants, trajectories, and kinetics of reactions involving soluble oxide precursors. In this chapter, we highlight those features that make reactions on extended oxide surfaces either similar to or dramatically different from the reactions documented in hydrolysis diagrams for each metal cation (see Chapter 5). We first describe oxide surface structures and then discuss how these structures impact both acid–base and ligand-exchange phenomena. In addition to dense oxides, we also introduce some of the chemistry associated with layered materials. Lamellar materials are important from both a fundamental and technological perspective, because water and ions can readily penetrate such structures and provide conditions under which almost every oxygen anion is at an oxide–water interface (see Chapter 10 and Chapter 11). This chapter focuses on oxides containing octahedral cations. The distinctive chemistry of oxides based on tetrahedral cations, including the clay minerals and the zeolites, are the focus of Part Five.

6.2 IDEAL OXIDE SURFACES IN THE ABSENCE OF WATER

The structures of bulk oxides were introduced in Chapter 2. However, for many oxides, the surface structures that interact with aqueous solutions are substantially different from structures found in the bulk. Here, we introduce the basic principles of oxide surfaces that make them chemically active. As a starting point, consider ideal oxide surfaces containing +2 octahedral cations (Fig. 6.1). Pristine oxide surfaces can be created by cleaving perfect crystals in an ultrahigh-vacuum environment. The creation of new surfaces requires an expenditure of energy corresponding to the cohesive energy of the solid, which in turn represents the energy required to break every bond along a given fracture plane. For MgO, the Mg—O bond energy is 380 kJ/mole [1]. Each surface created contains $1.4 \cdot 10^{19}$ oxygen atoms/m^2, or $2.4 \cdot 10^{-5}$ moles of bonds. Because two surfaces are created in the fracture event, the initial interfacial energy of each resulting MgO surface is $(1/2)(380\ \mathrm{kJ/mole})/(2.4 \cdot 10^{-5}\ \mathrm{mole/m^2}) = 4560\ \mathrm{mJ/m^2}$. To put this energy into perspective, the interfacial energies of representative soft and hard metals at room temperature [2] are 1100 mJ/m^2 for Al and 4400 mJ/m^2 for W. A particularly important interfacial energy is that of liquid water. The energy required to divide one droplet into two droplets in air [3] is only 72 mJ/m^2. However, this energy influences many chemical events that occur at the oxide–water interface.

Cleavage surfaces are so energetic that they reconstruct even as they are being formed. Pristine oxide surfaces pursue a number of strategies to reduce the enormous interfacial energy created as a result of rupturing bonds that exist in the bulk oxide. At an atomic level, bond distances readjust to reduce local charge

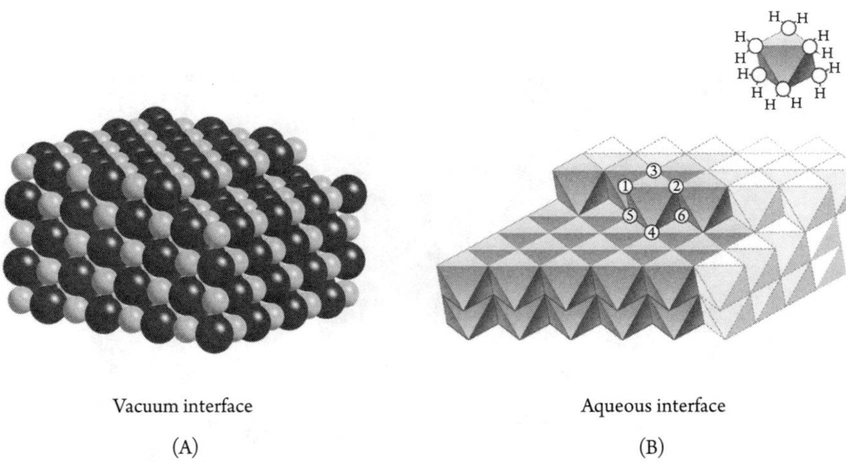

Figure 6.1 (A) Depiction of a fracture surface of Mg(100) showing exposed metal cations (small light-colored spheres) and underbonded anions (large dark-colored spheres) on terraces, corners, and edges. (B) A polyhedral representation of the Mg(100) surface after it is exposed to water, with an aquated cation shown as a reference species. The surface contains oxygen connectivity numbers ranging from one to five for the sites depicted. In contrast to the fracture surface (A), the Mg(II) sites in water all become fully coordinated to oxygens immediately. Note also that the Mg(II) coordination number is always 6 in the aquated surface (B), but the *oxygen* coordination numbers vary from 1 to 6, as shown by the numbers.

imbalances. If geometric constraints are favorable, adjacent cations and anions react with each other or pair up to eliminate unsatisfied bonds. For example, a nonbridging oxygen adjacent to an open cation site can shift to and fill the site and create a bridging oxygen. Although such bonding can repair surface damage and lower surface energies in part, the resulting bonds often contain residual strain and are more reactive than similar bonds in the bulk oxide (see Chapter 16).

Another strategy that surfaces use to reduce interfacial energy involves more macroscopic reconstructions. In principle, an infinite number of fracture surfaces could be created from a single oxide crystal by cleaving along different planes. In practice, surface science studies [4] show that surfaces reconstruct to minimize interfacial energy, creating terraces that are typically terminated by only a few low-Miller index planes. For example, the lowest energy plane for MgO is the electrically neutral and nonpolar (100) surface shown in Figure 6.1. In contrast, cleavage along the (111) plane creates surfaces that are terminated completely by either Mg^{2+} or O^{2-} ions, resulting in highly charged and polar surfaces. As a result, (111) surfaces are metastable and tend to facet to create new surfaces that are terminated by (100) planes [5] (Fig. 6.2). The relatively low interfacial energy of liquid water arises because the liquid restructures its hydrogen bonding distributions spontaneously at the air–water interface when a droplet is divided. Similarly, liquid water restructures when placed in contact with any solid surface.

Figure 6.2 Scanning electron micrograph of the MgO(111) surface after annealing at 1400°C for 1 minute, showing the effects of surface faceting.
SOURCE: Henrich, V. E. Thermal Faceting of (110) and (111) Surfaces of MgO. *Surf. Sci.*, **1976**, 57, 385–392, with permission.

6.3 INTERACTIONS BETWEEN PRISTINE SURFACES AND WATER MOLECULES

All processes described earlier contribute to a significant lowering of surface energies in a vacuum. For example, the experimentally measured interfacial energy [6] for MgO that is cleaved along the (100) plane is around 1200 mJ/m^2, or only 26% of the 4560 mJ/m^2 of all ruptured Mg—O bonds. However, the pristine surfaces are still quite energetic, containing 94 kJ/mole of energy per surface site. At this point, chemical reactions represent the primary mechanism for lowering the interfacial energy of pristine oxide surfaces further. The dominant environmental species that reacts with fresh oxide surfaces is the water molecule. Mechanisms for water adsorption involve: (1) Lewis acid–base interactions on exposed metal cation sites and (2) both Lewis and Brønsted interactions on undercoordinated O^{2-} sites.

On a conceptual level, the Lewis acidity of an exposed cation is proportional to its formal charge. To illustrate this Lewis acidity, consider the surface of MgO, where bond rupture during cleavage creates open coordination sites on Mg(II) centers. At an atomic level, the freshly cleaved surface not only contains ideal cleavage planes, but also steps, corners, and point defects (missing atoms) (Fig. 6.1). Even for a simple oxide such as MgO, the diversity of potential surface sites makes the surface chemistry of oxides diverse and complex. On vacuum-cleaved MgO, there can be Mg^{2+} sites where the cation is bonded to five, four, three, two, and even only one oxygen anion. From a formal charge perspective, these cationic sites exhibit formal charges of $+1/3$, $+2/3$, $+1$, $+4/3$, and $+5/3$, respectively. Even a site with a formal charge of only $+1/3$ is a Lewis acid that can bond to Lewis, bases, such as the electron lone pairs on a water molecule. Similarly, the O^{2-} sites of coordination numbers 5, 4, 3, 2, and 1 have formal charges of $-1/3$, $-2/3$, -1, $-4/3$, and $-5/3$ respectively, and are thus increasingly strong Lewis bases. The important point is that the surface energy of a freshly cleaved oxide is expressed in the creation of surface sites that span a wide range of acid–base activity. Some of these sites are so reactive that they do not survive except in ultrahigh vacuum environments.

Consider what happens when a freshly cleaved surface is exposed to even traces of water from the vapor phase. The extent of water adsorption and the strength of the acid–base interactions between water molecules and surface sites is typically quantified by performing temperature-programmed desorption (TPD) experiments on the oxide surface. In a TPD experiment, the surface is created (or cleaned) carefully in an ultrahigh-vacuum environment, cooled to a low temperature, and then dosed with known quantities of water. Next, the surface is gradually heated, and one uses a mass spectrometer to measure the amount of water that desorbs as a function of temperature. At the lowest dosing levels, the most reactive sites are the first to fill. With increased dosing levels, all active sites are eventually filled, followed by the completion of the first water monolayer, and finally by multiple water layers (i.e., ice layers) on top of the oxide surface. On heating, those molecules that form the weakest bonds to the surface come off first, followed by bound water molecules having higher bond strengths. TPD spectra [7–9] for representative oxides are shown in Figure 6.3.

Provided that the water molecule does not dissociate, there is a linear correlation between desorption temperature and the enthalpy of bond formation between the water molecule and specific surface sites (Fig. 6.4). That being said, TPD results alone are insufficient to characterize fully either the specific surface sites or the specific reactions occurring at those sites. This ambiguity arises because different bonding configurations on different sites can result in identical desorption temperatures. In addition, it can be extremely difficult to distinguish between distinct surface species such as $M-OH_2$ and $M-OH$ within an atomically thin surface layer on the basis of conventional spectroscopic techniques alone. (See Chapter 16 to see how vibrational spectroscopies on high-surface area powders can elucidate some of these structures and reactions.) For this reason, conclusions drawn from the extensive studies that have been performed on the interactions between water and oxide surfaces are often controversial (see [10] and [11] for excellent reviews). Our goal here is not to resolve these controversies, but to highlight phenomena that are important in the chemistry of water adsorption.

Oxide surfaces exhibit four different generic types of behavior when it comes to water adsorption. In order of increasing TPD desorption temperatures, these behaviors include (1) negligible adsorption, (2) hydrogen bonding to oxygen sites that function as Lewis bases, (3) adsorption on exposed cations (Lewis acid sites) to form $M-O$ bonds, and (4) water dissociation to form surface hydroxyl groups. Starting with the minimal adsorption case, water can be made to adsorb on almost any surface at sufficiently low temperatures. However, for temperatures below around 150 K, heats of adsorption must be estimated by the *adsorption* temperature rather than the *desorption* temperature because, when on the surface, water molecules hydrogen-bond to each other to form ice-like structures. When this happens, the minimum temperature for water desorption is around 160 K, which corresponds to the 42 kJ/mole of energy required to disrupt the hydrogen bonding network in ice (see desorption of water from FeO in Fig. 6.3). Hydrogen bonding to oxygen anions at the oxide surface can be somewhat stronger than this, resulting in desorption temperatures that typically range from 160 to 200 K (see MgO and ZnO in Fig. 6.3).

Water adsorption on Lewis acid sites is typically observed within the temperature window from 130 to 300 K, corresponding to interaction energies of roughly 30 to 70 kJ/mole. For cationic sites with a single open coordination site, these bond energies are comparable with the enthalpies of $M-O$ bond formation for

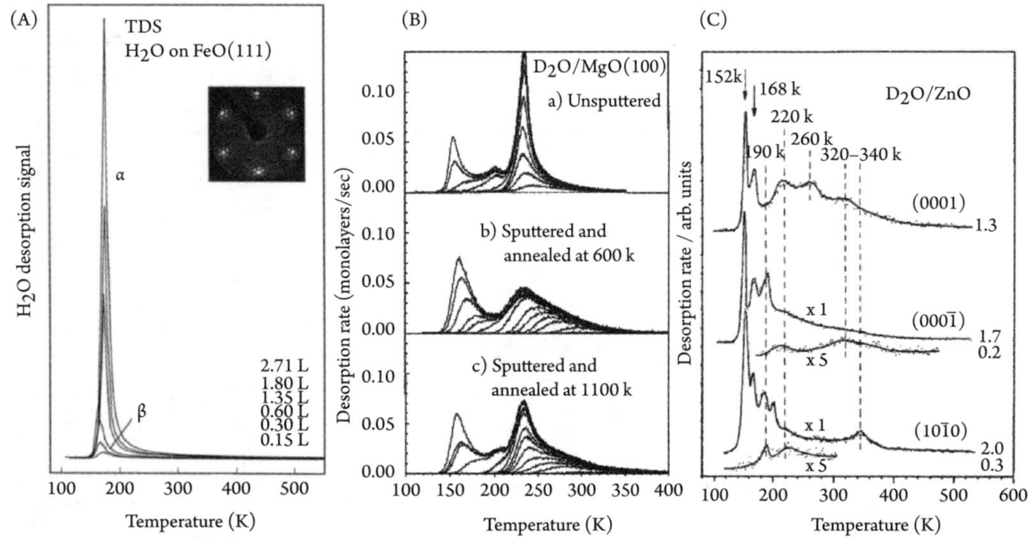

Figure 6.3 Temperature-programmed desorption (TPD) results obtained for MO oxides in which the dominant desorption features are representative of: (A) weak adsorption of molecular water on FeO (SOURCE: Joseph, Y.; Kuhrs, C.; Ranke, W.; Ritter, M.; Weiss, W. Adsorption of water on FeO(111) and Fe_3O_4(111): identification of active sites for dissociation. *Chem. Phys. Lett.*, **1999**, *314*, 195–202, with permission), (B) moderate molecular and/or dissociative adsorption on Lewis acid sites (on FeO) (SOURCE: Stirniman, M. J.; Huang, C.; Smith, R. S.; Joyce, S. A.; and Kay, B. D. The Adsorption and Desorption of Water on Single Crystal MgO(100): The Role of Surface Defects. *J. Chem. Phys.*, **1996**, *105*, 1295–1298, with permission), and (C) strong adsorption (with probable dissociation) on Lewis acid sites (on the (0001) surface of ZnO (SOURCE: Zwicker, G.; Jacobi, K. Site-specific Interaction of H_2O with ZnO Single-Crystal Surfaces Studied by Thermal Desorption and UV Photoelectron Spectroscopy. *Surf. Sci.*, **1983**, *131*, 179–194, with permission). On ZnO (C), note the different desorption patterns that are observed on different crystal faces reflecting different surface-site distributions.

the same cation dissolved in water (see Chapter 4 and Fig. 6.4). This similarity is not surprising because the formal charge donated by the cation (related to Lewis acidity) is identical in both cases. As in aqueous complexes, the $M-OH_2$ bond strength increases as the cation valence (and its effective formal charge) increases. The $M-OH_2$ bond strength also increases with the number of open coordination sites that are available around a given metal cation. For example, cations on the step edges of MgO have formal charges of $+2/3$, which are identical to those

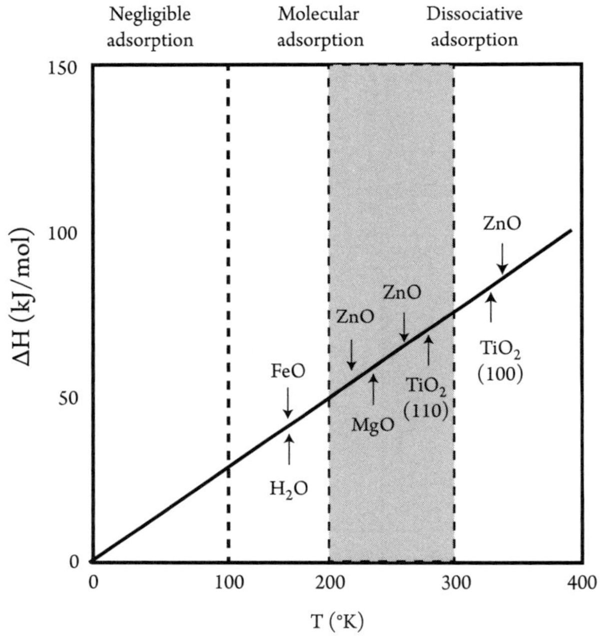

Figure 6.4 The heat of adsorption for water correlates with the temperature at which that water desorbs from the surface because both reflect the strength of bonds. The data used in this figure are contained in Figures 6.3 (for FeO, MgO, and several ZnO surfaces) and 6.7 (for the (110) and (100) surfaces of TiO_2). The primary desorption feature associated with each oxide is shown, with the exception of ZnO, where several desorption features are apparent, depending on the exposed surface. (Note that the peak on most oxides associated with water adsorbing on itself is not shown.) The designations at the top of the graph provide rough indications of desorption temperature ranges corresponding to surfaces with sites that are too weak to adsorb much water (negligible adsorption), sites that adsorb water reversibly as an intact molecule (molecular adsorption), and sites at which the water molecule is dissociated to form two surface hydroxyl groups (dissociative adsorption). The boundary between molecular and dissociative adsorption is indicated by a gray band instead of a line to indicate there is considerable overlap (and controversy) regarding the adsorption mechanism for desorption features that appear at intermediate temperatures (between roughly 200 and 300 K).

seen for bare cations in the middle of the (110) planes on TiO_2 surfaces. Because oxide surfaces are often highly defective, this means that more than one desorption state from Lewis acid sites can be seen frequently in TPD results.

On low-index or low-energy planes, it is rare to observe more than two distinct TPD features associated with Lewis acid sites on a given oxide surface. First, site populations decrease as the number of open coordination positions on the site (and site energies) increase. In other words, planar sites are more prevalent than step edges, which are more populous than corner sites and other defects. Second, just as in aqueous complexes (Chapter 4), as the charge donated by the cation (i.e., the Lewis acidity) increases, the strength of the $M{-}OH_2$ bond also increases. An interesting example that illustrates this point is provided by the TPD experiments performed on different cleavage planes of ZnO [9] (Fig. 6.3). The $(000\bar{1})$ surface is terminated by oxygen anions and therefore shows negligible TPD activity associated with Lewis acid sites. In contrast, the (0001) surface is terminated by Zn^{2+} cations, each of which has three adjacent open coordination sites (i.e., the entire surface is terminated by cationic "corner" sites). On this surface, the TPD feature associated with hydrogen bonding to surface oxygen anions (at 190 K) disappears, whereas three new features of equal intensity appear (at 220 K, 260 K, and 330 K). Because each Zn^{2+} site can adsorb up to three waters of hydration, it is probable that the three TPD features represent the sequential losses of one, two, and three waters of hydration from each site. As each water of hydration is removed, the Lewis acidity of the Zn^{2+} sites increases as the formal charge increases from $+1/3$ to $+2/3$ and finally to $+1$. As a cross-check to this hypothesis, desorption of molecular water from Ti^{4+} sites on the (110) surface of TiO_2 (with a formal charge of $+2/3$) occurs at 270 K compared with the 260-K peak on ZnO for sites that have already lost one water molecule, resulting in the same formal charge of $+2/3$.

Although logical, the assignments of the high-temperature desorption events on ZnO do not necessarily represent reactions involving molecular water. We saw in Chapter 4 that, as the $M{-}OH_2$ bond becomes stronger, the $O{-}H$ bonds in the coordinated water molecule become weaker. In other words, the coordinated water molecule becomes a stronger Brønsted acid. These acidic water molecules are increasingly eager to donate their protons to any oxygen anions in their immediate vicinity to create hydroxyl groups (Fig. 6.5) or, in extreme cases, even $M{-}O$ species. The dissociation of water to form two hydroxyl groups on the surface is called *hydroxylation*. As in aqueous complexes, these surface hydroxyl groups are bound more tightly to cationic sites than are coordinated water molecules as a result of simple electrostatics.

Dissociated water in the form of surface hydroxyl groups represents the most difficult form of water to remove from a surface. Removal involves *dehydroxylation* processes in which adjacent hydroxyl groups react with each other to reform a water

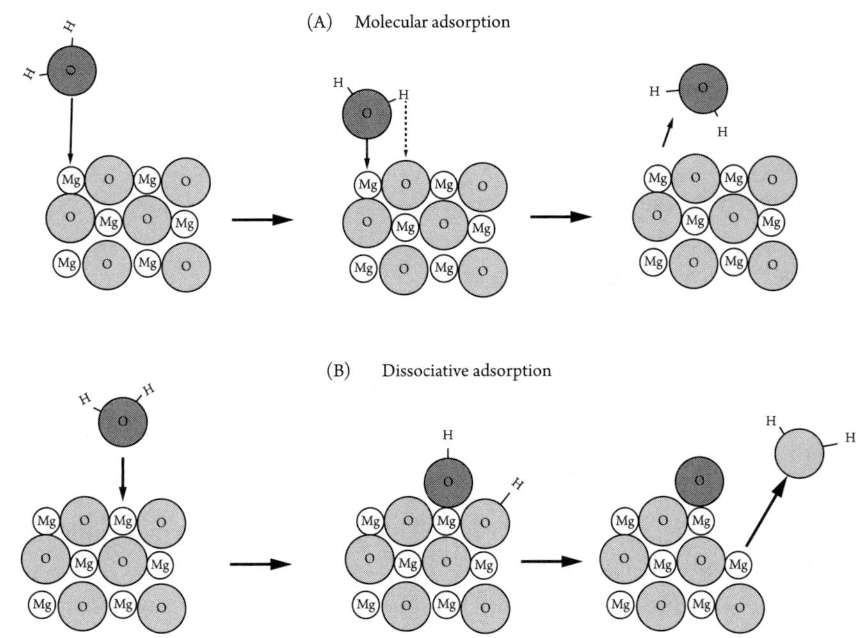

Figure 6.5 The two dominant adsorption modes for water on oxide surfaces (shown for MgO). (A) During molecular adsorption, the only interaction between water and the surface involves exposed metal cation sites. Here, desorption of the water molecule results in no oxygen exchange with the surface. (B) During dissociative adsorption, the proton on the incoming water molecule can interact sufficiently strongly with a surface oxygen site to dissociate the water molecule and form two surface hydroxyl groups. Here, water desorption requires the recombination of two hydroxyl groups. The hydroxyl group that captures a proton from its neighbor and desorbs as a water molecule can either represent the original water molecule or the hydroxyl group with the oxygen that was on the original surface. This results in a scrambling of oxygens between the adsorbed water and the surface, and is detectable via isotope-labeling studies. One possible explanation for the temperature-programmed desorption results seen on MgO(100) is that the feature at 200 K represents molecular water, whereas the dominant feature at 240 K represents water that has dissociated into hydroxyl groups (shown in the depiction). However, if dissociation occurs, experiments performed with $H_2{}^{18}O$ show the reaction is completely reversible, with negligible oxygen exchange involving the underlying oxide.

molecule that then desorbs from the surface. At high surface coverages, the rate-limiting step for desorption via dehydroxylation is often the disruption of the M—OH_2 bond, which means that the TPD desorption temperature is only slightly higher than that predicted for a given Lewis acid site had the water molecule not dissociated in the first place. For the case of MgO, extrapolation from the TPD results in Figure 6.4 to the strongest possible Lewis acid site (Mg^{2+} with only one bond to the

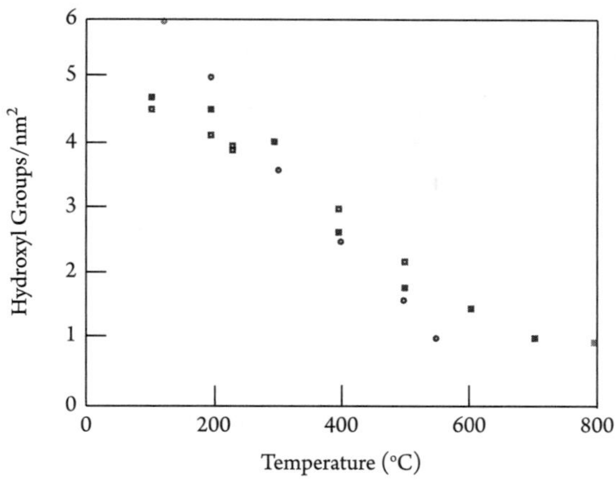

Figure 6.6 Surface hydroxyl group concentrations on SiO_2 as a function of temperature. All surface sites are occupied by hydroxyl groups at a coverage of 5 OH/nm^2. Almost 10% of the surface hydroxyl groups persist even at 1000°C. Adapted from Iler, R. K. *The Chemistry of Silica*; Wiley: New York, 1979, with permission.

surface and a formal charge of +5/3), leads to the prediction that complete dehydroxylation of MgO surfaces could require temperatures exceeding 500 K.

On many oxides, surface hydroxyl groups persist up to temperatures that exceed 1000°C. However, the reason for such high dehydroxylation temperatures often has as much to do with the geometry of the surface sites as it does with their Lewis acidity. For the case of fully hydroxylated silica surfaces (Fig. 6.6), dehydroxylation involves the recombination of two nonbridging silanol groups to form a bridging oxygen and a water molecule:

$$>\text{Si}-\text{OH}+\text{HO}-\text{Si}< \rightleftarrows >\text{Si}-\text{O}-\text{Si}<+\text{H}_2\text{O} \quad \textbf{(6.1)}$$

As more water is removed from the surface, the concentration of adjacent hydroxyl groups drops and it becomes increasingly difficult for two hydroxyl groups to encounter each other [12]. It eventually becomes necessary to break and rearrange bonds in the silicate network to promote the removal of the last vestiges of hydroxyl groups from the surface.

The previous example illustrates that reactions on an oxide surface can be substantially different from those of soluble hydrolysis products as a result of geometric effects. On a surface, the size, shape, and proximity to other surface species can vary significantly, depending on the local architecture of the oxide. Geometric effects are critical for hydroxylation, dehydroxylation, network hydrolysis and dissolution, catalytic reactions, as well as ion exchange (Chapter 10) and electrochemical (Chapter 11) processes. A specific example that illustrates the role of geometry on surface hydroxylation involves the reactions that occur between water molecules and titanium dioxide surfaces [13] (Figs. 6.7, 6.8). On the (110) TiO_2 surface, adsorption and desorption processes are limited to Lewis acid–base interactions involving molecular water and exposed Ti(IV) cations. However, on (100) surfaces, a proton on the adsorbed water molecule reacts with an adjacent Ti$-\mu_2$O$-$Ti bridge to form two surface hydroxyl groups (see Section 6.5). The geometric explanation for this observation is that, although the O$-$O separation distance between

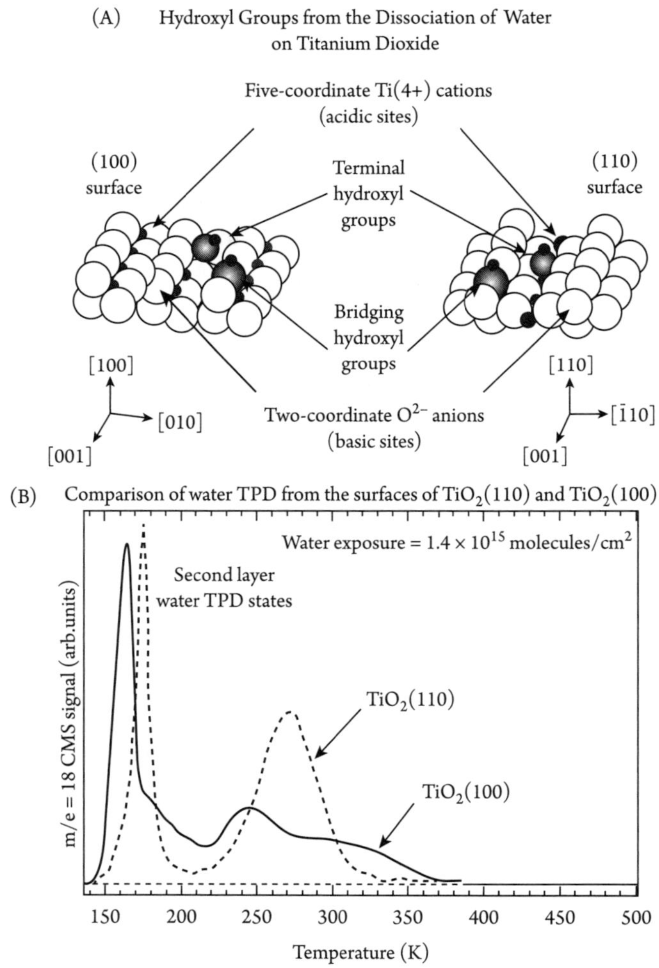

Figure 6.7 Geometric effects in water adsorption on TiO_2 surfaces. Temperature-programmed desorption results on TiO_2 surfaces highlight clear differences between the (110) surface, which is dominated by the adsorption and desorption of molecular water (at 270 K), and the (100) surface, on which water clearly dissociates (as indicated by the feature at 330 K).

SOURCE: Images and data from Henderson, M. A. Sensitivity in the Dissociation of Water on TiO_2 Single-Crystal Surfaces. *Langmuir*, **1996**, *12*, 5093, with permission.

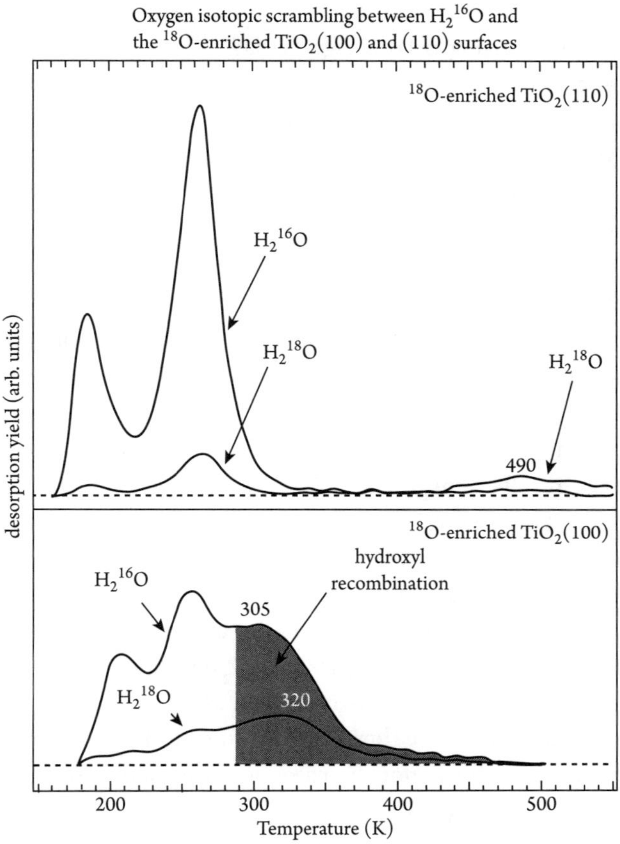

Figure 6.8 Clear evidence for the two reaction pathways shown in Figure 6.7 is provided by dosing experiments involving $H_2{}^{18}O$, which indicate that unlabeled bridging hydroxyl groups are protonated and released as water on the (100) but not the (110) surface.

SOURCE: From Henderson, M. A. Sensitivity in the Dissociation of Water on TiO_2 Single-Crystal Surfaces. *Langmuir,* **1996,** *12,* 5093, with permission.

potential reactants on the (100) surface (0.27 nm) is within range of hydrogen bond formation, the bound water and the adjacent Ti—O—Ti bridge are too far apart (0.33 nm) to react with each other on the (110) surface.

The hydroxyl groups formed via water adsorption do not necessarily have the same local bonding configurations as the equivalent bulk hydroxide phase, nor the same chemical stability as measured by dehydroxylation temperature. For example, $Mg(OH)_2$ contains neutral μ_3-OH bridges whereas the surface shown in Figure 6.5 contains nonbridging M—OH groups. The MgO surface sites shown can be dehydroxylated via the disruption of a single M—O bond, whereas removal of the μ_3-OH bridge from $Mg(OH)_2$ requires the disruption of three

M—O bonds. An examination of the array of different sites present in a typical defective MgO surface (Fig. 6.1B) suggests that it should be easier to dehydroxylate some sites on hydrated surfaces than to decompose stable hydroxide phases. As confirming evidence, the thermal decomposition of $Mg(OH)_2$ requires heating [14] to 350°C (623 K) compared with the more modest TPD feature at 240 K that some scientists have assigned to dehydroxylation of MgO defect sites [11].

Experimental results on MgO highlight the fact that reaction pathways for oxides in air can be substantially different from those that occur in liquid water. In liquid water, water molecules can hydrolyze not only the immediate surface, but also the underlying solid. Some of these reactions involve relative rapid proton transfer processes. Such reactions can result in either oxide dissolution or phase transformations, such as the conversion of MgO into $Mg(OH)_2$. However, the exposure of a perfect MgO [100] surface to limited amounts of water from the vapor phase results in negligible hydrolysis of the underlying oxide. In fact, isotopic labeling experiments [15, 16] show that if surface hydroxyl groups do indeed form on water dosing, the reaction is completely reversible with no exchange of oxygen anions between adsorbed water molecules and the surface. This is in contrast to the extensive oxygen exchange associated with hydroxyl recombination that occurs on the (100) surface of TiO_2 described earlier [13] (Fig. 6.8).

The previous result for MgO illustrates that, as some oxides are heated in air, water molecules desorb before they have a chance to hydrolyze the underlying oxide. In liquid water, this is not the case, because water molecules are always present and available for ligand exchange on available Mg(II) sites, provided the temperature is kept below the boiling point of water. It is usually impossible to maintain open coordination sites on metal surface sites immersed in liquid water because the sites are highly reactive and water concentrations are extremely high (~ 55 M). Liquid water serves as a solvent that can readily accommodate ionic species, which facilitates both hydrolysis and dissolution processes. In contrast, removal of ions from a solid oxide in air can require temperatures in excess of 1273 K or other extreme conditions such as electron, ion, or photon bombardment (see Chapter 13).

6.4 THE INTERFACE BETWEEN LIQUID WATER AND OXIDE SURFACES

It is now time to examine what happens when pristine oxide surfaces are immersed in liquid water. For the example of MgO, we saw that newly created oxide surfaces are highly energetic. Through the readjustment of surface bonds and reconstruction, these surfaces lower their interfacial energy from 4560 mJ/m^2 to an experimental value of around 1200 mJ/m^2. At this point, surface chemistry is from interactions with the solvent at the interface. In liquid water, every exposed cationic site is immediately occupied with an adsorbed water molecule. If the average $Mg-OH_2$ bond energy is the 50 kJ/mole indicated from TPD experiments (Fig. 6.4), water adsorption is expected to lower the surface energy by an additional 600 mJ/m^2. Because each adsorbed water molecule will hydrogen-bond to adjacent oxygen anions (or water molecules in the liquid), these hydrogen bonds should lower the interfacial energy by an addition 500 mJ/m^2. This means the remaining interfacial energy should be $(1200-600-500)=100\ mJ/m^2$. The experimental value for the interfacial energy of $Mg(OH)_2$ obtained from precipitation experiments conducted in water (see Chapter 7) is 115 mJ/m^2. This means we have now come close to quantifying the energy of the oxide–water interface based on the reactions that dominate the adsorption of molecular water from the vapor phase.

Although an interfacial energy of 115 mJ/m^2 is relatively low, it is still sufficient to drive additional chemical and physical phenomena. In fact, the oxide–water interface represents a very reactive and dynamic environment in which liquid water responds to the oxide and conversely the oxide responds to the presence of liquid water (Fig. 6.9). First, consider how liquid water responds to oxide

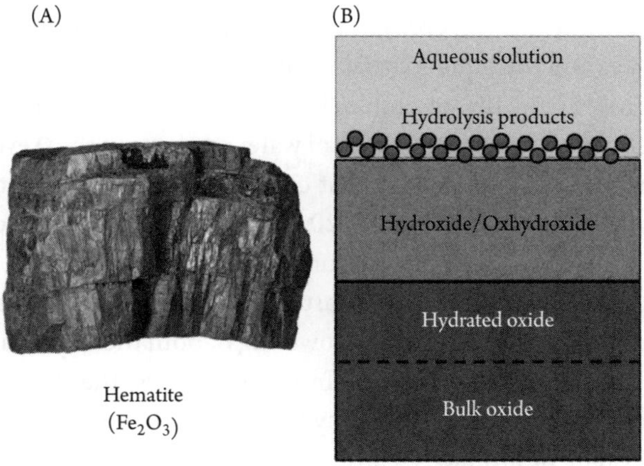

Figure 6.9 (A) When an anhydrous oxide such as hematite (Fe_2O_3) (SOURCE: SiimSepp/Shutterstock.com) is exposed to water, hydration and hydrolysis reactions can lead to the formation of complex surface alteration layers that cause some of the color changes observed in sample surfaces. (B) A schematic cross-section of the oxide–water interface shows some of the hydrated layers that can develop with depth, time, and/or degree of hydration. First, a layer of hydrated oxide forms. As the water content increases, this layer can be transformed into new phases such as oxyhydroxides (such as FeOOH) and hydroxides (such as $Fe(OH)_3$). Further hydrolysis structural breakdown forms a thin layer of adsorbed hydrolysis products at the immediate oxide–water interface. Desorption of these products results in oxide dissolution.

surfaces (see Chapter 8). In Chapters 3 and 4, we saw that the extensive hydrogen-bonding networks in water respond to the presence of foreign solutes. Hydrogen bonding at interfaces has several important consequences. First, hydrogen bonding controls whether water wets or spreads out on a surface. Wetting is often characterized by determining the contact angle between a liquid droplet and the surface [17]. On a superhydrophobic surface, where the hydrogen-bonding network runs orthogonal to the surface (as in clathrate hydrates), the contact angle is 180°. On a hydrophilic surface, where hydrogen bonding between water and surface sites is extensive, the contact angle is 0°. In general, bridging oxo-groups (which can terminate oxide surfaces containing higher valence cations) are hydrophobic relative to the surface hydroxyl groups. Also relatively hydrophobic are the terminal oxo-groups with multiple bonding character found for polyoxometalate ions and oxides of group V and group VI metals. The V_2O_5 sheets shown in Figure 4.2 with rows of highly coordinated oxygens having high bond orders exemplify such surfaces. Flat TiO_2 surfaces are reported to exhibit a contact angle [18] of 71° (compared with 108°, 91°, and 86° for paraffin, polystyrene, and graphite, respectively [17, 19]). In contrast, the contact angle on fully hydroxylated silica is 0°. However, dehydroxylation of silica at 1000°C, which replaces 90% of the surface hydroxyl groups by Si—O—Si groups, increases the contact angle to almost 60°.

If the oxide is heavily hydroxylated, extensive hydrogen bonding between water molecules and the liquid partially immobilizes water molecules at the surface, creating a structure that resembles ice. The anchoring of interfacial water has several consequences. First, a thin layer of water can form within several nanometers of the surface, with a viscosity that can be on the order of 1 to 10 million times greater than that of bulk water [20]. This layer is the source of a so-called repulsive *hydration force* between the surface and other objects [21] (see Chapter 8). Second, dielectric-relaxation measurements show that the decreased water mobility near the interface results in a lowering of both the dielectric constant of water and dielectric-relaxation times. In extreme cases, the dielectric constant can be lowered from the zero-frequency bulk value of 79 to the high-frequency dielectric constant of around 4 [22], where the frequency designation corresponds to that of the sinusoidal electic field applied to the sample in order to make the measurement.

The second attribute of a surface that has a major impact on the solution is surface charge. Under most solution conditions, oxide surfaces are charged as a result of the formal charge on exposed oxygen anions as mediated by the adsorption or desorption of protons (see Section 6.5). Dissolved cations and anions in solution are attracted to or repelled by this surface charge. The immediate surface charge plus the effective charge buildup in the adjacent solution creates a zone called the *electrical double layer* that can extend for microns into the solution phase. One consequence of the presence of the double layer is the generation of electrostatic forces between the surface and other objects (see Chapter 8). Another consequence of the double layer involves the creation of gradients in ionic concentrations near the oxide surface.

Both the electrical double layer and interfacial hydrogen bonding exert significant control over chemical equilibria at the interface. Because of the double layer, ionic concentrations at the surface are normally not the same as they are in the bulk solution. In addition, the lowering of the interfacial dielectric constant resulting from local hydrogen bonding means that liquid water at the interface is less effective at screening ionic charges from each other. In the region exhibiting a low dielectric constant, ions in solution are less stable and less soluble than would be indicated on the basis of bulk-solution equilibrium constants. To correct for nonideality in interfacial phenomena, equilibrium-constant expressions on oxide surfaces involve the use of activity coefficients [23] that separately take into account both electrostatic considerations and the dielectric constant of the solvent as ordered by the interface.

Now consider the solid side of the oxide–solution interface. Immediately after immersion in liquid water, oxide surface structures resemble bulk structures except that (1) each cation fills each broken M—O bond with a water molecule (see Fig. 6.1) and (2) each oxygen anion replaces its lost cation with a proton to create both bridging and nonbridging hydroxyl groups. In contrast to the vapor phase, so much water is present that even those cationic sites with a low formal charge and Lewis basicity are occupied. As in the case of solvated cations, coordinated water molecules on relatively weak Lewis acid sites are labile. As in solution, water-exchange rates scale with formal charge and correspond roughly to the rates seen for solvated cations (see Chapter 4). When a hydrated surface is established, coordinated water molecules and surface hydroxyl groups adjust their degree of protonation to meet local charge constraints. On a larger scale, reactions with water and the presence of the bulk fluid modify interfacial energies. Changes in interfacial energies can lead to surface reconstructions that favor a different set of low-energy lattice planes than those encountered on a freshly cleaved fracture surface.

The discussion thus far assumes that oxide–water reactions are confined to the immediate oxide surface, which is usually not the case (Fig. 6.9). Network hydrolysis can proceed to such an extent that oxide materials can be converted completely into hydroxides. The penetration depth for such reactions can range from a few nanometers to complete bulk transformations. Most commonly, hydration is confined to a thin layer near the surface, because extensive hydrolysis results in oxide dissolution. Within hydrated surface layers, network hydrolysis results in the formation of additional hydroxyl groups and coordinated water molecules, which decrease further the points of attachment between the anions and the host matrix. Hydrogen depth profiles (see Chapter 15) show there is usually a transition zone with a high degree of hydration at the immediate surface that decays with distance into the bulk. For thicker films, phase changes can occur within the hydrated layer (i.e., conversion of oxides into hydroxides above a thin transition zone). Regardless, the immediate surface often ends up resembling an adsorbed layer of hydrolysis products. The formation of hydration layers and oxide dissolution mechanisms are covered in detail in Chapter 15.

A final point to make regarding oxide surface structures is that solution species other than water can be incorporated into near-surface sites. These species range from simple cations or anions that are adsorbed to minimize local charge imbalances to foreign cations and/or ligands that can substitute for network cations or oxygen anions, respectively. Such substitutions are particularly important for the layered oxides and hydroxides, dictating critical ion-exchange properties (Chapter 10), surface charge and colloid chemistry (Chapter 8), electrochemical properties (Chapter 11), and environmental chemistry (Chapter 18). Such substitutions are also critical for the chemical behavior of open tetrahedral networks such as zeolites (see Chapter 14).

6.5 ACID–BASE REACTIONS ON OXIDE SURFACES IN LIQUID WATER

Acid–base reactions for oxides immersed in water are dominated by the Brønsted acidities (i.e., proton-transfer reactions) of exposed oxygen anions, because exposed Lewis acid sites on metal cations are invariably filled by water molecules or other ligands. The acidity of each oxygen anion is dictated by the local charge on that anion (see Chapters 4 and 5). Examinations of oxide surfaces (Figs. 6.1) indicate that even for simple, single-component octahedral oxides, the points of connection between an oxygen anion and the bulk can range from a high of one less than the bulk coordination number (e.g., 5 for MgO) down to one, which is the minimum required for attachment to the surface. Each oxygen anion site exhibits its own unique acid–base chemistry and pK_a value. Most oxide surfaces contain a distribution of sites and a distribution of pK_a values, ranging from highly acidic to highly basic. These site reactivities are also influenced by electrostatic charges at the interface (see Chapter 8) because it takes more energy to protonate a positively charged surface than a negatively charged surface, and vice versa. Ideally, if the distribution of oxygen anion sites is known, and the pK_a for each site is determined, the degree of protonation of the entire surface can be predicted as a function of solution pH.

For many oxide surfaces, quantitative surface site distributions are not known, and it is difficult to correlate specific pK_a values with specific sites. However, for crystalline, layered clay-like minerals, essentially all oxygen anions have access to the solution, site distributions are known, and sufficient numbers of sites are present to allow site behaviors to be investigated using acid–base titrations. When combined with formal-charge models, acid–base titrations of such materials allow for the assignment of specific pK_a values to specific sites. An example of such a material is $Na_2Ti_4O_9$, which is discussed in more detail in Chapter 10. Here, we focus on just the acid–base properties of this model material and the correlation between acid–base properties and formal charge.

A formal charge analysis of $Na_2Ti_4O_9$ indicates that the excess anionic network charge of -2.0 in this layered titanate is associated with one $Ti-\mu_2O-Ti$ bridging oxygen with a formal charge of -0.67 and one nonbridging oxygen with a formal charge of -1.33. Note that this "surface site" distribution is exactly the same as that found on the (110) surface of single-crystal TiO_2 (Fig. 6.8), although the two materials have substantially different crystal structures. Acid–base titrations of $Na_2Ti_4O_9$ from high pH to low pH [24] (Fig. 6.10) show that $Na_2Ti_4O_9$ can adsorb three equivalents of protons. The first acid–base reaction exhibits a pK_a value of around 11, and results in the removal of one equivalent of Na^+ from the structure. The second reaction, with a pK_a value near 7, results in the replacement of the second Na^+ equivalent by protons. The third reaction, with a pK_a value of around 2, results in the adsorption of one equivalent of a monovalent anion such as Cl^- onto the now sodium-free titanate to compensate for the excess positive charge associated with the adsorption of the final proton.

Based on formal-charge arguments, the three observed acid–base reactions correspond to the following:

$$\underset{(-1.33)}{\text{Ti-O-Na}^+} + \text{H}^+ \rightleftarrows \underset{(-0.33)}{\text{Ti-OH}} + \text{Na}^+ \qquad pK_3 = 11 \tag{6.2}$$

$$\underset{(-0.67)}{\text{Ti-O-Ti}(\text{Na}^+)} + \text{H}^+ \rightleftarrows \underset{(+0.33)}{\text{Ti-OH-Ti}} + \text{Na}^+ \qquad pK_2 = 7 \tag{6.3}$$

$$\underset{(-0.33)}{\text{Ti-OH}} + \text{H}^+ \rightleftarrows \underset{(+0.67)}{\text{Ti-OH}_2^+} \qquad pK_1 = 2 \tag{6.4}$$

In Eqs. 6.2 through 6.4, the formal oxygen charge is shown in parentheses below each oxygen site. Use these formal charges to assess the reaction charge balance. As in the case of cations and hydrolysis products dissolved in water (Figs. 4.4 and 5.4), there is a clear correlation between pK_a values and formal charge (Fig. 6.11).

The titration curve in Figure 6.10 (and Eqs. 6.2–6.4) highlights several important points regarding the acid–base chemistry of oxide surfaces. First, the oxygen anions in the surface of both TiO_2 and $Na_2Ti_4O_9$ exhibit a wide range of acid-base reactivities. In this case, the coordinated water molecule is a strong acid and

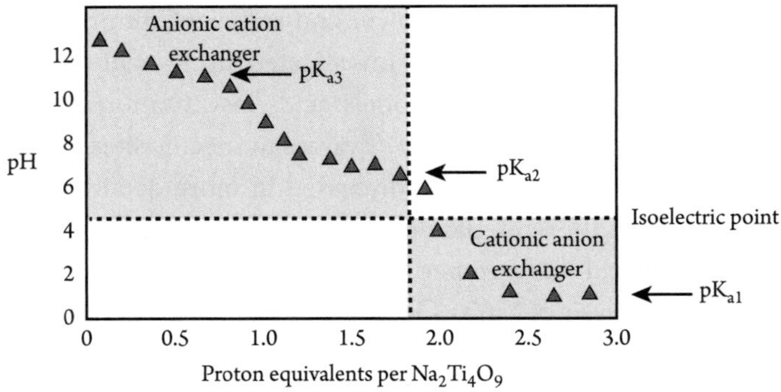

Figure 6.10 Acid–base titration results in a suspension of $Na_2Ti_4O_9$, indicating pK_a values, the isoelectric point of the solid, and the pH regimes in which the solid behaves as a cation and anion exchanger, respectively. Adapted from Sasaki, T.; Watanabe, M.; Komatsu, Y.; Fujiki, Y. Layered Hydrous Titanium Dioxide: Potassium Ion Exchange and Structural Characterization *Inorg. Chem.* **1985**, *24*, 2265–2271.

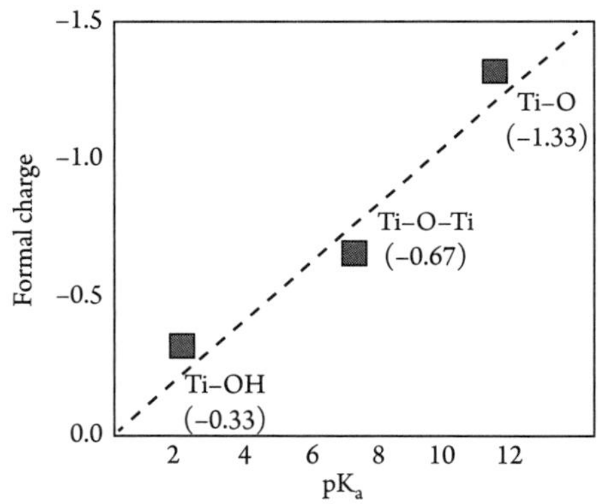

Figure 6.11 Formal charge correlates with the pKa values determined from acid–base titration experiments on the sodium titanate shown in Figure 6.10.

would like to donate its proton to form a terminal hydroxyl group, the completely deprotonated water molecule is a strong base and would like to seize a proton to make a terminal hydroxyl group, whereas the Ti-μ_2O-Ti bridge is weakly basic and can be protonated under neutral pH conditions to make a μ_2-OH bridge. Second, it is possible for a single site (in this case a coordinated water molecule) to exhibit two pK_a values within the normal pH range available in water. Third, the acid–base reactions are coupled to surface charge (see Chapters 8 and 10). The pH at which neither cation nor anion adsorption occurs (near pH 5 in Fig. 6.10) is the Point of Zero Charge (PZC) defined in Chapter 5 (see also Chapter 8) where the surface is electrically neutral.

Finally, if adjacent acid–base sites reside in a surface geometry where they are within range of hydrogen bond formation, sites with different pK_a values can react with each other to minimize surface charge imbalances. For example, the TPD experiments performed on single crystal TiO_2 surfaces exposed to water vapor (Fig. 6.8) show that water dissociates on (100) surfaces to form bound hydroxyl groups via the reaction:

$$\underset{(+0.67)}{\text{Ti-OH}_2} + \underset{(-0.67)}{\text{Ti-O-Ti}} \rightleftarrows \underset{(-0.33)}{\text{Ti-OH}} + \underset{(+0.33)}{\text{Ti-}\mu_2\text{OH-Ti}} \qquad K = 10^5 \tag{6.5}$$

Based on the solution titrations of $Na_2Ti_4O_9$, the equilibrium constant of 10^5 for this acid–base reaction is obtained by reversing Eq. 6.4 and adding it to Eq. 6.3. In other words, there is a strong chemical driving force for the surface to neutralize itself. In contrast, the application of the same analysis to the MgO (100) surface shown in Figure 6.6 suggests that water molecules adsorbed on Lewis acid sites within the plane should not dissociate:

$$\underset{(+0.33)}{\text{Mg-OH}_2} + \underset{(-0.33)}{\text{Mg-O-Mg}} \rightleftarrows \underset{(-0.67)}{\text{Mg-OH}} + \underset{(+0.67)}{\text{Mg-OH-Mg}} \qquad K = 10^{-5} \tag{6.6}$$

A similar analysis applied to ZnO (0001) surfaces suggests that bound water molecules should deprotonate in water, followed by the formation of bridging hydroxyl groups.

Acid–base titrations have also been performed [25] on high-surface area colloidal suspensions of crystalline rutile (TiO_2) (Fig. 6.12). It is interesting to note that rutile and $Na_2Ti_4O_9$ exhibit similar isoelectric points. However, their titration curves are not identical. The curve for TiO_2 is much smoother (it has fewer distinct inflection points) and the three distinct acid–base reactions seen for $Na_2Ti_4O_9$ are not apparent. There are at least three reasons for the observed differences. The first reason for smooth titration curves is that high-surface-area TiO_2 particles are terminated with a wide range of different crystal faces and local bonding configurations.

A second phenomenon leading to a smearing of titration curves is associated with surface electrostatics. Electrostatic factors are always important in chemical reactions. For species dissolved in solution, ionic repulsion and ion pairing are phenomena that influence the acid–base chemistry of the surface. For example,

$$\text{Al}\left(\text{H}_2\text{O}\right)_6^{3+}(\text{aq}) + \text{Cl}^-(\text{aq}) \rightleftarrows \left[\text{Al}\left(\text{H}_2\text{O}\right)_6\text{-Cl}^-\right]^{2+} \tag{6.7}$$

The separation of electrostatic energy can be handled explicitly (as in the Fuoss model for calculating equilibrium constants [26]). However, because each

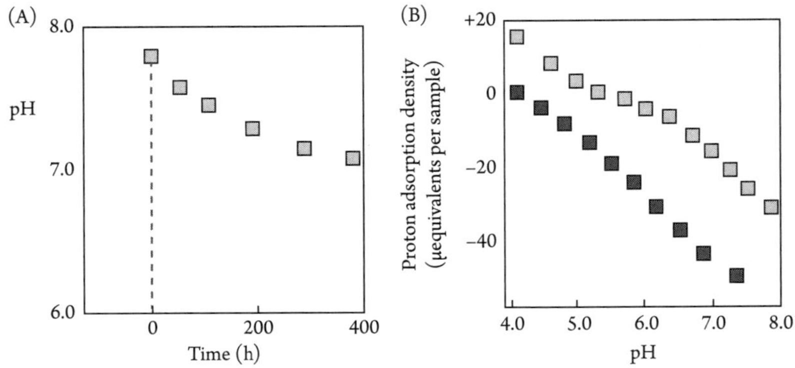

Figure 6.12 The evolution of TiO_2 surfaces immersed in water as indicated by acid–base reactions. (A) When high surface area suspensions of rutile (or almost any other high-surface area oxide) are immersed in aqueous solutions with a pH other than that of the point of zero charge (PZC) or isoelectric point, hydrolysis reactions on the oxide surface consume or release protons. This process can be observed by monitoring the drift in solution pH as a function of time. The dashed line indicates a rapid pH adjustment from the PZC, whereas the data points indicate the pH drift. These results are consistent with a slow hydrolysis of the oxide surface, resulting in the formation of a hydrated surface layer that is only a few nanometers thick. Similar results indicate that a similar protonated layer forms on Fe_2O_3 at a rate that is roughly 10 times faster. (B) Rapid acid–base titrations performed on pristine TiO_2 (top curve) as well as the same sample aged in basic solutions for 48 days (bottom curve) show that surfaces at pH values removed from the PZC contain a distributions of local oxygen-bonding configurations that differ from those of the parent oxide. In this case, the aged surface has a lower PZC than the native surface, indicating that at least some surface sites with high pK_a values have been eliminated. Adapted from Berube, Y. G.; de Bruyn, P. L. Adsorption at the Rutile-Solution Interface. I. Thermodynamic and Experimental Study. *J. Colloid Interface Sci.* **1968**, *27*, 305–318, with permission.

solution complex is an isolated entity, electrostatics are normally lumped in with all other factors that determine the equilibrium constant. In contrast, on extended surfaces, the charge experienced at one site is strongly influenced by the charge on all adjacent sites. This phenomenon is commonly addressed using a formalism [26] in which the equilibrium constant is expressed in terms of an intrinsic equilibrium constant (K_{int}) associated with an uncharged surface multiplied by an electrostatic term that accounts for the energy required to move the ion relative to the charged surface:

$$K = K_{int} \exp(-ze\Psi/kT) \tag{6.8}$$

In the electrostatic term, Ψ corresponds to the potential at the surface that arises from the electrostatic field and surface charge (see Chapter 8). To be rigorous and generally applicable, activity coefficients should also be used instead of molar concentrations [23].

The third important reason for the observance of smooth pH titration curves for many oxides is that oxide surfaces in solution are not immutable, but are constantly changing in response to changes in solution pH (see data for sodium titanate in Chapter 10 [27]). This phenomenon is readily apparent if one stops in the middle of a pH titration involving a high-surface area oxide [25]. As shown in Figure 6.12, if an oxide suspension is displaced from the isoelectric point, it invariably consumes or releases protons to shift the solution pH gradually back to the isoelectric point. The point-of-zero charge corresponds to the minimum energy surface, and is the pH where the solubility of the oxide is at a minimum. These slow proton capture or release processes are not the result of typical acid–base reactions (which are almost instantaneous), but are associated with reversible hydrolysis and condensation reactions that occur within the oxide surface. The coupling between surface protonation, surface-hydrolysis and surface-condensation reactions is exactly analogous to the coupling seen in the soluble hydrolysis products, although on surfaces such reactions tend to proceed at much slower rates. Once again, the importance of hydrolysis diagrams is readily apparent; species distributions as well as the acid–base properties of specific surface species are correlated strongly with the behaviors exhibited by the soluble hydrolysis products.

6.6 LIGAND-EXCHANGE REACTIONS INVOLVING WATER: OXIDE DISSOLUTION

The principles of ligand exchange reactions at surfaces can be highlighted by considering those reactions that lead either to the formation (see Chapter 7) or dissolution (see Chapter 15) of oxides in water. As for soluble hydrolysis products, the ligand-exchange properties of oxide surfaces are dominated by the Lewis acid–base interactions between the metal cations (the acids) and the oxygen anions (the bases). We have already seen that the coordination numbers for metal cations in bulk oxides and in solution are normally identical. For example, Mg^{2+} in both MgO(s) and $[Mg(OH_2)_6]^{2+}$ is coordinated to six oxygen anions. What changes dramatically during dissolution is the coordination numbers of the oxygen anions. *Oxide dissolution involves nothing more than a series of ligand exchange reactions in which bridging oxygens are replaced by nonbridging waters of hydration, nonbridging hydroxide or oxide ligands.* In Chapters 4 and 5, we highlighted that for octahedral cations, these exchange reactions typically involve a dissociative mechanism in which one of the oxygen anions leaves the inner coordination sphere and opens up a site that can be occupied by another nucleophile. In dissolution, this nucleophile is almost always a water molecule or hydroxide ion. Therefore, oxide dissolution kinetics should mirror trends seen for water-exchange rates around the same metal cations if the coordination geometry is preserved.

Strong correlations between dissolution rates and water-exchange rates are indeed observed in data for rock salt-structure MO oxides [28], orthosilicate minerals [29], and transition-metal oxides [30] (Fig. 6.13). These correlations highlight several important features of ligand-exchange reactions involving octahedral cations at surfaces. First, local cation bonding configurations control dissociation kinetics and dissolution rates. The almost quantitative correlation between dissolution rates and water-exchange rates is seen whether the ligand is a simple oxygen anion (O^{2-}) or the orthosilicate anion (SiO_4^{4-}). Second, for simple cations such as alkaline earth ions, the ratio of ion charge to ion size dictates both exchange and dissolution rates as predicted by formal charge and bond strength arguments. For cations with the same charge, the smaller the cation, the slower the exchange and dissolution rates. Third, although the transition metals do not follow a simple charge-to-radius correlation [30], dissolution rates still track ligand-exchange rates. Here, the rates scale with the number of d electrons and crystal field energies (see Chapter 4). For example, NiO(s) with a d^8 electronic configuration for Ni(II)

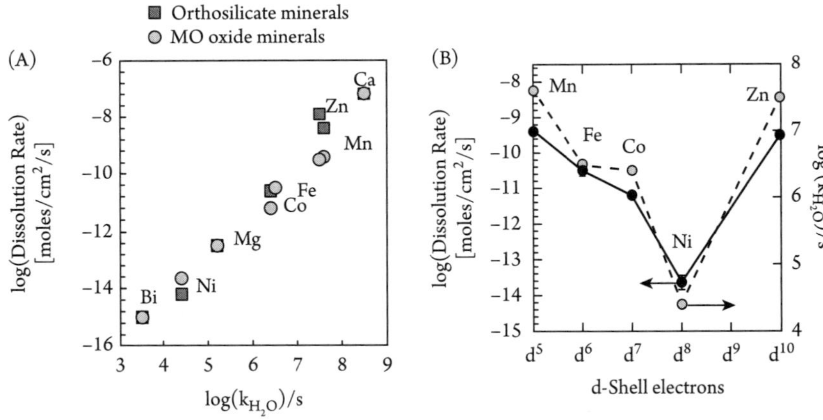

Figure 6.13 (A) Correlations between the dissolution rates for both MO oxides and divalent orthosilicate minerals, and water-exchange rates around the same divalent metal cations. (B) A similar correlation for data involving the dissolution behavior of transition-metal oxides. The correlation is nonlinear for solids containing first-row transition metals because of the influence of ligand or crystal-field stabilization energies on the kinetic transition. Adapted from Casey, W. H.; Westrich, H. R. Control of Dissolution Rates of Orthosilicate Minerals by Divalent Metal-Oxygen Bonds. *Nature*, **1992**, 355, 157–159, with permission.

reacts much more slowly than the isostructural MnO(s), which contains d^5 Mn(II) centers that are easier to distort and do not benefit from crystal-field stabilization energies. This nonlinear variation in rates with the number of valence d electrons is strong evidence that the bonding that effects ligand exchanges for soluble complexes also affects ligand-exchange reactions on solid surfaces.

Although the relative rates of dissolution and water exchange track one another, dissolution rates are almost always many orders of magnitude slower than water-exchange rates. For example, if the data presented in Figure 6.13 are used to calculate the rates of dissolution on a "per ligand exchange event" basis, a dissolution rate of around $1 \cdot 10^{-8}$ mole cm^{-2} s^{-1} (on the y-axes) is roughly equivalent to an exchange rate of 1/second (on the x-axes). If we compare these exchange rates with the water-exchange rates for dissolved cations (Fig. 4.7), this means that, at an atomic level, the dissolution rates shown in Figures 6.10 and 6.11 are between 10 million and 100 million times slower than water-exchange rates.

The primary reason for relatively slow dissolution rates is the chelate effect described in Chapter 5. To restate this effect, most hydrolysis and ligand-exchange reactions are reversible, so disrupted bonds can often reform before all other bonds connecting the complex to the surface are broken. For this reason, dissolution rates on specific sites tend to scale with site connectivity by factors resembling those described for chelating ligands in Chapter 5. For the MgO(s) surface shown in Figure 6.1, oxygen reactivity versus connectivity should follow the trend $5<4<3<2<1$. This does not mean that high-connectivity sites are inert to ligand exchange, but they must undergo a series of depolymerization reactions that reduce connectivity to the solid in a stepwise fashion before a free metal complex can detach from the surface.

As detailed in the discussion of oxygen exchange in polyoxometalates in Chapter 5, it is clear that the hydrolysis of low-connectivity sites can create fluxional reactive intermediates in which the concerted motions of many ions result in extensive isotopic scrambling and even phase changes (e.g., see Chapter 10 for transformations involving titanate ion exchangers [27]). For example, surface science studies on solid α-Fe_2O_3 [31] clearly show that reversible hydrolysis triggers isotopic mixing between terminal and μ_3-oxo surface sites.

In heterogeneous mixed-oxide structures such as glasses and clays, individual cations can be attacked and released via highly localized ligand exchange processes, resulting in what is called *selective leaching* (see Chapter 15). Here, one set of cations is released preferentially from oxide surfaces relative to another set. Dense mixed-metal oxides can also exhibit hydrolysis and dissolution behaviors that are better described as averages of the oxide's constituents. For example, the dissolution rate of Ca_2SiO_4(s) is many orders of magnitude faster than that of

Mg_2SiO_4. However, when one dissolves $CaMgSiO_4$(s), the Ca(II) is not leached away faster than Mg(II). Solid $CaMgSiO_4$(s) dissolves uniformly at a rate that is intermediate between both end members [32]. This averaging of reactivity arises because all anions in the structure are shared by both Mg(II) and Ca(II), resulting in bridging oxygen sites with acid–base and ligand exchange properties that are dictated by both cations. This example illustrates that predicting the dissolution behavior of multicomponent oxides can be complex and requires an in-depth understanding of the specific bond distributions and relative site reactivity patterns in a given structure.

After composition, pH is the most important parameter influencing the ligand exchange and dissolution behavior of extended oxide surfaces. As with surface acidity, hydrolysis diagrams (Chapter 5) represent a good starting point for predicting this ligand exchange behavior. First, the solubility of an oxide immersed in water is determined by the sum of the equilibrium concentrations of all hydrolysis products of that oxide as depicted in standard hydrolysis diagrams. For pH values and concentrations that fall below this solubility limit, the oxide surface dissolves (see Chapter 15), whereas above the solubility limit, hydroxides and oxides nucleate and grow from solution (see Chapter 7). The kinetics of reversible hydrolysis and ligand exchange reactions typically reflect how far removed from the equilibrium solubility curve a given contacting solution is.

The pH dependence of ligand exchange on surfaces also reflects the basic ligand exchange mechanisms observed for soluble species containing the same cations (see Chapter 5). Once again, protonation of bridging oxygens removes electron density from bridging bonds and makes them more susceptible to dissociative ligand-exchange processes. In basic solutions, increasing the concentration of nucleophilic hydroxide ions accelerates the rate at which bridging bonds are disrupted by associative ligand-exchange processes.

Because the localized bonding and acid–base properties of specific oxygen sites in hydrolysis products and oxide surfaces tend to be similar, the pH dependence of oxide dissolution mirrors the species distributions seen in hydrolysis diagrams. The net result is that most oxide surfaces are amphoteric. Dissolution rates accelerate as pH either decreases or increases on either side of a zone in which dissolution rates are slowest and unaffected by pH. The pH at which minimum dissolution rates are observed usually coincides with the point of zero charge indicated in the hydrolysis diagram for the same oxide.

Exceptions to this trend indicate there is a structural dissimilarity between local surface structures and the corresponding hydrolysis products in the aqueous phase. For example, the balance between octahedral and tetrahedral coordination in Al(III)-containing solids and soluble solution species can be substantially different as a function of pH, leading to a disparity in the pH of minimum solubility (which is often near pH 6) and the PZC (which is often closer to pH 9). An electrochemical example of structural dissimilarity is provided by the reductive dissolution of Fe_2O_3, in a case where surfaces are reduced only in part, yet the aqueous solution is dominated by the hydrolysis products of Fe(II).

6.7 REACTIONS INVOLVING FOREIGN LIGANDS

At oxide surfaces, the O^{2-} anion can be replaced by foreign ligands in analogy to the replacement of water molecules within the inner solvation shell of solvated cations (see Chapter 4). Illustrative examples (where the ">" denotes the surface) include the following:

$$>\text{Al-OH} + \text{F}^-(\text{aq}) \rightleftarrows\ >\text{Al-F} + \text{OH}^- \qquad \textbf{(6.9)}$$

$$>\text{Ti-(OH)}_2 + \text{CH}_3\text{COOH} \rightleftarrows\ >\text{Ti-}\left(\text{OOCCH}_3\right) + \text{H}_2\text{O} \qquad \textbf{(6.10)}$$

$$>\text{Al-OH} + \text{Pb}\left(\text{H}_2\text{O}\right)_6^{2+} \rightleftarrows\ >\text{Al-O-Pb}\left(\text{H}_2\text{O}\right)_5^{+} + \text{H}_3\text{O}^+ \qquad \textbf{(6.11)}$$

Eq. 6.9 illustrates that anionic substitutions can modify the protonation state of the surface. Eq. 6.10 illustrates that some ligands are capable of binding to multiple surface sites. Such ligands can either form chelates with a single metal center (which might destabilize the surface and/or promote dissolution) or bind to multiple metal centers (which might increase rather than decrease surface stability). The latter bonding mode is illustrated in Figure 6.14 for the binding of formic acid to TiO_2 (110) surfaces [33]. Increases in dissolution rates often track the rates of ligand exchange for structurally similar soluble metal complexes. Amine ligands, for example, are known to labilize the waters in Ni(II) ions in solution (see Chapter 4). Similar amines labilize the surface of NiO(s) to dissolution [34].

Eq. 6.11 illustrates that the adsorption of many cations onto oxide surfaces actually involves hydrolysis and condensation reactions between soluble and surface-bound species. In Eq. 6.11, this condensation forms a μ_2-oxo-bridge in $>\text{Al-O-Pb}\left(\text{H}_2\text{O}\right)_5^{+}$, linking the lead-containing species to the surface. At high adsorbate loadings, such reactions lead to the nucleation and growth of oxide films (see Chapter 7). However, although an ion in solution is often known to have a fixed composition and structure based on in-depth spectroscopic characterization, complexes at the surface of an oxide are typically depicted generically, because it is exceedingly difficult to characterize complex monolayers in the surface of a solid oxide. For this reason, much of the following discussion is based on inferences drawn from solution complexes as well as from surface-science studies performed under high-vacuum conditions.

The starting point for evaluating surface complexation involves examining modified hydrolysis diagrams in which the equilibrium constants for ligand binding to

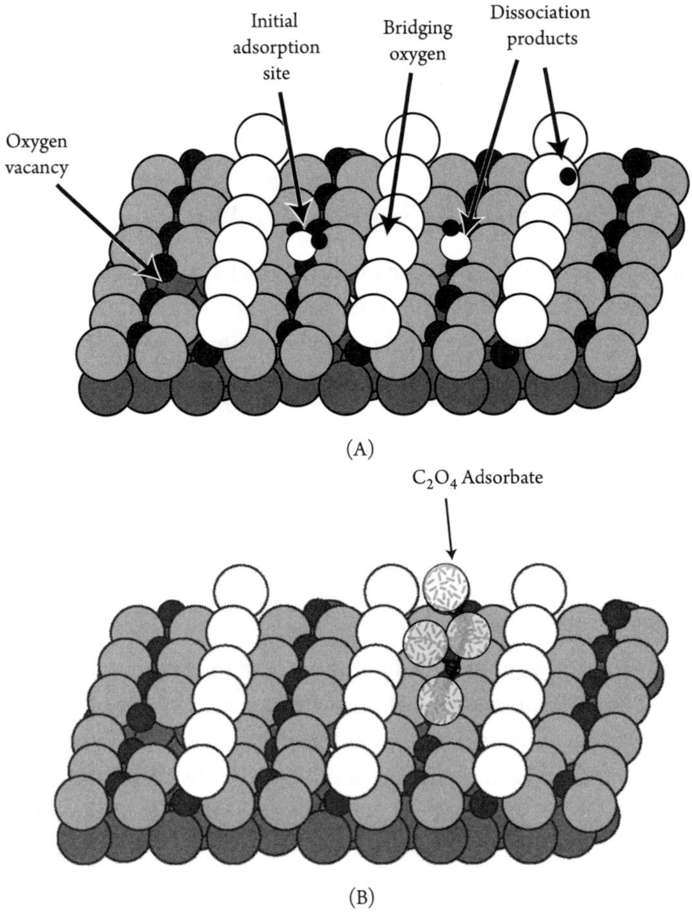

Figure 6.14 Adsorption on the [110] TiO_2 surface often involves interactions between Lewis base sites in the adsorbate and Lewis acid sites, here represented by exposed titanium metal centers. (A) In water adsorption a single Lewis base interacts with a single metal center. (B) Molecules such as oxalate contain two Lewis bases (oxygens as patterned spheres) that can bind to adjacent Lewis acids, or exposed metal centers. Adapted from results on formic acid from Henderson, M. A. Complexity in the Decomposition of Formic Acid on the TiO_2(110) Surface *J. Phys. Chem.* **1997**, *101*, 221–229, with permission.

soluble species are considered (see Fig. 5.4). Such hydrolysis diagrams take into account the affinity of hydrated metal cations for ligands having a nucleophilic character (i.e., Lewis basicity) that differs from that of a hydroxide ion, as well as how the Brønsted acidity of both the ligand and the hydrated cation vary with solution pH. Generally speaking, there is a close correlation between the net extent of ligand adsorption on oxide surfaces and complexation diagrams [35] (Fig. 6.15).

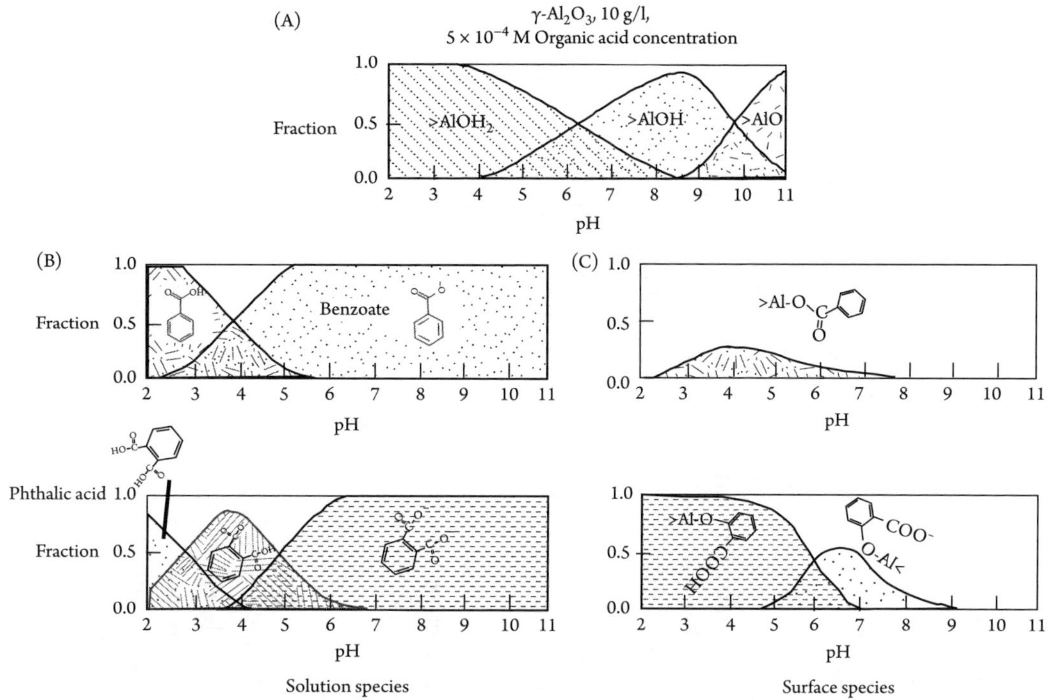

Figure 6.15 The extent of adsorption of aromatic acids on *α*-Al_2O_3 as a function of pH. (A) The distribution of acid–base sites on the alumina surface as a function of pH. (B) The acid–base properties of the ligands benzoic acid and phthalic acid, as indicated by the fraction of a given ionic state as a function of solution pH. (C) The extent of adsorption of benzoic acid and phthalic acid on alumina as a function of pH, as indicated by the fraction of sites on the surface occupied as a function of pH. Note that benzoate species are adsorbed within a pH window where the ligand is anionic and the surface is cationic, resulting in the formation of a neutral surface. Phthalic acid adsorption follows similar trends. However, in this case, the uncoordinated carboxyl group that is left over after adsorption imparts a negative charge to the surface between pH 6 and pH 8.5, where the native surface is normally cationic. Adapted from Kummert, R.; Stumm, W. The Surface Complexation of Organic Acids on Hydrous *γ*-Al_2O_3 *J. Colloid Interface Sci.* **1980**, 75, 373–385, with permission.

General trends seen as a function of solution pH for ligand binding to surfaces (Eqs. 6.9–6.11) include the following

- Most ligands cannot compete with hydroxide ions in basic solutions (i.e., for pH values above the isoelectric point). This unfavorable competition arises because OH^- is a strong nucleophile and hydroxide ions are present in high concentrations in basic solutions (see Chapter 4).
- Protonated ligands that are present below the pK_a of the ligand tend to be weakly adsorbed by oxide surfaces because surface metals have difficulty in competing with protons for binding to the basic sites on the ligand.
- Ligand binding is often maximized for pH values between the pK_a of the ligand and the isoelectric point of the oxide. Here, anionic ligands often outcompete either neutral or cationic oxygen-containing species by virtue of their stronger Lewis basicity.
- Soluble hydrolysis products can themselves function as ligands and are readily incorporated into oxide surfaces. Just as for aquated cations, condensation reactions between soluble species and surfaces are often triggered by the hydrolysis of the soluble species. For example, the adsorption of metal cations by silica surfaces (Fig. 6.16) tends to occur at the pH where the aquated cation is hydrolyzed [36]. In this instance, the silanol groups in the silica surface represent a ligand that exchanges for the inner coordination sphere of the soluble cation. The inverse reaction can also occur, in which the hydrolysis product serves as an incoming ligand to the surface. The reaction pathway typically depends on which moiety (adsorbate or substrate) has the fastest rate of ligand exchange. For example, simulations of the rates of water losses from surface sites on Al(III) minerals indicate they can be many orders of magnitude more labile than for $Al^{3+}(aq)$ [37], because of the lower charge density and surface relaxation of the bonds. These rates approach those of the most labile aqueous complexes (see Fig. 4.8).
- As in most hydrolysis reactions, rates of ligand exchange on surface sites generally increase as surface connectivity decreases. This means that rates of ligand exchange are more pronounced on sites with low connectivity to the surface, such as corners and edges of monomolecular steps, where there are high concentrations of nonbridging oxygens. When the incoming or outgoing ligand is a hydrolysis product, reactions leading either to the growth or the dissolution of oxides are most rapid on steps, edges, and other surface defects.

It is important to point out that the structures and chemistry of ligand adsorbates on surfaces are not always the same as those encountered in solution. This

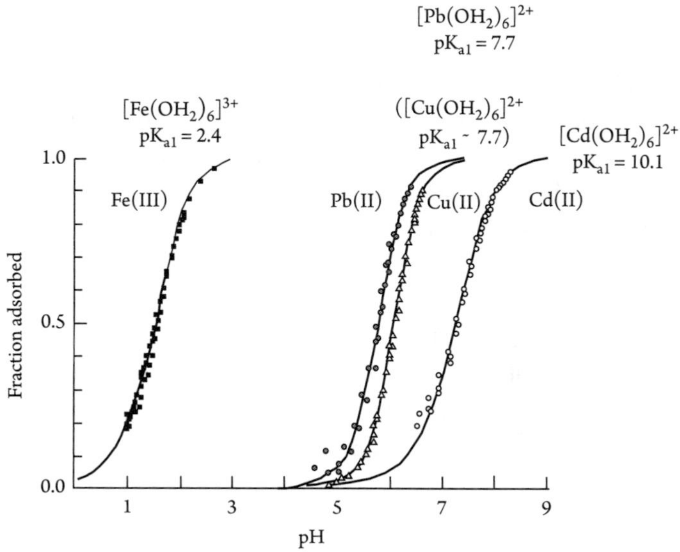

Figure 6.16 The extent of adsorption of dissolved metal cations onto high-surface area silica as a function of solution pH showing the correlation between initial cation hydrolysis (i.e., the pK_a for $[M(OH_2)_6]^{2+} = [M(OH_2)_5OH]^+ + H^+$) and cation adsorption. Adapted from Schindler, P. W.; Fürst, B.; Dick, R.; and P.U. Wolf Ligand Properties of Surface Silanol Groups. I. Surface Complex Formation with Fe^{3+}, Cu^{2+}, Cd^{2+}, and Pb^{2+} *J. Colloid Interface Sci.* **1976**, 55, 469–475, with permission.

is especially true for multidentate ligands. For example, carboxylic acid groups in compounds such as citric acid are known to bind to many metal cations in a bidentate fashion, as in the monomeric tricitrato complexes of Ti(IV) [38] (Fig. 6.16). In contrast, surface-science studies show [34] that the two oxygen atoms in formic acid on the (110) surface of single-crystal TiO_2 bind to two adjacent Ti(IV) sites in a monodentate fashion (Fig. 6.14). Equilibrium constants for these two distinct binding motifs are substantially different from one another. As this example illustrates, it is not surprising that surfaces can exhibit different reactivity patterns than aqueous complexes. What is stunning is the pervasive extent to which the solution chemistry of aqueous complexes can guide an understanding of the aqueous–oxide interface.

Ligand binding at surfaces can have several important consequences for the underlying oxide. The new ligands can exert a strong influence on oxide surface charge and the underlying space-charge layer. As examples, adsorption of citric acid (a tricarboxylic acid) onto alumina can reverse the surface charge from positive to negative for pH values above the pK_a of the citrate and below the point of zero charge [35]. Here, dangling carboxylic acid groups terminate the surface and dictate the surface charge rather than the underlying hydroxyl groups.

Conversely, the adsorption of hydrolyzed cations onto anionic oxide surfaces can reverse the surface charge from positive to negative, reflecting the acid–base chemistry of the adsorbed hydrolysis products [39] (Fig. 6.16).

Finally, oxide dissolution rates are strongly influenced by the presence of foreign ligands, mirroring the ligand exchange kinetics of soluble species. Ammine and amine ligands, for example, are known to labilize the rates of dissociation of Ni(II)$-OH_2$ bonds in nickel ions in solution [34]. These ligands are particularly effective at inducing the rapid dissolution of NiO(s).

Hard Lewis bases such as carboxylate and phenolic ligands that form strong complexes with aquated Al(III) also promote the rapid dissolution of aluminum oxide and hydroxide solids. For natural chemists, this is important because such ligands represent the natural breakdown products of cellulose and lignin in soils. These ligands have a dramatic impact on the weathering of soil minerals and the cycling of aluminum in natural environments (see Chapter 18). Although HF is a relatively weak acid (with a pK_a of 3.45), it is particularly effective in attacking and dissolving many oxides. The fluoride ion often promotes oxide dissolution because it is a strong nucleophile that readily displaces oxygen anions. In addition, because F^- has only one valence electron to share, it rarely forms bridges between metal cations. Thus, the products of fluoride-enhanced dissolution tend to be monomeric soluble complexes rather than polymeric complexes. The fluoride ion has been shown to accelerate dramatically the dissociation of the aluminum hydrolysis product $[AlO_4Al_{12}(OH)_{24}(H_2O)_{12}]^{7+}$ under solution conditions that are almost identical to the conditions used to accelerate the dissolution of $Al(OH)_3$ resulting from fluoridation [40–42]. Even in the case of solids, where bridging fluoride ions have been identified spectroscopically, fluoride ions are still particularly destabilized to many oxide surfaces.

References

1. Weast, R. C. Ed. *CRC Handbook of Chemistry and Physics, 58th ed.*; CRC Press: Cleveland, OH, 1977; p F223.
2. Israelachvili, J. *Intermolecular and Surface Forces*; Academic Press: San Diego, 1992; p 205.
3. Israelachvili, J. *Intermolecular and Surface Forces*; Academic Press: San Diego, 1992; p 315.
4. Henrich, V. E.; Cox, P. A. *The Surface Science of Metal Oxides*; Cambridge University Press: Cambridge, UK, 1996.
5. Henrich, V. E. Thermal Faceting of (110) and (111) Surfaces of MgO. *Surf. Sci.* **1976**, *57*, 385–392.
6. Adamson, A. W. *Physical Chemistry of Surfaces*; Wiley: New York, 1976; p 269.
7. Joseph, Y.; Kuhrs, C.; Ranke, W.; Ritter, M.; Weiss, W. Adsorption of Water on FeO(111) and Fe_3O_4(111): Identification of Active Sites for Dissociation. *Chem. Phys. Lett.* **1999**, *314*, 195–202.

8. Stirniman, M. J.; Huang, C.; Smith, R. S.; Joyce, S. A.; Kay, B. D. The Adsorption and Desorption of Water on Single Crystal MgO(100): The Role of Surface Defects. *J. Chem. Phys.* **1996**, *105*, 1295–1298.
9. Zwicker, G.; Jacobi, K. Site Specific Interaction of H_2O with ZnO Single Crystal Surfaces Studied by Thermal Desorption and UV Photoelectron Spectroscopy. *Surf. Sci.* **1983**, *131*, 179–194.
10. Thiel, P. A.; Madey, T. E. The Interaction of Water with Solid Surfaces: Fundamental Aspects. *Surf. Sci. Rep.* **1987**, *7*, 211–395.
11. Henderson, M. A. The Interaction of Water with Solid Surfaces: Fundamental Aspects Revisited. *Surf. Sci. Rep.* **2002**, *46*, 1–308.
12. Iler, R. K. *The Chemistry of Silica*; Wiley: New York, 1979; p 635.
13. Henderson, M. A. Structural Sensitivity in the Dissociation of Water on TiO_2 Single-Crystal Surfaces. *Langmuir* **1996**, *12*, 5093–5098.
14. Weast, R. C., Ed. *CRC Handbook of Chemistry and Physics, 58th ed.*; CRC Press: Cleveland, OH, 1977; p B128.
15. Stirniman, M. J.; Huang, C.; Smith, R. S.; Joyce, S. A.; Kay, B. D. The Adsorption and Desorption of Water on Single Crystal MgO(100): The Role of Surface Defects. *J. Chem. Phys.* **1996**, *105*, 1295–1298.
16. Yu, Y.; Guo, Q.; Liu, S.; Wang, E. Partial Dissociation of Water on a MgO(100) Film. *Phys. Rev. B* **2003**, *68*, 115414-1-115414-4.
17. Kwok, D. Y.; Neumann, A. W. Contact Angle Interpretation: Re-evaluation of Existing Contact Angle Data. *Colloids Surfaces A* **2000**, *161*, 49–62.
18. Feng, A.; McCoy, B. J.; Munir, Z. A.; Cagliostro, D. Wettability of Transition Metal Oxide Surfaces. *Mater. Sci. Eng.* **1998**, *A242*, 50–56.
19. Adamson, A. W. *Physical Chemistry of Surfaces*; Wiley: New York, 1976; p 352.
20. Kim, H. I.; Kushmerick, J. G.; Houston, J. E.; Bunker, B. C. Viscous "Interphase" Water Adjacent to Oligo(ethylene glycol)-Terminated Monolayers. *Langmuir* **2003**, *19*, 9271–9275.
21. Israelachvili, J. *Intermolecular and Surface Forces*; Academic Press: San Diego, 1992.
22. Conway, B. E.; Bockris, J. O'M.; Ammar, I. A. The Dielectric Constant of the Solution in the Diffuse and Helmholtz Double Layers at a Charged Interface in Aqueous Solution. *Trans. Farad. Soc.* **1951**, *47*, 756–766.
23. Sverjensky, D. A. Standard States for the Activities of Mineral Surface Sites and Species. *Geochim. Cosmochim. Acta* **2003**, *67*, 17–28.
24. Sasaki, T.; Watanabe, M.; Komatsu, Y.; Fujiki, Y. Layered Hydrous Titanium Dioxide: Potassium Ion Exchange and Structural Characterization. *Inorg. Chem.* **1985**, *24*, 2265–2271.
25. Berube, Y. G.; de Bruyn, P. L. Adsorption at the Rutile–Solution Interface. *J. Colloid Interface Sci.* **1968**, *27*, 305–318.
26. Morel, F. M. M.; Hering, J. G. *Principles and Applications of Aquatic Chemistry*; Wiley-Interscience: New York, 1993.
27. Bunker, B. C.; Peden, C. H. F.; Kirkpatrick, R. J.; Turner, G. L. ^{17}O NMR Studies of Titanate Repolymerization in Water. *Mater. Res. Soc. Symp. Proc.* **1997**, *432*, 39–43.
28. Ohlin, C. A.; Villa, E. M.; Rustad, J. R.; and Casey, W. H. The Dissolution of Insulating Oxides at the Molecular Scale. *Nat. Mater.* **2010**, *9*, 11–19.

29. Casey, W. H.; Westrich, H. R. Control of Dissolution Rates of Orthosilicate Minerals by Divalent Metal–Oxygen Bonds. *Nature* **1992**, *355*, 157–159.
30. Basolo, F.; Pearson, R. G. *Mechanisms of Inorganic Reactions, 2nd ed.*; Wiley: New York, 1967.
31. Henderson, M. A.; Joyce, S. A.; Rustad, J.R. Interaction of Water with the (1*1) and (2*1) Surfaces of α-Fe_2O_3(012). *Surf. Sci.* **1998**, *417*, 66–81.
32. Westrich, H. R.; Cygan, R. T.; Casey, W. H.; Zemitis, C.; and Arnold, G. W. The Dissolution Kinetics of Mixed-Cation Orthosilicate Minerals. *Am. J. Sci.* **1993**, *293*, 869–893.
33. Henderson, M. A. Complexity in the Decomposition of Formic Acid on the TiO_2(110) Surface. *J. Phys. Chem.* **1997**, *101*, 221–229.
34. Ludwig, C.; Devidal, J.-L, and Casey, W. H. The Effect of Different Functional Groups on the Ligand-Promoted Dissolution of NiO and Other Oxide Minerals. *Geochim. Cosmochim. Acta* **1996**, *60*, 213–224.
35. Kummert, R.; Stumm, W. The Surface Complexation of Organic Acids on Hydrous α-Al_2O_3. *J. Colloid Interface Sci.* **1980**, *75*, 373–385.
36. Schindler, P. W.; Furst, B.; Dick, R.; Wolf, P. U. Ligand Properties of Surface Silanol Groups. *J. Colloid Interface Sci.* **1976**, *55*, 469–475.
37. Wang, J.; Rustad, J. R.; and Casey, W. H. Calculation of Water-Exchange Rates on Aqueous Polynuclear Clusters and at Oxide–Water Interfaces. *Inorg. Chem.* **2007**, *46*, 2962–2964.
38. Deng, Y.- F.; Zhou, Z.- H.; Wan, H.- L. pH-Dependent Isolations and Spectroscopic, Structural, and Thermal Studies of Titanium Citrate Complexes. *Inorg. Chem.* **2004**, *43*, 6266–6273.
39. Tewari, P. H.; Lee, W. Adsorption of Co(II) at the Oxide–Water Interface. *J. Colloid Interface Sci.* **1975**, *52*, 77–88.
40. Nordin, J. P.; Sullivan, D. J.; Phillips, B. L.; and Casey, W. H. Mechanisms for Fluoride-Promoted Dissolution of Bayerite [β-$Al(OH)_3$(s)] and Boehmite [γ-AlOOH]: ^{19}F-NMR Spectroscopy and Aqueous Surface Chemistry. *Geochim. Cosmochim. Acta* **1999**, *63*, 3513–3524.
41. Yu, P.; Lee, A. P.; Phillips, B. L.; and Casey, W. H. Potentiometric and ^{19}F Nuclear Magnetic Resonance Spectroscopy Study of Fluorine Substitution in the $GaAl_{12}$ Polyoxocation: Implications for Aluminum (Hydr)oxide Mineral Surfaces. *Geochim. Cosmochim. Acta* **2001**, *67*, 1065–1080.
42. Yu, P.; Phillips, B. L.; and Casey, W. H. Water Exchange in Fluoroaluminate Complexes in Aqueous Solution: A Variable Temperature Multinuclear NMR Study. *Inorg. Chem.* **2001**, *40*, 4750–4754.

PART THREE

The Aqueous Synthesis and Processing of Oxides

7

Nucleation and Growth of Solid Oxide and Hydroxide Phases

7.1 INTRODUCTION

In this chapter, we consider what happens when solids begin to form from solution. To grow solids from solution, solution conditions are changed from a condition in which all species are completely soluble to a condition in which they are insoluble. In the context of hydrolysis diagrams (see Figs. 5.2 and 5.3), the solution composition moves in pH and total dissolved metal concentration from a regime below a solubility or saturation limit (given by the bold solid line in Figs. 5.2 and 5.3) to a regime above this limit where the solution is *supersaturated*.

Supersaturated solutions are inherently unstable and have the potential to generate hydroxide or oxide solids. Sometimes these solutions can be maintained in a metastable state in which precipitation does not occur immediately. However, Mother Nature eventually reduces the energy of the solution by forming a stable mixture of solids plus solution species. As solids form, soluble complexes are removed from solution until concentrations drop back to the solubility limit.

The precipitation of a solid from an aqueous solution is a surprisingly complex process, involving nucleation and growth phenomena that occur at nanometer-length scales. *Nucleation* involves reactions between oligomers to form new clusters or particles that are sufficiently large that they do not redissolve spontaneously via the reversible reactions denoted in hydrolysis diagrams. *Homogeneous* and *heterogeneous* nucleation processes represent events that occur within the bulk solution or at the interface of another phase, respectively. *Growth* involves the addition of monomers to clusters in solution or oligomers to existing particles or surfaces.

The combination of nucleation and growth phenomena can lead to oxides exhibiting a bewildering range of sizes, shapes, and crystal structures [1] (Fig. 7.1). How do metal complexes decide whether to form a new particle or add to an existing particle? What determines the size, shape, and crystal structure of evolving particles? Do the particles aggregate with one another in an organized fashion? Because nucleation typically involves extremely rapid (<1 millisecond) events involving objects that are extremely small (on the order of a nanometer), it is difficult to probe such phenomena at a molecular level. Therefore, this chapter focuses on well-established thermodynamic descriptions of precipitation phenomena. Both the blessing and curse of thermodynamic descriptions is that

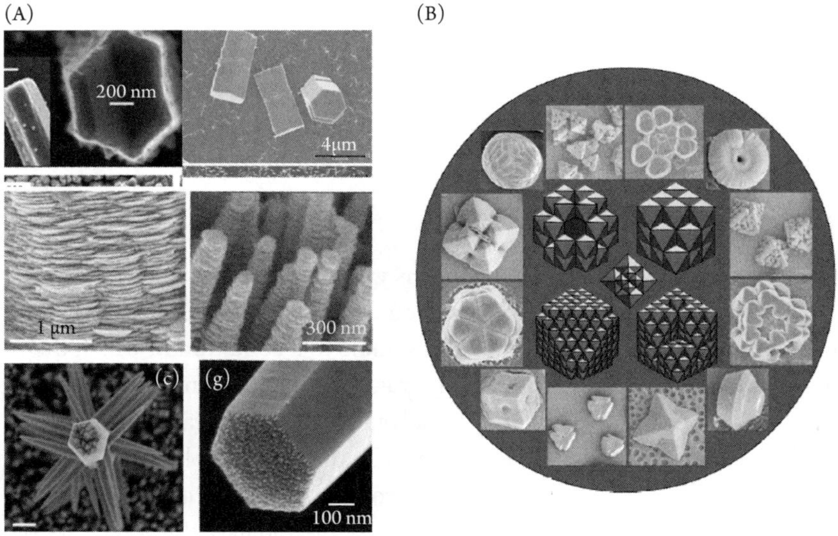

Figure 7.1 (A) The morphologies of ZnO crystals that have been produced via controlled nucleation and growth in aqueous solutions. Primary nucleation followed by growth can produce morphologies ranging from long needles to flat plates. The other morphologies shown were produced via heterogeneous nucleation and growth on existing ZnO crystals. (SOURCE: Reproduced with permission from Sounart, T. L, Liu, J. Voigt, J. A.; Huo, M.; Spoerke, E. D.; and McKenzie, B. Secondary Nucleation and Growth of ZnO *J. Am. Chem. Soc.*, **2007**, *129*, 15786–15793, with permission.) (B) Amazingly, crystalline objects can also be produced via similar nucleation and growth phenomena using submicrometer oxide clusters such as the mesoporous silicate structures shown. (SOURCE: Reproduced with permission from Tian, Z. R.; Liu, J.; Voigt, J. A.; McKenzie, B.; Xu, H.; Hierarchical and Self-Similar Growth of Self-Assembled Crystals *Angew. Chem.*, **2003**, *115*, 430–433.)

they give no information about reaction pathways. Molecular bonding and reactivity are not captured in these thermodynamic descriptions.

Scientists are beginning to appreciate the important roles that the structure and chemistry of preassembled metal–hydroxide oligomers play in determining the transition between solutions and solid phases. These *aggregated growth models* are enhancing the classic and geometric view of nucleation. When appropriate, we insert molecular-level models for nucleation and growth into the discussion of thermodynamics based on the chemistry and structure of the hydrolysis products (Chapter 5) and oxide surfaces (Chapter 6).

7.2 CLASSIC NUCLEATION THEORY

Let us illustrate the differences between nucleation and growth using a "wet" example that everyone can relate to: beer! Beer is a liquid that is supersaturated with respect to CO_2 gas if exposed to atmospheric pressures. Once opened and poured, beer reestablishes chemical equilibrium by *nucleating* and then *growing* CO_2 bubbles that are released into the atmosphere as CO_2 partial pressures equilibrate. When gazing into a glass of beer, one notices that some bubbles form within the bulk liquid. However, the rate and extent of nucleation is almost always more pronounced on the sides of the beer mug (or, for other carbonated beverages, on ice cubes). This is because heterogeneous nucleation events exploit solid surfaces to reduce the energy costs associated with making new phases and interfaces. Next, we examine the thermodynamics of both homogeneous and heterogeneous nucleation processes.

7.2.1 Supersaturation: The Driving Force for Nucleation

Thermodynamic models [2–4] can be used to calculate where precipitates form, how fast they form, how many form, and how large the nucleated objects get. As a starting point, consider homogeneous nucleation, in which our previous example is small bubbles forming in beer. The driving force for nucleation is provided by the degree of supersaturation (S). The term *S* represents the ratio of the equilibrium constant expression containing actual solution activities divided by the same expression containing concentrations that would be present at equilibrium. For example, consider a precipitation reaction:

$$Mg^{2+}(aq) + 2OH^-(aq) \rightleftarrows Mg(OH)_2(s) \tag{7.1}$$

$$K_{eq} = \frac{(Mg(OH)_2(s))}{(Mg^{2+})(OH^-)^2} \tag{7.2}$$

Here, the parentheses ( ) identify thermodynamic activities relative to well-defined reference states. The activity of solid brucite ($Mg(OH)_2$) is equal to unity if it is a coarse, pure, and well-crystallized material. Activities of the solutes are normally defined relative to a one molar solution having the properties of infinite dilution. More detailed absolute activity coefficients can be calculated using standard methods such as the Debye-Hückel equations [5]. However, these corrections are usually small and are henceforth ignored. Using the activities defined previously, the equilibrium constant for Eq. 7.2 is approximated as

$$K_{eq} = \frac{1}{([Mg^{2+}][OH^-]^2)_{eq}} \tag{7.3}$$

where the brackets indicate molal concentrations at equilibrium (i.e., activity coefficients are set to unity). For the actual supersaturated solution, potentially at disequilibrium, measured concentrations are used to determine an equivalent expression called *Q*:

$$Q = \frac{1}{([Mg^{2+}][OH^-]^2)_{actual}} \tag{7.4}$$

A parameter S, representing the degree of supersaturation, is then defined as

$$S = \frac{K_{eq}}{Q} \tag{7.5}$$

Supersaturation levels provide a measure of the amount of chemical energy that can be exploited to induce the formation of solid phases. (We define 'S' to be consistent with formalism of crystal-growth kinetics, below. It is distinct from the conventional 'Saturation Index (SI)' defined as: $SI = \log\left(\frac{Q}{K_{eq}}\right)$ that is familiar in geochemistry and in solution chemistry. Because Eqn. 7.1 is written as precipitation, and the solid activities are unity, S becomes the ion product at disequilibrium divided by that at equilibrium.)

7.2.2 The Free Energy for Homogeneous Nucleation

Having identified supersaturation levels as the driving force for nucleation, we are now in a position to examine the thermodynamic expressions that are used to understand nucleation. The free energy for forming a stable particle contains two terms:

$$\Delta G_{net} = -R \cdot T \cdot \ln(S) + \sigma \cdot A \tag{7.6}$$

Here, S is the supersaturation level, T is temperature, σ is the solution–particle interfacial energy introduced in Chapter 6, and A is the particle surface area. Eq. 7.6 describes the chemical energy available to change the solution composition by making solids.

The first term in Eq. 7.6 captures the chemical driving force for forming a particle, which for oxides involves the free energy of hydrolysis and condensation reactions that are driven by supersaturation. As the energy released via the formation of metal oxide and/or hydroxide bonds exceeds that associated with the loss of waters of hydration, this term is negative or favorable. For example, if the concentrations of both $[Mg]^{2+}$ and $[OH]^{-}$ are twice those found at the solubility limit, S would be equal to 8 and $-RT\ln(S) = -5$ kJ/mol. The negative energy reflects the fact that the solution could spontaneously form the solid because it is supersaturated. However, the second term in Eqn. 7.6, the $\sigma.A$ term, captures the fact that there is an unfavorable energy penalty in making new solid-liquid interfaces from a homogeneous fluid because work must be performed. This work is the interfacial energy discussed in Chapter 6. For example, the interfacial energy for silica in water [6] is estimated to be around 100 mJ/m², whereas that for MgO [7] is 115 mJ/m².

The balance between the opposing bulk and surface energies in Eq. 7.6 depends on the size and shape of the particle (Fig. 7.2). For extremely small particles, almost all atoms in the system are surface atoms, and the unfavorable surface-energy term dominates. It can be so large that it prevents nucleation. It is only when supersaturation levels force bulk bonds to form faster than the incipient cluster can redissolve that nuclei become stable. For spherical particles, the net free energy is given by

$$\Delta G_{net} = -\frac{4}{3}\pi \cdot r^3 \Delta G_{bulk} + 4 \cdot \pi \cdot r^2 \sigma \tag{7.7}$$

In Eq. 7.7, ΔG_{bulk} is related to the supersaturation term in Eq. 7.6. As shown in Fig. 7.2, as the radius, r, of a spherical particle increases, the free energy starts out positive, goes through a maximum at a critical nucleus size r*, and then decreases until the net energy becomes negative and a stable phase is formed. The maximum in the free energy-versus-radius curve is equivalent to an energy barrier for nucleation (ΔG*). The value of ΔG* is controlled by S and σ. For a spherical particle, it is

$$\Delta G^* = \frac{16 \cdot \pi \cdot \sigma^3 \cdot V^2}{3 \cdot RT \ln(S)} \tag{7.8}$$

In Eq. 7.8, V is the molar volume of the precipitating phase. As is shown later in this chapter, expressions such as Eq. 7.8 can be used to predict the size of nucleating particles.

In examining Eq. 7.6, it soon becomes apparent that nucleation events involve small changes in energy when viewed at a chemical level. The condensation

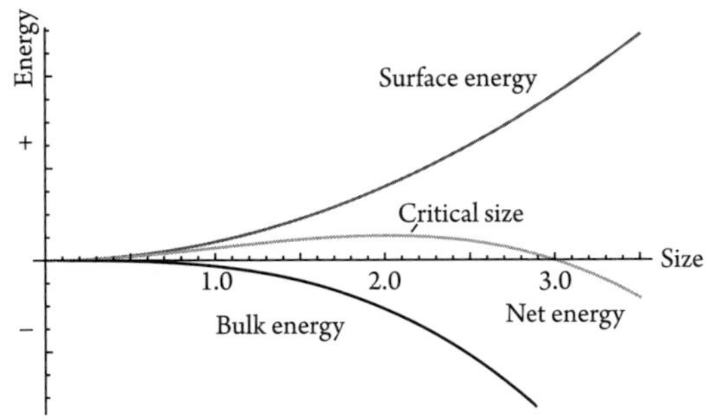

Figure 7.2 Generic curves showing how both the surface and bulk energy of a spherical particle vary with particle size. The sum of these two energies has a peak corresponding to the activation energy for particle nucleation.

reactions that produce the solid phase (as captured by the first term) tend to have small free energies on a per-bond basis. For example, for silica, the free energy for the reaction $>\mathrm{Si-OH+HO-Si}< \rightleftarrows >\mathrm{Si-O-Si}<+\mathrm{H_2O}$ is estimated to be around 7 kJ/mole [6]. Similarly, the value of σ for the silica–water interface of around 100 mJ/m^2 is equivalent to an interfacial energy per surface site of 12 kJ/mole. Both these energies are comparable with that of a hydrogen bond, which means that subtle changes in interfacial energies can have a dramatic impact on particle nucleation rates.

7.2.3 The Free Energy of Heterogeneous Nucleation

As we notice when gazing into a beer mug, bubbles tend to originate on the sides of the mug rather than in solution. An existing surface often profoundly influences the interfacial energy associated with nucleation. In other words, *heterogeneous* nucleation on existing surfaces is normally preferred to *homogeneous* nucleation in the bulk fluid. In heterogeneous nucleation, the surface free energy is described by three terms:

$$\Delta G = -RT\ln S + \sigma_{cl}A_{cl} + (\sigma_{cs} - \sigma_{sl})A_{cs} \tag{7.9}$$

Here, the subscripts c, s, and l refer to interfaces involving the crystalline particle (c), the substrate (s), and the liquid (l). In contrast to homogeneous nucleation, when the only interface to consider is that between the particle and the solution (σ_{cl}), two additional interfaces must be considered: the interface between the precipitant and the solid surface (σ_{cs}) and the interface between the solid surface and the liquid (σ_{sl}). The latter surface is being covered up by the growing film, which leads to the negative sign in Eq. 7.9.

As a simple case, consider the free energy for nucleating a six-sided cube on an existing oxide surface having the same structure and composition. For this geometry, σ_{cs} is equal to zero, σ_{sl} is equal to σ_{cl}, A_{cl} is equal to five-sixths of the total particle area, and A_{cs} represents the remaining one-sixth (i.e., the side of the cube in contact with the surface). The net result is that, although the surface energy term for homogeneous nucleation is $\sigma \cdot A$, the net surface energy term for heterogeneous nucleation is $(5/6 - 1/6)$, or only two-thirds of the barrier to nucleation in the bulk solution. Solid-solution interfaces almost always have a higher energy than the interface between a precipitating phase and another solid substrate. An important consequence of this observation is that heterogeneous nucleation is almost always a lower energy process than homogeneous nucleation, and it represents the preferred pathway for relieving a condition of supersaturation.

The supersaturation conditions required to induce both homogeneous precipitation and heterogeneous film formation are indicated in terms of a generic hydrolysis diagram in Fig. 7.3. The diagram indicates that heterogeneous film formation can occur at low supersaturation levels, whereas precipitation dominates only when the solution is pushed even further out of equilibrium. For processes in which precipitation is exclusively desired, surfaces must normally be treated to inhibit heterogeneous nucleation (see Chapter 9).

Free energy relationships have been used to derive equations for all the critical parameters associated with the nucleation of solid phases, including particle size,

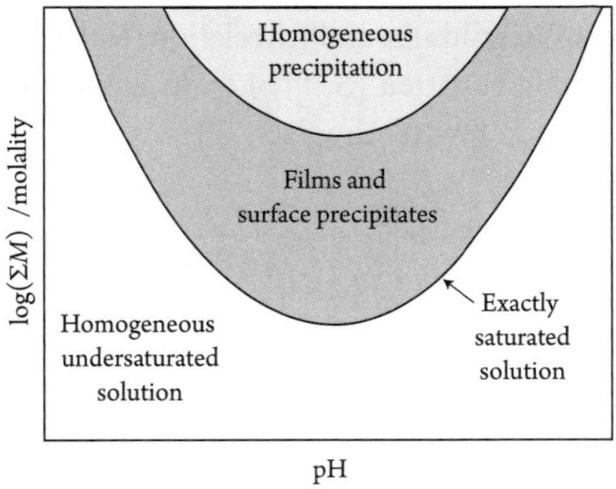

Figure 7.3 A generic diagram for an amphoteric solid showing how supersaturation levels for amphoteric solids, as defined by solution pH and hydrolysis product concentrations, influence both heterogeneous and homogeneous nucleation processes. The lower solid curve represents the solubility limit for the precipitating phase whereas the upper solid curve indicates the level of supersaturation required to stimulate homogeneous precipitation within the bulk solution. Adapted from Bunker, B. C.; Rieke, P. C.; Tarasevich, B. J.; Campbell, A. A.; Fryxell, G. E.; Graff, G. L.; Song, L.; Virden, J. W.; and McVay, G. L. Ceramic Thin Film Formation on Functionalized Interfaces Through Biomimetic Processing. *Science* **1994**, *264*, 48–55, with permission.

the number of nuclei formed, and the induction time for particle formation. See [3] for derivations of the simplified equations used in the following discussion.

7.2.4 Factors Influencing the Sizes of Nucleated Particles

Starting with particle size, we have already seen (Fig. 7.2) that the critical nucleus size r* depends on both supersaturation levels and interfacial energy, as captured in the expression

$$r^* = \frac{2 \cdot \sigma \cdot V}{R \cdot T \cdot ln(S)} \tag{7.10}$$

Here, V is the molar volume, which for silica (SiO_2) is 30 cm^3/mole. For $r < r^*$, nuclei are unstable and redissolve. For $r > r^*$, stable nuclei form and can continue to grow. For the example of silica at a supersaturation level of 10, and assuming that $\sigma = 100$ mJ/m^2, r* is equal to 1.2 nm, which represents a spherical cluster containing on the order of 140 silicon atoms. Increasing S, the saturation state, lowers both the activation energy for nucleation and the size of the critical nucleus, whereas increasing σ, increases both parameters. The dependence of nucleation on S is manifested in everyday chemistry. Systems in a high state of disequilibrium tend to nucleate large numbers of small particles whereas systems near equilibrium make a few large crystals.

Although solubility influences r*, the converse is also true: the radius of curvature of a particle can influence its solubility. The thermodynamics of this relationship is captured in the Ostwald–Freundlich relation [8–10], which is an extension of the Gibbs–Thompson equation [5, 11, 12] developed to describe the vapor pressure in equilibrium with water droplets:

$$ln\left(\frac{S_r}{S_f}\right) = \frac{2 \cdot \sigma \cdot V}{r(R \cdot T)} \tag{7.11}$$

In Eq. 7.11, S_r and S_f refer to the solubilities of substances with radius of curvature 'r', and a 'flat' surface, respectively, and V is the molar volume. Although there is debate regarding the quantitative accuracy of this expression, the functional form of the expression is consistent with measured solubility measurements (Fig. 7.4). An important consequence of Eq. 7.6 is that if one starts with a mixture of large and small particles, the small particles eventually dissolve and are consumed by the larger particles. Conversely, regions with a high negative radius of curvature, such as pores or the contact points between particles, eventually fill in.

7.2.5 Parameters Influencing the Number of Nuclei Formed

The number of particles (N) that precipitate is also a strong function of supersaturation. Larger S values result in the nucleation of more particles. There are two

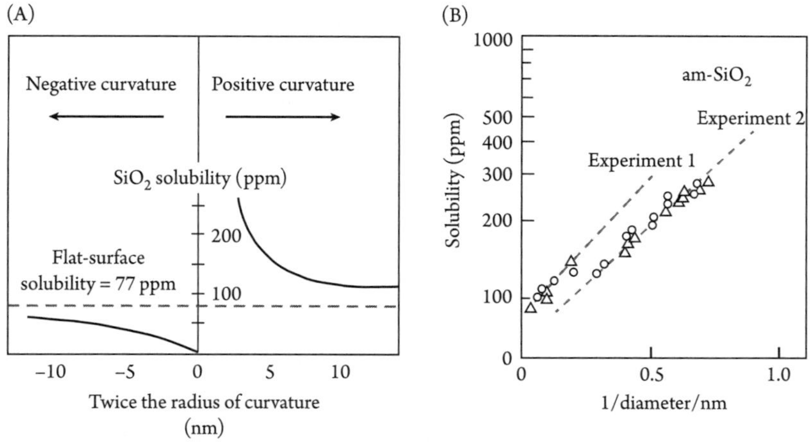

Figure 7.4 The solubility of amorphous silica as a function of radius of curvature. (A) Solubility versus curvature for both pores and particles. (B) Experimental data for the solubility of silica particles as a function of particle diameter. Adapted from Iler, R. K. *The Chemistry of Silica*; Wiley: New York, 1979, with permission.

reasons for this. First, increasing S lowers the barrier to nucleation, so many more incipient nuclei can surmount the activation barrier and continue to grow. Second, there is a finite amount of dissolved material in a supersaturated solution that is available for precipitation. If r^* is small, each nucleated particle consumes less mass, so many more particles can be created. Although a range of thermodynamic expressions have been derived to describe how N depends on S and σ, these expressions all have a similar functional form:

$$ln(N) = C_1 - \frac{C_2 \cdot \sigma^3}{R \cdot T^3 \cdot (\ln(S))^2} \quad (7.12)$$

Here, C_1 and C_2 are constants. These equations capture the fact that N is *extremely* sensitive to both S and σ. For example, simple math shows that a factor of 10 increase in S at room temperature corresponds to an increase in Gibbs energy of RT ln(10) = 5.7 kJ/mole. As a consequence, ln(N) is increased by a factor of $(5.7)^2$, or 32.5, which means that N increases by a factor of more than e^{32}, or around 10^{14}. *Manipulating S can drive N values to exceed 100 trillion particles per liter of water.* The sensitivity of N to σ can be seen in comparing nucleation densities for a range of different precipitating phases (Fig. 7.5). Substances such as $Ca(OH)_2$, with hydroxyl bridges and a high surface concentration of hydrogen bonds, have a low interfacial energy with water and thus generate the largest numbers of nuclei. Substances such as $BaSO_4$, which have higher interfacial energies, require much higher supersaturation levels to induce precipitation, and exhibit far lower nucleation densities.

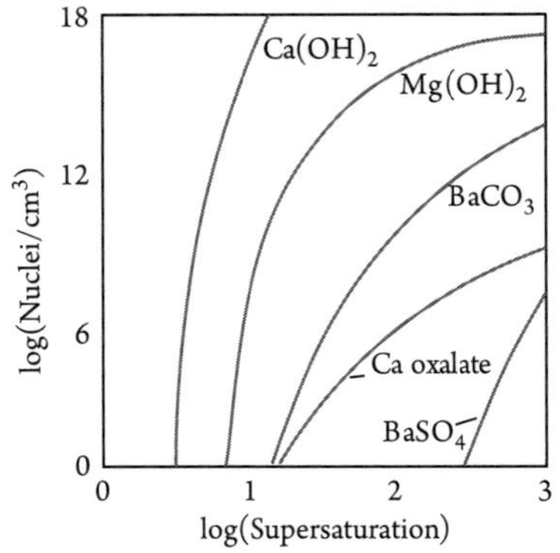

Figure 7.5 Experimental nucleation densities obtained for different oxides and ionic crystals as a function of supersaturation levels. Note that nucleation densities can vary by more than 18 orders of magnitude. Adapted from data supplied in Mohanty, R.; Bhandarkar, S.; Estrin, J. Kinetics of Nucleation from Aqueous Solution. *A. I. Ch. E. J.*, **1990**, *36*, 1536–1544, with permission.

The lowering of the interfacial energy in moving from *homogeneous* to *heterogeneous* nucleation allows more nuclei to form at lower supersaturation levels. An examination of N versus S with and without the presence of existing surfaces (Fig. 7.6) indicates that one consequence of interfacial energy effects is that three different solution regimes can be identified based on solution saturation levels. At low supersaturation levels, no nucleation occurs (at least on reasonable timescales); at intermediate supersaturation levels, heterogeneous film formation can be induced without promoting precipitation; and at the highest supersaturation levels, homogeneous precipitation dominates.

7.2.6 Parameters Influencing Nucleation Induction Times

The third important parameter that needs to be controlled in the precipitation of oxide and hydroxide phases is the induction time (τ) for nucleation. Just because a solution is supersaturated does not guarantee that precipitates will form. Honey, for example, is a supersaturated solution of sugar that can remain a homogeneous solution for months. In the honey case, nucleation is inhibited by the high viscosity of the fluid. However, in most aqueous solutions, induction times are controlled via the activation barrier to nucleation. Thermodynamic expressions for induction times take the form

$$ln\left(\frac{1}{\tau}\right) = C_1 - \frac{C_2 \cdot \sigma^2}{R \cdot T^3 \cdot (\ln(S))^2} \tag{7.13}$$

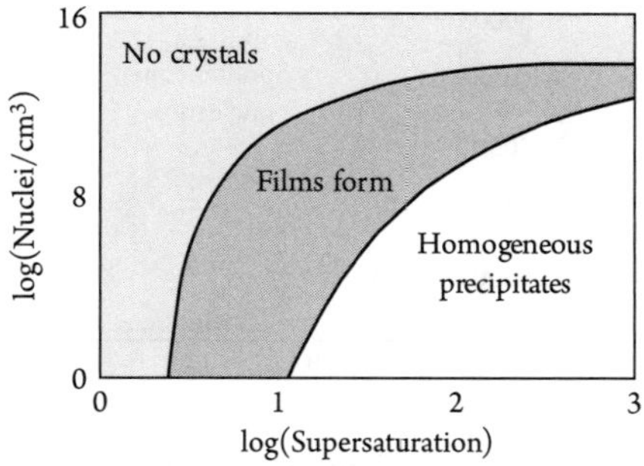

Figure 7.6 Nucleation densities calculated using Eq. 7.7 for a precipitating phase that has access to both heterogeneous (top curve) and homogeneous (bottom curve) reaction pathways. The top curve is calculated based on the assumption that the net interfacial energy for heterogeneous nucleation is half that for homogeneous nucleation. The two curves define three distinct regimes of nucleation behavior as a function of solution saturation, including no nucleation and growth (far left), exclusive heterogeneous nucleation (between curves 1 and 2), and a regime to the far right in which homogeneous precipitation is predominant. Adapted from Bunker, B. C.; Rieke, P. C.; Tarasevich, B. J.; Campbell, A. A.; Fryxell, G. E.; Graff, G. L.; Song, L.; Virden, J. W.; and McVay, G. L. Ceramic Thin Film Formation on Functionalized Interfaces Through Biomimetic Processing. *Science* **1994**, *264*, 48–55, with permission.

The impact of supersaturation on the induction times for both homogeneous and heterogeneous nucleation is illustrated for the case of a calcium-oxalate hydrate (a solid that affects millions of people as kidney stones) forming in solution as well as on a silicon-dioxide surface [13] (Fig. 7.7). Issues addressed in this example are the same as those associated with the heterogeneous nucleation of oxides onto calcium-oxalate substrates. Induction times for homogeneous nucleation of the oxalate solid can be reduced from seconds to virtually never by lowering supersaturation levels by less than a factor of two (i.e., from 4 to 2.5). When existing surfaces are present, the three regimes seen in the plot of N versus S (Fig. 7.6) are again observed; at low values of S (in this case, $S < 2$), no nucleation is observed; at intermediate values of $S (2.4 < S < 2.8)$, heterogeneous nucleation dominates; whereas for high S values ($S > 4$), homogeneous nucleation prevails.

In the discussions so far, the interfacial energy between the substrate and the nucleating crystal (σ_{cs} in Eq. 7.9) responsible for lowering the barrier to nucleation has been discussed as an empirical quantity. However, the chemical and/or atomic-level meaning of this energy is somewhat mysterious. As a starting point, consider the nucleation of a perfectly flat crystal on a perfectly flat substrate. When nucleation occurs, the interfacial energy is defined by the interfacial strain energy between the two contacting phases. This strain energy can be calculated based on the degree of lattice mismatch between the two phases and their elastic constants

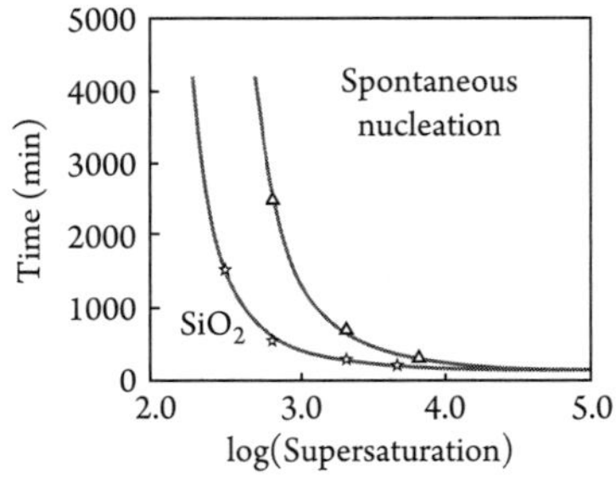

Figure 7.7 Experimental results on induction times for the nucleation of calcium oxalate crystals in a bulk solution and on silica surfaces as a function of supersaturation levels. The curves represent fits to the experimental data based on classic nucleation and growth theories. Adapted from Bunker, B. C.; Rieke, P. C.; Tarasevich, B. J.; Campbell, A. A.; Fryxell, G. E.; Graff, G. L.; Song, L.; Virden, J. W.; and McVay, G. L. Ceramic Thin Film Formation on Functionalized Interfaces Through Biomimetic Processing. *Science* **1994**, *264*, 48–55, with permission.

[14]. In other words, this energy corresponds to how hard it is to deform each phase to get atoms to match up at the interface. If the ultimate structure of the crystal–substrate interface was all that mattered, selecting conditions to induce heterogeneous nucleation would be straightforward: (1) select a substrate with a good epitaxial match to the crystal and (2) induce an appropriate level of supersaturation within the contacting solution. According to Eq. 7.9, so long as $\sigma_{cs} < \sigma_{sl}$, the free energy for heterogeneous nucleation should be less than that for homogeneous nucleation, allowing for film formation at intermediate S values. This condition is met for the case of weddellite (calcium-oxalate hydrate) mineral solids on silica, where $\sigma_{cs} = 40$ mJ/m^2, $\sigma_{sl} = 100$ mJ/m^2, and the net experimental value for heterogeneous nucleation is $\sigma = 24$ mJ/m^2 (compared with 70 mJ/m^2 for homogeneous precipitation).

Unfortunately, the simple epitaxial matching model for σ_{cs} breaks down upon detailed inspection. First, heterogeneous nucleation can occur on substrates with a poor lattice match with the nucleating phase (such as TiO_2, where $\sigma_{cs} =$ 120 mJ/m^2). In fact, nucleation occurs on substrates where $\sigma_{cs} < \sigma_{sl}$. Second, this model predicts that the surface chemistry of the substrate–solution interface should be irrelevant. Experiments involving the calcium-oxalate solid show [15] this is clearly not the case. Calcium oxalate represents an ideal model for a nucleating system because its solubility is constant between pH 4 and pH 11. The protonation state of the oxalate is constant and the metal ion does not hydrolyze, which means that nucleation on oxide particles can be examined as a function of pH, allowing researchers to vary the degree of protonation and the charge on the oxide surface (see Chapter 8). Conversely, oxides with different surface charge and protonation characteristics can be examined at the same pH.

Experiments have been conducted on SiO_2, TiO_2, and Al_2O_3 (with isoelectric points of 2, 6, and 9, respectively) at pH 6.5 (Fig. 7.8). Results indicate that nucleation of solid calcium oxalate is best promoted on anionic SiO_2, followed by neutral TiO_2, whereas the cationic Al_2O_3 surface is least effective. This disparity is not a result of differences in epitaxial matching, because TiO_2 surfaces can be made as active as SiO_2 by going to a higher pH (where TiO_2 is anionic), or can be relatively inactive (such as Al_2O_3) by going to a lower pH (where TiO_2 is cationic). In fact, at pH 6.5, where Al_2O_3 surfaces are normally cationic, nucleation can be enhanced by first adsorbing the anionic surfactant dextran sulfate onto the Al_2O_3 surface.

These results emphasize the obvious point that surface complexation plays a key role in heterogeneous nucleation. Although more studies are required to quantify this effect, it appears the interfacial energy that matters most for heterogeneous nucleation is the energy of the hydrated substrate relative to the energy of the surface after complexation has occurred (i.e., *not* the energy of the ultimate postnucleation interface).

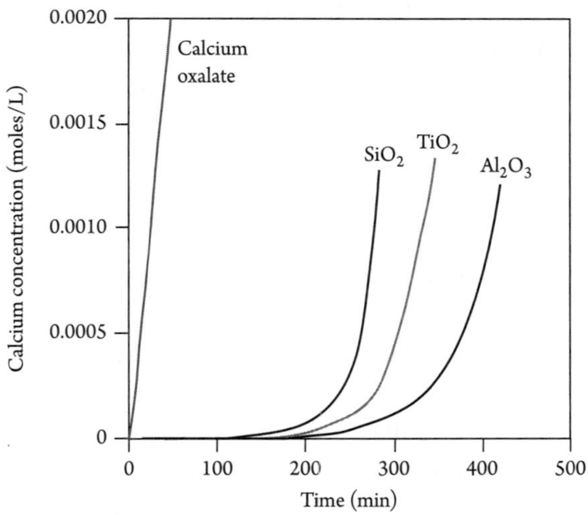

Figure 7.8 Experimental uptake of Ca^{2+} per unit surface area by high-surface area colloidal oxides as a function of time from supersaturated calcium oxalate solutions ($[Ca^{2+}] = [Ox^{2-}] = 2 \cdot 10^{-4}$ M, pH = 6.5, supersaturation level S = 3.3). Solution concentrations are held constant using an automatic burette system in conjunction with a calcium ion electrode (called a *constant composition experiment*). Adapted from Bunker, B. C.; Rieke, P. C.; Tarasevich, B. J.; Campbell, A. A.; Fryxell, G. E.; Graff, G. L.; Song, L.; Virden, J. W.; and McVay, G. L. Ceramic Thin Film Formation on Functionalized Interfaces Through Biomimetic Processing. *Science* **1994**, *264*, 48–55, with permission.

7.2.7 The Nature of the Precipitating Phase

The final parameter of interest in nucleation involves the phase that is nucleated. For the case of solutions containing the hydrolysis products of iron, more than a dozen compounds can form [13], including hydroxides such as ferrihydrite ($Fe(OH)_3$), oxyhydroxides such as goethite or akageneite (FeOOH), and oxides such as hematite (α-Fe_2O_3). How does a supersaturated solution "decide" on which of these 16 solid materials to precipitate? Once again, a creative tension exists between supersaturation and surface energy, leading to one dominant pathway. In terms of solubility, the least soluble and most thermodynamically stable phases tend to be those that contain the fewest number of protons, the highest number of bridging oxo bonds, and the highest symmetry. Dense oxides are almost always less soluble than oxyhydroxides, which in turn are less soluble than the hydroxides. For example, at pH 7, layered $Zn(OH)_2$ has a solubility of $5\cdot10^{-2}$ M, which is 40 times greater than the solubility of cubic ZnO ($1.3\cdot10^{-3}$ M). The most soluble precipitates are amorphous materials formed by linking oligomers together randomly, such as the aluminum-hydroxide flocs that form in streams.

Based on solubility alone, one might expect that the least-soluble phases should always precipitate first, because solutions are the most supersaturated with respect to the least-soluble phase. Instead, the exact opposite is almost always observed, because the energy penalty associated with creating a solid–liquid interface is the lowest for those phases that come the closest to resembling solvated ions. The interfacial energy for solid hydroxides is lower than that for oxyhydroxides, which is lower than that for oxides, which often means that the activation barrier for the nucleation of hydroxides is lower than that for oxyhydroxides and oxides. Because nucleation is an extremely rapid event controlled by kinetic factors rather than thermodynamic equilibrium, disordered amorphous hydroxides are almost always the phases that nucleate initially. In other words, oxide nucleation often involves the rapid and relatively random polymerization of soluble hydrolysis products. This does not mean that amorphous hydroxides are invariably the only materials that can form in aqueous solutions. Instead it means that amorphous hydroxides are the most likely materials to be formed in the instant of precipitation.

Suspensions containing amorphous hydroxides continue to evolve with time if left in solution via transformations that involve reversible hydrolysis and condensation reactions. The sequence of solid phases produced generally proceeds from the most-soluble to the least-soluble structure until the most thermodynamically stable phase is produced. This general phenomenon is referred to as the *Ostwald Law of Stages* [17, 18] and is wholly empirical. For the example of Al(III) precipitation, the sequence of phases observed evolves from amorphous to pseudo-boehmite to bayerite to gibbsite, and finally (under hydrothermal conditions) to boehmite. Because both the precipitation process and the

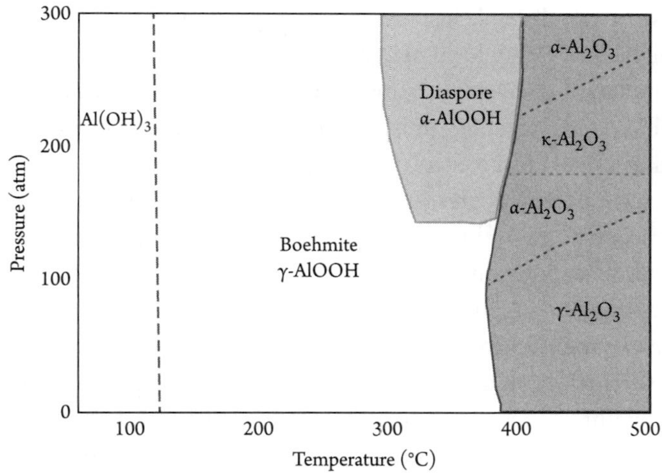

Figure 7.9 An example of the role of temperature in controlling oxide phases during precipitation from aqueous media. Shown is a phase diagram depicting the hydrothermal conditions required to produce hydroxides ($Al(OH)_3$), oxyhydroxides (boehmite and diaspore), and oxides of aluminum. Adapted from Levin, E. M.; Robbins; C.R. McMurdie, H. F. *Phase Diagrams for Ceramists, 1969 Supplement*; The American Ceramic Society: Columbus, OH, 1969.

subsequent phase transformations are activated processes, the thermal barriers to the formation of more stable phases can often be overcome by increasing the temperature and by judicious choice of nucleating surfaces with which the solution can interact. The hydrothermal conditions required to produce various aluminum hydroxides, oxyhydroxides, and oxides are indicated in Figure 7.9 [19].

7.3 CHEMICAL CONTROL OF OXIDE NUCLEATION

Thermodynamic expressions for nucleation clearly indicate that chemical supersaturation and interfacial energy are the two critical parameters that control the nucleation of oxides from aqueous solutions. Chemical strategies are commonly used to manipulate these two parameters in industrial precipitation processes. Starting with supersaturation, hydrolysis diagrams clearly indicate that pH control is of the utmost importance for oxide precipitation. Earlier, we saw that increasing the saturation term S by a factor of 10 can increase nucleation densities by a factor of ~10^{14}. For many dissolved cations, this increase in S can be accomplished by changing the solution pH by 1 unit or less. In the extreme case of acidic Zr(IV) solutions, S can change by a factor of 100 million by changing pH by 1 unit, or $\Delta pH = 1$. The easiest way to change pH involves titrating the solution with a strong acid such as HCl or a strong base such as KOH. However, droplets of added acid or base do not disperse instantaneously regardless of mixing. As it disperses, each droplet creates large local pH gradients, resulting in large gradients in S, N, and in induction times. A fast titration occurs in seconds, but nucleation can occur within milliseconds, resulting in huge variations in nucleation and growth rates. The resulting lack of local control over S can result in a heterogeneous distribution of particle sizes, morphologies, and even compositions. The challenge in producing monodisperse and/or high-quality powders is to move from point A to point B in the hydrolysis diagram in such a way that supersaturation levels are changed in a more homogeneous and/or controlled fashion.

One simple method for inducing homogeneous changes in supersaturation levels and/or nucleation tendencies involves manipulating the solution temperature. Temperature appears in all the thermodynamic expressions for nucleation, and clearly plays a major role in inducing precipitation. Increasing the temperature allows more incipient nuclei to surmount the activation barrier to nucleation. This increase (and the lower interfacial energy associated with heterogeneous nucleation) is one reason why scale deposits can be such a problem in a hot-water heater. However, an equally important, although perhaps less obvious, effect is that temperature changes can alter all the chemical equilibrium constants that control speciation, pH, solubility, and, ultimately, supersaturation levels. For solid phases with a positive enthalpy for precipitation (and a negligible entropy term), S increases with heating, whereas for phases with a negative enthalpy for precipitation, S is reduced by heating. For the example of Al(III) solutions (Fig. 7.10), temperature increases shift the entire solubility curve to lower pH values and induce a slight increase in the minimum solubility value [20]. At room temperature, this minimum occurs near pH 7. By 90°C, the minimum has moved to pH 5.3, whereas at 300°C, the minimum is pushed down to almost pH 3. Considering a specific example (Fig. 7.10), at room temperature, a solution of pH 4 containing a total concentration of dissolved aluminum (ΣAl(III)) $= 10^{-3}$ M is near the solubility limit. However, at 90°C, the same pH 4, ΣAl(III) $= 10^{-3}$ M solution is supersaturated by a factor of almost 1000. In practice, such large gains in S are not achieved via simple heating because the thermally activated hydrolysis of dissolved aluminum species releases protons and lowers the pH below 4. However, the net increases in S are still sufficient to lead to more controlled particle nucleation phenomena.

A second strategy for manipulating S involves changing the pH in a homogeneous fashion via the decomposition of organic additives. The classic example of such a strategy involves the thermal decomposition of urea:

$$(NH_2)_2CO + H_2O \rightleftarrows 2NH_3 + CO_2(g) \tag{7.14}$$

$$NH_3 + H_2O \rightleftarrows NH_4^+ + OH^-,\ K_b = 1.8 \cdot 10^{-5} \tag{7.15}$$

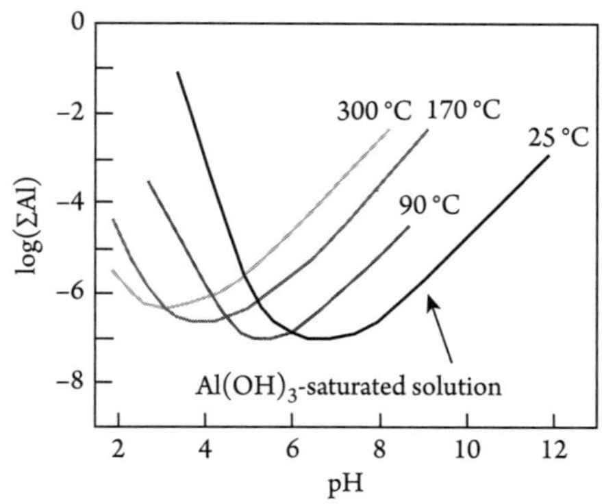

Figure 7.10 Solubility diagrams for dissolved aluminum hydrolysis products in equilibrium with AlOOH as a function of solution temperature. Note how the entire hydrolysis diagram and solubility curve shifts to lower pH as the temperature is increased. Adapted from Castet, S.;, Dandurand, J.-L.; Schott, J.; and Gout, R. Boehmite Solubility and Aqueous Aluminum Speciation in Hydrothermal Solutions (90–350°C) : Experimental Study and Modeling *Geochim. Cosmochim. Acta* **1993**, 57, 4869–4884, with permission.

The decomposition reaction results in controlled release of ammonia into the solution, which makes the solution pH more basic in a slow homogeneous fashion. In some instances, the products released ($[NH_3]^\circ$ and, in more basic solutions, $[HCO_3]^-$) can also form complexes with dissolved metal cations, which impacts solubility further. An example in which decomposition results in a lowering of the solution pH involves the slow conversion of chloroacetate into the glycolate anion:

$$ClCH_2COO^- + H_2O \rightleftarrows HOCH_2COO^- + H^+ + Cl^- \qquad \textbf{(7.16)}$$

In acidic solutions, the acid–base chemistry of the carboxylate anions can also contribute to pH changes, because glycolic acid ($pK_a = 3.8$) is a weaker acid than chloroacetic acid ($pK_a = 2.8$).

A third strategy for improving the control of supersaturation conditions involves manipulating hydrolysis product distributions and solubility by adding other complexing ligands to the solution. This strategy is used for oxides that exhibit extremely steep solubility-versus-pH curves (such as TiO_2, SnO_2, or ZrO_2) or for the precipitation of multicomponent oxides. For example [21], the addition of the oxalate anion to Sn(IV) solutions not only lowers the pH sensitivity (noting that S increases by 10^4 for $\Delta pH = +1$ without oxalate), but also reduces the severity of reaction conditions (from pH 0.5 without oxalate to pH 5 with oxalate [Fig. 7.11]).

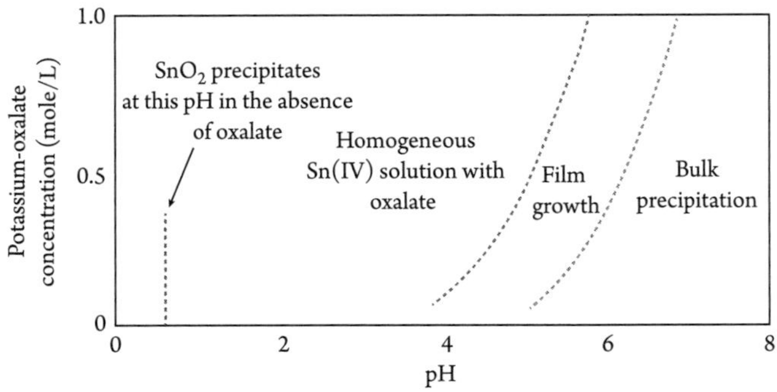

Figure 7.11 Experimental results for the growth of tin oxide films as a function of solution pH and the presence of the oxalate anion as a complexant. Without oxalate, uncontrolled precipitation occurs at pH 0.5. With oxalate, controlled film growth can be achieved between the two curves shown (between roughly pH 4 and 6). Bulk precipitation is observed only for solutions with higher pH values.
SOURCE: Graff, G. L.; Bunker, B. C. Pacific Northwest National Laboratory, unpubl. results.

For multicomponent systems, it is desirable to try to match the solubilities of several components to induce simultaneous precipitation of all desired constituents. A sampling of Al(III) complexes (Fig. 5.4) illustrates the means by which cation solubilities can be manipulated to achieve such matching. Addition of fluoride ions forms new soluble species that consume a fraction of the existing hydrolysis products. In this case, the slope of the solubility curve in acidic media is unchanged, but the $\log[M] = 0$ intercept increases by roughly 1 pH unit for each 10-fold increase in $[F]^-$. In the acidic regime, this results in a corresponding 10-fold increase in solubility at a given pH. For chelating ligands such as ethylenediamine tetraacetic acid (*EDTA*), ligand binding can result in changes in both the slope and intercept of the solubility curve. For the case of acetyl acetonate, a pH window exists in which the solubility is invariant with solution pH. There are even some complexing agents (such as Tiron) that can create solubility curves with multiple local solubility minima.

A specific example in which complexation has been used to match solubilities is provided by the process used to precipitate powders used in the synthesis of Y–Ba–Cu oxide superconductors [22]. Although the hydrolysis products of Ba(II), Cu(II), and Y(III) are all very soluble in acidic precursor solutions, by pH 7 the solubilities are strongly divergent, with Cu(II) being the least soluble (10^{-5} M), followed by Y(III) (10^{-2} M), and Ba(II) (which is extremely soluble as a carbonate or sulfate mineral). Simultaneous precipitation is achieved via extremely rapid mixing of the acidic soluble precursor solutions with basic solutions containing carbonate ions, primarily to form insoluble carbonate complexes with

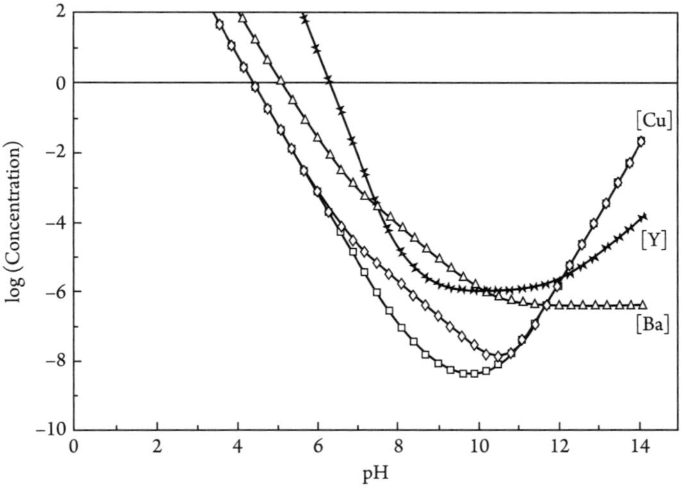

Figure 7.12 Solubility curves for Y^{3+}, Ba^{2+}, and Cu^{2+} solutions used to prepare Y-Ba-Cu oxide superconductors. The solubility of Ba^{2+} in particular has been lowered via the addition of carbonate ions to the solution. The optimum pH window for the simultaneous precipitation of all three components is pH 11 to 12, which is achieved by rapid additions of appropriate amounts of KOH to acidic (pH < 4) precursor solutions. The resulting amorphous precursor powders have compositions that are then appropriate for calcining and sintering to produce oxide superconductors.
SOURCE: Bunker, B. C., et al.; in Hatfield, W. E.; Miller, Jr.; J.H., Eds., *High-Temperature Superconducting Materials: Preparations, Properties, and Properties*; Marcel Dekker, Inc.: New York, 1988, with permission.

Ba(II). By tailoring feed solutions such that the final pH of the mixture is between pH 11 and 12, precipitation can occur within a solution in which all components have a solubility of around 10^{-6} M (Fig. 7.12). Although the resulting nanoparticle precipitate is an oxycarbonate rather than a pure oxide, the powder can readily be calcined and sintered to produce phase-pure oxide superconductors.

Another way to control the monospecificity of size distributions of nuclei is to physically isolate the particles in micelles (see Chapter 9) or by using nonaqueous solvents to simplify speciation. Because this is a book on aqueous oxide chemistry, we do not discuss nonaqueous strategies, although they have become highly refined for making nanoparticles of uniformly small sizes.

7.4 CLASSIC GROWTH THEORY

Nucleation is only the first step in a typical precipitation process (Fig. 7.13). Immediately after solutions become supersaturated, there is a burst of nucleation until supersaturation levels are no longer sufficient to allow incipient nuclei to

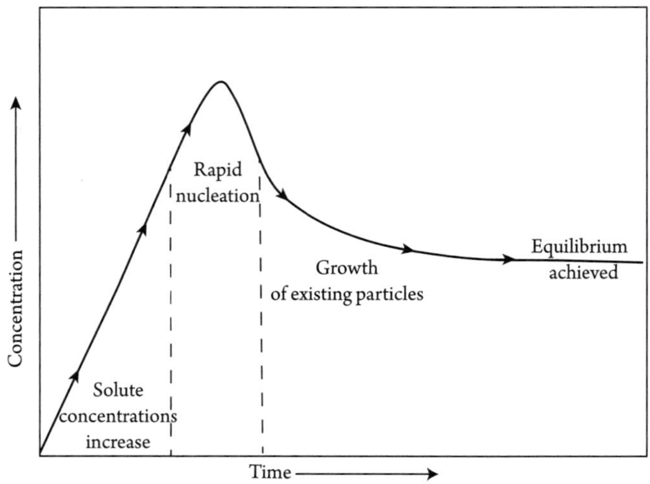

Figure 7.13 A schematic curve indicating solution concentrations relative to supersaturation as a function of time during nucleation and growth from aqueous solutions. In practice, concentrations are typically held fixed, whereas solution conditions (such as pH) are changed to promote supersaturation and induce nucleation in the initial part of the curve. The peak in the curve corresponds to a burst of nucleation, after which solution concentrations drop gradually to the solubility limit as nucleated particles grow.

surmount the free energy barrier. At this point, there is a transition to lower energy phenomena associated with particle growth. Growth typically continues to consume dissolved species until the system reestablishes chemical equilibrium at the solubility limit.

The classic thermodynamics of particle growth, as described by Burton–Cabrera–Frank theories [19], are dictated by similar parameters as those associated with nucleation, including supersaturation levels, surface energies, as well as surface topologies and defect states. Material can add to surfaces via three distinct mechanisms in order of decreasing energy: the creation of new active sites for growth (called *birth* in birth-and-spread models), growth of material at site or step edges (called *spread*), and growth from screw dislocations. First, consider an ideal oxide surface that is perfectly flat and contains no defects. The "birth" step involves converting soluble species that adsorb to the surface and diffuse on this surface until they reach heterogeneous nuclei or "islands." The edges of these islands are more reactive (i.e., have a higher free energy) than the ideal surface because there are more terminal ligands for replacement. From a chemical perspective, we have already seen (Chapter 6) that these energetic step edges contain sites that (1) are stronger Lewis acids and bases than sites on the flat surface and (2) have more open coordination sites that are available for polymerization via ligand exchange for coordinated water molecules. This means that the growth via the spreading of islands occurs at step edges. In growth models, it is often assumed that island edges are unit-cell-height steps, called *elementary steps*, that propagate across the original surface as two-dimensional sheets.

The rate-limiting step in the birth-and-spread mechanism is the heterogeneous nucleation of a stable island on a perfectly smooth surface. We have already seen that island nucleation requires high supersaturation levels to overcome that component of the nucleation barrier associated with interfacial energy effects (Eq. 7.6). However, after a new island forms, growth can continue via the addition of adatoms to the edge of the island. The free energy for step edge growth is given by

$$\Delta G = -(1/a)^2 (RT \cdot ln(S)) + 4 \cdot l \cdot \sigma_{edge} \quad \textbf{(7.17)}$$

Here, l is the length of an elementary step, a is the distance between adatoms, and σ_{edge} is the step-edge free energy, which is analogous to the interfacial free energy in Eq. 6.6. As in the case of nucleation from solution, there exists a critical size for the edge length above which the step grows but below which the step erodes. In highly supersaturated solutions, new nuclei form on surfaces faster than the steps can grow laterally, resulting in a rough surface. Conversely, growth shuts down at low supersaturation levels because the barrier to form new nuclei is too high to allow for new island formation. When existing islands or steps grow until they cover the entire surface, the steps disappear and the lower energy step-edge pathway for relieving the condition of supersaturation is eliminated.

Crystals continue to grow in slightly supersaturated solutions even after the birth-and-spread and step-edge mechanisms have stopped, as a result of an additional low-energy pathway based on screw dislocations [24]. Finite populations of screw dislocations are commonly found on crystal surfaces. As mass adds to the edges emanating from a screw dislocation, the new edge surface never extinguishes. Instead, it rotates around the dislocation in a helical fashion (Fig. 7.14). Here, pyramidal facets form on the surface as sets of steps radiate outward from the screw dislocation. These often truncate in highly reactive kink sites that are stabilized by the addition of adatoms to the surface.

As in both homogeneous and heterogeneous nucleation, the presence of foreign ligands can have a profound impact on oxide growth phenomena. Schoolchildren know that apparently innocuous additives to a solution can cause crystals to change morphology (e.g., from prismatic to tabular, see Fig. 7.1), although the impurity is not incorporated into the bulk crystal structure. Addition of millimolar concentrations of ferricyanide $[Fe(CN)_6]^{4-}$ to a saturated NaCl solution (~6 molal) causes the crystals to change from the simple NaCl(s) cubes that grow naturally to skeletal and spiral-form crystals. Many of the observed changes can be

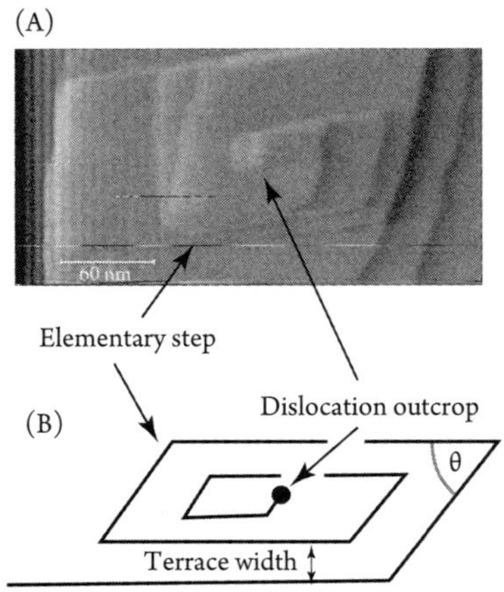

Figure 7.14 (A, B) A schematic (B) and an experimental atomic microscope image (A) showing a growing screw dislocation on calcium-oxalate monohydrate.
SOURCE: From Friddle, R. W.; Weaver, M.L.; Qiu, S. R.; Wierzbicki, A.; Casey, W. H.; and De Yoreo, J. J. Sub-nm AFM Imaging Reveals Source of Acceleration and Hysteresis During Peptide-Directed Crystallization. *Proc. Nat. Acad. Sci.*, **2010**, *107*, 11–15, with permission.

rationalized on the basis of the thermodynamic expressions that were introduced earlier [25]. First, ligand complexation in solution impacts the solubility of oxide and hydroxide phases, which in turn dictates the supersaturation level S. Second, ligands can dictate the magnitude of the net interfacial energy σ for heterogeneous nucleation (see Eq. 7.6). Ligands can modify both the epitaxial matching between the nucleating phase and the substrate (changing σ_{cs}) and the "mismatch" between the substrate and the aqueous phase (changing σ_{sl}). Ironically, functionalization with sulfates can sometimes enhance surface nucleation by *increasing* σ_{sl}. Enhancement is achieved by covering up an energetically unfavored sulfate surface with a nucleated hydroxide phase, which has many sites for hydrogen-bonding to water and is a better match to the adjacent aqueous solution phase.

Ligand adsorption densities and their impact on interfacial energies both depend on the crystal lattice plane exposed to the solution. Ligand adsorption can promote the growth of specific crystal faces relative to other faces, which dictates the morphology of the growing crystal. Third, ligand adsorption can have highly localized effects on growth, because ligand adsorption can occur preferentially on step edges and point defects (see Chapter 6). This preferential adsorption probably accounts for the effect of ferricyanide $[Fe(CN)_6]^{4-}$ on NaCl crystal morphologies mentioned earlier.

Even at impurity levels, foreign adsorbates can poison surfaces by pinning steps and preventing them from moving laterally [26]. If the spacing between pinning sites drops below the critical length for step growth and/or increase step energies, steps can be induced to retreat rather than grow. Once again, each crystal face can exhibit its own unique step and defect chemistries, leading to crystal growth habits that depend on the nature of the adsorbate. Foreign adsorbates can also be used to suppress aggregation and interactions between colliding nuclei, aiding in the generation of monodisperse particles (see Chapter 8). The surface adsorbates that are used to mediate surface nucleation, growth, and aggregation phenomena are called *surfactants*. In Chapter 9, we provide examples of the impact of surfactants on biomineralization and the biomimetic synthesis of complex oxide architectures.

Finally, there is increased appreciation for the growth of minerals via nonclassic pathways, such as oriented aggregation of nanoparticles and liquid-like prenucleation clusters [27–29]. These particles are so small they move relatively quickly in solution and can reorient when attached to one another. Growth of crystallites can thus proceed not by the birth-and-spread or step-flow models, but by the accumulation of nanoparticles that attach to each another and the substrate in a way that minimizes the lattice mismatch at the interface. These growth mechanisms extend well beyond classic treatments based solely on thermodynamics.

References

1. Sounart, T. L.; Liu, J.; Voigt, J. A.; Huo, M.; Spoerke, E. D.; McKenzie, B. Secondary Nucleation and Growth of ZnO. *J. Am. Chem. Soc.* **2007**, *129*, 15786–15793.
2. Mohanty, R.; Bhandarkar, S.; Estrin, J. Kinetics of Nucleation from Aqueous Solution. *A. I. Ch. E. J.* **1990**, *36*, 1536–1544.
3. De Yoreo, J. J; Vekilov. P. G. Principles of Crystal Nucleation and Growth. In *Biomineralization: Reviews in Mineralogy and Geochemistry*; Dove, P. M.; De Yoreo, J. J.; Weiner, S., Eds.; 2003; pp 57–94.
4. Burton, W. K.; Cabrera, N.; Frank, F.C. Dislocations in Crystal Growth. *Nature* **1949**, *163*, 398–399.
5. Schindler, P. W. Heterogeneous Equilibria Involving Oxides, Hydroxides, Carbonates, and Hydroxide Carbonates. In *Equilibrium Concepts in Natural Water Systems*; Stumm, W., Ed.; Advances in Chemistry Series; American Chemical Society: Washington, DC, 1968; pp 196–221.
6. Iler, R. K. *The Chemistry of Silica*; Wiley: New York, 1979.
7. Klein, D. H.; Smith, M. D.; Driy, J. A. Homogeneous Nucleation of Magnesium Oxide. *Talanta* **1967**, *14*, 937–940.
8. Thomson, W. On the Equilibrium of Vapour at Curved Surface of Liquid. *Phil. Mag. Ser. 4*, **1871**, *42*, 448–452.
9. Ostwald, W. Uber die Vermeintliche Isomerie des Roten und Gelben Quecksilbersoxyds und die Oberflachenspanning Fester Korper. *Zeit. Phys. Chem.* **1900**, *34*, 495–503.
10. Meldrum, F. C.; Colfen, H. Controlling Mineral Morphologies and Structures in Biological and Synthetic Systems. *Chem. Rev.* **2008**, *108*, 4332–4432.

11. Dundon, M. L.; Mack, E., Jr. The Solubility and Surface Energy of Calcium Sulfate. *J. Am. Chem. Soc.* **1923**, *45*, 2479–2485.
12. Johnson, G. Generalization of the Gibbs–Thomson Equation. *Surf. Sci.* **1965**, *3*, 429–444.
13. Bunker, B. C.; Rieke, P. C.; Tarasevich, B. J.; Campbell, A. A.; Fryxell, G. E.; Graff, G. L.; Song, L.; Virden, J. W. ; and McVay, G. L. Ceramic Thin Film Formation on Functionalized Interfaces Through Biomimetic Processing. *Science* **1994**, *264*, 48–55.
14. van der Merwe, J. H. In *Treatise on Materials Science and Technology*; Herman, H., Ed.; Academic Press: New York, 1973; p 1.
15. Song, L.; Pattillo, M. J.; Graff, G. L.; Campbell, A. A.; Bunker, B.C. Heterogeneous Nucleation of Calcium Oxalate on Native Oxide Surfaces. *Mater. Res. Soc. Symp. Proc.* **1994**, *346*, 223–228.
16. Cornell, R. M.; Schwertmann, U. *The Iron Oxides*; Wiley-VCH: Bern, Switzerland, 2003.
17. Ostwald, W. Studien uber die Bildung und Umwandlung fester Körper. *Z. Phys. Chem.* **1897**, *22*, 289–330.
18. ten Wolde, P. R.; Frenkel, D. Homogeneous Nucleation and the Ostwald Step Rule. *Phys. Chem. Chem. Phys.* **1999**, *1*, 2191–2196.
19. Levin, E. M.; Robbins, C. R.; McMurdie, H. F. *Phase Diagrams for Ceramists 1969 Supplement*; The American Ceramic Society: Columbus, OH, 1969; p 523.
20. Castet, S.; Dandurand, J.- L.; Schott, J.; Gout, R. Boehmite Solubility and Aqueous Aluminum Speciation in Hydrothermal Solutions (90–350°C): Experimental Study and Modeling. *Geochim. Cosmochim. Acta* **1993**, *57*, 4869–4884.
21. Graff, G. L. Pacific Northwest National Laboratory. Unpublished results.
22. Bunker, B. C.; Voigt, J. A.; Doughty, D. H.; Lamppa, D. L.; Kimball, K. M., In *High-Temperature Superconducting Materials: Preparations, Properties, and Processing*; Hatfield, W. E.; Miller, J. H., Jr., Eds.; Marcel Dekker: New York, 1988.
23. Burton, W. K.; Cabrera, N.; Frank, F. C. The Growth of Crystals and the Equilibrium Structure of Their Surfaces. *R. Soc. Lond. Phil. Trans.* **1951**, *A243*, 299–358.
24. Frank, F. C. The Influence of Dislocations on Crystal Growth. *Disc. Farad. Soc.* **1949**, *5*, 48–54.
25. Cabrera, N.; Vermilyea, D. A. *Growth and Perfection of Crystals*; Chapman and Hall: New York, 1958.
26. Friddle, R. W.; Weaver, M. L.; Qiu, S. R.; Wierzbicki, A.; Casey, W.H.; De Yoreo, J. J. Subnanometer Atomic Force Microscopy of Peptide–Mineral Interactions Links Clustering and Competition to Acceleration and Catastrophe. *Proc. Nat. Acad. Sci. USA* **2010**, *107*, 11–15.
27. Gebauer, D.; Voekel, A.; Coelfen, H. Stable Prenucleation Calcium Carbonate Clusters. *Science* **2008**, *322*, 1819–1822.
28. Li, D.; Nielsen, M. H.; Lee, J. R. I.; Frandsen, C.; Banfield, J. F.; De Yoreo, J. J. Direction-Specific Interactions Control Crystal Growth by Oriented Attachment. *Science* **2012**, *336*, 1014–1018.
29. Banfield, J. F.; Welch, S. A.; Zhang, H.; Ebert, T. T.; Penn, R.L. Aggregation-Based Crystal Growth and Microstructure Development in Natural Iron Oxyhydroxide Biomineralization Products. *Science* **2000**, *289*, 751–754.

8

The Colloidal Chemistry of Oxides

8.1 INTRODUCTION

Colloids are defined as suspensions of finely divided particles in a continuous medium that do not settle rapidly and are not readily filtered. To be more specific, the International Union of Pure and Applied Chemistry defines a colloid as any material for which one or more of its three dimensions lies within the size range of 1 to 1000 nm [1]. As the nucleation and growth of oxides from aqueous solutions almost always produces suspensions containing submicron particles (see Chapter 7), typical oxide suspensions fall squarely within the colloidal domain. In this book, we consider colloidal particles to represent oxides or hydroxides that are small enough to stay in aqueous suspensions for more than a few hours, yet are larger and lacking in the specific molecular structures of typical hydrolysis products (see Chapter 5). Given the density range of most oxides (from around 2–10 g/cm^3), the sizes of most colloidal oxides fall within the limits of the International Union of Pure and Applied Chemistry (see Section 8.4.5).

Colloidal oxide particles suspended in water represent a complex chemical environment. At the molecular level, protons, ions, small molecules, and polymeric species interact with particle surfaces to create charged surface sites and promote adsorption and desorption phenomena (see Chapter 6). These modified surfaces perturb the adjacent liquid, creating ordered solvent layers and strong concentration gradients in ions and other dissolved species. These interfacial phenomena generate a range of forces called *interaction potentials.* Such forces determine whether particles repel each other (leading to stable suspensions) or are attracted to one another, resulting in agglomeration and sedimentation phenomena.

The length scales of those components of the oxide–water interface that influence the interaction potentials to be discussed in this chapter are introduced in Figure 8.1. At the subatomic level, the correlated polarization of electron clouds gives rise to dispersion forces described by quantum mechanics that contribute to van der Waals interactions. At the atomic level, the inherent charge on each exposed oxygen anion that terminates the oxide surface is controlled by local chemical bonds to adjacent cations (see Chapter 6). This inherent charge, as mediated by the adsorption and/or desorption of protons from solution, generates electrostatic interactions.

At nanometer-length scales, the water molecules that solvate the immediate oxide surface are relatively immobile and define a zone that is called the *Stern layer*. Hydration forces originate within this layer. Ions from solution can enter the Stern layer. The boundary between the Stern layer and the bulk solution is

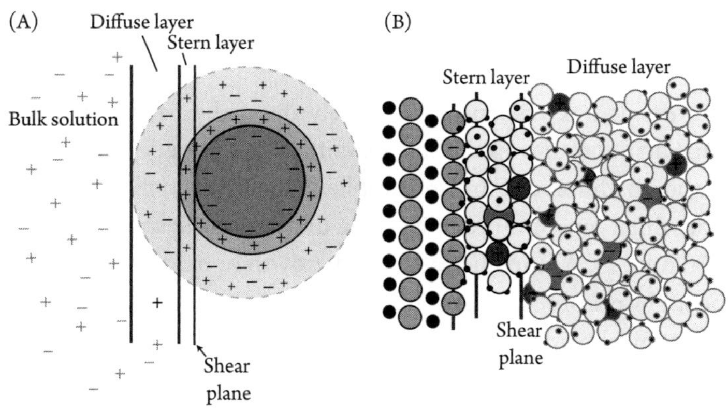

Figure 8.1 An image depicting the interface between an oxide and an aqueous solution indicating the origins of primary colloidal forces. (A) A generic view of an anionic oxide particle in water, showing the fixed charges in the immediate oxide surface. These anionic charges are partially neutralized by cations that enter the Stern layer of immobile water adjacent to the immediate surface that represents the immobile portion of the electrical double layer. The readjustment of local concentrations of both cations and anions within the mobile fluid representing the diffuse portion of the double layer complete the charge neutralization process. (Adapted from Hirtzel, C. S.; Rajagopalan, R.; *Colloidal Phenomena: Advanced Topics*; Noyes Publications: Park Ridge, NJ, 1985; Figure 3.A.1, p 33, with permission.) (B) An atomic-level view of the same interface, showing the positions of the fixed charges and ions.

called the *outer Helmholtz* or *slipping plane*. The outer Helmholtz plane represents the boundary between two aqueous zones: a high-viscosity inner (fixed) layer and a low-viscosity outer (or diffuse) region of the electrical double layer. The *electrical double layer* is that region within which the solution concentrations of ions are perturbed by the fixed charges present in the oxide surface. The diffuse portion of the double layer can extend for distances exceeding a micron from the surface in nearly pure water. The electrical double layer controls the electrostatic interactions between oxide surfaces in water.

An understanding of interaction potentials is critical for understanding how oxide surfaces interact with their local environments. For oxide suspensions, it is critical to control how suspended particles interact with each other. The structures created by such interactions also involve multiple length scales, as illustrated by the aggregated colloids that form the building blocks of soils (Fig. 8.2). When particles repel each other, it is sometimes possible to create suspensions containing isolated particles. More commonly, suspensions contain agglomerates that incorporate large numbers of primary particles. These agglomerates can then interact with each other to form larger flocs, or even semisolid phases called *gels* that span and entrain the entire solution volume. Depending on the nature and extent of agglomeration, particles either stay in suspension or settle out, and can create slurries that exhibit high or low viscosities.

In this chapter, we begin our discussion of colloidal oxides by describing the fundamental interactions between oxide particles in aqueous solutions. Next, we focus on how the solution and surface chemistry of oxides determine those colloidal interactions. We then discuss how colloidal interactions control aggregate

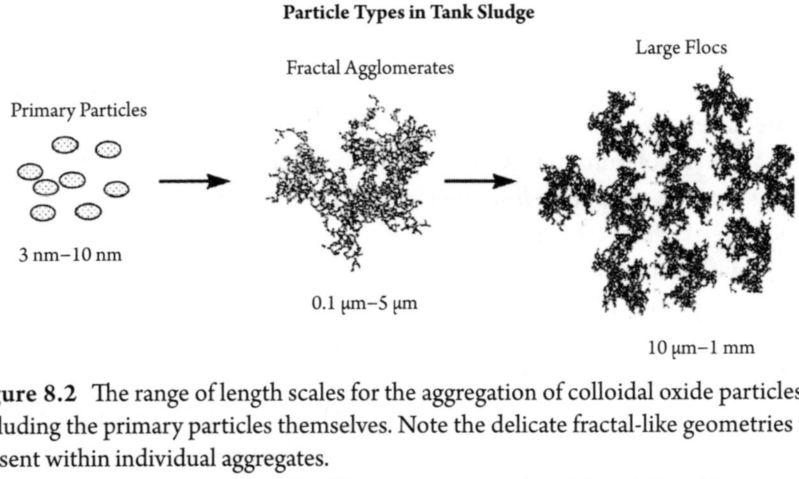

Figure 8.2 The range of length scales for the aggregation of colloidal oxide particles, including the primary particles themselves. Note the delicate fractal-like geometries that present within individual aggregates.
SOURCE: SEM images of colloidal gold aggregates reproduced from Weitz, D. A.; Oliveria, M.; *Phys. Rev. Lett.*, **1984**, *52*, 1433–1436, with permission.

structures, which in turn control the physical properties of aqueous suspensions. Finally, we describe the processing techniques that are used by engineers to convert oxide slurries into particle compacts called *green bodies* that are subsequently heat-treated to form ceramic materials. The environmental impact of colloidal interactions on clays and other oxide suspensions is deferred until Chapter 18.

8.2 FUNDAMENTAL COLLOIDAL INTERACTIONS

The oxide–water interface is a phase discontinuity. As discussed in Chapter 6, the structure, bonding, electronic configurations, and ionic distributions in both the solid and the liquid phases readjust themselves in the vicinity of the interface to minimize the energy difference between the two phases. These adjustments create both physical and chemical gradients that give rise to the forces that determine how particles respond to each other and their local environments. Interaction potentials are so important, and involve so many areas of science and technology, that numerous books and reviews have been written on the subject [1–3]. Here, we highlight key features of the major interaction potentials that correspond to the zones depicted in Figure 8.1. These potentials include hydration forces, van der Waals interactions, and electrostatic forces.

8.2.1 Hydration Forces

The immobile water molecules in the Stern layer are held in place by relatively strong ion–dipole, dipole–dipole, and hydrogen bonding interactions involving both surface sites and adjacent water molecules. The minimum thickness this immobile layer could have corresponds to the size of one water molecule (0.14 nm), which is equivalent to the first solvation shell of an ion in solution. However, on surfaces such as oxides, the ordering of water can extend out to 5 nm, or almost 30 water layers [2,4].

The viscosity of immobile water within the Stern layer is responsible for generating short-range *repulsive hydration forces*. When approaching objects encounter the outer Helmholtz plane, the viscous inner water layer must be moved out of the way for the objects to get any closer to each other. The thickness and mechanical properties of the ordered water layer have been measured directly by approaching surfaces with a probe tip in a scanning probe instrument called the *interfacial force microscope*. Measurements performed on polyethylene oxide-based self-assembled monolayers [4] (Fig. 8.3) show that the hydration force is unusual in that it depends on the speed at which the tip approaches the surface. Faster approach speeds result in stronger hydration forces. This approach speed dependence is just what is expected for the displacement of a viscous liquid.

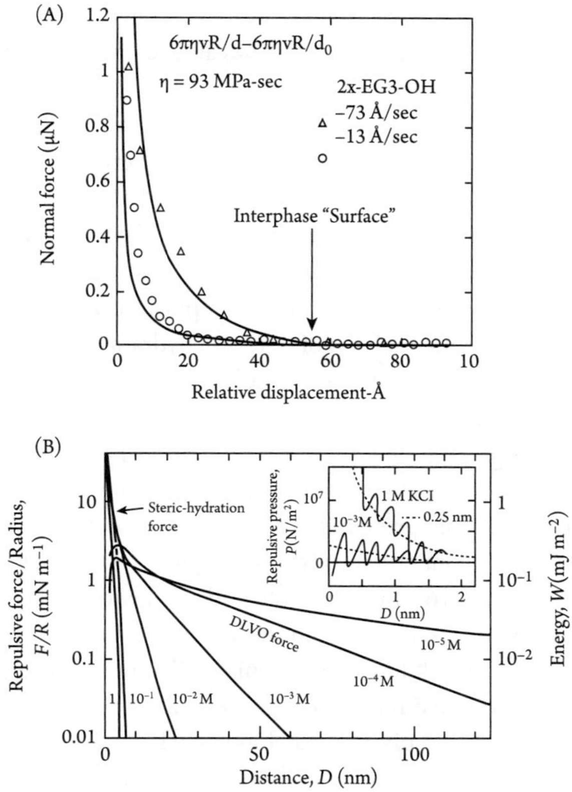

Figure 8.3 Repulsive hydration forces. (A) Force profiles between a tungsten tip and a surface functionalized with a short-chain polyethylene oxide monolayer at two different tip speeds. The speed dependence of the measured hydration force is indicative of the displacement of viscous water within the immobile Stern layer. (Adapted from Kim, H.I.; Kushmerick, J.G.; Houston, J.E.; and Bunker, B.C Viscous "Interphase" Water Adjacent to Oligo(ethylene glycol)-Terminated Monolayers. *Langmuir* **2003**, *19*, 9271–9275, with permission.) (B) Force curves between two mica plates interacting in 10^{-3} M KCl electrolyte. The inset shows theoretical computations for the same system. (SOURCE: Reproduced from Israelachvili, J.; *Intermolecular and Surface Forces, 2nd Ed.*; Academic Press: New York, 1992, Fig. 13.10, p. 279, with permission.)

The difficulty of draining a high-viscosity liquid between a scanning probe tip and a surface has been characterized using a modified Stokes equation:

$$F_{\eta} = \frac{6\pi\eta\upsilon R^2}{D} \tag{8.1}$$

In Eq. 8.1, F is the force experienced between the objects, η is the viscosity of the water in the Stern layer, υ is the approach speed, R is the tip radius, and D is the tip–substrate separation distance. On polyethylene oxide functionalized surfaces, the magnitude of the repulsive hydration force is consistent with an ordered interfacial water layer that is 4 to 5 nm thick and has an effective viscosity that is 10^6 to 10^7 times that of bulk water (Fig. 8.3).

The connection between surface solvation and hydration forces is reinforced by studies performed in the surface-force apparatus between anionic mica surfaces separated by concentrated aqueous salt solutions [5]. At sufficiently high concentrations, the Stern layer no longer represents a layer of ordered water, but a layer of hydrated cations. Under these conditions, hydration forces now exhibit strong oscillations between net attraction and repulsion. These oscillations have been attributed to the sequential loss of waters of hydration around the adsorbed cations followed by the reorganization of solvation shells as the opposing mica surfaces in the apparatus are forced together.

As a final note regarding hydration forces, it appears that surface geometry plays a role in the extent and strength of water ordering at the interface. Such geometric effects have been probed in infrared studies of water ordering on self-assembled monolayers containing short polyethylene oxide chains adsorbed on gold and silver substrates [6]. Both the gold and silver surfaces are terminated by identical functional groups. However, ordering of water is detected on the gold substrate, but not on silver. It has been speculated that this is because the spacing of functional groups on gold represent a good lattice match for ice, whereas the matching on silver is poor. Epitaxial matching may be one reason why mica surfaces are effective at ordering water in both surface force experiments and in cloud seeding (see Chapter 18).

8.2.2 Van der Waals Interactions

The most common short-range interactions between all substances regardless of size are the *van der Waals* forces. Three separate interactions give rise to van der Waals forces [2]: (1) the normally dominant *dispersion* or *London forces* associated with transient polarization of electron clouds present in all atoms, (2) the polarization of electron clouds in one atom or molecule in response to a fixed dipole in another (corresponding to *induction* or *Debye interactions*), and (3) the attraction and reorientation of two permanent dipoles with respect to one another, which are called *orientation* or *Keesom interactions*. All three forces are generally attractive. Because all three forces between atoms or molecules follow a $1/r^6$ distance dependence, the three contributions are readily combined to describe the van der Waals force. The $1/r^6$ distance also explains why van der Waals interactions are only effective at short separation distances (0.2–10 nm).

Global equations for van der Waals interactions have been derived that describe the attraction between any two bodies regardless of size or shape by integrating over all interactions between all atoms within the object. For example, the potential energy, W, between two spheres with different radii resulting from van der Waals forces is given by [2]

$$W = \frac{-A}{6D} \cdot \frac{r_1 r_2}{(r_1 + r_2)} \quad \textbf{(8.2)}$$

Here, r_i are the sphere's radii, A is an important empirical coefficient, called the Hamaker constant (which has units of energy), and D is the interparticle distance. Hamaker coefficients do not depend strongly on the geometry of the system and range from around $1.4 \cdot 10^{-19}$ J for Al_2O_3 to $4.3 \cdot 10^{-19}$ J for the more polarizable TiO_2. Similar equations can be derived for slurries of colloids with shapes more complicated than spheres, or for interactions between colloids and large-scale surfaces.

8.2.3 Electrostatic Interactions

Oxide surfaces almost always have either a positive or negative surface charge. As a result, electrostatic forces often play a dominant role in the interactions between oxides immersed in water. Species of opposite and like charges attract or repel each other, respectively. Charged oxide surfaces can attract objects ranging in size from simple ions to colloidal particles to macroscopic objects. At the smallest scale, two point charges experience forces between each other such that the interaction energies scale as a function of $1/D$ compared with the $1/D^6$ dependence seen for van der Waals forces. This means that electrostatic forces are felt at much longer distances than van der Waals forces.

If the only relevant charges mediating the interactions between particles are the fixed charges present in the native oxide surface, electrostatic forces are maximized. However, if the solution contains ions, these ions sense the surface and arrange themselves in such a way as to try to neutralize that charge. Some ions can enter the immobile Stern layer, resulting in a reduction in the number of fixed surface charges present. Above the outer Helmholtz plane, in the diffuse part of the electrical double layer, ions are free to move and arrange themselves in such a way that ionic concentration gradients are established in solution. For example, if the oxide surface is anionic, the concentrations of soluble cations are greatest near the surface and decay into solution, whereas the concentrations of soluble anions are lowest at the surface and increase with distance from the surface. Although these ionic clouds neutralize some of the fixed surface charge, the concentration gradients within the clouds create charge imbalances that also generate electrostatic forces. Ionic concentration gradients in the diffuse double layer create electrostatic potentials that extend as far from the surface as the gradients.

As with van der Waals forces, the net interaction potential and force experienced between two objects represents the average overall charges that are present. Once again, these forces depend on the sizes and shapes of the interacting objects. For example, a continuum equation that describes the electrostatic energy (W) between two identical spherical particles of radius *r* at low surface potentials is given by [2]

$$W = \frac{2\pi r\sigma^2 \exp(-\kappa D)}{\kappa^2 \varepsilon\varepsilon_o} \tag{8.3}$$

Here, the surface charge density, σ, is in units of coulombs (C) per square meter. This is the electrostatic equivalent of Eq. 8.1 that describes van der Waals interactions between spherical particles. It relates to the electrochemical potential in volts at the immediate surface, ψ_o, via $\sigma = \varepsilon\varepsilon_o\kappa\Psi_o$. In these relations, κ is a decay constant with units of inverse length, D is the particle–particle separation distance, ε is the dielectric constant of the fluid, and ε_o is the permittivity of free space.

Although an approximation, Eq. 8.3 embodies several key features regarding electrostatic interactions involving all oxide surfaces in aqueous electrolytes, regardless of size or shape. First, the energy of interaction depends on the charge density on the surface. The higher the charge density, the greater the electrostatic force. Second, the energy between the particles decays exponentially away from a charged shape. The decay constant, κ, has units of inverse distance and is referred to as the *Debye length* when expressed as $1/\kappa$. The Debye length corresponds to the distance over which ionic concentration gradients are produced in solution, allowing charged objects to sense one another. If spheres are separated to $D >> 1/\kappa$, they neither repel nor attract one another because of electrostatics.

In aqueous electrolyte solutions, $1/\kappa$ varies with ion concentrations, or ionic strength. The Debye length depends on the square root of the ionic strength because each ion helps to screen the presence of other charged objects in solution. For an aqueous electrolyte containing monovalent cations and anions (such as NaCl), $1/\kappa = 0.304/[\text{NaCl}]^{1/2}$, where the units in $1/\kappa$ are in nanometers [2]. In a very dilute solution, the Debye length can extend many nanometers into the solution (for $[\text{NaCl}] = 10^{-5}$ M, $1/\kappa = 90$ nm), but it collapses when the solution becomes salty (for $[\text{NaCl}] = 10^{-1}$ M, $1/\kappa = 0.9$ nm).

8.2.4 Net Colloidal Interactions and the DLVO Theory

Under many conditions, the net interaction potential between two colloidal oxide particles is the sum of the electrostatic interaction potential and the van der Waals force (ignoring hydration forces). The interplay and manipulation of these attractive and repulsive forces to stabilize colloids is at the heart of the famous DLVO theory, which is named for its authors: Derjagian, Landau, Verwey, and Overbeek [7, 8]. Particles of like charge repel each other at large

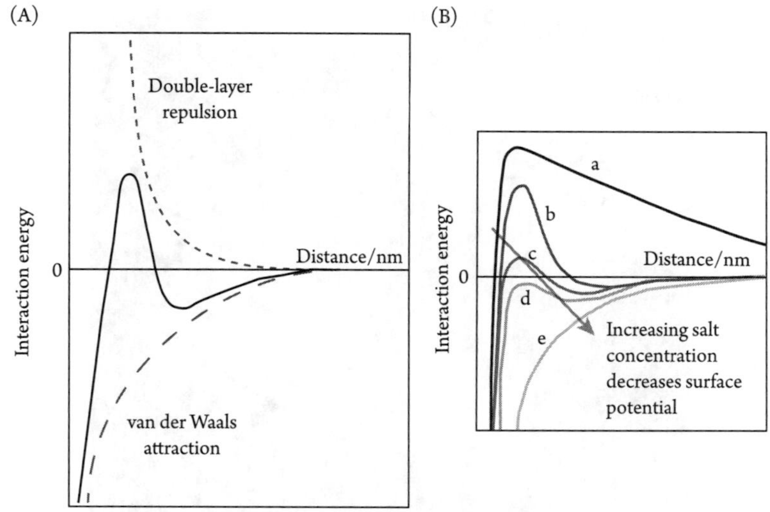

Figure 8.4 The electrical double-layer interaction energy as calculated via the Derjagian, Landau, Verwey, and Overbeek (DLVO) theory. (A) The graph shows how a complex interaction potential is generated as a sum of van der Waals attractions and electrostatic repulsions. (B) The curves show how the DLVO potential varies as a function of increasing the salt content of the electrolyte or decreasing the surface charge. Note the potential can vary from being strictly repulsive at a low salt content to strictly attractive at a high salt content. Adapted from Israelachvili, J. *Intermolecular and Surface Forces, 2nd ed.*; Academic Press: New York, 1992; Figure 12.12, p 248, with permission.

separation distances (where the electrostatic force is dominant) and are attracted to each other at very close separation distances (where van der Waals forces dominate). At intermediate separation distances, the sum of the forces can exhibit a complex distance dependence that, in some cases, exhibits a secondary attractive minimum (Fig. 8.4). This minimum is encountered at high ionic strength and leads to particle aggregation. Interactions between particles can either be attractive or repulsive depending on the surface charge (e.g., opposite charges attract whereas like charges repel each other) and electrolyte concentrations.

Changes in the Debye length with salt concentration lead to colloid chemistry that occurs on such a large scale that it can be observed by satellites [9], representing some of the most globally conspicuous chemistry on Earth (Fig. 8.5, see also Plate 14). Most of us have seen the enormous sediment deltas that accumulate at the mouths of rivers such as the Mississippi, the Nile, or the Amazon. These rivers are made up of snowmelt and rainwater and have very low salt concentrations. As a consequence, the Debye length on charged particles such as clays is long. Because the particles have similar charges, electrostatic repulsions

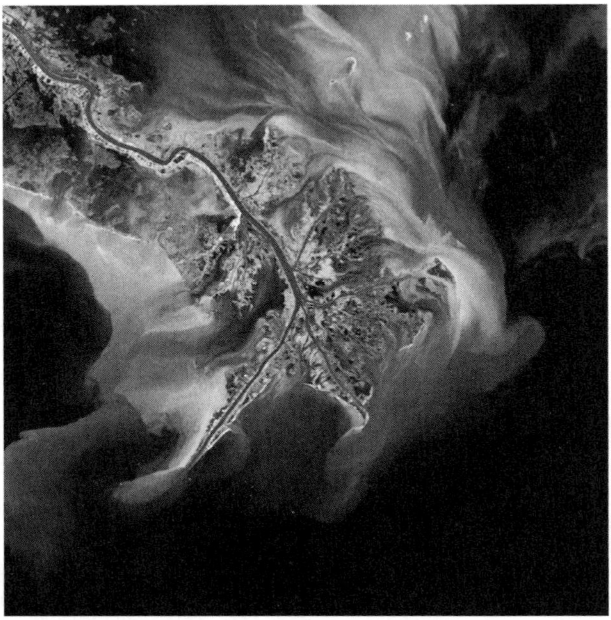

Figure 8.5 A Landsat 7 satellite view of the mouth of the Mississippi River showing the massive aggregation and sedimentation of suspended clay particles that occurs when colloids suspended in rivers with a low-salt content solutions encounter the ocean, which is a solution with a high salt content. The salt concentrations cause the double layer to shrink sufficiently for attractive van der Waals forces to flocculate the clays, causing them to settle. (see Plate 14)
SOURCE: NASA/Science Source with permission.

between colloidal clay particles are extensive, and the colloids stay in suspension. Suspended oxide particles are the reason why the Mississippi River is called "The Big Muddy." However, when these muddy colloidal suspensions reach the ocean, salt concentrations increase to around 0.7 M. The increased salinity causes the repulsive double layers to collapse, allowing the suspended particles to agglomerate as a result of attractive van der Waals interactions. This flocculation process leads to the precipitation of the semisolid masses that form extensive delta sediments (see Chapter 18).

8.3 THE CHEMICAL ORIGIN OF SURFACE CHARGE

The colloidal interactions that are sensitive to oxide structure are dominated by electrostatics. Electrostatics, in turn, are controlled by charged surface sites as mediated by dissolved ions that interact at the oxide surface. Three classes of ion interactions affect oxide surface charge and influence the structure and properties of the electrical double layer: (1) the adsorption of potential-determining ions, (2) the influence of indifferent electrolytes, and (3) the presence of specifically adsorbed ions.

8.3.1 The Interplay between Surface Sites and Potential-Determining Ions

The most important ions in the electrical double layer are the *potential-determining ions* that dictate local charge distributions at the immediate oxide surface. The potential-determining ions for oxides are protons [10], because native oxide surfaces are terminated with Brønsted bases and acids, as anionic oxygen sites, hydroxyl groups, and coordinated water molecules. The charge on each of these sites (based on local bonding configurations [see Chapters 4–6]) is changed by 1 unit in response to the adsorption or desorption of a proton. The net surface charge is the sum of the charges on all exposed surface sites, taking into account the degree of site protonation. The protonation state of each site is determined by the pK_a of the oxygen anion and the solution pH. However the surface equilibrium constants differ from those in solution in that they vary with the magnitude of the surface charge (see Chapter 6 for details). For example, it takes more energy to place a proton on a positively charged surface than a negatively charged surface. Thus, equilibrium constants reported in the colloid chemistry literature and those discussed here are often defined for uncharged surfaces.

Most oxide surfaces contain distributions of local bonding configurations for oxygen anions at the oxide–solution interface, resulting in a broad range of pK_a values as well as relatively smooth surface charge versus pH curves (see Chapters 6 and 10). Here, we provide a brief example to illustrate the coupling between site structure, site protonation, and surface charge. Imagine a hypothetical aluminosilicate surface containing two types of anionic surface sites, each with a surface concentration of 4 sites/nm^2: (1) $>Si-O^-$ sites (η^1-O, each with a formal charge of -1) and (2) $>Si-O-Al<$ sites (μ_2-O, each with a formal charge of -0.25). In strongly basic solutions, all sites are deprotonated, resulting in a net surface charge of $-5/nm^2$ (i.e., -4 for the four $>Si-O^-$ sites and -1 for the four $>Si-O-Al<$ sites, each with a charge of -0.25). This surface charge corresponds to a surface charge density σ in Eq. 8.3 of 0.8 C/m^2. When the solution pH is lowered to mildly basic or neutral conditions, all the $>Si-O^-$ sites become protonated to form neutral silanol ($>SiOH$) groups. The remaining anionic sites are the $>Si-O-Al<$ sites with a net surface charge density of $-1/nm^2$. Under mildly acidic conditions, it becomes possible to start protonating the anionic $>Si-O-Al<$ sites and converting them into $>Si-OH-Al<$ sites with a formal charge of $+0.75$.

At the point where one-quarter of the >Si–O–Al< sites are protonated, the net charge on the surface becomes zero (i.e., $[(3)(-0.25)+(1)(+0.75)]/\text{nm}^2$. This pH is defined as the point of zero charge, or PZC (see Chapters 5 and 6) [11]. As the pH is lowered below the PZC, the surface eventually assumes a net positive charge of $+3/\text{nm}^2$ when all the >Si–O–Al< sites are protonated. In strongly acidic solutions, it is possible to protonate the silanol groups further to create bound nonbridging waters: $>Si(OH_2)$ groups, each with a formal charge of +1. This results in the maximum possible positive charge density for our hypothetical surface of $+7/\text{nm}^2$. In practice, even higher surface charge densities can be generated if the oxide reacts with water to create a hydrated surface layer of finite thickness containing charged sites similar to those found at the immediate surface.

The PZC marks the pH boundary between solution environments in which the oxide is charged either positively or negatively. For pH values below the PZC, the surface is charged positively, whereas above the PZC, the surface is anionic. As shown earlier, although the net charge is zero at the PZC, both cationic and anionic sites may both be present. Another critical feature of the PZC is that it is largely independent of the ionic composition of the contacting solution (with the exception of pH). The PZC can often be inferred by performing potentiometric acid–base titrations on high-surface area oxide powders [12, 13].

A wide range of surface charge measurements have been performed on aqueous slurries containing all the major oxide and hydroxide phases. An excellent compilation of these measurements [10] (condensed in Table 8.1) confirms that, in most cases, (1) the PZC corresponds to the pH of minimum solubility for a given oxide and (2) the distribution of charged sites on the surface is similar to the distribution of charges in the soluble hydrolysis products at the same pH. The correlation between the PZC and hydrolysis product chemistry arises as a result of the structural and chemical similarities between the hydrolysis products and specific sites in the metal oxide surface. Trends in PZC for oxides containing different cations mirror those seen in their corresponding hydrolysis products. Oxides containing low-valence (i.e., +2) cations have high PZC's and have cationic surfaces except at high pH. As the cation valence increases, the PZC drops. Oxides containing +5 and +6 cations tend to be anionic except in the most acidic solutions. When there is a divergence between the PZC and the pH of minimum solubility, the difference is often a result of variations in the degree of hydration or impurity levels in the oxide. Structure also plays an important role. For example, in γ-Al_2O_3, this divergence may be associated with the fact that Al(III) metals in the oxide surface exist in both octahedral and tetrahedral coordination geometries to oxygens, but the soluble hydrolysis products are dominated by octahedrally coordinated metals.

Table 8-1. Isoelectric Points of Representative Oxides

Phase	pH of Minimum Solubility	Isoelectric Point
+2		
MgO	>12	12.4
$Mg(OH)_2$	>12	12
$Fe(OH)_2$	10.5	12
$Co(OH)_2$	11	11.4
NiO	10	10.3
$Ni(OH)_2$	10	11.1, 12
CdO	11.5	10.4
$Cd(OH)_2$	11.5	>10.5
BeO	—	10.2
$Pb(OH)_2$	10.7	9.8, 11
CuO	9.7	9.5
$Cu(OH)_2$	9.7	9.4, 7.6
ZnO	9.5	8.7–9.3
HgO	8	7.3
$Mn(OH)_2$	12.5	7
SnO	6.4	6.6
+3		
La_2O_3	—	10.4
Y_2O_3	—	9.0
Cr_2O_3	8.5	7
Fe_2O_3	8	5–8.6
FeOOH	8	6.7–7.4
$Fe(OH)_3$	8	8.5
Al_2O_3	6.5	5–9
AlOOH	6.5	8–9
$Al(OH)_3$	6.5	5–9
+4		
ThO_2	>6	9
UO_2	5	5–6.6
TiO_2	—	4.7
SnO_2	—	4.5, 5.5, 7.3
ZrO_2	5	4–6.7
MnO_2	—	4–4.5
SiO_2	—	2
+5		
Sb_2O_5	—	<0.4
+6		
WO_3	3.5	0.5

8.3.2 Surface Charge Measurements in Indifferent Electrolytes

Although it is relatively easy to monitor the protonation and deprotonation of oxide surfaces via acid–base titrations, it is not easy to measure the resulting charge state of either specific acid–base sites or the net charge at the immediate oxide surface. Most techniques for measuring surface charge rely on the coupling between electric fields and motion. For example, electrophoretic mobility [14] and electrosonic amplitude measurements [15,16] detect the velocity of colloidal particles under the influence of static or oscillating electric fields, respectively. Streaming-potential measurements [17] determine the voltages generated by moving fluids relative to charged surfaces. In all these cases, the parameter measured is the ζ (zeta) potential [11, 18] induced when the fixed portion of the electrical double layer is moved relative to the diffuse part that extends into the bulk fluid. For oxides, the ζ potential varies with solution pH just as the actual surface charge does. The ζ potential assumes a value of zero at a pH called the *isoelectric point* (iep) when the charge density at the slipping plane is zero [15, 18]. For example, electrophoretic mobility measurements performed on α-Al_2O_3 particle suspensions indicate that the iep for this particular slurry is around pH 8 (Fig. 8.6).

Although a useful parameter, the ξ potential is not identical to the potential generated between the actual oxide surface and the solution. It is a hydrodynamic parameter. Not only is the ξ potential measured at a stand-off distance from the actual surface (i.e., at the slipping plane), but also it can be biased by the adsorption of ions into the immobile Stern layer. For monovalent ions, the interactions between dissolved cations and anionic surface sites can be sufficiently weak that ionic concentrations within the Stern layer are largely independent of the ionic

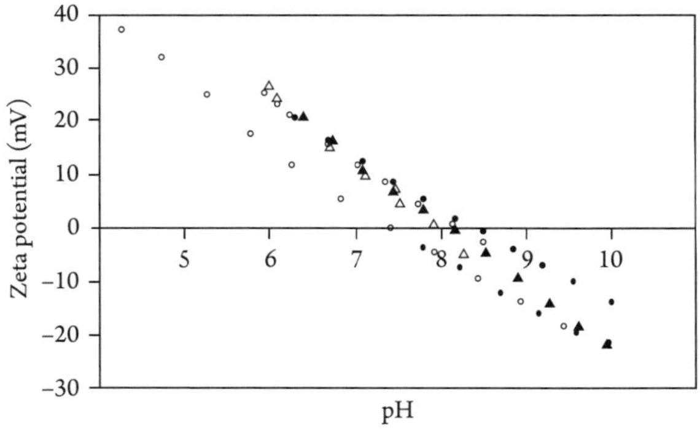

Figure 8.6 The zeta potential of alumina as a function of pH. The symbols refer to different total NaCl concentrations in solution. A positive zeta potential is indicative of a positive surface charge. The isoelectric point of the suspension is around pH 8. Adapted from Yopps, J. A.; Fuerstenau, D. W. The Zero Point of Charge of Alpha-Alumina. *J. Colloid Sci.* **1964**, *19*, 61–71, with permission.

strength of the solution. Solutions containing such ions are called *indifferent electrolytes*. For indifferent electrolytes, the iep shows little dependence on electrolyte concentration, and the iep and PZC are identical. However, it is important to note that these indifferent electrolytes typically dominate charge distributions and potentials involving the diffuse part of the electrical double layer.

8.3.3 The Role of Specifically Adsorbed Ions on Surface Charge

Some ions, called *specifically adsorbed ions*, can penetrate the Stern layer, interact with ion-exchange sites in the oxide surface, and partially neutralize the net charge as sensed at the outer Helmholtz plane. Although such interactions are often weaker than those involving protons, such ion-exchange reactions (see Chapter 10) are sufficient to influence the surface charge. Ions within the Stern layer are in equilibrium with the bulk solution. This means that as the ionic concentration in solution increases, more ions present in the immobile Stern layer. Correspondingly, the effective surface charge decreases, and the iep is shifted to higher pH values. In contrast to the behavior observed in indifferent electrolytes, this means that the iep can exhibit a significant dependence on ionic strength. It also means that the iep is no longer identical to the PZC.

Specifically adsorbed divalent cations can have a much greater impact on surface charge than monovalent cations. For example, it has been shown that on oxides such as silica, the addition of divalent cations such as Ca^{2+} can neutralize the anionic surface charge completely and induce coagulation [19]. It appears that a single Ca^{2+} ion interacts only with a single nonbridging oxygen anion in the silica surface, resulting in a charge reversal on the site from -1 to $+1$. Similar charge reversals are seen as a result of Ca^{2+} adsorption on Al_2O_3 [20] (Fig. 8.7).

8.3.4 The Impact of Hydrolysis Products on Surface Charge

Charge reversals and large shifts in the iep are even more pronounced for adsorbed cations that can undergo hydrolysis reactions. The complexities in surface charge behavior associated with cation hydrolysis can be predicted on the basis of hydrolysis diagrams. For example, consider the behavior of colloidal silica suspended in Co(II) solutions [21] (Fig. 8.7). In the absence of Co(II), silica has an iep at pH 2 and is anionic throughout most of the normal pH range of water. However, in the presence of Co(II), silica exhibits wild swings in surface charge behavior. Under acidic to neutral conditions, silica retains its anionic surface charge, although this charge is reduced slightly via ion pairing with $[Co(H_2O)_6]^{2+}$. However, when the pH is increased to pH 8, where the (1,0) species of Co(II) (corresponding to $[Co(H_2O)_6]^{2+}$) is hydrolyzed to (1,1) (corresponding to $[CoOH(H_2O)_5]^{+}$ see Chapter 5), the dissolved Co(II) condenses with surface

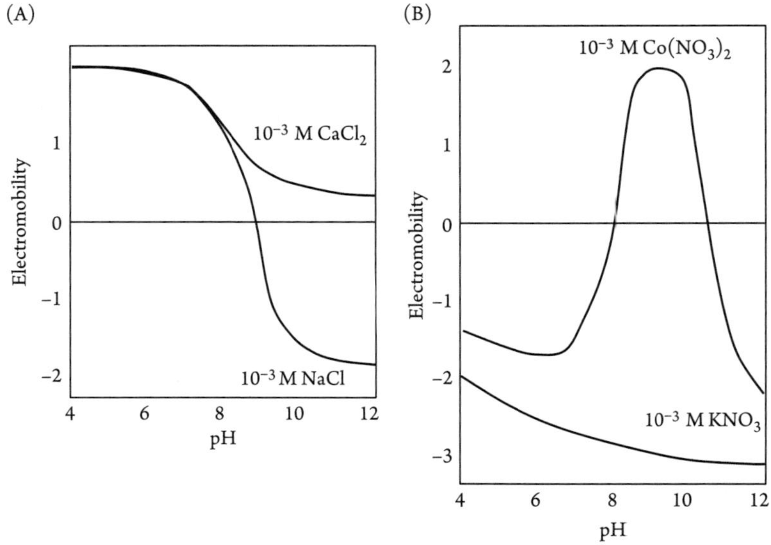

Figure 8.7 (A) The surface charge on alumina particles as a function of pH in the presence of an indifferent electrolyte (NaCl) and an electrolyte containing specifically adsorbed cations (i.e., the Ca^{2+} in $CaCl_2$). Although native alumina exhibits a point of zero charge, PZC, of around pH 9, Ca^{2+} adsorption results in a charge reversal, allowing alumina to exhibit a positive surface charge even above pH 12. (Adapted with permission from Huang, C.- P.; Stumm, W. Specific Adsorption of Cations onto Hydrous γ-Al_2O_3 Surface. *J. Colloid Interface Sci.* **1973**, *43*, 416.) (B) The surface charge on silica particles in an indifferent electrolyte (KNO_3) compared with the charge observed in the presence of a hydrolysable cation (Co^{2+} in $Co(NO_3)_2$). The dramatic shifts in surface charge for Co^{2+} are all associated with the charge generated on adsorbed hydrolysis products. (Adapted with permission from James, R. O.; Healey, T. W. Adsorption of Hydrolyzable Metal Ions at the Oxide—Water Interface. I. Co(II) Adsorption on SiO_2 and TiO_2 as Model Systems. *J. Colloid Interface Sci.* **1972**, *40*, 42–52.

silanol groups. When this happens, the silica surface is terminated by a layer of cobalt hydrolysis products rather than by silanol groups. The covered silica surface now exhibits surface charge characteristics that are representative of $Co(OH)_2$, which has an iep of pH 11. This means that at pH 9, the surface has a large positive surface charge rather than the large negative charge found on native silica surfaces. However, when the solution pH exceeds the iep of $Co(OH)_2$, deprotonation of the adsorbed hydrolysis products reestablishes an anionic surface charge.

8.3.5 The Influence of Anionic Ligands on Oxide Surface Charge

The impact of multivalent anion adsorption on surface charge mirrors that exhibited by multivalent cations, with a few minor variations. Multivalent anions can reverse the surface charge via physisorption. However, such anions are often quite nucleophilic and can sometimes replace oxygen anions at the surface via ligand exchange reactions (Chapter 6). Regardless, the acid–base chemistry of free functional groups on the adsorbed ligands can define the surface charge (Fig. 6.17) [22]. For example, adsorption of dicarboxylates such as the oxalate ion [23] (Fig. 8.8) can shift the iep of Al_2O_3 from around pH 9 down to around pH 2. Here, one of the carboxyl groups binds to the surface, whereas the other now terminates the surface and defines its acid–base chemistry and surface charge characteristics. However, although pK_a values for adsorbed ligands can be used as guidelines for predicting when adsorption occurs and what the resulting pK_a of the surface is, direct coordination to metal cations within the oxide can shift the pK_a values of the adsorbate every bit as much as coordination to cations influences the pK_a values of oxygen anions (see Chapter 4). For example, in solution, pK_1 and pK_2 for oxalic acid are 1.0 and 3.8, respectively, whereas the apparent pK_a value for the pendant carboxyl group of oxalic acid adsorbed on alumina is ~2.

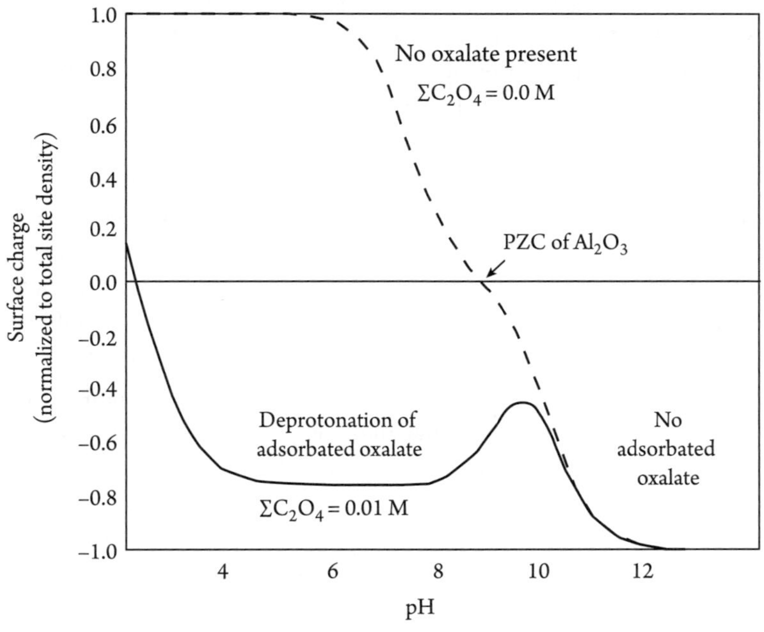

Figure 8.8 The surface charge on Al_2O_3 particles in the absence (dashed curve) and presence (solid curve) of oxalic acid and deprotonated oxalate anions as a function of solution pH. Above the pK_a of the oxalic acid, oxalate adsorption renders the surface anionic.

SOURCE: Bunker, B. C. Unpublished results Pacific Northwest National Laboratories, *ca* 1995.

8.4 THE IMPACT OF AGGREGATION ON THE PROPERTIES OF COLLOIDAL SLURRIES

The magnitude and sign of the surface charge are critical parameters that control the nature and extent of particulate agglomeration within typical oxide slurries. In turn, the agglomeration state dictates many key slurry properties, including viscosity and sedimentation. Such properties not only determine how slurries need to be handled, but also establish the solution conditions that must be maintained in practical applications, such as in ceramic processing. Here, we first describe the behavior of radioactive colloids in nuclear tank wastes and in the environment to illustrate why understanding colloidal aggregation is so important. We then highlight how agglomerates form, what structures they have, and how such structures influence specific properties using single-component tank waste simulants such as boehmite and various phases of $Fe(OH)_3$.

8.4.1 The Ultimate Challenge in Colloidal Oxide Processing: Nuclear Tank Waste

The importance of aggregation is highlighted by examining what is required to process the millions of gallons of radioactive wastes stored at reactor fuel-rod processing facilities such as the Hanford site in Richland, Washington, or Savannah River in South Carolina. At the Hanford site alone [24, 25], 94 million gallons of wastes are stored in more than 150 large tanks, many with a capacity of 1 million gallons. The total radioactive content of these tanks is stunning, amounting to $\sim 2 \cdot 10^8$ Ci at Hanford and $\sim 4 \cdot 10^8$ Ci at Savannah River. Some tanks are so radioactive that tank inspections must be performed using robots, and solution components are radiolyzing into gases such as H_2. The contents of each tank started their existence as aqueous solutions containing soluble species from dissolved fuel rods. However, as a result of evaporative water losses, as well as changes in temperature, pH, and species concentrations, each tank has become supersaturated with respect to a range of oxide phases, resulting in massive precipitation (Fig. 8.9, see Plate 15). The resulting solutions contain high concentrations of radioactive colloidal particles made from metal hydrolysis products suspended in the brine, along with dissolved radioactive fission products.

Transmission electron micrographs [26] (Fig. 8.10) show that the colloids in tank wastes are diverse in terms of both composition and particle size. Aluminum-containing precipitates represent the largest fraction of insoluble material present, including boehmite (AlOOH) and gibbsite ($Al(OH)_3$. Other major phases present include hydrated oxides of uranium, zirconium, iron, manganese, chromium, nickel, and titanium, as well as zeolites and clay minerals. Primary particle and agglomerate sizes span five orders of magnitude. The smallest primary particles, such as ZrO_2 and $Fe(OH)_3$, are only 3 to 6 nm in diameter,

Figure 8.9 Photographs taken inside two different tanks of radioactive waste at the Hanford site in Richland, Washington. Radioactivity levels are so high in many tanks that images such as these are usually taken robotically. (see Plate 15)

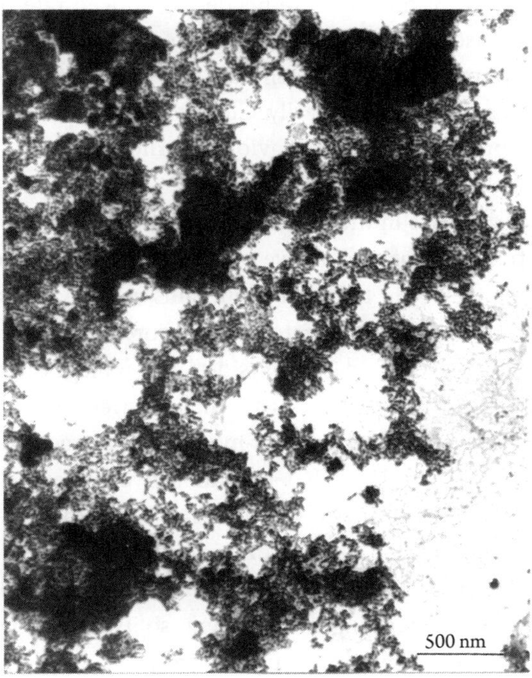

Figure 8.10 A transmission electron micrograph of sludge taken from tank 105-AW at the Hanford site in Richland, Washington. The primary cations present in the oxide particles shown include Al, Zr, Si, Fe, Cr, Ca, and U. Note that many of the primary particles in this sludge have a long dimension of around 25 nm.
SOURCE: LaFemina, J. P. *Tank Waste Treatment Science: Report for the First Quarter FY 1994*, Report TWRSPP-94-00; Pacific Northwest National Laboratory: Richland, WA, 1995; p. 6.)

followed by 100 nm to 1 μm plates and or needles of AlOOH, and large (20 μm) primary particles such as gibbsite. Agglomerates are typically 1 to 10 μm in diameter, but can exceed 100 μm in size. Eventually, slurries containing all these particles have to be removed from the tanks and pumped to central processing facilities for conversion into glass or ceramic waste forms.

Colloidal particles also form naturally in fuel-rod storage facilities such as the Hanford site's K-Basins [27]. Any work done to manipulate the now-decomposing fuel rods in these basins disturbs the reactive colloidal sediments that have formed via fuel rod corrosion to produce highly opaque clouds of suspended particles that reduce the visibility to the point where additional work cannot be done for days or even weeks. At the other extreme, minor changes in solution conditions can convert diluted tank waste slurries that have a viscosity similar to that of water into highly viscous fluids with mechanical properties comparable with those of toothpaste. When such a transition occurs while pumping slurries from one area to another, miles of pipeline can become completely clogged with highly radioactive materials. At processing facilities, solid–liquid separations such as sedimentation or filtration are needed to concentrate the waste before high-temperature waste conversions. However, some sludges do not settle at all, and all water is entrapped within the colloidal mass, rather than forming a supernatant liquid that can be removed easily. Clearly, waste engineers need to have a scientific basis for predicting slurry behavior as a function of factors such as solids loadings, pH, and other processing parameters.

Environmental disasters in the former Soviet Union bring home the need to understand the transport of radioactive colloids in natural waters. The Mayak fuel rod processing facility in the Ural Mountains experienced a series of catastrophic accidents and explosions that released more than 100 million Ci of radiation into the local rivers and countryside [28]. Villages near the site were abandoned and are still unoccupied, as are major agricultural and industrial facilities. Lake Karachai, a small, open, man-made waste storage pond, was eventually filled with more than 120 MCi of radioactive materials, exceeding the combined inventory of storage tanks at both the Hanford and Savannah River sites in the United States [29]. At one point in its history, the lake went dry, and radioactive dust was distributed by the wind. Even when wet, Lake Karachai was not isolated from the surrounding environment, so its radioactive inventory entered the groundwater and migrated outward from the site. In such events, the radioactive species of greatest concern are extremely fine and dispersed colloidal particles of uranium and plutonium oxides [30–33], which are not readily impeded by surrounding soils and that have been dispersed by natural forces. Next, we highlight a few of the key factors that lead to aggregation or dispersion phenomena of such colloidal oxides and discuss how such interactions mediate waste properties.

8.4.2 The Roles of Size and Surface Charge on Agglomeration

In the absence of repulsive forces, colloidal particles aggregate as a result of attractive van der Waals forces. Large particles tend to move independently because they have sufficient inertia to break up weak particle-particle interactions at contact points. In contrast, smaller primary particles agglomerate extensively. For example, light-scattering measurements performed on suspensions containing $Al(OH)_3$ (gibbsite) or AlOOH (boehmite) particles [34] (Fig. 8.11) show that when 10 μm particles are present, the primary particle radius, r, and the agglomerate radius, R, are identical. However, for the smallest particles shown, $R/r = 30$. This means that the average agglomerate is almost 4 μm in diameter and can contain more than 30,000 primary particles. For clay and other soil minerals (as well as tank wastes), these small agglomerates aggregate to form even larger porous arrays ranging from micrometers to millimeters and even tens of centimeters. Let us now focus on the aggregation of these submicron particles.

A calculated correlation between surface charge and agglomeration state [35] is shown in a "stability diagram" for colloidal AlOOH as a function of both pH and ionic strength in Figure 8.12. The curve defines the boundary between solution regimes in which aggregation should be either suppressed (labeled *stable*) or promoted (labeled *unstable*). The absolute minimum stability and maximum aggregation is expected at the iep (pH 8–9), where there are no electrostatic repulsions to counteract attractive van der Waals interactions. Similarly, suspensions are expected to be unstable and to flocculate when the ionic strength of the solution exceeds around 0.1 M as a result of the collapse of the electrical double layer. Although the boundary between "stable" and "unstable" solutions is more diffuse than indicated in Figure 8.12, it does provide useful guidelines for predicting

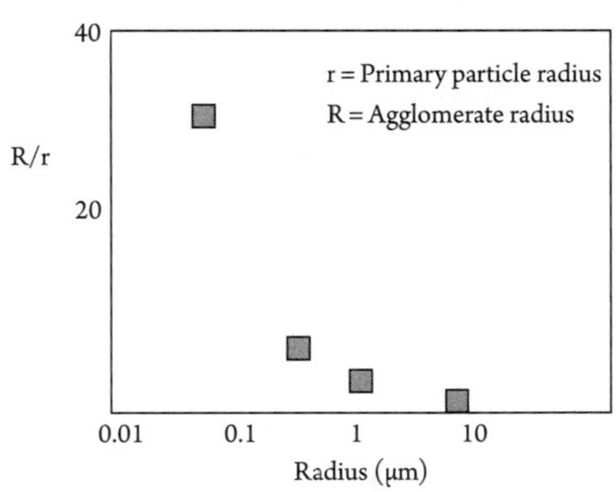

Figure 8.11 Agglomerate sizes as a function of primary particle size for suspensions containing either gibbsite or boehmite particles (Bunker, B. C. Unpublished results, Pacific Northwest National Laboratories, *ca* 1995).

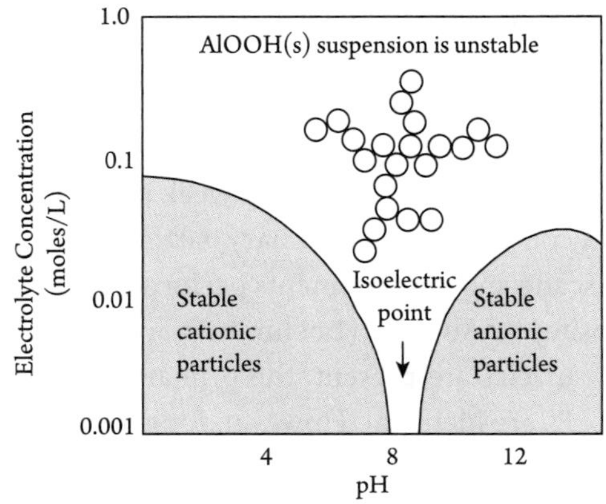

Figure 8.12 The consequences predicted for how the surface charge and salt concentration on boehmite should influence colloidal stability. Regions in the diagram labeled "stable" and "unstable" represent conditions under which particles are expected to be dispersed or agglomerated, respectively. Adapted from LaFemina, J. P. *Tank Waste Treatment Science: Report for the Second Quarter FY 1995*, Report *TWRSPP-95-008*; Pacific Northwest National Laboratory: Richland, WA, 1995; Figure 28, p. 66.

when agglomeration will occur. For example, light-scattering measurements done on 1 vol% suspensions of 50 nm boehmite particles [34] indicate that between pH 3 and 4, the suspensions are largely dispersed (with the agglomerate radius R corresponding to the primary particle radius r of 50 nm). At pH 7 and 8, huge agglomerates (70 μm) form. At pH 13, all particles have a large negative charge, but because a pH-13 solution corresponds to an ionic strength of 0.1 M, aggregation is still observed (R = 1.5 μm).

8.4.3 Agglomerate Structures: The Importance of Fractal Geometry

Size is not the only critical parameter dictating the properties of colloidal suspensions. The *nature* of such agglomerates can be completely different depending on the magnitude of the interactions between the particles. In most cases, including in soils, the aggregates are not compact, close-packed structures; they form what are called *fractal objects* [35]. The mass (M) of a fractal aggregate scales with the aggregate radius of gyration, r, according to

$$M = k \cdot r^{\gamma} \tag{8.4}$$

Here, the exponent, γ, is defined as the fractal dimension.

Agglomerate structures fall between two extremes, each with its own theoretical values for γ [36–40]. In *reaction-limited aggregation,* particles are weakly associated with each other and can rearrange into denser structures after the initial agglomerate forms. In the ideal case of spheres that rearrange into in a perfect hexagonal close-packed array, $\gamma = 3$. In other words, the volume occupied by the agglomerate is proportional to R^3 just as it is for a completely dense sphere. However, perfect packing is almost never achieved, resulting in $\gamma < 3$. The other extreme of aggregation is called *diffusion-limited aggregation*. Here, particles interact so strongly with one another they stick irreversibly wherever they land. No particle rearrangements occur within the growing aggregate. Diffusion-limited aggregates exhibit wispy, open structures $(\gamma \sim 1.7 - 1.8)$ that resemble snowflakes, soot, or the wispy macromolecular aggregates of organic acids and phenolic compounds found in soils (see Chapter 18).

The physical properties of colloidal suspensions are controlled by both the sizes and fractal dimensions of suspended aggregates. The critical parameter that controls most properties is the volume occupied by particles within a given suspension. The volume fraction Φ_a occupied by spherical particles within a single agglomerate is given by the expression [39, 40]

$$\Phi_a = \left(\frac{R}{r}\right)^{\gamma-3} \tag{8.5}$$

Here, R is the agglomerate radius, r is the primary particle radius, and γ is the fractal dimension. This relationship is represented graphically in Figure 8.13. For example, if an oxide precipitate with a primary particle radius of 10 nm forms diffusion-limited aggregates having radii of 1 μm, $R/r = 100$, and $\gamma = 1.8$, resulting in $F_a = 0.004$. The volume occupied per unit mass of oxide for these aggregates is $1/\Phi_a$ relative to a completely dense spherical particle. This means our fractal aggregate of small primary particles occupies 250 times more space and has an effective density that is roughly 250 times less than that of the dense particle. The importance of both the low density and high volume of these fractal agglomerates is immediately apparent when one examines the parameters that control slurry properties.

8.4.4 Slurry Viscosity and Gelation

Although the viscosity of water is relatively high for a liquid (1 centipoise [cp] at room temperature), the addition of particles to water can increase this viscosity by many orders of magnitude. At low solids loadings, particles are free to move past one another, the presence of particles has a relatively small impact on viscosity, and the fluid exhibits ideal Newtonian fluid flow behavior. However, as more particles are added, the motion of one particle is constrained by other particles, resulting in non-Newtonian flow. Eventually, at a point called the *percolation threshold* (or *gel point*), particles form a continuous network that spans the entire solution volume

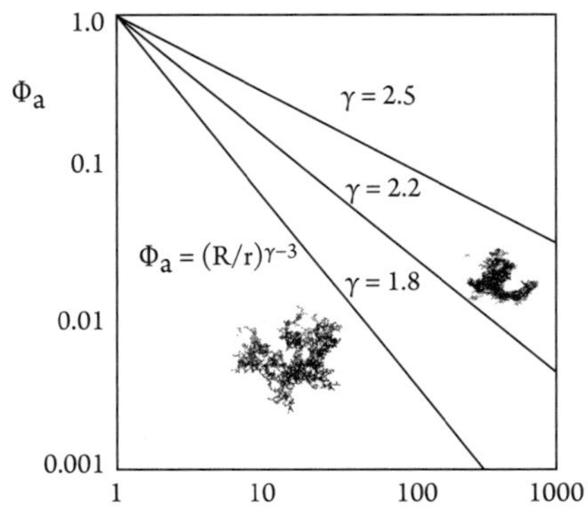

Figure 8.13 The calculated volume fraction occupied by particles within fractal agglomerates as a function of fractal dimension. For a fully dense particle, the occupied volume fraction is one.
SOURCE: SEM images of colloidal gold aggregates reproduced from Weitz, D. A.; Oliveria, M.; Fractal Structures Formed by Kinetic Aggregation of Aqueous Gold Colloids. *Phys. Rev. Lett.*, **1984**, *52*, 1433–1436; Weitz, D. A.; Huang, J. S.; Lin, M. Y.; and Sung, J. Limits of the Fractal Dimension for Irreversible Kinetic Aggregation of Gold Colloids *Phys. Rev. Lett.*, **1985**, *54*, 1416–1419, with permissions.)

[41]. For hard spheres, this percolation threshold occurs at a solids loading of 16 vol%. Above the percolation threshold, particle motions require displacements and rearrangements within this particle network. Eventually, all available space is occupied by particles to form a semi-solid, random close-packed array, and viscosities become very high indeed.

The relationship between solids loading and viscosity for "noninteracting" particles [43, 43] is shown in Figure 8.14 for both spheres and plates (represented by boehmite):

$$\eta = \eta_o \left(1 - \frac{\Phi}{\Phi_m}\right)^{-[n]} \Phi^m \tag{8.6}$$

Here, η_o is the viscosity of the bulk fluid, Φ is the volume fraction occupied by particles, Φ_m is the maximum volume fraction that can be occupied by the particles (64% for a random array of close-packed spheres), and [n] is a geometric factor that accounts for the particle shape. As shown in Figure 8.12, large viscosity excursions are not expected for spheres until the volume fraction exceeds around 40%. For the thin plates representing layered hydroxides and clay minerals, the viscosity is expected to rise more quickly (at around 10 vol%), simply because thin sheets or rods form continuous networks at lower solids loadings than spheres.

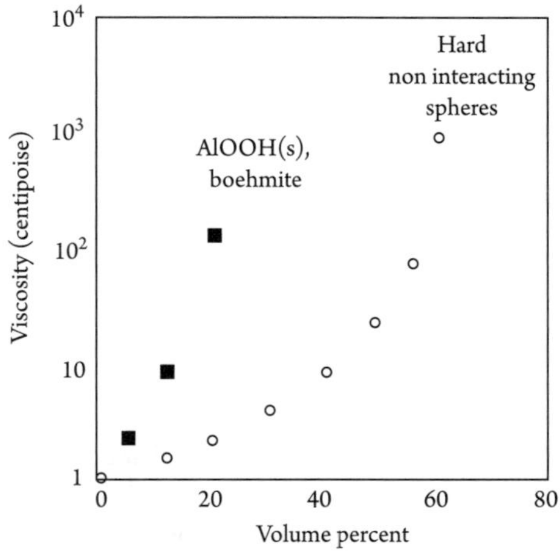

Figure 8.14 Calculated viscosity curves for both spherical and plate-like particles as a function of the volume fraction occupied by noninteracting particles. In the absence of agglomeration, the viscosity of slurries containing spherical particles is actually quite low until the load of solids exceeds 50 vol%. Plate-like particles form interacting networks at lower solids loading values.
SOURCE: Rector, D. R. Pacific Northwest National Laboratory, Richland, WA. Unpublished results, *ca* 1995.

If spherical suspensions do not start exhibiting high viscosities until solids loadings exceed 40 vol%, then why is the viscosity of oxide suspensions of concern? The reason is that oxide slurries containing submicron particles almost never contain noninteracting spheres, but do contain highly interacting fractal agglomerates. Here, the gel point depends not on the packing of ideal spheres, but on the packing of fractal objects, where the primary particle size, agglomerate size, and fractal dimension are all critical parameters (Table 8.2). The lowest gel points are encountered for large aggregates containing small primary particles with low fractal dimensions. Although a suspension of dense spheres with a diameter of 100 μm is expected to exhibit a gel point at 16 vol%, spherical 100-μm diffusion-limited aggregates containing 10-nm primary particles reach the gel point at only 0.32 vol%. This compresses the x-axis shown in Figure 8.14 by a factor of 50, which means that enormous viscosities can be experienced at solids loadings of only 1 vol%.

It now becomes apparent why surface chemistry and chemical interactions between particles as expressed via DLVO theory are so critical for colloidal suspensions. If particle interactions are dominated by electrostatic repulsions, agglomeration is suppressed because the particles cannot get close enough to each other to be within range of attractive van der Waals interactions. Here, viscosities can be maintained at low values up to relatively high solids loadings.

Table 8-2. DIMENSIONS AND PROPERTIES OF FRACTAL AGGLOMERATES

d (μm)	D (μm)	Fractal Dimension	Gel Point (vol%)	Settling Rate (cm/h)
0.01	100	1.8	0.32	0.07
1	100	1.8	2.7	18
0.01	100	2.5	3	72
1	100	2.5	10	3000
0.01	1	2.5	10	0.036
0.01	10	2.5	7	1.4
0.01	100	2.5	3	72
100	(100)	316	6000	

Shown are the diameters (d) of the primary particles that comprise the agglomerate, the agglomerate diameters (D), and their fractal dimension. The resulting physical properties of the agglomerates include the gel point for forming a continuous particulate network (in vol%) and the rate at which the agglomerates are predicted to settle (assuming a primary particle density of 2 g/cm^3). The last entry represents a single 100-μm particle for comparison. Note how much more difficult colloidal slurries are to handle as the primary particle decreases, the agglomerate size increases, and the fractal dimension decreases.

However, if particle interactions are attractive, which occurs under conditions of high salinity or when the iep is approached, huge slurry viscosities can result.

The effect of interparticle interactions on viscosity is illustrated by the impact of pH on the behavior of boehmite suspensions [34] (Fig. 8.15). When boehmite particles are positively charged in dilute electrolyte solutions (between pH 3 and 6), they repel each other and undergo minimum aggregation (or flocculation). Even at 10 vol%, such suspensions exhibit a viscosity of less than 10 cp. Increasing the pH to that of the iep (near pH 9) eliminates the electrostatic repulsion and allows extensive aggregation to occur. This aggregation increases the slurry viscosity from a few cp to more than 20,000 cp. Above the iep, where the particles assume a negative charge, agglomerates are disrupted and the viscosity drops. Notice that the extent of agglomeration and the viscosity are both high under extreme pH conditions. This result is because electrolyte concentrations below pH 2 and above pH 12 are sufficiently high to collapse the electrical double layer, just as we saw in Fig. 8.5 for river deltas (see Plate 14).

The retention of high slurry viscosities depends on the mechanical stability of the suspended fractal aggregates. Because oxide aggregates tend to have small primary particles and low fractal dimensions, they tend to be fragile. Models for the breakup of flocs in moving fluids [44] suggest that the stress dependence for the steady-state floc size is of the form

$$R = b \cdot \alpha^{(\gamma-1)/2p(\gamma-3)} \tag{8.7}$$

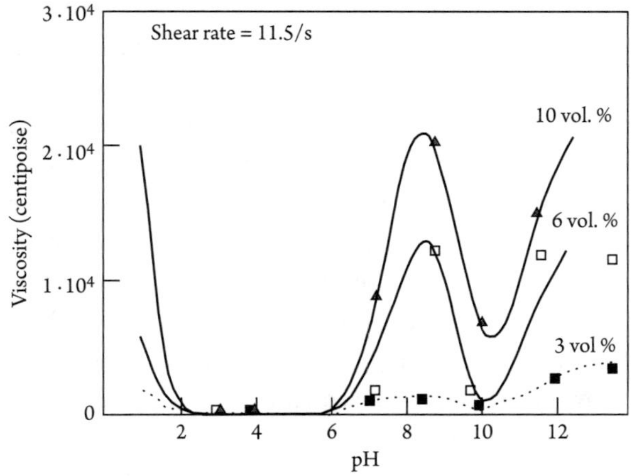

Figure 8.15 Measured viscosities for boehmite slurries as a function of solids loading (volume percent) and pH. When particles are dispersed, viscosities are low. However, when agglomeration occurs (at the isoelectric point near pH 9 and at high electrolyte concentrations (i.e., at high and low pH), viscosities increase by a factor of as much as 20,000. Adapted from LaFemina, J. P. *Tank Waste Treatment Science: Report for the Second Quarter FY 1995, Report TWRSPP-95-008*; Pacific Northwest National Laboratory: Richland, WA, 1995; Figure 34, p 77.

Here, α is the shear rate (per second), b is a constant, p is a constant that is inversely proportional to the agglomerate's shear strength (where the moduli and critical stress leading to agglomerate rupture vary with volume fraction), and γ is the fractal dimension. As a result of stress-induced aggregate disruption, oxide suspensions almost always exhibit *shear-thinning behavior* [45] (Fig. 8.16). This behavior means that as shear rates increase, slurry viscosities decrease as a result

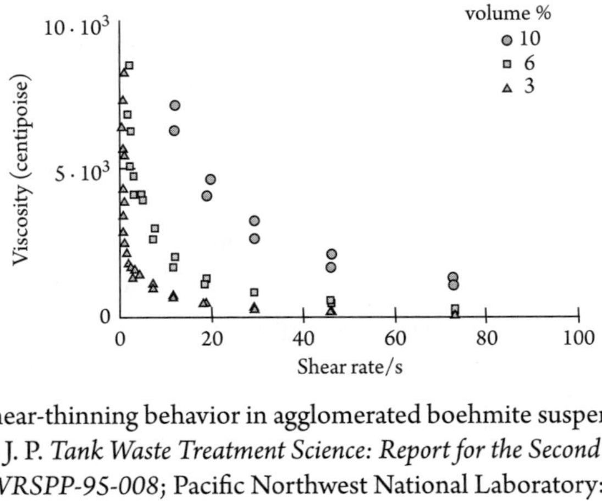

Figure 8.16 Shear-thinning behavior in agglomerated boehmite suspensions. Adapted from LaFemina, J. P. *Tank Waste Treatment Science: Report for the Second Quarter FY 1995, Report TWRSPP-95-008*; Pacific Northwest National Laboratory: Richland, WA, 1995; Figure 37, p 81.

of the motion-induced disruption of aggregate structures. For model boehmite suspensions, viscosities can be lowered by at least an order of magnitude just by increasing the shear rate. However, high shear rates must be maintained to break up the agglomerates faster than they reform. For the boehmite example, the viscosities achieved by pumping at moderate shear rates reduce aggregate sizes from 10 μm down to 2 μm in less than 1 hour. However, these aggregates regrow back to 10 μm if allowed to sit for 3 hours. If the desire is to maintain low-viscosity solutions, oxide suspensions must be kept moving. In stagnant solutions, gel networks can form that eventually generate such high viscosities that stirring or agitation cannot be reinitiated. In the case of nuclear waste slurries, such gel formation can have catastrophic consequences.

The disruption of oxide agglomerates can also be catastrophic. In Nature, the horrors associated with earthquakes and massive volcanic ash flows called *lahars* are amplified when agglomerated particle networks are disrupted. The shaking of wet soil leads to *thixotropic*, or liquefaction, behavior during which the soil viscosity drops precipitously as a result of shear thinning. For example, soil liquefaction can cause the support pillars that hold up a freeway to sink by several feet in a matter of seconds. In the 1989 Loma Prieto earthquake, soil liquefaction caused the collapse of many homes in the fashionable districts of San Francisco because they were built on wet, loosely compacted fill from San Francisco Bay [46]. Liquefaction also caused buried rubbish to pop to the surface when it suddenly became buoyant. Ironically, this newly exposed rubbish was debris that was originally buried during the famous 1906 San Francisco earthquake. This liquefaction alone caused ~2% of the total earthquake damage [47].

8.4.5 Sedimentation Behavior

In colloidal processing, the desire is either to keep 100% of the particles in suspension indefinitely or to remove all particles completely from solution as quickly and efficiently as possible. The cheapest and easiest removal strategy often involves sedimentation, during which gravity performs all the work. In the removal of particles from solution via sedimentation, the two parameters of interest are the sedimentation rate and the volume the sediment cake occupies after sedimentation is complete (which is related to sediment density).

The fractal nature of oxide aggregates is a critical factor controlling the rate and extent of sedimentation. The common expression used to predict the sedimentation rate V for a particle in a fluid under the force of gravity or centrifugation is Stokes' Law [48]:

$$V = \frac{2gr^2\left(d_p - d_l\right)}{9\eta} \tag{8.8}$$

Here, g is the force of gravity (or a larger constant associated with centrifugal acceleration), r is the particle radius, d_p is the particle density, d_l is the density of the fluid, and η is the fluid viscosity (Table 8.2). This expression shows that dense, sand-size (100 μm) particles should fall through aqueous solutions at a rapid rate. Smaller particles settle more slowly. Sedimentation rates also depend on density. The more rapid sedimentation rates of dense minerals such as gold particles relative to most oxides explain why there is a happy concentration of pay dirt in a miner's pan or sluice. Lower density fractal objects settle at lower rates than dense particles (Fig. 8.17). In fact, a 100 μm diffusion-limited boehmite aggregate containing 0.1 μm primary particles settles at a rate that is almost 10,000 times slower than a dense 100 μm particle (see Table 8.2). These sedimentation behaviors lead to the sorting of particles observed at the mouth of a river, where coarse sand is always deposited upstream of finer silt and clay particles.

As with viscosity, sedimentation rates depend on whether single particles or aggregates encounter each other. As particles sink to the bottom, they become more concentrated and the effective fluid density increases. Sedimentation rates drop as a result of hydrodynamic interactions, osmotic pressure effects, and, last, interparticle contact. For example, Figure 8.18 shows the sedimentation behavior of $Fe(OH)_3$ suspensions as a function of initial solids loading [34]. The 8 wt% solution is 2.4 vol% in $Fe(OH)_3$. This solution exhibits almost no settling because it starts out containing a dense particulate network that is above the percolation threshold. After settling, there is essentially no supernatant liquid to remove, and the efficiency of a solid–liquid separation based on sedimentation

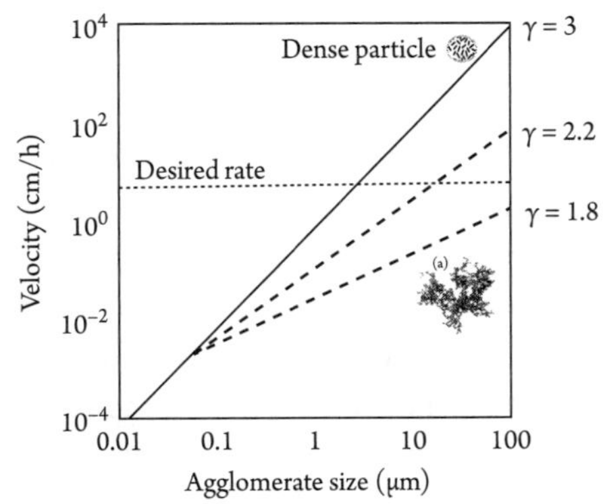

Figure 8.17 Calculated sedimentation velocities for suspensions as a function of agglomerate size and fractal dimension. Adapted from LaFemina, J. P. *Tank Waste Treatment Science: Report for the Second Quarter FY 1995, Report TWRSPP-95-008*; Pacific Northwest National Laboratory: Richland, WA, 1995; Figure 30, p 70. SOURCE: SEM images of colloidal gold aggregates reproduced from Weitz, D. A.; Oliveria, M.; Fractal Structures Formed by Kinetic Aggregation of Aqueous Gold Colloids. *Phys. Rev. Lett.*, **1984**, 52, 1433–1436, with permission.

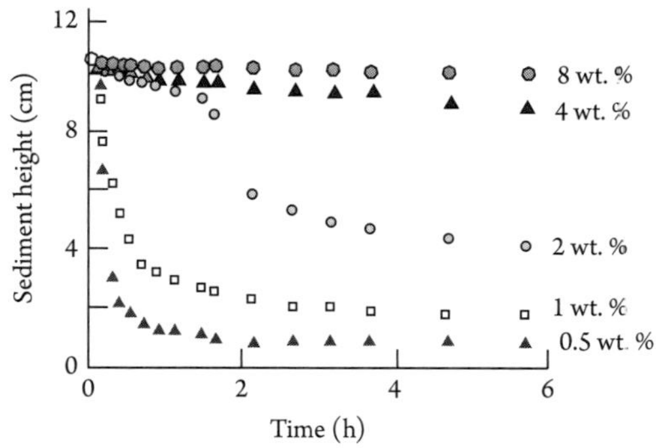

Figure 8.18 Observed sedimentation rates for $Fe(OH)_3$ suspensions in pH 12 aqueous solutions as a function of solids loadings (in weight percent). Adapted from LaFemina, J. P. *Tank Waste Treatment Science: Report for the Second Quarter FY 1995, Report TWRSPP-95-008*; Pacific Northwest National Laboratory: Richland, WA, 1995; Figure 43, p. 90.

is negligible. At the other extreme (below the percolation threshold), individual agglomerates each settle at their own rate, leaving behind a large fraction of suspended particles after the larger agglomerates settle out. In this case, neither extreme is desired or results in efficient solid–liquid separations, illustrating the challenges associated with handling submicron colloidal oxide suspensions.

When continuous particle networks form, further sedimentation is possible only if the forces exerted on the networks (by gravity or centrifugation) exceed the yield stress $P_y(\Phi)$ of the particle network [49]:

$$P_y(\Phi) = c\left[\left(\frac{\Phi}{\Phi_p}\right) - 1\right]^n \quad \textbf{(8.9)}$$

Here, c is the compressive yield strength of the network, Φ is the volume fraction of particles in the sediment at equilibrium Φ_p is the volume fraction corresponding to the percolation threshold, and n is a constant (normally set equal to 4). When Φ_p and c are known, sediment packing densities can be predicted for a wide range of processing conditions. For example, a sediment with $c = 0.3$ gm/cm^2 and $\Phi_p = 1$ vol% (representative of $Fe(OH)_3$ agglomerates in pH-12 water) should exhibit equilibrium sediment densities of only 1, 4.3, 7.7, and 8.6 vol% when subjected to pressures representative of lab tests (i.e., 0.3 g/cm^2 or the weight resulting from gravity in the tests shown in Fig. 8.18), sedimentation in a Hanford waste tank (30 g/cm^2), centrifugation at 20,000 *g* (600 g/cm^2), or pressure filtration (980 g/cm^2), respectively. Even the densest of these sediment compacts is more than 90 vol% water. These results illustrate that for any solid–liquid separation or compaction process involving submicron oxide particles, it is absolutely critical to suppress particle aggregation and agglomeration phenomena.

8.5 CERAMIC PROCESSING: THE ROLE OF ORGANIC ADDITIVES

Colloidal oxides are used in an enormous range of commercial products, including inks, paints, protective coatings, and even pharmaceutical materials. In all these applications, the processing and properties of oxide slurries are controlled by colloidal interactions and aggregation phenomena. Controlling these interactions is so important to so many technologies that entire industries are based on producing specific additives to manipulate slurry properties, including (1) flotation agents that are used to separate specific minerals from raw milled ore [50], (2) flocculating agents to promote aggregation and sedimentation [51], (3) dispersants to lower slurry viscosities and maximize particle packing in processes such as slip casting [52], and (4) binders to assist in ceramic processes such as extrusion, tape casting, thick-film printing, and injection molding [53].

Next, we discuss how colloidal interactions are manipulated in the processing of ceramic materials (see [54,55] for comprehensive discussions of ceramic processing). The first step in the processing of many ceramics involves the production of oxide particles via the nucleation and growth processes described in Chapter 7. The resulting colloidal slurries can sometimes be used directly in subsequent processes. More commonly, the oxides are removed from solution via processes such as sedimentation and filtration to create an oxide powder. This powder is redispersed in an aqueous solution that serves as a processing fluid used to form relatively dense and low-water content powder compacts called *green bodies*. The green body has the desired shape for the ultimate ceramic and sufficient structural integrity to survive subsequent handling. The green body is later calcined, which involves heating to remove all remaining water, organic materials, and other volatile impurities. Finally, the calcined material is sintered or hot-pressed at temperatures approaching the oxide's melting point to fuse the oxide particles together, remove porosity, and produce a dense ceramic product.

Colloidal chemistry is of critical importance in the formation of all green bodies. Dilute (typically less than 5 vol% solids) suspensions are used to form green bodies using techniques including sedimentation, centrifugation, pressure filtration, and slip casting. More concentrated high-viscosity suspensions are used in processes such as tape casting, extrusion, and injection molding. Many of these processes involve the use of organic additives. Next, we describe the interactions of organic additives with oxide surfaces, followed by detailing some highlights of how such additives are used in specific ceramic processing operations.

8.5.1 Surfactants and Other Organic Additives

The most extensive classes of molecular species in both technology and Nature that impact interparticle interactions are organic molecules and polymeric species called *surfactants* [56]. Every day, we use surfactants called soaps to remove dirt in the form of colloidal oxides from our skin and clothes. Typical surfactant molecules contain both hydrophilic ionic groups that can bind to oxide surfaces via either electrostatic or ligand exchange interactions, and hydrophobic groups such as hydrocarbon chains. These hydrophobic groups sometimes promote surfactant adsorption resulting from solvation effects (i.e., the hydrophobic groups are insoluble in water). A more detailed discussion of surfactants is provided in Chapter 9, where surfactants are major tools in the arsenal that organisms use to create complex oxide nanostructures.

Engineers often use surfactants both to stabilize and aggregate colloidal particles [57]. In addition to electrical double-layer forces, additional forces that can be manipulated in water include (1) attractive solvation forces that arise when hydrophobic surfaces are attracted to one another as a means for escaping from water and (2) entropic forces [2]. Entropic forces arise when approaching surfaces begin to constrain the equilibrium conformations of polymer chains within confined spaces. Tethered chains that are being compressed or forced to interpenetrate resist approach, resulting in what is called *steric stabilization*. Untethered chains in the solution phase can retain chain conformations by leaving the confining space between approaching surfaces. The resulting polymer concentration gradients and osmotic pressure can generate *attractive depletion forces*. Unfortunately, the incredible diversity of surfactants and their associated interaction potentials cannot be used in conjunction with the radioactive tank wastes discussed earlier because most organic compounds experience radiation damage (see Chapter 13), which not only destroys the surfactant, but also generates potentially explosive hydrogen gas.

Depending on surfactant composition, as well as which components of a particular surfactant bind to the surface, is it possible to create oxide surfaces that express almost any desired surface functional group. Surfactant adsorption can (1) reverse the charge on oxide surfaces from positive to negative or negative to positive, or (2) transform hydrophilic (water-loving) oxide surfaces into surfaces that are hydrophobic and repel water enthusiastically. Another advantage of the use of surfactants is that they can interact with multiple components in the same slurry and make particles with dissimilar surface chemistries appear to be identical.

An example illustrating the complexities of surfactant–oxide interactions is provided by the adsorption of polyacrylic acid (PAA) (a polymer containing pendent carboxyl groups) on ZrO_2 [58]. Electrosonic amplitude measurements show that native ZrO_2 has an iep of around pH 6. Native PAA is anionic above its pK_a value of around 5, but represents a neutral, hydrophobic molecule when all its carboxyl groups have been protonated (below pH 3). At low pH, the hydrophobic nature of PAA promotes the adsorption of the surfactant onto ZrO_2 surfaces, rendering the ZrO_2 surfaces hydrophobic as well. Adsorption also shifts the effective pK_a of the carboxyl groups within PAA from 5 down to around 3. The presence of anionic carboxyl groups imparts a negative charge to the ZrO_2 surface above pH 3. Because of the mutual repulsion of the carboxyl groups, the PAA also swells to generate enhanced steric repulsion between particles. The combination of steric and electrostatic repulsion is called *electrosteric stabilization*. Above the iep of native ZrO_2 (at pH 6), the underlying oxide becomes anionic and starts to repel the PAA overlayer. This repulsion eventually stimulates the PAA to desorb, reestablishing the native anionic ZrO_2 surface.

Surfactant–oxide interactions have key roles in agriculture. Polymers such as polyacrylamide gels are used to retain water in soil and reduce mineral losses during irrigation. A tiny amount of soluble polymer surfactant added to flowing irrigation waters flocculates clays and prevents their loss by entrainment. These molecules commonly adsorb to the surface of the oxide minerals via their hydrophobic tails, sometimes with the aid of $[Ca(OH_2)_6]^{2+}$ as a cationic bridge to the mineral. The molecules have several hydrophilic functional groups that bind to the surfaces of multiple particles. The resulting tangle of linked clays settles quickly rather than being carried off by flowing water. Such polymers are the artificial equivalents of naturally occurring organic matter that plays key roles in the structure and chemistry of soil (see Chapter 18).

8.5.2 Mineral Flotation: The Role of Classic Surfactants

The simplest surfactants contain a hydrophilic, ionic head group that is attached to one or more hydrocarbon tails. Such surfactants can be used to separate one class of colloidal particles from another. A classic example of such separations involves the "froth flotation" methods used to extract minerals safely from milled ore on a massive scale. Here, head groups in surfactants such as xanthates, dithiophosphates, petroleum sulfonates, and fatty amines are designed specifically to adsorb onto the surfaces of colloidal sulfide minerals, whereas the tails render the resulting surfaces slightly hydrophobic. The particles are then suspended in a solution agitated by bubbles of air. The hydrophobic particles escape their incompatible aqueous environment by migrating to air–water interfaces at the surfaces of the bubbles, where they are trapped. The buoyant bubbles rise to the tank surface, carrying the desired sulfide minerals with them. The floating sulfide ore is then skimmed away from the undesired waste rock. The froth flotation process is much safer and less expensive than alternative processes that involve roasting the ore followed by leaching with dangerous and toxic chemicals such cyanide.

8.5.3 Powder Separations via Sedimentation: The Role of Flocculating Agents

Sedimentation and centrifugation processes represent simple methods for performing crude particle separations based on particle size or surface chemistry. Simple flocculation can be achieved in single-component systems by adjusting the solution pH to the iep. Here, strong van der Waals interactions dominate, resulting in the formation of diffusion-limited aggregates with a low fractal dimension. Such aggregates are ideal for the isolation of powders, because they are redispersed readily. However, such aggregates do not pack to form dense structures and are not recommended for the formation of defect-free green bodies. Multicomponent systems can exhibit more complex sedimentation behaviors. For example, for two oxides having well-separated iep's, it is possible to select pH conditions where some particles are positively charged whereas others are negative. Under such pH conditions, one component can coat another via electrostatic attraction or can form particulate bridges that induce multicomponent flocculation.

The simplest means of ensuring that all components undergo flocculation is to add salts such as NaCl to the solution to collapse all electrical double layers, as often happens in natural environments (see Fig. 8.5). However, many inorganic salts used for flocculation cannot be pyrolyzed and add impurities to the resulting sedimented powders. Salt addition typically results in the formation of denser reaction-limited aggregates than those produced at the iep because they arise from attractions in the weak DLVO secondary minimum (see Fig. 8.4). Here, sedimentation tends to form dense bodies that are difficult to redisperse. This dense sedimentation is preferred for green-body formation, but creates problems in the production of oxide powders.

Sedimentation processes are often facilitated by the use of organic flocculating agents. The most effective organic flocculating agents are high-molecular weight (i.e., 10^6-10^7 g/mole) polymeric species containing functional groups capable of forming strong bonds to the oxide of interest via either ligand exchange or electrostatic interactions. For example, anionic surfactants such as polyacrylamides are effective when used in conjunction with oxides with a positive surface charge. Flocculating agents are typically used at low concentrations (< 0.5 wt %) that facilitate what is called *bridging flocculation*. Here, single chains form robust connections between two or more particles. Low concentrations are required to ensure that the polymer coverage on oxide surfaces is sufficiently low to provide bare patches available for bridge formation.

8.5.4 Slip Casting and Filter Pressing: The Role of Dispersants

A green body is a stable particle compact that is a precursor to a fully fired, dense ceramic material. Green bodies are often produced by pouring oxide

Figure 8.19 Scanning electron micrographs of monodisperse silica particles that were assembled into a dense green body in the presence of dispersants.
SOURCE: Reproduced from Xia, Y.; Gates, B.; Yin, Y.; Lu, Y. Monodispersed Colloidal Spheres: Old Materials with New Applications *Adv. Mater.*, **2000**, *12*, 693–713, with permission.

slurries into porous molds and then wicking away as much water as possible through the mold. The simplest of these techniques is called *slip casting*, where the only source of pressure to induce water removal is the weight of the fluid contained in the mold. More commonly, external pressure sources are applied, including a vacuum (as in filtration), or a positive pressure, as in pressure filtration or filter pressing. In slip casting and related techniques, the goal is to maximize particle packing to create a green body that is as dense and defect free as possible. Particle compacts approaching theoretical packing densities can be achieved from slurries containing noninteracting particles, as illustrated by the production of *colloidal crystals*. Monodisperse particles can exhibit perfect packing, providing agglomeration is inhibited. For example, spherical particles of silica can be assembled into hexagonal close-packed arrays for optical applications [59] (Fig. 8.19).

The densest green bodies are most often obtained using suspensions in which particles repel each other and are totally dispersed. Although pH conditions can sometimes be selected to achieve such dispersion based on oxide surface charge alone, it is more common to achieve high packing densities using *dispersants*. The most effective organic dispersants for use in aqueous systems rely on either steric or electrosteric stabilization. Ironically, some of the best dispersants can also serve as flocculating agents. The transition between dispersion and flocculation behaviors is normally determined by the ratio of the solution concentrations of the polymer compared with the surface area of the oxide powder. At low concentrations, bridges can form. At high concentrations, oxide surfaces are saturated with the polymer and no space is left for bridge formation.

8.5.5 Tape Casting, Extrusion, and Injection Molding: The Role of Binders

Several forms of ceramic processing rely on slurries with a high solids loading (typically 10%–50%). High solids loadings minimize the amount of fluid removal required after casting. More important, solids loadings are optimized to create the slurry viscosity required to form desired green-body shapes. In tape casting, the oxide slurry is placed on a moving belt that passes under a stationary doctor blade. The result is the formation of a sheet or tape with a constant thickness. After casting and drying, the tape can be cut to near-final shape. Tape casting is often used in electronic ceramics to make multilayer structures, where some of the layers contain metal particles to generate electrode arrays within the stack (Fig. 8.20). In extrusion molding, pressure is used to force the slurry through a complex die, like

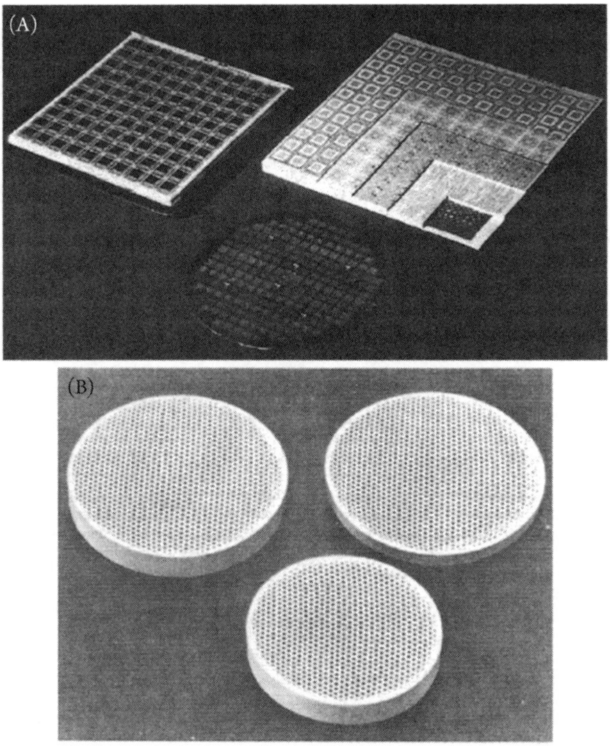

Figure 8.20 (A) Photograph of a multilayer ceramic module produced via tape casting and used in the packaging of microelectronic components. (B) Photograph of ceramic air distributors for heaters and hair dryers produced via extrusion of oxide powder and binder mixtures.
SOURCE: Images reproduced from Buchanen, R. C., Ed.; *Ceramic Materials for Electronics*; Marcel Dekker, Inc.: New York, 1986, pp. 343 and 406, with permission.

toothpaste being forced out of a tube. The extruded column has a cross-sectional structure that replicates the die. The resulting green body can then be cut into slices to generate the final product (Fig. 8.20). Injection molding involves using pressure to force an oxide paste into an impermeable mold. Removal of the mold leaves behind a complex three-dimensional oxide structure.

The optimization of slurry viscosity in green-body formation involves controlling both the oxide solids loading and the use of organic additives called *binders*. In the absence of binders, the mechanical properties of oxide slurries progress from low-viscosity Newtonian fluids to viscous non-Newtonian fluids to viscoelastic solids with increasing solids loadings. Binders are used to fine-tune these transitions at lower solids loadings. Binders are typically polymeric chains with a high molecular weight. These binders can form extended gel networks on their own at low (typically < 1 vol%) solution concentrations. However, as their name implies, most binders can also attach themselves in an irreversible fashion to oxide surfaces, forming strong bridges between adjacent particles. These bridges mean that binders form hybrid networks that incorporate both oxide particles and polymeric species. Artificial binders include polyvinyl alcohol, PAA, polyethylene glycol, and polyethylenimine. Many naturally occurring molecules found in soils and plants also function as binders (see Chapter 18). These molecules include methylcellulose, sodium carboxymethylcellulose, starches, dextrins, and sodium alginate.

References

1. Hirtzel, C. S.; Rajagopalan, R. *Colloidal Phenomena: Advanced Topics*; Noyes Publications: Park Ridge, NJ, 1985.
2. Israelachvili, J. *Intermolecular and Surface Forces*; Academic Press: London, 1992.
3. Evans, D. F.; Wennerstrom, H. *The Colloidal Domain Where Physics Chemistry, Biology, and Technology Meet*; VCH Publishers: New York, 1994.
4. Kim, H. I.; Kushmerick, J. G.; Houston, J. E.; Bunker, B. C. Viscous "Interphase" Water Adjacent to Oligo(ethylene glycol)-Terminated Monolayers. *Langmuir* **2003**, *19*, 9271–9275.
5. Pashley, R. M.; Israelachvili, J. N. Molecular Layering of Water in Thin Films between Mica Surfaces and Its Relation to Hydration Forces. *J. Colloid Interface Sci.* **1984**, *101*, 511–523.
6. Harder, P.; Grunze, M.; Dahint, R.; Whitesides, G. M.; and Laibinis, P. E. Molecular Conformation in Oligo(ethylene glycol)-Terminated Self-Assembled Monolayers on Gold and Silver Surfaces Determines Their Ability To Resist Protein Adsorption. *The Journal of Physical Chemistry B*, **1998**, 102(2), 426–436.
7. Derjaguin, B. V.; Landau, L. Theory of the Stability of Strongly Charged Lyophobic Sols and of the Adhesion of Strongly Charged Particles in Solution of Electrolytes *Acta Physiocochim URSS* **1941**, *14*, 633–662.
8. Verwey, E. J. W.; Overbeek, J. T. G. *Theory of Stability of Lyophobic Colloids*; Elsevier: Amsterdam, 1948.

9. Evans, D. F. and Wennerström, H. *The Colloidal Domain: Where Physics, Chemistry, Biology, and Technology Meet, 2nd Edition*, 672 pp. Wiley-VCH, New York.
10. Parks, G. A. The Isoelectric Points of Solid Oxides, Solid Hydroxides, and Aqueous Hydroxo Complex Systems. *Chem. Rev.*, **1965**, *65*, 177–198.
11. Hunter, R. J. *Zeta Potential in Colloid Science: Principles and Applications*; Academic Press: London, 1981.
12. Lodge, K. B. Techniques for the Measurement of Forces between Solids. *Adv. Colloid Interface Sci.* **1983**, *19*, 27–73.
13. Onoda, G. Y.; de Bruyn, P. L. Proton Adsorption at the Ferric Oxide/Aqueous Solution Interface I: A Kinetic Study of Adsorption. *Surf. Sci.* **1966**, *4*, 48–63.
14. Yopps, J. A.; Fuerstenau, D. W. The Zero Point of Charge of Alpha-Alumina. *J. Colloid Sci.*, **1964**, *19*, 61–71.
15. Greenwood, R. Review of the Measurement of Zeta Potentials in Concentrated Aqueous Suspensions Using Electroacoustics. *Adv. Colloid Interface Sci.* **2003**, *106*, 55–81.
16. O'Brien, R. W.; Midmore, B. R.; Lamb, A.; Hunter, R. J. Electroacoustic Studies of Moderately Concentrated Colloidal Suspensions. *Farad. Disc. Chem. Soc.* **1990**, *90*, 1–12.
17. Li, H. C.; de Bruyn, P. L. Electrokinetic and Adsorption Studies on Quartz. *Surf. Sci.* **1996**, *5*, 203–220.
18. Delgado, A. V.; Gonzolex-Caballero, F.; Hunter, R. J.; Koopal, L. K.; Lyklema, J. Measurement and Interpretation of Electrokinetic Phenomena. *J. Colloid Interface Sci.* **2007**, *309*, 194–224.
19. Iler, R. K. Coagulation of Colloidal Silica by Calcium Ions, Mechanism, and the Effect of Particle Size. *J. Colloid Interface Sci.* **1975**, *53*, 476–488.
20. Huang, C.- P.; Stumm, W. Specific Adsorption of Cations on Hydrous α-Al_2O_3. *J. Colloid Interface Sci.* **1973**, *43*, 409–420.
21. James, R. O.; Healey, T. W. Adsorption of Hydrolyzable Metal Ions at the Oxide–Water Interface. *J. Colloid Interface Sci.* **1972**, *40*, 42–52.
22. Kummert, R.; Stumm, J. The Surface Complexation of Organic Acids on Hydrous γ-Al_2O_3. *J. Colloid Interface Sci.* **1980**, *75*, 373–385.
23. Yoon, T. H.; Johnson, S. B.; Musgrave, C. B.; Brown Jr, G. E., *Adsorption of Organic Matter at Mineral/Water Interfaces: I. ATR-FTIR Spectroscopic and Quantum Chemical Study of Oxalate Adsorbed at Boehmite/Water and Corundum/Water Interfaces.* Geochimica et Cosmochimica Acta, 2004. **68**(22): p. 4505–4518.
24. Zorpette, G. Confronting the Nuclear Legacy: Part II. *Sci. Am.* **1996**, *274*, 88–97.
25. Gephart, R. E.; Lundgren, R. E. *Hanford Tank Clean Up: A Guide to Understanding the Technical Issues*, PNNL Report PNL-10773; Pacific Northwest National Laboratory: Richland, WA, 1997.
26. Liu, J.; Lumetta, G. J.; Virden, J. W.; Graff, G. L.; Colton, N. G.; Bunker, B. C.; Krumhansl, J. L. Characterization of Colloidal Phases in Tank Wastes. *Environ. Issues Waste Manage. Technol.* **1999**, *V*, 107–114.
27. Baker, R. B. *Summary Status of K Basins Sludge Characterization*, Report WHC-SD-SNF-TI-006; Westinghouse Hanford Company: Richland, WA, 1995.
28. Standring, W. J. F.; Dowdall, M.; Strand, P. Review of Dose Assessment Developments and the Health of Riverside Residents Close to the "Mayak" PA Facilities, Russia. *Int. J. Environ. Res. Public Heath*, **2009**, *6*, 174–199.

29. Goldman, M. The Russian Radiation Legacy: Its Integrated Impact and Lessons. *Environ. Health Perspect.* **1997**, *105*, 1385–1391.
30. Navikov, A. P.; Kalmykov, S. N.; Utsunomiya, S.; Ewing, R. C.; Horreard, F.; Merkulov, A.; Clark, S.; Tkachev, V. V.; Myasoedov, B. F. Colloid Transport of Plutonium in the Far-Field of the Mayak Production Association, Russia. *Science* **2006**, *314*, 638–641.
31. Lindenbaum, A.;, Westfall, W. Colloidal Properties of Plutonium in Dilute Aqueous Solution. *Int. J. Appl. Radiat. Isotopes* **1963**, *16*, 545–553.
32. Cleveland, J. M. Critical Review of Plutonium Equilibria of Environmental Concern. *Am. Chem. Soc. Symp. Ser.* **1979**, *93*, 321–338; Jenne, C. A., Ed.;
33. Giblin, A. M.; Bates, B. D.; Swaine, D. J. Laboratory Simulation Studies of Uranium Mobility in Natural Waters. *Geochim. Cosmochim. Acta* **1981**, *45*, 699–709.
34. (a) LaFemina, J. P. *Tank Waste Treatment Science: Report for the First Quarter FY 1995*, Report TWRSPP-94-118; Pacific Northwest National Laboratory: Richland, WA, 1995; (b) LaFemina, J. P. *Tank Waste Treatment Science: Report for the Second Quarter FY 1995*, Report TWRSPP-95-008; Pacific Northwest National Laboratory: Richland, WA, 1995.
35. Shih, W. H.; Shih, W. Y.; Kim, S. I.; Liu, J.; Aksay, I. A. Scaling Behavior of the Elastic Properties of Colloidal Gels. *Phys. Rev. A* **1990**, *42*, 4772–4779.
36. Meakin, P. Fractal Aggregates. *Adv. Colloid Interface Sci.*, **1988**, *28*, 249–331.
37. Schaefer, D. W.; Martin, J. E.; Wiltzius, P.; Cannell, D. S. Fractal Geometry of Colloidal Aggregates. *Phys. Rev. Lett.* **1984**, *52*, 2371–2374.
38. Schaefer, D. W.; Keefer, K. D. Structure of Random Porous Materials: Silica Aerogel. *Phys. Rev. Lett.* **1986**, *56*, 2199–2202.
39. Weitz, D. A.; Oliveria, M. Fractal Structures Formed by Kinetic Aggregation of Aqueous Gold Colloids. *Phys. Rev. Lett.* **1984**, *52*, 1433–1436.
40. Weitz, D. A.; Huang, J. S.; Lin, M. Y.; Sung, J. Limits of the Fractal Dimension for Irreversible Kinetic Aggregation of Gold Colloids *Phys. Rev. Lett.* **1985**, *54*, 1416–1419.
41. Brinker, C. J.; Scherer, G. W. *Sol-Gel Science: The Physics and Chemistry of Sol-Gel Processing*; Academic Press: San Diego, 1990.
42. Krieger, I. M.; Dougherty, T. J. A Mechanism for Non-Newtonian Flow In Suspensions of Rigid Spheres. *Trans. Soc. Rheol.* **1959**, *3*, 137.
43. Woods, M. E.; Krieger, I. M. Rheological Studies on Dispersions of Uniform Colloidal Spheres: I. Aqueous Dispersions in Steady Shear Flow. *J. Colloid Interface Sci.* **1970**, *34*, 91–99.
44. (a) Sontagg, R. S.; Russel, W. B. Structure and Breakup of Flocs Subject to Fluid Stresses: I. Shear Experiments. *J. Colloid Interface Sci.* **1986**, *113*, 399–413; (b) Sontagg, R. S.; Russel, W. B. Structure and Breakup of Flocs Subject to Fluid Stresses: II. Theory. *J. Colloid Interface Sci.* **1987**, *115*, 378–389.
45. Barnes, H. A. Thixotropy: A Review. *J. Non-Newtonian Fluid Mech.* **1997**, *70*, 1–33.
46. Borcherdt, R. D.; Glassmoyer, G. On the Characteristics of Local Geology and Their Influence on Ground Motions Generated by the Loma Prieta Earthquake in the San Francisco Bay Region, California. *Bull. Seismol. Soc. Am.* **1992**, *82*, 603–641.
47. Holzer, T. L. Loma Prieta Damage Largely Attributed to Enhanced Ground Shaking. *EOS* **1994**, *75*, 299–301.

48. Buscall, R.; White, L. R. The Consolidation of Concentrated Suspensions. Part 1: The Theory of Sedimentation. *J. Chem. Soc. Farad. Trans. I* **1987**, *83*, 873–891.
49. Landman, K. A.; White, L. R.; Buscall, R. The Continuous-Flow Gravity Thickener: Steady State Behavior. *AIChEJ*, **1988**, *34*, 239–252.
50. Nutt, C. W. Froth Flotation: The Adhesion of Solid Particles to Flat Interfaces and Bubbles. *Chem. Eng. Sci.* **1960**, *12*, 136–141.
51. Rumpf, H.; Schubert, H. Adhesion Forces in Agglomeration Processes. In *Ceramic Processing before Firing*; Onoda, G. Y.; Hench, L. L., Eds.; Wiley-Interscience: New York, 1978; pp 357–376.
52. Ren, T.; Aksay, I. A.; Yasrebi, M.; Pellerin, N. B.; Staley, J. T. Biogenic Hydroxylated Carboxylic Acid Monomers Serve as Dispersants for Ceramic Particles. Mat. Res. Soc. Symp. Proc. v. 330, pp. 113–118.
53. Onoda, G. Y. The Rheology of Organic Binder Solutions. In *Ceramic Processing Before Firing*; Onoda, G. Y.; Hench, L. L., Eds.; Wiley-Interscience: New York, 1978; pp 235–251.
54. Ring, T. A. *Fundamentals of Ceramic Powder Processing and Synthesis*; Academic Press: San Diego, 1996.
55. Reed, J. S. *Introduction to the Principles of Ceramic Processing*; Wiley: New York, 1988.
56. Fuerstenau, D. W. Equilibrium and Nonequilibrium Phenomena Associated with the Adsorption of Ionic Surfactants at Solid-=–Water Interfaces. *J. Colloid Interface Sci.* **2002**, *256*, 79–90.
57. Horn, R. G. Surface Forces and Their Action in Ceramic Materials. *J. Am. Ceram. Soc.* **1990**, *73*, 1117–1135.
58. Bunker, B. C. Unpublished results Pacific Northwest National Laboratories, *ca* 1995.
59. Xia, Y.; Gates, B.; Yin, Y.; Lu, Y. Monodispersed Colloidal Spheres: Old Materials with New Uses. *Adv. Mater.* **2000**, *12*, 693–713.

9

Bio-inspired Synthesis of Oxide Nanostructures

9.1 INTRODUCTION

Nature is capable of building magnificently intricate and detailed structures out of otherwise boring materials such as calcium carbonate and silica. Anyone who has taken their children to see dinosaurs at a Natural History museum or visited natural wonders such as the Petrified Forest in Arizona are familiar with the natural process called *fossilization* by which the tissues of dead organisms are eventually replicated by objects of stone (Fig. 9.1). Most living organisms (including humans) are critically dependent on more deliberate and controlled *biomineralization* [1, 2] phenomena that lead to the production of all hard tissues, including our teeth and bones, seashells and diatom skeletons, egg shells, and the magnetic nanoparticles that provide homing devices from bacteria to birds (Fig. 9.1). All these processes are nothing more than specific examples of highly controlled nucleation and growth phenomena such as those described in generic terms in Chapter 7. At a molecular level, these processes are controlled by the same reaction mechanisms involving oxide surfaces, which were outlined in Chapter 6. However, biomineralization is orders of magnitude more sophisticated than standard nucleation and growth processes. The unique features of biomineralization involve the interplay between organic biomolecules and the nucleation and growth of inorganic phases such as oxides. This interplay is of critical importance

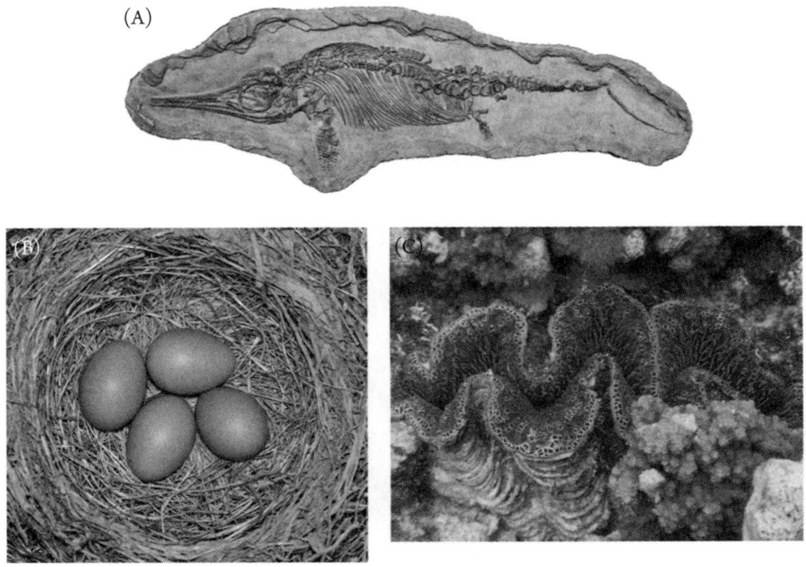

Figure 9.1 Examples of biomineralization. (A) The fossilized skeleton of an *Ichthyosaur* cub skeleton, illustrating both biomineralization and fossilization phenomena. (SOURCE: Sombra/Shutterstock.com). (B) American robin eggs in a nest, where the egg shells are composed of calcium-carbonate minerals. (SOURCE: StevenRussellSmithPhotos/Shutterstock.com). (C) A giant clam with a membrane that performs biomineralization functions, next to coral, which is also formed via biomineralization (SOURCE: Moize nicolas/Shutterstock.com).

in both biology and emerging nanotechnologies, providing specific examples that illustrate many of the concepts of oxide chemistry introduced in Chapters 5 through 7.

In this chapter, we highlight the key concepts of biomineralization and provide examples of how researchers can now produce complex nanostructured oxides via biomimetic nucleation and growth strategies that replicate some of the key features used to make hard tissues in living systems. These strategies include the use of (1) molecular complexation and compartmentalization to control supersaturation levels, (2) specific ligands and surface structures to mediate nucleation phenomena, (3) hierarchical self-assembled organic architectures as templates for oxide formation, (4) functionalization to stimulate desired heterogeneous nucleation and growth processes on those templates, and (5) organic surfactants to manipulate both crystal-phase preferences and growth habits.

Biomineralization processes involve the use of several of these strategies in either a sequential or simultaneous fashion, resulting in highly complex mineralization phenomena. A classic example illustrating an integrated biomineralization process involves the formation of the iridescent nacre [2] or "mother

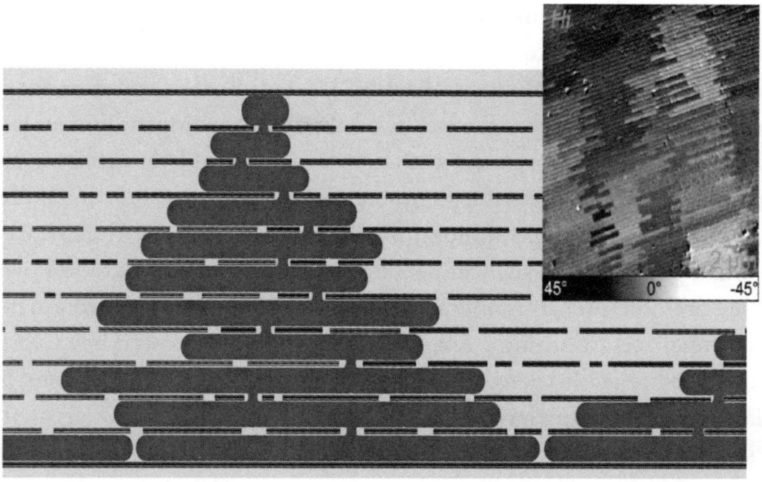

Figure 9.2 Crystals of mollusk nacre growing between predeposited matrix sheets. The sheets (black) have a central layer of core fibers covered with nucleation that promotes acidic protein layers. Each layer is pierced with small pores that allow soluble species to be supplied to growing crystals of aragonite (gray). The pores also allow for the formation of mineral bridges that ensure all crystals assume the same crystalline orientation. The inset shows the resulting aligned crystals in the shell of a blackfoot abalone *Haliotis iris*. SOURCE: Reproduced from Addadi, L.; Weiner, S.; Biomineralization: A pavement of pearl. *Nature*, 1997, *389*, 912–915; with permission. see also Olson, I. C.; Kozdon, R.; Valley, J. W.; Gilbert, P. U. P. A. Mollusk Shell Nacre Ultrastructure Correlates with Environmental Temperature and Pressure. *J. Am. Chem. Soc.* **2012**, *134*, 7351–7358.

of pearl" on the interior of some seashells (Fig. 9.2). Nacre is a composite material consisting of continuous $CaCO_3$ plates that are separated by a compliant, layered organic template. The interdigitation of the inorganic and organic phases results in a material that is much stronger than either $CaCO_3$ or the matrix alone. The organic template consists of layers of chitin [3] (the polysaccharide that comprises the exoskeletons of insects) and/or glycine- and alanine-rich proteins in an antiparallel β-sheet configuration. The layer spacing controls the dimensions of the depositing $CaCO_3$ crystals. Pores in the layers allow ions to flow into each compartment to maintain sufficient supersaturation to support growth and to make continuous crystals of highly regular morphology. Finally, layer walls are coated with self-assembled layers of acidic proteins that control the nucleation, growth, and crystal phase (in this case, aragonite) of $CaCO_3$.

The complexities of biomineralization involve many topics on the forefront of both biology and materials science. As such, the field of biomineralization has taken enormous strides even during the writing of this book. Here, we provide examples that highlight each of the primary strategies encountered in both biomineralization and biomimetic oxide synthesis (see Chapter 11 for additional examples). Those of you seeking more details are referred to several exceptional reviews on biomineralization [1, 2, 4–8] that cover recent developments in this rapidly changing field. Rather than probing individual biomineralization processes from start to finish, we organized this chapter around specific steps involving specific phenomena. First, we introduce the molecular species that form the basis for biomineralization phenomena. The first two sections highlight (1) how such molecules self-assemble to create templates as exquisite architectures for mineralization and (2) how complex templates can be functionalized to promote specific nucleation and growth behaviors. The next three sections illustrate how biomolecules and templates are used to manipulate supersaturation levels, nucleation phenomena, and oxide growth. The final section deals with energy-consuming molecules and architectures that can be used to create complicated and sometimes dynamic nonequilibrium structures using active assembly processes.

9.2 HIERARCHICAL SELF-ASSEMBLY: FROM MOLECULES TO TEMPLATE ARCHITECTURES

The optimization of nucleation and growth conditions alone is insufficient to produce any of the complex structures shown in Figure 9.1. As described in Chapter 7, classical nucleation and growth lead to the formation of particles, aggregates, and films, but nothing approaching the intricate architectures of diatom skeletons. Complex oxide structures are produced only when nucleation and growth are performed on a preexisting organic template. Organisms are capable of creating a bewildering array of organic scaffolds to support heterogeneous nucleation and growth processes, including lipid vesicles, cellular membranes, fibrous networks, and hierarchical assemblies spanning length scales from nanometers to meters. A critical attribute of all templates for biomineralization is that they must be relatively easy to prepare. Surprisingly, many of the complex templates used by organisms form spontaneously under ambient conditions by processes that fall under the heading of *hierarchical self-assembly.* We now work our way up the hierarchy of such self-assembly, starting with molecules and moving up to structures that are as large as an elephant.

At the smallest length scale are the primary classes of molecular species that form the basis for most higher levels of biological assemblies, including (Fig. 9.3) (1) fatty acid and phospholipid surfactants, which are used to generate membranes; (2) amino acids, which form the basis for proteins [9]; and (3) saccharides (or sugars), which ultimately form the cellulose scaffolds of plants and the chitin for insect exoskeletons. All these molecules have the ability to assemble into larger and often highly complex structures. However, although the saccharides are

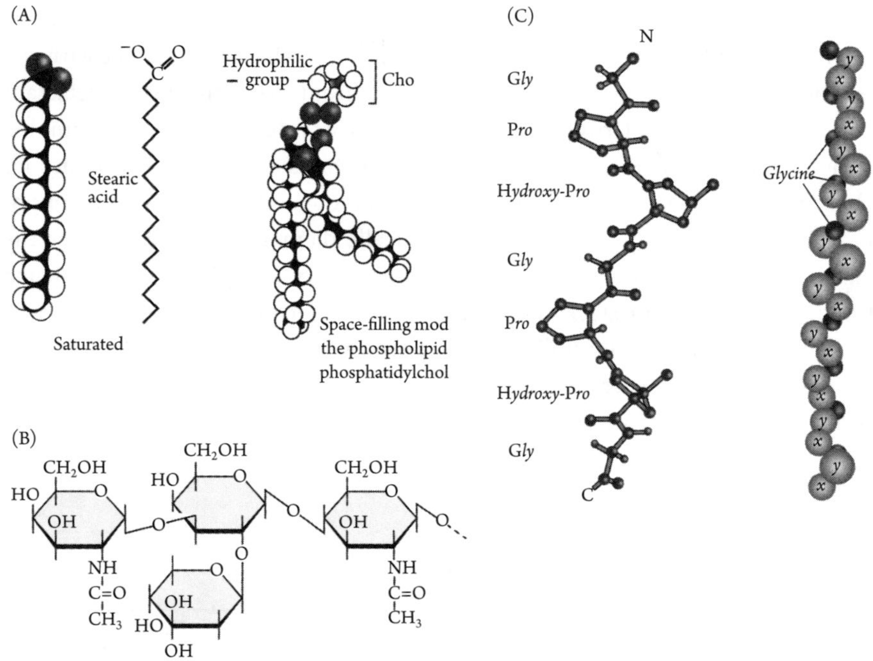

Figure 9.3 The primary classes of molecular species used by organisms for biomineralization. (A) Surfactant molecules including a fatty acid and a phospholipid. (SOURCE: Reproduced with permission from Alberts, B.; Bray, D.; Hopkin, K.; Johnson, A.; Lewis, J.; Raff, M.; Roberts, K.; and Walter, P. *Essential Cell Biology*; Garland Publishing: New York, 1998, p. 58.) (B) A functionalized polysaccharide. (SOURCE: Reproduced from Alberts, B.; Bray, D.; Hopkin, K.; Johnson, A.; Lewis, J.; Raff, M.; Roberts, K.; and Walter, P. *Essential Cell Biology*; Garland Publishing: New York, 1998, p. 58.) (C) Depictions of the primary protein strand found in collagen. (SOURCE: Reproduced from Branden, C.; Tooze, J.; *Introduction to Protein Structure*; Garland Publishing: New York, 1999, p. 284.)

clearly active (as demonstrated by petrified wood), they are not discussed further because many of their key attributes mirror those of the protein families that are most active in biomineralization processes.

9.2.1 Surfactant Templates

Surfactant molecules of relevance for biomineralization consist of a polar or ionic head group attached to a hydrocarbon tail. The hydrophilic head group provides sites that can stimulate oxide formation whereas the hydrophobic tail provides a major driving force for molecular self-assembly processes. When surfactants such as lipids are placed in a polar solvent such as water, they immediately organize themselves into phase assemblies in which the hydrocarbon tails are buried and screened beneath a continuous layer of head groups that is in contact with the solution. Depending on factors including head group size, tail length, surfactant concentration, and polarity of added cosolvents, a wide range of self-assembled structures can form (Fig. 9.4). The simplest structures include spherical micelles (heads out and tails in) and inverse micelles (tails out and heads in within nonpolar solvents) that contain no water. The micelle diameter is equal to twice the chain length (typically 1–10 nm). Next in complexity are lipid bilayers, in which two surfactant films are stacked back to back, with head groups pointing outward. In the presence of a hydrophilic surface, one layer of head groups interacts with the surface and the other head group layer contacts the solution. In the presence of a hydrophobic surface, surfactants can form a *self-assembled monolayer* (SAM)

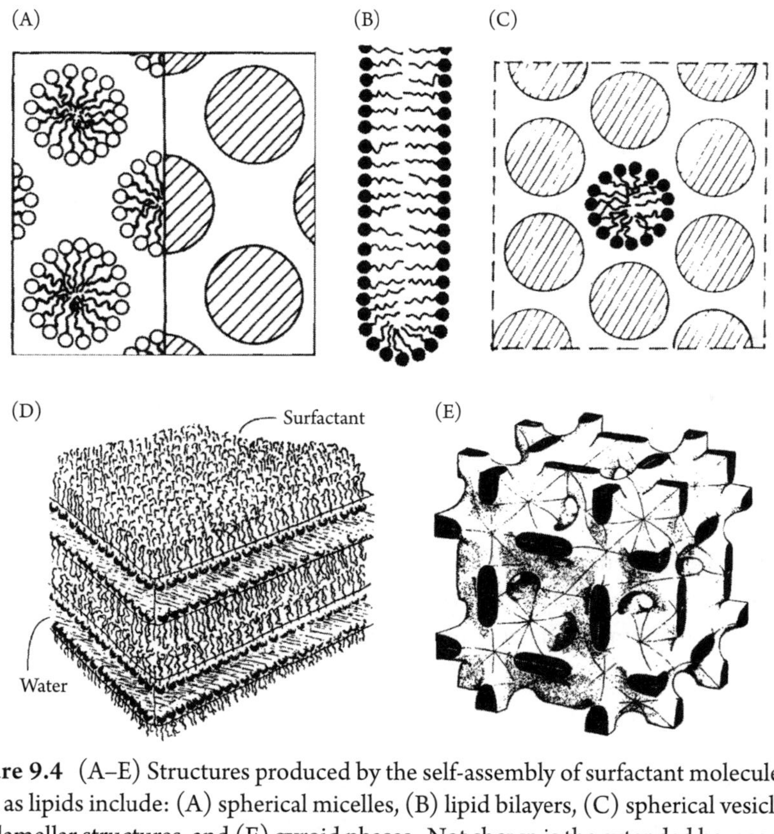

Figure 9.4 (A–E) Structures produced by the self-assembly of surfactant molecules such as lipids include: (A) spherical micelles, (B) lipid bilayers, (C) spherical vesicles, (D) lamellar structures, and (E) gyroid phases. Not shown is the extended hexagonal phase that is formed by the hexagonal closest packing of lipid cylinders. (SOURCE: Reproduced with permission from Tiddy, G. J. T.; Surfactant-water liquid crystal phases. Physics Reports, *Phys. Reports*, **1980**, 57, 1–46, pgs. 6, 7).

in which the tails are next to the surface and the head groups contact the solution. Such monolayers are a primary tool used in surface functionalization. In solution, lipid bilayers can curve back on themselves to form spherical vesicles and cells that encapsulate a fixed solution volume. These water-filled "balloons" can range in diameter from around 10 nm up to more than 1 μm, forming the basis for cells and many subcellular structures, including lipid nanotubes.

Ironically, materials scientists have used lipids to a greater extent than organisms when it comes to creating complex templates for biomineralization. The addition of less polar cosolvents (not found in most organisms) to aqueous surfactant solutions results in the formation of extended and periodic structures called *lyotropic liquid crystals* [10] (Fig. 9.4). Such extended structures include lamellar, hexagonal, cubic, and gyroid phases. In 1992, lyotropic liquid crystals were first reported as templates for the preparation of mesoporous ceramics, including amorphous silica and aluminosilicates [11]. The synthesis of mesoporous ceramics involves adding soluble precursors such as tetraethylorthosilicate to lyotropic liquid crystal solutions. The most effective surfactants for forming mesoporous aluminosilicates contain head groups such as the quaternary ammonium group. It is thought that these cationic head groups help attract and organize hydrolyzed silicate and aluminate anions to the template, which then condense to create a continuous oxide phase. The surfactant is later removed via low-temperature pyrolysis, leaving behind a silica matrix containing ordered porosity (Fig. 9.5).

More traditional zeolites are based on the polymerization of aluminosilicates around molecular templates [12] (Chapter 14). Zeolites exhibit pore sizes ranging from 0.3 to a maximum of 1.4 nm. In contrast, the mesoporous ceramics exhibit pore sizes that can be engineered from 1.5 to 10 nm. The pore size corresponds to roughly twice the length of the surfactant's hydrocarbon tail. The upper limit on pore size is determined by the maximum chain length that can be dispersed in an aqueous solution and still form the liquid-crystal template. More recently, it has been shown that di- and triblock copolymers can be used to extend the pore size range from 4.6 to 30 nm, with silica wall thicknesses ranging from 3 to 6 nm. These block copolymers form phases that are analogous to the lyophobic liquid crystals, where hydrophobic and hydrophilic polymer blocks play the roles of the surfactant's hydrocarbon tail and ionic head group. The pores in the resulting ceramics can be made as large as the "blocks" in these phase-separated polymer systems. Block copolymer templates have been exploited for the nucleation of calcite [13], resulting in materials that rival the nacre created when aragonite is nucleated and grown within layered proteinaceous material [14, 15].

It is much easier to form mesoporous ceramics using oxides based on tetrahedral rather than octahedral cations. Tetrahedral hydrolysis products polymerize

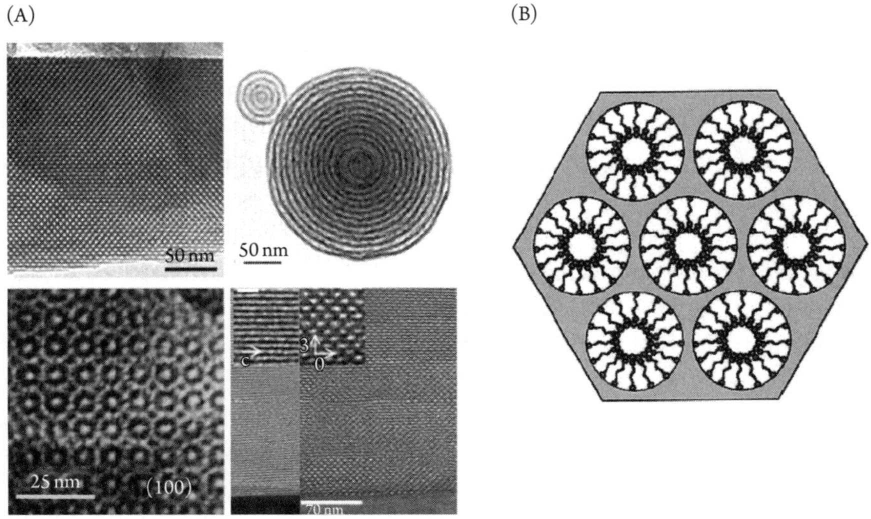

Figure 9.5 (A) Transmission electron micrographs of several of the unique phases that can be generated by amorphous silica grown in the presence of lyotropic liquid crystals. (SOURCE: Unpublished montage, Sandia National Laboratories, *ca* 1999.). (B) When the surfactant template is removed via pyrolysis, it is possible to reintroduce shorter chain surfactants with any desired head group to create a functionalized mesoporous material. In such materials, the silica mimics the role of the framework protein, whereas the self-assembled monolayer mimics the acidic macromolecules that are found in many two-component biomineralization systems.
SOURCE: Reproduced with permission from Liu, J.; Feng, X.; Fryxell, G.E.; Wang, L.Q.; Kim, A.Y.; Liu, J.; and Kemner, K.M., 1997. Functionalized Monolayers on Ordered Mesoporous Supports. Science, 276(5314): 923–926.

to form amorphous structures that can readily conform to almost any template. In contrast, oxides based on octahedral cations are almost always crystalline and form regular, nonconformal solids. Because the oxide is much more rigid than the surfactant array, the nucleated oxide imposes its structure on the template rather than the other way around, so no mesoporous materials are formed. This shortcoming can be overcome by strategies such as crosslinking the surfactant array before mineralization, or using hydrolysis products as head groups [16]. In the latter strategy, the hydrolysis products are made to condense with each other immediately after the lyotropic liquid crystal array is formed, creating a cross-linked head group array that locks the structure into place. This head group monolayer has the added advantage that it strongly resembles the oxide that it is intended to nucleate. When these hybrid surfactant architectures are exposed to supersaturated solutions of hydrolysis products, facile oxide formation occurs that replicates the template.

9.2.2 Protein Templates

The range of hierarchical assembly motifs that are available to proteins [9] are much more diverse than those associated with lipids and other surfactants. Proteins are constructed from 20 standard amino-acid molecules. These amino acids can have cationic, anionic, polar, or nonpolar side chains. At the molecular level, amino acids can condense with each other through amide linkages to form chains. These chains can contain a limitless number of different amino-acid sequences. At a local level, the chains self-assemble as a result of a balance of ionic, hydrophilic–hydrophobic, and hydration forces into secondary structures such as the α-helix and the β-sheet. The secondary structures then fold into more compact tertiary structures. These tertiary structures can then self-assemble with other folded proteins to form even larger quaternary structures.

The complexity of protein assemblies is illustrated by relatively simple collagen proteins [3, 9] (Fig. 9.6). These proteins are the most prevalent proteins in mammals (25% by mass), and form the basis for all connective tissues, including skin, tendons, cartilage, and the biomineralization matrix for egg shells and bone. In collagen, amino acids assemble into single, helical polyproline type II chains. Three of these helical chains supercoil around a common axis to form a "coiled-coil" structure with a diameter of 1.5 nm. The coiled coils assemble into fibrils (10–300 nm in diameter) that in turn assemble into collagen fibers ($0.5 - 3\ \mu$m in diameter and many microns in length). These fibers are woven into tissues that can be as large as an elephant's skin. Collagen can also coassemble with SAMs that

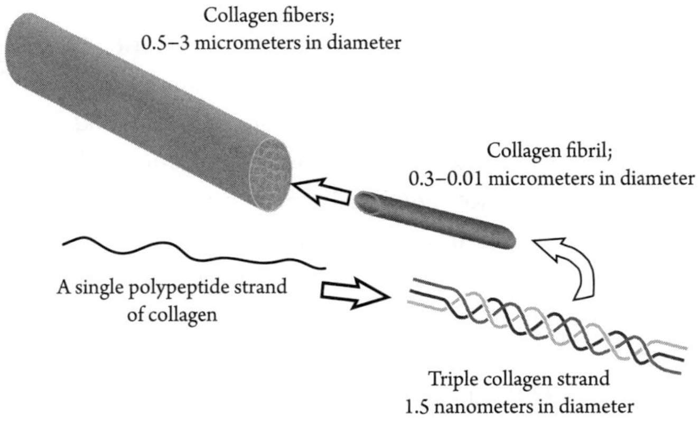

Figure 9.6 The hierarchical assembly of collagen fibers from (counterclockwise from bottom) single protein strands to a coiled coil to collagen fibrils to collagen fibers. The fibers assemble into collagen tissues.
SOURCE: Adapted from Alberts, B.; Bray, D.; Hopkin, K.; Johnson, A.; Lewis, J.; Raff, M.; Roberts, K.; and Walter, P. *Essential Cell Biology*; Garland Publishing: New York, 1998, p. 601.

promote biomineralization processes. Researchers are just beginning to unravel the mechanisms underlying this multiscale, hierarchical assembly. Materials scientists are now using these and many other biological constructs as templates for oxide growth, including cytoskeletal fibers, viruses [17, 18], and even living cells. The formation of our own bones involves the deposition by osteoblast cells of osteoid organic material that is then mineralized with calcium phosphate [19].

9.3 TEMPLATE FUNCTIONALIZATION: THE ROLE OF SELF-ASSEMBLED MONOLAYERS

Although templates can be prepared with complex architectures, the physical structure of the scaffold is insufficient to induce the nucleation and growth of mineral phases from solution. A key component of biomineralization involves modifying the scaffold deliberately to stimulate the formation of oxides on exposed surfaces. For example, the collagen described earlier is a relatively insoluble, hydrophobic protein that is normally ineffective at stimulating nucleation and growth processes. This is fortunate, because otherwise our skin, hair, tendons, and even our eyeballs would all become fossilized. Yet this same collagen protein provides the scaffold for our bones. What allows collagen to stimulate biomineralization in some instances and not in others? The answer is *functionalization*.

Several key factors can be manipulated via functionalization to stimulate the heterogeneous nucleation of oxides on template surfaces. First, any surface with a charge or net polarity can generate an electrostatic driving force for surface nucleation. For example, negatively charged surfaces will attract cationic hydrolysis products within their electrical double layers (see Chapter 8) to create a condition of local supersaturation. Second, ligands can bind to specific hydrolysis products, which also leads to high surface concentrations. Ligands can be selected for binding to specific hydrolysis products based on solution speciation diagrams (see Chapter 5). Third, surface functional groups can be arranged such that they form a local symmetry, geometric relationship, or epitaxial match to specific crystal faces of the oxide to be nucleated. Here, the surface layer has the potential not only to nucleate an oxide, but also to bias the crystalline phase, orientation, and growth habit.

Surface functionalization is as vital for the biomimetic processing of oxide films as it is in living systems. Some substrates, including oxides or metals passivated by native oxides (see Chapter 12), have surfaces covered with hydroxyl groups that are capable of activating nucleation. In contrast, most commercial plastics, including polystyrene, polycarbonate, and polyethylene, have hydrophobic surfaces that behave like native collagen and inhibit heterogeneous nucleation. However, these plastics can be modified chemically to produce hydrophilic and/or ionic surfaces that are much more stimulating [20, 21]. Examples of such modifications include the sulfonation of polystyrene and polycarbonate, and the introduction of carboxylate, amino, hydroxyl, and phosphate groups into the surfaces of polyethylene. Such functionalization makes the polymer chains hydrophilic and induces the modified polymer surfaces to swell in water. Swelling allows functionalization to occur beneath the immediate polymer–water interface, which means that mineralization occurs within a functionalized zone, not just the immediate surface [22] (Fig. 9.7). The presence of an interlocked polymer–oxide composite within the near-surface zone enhances film stability, in analogy to the structure of nacre.

In addition to robustness, advantages of biomimetic processing on plastics relative to standard ceramic thin-film processing methods include the following: (1) processing occurs at room temperature in contrast to methods such as sol–gel techniques that require temperatures that exceed 400°C and incinerate most plastics; (2) processing is more environmentally friendly than many deposition techniques, which require toxic chemicals and/or solvents; (3) coatings are conformal, allowing for the deposition of uniform coatings onto complex shapes and porous materials (Fig. 9.7); and (4) patterned films can be produced with micron-scale resolution when used in conjunction with techniques such as photolithography.

Some biological templates such as ferritin and the silicateins are known to contain functional groups in their native surfaces that catalyze the heterogeneous nucleation of oxides and other minerals. However, most biological templates typically contain not one, but two, key components [6]. Relatively inert templates such as collagen and cellulose are called *framework proteins*. The framework proteins only become activated to stimulate nucleation and growth when they are covered with a thin layer of more ionic water-soluble, low-molecular weight proteins that include the so-called *acidic macromolecules*. Overlayers of acidic macromolecules decorate template surfaces with amino acids having ionic side chains, such as the carboxylic acids (in aspartic and glutamic acid), hydroxyl groups (as in serine and threonine), and organic phosphates. The acidic macromolecules are deposited onto templates at specific times and at specific locations for the purpose of inducing biomineralization in a highly controlled fashion.

Although many oxides and plastics can contain functional groups that stimulate heterogeneous nucleation and growth, the most flexible biomineralization methods involve the deposition of desired functional groups on top of the substrate before mineralization. In artificial systems, the role of biology's acidic macromolecules is typically filled by SAMs [23] (Figs. 9.8). Molecules used in SAMs contain three key components: a head group that defines surface functionality, a hydrophobic hydrocarbon chain that drives the self-assembly process (as in lipid bilayers), and a tail group that is used to anchor the SAM to the desired surface. For oxide substrates, the tail group is typically a silane coupling

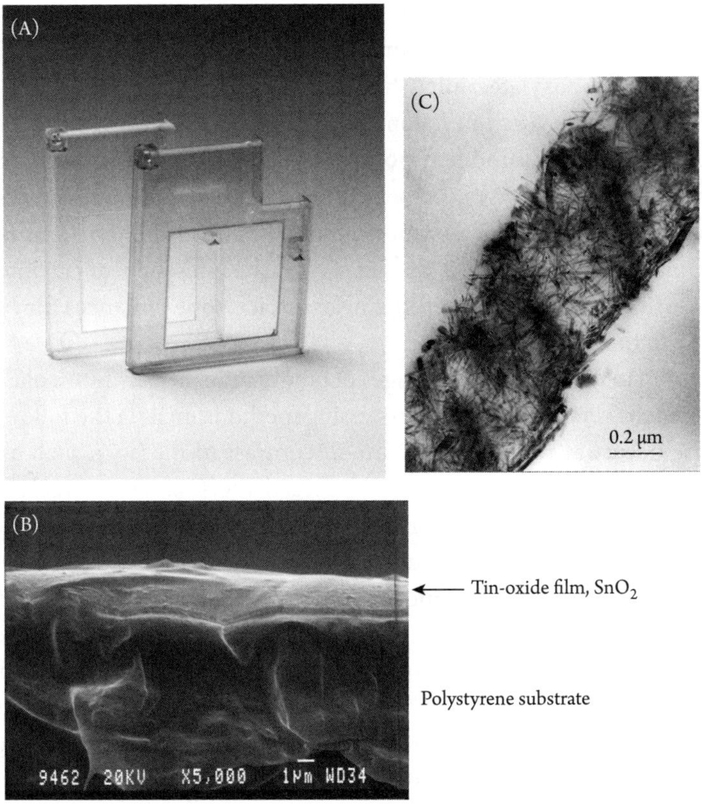

Figure 9.7 Wear-resistant oxide coatings formed on functionalized plastic viewed at increasing levels of magnification (clockwise from top right). (A) A complex sulfonated polystyrene part before and after exposure to a supersaturated solution of iron hydrolysis products. Spontaneous formation of an FeOOH film imparts a reddish tint to the part. (B) A scanning electron micrograph of the cross-section of a similar sulfonated polystyrene part after exposure to an aqueous solution containing soluble Sn(IV) hydrolysis products. The SnO_2 film is dense, completely conformal, and has a thickness of 0.4 μm. (C) A transmission electron micrograph of an FeOOH film formed on sulfonated polystyrene. In this case, the bulk of the observed nucleation and growth occurs within a swollen sulfonated surface gel on the plastic, resulting in the formation of a FeOOH–polystyrene composite with a random array of FeOOH crystallites. This film morphology is in sharp contrast to what is observed when films nucleate on top of self-assembled monolayers, where a dense film forms in which all crystalline needles are oriented perpendicular to the surface. In many biominerals such as nacre, the benefits of crystalline orientation and composite formation are both achieved.

SOURCE: All images reproduced with permission from Bunker, B. C.; Rieke, P. C.; Tarasevich, B. J.; Campbell, A. A.; Fryxell, G. E.; Graff, G. L.; Song, L.; Virden, J. W.; and McVay, G. L. Ceramic Thin Film Formation on Functionalized Interfaces Through Biomimetic Processing. *Science* **1994**, *264*, 48–55.

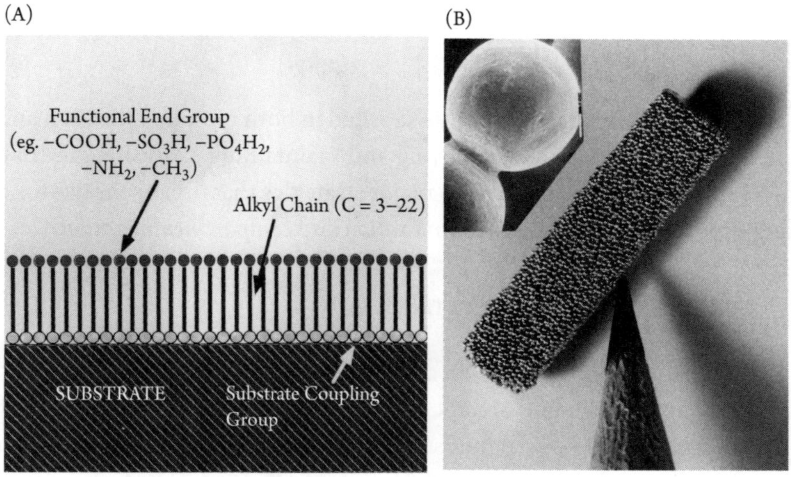

Figure 9.8 (A) A depiction of a self-assembled monolayer (SAM) on a substrate surface, showing head groups adjacent to the solution, the organizing hydrocarbon tails in the middle, and the functionalized tail groups that anchor the SAM to the surface, enhancing SAM stability. Such anchor groups are absent in biological lipids. Common anchor groups for oxide groups are silane coupling agents that hydrolyze and condense on the surface. The hydrocarbon chains typically contain from 6 to 22 $-CH_2-$ groups. Typical head groups to help induce mineralization include cationic amines, as well as groups that can be deprotonated to form anionic surfaces such as $-COOH$, $-SO_3H$, and $-PO_4H_2$. (B) A picture of a complex titanium substrate for a bone implant that was functionalized to nucleate and grow a conformal film of octacalcium phosphate (inset) to stimulate bone formation.
SOURCE: All images reproduced with permission from Bunker, B. C.; Rieke, P. C.; Tarasevich, B. J.; Campbell, A. A.; Fryxell, G. E.; Graff, G. L.; Song, L.; Virden, J. W.; and McVay, G. L. Ceramic Thin Film Formation on Functionalized Interfaces Through Biomimetic Processing. *Science* **1994**, *264*, 48–55.

agent that condenses with surface hydroxyl groups and adjacent molecules to form robust covalent $>M-O-Si<$ or $>Si-O-Si<$ linkages. However, other tail groups including carboxylates, phosphonates, and thiols (on Au° or Ag° substrates) are also popular. A wide range of ionic head groups can be purchased or prepared, including the cationic amine groups and anionic carboxylate, sulfonate, and phosphate groups found in both natural lipids and modified plastic surfaces. A major advantage of SAMs is that they can be used to coat almost any template, regardless of template complexity (Figs. 9.5 and 9.8). In this respect, SAMs mirror the versatility of the acidic macromolecules organisms use in biomineralization. Next, we describe how functionalized surfaces promote biomineralization processes by influencing local supersaturation levels, as well as by lowering the activation barriers to both nucleation and growth.

9.4 CONTROLLING SUPERSATURATION LEVELS: COMPLEXATION AND COMPARTMENTALIZATION

The minimum condition that must be satisfied in both biomineralization and biomimetic processing involves achieving and maintaining exquisite control over solution supersaturation levels. Two major strategies that organisms use for controlling supersaturation include *complexation* and *compartmentalization*.

9.4.1 Complexation in Biomineralization

The general use of complexation in artificial oxide precipitation schemes is discussed in Chapter 7. Simple organic ligands such as citrate and pyrophosphate ions are common in biological fluids and are used to manipulate species solubility. However, in organisms, the most important (and prevalent) complexes involve more complicated metalloproteins of Fe, Cu, and Zn, followed by those incorporating Si, V, Cr, Mn, Co, Ni, Mo, and W. Many of these metalloenzymes provide critical storage, transport, and catalytic functions, including (1) iron storage and transport (e.g., by the ferritin (Fig. 9.9) described later), (2) oxygen transport (by Fe-containing hemoglobin), (3) CO_2 sequestration (by Zn-containing carbonic anhydrase), and (4) light harvesting and electron-transfer

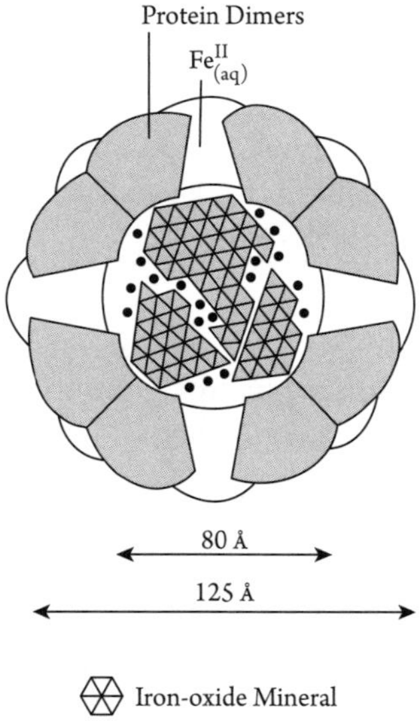

Figure 9.9 A depiction of the porous apoferritin protein shell in iron-carrying ferritin proteins showing the interior space available for the nucleation and growth of iron oxide.

SOURCE: Image reproduced from Mann, S.; Biomineralization: A New Horizon in Bioinorganic Chemistry. *Chemie in Unserer Zeit*, **1986**, *20*, 69–76, with permission.

mediators important in photosynthesis (such as Cu-containing azurin). Because of their importance in energy management and metabolic cycles, the structures and properties of the metalloenzymes have been studied more extensively than most of the proteins that mediate oxide mineralization.

One example of a protein that serves as a complexing agent for hydrolysis products involves the molybdenum storage protein MoSto [24] (Fig. 9.10). This protein can encapsulate polyoxometalate structures containing up to 100 molybdenum or tungsten atoms and demonstrates a clear link between biochemistry and the principles of hydrolytic coupling discussed in Chapter 5. Here, a primary factor that prevents these polyoxometalate clusters from dissipating under physiological pH conditions (where they are normally very unstable) is bonding to the amino acids that line the interior wall of the cluster.

9.4.2 The Role of Compartmentalization in Mediating Supersaturation

Ferritin (Fig. 9.9) and the MoSto protein (Fig. 9.10) both illustrate a second critical biomineralization strategy, which is *compartmentalization*. The classic compartment in biology is the cell. The lipid bilayers that define the cell membrane also define compartments on smaller length scales, including organelles and vesicles. The sizes of biological compartments range from several microns down to the nanometer-scale cavities found at the interior of many mineralization proteins. A similar range of compartment sizes has been achieved in man-made systems, including artificial vesicles and emulsions. Key features of membrane compartments include (1) the compartments are monodisperse (i.e., all have the same size), defining a specific solution volume within which nucleation and growth occurs; (2) compartments define carefully controlled environments with limited communication with the outside world; and (3) membranes can be made to be permeable to desired chemicals. This permeability allows the environment within the compartment to be changed in a carefully controlled manner. In mineralization, selective permeability is used to change the compartment's environment from an undersaturated to a supersaturated condition, promoting the precipitation phenomena discussed in Chapter 7. Parameters that can be carefully controlled to mediate supersaturation levels include (1) compartment pH via the use of proton-selective ion channels and pumps, (2) compartment redox potential, E_h, via the controlled admittance of O_2 or other redox-active agents, and (3) ionic concentrations via the use of ion channels, ion pumps, and concentration gradients that are established across the membrane. Diatoms have even evolved to create a transmembrane pump that is selective for $Si(OH)_4$.

The classic example of a biological compartment for oxides is the protein ferritin [25–27] (Fig. 9.9), which is the major construct used to store and transport iron in organisms. The compartment in ferritin is a self-assembled polypeptide

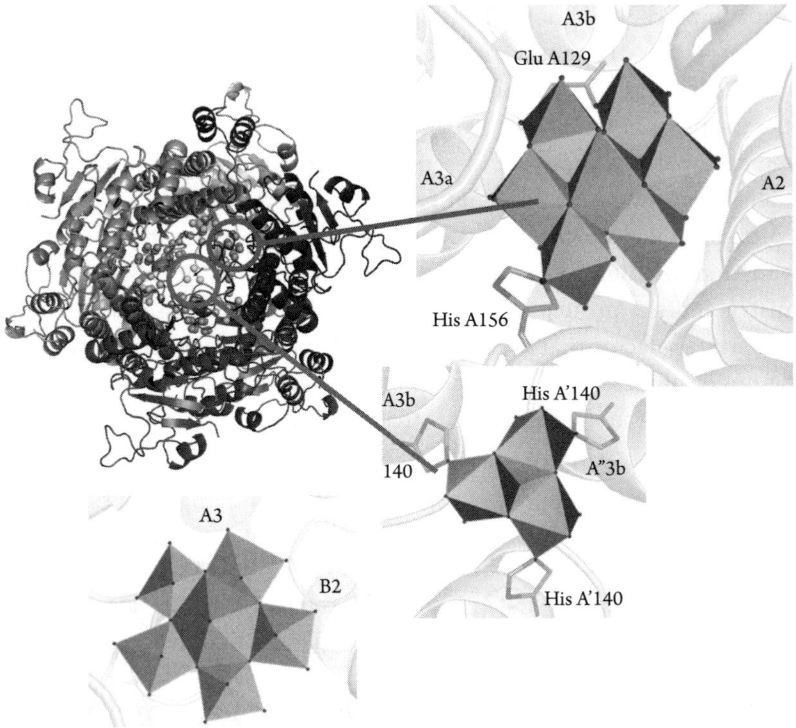

Figure 9.10 A depiction of the molybdenum-storage protein MoSto, from *Azobacter vinlandii.* The molybdate clusters in the core of this protein are nearly identical to those that are created abiotically. However, the unique environment within the protein core stabilizes the clusters to survive under physiological conditions that would normally cause them to dissociate spontaneously. Note that the core of the pentameric cluster (bottom) is the building block for the enormous Keplerate polyoxometalates described in Chapter 5.

SOURCE: Image reproduced from Kowalewski, B.r.; Poppe, J.; Demmer, U.; Warkentin, E.; Dierks, T.; Ermler, U.; and Schneider, K. Nature's Polyoxometalate Chemistry: X-ray Structure of the Mo Storage Protein Loaded with Discrete Polynuclear Mo-O Clusters. *J. Am. Chem. Soc.*, **2012**, *134*, pgs. 9768, 9770, with permission.

capsule called *apoferritin* [6, 25, 26] that surrounds a spherical, solution-filled cavity with a diameter of 8 nm. Apoferritin is permeable to both soluble Fe(III) complexes such as iron citrate and soluble Fe(II) hydrolysis products as a result of the presence of eight 0.3-nm-wide hydrophilic molecular channels that penetrate the protein shell. During loading at around pH 7, the complexes move down concentration gradients from the outside to the inside of the apoferritin; the complexes are consumed after they penetrate the membrane. For example, soluble Fe(II)(aq) is oxidized to form much less soluble Fe(III) hydrolysis products by a catalytic site in the inner protein wall called the *ferroxidase center*. All Fe(III) in the cavity is then bound at a single, specific nucleation site, resulting in the growth of a single, encapsulated ferrihydrite (FeOOH) nanoparticle. The net reaction in the case of Fe(II) corresponds to Eqs. 9.1 and 9.2 for iron concentrations of less than 30 Fe(II) per protein and more than 250 Fe(II) per protein, respectively:

$$2Fe^{2+}(aq) + O_2(aq) + 4H_2O(l) \rightleftarrows 2FeOOH(s) + H_2O_2(aq) + 4H^+(aq) \quad \textbf{(9.1)}$$

$$4Fe^{2+}(aq) + O_2(aq) + 6H_2O(l) \rightleftarrows 4FeOOH(s) + 8H^+(aq) \quad \textbf{(9.2)}$$

These reactions continue until the central compartment is filled with an amorphous ferrihydrite (FeOOH) particle with a diameter of around 8 nm. The protective ferritin coat prevents the FeOOH nanoparticles from aggregating. When the ferritin is transported to desired locations, a lowering of the pH to 5.5 or, more commonly, a reduction of Fe(III) to Fe(II) is sufficient to redissolve the FeOOH core. Dissolution of the core can release up to 4500 iron ions at the destination point. This is much more efficient than transport by conventional proteins, which typically only contain a few complexed iron centers. Materials scientists have shown that apoferritin can be used to create monodisperse nanoparticles other than FeOOH [28–31], including magnetite (Fe_3O_4), iron sulfide, and iron-free compounds such as CoOOH, MnOOH, uranium oxide, and even sulfide and selenide quantum dots.

The biomimetic *in vitro* synthesis of oxide nanoparticles can also involve the use of abiotic vesicles or emulsions rather than frameworks as sophisticated as apoferritin. However, the basic synthetic principles associated with compartmentalization still apply. Solutions containing monodisperse fluid-filled vesicles ranging in size from around 10 nm to 10 μm are formed via the self-assembly of lipid bilayer membranes. Although the lipid bilayers are continuous, they are still permeable to simple ions, protons, and hydroxide ions. However, these membranes are normally impermeable to larger species such as the hydrolysis products of metal cations.

The strategies used to synthesize oxide nanoparticles within vesicles start by forming the vesicles in the presence of soluble hydrolysis products to encapsulate these oxide precursors. Next, the solution environment exterior to the vesicles is modified with species that can diffuse through the lipid bilayer to induce a condition of supersaturation for the encapsulated species via pH or E_h changes. For example, OH^- ions outside can be induced to exchange with NO_3^- ions inside in response to concentration gradients across the vesicle membrane, allowing the pH inside to increase while maintaining a constant ionic strength. Conversely, the exchange of encapsulated Na^+ by H^+ can be used to lower the pH inside the vesicle. Such exchanges can be used to change the interior pH by 3 to even 6 units, depending on whether monovalent or divalent ions are exchanged for the protons or hydroxide ions. Such pH changes can induce enormous increases in the supersaturation levels that trigger particle nucleation (see Chapter 7).

Vesicle-assisted processes, including apoferritin syntheses, have three advantages relative to normal homogeneous precipitation: (1) the nanoparticle size can be controlled with precision based on the vesicle size and/or the concentrations of the encapsulated ions, (2) changes in solution conditions leading to supersaturation are more gradual and controlled than those achieved by adding acids and bases to a bulk solution, and (3) the vesicle coating prevents particle aggregation, which aids in the production of ideal monodisperse particles. Controlled-diameter nanoparticles of almost any oxide or hydroxide can be prepared via this technique [32, 33].

9.5 NUCLEATION PROMOTERS AND INHIBITORS: CATALYTIC AND EPITAXIAL MANIPULATIONS

Although controlling supersaturation levels is critical for biomineralization, it is insufficient to ensure that solid oxide phases nucleate in desired locations. As discussed in Chapter 6, soluble solution species must overcome an activation barrier for nucleation to occur. Organisms have developed two distinct strategies that enable them to lower or raise this activation barrier to promote or inhibit nucleation events, respectively. These strategies involve *catalysis* and *epitaxy*.

9.5.1 Catalysis of Hydrolysis and Condensation Reactions

Organisms have developed proteins that are capable of catalyzing hydrolysis and condensation reactions. Mineralization proteins often contain active sites that are capable of (1) binding soluble species at a common location to increase local supersaturation levels greatly, (2) orienting reactants such that they approach each other along low-energy reaction trajectories, (3) activating reactants by shifting proton and/or electron densities, and (4) releasing products from the active site, which resets to repeat the cycle as many times as required. For example, diatoms and sponges have developed proteins called *silicateins* [34–36] (for silica proteins) or silaffins (named for their affinity for silica) to catalyze the nucleation of spherical silica nanoparticles (Fig. 9.11). Scientists have learned how to deploy the silaffins in artificial settings to create silica-based nanomaterials [34, 35, 37] as well as Ga(III) oxides [38], titanate [39], and titanium-phosphate [40] nanostructures. The active site in these small (6–17-kDa or 1–2-nm) proteins consists of a pocket containing three critical amino acids: aspartine, histidine, and serine [41]. All three amino acids promote the binding of silicic acid to the active

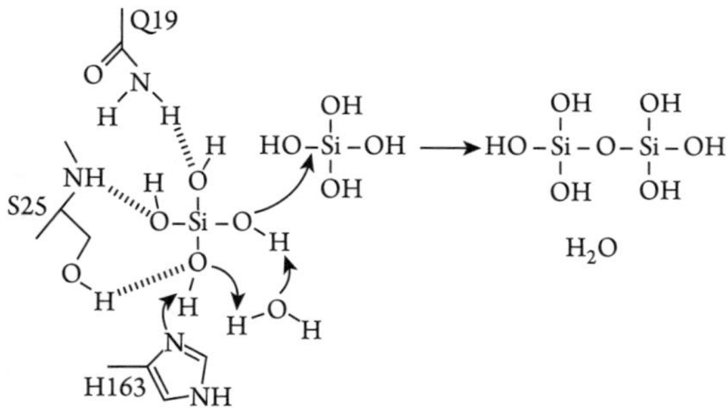

Figure 9.11 The active site in a silicatein protein that catalyzes the polymerization of amorphous SiO_2. Nitrogen sites in the active pocket help to bind $Si(OH)_4$ in a preferred orientation. The histidine site H163 is proposed to assist in the deprotonation of the $Si(OH)_4$, which allows the hydroxyl group on the serine residue S25 to attack and activate further the silicic acid to form Si–O–Si bonds with other silicate species.
SOURCE: Image reproduced from Fairhead, M.; Johnson, K.A.; Kowatz, T.; McMahon, S.A.; Carter, L.G.; Oke, M.; Liu, H.; Naismith, J.H.; and van der Walle, C.F. Crystal Structure and Silica Condensing Activities of Silicatein alpha-Cathepsin L Chimeras. *Chem. Comm.*, **2008**, 1765–1767, with permission.

site. The basic residue on the histidine promotes deprotonation and activation of the silicic acid. The hydroxyl group on the serine performs a nucleophilic S_N2 attack on the tetrahedral Si(IV) site (see Chapter 14) increasing the silicon coordination number temporarily and activating polymerization reactions further. Another example of a catalytic site that promotes oxide polymerization is the redox-active ferroxidase center in the ferritin protein discussed earlier.

9.5.2 The Role of Epitaxy in Lowering Nucleation Barriers

In Chapter 7, it was shown that the activation barrier to heterogeneous nucleation is almost always lower than that for homogeneous nucleation, and that the closer the nucleating surface comes to resembling the nucleating phase, the lower the barrier to nucleation. Most biomediated nucleation processes involve one of two strategies for manipulating the barriers to heterogeneous nucleation. First, the nucleating surface often contains functional groups that bind to the soluble species that form the nucleating phase. Second, the barrier to nucleation is the lowest when there is *epitaxial matching* between the functional groups on the substrate and the nucleating phase. Epitaxial matching means the surface of the nucleating agent exhibits a symmetry and lattice constants that are as close as possible to that of the phase being nucleated. The barrier to nucleation is at an absolute minimum when an oxide nucleates on itself.

The principles of epitaxial matching are best illustrated by examining the behavior of materials that are used to nucleate ice for cloud seeding [42]. The most effective nucleating agents for cloud seeding are inorganic nanoparticles, many clay minerals, and even silica. The most effective agent is AgI, with a lattice mismatch to ice of only 1%. Micas and clays exhibit hexagonal ring structures that are identical to those found in ice, although with a larger lattice mismatch (13%). Ice nucleation and growth on freshly cleaved and atomically flat mica surfaces have been imaged using the atomic-force microscope (AFM). Although complex nucleation behaviors are observed, the AFM results [43] show that, under certain humidity conditions, ice bilayers nucleate and grow to form islands that replicate the rings in the underlying mica structure (Fig. 9.12).

One advantage of nucleation on biological or organic interfaces is that these surfaces can be soft and compliant, which means they can deform and rearrange themselves easily to accommodate the crystal lattice of the precipitating phase. Soft templates are often more effective in promoting heterogeneous nucleation than rigid interfaces. However, the mesoporous ceramics illustrate that if the organic architecture is too soft, the precipitating phase can dictate the structure of the template, which can modify or destroy desired structures. This is why two-component systems consisting of rigid templates covered by a thin functionalized overlayers are often most effective in promoting biomineralization.

Next, let us consider biological species that stimulate nucleation within the bulk of an aqueous solution. Although such nucleation may resemble homogeneous nucleation, it is really heterogeneous nucleation at the nanoscale. Nucleation promoters are typically proteins that resemble critical nuclei (see Chapter 7) in terms of size, shape, and surface functionality. In fact, as far as the desired precipitating phase is concerned, these promoters *are* nuclei. This means the primary barrier to oxide formation is associated with growth rather than nucleation.

The model oxide with nucleation and growth behavior on proteins that has been studied in greatest detail is ice. The *ice nucleating proteins* (INPs) are among the most extensively studied biological agents for stimulating both homogeneous and heterogeneous nucleation. INPs typically have molecular weights in the range of 150 kDa, and are 5 to 7 nm in diameter, which is larger than a typical oxide's critical nucleus diameter of around 2 nm. The INPs have been studied in depth because of their enormous economic impact. A common bacterium, *Pseudomonas syringae*, has an INP on its outer cell wall [44], which is used by the bacterium to nucleate ice on adjacent plant cells, which assists the bacteria in breaking down those cells. *P. syringae* and other INPs are implicated in billions of dollars of frost damage to our crops each year. To overcome this problem, scientists identified the genes responsible for INP production [44], expressed the

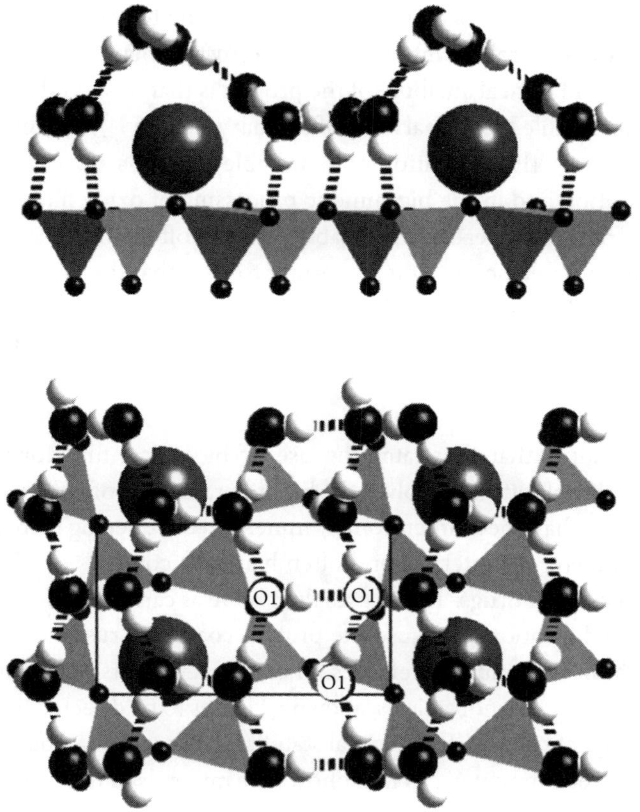

Figure 9.12 Two views of the optimized structure calculated for the first monolayer of water adsorbed on mica. Silicate tetrahedra in the mica surface are depicted as interconnected triangles. The large circles represent potassium ions, while the smaller circles represent the oxygen atoms (dark) and protons (white) in water molecules. Top – A side view of the adsorbed water molecules showing that hydrogen bonding to the surface creates the template for the formation of the hexagonal ring structures in ice. Bottom – A top view of the water layer shows that that while the hexagonal rings in the water layer are offset from the rings in the underlying mica, the mica surface controls the crystal orientation and water molecule spacing in the adsorbed layer.
SOURCE: Odelius, M.; Bernasconi, M.; Parrinello, M. Two Dimensional Ice Adsorbed on Mica Surface. *Phys. Rev. Lett.*, **1997**, *78*, 2855–2858, Fig. 2 on p. 2857, reproduced with permission.)

protein to determine its structure, and created a genetically mutated version of *P. syringae* that has no INP on its cell walls. Use of this "ice-free" bacterium to inhibit frost damage was the first unconstrained release of genetically modified bacteria into the environment.

The *P. syringae* INP illustrates several key features associated with proteins that can lower the barrier to nucleation. First, the protein contains hydrogen bond donors and acceptors that are ideal for bonding to water and ice. Second, its peptide chains are folded into a β-strand array that matches a crystal surface of ice. A third and critical attribute of the protein is that it is not designed to nucleate the most stable hexagonal form of ordinary ice (ice I_h), but the metastable cubic-phase ice, I_c. The nucleation of metastable phases is a common theme in biomineralization and in the biomimetic processing of oxide films because the activation barrier to nucleating metastable phases is often lower than that associated with more stable phases. It is important to note that the protein represents a soft and compliant substrate. Not only does this assist in the alignment and matching of functional groups and the ice during nucleation, but it also allows the protein surface to accommodate the subsequent transformation of ice I_c into more stable ice I_h.

A final example that illustrates the use of biological functionalization by humans in artificial settings involves the biomineralization of virus particles [17, 18]. Viruses that have been intentionally mineralized are seeing increasing use in biomedicine. Here, the interior (and often harmful) contents of viruses are replaced by vaccines or drugs. The viruses then serve as carriers to deliver the medicine to desired locations, because the protein coats of viruses are designed to pass through most immunity defenses. These protein coats can be mineralized with magnetic Fe_3O_4 nanoparticles, allowing viruses to be directed to desired tissues using magnetic fields. The thermal stability of the virus can be improved by inducing the precipitation of calcium phosphate minerals onto the exterior of the virus capsid. In artificial systems, it has been shown that viruses can be mineralized with amorphous lithium-iron phosphate, and can be used to establish extremely fine-scale connections within lithium-ion batteries [15].

9.6 MEDIATION OF CRYSTAL GROWTH: THE ROLE OF SITE BLOCKERS

Even if supersaturation levels, nucleation conditions, template architectures, and surface functionalization are just right, none of these factors guarantee the resulting oxide has the right crystalline phase and/or orientation to achieve a desired function. Magnetotactic bacteria provide examples of what materials scientists are striving for in terms of crystal phase, orientation, and growth habit [6, 27] (Fig. 9.13). These bacteria produce magnetic magnetite (Fe_3O_4) with exactly the right size (i.e., the 40–50 nm of a single magnetic domain) and a controlled octahedral morphology within linear chains of vesicles. Each crystal has exactly the same [111] orientation running parallel to the vesicle chain axis. All these factors result in the maximum magnetization in the desired direction with the minimum use of iron. Unfortunately, the specific mechanisms leading to the observed nucleation and growth phenomena are not well understood. However, it appears

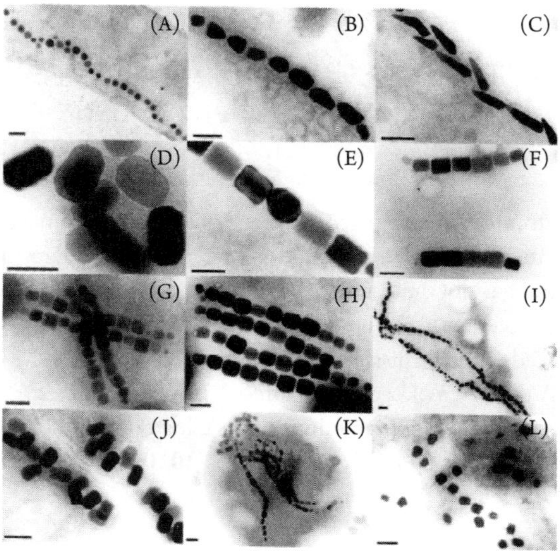

Figure 9.13 (A–L) Transmission electron micrographs of the chains of oriented single-domain magnetite particles (called magnetosomes) that are produced by controlled nucleation and growth within magnetotactic bacteria, or oriented aggregation of nanoparticles. The scale bars correspond to 0.1 μm.
SOURCE: Images reproduced with permission from Schuler, D.; Formation of Magnetosomes in Magnetotactic Bacteria *J. Molec. Microbiol. BioTechn.*, **1999**, *1*, 79–86.

that solution species that are bound to specific sites or surfaces can serve as *site blockers* to bias the growth of specific crystalline faces or induce the growth of specific phases. The roles of biological adsorbates are similar to those highlighted in Chapter 7 regarding the generation of ZnO crystals with a wide range of crystal growth habits.

Active sites for crystal growth are usually at the step edges (see Chapter 7). Site blockers adsorb to specific crystal faces. When a growing step translating across a crystal face encounters site blockers, growth on that face is inhibited because the movement of growth steps is pinned. A biologically relevant example of the use of site blockers involves AFM studies of the growth of the calcium oxalate mineral whewellite ($CaC_2O_4 \cdot H_2O(s)$) [46], a major constituent of kidney stones. The undesired growth of kidney stones that nucleate in urine can have highly painful and even life-threatening consequences. It has long been known that proteins rich in aspartic acid (an amino acid with a pendant carboxyl group) have a high affinity for calcite ($CaCO_3$) surfaces, and can have a dramatic effect on calcite growth rates and crystal growth habits. Figure 9.14 shows that an engineered protein rich in aspartic acid is readily adsorbed by calcium oxalate surfaces, where it interferes with mineral growth. Such proteins could be enormously beneficial in the treatment of kidney stones.

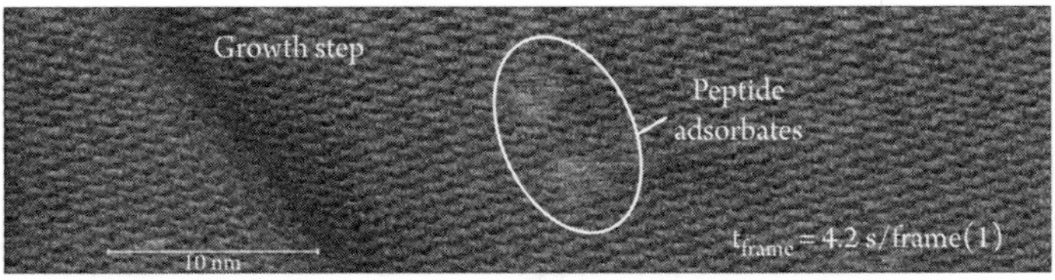

Figure 9.14 An atomic force microscopic image of step growth on a crystal of calcium oxalate monohydrate in the presence of an engineered peptide. Here, the step is growing from left to right and has not quite reached the adsorbed proteins (corresponding to the diffuse blobs near the center of the image).
SOURCE: Reproduced with permission from Friddle, R.W.; Weaver, M.L.; Qiu, S.R.; Wierzbicki, A.; Casey, W.H.; and De Yoreo, J. J. Subnanometer Atomic Force Microscopy of Peptide-Mineral Interactions Links Clustering and Competition to Acceleration and Catastrophe. *Proc. Natl. Acad. Sci.*, **2010**, *107*, 11–15.

One of the more extensively studied oxide systems in which biological site blockers are known to be effective involve the role of the *antifreeze proteins* in altering the crystal growth habits of ice [47, 48]. The nucleation and growth of dendritic ice crystals can destroy cellular structures. To prevent death by freezing, fish, insects, and even bacteria that are exposed to subzero temperatures in the Arctic and Antarctic have evolved within the past million years to produce antifreeze proteins. Because ice nucleation can also be induced at high pressures (see Chapter 3), such proteins also enhance the viability of bacteria in high-pressure water that they can encounter in deep-ocean or subsurface environments [49–51]. Early researchers thought that antifreeze proteins functioned like the antifreeze in a car's radiator or salt used to deice roads. Both substances lower the freezing temperature of water via the colligative property of freezing-point depression. However, it has since been determined that the antifreeze proteins do not inhibit the nucleation of ice particles. Instead, they adhere strongly to the surfaces of freshly formed nuclei to inhibit their subsequent growth.

As mentioned earlier, ice typically crystallizes as the hexagonal I_h phase, where the most extensive and damaging crystal growth involves dendrites that form on prismatic faces [47]. Scientists believe that ice-nucleating proteins bind strongly and extensively to the prismatic faces, but not to the basal planes that grow along the c-axis of a nucleated ice crystal [48] (Fig. 9.15). As each new layer grows along the c-axis, more antifreeze protein (AFP) is adsorbed on the newly formed prismatic phases, resulting in smaller and smaller basal planes in each successive layer. Eventually, the ice crystal pinches off into bipyramidal ice crystals that are coated completely with AFP molecules, which terminates the crystal-growth process. The exact mechanism for the preferential adsorption of AFP molecules on the prismatic faces is still unknown. However, it is speculated that the repeat distance in the

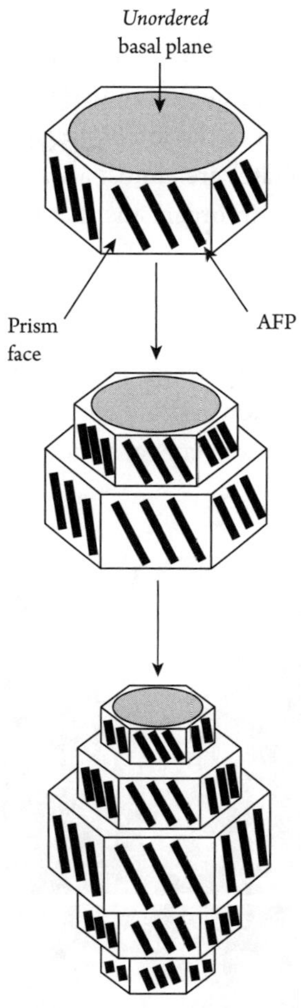

Figure 9.15 Antifreeze proteins play an important role in inhibiting crystal growth. In growth suppression, antifreeze protein is adsorbed preferentially on the prismatic faces of ice, leaving the basal planes largely untouched. Because growth is thus more rapid on basal planes, the crystal eventually pinches off to form a hexagonal bipyramid (bottom) in which only slow-growing prism faces are present.
SOURCE: Images reproduced with permission from Hew, C. L.; and Yang, D. S. C.; Protein Interaction with Ice. *Eur. J. Biochem.*, **1992**, *203*, 33–42 (p. 36, Fig. 3).

α- helix of some AFPs are an exact match for specific planes on the ice crystal [52]. Binding is thought to arise as a result of hydrogen bond formation between the side chains of a three-amino-acid repeat unit (threonine, aspartic acid, and asparagine) and the oxygen atoms of the ice crystal surface. Differences in polarity between the basal and prismatic planes are also thought to bias AFP adsorption.

Biological adsorbates can also determine which phases of a given oxide are expressed. For example, organisms are able to produce selectively all the oxides, oxyhydroxides, and hydroxides of iron, with the exception of hematite. The mechanisms for such phase control remain murky. It appears that, in many instances, these adsorbates do not nucleate the desired phase, but are involved in the subsequent transformation of the initial precipitates into the desired phase. For oxides, the initial precipitates are almost always the most soluble (and often amorphous) hydrated phases. In organisms that form iron oxides, this phase is usually amorphous ferrihydrite. The initial precipitates are then transformed into progressively less soluble phases via reversible hydrolysis and condensation reactions as described by the Ostwald Law of Stages (see Chapter 7). The role of biomolecules in this process appears to involve directing this progression via phenomena such as complexation and site blocking, and lowering the activation barriers to the formation of desired crystal structures.

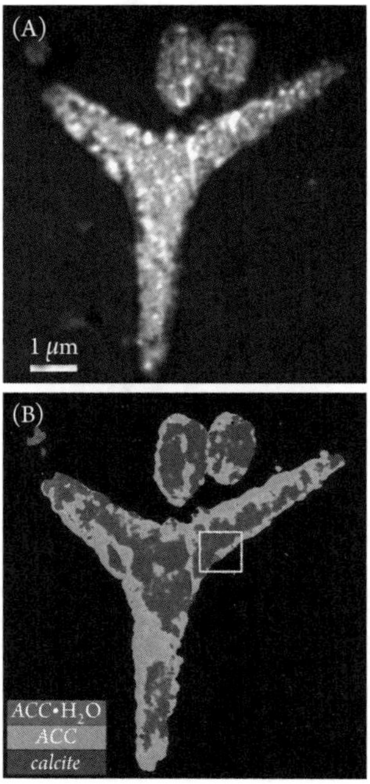

Figure 9.16 The evolution of calcium carbonate phases in sea urchin (*Strongylocentrotus pupuratus*) spicules as determined by X-ray structural mapping (see Plate 16). Sea urchin larvae make a set of spicules that begin as a trigonal form. This trigonal template is first filled with amorphous, hydrated calcium carbonate (ACC) that eventually converts into a single crystal of calcite. For the intermediate stage shown, the gray scales in the inset correspond to different phases (hydrated ACC, dehydrated ACC, and calcite).
SOURCE: Images reproduced with permission from Gong, Y. U. T.; Killian, C. E.; Olson, I. C.; Appathurai, N. P.; Amasino, A. L.; Martin, M. C.; Holt, L. J.; Wilt, F. H.; Gilbert, P. U. P. A. Phase Transitions in Biogenic Amorphous Calcium Carbonate. *Proc. Nat. Acad. Sci.*, **2012**, *109*, 6088–6093. (see Plate 16)

The phase transformation sequence in a related biomineralization process involving $CaCO_3$ deposition in sea urchin larvae has been observed *in situ* at high magnification using high-fluence X-ray structural mapping [53] (Fig. 9.16, Plate 16). In sea urchins, it is thought that the first step in $CaCO_3$ spicule formation involves the nucleation and growth of hydrated, amorphous, calcium-carbonate (ACC) nanoparticles. Next, a membrane template is used to bind and organize numerous nuclei of ACC within a trigonal array. The third and final step involves the transformation of the bound ACC into more stable crystalline forms. In general, ACC can be induced to dehydrate and transform into various $CaCO_3$ phases, including vaterite, the aragonite in nacre, and calcite. However, the *in situ* images in Figure 9.16 suggest that in sea urchins, the reaction sequence involves the transformation of hydrated ACC to dehydrated ACC, which is then transformed directly into calcite. The final product is a desirable trigonal spicule architecture in which most of the initial nuclei merge to form a single calcite crystal. The use of amorphous precursors that transform into desired crystalline phases has been adapted to assist in the repair of bone. For more than a decade, amorphous calcium phosphate has been available in kits that start making apatite-like bone material within a few days after application in surgery [54].

9.7 ACTIVE ASSEMBLY USING MICROTUBULES AND MOTOR PROTEINS

Although complex architectures can be clearly created using self-assembly, self-assembly alone is unable to generate structures that are as complex as bone or a diatom skeleton. The creation of such structures requires nonthermodynamic processing strategies called *active assembly*. Active assembly involves the use of energy-consuming proteins to transport and assemble components. The classic active assembly system in eukaryotic cells involves two classes of energy-consuming proteins. The first class involves the intricate networks of dynamic protein filaments that comprise the cytoskeleton, including microtubules, actin filaments, and intermediate filaments. The second class of proteins involves energy-consuming nanorobots called *motor proteins*, which are programmed to walk along cytoskeletal fibers carrying desired cargo to desired locations. Because the combination of microtubules and the motor protein kinesin have seen the most use in artificial systems, we highlight the use of these two proteins in oxide synthesis here. Those of you interested in more detail regarding the structures, properties, and functions of the motor proteins and cytoskeletal fibers are directed to other sources [55–57].

Microtubules are polar hollow protein fibers that are 25 nm in diameter and can be more than 1 mm in length (Figs. 9.17). Microtubules resemble collagen fibers in that they exhibit hierarchical assembly. Microtubules are formed via the

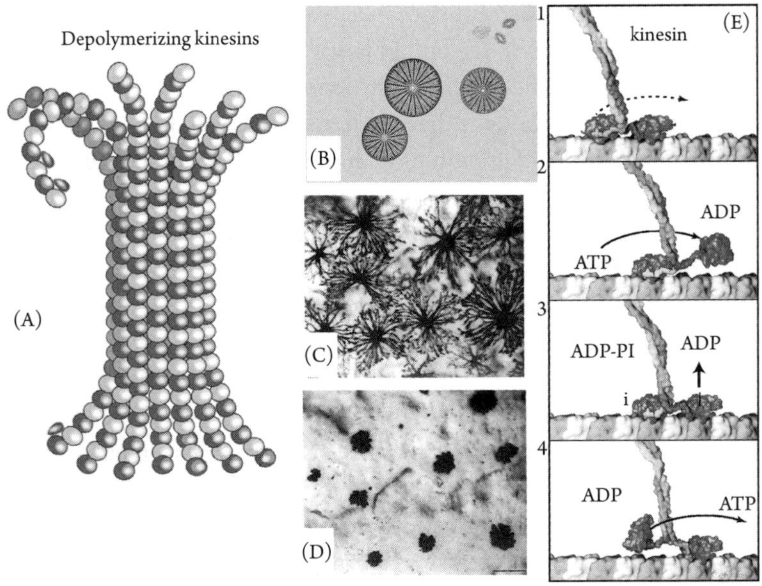

Figure 9.17 Components used in the "active" assembly of oxides. (A) – A depiction of a tubulin fiber during the catastrophic disassembly stage of dynamic instability. (Image reproduced with permission from Amos, L. A.; Microtubule Structure and Stabilization. *Org. Biomol. Chem.*, **2004**, *2*, 2153–2160.) (B)-(D) – Passive and active constructs organized around microtubule organizing centers (MOCs). In all cases, microtubules radiate outward from the central MOC, with kinesin motors transporting materials along the microtubule networks. (B) – The transport of vesicles and/or nanoparticles along fiber networks eventually results in the formation of rigid diatom skeletons composed of silica. (SOURCE: The round diatom *Arachnoidiscus. Jubal* Harshaw/Shutterstock.com with permission.) (C)-(D) – Dye particles carried along polar fiber networks are used to define two distinct states in the reversible color changing systems of fish (shown) and chameleons. The fiber network clearly appears in the "extended" state shown in (C). (Images reproduced with permission from Haimo, I. T.; Thaler, C. D. Regulation of Organelle Transport: Lessons from Colour Change in Fish. *BioEssay*, **1994**, *16*, 727.) (E) – A sequence of images showing kinesin motors "walking" along a microtubule by taking 8 nm steps corresponding to the length of the tubulin dimer. Motors can walk as speeds exceeding 1 μm/sec. (Images reproduced with permission from Vale, R. D.; Milligan, R. A.; The Way Things Move: Looking Under the Hood of Molecular Motor Proteins. *Science*, **2000**, *288*, 88–95.)

polymerization of tubulin dimers. These dimers form linear chains called *protofilaments*. Thirteen protofilaments commonly self-assemble to create a single microtubule. Microtubules are sometimes arranged in a centro-symmetric polar fashion around particles called *microtubule-organizing centers* (corresponding to the centrosomes in animal cells). However, although microtubules undergo hierarchical assembly like collagen, microtubules exhibit one property that collagen does not. Microtubules can assemble and disassemble on command via a process called *dynamic instability*.

Dynamic instability involves two key steps: assembly and disassembly. During the assembly step, the ends of tubulin dimers are glued together via interactions with a triphosphate molecule called *guanosine triphosphate* (GTP), which leads to the growth of the microtubule. In the disassembly step, an enzymatic site in the tubulin hydrolyzes GTP to the diphosphate molecule guanosine diphosphate. GTP hydrolysis releases energy in the form of protein reconfiguration. This conformation change, and the fact that guanosine diphosphate is a much weaker glue than GTP, triggers a catastrophic depolymerization or shrinkage of the microtubule, indicating that microtubules are dynamic fibers that are in sharp contrast to static collagen fibers. More important, cells can control this dynamic behavior.

Growing and shrinking microtubules are used to push or pull on other objects for processes ranging from cell division to the crawling motion of the amoeba. For biomineralization, organisms such as diatoms use dynamic instability to grow microtubule architectures with the exact size, shape, and internal structure needed as templates for oxide growth. These architectures often involve other cellular components such as microtubule organizing centers, lipid membranes, or protein fibers that the microtubules help push or pull into place. When the template network is optimized, dynamic instability is turned off and the template is stabilized for subsequent oxide deposition processes.

Motor proteins such as kinesin (Fig. 9.17) provide an additional mechanism that assists in the transport and assembly of nanomaterials. Kinesin motors are adsorbed onto microtubule surfaces by two "feet." Kinesin hydrolyzes adenosine triphosphate (ATP) as a fuel that powers the motor protein to walk along the microtubule in a specific polar direction. The tail of the motor protein is functionalized to carry specific cargo such as vesicles, organelles, or even other fibers. For example, the motor protein myosin interacting with multiple actin filaments is responsible for muscle contraction. In cells, the moving objects in a given transportation system are almost always motors that walk along microtubules in what is called *normal motility* [57] (Fig. 9.18). On substrate surfaces in artificial systems it is often more convenient (and certainly more robust) to create transportation systems based on so-called *inverted motility* [58] (Fig. 9.18). Here, motor tails are inserted into surface-bound SAMs and/or protein arrays, placing the mobile feet at the immediate solid–solution interface. Short segments of microtubules that have either been stabilized or crosslinked to prevent dynamic instability are then adsorbed onto multiple motors protruding from the monolayer surface. In the presence of adenosine triphosphate fuel, each microtubule is propelled across the surface. During inverted motility, functionalized shuttles are propelled by thousands of motors rather than the single motors deployed in

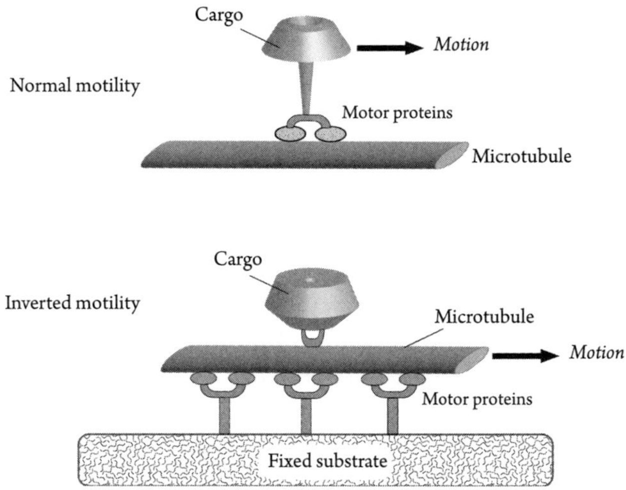

Figure 9.18 Cargo-carrying modes for active and/or dynamic assembly of oxides and other species. (Top) Single kinesin motors carry cargo by walking along microtubules in a normal motility geometry. Single motors can carry objects exceeding a micron in diameter. (Bottom) In artificial systems, transport is often provided using inverted motility. Here, many motors are anchored to substrate surfaces by their tails within self-assembled monolayers that are often patterned using lithographic techniques. Thousands of motors propel short (10 μm) segments of stabilized, cargo-carrying microtubules across the surface.

normal motility. These microtubule shuttles then serve as the vehicles that transport and assemble desired cargo such as oxide nanoparticles.

The role of energy-consuming proteins in the active assembly of oxides is illustrated by the creation of the silica skeleton of diatoms, radiolarians, and chrysophytes [59]. These silica-containing photoplankton may be small, but they represent one of the most pervasive and important ecosystems on our planet. These mostly single-celled creatures fix more organic carbon (105 Pg C/y) than the rain forests and savannas of Earth combined (17.8 Pg C/y and 16.8 Pg C/y, respectively), as well as generate one fifth of the oxygen we breathe. The skeletons of diatoms (Fig. 9.17) form huge and highly useful silica deposits called *diatomaceous earth*.

The diatom creates the architecture desired for its skeleton by controlling the dynamic instability of an axially symmetric polar microtubule array defined by a central microtubule organizing center. These structures have been replicated in artificial systems by attaching microtubules containing biotin-decorated segments onto streptavidin-coated nanoparticles. Next, vesicle-loving motor proteins attach themselves to microtubule surfaces and capture the desired vesicles as cargo. These vesicles are associated with proteins such as the silaffins, which promote the condensation and polymerization of soluble silicate monomers. In some instances, an amorphous silica particle forms within each vesicle, which is then transported by the motor protein to desired locations on the template surface. However, vesicle transport often occurs before mineralization. In addition, interactions involving either multiple motors or the microtubules themselves can distort the vesicles into nonspherical shapes, including sausage-like structures, linear chains of interconnected strings-of-pearls, flattened balloons, and even long lipid nanotubes [60]. In artificial systems, it has been shown that microtubule shuttles deployed on substrates in the inverted motility geometry can pull out extended networks of lipid nanotubes from large vesicles that are adsorbed to the surface [61]. In the diatom skeleton, controlled-porosity motifs can be created by having motor proteins deliver controlled mixtures of distorted silica deposition vesicles within close-packed arrays of larger normal vesicles. When the array is assembled, mineralization commences. Distorted vesicles form the basis for mineralized structures including curved shells, sheets, and even tangential rods.

Although materials scientists have yet to replicate the synthesis of structure as beautiful or complex as a diatom skeleton, there have been more simple demonstrations of the use of microtubules and motor proteins in the active assembly of oxide structures in artificial systems. The simplest strategy for exploiting microtubules involves (1) growing microtubules in desired locations via control of local temperatures (e.g., heating induces microtubule polymerization) and then (2) mineralizing the native microtubules in place. Using this strategy, a range of mineralized microtubules has been formed (Fig. 9.19), including FeOOH [62]; ZnO, ZnS, and CdS [63]; as well as Ag and Au. In the case of Au°, microtubules have been used to form conductive interconnects between electrodes deployed

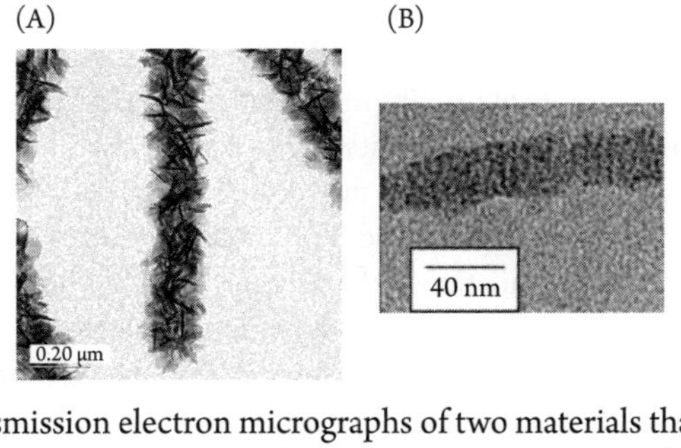

Figure 9.19 Transmission electron micrographs of two materials that have been nucleated and grown on microtubules. (A) Lepidocrocite (FeOOH). (B) CdS quantum dots. For CdS, one quantum dot appears to be nucleated per tubulin unit, replicating the exact structure of the underlying microtubule.

SOURCE: Images reproduced with permission from Boal, A. K.; Headley, T. J.; Tissot, R. G.; Bunker, B. C.; Microtubule-Templated Biomimetic Mineralization of Lepidocrocite *Adv. Funct. Mater.*, **2004**, *14*, 19–24.

in microfluidic systems [64]. A slightly more complex strategy involves functionalizing tubulin monomers to bind specific oxide nanoparticles. For example, tubulin functionalized with biotin is available commercially. Because biotin forms a strong complex with streptavidin, streptavidin-coated nanoparticles (many of which are also available commercially) are readily attached to tubulin and/or prepolymerized microtubules. This approach does not produce the continuous coatings that result from in situ mineralization of microtubules. However, mineralization of microtubules is irreversible because the mineralized coating forms a hard, continuous shell. In contrast, particle-laden microtubules can be depolymerized, resulting in programmable nanoparticle arrays.

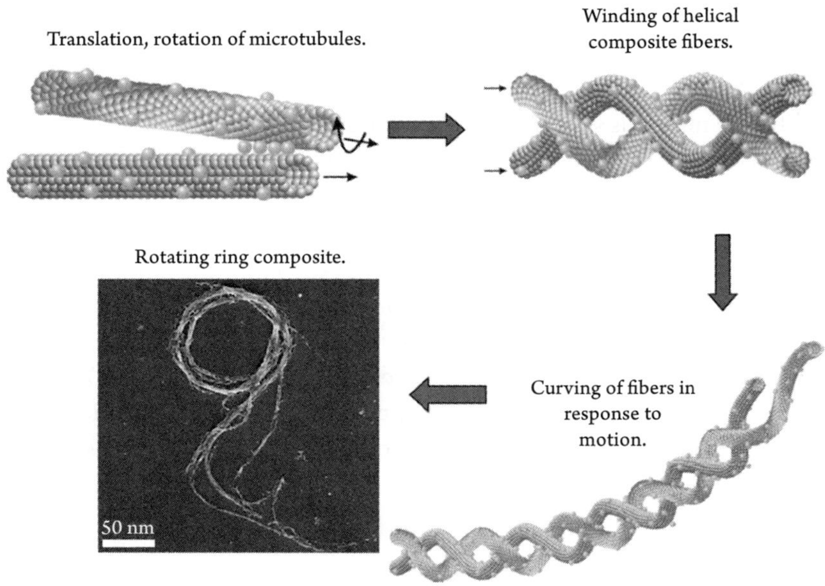

Figure 9.20 Active assembly of rotating-ring nanocomposites (clockwise from top left). Motor proteins in an inverted motility assay propel microtubules and their nanoparticle cargo across a surface. If the microtubule is helical (i.e., contains 12 or 14 protofilaments), it rotates as it is propelled. When two particle-laden microtubules collide with one another, the functionalized particles glue the shuttles together. The rotation of helical shuttles winds the two together to form a double helix. Translation of the helix induces it to curve back on itself to form a ring. The ring then rotates around a central axis to wind up all microtubules that encounter it much as a spool winds up thread. The assembled rings continue to rotate so long as adenosine triphosphate fuel is available to the motors.

SOURCE: Images reproduced with permission from Liu, H.; Spoerke, E.D.; Bachand, M.; Koch, S.J.; Bunker, B.C.; and Bachand, G.D. Biomolecular Motor-Powered Self-Assembly of Dissipative Nanocomposite Rings. *Adv. Mater.*, **2008**, *20*, 4476–4481.

The final and most complex strategies for creating active oxide arrays involve attaching oxide nanoparticles to mobile components. This strategy resembles what happens during the assembly of the diatom skeleton, except that instead of transporting vesicles, active components carry functionalized nanoparticles [65]. It was recently discovered that, under certain conditions, microtubules laden with nanoparticles and propelled in the inverted motility geometry stick to each other. Constructs containing multiple microtubules can then form coiled-coil structures that bend back on themselves to form rotating-ring microtubule–nanoparticle composites [66]. Images of these composites and details regarding the mechanisms of their formation are depicted in Figure 9.20 for the specific case when the particles are optically active quantum dots. Such rotating rings have never been observed in cellular systems. There is no telling how many other new forms of active oxide nanocomposites await to be discovered in artificial systems.

References

1. Cuif.; J.- P.; Dauphin, Y.; Sorauf, J. E. *Biominerals and Fossils through Time.* Cambridge University Press: Cambridge, 2011.
2. Addadi, L.; Weiner, S. Biomineralization: A Pavement of Pearl. *Nature* **1997**, *389,* 912–915.
3. Alberts, B.; Bray, D.; Johnson, A.; Lewis, J.; Raff, M.; Roberts, K.; Walter, P. *Essential Cell Biology*; Garland Publishing: New York, 1998.
4. Mueller, W. E. G. *Molecular Biomineralization: Aquatic Organisms Forming Extraordinary Materials*; Springer: New York, 2011.
5. Lowenstam, H. A.; Weiner, S. *On Biomineralization.* Oxford: New York, 1989.
6. Mann, S. *Biomineralization: Principles and Concepts in Bioinorganic Materials Chemistry*; Oxford University Press: Oxford, 2001.
7. Mann, S.; Webb, J.; Williams, R. J. P. *Biomineralization: Chemical and Biochemical Perspectives*; VCH: New York, 1989.
8. Bauerlein, E.; Behrens, P.; Epple, M. *Handbook of Biomineralization: Biomimetic and Bioinspired Chemistry.* Wiley-VCH: New York, 2009.
9. Branden, C.; Tooze, J. *Introduction to Protein Structure, 2nd ed.*; Garland Publishing: New York, 1999.
10. Tiddy, G. J. T. Surfactant–Water Liquid Crystal Phases. *Phys. Rep.* **1980**, *57*, 1–46.
11. Beck, J. S.; Vartuli, J. C.; Roth, W. J.; Leonowicz, M. E.; Kresge, C. T.; Schmitt, K. D.; Chu, C. T. W.; Olson, D. H. Sheppard, E. W. A New Family of Mesoporous Molecular Sieves Prepared with Liquid Crystal Templates. *J. Am. Chem. Soc.* **1992**, *114,* 10834–10843.
12. Barrer, R. M. *Hydrothermal Chemistry of Zeolites*; Academic Press: New York, 1982.
13. Mayer, G. Rigid Biological Systems as Models for Synthetic Composites. *Science* **2005**, *310*, 1144–1147.
14. Kim, Y.-Y.; Ganesan, K.; Yang, P.; Kulak, A.N.; Borukhin, S.; Pechook, S.; Ribeiro, L.; Kröger, R.; Eichhorn, S. J.; Armes, S. P.; Pokroy, B.; and Meldrum, F. C. An Artificial Biomineral Formed by Incorporation of Copolymer Micelles in Calcite Crystals. *Nat. Mater.* **2011**, *10*, 890–896.

15. Estroff, L. A.; Cohen, I. Biomineralization: Micelles in a Crystal. *Nat. Mater.* **2011**, *10*, 810–811.
16. Antonelli, D. M.; Ying, J. Y. Synthesis of a Stable Hexagonally Packed Mesoporous Niobium Oxide Molecular Sieve Through a Novel Ligand-Assisted Templating Mechanism. *Angew. Chem. Int. Ed. Engl.* **1996**, *35*, 426–430.
17. Lee, Y.J.; Yi, H.; Kim, W.-J.; Kang, K.; Yun, D. S.; Strano, M. S.; Ceder, G.; and Belcher, A.M. Fabricating Genetically Engineered High-Power Lithium-Ion Batteries Using Multiple Virus Genes. *Science* **2009**, *324*, 1051–1055.
18. Wang, G.; Cao, R.-Y.; Chen, R.; Mo, L.; Han, J.-F.; Wang, X.; Xu, X.; Jiang, T.; Deng, Y.-Q.; Lyu, K.; Zhu, S.-Y.; Qin, E. D.; Tang, R.; and Qin, C.-F. Rational Design of Thermostable Vaccines by Engineered Peptide-Induced Virus Self-Biomineralization under Physiological Conditions. *Proc. Nat. Acad. Sci. USA* **2013**, *110*, 7619–7624.
19. Urist, M. R. Bone Formation by Autoinduction. *Science* **1965**, *150*, 893–899.
20. Dalas, E. Crystallization of Sparingly Soluble Salts on Functionalized Polymers. *J. Mater. Chem.* **1991**, *1*, 473–474.
21. Dalas, E.; Kallitsis, J.; Koutsoukos, P. G. The Crystallization of Calcium Carbonate on Polymeric Substrates. *J. Cryst. Growth* **1988**, *89*, 287–294.
22. Bunker, B. C.; Rieke, P. C.; Tarasevich, B. J.; Campbell, A. A.; Fryxell, G. E.; Graff, G. L.; Song, L.; Liu, J.; Virden, J. W.; McVay, G. L. Ceramic Thin-Film Formation on Functionalized Interfaces through Biomimetic Processing. *Science* **1994**, *264*, 48–55.
23. Ulman, A. *An Introduction to Ultrathin Organic Films*; Academic Press: Boston, 1991.
24. Kowalewski, B. R.; Poppe, J.; Demmer, U.; Warkentin, E.; Dierks, T.; Ermler, U.; Schneider, K. Nature's Polyoxometalate Chemistry: X-ray Structure of the Mo Storage Protein Loaded with Discrete Polynuclear Mo-O Clusters. *J. Am. Chem. Soc.* **2012**, *134*, 9768–9774.
25. Powell, A. K. Ferritin: Its Mineralization. *Metal Ions Biol. Syst.* **1998**, *35*, 515–561.
26. Powell, A. K. Ferritins. *Comprehens. Coord. Chem. II* **2004**, *8*, 169–194.
27. Cornell, R. M.; Schwertmann, U. *The Iron Oxides*; Wiley-VCH: Weinheim, Germany, **2003**; p. 17.
28. Douglas, T.' Stark, V. T. Nanophase Cobalt Oxyhydroxide Mineral Synthesized within the Protein Cage of Ferritin. *Inorg. Chem.*, **2000**, *39*, 1828–1830.
29. Iwahori, K.; Yoshizawa, K.; Muraoka, M.; and Yamashita, I. Fabrication of ZnSe Nanoparticles in the Apoferritin Cavity by Designing a Slow Chemical Reaction System. *Inorg. Chem.* **2005**, *44*, 6393–6400.
30. Matijevic, E.; Melman, A. Synthesis of Inorganic Nanoparticles Using Protein Templates. In *Fine Particles in Medicine and Pharmacy*; Springer: New York, 2012; pp 195–221.
31. Meldrum, F. C.; Douglas, T.; Levi, S.; Arosio, P.; and Mann, S. Reconstitution of Manganese Oxide Cores in Horse Spleen and Recombinant Ferritins. *J. Inorg. Biochem.* **1995**, *58*, 59–68.
32. Mann, S.; Hannington, J. P.; Williams, R. J. P. Phospholipid Vesicles as a Model System for Biomineralization. *Nature* **1986**, *324*, 565–567.
33. Bhandarkar, S.; Bose, A. Synthesis of Submicrometer Crystals of Aluminum Oxide by Aqueous Intravesicular Precipitation. *J. Colloid Interface Sci*. **1990**, *135*, 531–538.
34. Shimizu, K.; Cha, J.; Stucky, G. D.; Morse, D. E. Silicatein: A Cathepsin L-Like Protein in Sponge Biosilica. *Proc. Nat. Acad. Sci. USA* **1998**, *95*, 6234–6238.

35. Tomczak, M. M.; Glawe, D. D.; Drummy, L. F.; Lawrence, C. G.; Stone, M. O.; Perry, C. C.; Pochan, D. J.; Deming, T. J.; and Naik, R. R. Polypeptide-Templated Synthesis of Hexagonal Silica Platelets. *J. Am. Chem. Soc.* **2005**, *127*, 12577–12582.
36. Fairhead, M.; Johnson, K. A.; Kowatz, T.; McMahon, S. A.; Carter, L. G.; Oke, M.; Liu, H.; Naismith, J. H.; van der Walle, C. F. Crystal Structure and Silica Condensing Activities of Silicatein alpha-Cathepsin L Chimeras. *Chem. Comm.* **2008**, 1765–1767.
37. Jensen, M.; Keding, R.; Hoêche, T.; Yue, Y. Biologically Formed Mesoporous Amorphous Silica. *J. Am. Chem. Soc.* **2009**, *131*, 2717–2721.
38. Kisailus, D.; Truong, Q.; Amemiya, Y.; Weaver, J. C.; Morse, D. E. Self-Assembled Bifunctional Surface Mimics an Enzymatic and Templating Protein for the Synthesis of a Metal Oxide Semiconductor. *Proc. Nat. Acad. Sci. USA* **2006**, *103*, 5652–5657.
39. Brutchey, R. L.; Yoo, E. S.; Morse, D. E. Biocatalytic Synthesis of a Nanostructured and Crystalline Bimetallic Perovskite-like Barium Oxofluorotitanate at Low Temperature. *J. Am. Chem. Soc.* **2006**, *128*, 10288–10294.
40. Curnow, P.; Bessette, P. H.; Kisailus, D.; Murr, M. M.; Daugherty, P. S.; Morse, D. E. Enzymatic Synthesis of Layered Titanium Phosphates at Low Temperature and Neutral pH by Cell-Surface Display of Silicatein. *J. Am. Chem. Soc.* **2005**, *127*, 15749–15755.
41. Cha, J. N.; Shimizu, K.; Zhou, Y.; Christiansen, S. C.; Chmelka, B. F.; Stucky, G. D.; Morse, D. E. Silicatein Filaments and Subunits from a Marine Sponge Direct the Polymerization of Silica and Silicones in Vitro. *Proc. Nat. Acad. Sci. USA* **1999**, *96*, 361–365.
42. Vonnegut, B.; Chessin, H. Ice Nucleation by Coprecipitated Silver Iodide and Silver Bromide. *Science* **1971**, *174*, 945–946.
43. Hu, J.: Xiao, X.- D.; Ogletree, D. F.; Salmeron, M. The Structure of Molecularly Thin Films of Water on Mica in Humid Environments. *Surf. Sci.* **1995**, *344*, 221–236.
44. Lindow, S. E.; Amy, D. C.; Upper, C. D. Bacterial Ice Nucleation: A Factor in Frost Injury to Plants. *Plant. Physiol.* **1982**, *70*, 1084–1089.
45. Kajava, A. V.; Lindow, S. E. A Model of the Three-Dimensional Structure of Ice Nucleation Proteins. *J. Mol. Biol.* **1993**, *232*, 709–717.
46. Friddle, R. W.; Weaver, M. L.; Qiu, S. R.; Wierzbicki, A.; Casey, W. H.; De Yoreo, J. J. Subnanometer Atomic Force Microscopy of Peptide–Mineral Interactions Links Clustering and Competition to Acceleration and Catastrophe. *Proc. Natl. Acad. Sci. USA* **2010**, *107*, 11–15.
47. Hew, C. L.; Yang, D. S. C. Protein Interaction with Ice. *Eur. J. Biochem.* **1992**, *203*, 33–42.
48. Yeh, Y.; Feeney, R. E. Antifreeze Proteins: Structures and Mechanisms of Function. *Chem. Rev.* **1996**, *96*, 601–617.
49. Vanlint, D.; Mitchell, R.; Bailey, E.; Meersman, F.; McMillan, P. F.; Michiels, C. W.; Aertsen, A. Rapid Acquisition of Gigapascal-High-Pressure Resistance by *Escherichia coli*. *mBio*, **2011**, *2*, 1–3.
50. Picard, A.; Testemale, D.; Hazemann, J.-L.; Daniel, I. The Influence of High Hydrostatic Pressure on Bacterial Dissimilatory Iron Reduction. *Geochim. Cosmochim Acta* **2012**, *88*, 120–129.

51. Sharma, A.; Scott, J. H.; Cody, G. D.; Fogel, M. L.; Hazen, R. M.; Hemley, R. J.; Huntress, W. T. Microbial Activity at Gigapascal Pressures. *Science* **2002**, *295*, 1514–1516.
52. Jia, Z.; Davies, P. L. Antifreeze Proteins: An Unusual Receptor–Ligand Interaction. *Trends Biochem. Sci.* **2002**, *27*, 101–106.
53. Gong, Y. U. T.; Killian, C. E.; Olson, I. C.; Appathurai, N. P.; Amasino, A. L.; Martin, M. C.; Holt, L. J.; Wilt, F. H.; Gilbert, P. U. P. A. Phase Transitions in Biogenic Amorphous Calcium Carbonate. *Proc. Nat. Acad. Sci. USA* **2012**, *109*, 6088–6093.
54. Fulmer, M. T.; Ison, I. C.; Hankermayer, C. R.; Constantz, B. R.; Ross, J. Measurements of the Solubilities and Dissolution Rates of Several Hydroxyapatites. *Biomaterials* **2002**, *23*, 751–755.
55. Howard, J. *Mechanics of Motor Proteins and the Cytoskeleton*; Sinauer Associates: Sunderland, MA, 2001.
56. Hess, H.; Vogel, V. Molecular Shuttles Based on Motor Proteins: Active Transport in Synthetic Environments. *Rev. Mol. Biotechnol.* **2001**, *82*, 67–85.
57. Limberis, L.; Stewart, R. J. Toward Kinesin-Powered Devices. *Nanotechnology* **2000**, *11*, 47–51.
58. Howard, J.; Hudspeth, A. J.; Vale, R. D. Movement of Microtubules by Single Kinesin Motors. *Nature* **1989**, *342*, 154–158.
59. Parkinson, J.; Brechet, Y.; Gordon, R. Centric Diatom Morphogenesis: A Model Based on a DLA Algorithm Investigating the Potential Role of Microtubules. *Biochim. Biophys. Acta* **1999**, *1452*, 89–102.
60. Emsellem, V.; Cardoso, O.; Tabeling, P. Vesicle Deformation by Microtubules: A Phase Diagram. *Phys. Rev. E* **1998**, *58*, 4807–4810.
61. Bouxsein, N. F.; Carroll-Portillo, A.; Bachand, M.; Sasaki, D. Y.; Bachand, G. D. A Continuous Network of Lipid Nanotube Fabricated from the Gliding Motility of Kinesin Powered Microtubule Filaments. *Langmuir* **2013**, *29*, 2992–2999.
62. Boal, A. K.; Headley, T. J.; Tissot, R. G.; Bunker, B. C. Microtubule-Templated Biomimetic Mineralization of Lepidocrocite. *Adv. Funct. Mater.* **2004**, *14*, 19–24.
63. Spoerke, E. D.; Connor, B. A.; Gough, D. V.; McKenzie, B. B.; Bachand, G. D. Microtubule-Templated Cadmium Sulfide Nanotube Assemblies. *Part. Part. Syst. Charac.* **2014**, *31*, 863–870.
64. Spoerke, E. D.; Hendricks, J. K.; Bachand, G. D.; Boal, A. K.; Liu, H.; Warrender, C.; Bouchard, A.; Osbourn, G.; Bunker, B. C. Raiding Nature's Toolbox: Adapting Microtubules and Motor Proteins for Nanoscale Materials Assembly. *Found. Nanosci.* **1998**, *FNANO08*, 129–133.
65. Boal, A. K.; Bachand, G. D.; Rivera, S. B.; Bunker, B. C. Interactions Between Cargo-Carrying Biomolecular Shuttles. *Nanotechnology* **2006**, *17*, 349–354.
66. Liu, H.; Spoerke, E. D.; Bachand, M.; Koch, S. J.; Bunker, B. C.; Bachand, G. D. Biomolecular Motor-Powered Self-Assembly of Dissipative Nanocomposite Rings. *Adv. Mater.* **2008**, *20*, 4476–4481.

PART FOUR

Technologically Important Oxide Reactions

10

The Ion Exchange Behavior of Oxides

10.1 INTRODUCTION

Oxides comprise the most common ion-exchange materials on our planet, with the clay minerals alone, formed by the weathering of rock, having a total mass of around 10^{25} g. This mass represents almost one-third of the total mass of Earth's crust and is more than six times the mass of Earth's oceans [1, 2]. These fine-grained ion exchange materials play a major role in mediating the concentrations of ionic species found in freshwater, groundwater, and our oceans (see Chapter 18). Oxide ion exchangers are also of critical importance in removing contaminants from the environment. Nowhere is this role more apparent than in the removal and sequestration of radioactive elements such as ^{137}Cs, ^{90}Sr, and ^{99}Tc, which are serious hazards present in nuclear wastes (Fig. 10.1).

Oxide ion exchangers exhibit several properties that make them materials of choice for treating nuclear wastes, including high selectivity, enhanced stability to radiation damage relative to organic exchangers, and the potential as materials to be condensed further into solid waste after they are loaded with radioactive species. Oxide exchangers are extremely useful for extracting valuable cations from complex fluids, such as the lithium used in our highest energy density batteries. Ion exchange also represents a pathway for creating unique nanomaterials, with applications including battery separators, catalysts, optical materials, magnets, and materials for drug delivery.

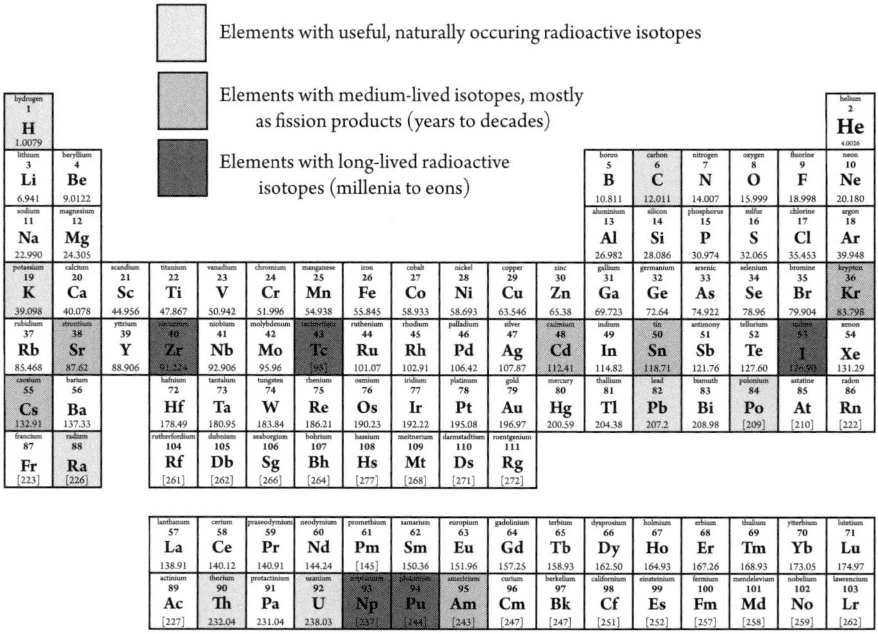

Figure 10.1 A Periodic Table highlighting some of the major radioactive elements that are either of concern in nuclear waste disposal or environmental chemistry. For reactor accidents and nuclear waste disposal, $^{134,137}Cs$, ^{90}Sr, and ^{99}Tc are particularly troublesome.

Oxides materials can exhibit exceptional properties as both cation and anion exchangers for a wide range of separation and water treatment technologies. Although the total ion-exchange capacity of an oxide is important for some applications, such as the deionization of water, separations require the use of oxides and hydroxides having the highest degree of *ion-exchange selectivity*. For selectivity, oxides must be designed with specific sites that exhibit a much higher affinity for one ion than any other, which requires much more sophistication than just generating a net charge. Here, we describe the key factors that control both the capacity and selectivity of inorganic ion exchangers, including (1) the role of acid–base reactions in controlling surface charge and ion-exchange capacity, (2) the role of local charge distributions in determining ion-exchange selectivity, and (3) the effect of shape and selective solvation on enhancing that selectivity. We then provide specific examples of oxide ion exchangers to highlight the critical issues associated with understanding and using their unique properties. Those of you interested in more comprehensive discussions of the basic principles and applications of ion exchange are referred to several books and reviews (such as [3]).

10.2 ION EXCHANGE CAPACITY: THE ROLE OF POTENTIAL-DETERMINING IONS

All oxide and hydroxide surfaces normally exhibit a surface charge when immersed in water (see Chapter 6). This charge arises because of charge imbalances in the bulk material that come from heterometal substitutions (such as Al(III) for Si(IV)) and from acid–base reactions at the oxide surface. The charge is associated with the formal charge on each oxygen surface site as mediated by the adsorption or desorption of protons (i.e., the potential-determining ions, see Chapter 8). At a pH below the point of zero charge (PZC), protonation of the surface creates an excess positive charge via Brønsted acid–base reactions (see Chapter 5). Neutralization of this positive charge usually involves the adsorption of soluble anions from solution. However, the charge can also be mediated by ligand-exchange reactions involving the replacement of oxygen anions by foreign anions (e.g., carbonate, phosphate, or sulfide ions). These reactions are analogous to anion uptake by solutes in homogeneous solutions, such as $[Al(OH_2)_6]^{3+}$ converting to $[AlF(OH_2)_5]^{2+}$ by F^- uptake in water. Anions in solution compete with one another for the privilege of occupying cationic surface sites based primarily on electrostatic attractions. Below the PZC, when the surfaces are positively charged, these oxide surfaces function as *anion exchangers*. The *anion exchange capacity* of such surfaces is equal to the sum of charges associated with all cationic surface sites.

As the pH increases toward the PZC, an oxide's anion-exchange capacity decreases, because proton desorption decreases the net positive surface charge on the solid. For pH values well above the PZC, oxide surfaces are largely deprotonated and are negatively charged. The net charge on these anionic surfaces must be compensated by cations. At pH values above the PZC, an oxide that exhibited anion-exchange properties at low pH now functions as a *cation exchanger*. In the basic pH regime, lowering the solution pH lowers the surface charge and, correspondingly, the cation-exchange capacity of the oxide surface. At the PZC, the net surface charge is zero. However, this does not necessarily mean the surface has absolutely no ion-exchange activity, only that concentrations of both cationic and anionic sites are equal.

There is similarity between the charge distributions on a solid surface and the distribution of aqueous solutes and complexes shown in a hydrolysis diagram. As discussed in Chapter 5 (see Fig. 5.3), the ensemble of metal hydrolysis complexes in solution can have a net positive or negative charge depending on the solution pH. For example, in Figure 5.3 one can see that the ensemble of aqueous Fe(III) species is positive at pH less than 6 (e.g., $[Fe(OH_2)_6]^{3+}$ and

$[FeOH(OH_2)_5]^{2+}$) because there are negligible anionic iron species present. In contrast, the ensemble of iron species has a net negative charge at pH more than 10, where the $[Fe(OH)_4]^-$ ion dominates. In the pH region $7 < pH < 9$, the net charge of all aqueous Fe(III) hydrolysis species is near zero. This point is important. Even for solutes, there is a pH region where the sum total charge of all hydrolysis species is near zero, just as it is for solid surfaces. Thus, hydrolysis diagrams can be an important starting point for predicting the pH conditions under which a given oxide surface is likely to function as either a cation or anion exchange material.

One of the critical factors in selecting an oxide for a specific application involves matching the pK_a values of active surface sites to the pH of the target solution. The ion-exchange capacity of materials is often designed to be extremely high. For example, $Na_2Ti_3O_9$ has a cation-exchange capacity per Ti(IV) of 0.67. However, this high exchange capacity is achieved only if the potential-determining ions (protons or hydroxide ions) are unable to compete with the target ions for ion-exchange sites. For cation exchange, this means the pK_a of the important anionic sites in the structure must be at least 1 unit less than the solution pH at which the exchange process is to be carried out. Exploiting the full cation-exchange capacity of $Na_2Ti_4O_9$ requires pH values greater than 11 (see Fig. 6.6) [4]. Similarly, the same material could be exploited as an anion exchanger, but only at pH values less than 2. Above this pH, exchange sites are anionic and repel anionic solution species. Fortunately, only a few oxygen site types on the surface of a material usually dominate the acid–base behavior, allowing the ion-exchange properties of oxides to be predicted and controlled (see Chapter 5). However, pH also influences cation speciation and the charge of the exchanging ion. Many dissolved metal ions are hydrolysis products (see Chapter 5) with a charge that varies with pH. Thus, both the ion exchange and solute chemistry must be matched.

Although the electrostatic energy is important, adsorption of some cations onto oxide surfaces is not a simple Coulombic ion-exchange process, but proceeds simultaneously with hydrolysis and condensation reactions (see Figs. 6.18 and 8.11). These Lewis acid–base reactions depend on pH and ligand concentrations. Of course, hydrolysis and condensation reactions can also be triggered within the oxide in response to changes in pH. In some instances, such reactions destabilize the ion exchange material by proton-enhanced dissolution or hydrolysis by hydroxide ion. Therefore, solubility and hydrolysis diagrams for both the solution species and the exchanger should always be consulted before selecting the optimum conditions for removing an ionic species from solution. Similarly, competition exists between the surface sites and ligands in solution (e.g., fluoride), and all aqueous species must be considered.

10.3 ION EXCHANGE EQUILIBRIA: THE IMPORTANCE OF SELECTIVITY

For most applications, the selectivity of an exchanger is even more important than its ion-exchange capacity. For the example of alkali–metal cations, some applications may require the exchanger to be much more selective for a small, hard cation such as Li^+ than for Na^+, whereas for other applications, such as mitigating radioactive contamination, a higher affinity may be desired for a large, soft cation such as Cs^+. Selectivities are described by ion exchange equilibrium constants (K_I) for reactions such as

$$[>Ti-O]^- \cdot Na^+ + Cs^+(aq) \rightleftarrows [>Ti-O]^- \cdot Cs^+ + Na^+(aq) \quad \textbf{(10.1)}$$

$$K_I = \frac{[Na^+]\cdot X_{[>Ti-O]^- \cdot Cs^+}}{[Cs^+]\cdot X_{[>Ti-O]^- \cdot Na^+}} \quad \textbf{(10.2)}$$

where > denotes the surface, the square brackets indicate molarity of the ions, and X_i indicates the mole fraction of the ion-exchange sites occupied by a particular ion. This choice of concentration units to describe the ion-exchange equilibria is called the *Gapon convention*. Here, ion-exchange equilibrium constants, thermodynamic energies, and activities are handled just like conventional equilibrium constants for solutes [5]. Using this convention, adjustments can be made for heterovalent ion-exchange reactions, for ionic strengths, and for competing ligands and metals in solution [5]. However, in the interest of simplicity, approximate equilibrium expressions such as Eqs. 10.1 and 10.2 are used in the remainder of this chapter.

A practical approach that engineers often use to predict ion-exchange behavior involves expressing the ion exchange site concentrations in terms of suspended masses, not mole fractions of sites. If the goal is to remove a particular species from a specific solution, the parameter of interest to engineers is the distribution coefficient (K_d), which defines how a given species of interest is partitioned between the liquid and a given mass of the solid sorbent:

$$K_d = C_s / C_l \quad \textbf{(10.3)}$$

which is measured in units of milliliters per gram. Here, C_s is the concentration of the species exchanged onto the solid phase (measured in moles per gram) and C_l is the concentration of the species remaining in solution after batch contact (measured in moles per milliliter). For the exchange shown in Eq. 10.1, K_d is proportional to the cesium half of the equilibrium constant expression shown in Eq. 10.2 (i.e., $K_d = [>Ti-O-Cs^+]/[Cs^+]$), where the square brackets again indicate mass concentrations. Notice that, although K_d is related to the ion-exchange equilibrium constant, a given K_d value applies only for a specific solution composition, feed concentration (e.g., $[Na^+]$) and exchanger mass, whereas K_I is consistent with familiar solution thermodynamics. Competition among the cations for sites on the exchanger can be treated using a multicomponent speciation model and a mass balance that keeps track of the loss or creation of surface sites. A related parameter of interest is λ, which is defined as the number of exchanger bed volumes of a specific feed that can be loaded onto an exchanger (required column volumes):

$$\lambda = K_d \cdot \rho_b \qquad \textbf{(10.4)}$$

where ρ_b is the bed density of the exchanger.

For separations, the desire is to make K_I and thus K_d as large as possible. One example for which a high K_d is critical is the separation of radioactive cesium (^{137}Cs and ^{134}Cs) from millions of gallons of nuclear tank-waste solutions [6]. Nuclear-waste streams can contain high sodium concentrations (e.g., $[Na^+] \geq 3\ M$) and low cesium concentrations (e.g., $[Cs^+] \leq 10^{-4}\ M$), thus presenting a profound challenge to chemists. The desire is to lower $[Cs^+]$ in the liquid down to 10^{-7} M, for a decontamination factor of 1000. For a typical ion-exchange material (with an exchange capacity of 5 meq/g and a bed density of 0.6 g/mL), a key question is: How much solid waste (corresponding to the loaded exchanger) is generated in treating the contents of a 1 million-gal waste tank ($\sim 3.8 \cdot 10^6$ liters)? If $K_I = 10^7$ for Cs^+ relative to Na^+, 100% of the active sites on the exchanger will be occupied with Cs^+ after the exchange, resulting in the creation of less than 200 lb (~ 91 kg) of radioactive solid waste. This mass is much smaller than the original enormous volume of contaminated water. On the other hand, if K_I reflects only a 10-fold preference for Cs^+ relative to Na^+, only 0.0002% of the cation exchange sites will be occupied by Cs^+ for the specified tank-waste feed, generating more than 50,000 tons of contaminated solid waste (i.e., more than half the volume of the original tank) requiring disposal in a waste repository. Clearly, selectivity matters here.

10.4 THE ROLE OF LOCAL CHARGE DISTRIBUTIONS IN OPTIMIZING ION-EXCHANGE SELECTIVITY

The first parameter to consider in achieving ion-exchange selectivity involves understanding the effects of local charge distributions. Oxygen anions in an oxide or hydroxide lattice can adsorb metal cations readily to achieve local charge neutrality. However, what determines the extent to which a local site prefers cation A to cation B? One general rule for predicting selectivity is that exchange sites prefer to adsorb ions that match their local charge distributions. For cations, the

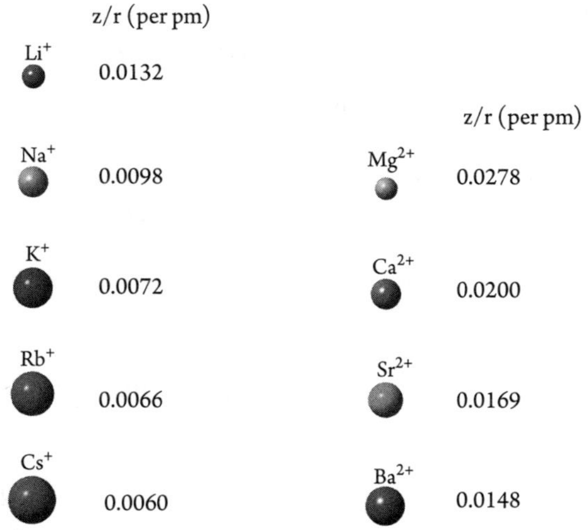

Figure 10.2 The sizes, charges, and charge densities of the alkali metal and alkaline earth cations. In terms of oxygen site selectivity, the most important parameter is the charge density (z/r), whereas for steric selectivity, the sizes of both bare and hydrated ions are most critical. From data compiled in Helferrich, F. *Ion Exchange*; McGraw Hill: New York, 1962.

local charge density is given by the parameter z/r, or the cation charge divided by its ionic radius in a given coordination geometry (Fig. 10.2). Large z/r values correspond to localized positive charge (Table 10.1). The charge density on four-coordinated Li^+, with an ionic radius of 0.6 Å is almost three times more localized than that on six-coordinated Cs^+, with an ionic radius of 1.67 Å. Although K^+ ($r = 1.38$ Å) and Ba^{2+} ($r = 1.35$ Å) are similar in size when similarly coordinated, the charge density on Ba^{2+} (1.48/Å) is twice as high as that found on K^+ (0.72/Å) because of the difference in ion charge (Table 10.1).

For the anionic surface sites on the exchanger, the charge density on each oxygen-anion surface site is defined by its formal charge (Fig. 10.3). As discussed in Chapter 4, formal charges are determined by the valence and local coordination geometry of adjacent network-forming cations, and the presence or absence of adsorbed protons.

When charge densities are known, predictions can be made regarding selectivity based on charge matching. Oxygen anion sites with a high site charge, such the $>Si-O^-$ site, prefer to associate with cations that also have a high charge density, or high z/r value (i.e., they prefer Li^+ or Mg^{2+} to Rb^+ or Ba^{2+}). Conversely, if the oxide charge is spread over several anions, such as the ring system shown in Figure 10.3, the material will have a higher relative affinity for large cations and/or cations with a lower net charge (i.e., will prefer K^+ to Li^+.) In the four-membered ring system shown in Figure 10.3, tetrahedral metals are linked

Table 10-1. Shannon-Prewitt Ionic Radii for Selected Elements and Their Ionic Potentials (z/r)

	Ionic Radius pm	Coordination	z/r (per pm)		Ionic Radius pm	Coordination	z/r (per pm)
Li(I)	59	4	0.0169	Be(II)	16	3	0.1250
	76	6	0.0132		27	4	0.0741
	92	8	0.0109		45	6	0.0444
Na(I)	99	4	0.0101	Mg(II)	57	4	0.0351
	102	6	0.0098		72	6	0.0278
	118	8	0.0085		89	8	0.0225
	139	12	0.0072	Ca(II)	100	6	0.0200
K(I)	137	4	0.0073		112	8	0.0179
	138	6	0.0072		123	10	0.0163
	151	8	0.0066		134	12	0.0149
	164	12	0.0061	Sr(II)	118	6	0.0169
Rb(I)	152	6	0.0066		126	8	0.0159
	161	8	0.0062		136	10	0.0147
	166	10	0.0060		144	12	0.0139
	172	12	0.0058	Ba(II)	135	6	0.0148
Cs(I)	167	6	0.0060		142	8	0.0141
	174	8	0.0057		152	10	0.0132
	181	10	0.0055		161	12	0.0124
	188	12	0.0053	Ti(IV)	42	4	0.0952
Al(III)	39	4	0.0769		60.5	6	0.0331
	48	5	0.0625				
	53.5	6	0.0561				
Si(IV)	26	4	0.1538				
	40	6	0.1000				

SOURCE: Shannon. R. D. Revised Effective Ionic Radii and Systematic Studies of Interatomic Distances in Halides and Chalcogenides. *Acta Crystallogr. A* **1976**, *32*, 751–767.

via μ_2-oxo-bridges so that each bridging oxygen has a formal charge of $-0.25(+[4/4]+[3/4]-2=-0.25)$, making an anionic ring. This contrasts with the nonbridging $>Si-O^-$ site with a formal charge on the oxygen of -1. Here, we show this relatively nonselective site bonded to Na^+ as a charge-compensating cation.

It is critical to note that the affinity sequences discussed earlier (Table 10.1) apply only for hydrated cations that have unobstructed access to oxygen anions. Free access is available for most exposed oxide surfaces and for some layered oxide and hydroxide materials that are able to swell to accommodate the waters of hydration associated with the exchanging cation. Obviously the effective radii

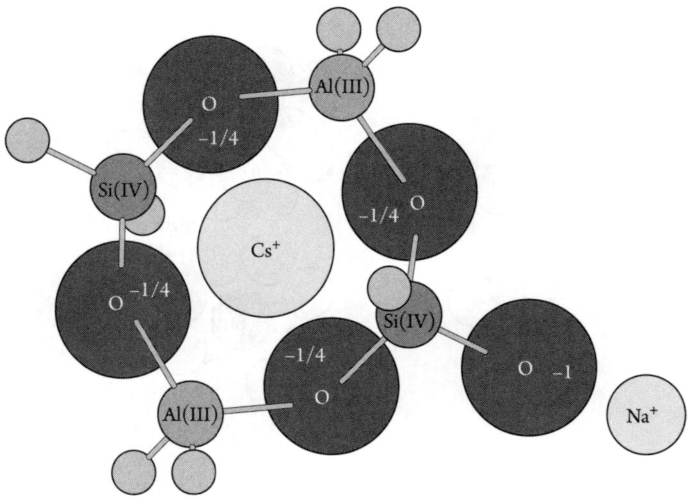

Figure 10.3 A depiction of an aluminosilicate structure, indicating oxygen formal charges. Cations with a large size and small charge (i.e., with a small z/r value) prefer to occupy regions of diffuse charge (such as the four $Si-\mu_2O-Al$ sites, each with a formal charge of $-1/4$), whereas cations with a large z/r value prefer to occupy sites where the anionic charge is more localized (such as a silicate nonbridging oxygen with a formal charge of -1).

of hydrated cations are much larger than the ionic radii shown in Table 10.1. Using such materials, it is difficult to separate cations from one another that exhibit similar z/r values such as Li^+ and Sr^{2+}.

10.5 THE ROLES OF STERIC EFFECTS AND SELECTIVE SOLVATION ON ION-EXCHANGE SELECTIVITY

The classic materials for that select on the basis of size are the zeolites. Most zeolites are aluminosilicate minerals with three-dimensional frameworks built around well-defined and interconnected pores (see Chapters 2 and 15). Pore diameters are defined by interconnected rings of SiO_4 and/or AlO_4 tetrahedra, with ring diameters being determined by the numbers of tetrahedra (n) in each ring ($n = 4$, 5, 6, and 8 are common) (Fig 10.4) [7]. Although almost all the charge in most zeolites is associated with $>Al-\mu_2O-Si<$ linkages (each of these oxygens with a formal charge of -0.25), some zeolites exhibit a strong preference for Li^+ whereas others exhibit a strong preference for K^+. How can this be?

For zeolites and other oxide materials that exhibit fixed nanoscale porosity (or layering), selectivities can be understood by considering cation size. In some instances, the ionic radius of the bare cation is the only thing that matters. For example, Cs^+ (with an ionic diameter of 3.4 Å) is too large to squeeze through the rings defined by six interconnected tetrahedra (with a pore opening of 2.7 Å).

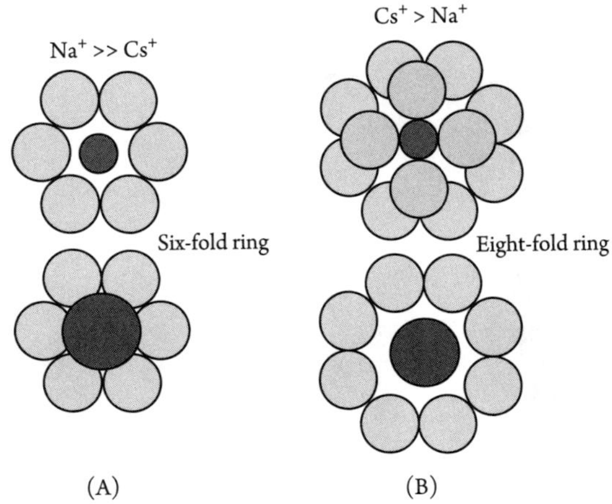

Figure 10.4 An illustration of the sizes of bare Na^+, bare Cs^+, and hydrated Na^+ relative to the openings in six- and eight-membered silicate rings such as those found in zeolites. (A) A bare Na^+ ion can fit easily through a six-membered ring (top), whereas Cs^+ is too large to squeeze through (bottom). For this reason, zeolites containing six-membered rings are highly selective for Na^+ relative to Cs^+. (B) The opposite selectivity trend is observed for zeolites containing eight-membered rings. A hydrated Na^+ ion is too large to enter the ring (top). In contrast (bottom), Cs^+ can shed its waters of hydration readily, allowing it to enter the zeolite in preference to Na^+.

Zeolites such as analcime that contain six-membered rings are much more selective for Na^+ than Cs^+, and in fact exclude Cs^+ completely.

In many instances, the size of the hydrated cation is what dictates selectivity (Fig. 10.4). Hydrated cation sizes are largely defined by the space occupied by waters of hydration (see Chapter 4). Considering the inner coordination sphere, hydrated monomeric cation sizes range from 5.5 to 7.0 Å. Hydrated Na^+ is too large to pass through a six-membered aluminosilicate ring or even an eight-membered ring. In fact, hydrated Na^+ can barely fit through a 10-membered ring. However, this does not mean that Na^+ cannot enter zeolites with 6-, 8-, and 10-membered rings. What it does mean is that Na^+ or any other hydrated cation must shed a substantial number of its waters of hydration to do so. This ion dehydration requires a substantial energy penalty that depends on the charge and size of the bare ion. The magnitude of this penalty can be appreciated by examining the heats of hydration for different cations, where heats of hydrations scale with z/r. For example, in moving from Li^+ to Na^+ to K^+ to Cs^+, heats of hydration drop from –540 kJ/mol to –427 kJ/mol to –342 kJ/mol to –297 kJ/mol [8]. Because it is much easier to dehydrate Cs^+ than Na^+, zeolites with 8- and 10-membered rings (which require dehydration for entry) exhibit a substantial preference for Cs^+ relative to Na^+.

When a cation and any of its associated water molecules enters an oxide exchanger such as a zeolite, it now encounters rigid local environments that are substantially different than bulk water. Interactions with the interior of the exchanger have a significant impact on exchanger selectivity. First, both cations and water molecules sense the acid–base sites of the framework, resulting in competitive solvation between the cations and exchanger walls. One consequence of these interactions is that, although measured heats of hydration within zeolites track cation z/r values as in the bulk solution, hydration enthalpies within zeolites are typically low (e.g., 2–12 kJ/mole for zeolites containing interconnected 12-membered rings that define pores so large that all hydrated ions can enter [Fig. 10.5]) [9–11].

Both cations and water molecules compete for space within the available pore volume. As indicated in Figure 10.6, up to half the available volume in a zeolite can be occupied by water molecules. However, as more cations are added (as reflected in the density of anionic Al sites), the water content drops. Regardless of water content, the structure of water within a zeolite is substantially different

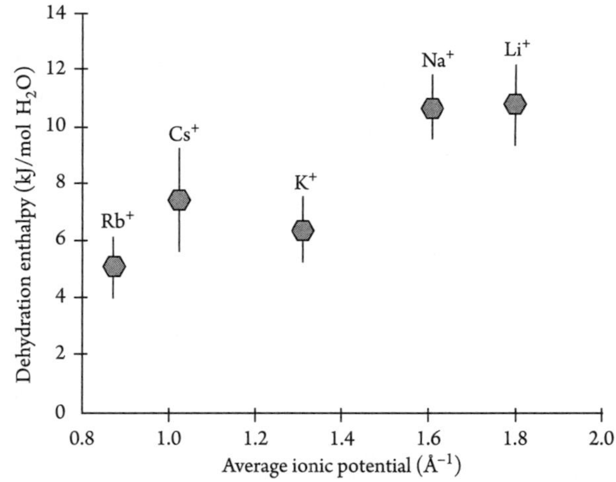

Figure 10.5 The dehydration enthalpy of ion-exchanged zeolite BEA per mole of water. As in bulk water, smaller ions interact more strongly with their waters of hydration than larger ions. However, the magnitude of the ion-size effect is much smaller than in the bulk solution. Created using data from the following sources: Barrer, R. M. *Hydrothermal Chemistry of Zeolites*; Academic Press: London, 1982. Shim, S.-H.; Navrotsky, A.; Gaffney, T. R.; and MacDougall, J. E. Chabazite: Energetics of Hydration, Enthalpy of Formation and Effects of Cations on Stability. *Am. Miner.* **1991**, *84*, 1870–1882. Sun, P.; Navrotsky, A. Enthalpy of Formation and Dehydration of Alkaline Earth Cation Exchanged Zeolite Beta. *Micropor. Mesopor. Mater.*, **2008**, *109*, 147–155. Sun, P.; Deore, S.; Navrotsky, A. Formation and Dehydration Enthalpy of Ion Exchanged Zeolite Beta. *Micropor. Mesopor. Mater.*, **2006**, *91*, 15–22.

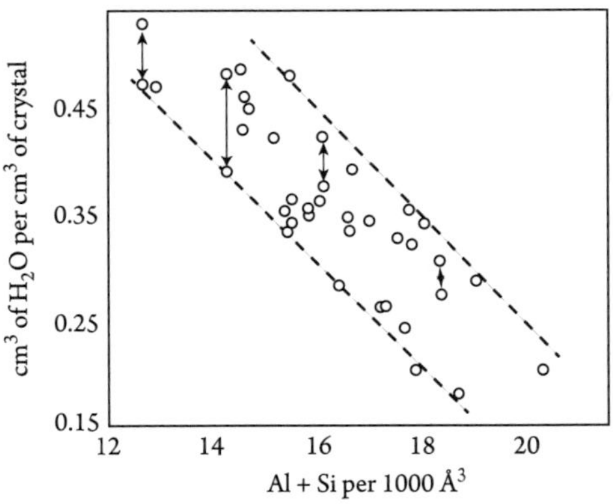

Figure 10.6 The volume of water contained within a range of zeolites as a function of the anionic charge density (and thus cation density) within the zeolitic structure. Because there is a finite amount of space available within the zeolite, the addition of more cations results in the presence of less water within the structure. Reproduced from Barrer, R. M.; *Hydrothermal Chemistry of Zeolites*; Academic Press: London, 1982, p. 21.

from that found in bulk water. This difference means that solvation of both cations and the zeolite network can vary considerably, both in terms of solvation number and heat of hydration. Although it can be difficult to quantify the net result of steric and competitive solvation effects, such effects provide a qualitative framework for selecting ring topologies and layer spacings to optimize ion-exchange selectivities. The remainder of this chapter is devoted to specific examples that illustrate how oxides are selected and designed for selectivity with regard to both cations and anions.

10.6 SODIUM TITANATES FOR $^{90}Sr^{2+}$ REMEDIATION

Sodium titanates represent some of the most attractive candidates for the removal of $^{90}Sr^{2+}$ from radioactive wastes [12–15] held as solutions in tanks at the Hanford site in Washington and at the Savannah River site in South Carolina. Derivative structures and peroxide-treated titanates also exhibit exceptional performance for $^{90}Sr^{2+}$ removal [13–14]. Here, we examine $Na_2Ti_4O_9$ as a model system to highlight how solution pH influences the performance of oxide materials both in terms of ion exchange selectivity and chemical stability. The structure of $Na_2Ti_4O_9$ consists of clay-like titanate sheets (see Fig. 2.4) [16, 17]. Two anionic sites are present in the sheets: $>Ti-O^-$ nonbridging oxygens, each with a formal charge of -1.33, and $>Ti\text{-}\mu_2O\text{-}Ti<$, each with a formal charge of -0.67. Sodium ions are located between the titanate sheets where they compensate for the structural charge of the anionic titanate network. The acid–base chemistry of the two anionic sites is described in Chapter 6 [18, 19]. The nonbridging oxygen sites have an effective pK_a for converting $>Ti-O^-$ to $>Ti-OH$ of around 11, and can be protonated further to create $>Ti-OH_2$ at around pH 2 to create cationic anion-exchange sites. The μ_2O exhibits a pK_a value of around 7; below pH 7, it is protonated to form μ_2OH.

In basic solutions, two different anionic sites are available for the exchange of Na^+ by Sr^{2+}. How selective are each of these sites, and under what solution conditions is this selectivity optimized? These questions are answered by examining the distribution coefficient for Sr^{2+} uptake by $Na_2Ti_4O_9$ as a function of solution pH (Fig. 10.7) [20]. In strongly basic solutions, the $Na_2Ti_4O_9$ solid has an exceptional affinity for Sr^{2+}, with a distribution coefficient in high-sodium waste simulants approaching 10^5. However, this exceptional performance is lost as the solution pH is lowered. After the nonbridging oxygen sites have been protonated, the remaining bridging oxygen sites exhibit a distribution coefficient for Sr^{2+} that decreases with pH. At very low pH, the cation exchange properties of the material are completely lost as protons outcompete Sr^{2+} for sites on the exchanger.

These observations are all consistent with predictions based on local charge distributions. Anionic sites with a high local charge, such as the nonbridging oxygen site, are expected (1) to have a strong preference for cations exhibiting a high charge-to-radius ratio (i.e., to prefer Sr^{2+} with $z/r=1.77$ to Na^+ with $z/r=1.05$) and (2) to exhibit a high pK_a value (i.e., requiring high pH values for deprotonation and site activation). The anionic bridging-oxygen site, with its formal charge of -0.67 relative to -1.33, is expected to have a lower affinity for Sr(II) and a lower pK_a, in agreement with the experimental observations. As a point of comparison, the nonbridging silanol groups ($>Si-OH$) present in

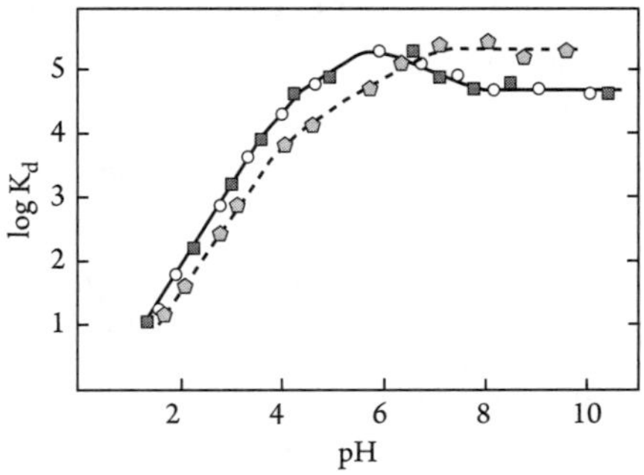

Figure 10.7 K_d values for Sr^{2+} adsorption onto $Na_4Ti_9O_{20} \cdot xH_2O$ as a function of solution pH. The plus sign, circles, and squares correspond to $[Sr(II)] = 10^{-3}M$, $10^{-5}M$, and $10^{-7}M$, respectively. Adapted from Lehto, J.; Harjula, R.; Girard, A. The Equilibrium of Strontium Ion Exchange on Sodium Titanate, $Na_4Ti_9O_{20}]xH_2O$. *J. Chem. Soc. Dalton Trans.* **1989**, *1989*, 101–103.

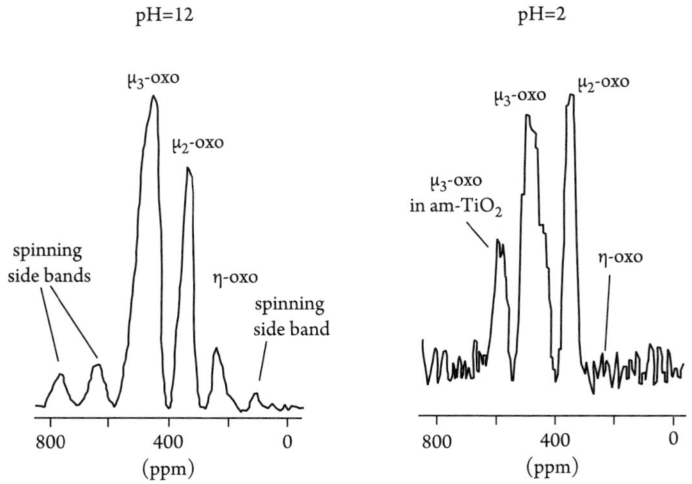

Figure 10.8 ^{17}O nuclear magnetic resonance spectra for amorphous $Na_2Ti_4O_9$ in a pH 12 and a pH 2 solution. The resonance associated with nonbridging oxygen sites largely disappears when sodium titanates are stored under acidic conditions, whereas the population of a new site with the same chemical shift as the μ_3-oxo-bridge in TiO_2 appears. Both observations indicate that protonation in acid triggers the polymerization of the titanate network to form amorphous TiO_2 and, ultimately, to eliminate most ion exchange activity. Reprinted with permission from Bunker, B. C.; Peden, C. H. F.; Kirkpatrick, R. J.; Turner, G. L.; ^{17}O NMR Studies of Titanate Repolymerization in Water. *Mat. Res. Soc. Symp. Proc.*, **1997**, *432*, 39–43.

many silicates have a formal charge of −1. As expected, their pK_a and Sr(II) selectivity are both intermediate between those of the two distinct titanate sites.

A second important observation regarding $Na_2Ti_4O_9$ has to do with its chemical stability. As pointed out in Chapter 6, oxide ion exchangers have a tendency to undergo hydrolysis and condensation reactions for pH values that are far removed from the PZC. These reactions can sometimes change or even eliminate ion exchange activity. For $Na_2Ti_4O_9$, the impact of these reactions is clearly demonstrated by exposing the oxide ion exchange material to low-pH solutions, where both bridging and nonbridging oxygens are protonated. Although titanates are relatively insoluble even under these conditions, ^{17}O nuclear magnetic resonance labeling studies [21] (Fig. 10.8) clearly show condensation reactions occur that result in a polymerization of the oxide structure via reactions such as

$$\begin{array}{l} >\text{TiOH} + >\underset{\displaystyle \underset{\text{H}}{|}}{\text{Ti-O-Ti}}< \rightleftarrows >\underset{\displaystyle \underset{\underset{\wedge}{\text{Ti}}}{|}}{\text{Ti-O-Ti}}< + H_2O \end{array} \qquad \textbf{(10.5)}$$

The net effect of such reactions is to convert $Na_2Ti_4O_9$ into amorphous TiO_2 particles. This transformation is kinetically irreversible, which is why the hydrated form of $Na_2Ti_4O_9$ loses substantial ion-exchange activity if stored under low pH conditions.

10.7 SILICOTITANATE ION EXCHANGERS FOR Cs^+ REMEDIATION

The water-soluble cations of the fission products ^{134}Cs and ^{137}Cs represent a major environmental threat from reactor accidents because these nuclei prevent people from returning to their homes. The ^{137}Cs isotope has a half-life of 30 years, which means that it is highly radioactive for several human generations. To capture $^{137}Cs^+$, the highest ionic selectivity is achieved within materials having structures that exploit both local charge distributions and steric effects. High selectivity is provided by oxides such as the silicotitanates that have been developed to extract $^{137}Cs^+$ from Na^+-rich radioactive tank-waste solutions [22]. Under optimum conditions, these silicotitanates can exhibit distribution coefficients for Cs^+ relative to Na^+ in sodium-rich solutions that are as high as 100,000 [23] (Fig. 10.9). Here, we describe design features that lead to these high selectivity values, as well as how silicotitanates can be converted into solid waste forms to ensure the extracted Cs^+ is never released back into the environment.

The first parameter used to promote high selectivities for Cs^+ relative to Na^+ involves steric selectivity, as illustrated by aluminosilicate zeolites. By tuning the ring openings in the zeolites in the range of 6.0 to 7.7 Å (corresponding to rings containing 8–10 interconnected tetrahedra), zeolites can be designed to admit bare Cs^+ while excluding hydrated Na^+ (Fig. 10.3). A zeolite named *IE-96*, which exhibits the highest known selectivity of any known zeolite for Cs^+ relative to

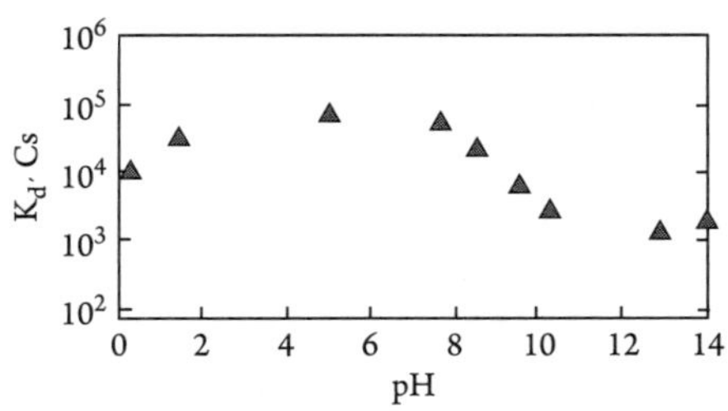

Figure 10.9 The distribution coefficient for Cs^+ as a function of solution pH for a proprietary silicotitanate material immersed in simulated tank-waste solutions. Here, selectivity relative to Na^+ is maximized within a pH regime where the predominant anionic species within the exchanger are $Si-\mu_2O-Ti$. Adapted from Anthony, R. G.; Philip, C. V.; Dosch, R. G. Selective Adsorption and Ion Exchange of Metal Cations and Anions with Silico-titanates and Layered Titanates *Waste Manage.* **1993**, *13*, 503–512, with permission.

Na^+ [24], exploits this steric selectivity mechanism. However, although *IE-96* is highly selective, the concentrations of Na^+ in tank waste are so high (e.g., $[Na^+] = 4$ M) and Cs^+ concentrations are so low (e.g., $[Cs^+] = 10^{-4}$ M) that *IE-96* exhibits a K_d value of only 120. This low K_d value is insufficient for achieving desired decontamination factors and would lead to the generation of large volumes of contaminated exchange material requiring disposal in a nuclear-waste repository. In addition, zeolites are relatively unstable except under near-neutral pH conditions and are not appropriate for treating native tank wastes, the pH of which can exceed 14.

Achieving higher K_d values requires the incorporation of cation-exchange sites that are intrinsically more selective for Cs^+ and more stable than the anionic $> Si - \mu_2 O - Al <$ sites present in a standard zeolite. Enhancing site selectivity for Cs^+ relative to Na^+ entails distributing single network charges over a greater number of oxygen anions than the four $> Al - \mu_2 O - Si <$ bonds that produce a charge of -1 in most zeolites.

Formal charge analyses suggest that sites with lower negative charges are possible for certain silicotitanate compositions, depending on the titanium coordination number. All known crystalline silicotitanates contain Si(IV), which is coordinated tetrahedrally to oxygen anions. However, Ti(IV) can be coordinated to four, five, or six oxygen anions (designated Ti(4), Ti(5), and Ti(6)). Formal charges for the most likely oxygen anion-bonding configurations for each of these titanium coordination numbers are provided in Table 10.2. As indicated, Ti(4) is equivalent to Si(4) in that all bridging oxygens are neutral and should exhibit no cation exchange activity. In addition, Ti(4) is found only in zeolitic structures at low concentrations as isolated tetrahedra associated with four

Table 10-2. Oxygen Formal Charges for Sites in Silicotitanates

Site Type	Nonbridging hydroxyl	μ_2-Bridge	Nonbridging oxygen
Silicate	0	0	−1
Titanate (4)	0	0	−1
Titanate (5)	−0.2	−0.4	−1.2
Titanate (6)	−0.33	−0.67	−1.33
Si−O−Ti(4)		0	
Si−O−Ti(5)		−0.2	
Si−O−Ti(6)		−0.33	

See Chapter 2 regarding the calculation of formal charges. Si(IV) cations are always tetrahedral. The coordination numbers for Ti(IV) cations are indicated in the left column in parentheses. Both cations in titanate μ_2-bridges have the same coordination number for entries in this table.

$>Si-\mu_2O-Ti<$ bonds. Therefore, Ti(4) can be eliminated from further consideration for active cation exchange within silicotitanates.

Silicotitanates containing Ti(6) are most common, because titanium prefers octahedral over tetrahedral coordination geometries [25]. However, when Ti(6) is present, the formal charge on oxygens in $Si-\mu_2O-Ti$ bonds is -0.33, slightly higher than the -0.25 charge on $>Si-\mu_2O-Al<$ bonds (Fig. 10.3). Here, any increase in Cs^+ selectivity relative to aluminosilicates would have to arise from subtle changes in ring topologies rather than from local charge distributions (Fig. 10.10). The most attractive Ti(5) geometry is encountered within Si-rich compositional ranges because it is easier to substitute a Ti(5) site into a tetrahedral network than it is to accommodate Ti(6), unless structures such as Ti(6) chains are present. The formal charge on the $>Si-\mu_2O-Ti(5)<$ bond of -0.2 is less than that found on an $>Si-\mu_2O-Al<$ which means that silicotitanates containing Ti(5) are predicted to exhibit the highest K_d values for Cs^+ based on local charge distributions.

If the titanium coordination number is known, formal charge models can be used to predict the most probable distribution of the anionic oxygens shown in Table 10.2 as a function of composition. As introduced in Chapter 2, formal charge models predict that the most stable, local bonding configurations are those that minimize anionic charge on the maximum number of oxygen anion sites. For octahedral titanium (Ti(6)), the relative ease of formation of anionic sites in proton-free material should follow the trend $Si-\mu_2O-Ti$ less than $Ti-\mu_2O-Ti$ less than $Si-\eta O$ less than $Ti-\eta O$. For the composition

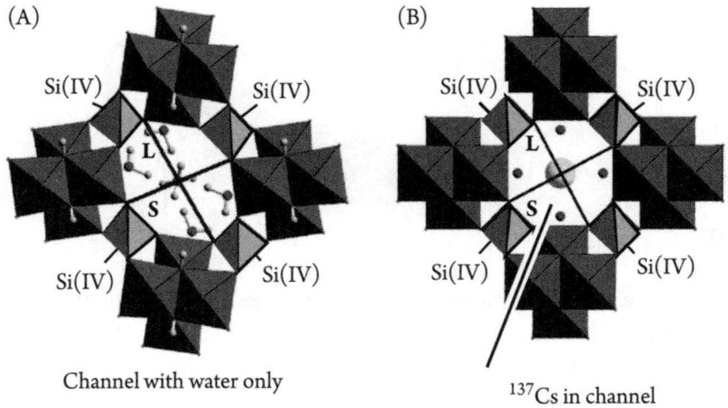

Figure 10.10 (A, B) Polyhedral representations of the structure of a crystalline silicotitanate before (A) and after (B) ion exchange for cesium. Note that the exchange process distorts the ring openings in the primary ion exchange channels, which amplifies selectivity toward Cs^+. Adapted from Celestian, A. J.; Kubicki, J. D.; Hanson, J.; Clearfield, A.; and Parise, J. B. 2008. The Mechanism Responsible for Extraordinary Cs Ion Selectivity in Crystalline Silicotitanate. *J. Am. Chem. Soc.* **2008**, *130*, 11689–11694, with permission.

$CaO-TiO_2-SiO_2$ (or $CaTiSiO_5$ corresponding to the mineral sphene or titanite [26]), there are five oxygen anions in the formula unit. The formula unit and cation coordination numbers dictate that these oxygens participate in a total of 10 bonds (six to Ti and four to Si). The formula also dictates that the sum of the formal charges on all oxygen sites be equal to -2 (to compensate the charge of Ca(II)).

Given these constraints, a wide range of oxygen-site distributions is possible (Table 10.3). However, the combination that generates the minimum charge on the maximum number of oxygen anions is $4 >Si-\mu_2O-Ti<+>Ti-\mu_2O-Ti<$, corresponding to the known crystal structure of titanite. This example illustrates that nonbridging oxygens are not favored in zeolitic silicotitanates because such oxygens carry a large negative charge. The same formal charge considerations explain why aluminosilicate minerals including the zeolites exhibit a strong preference for forming $>Al-\mu_2O-Si<$ bonds instead of nonbridging oxygens. Similar trends are observed in zeolitic

Table 10-3. TEN OF THE POSSIBLE DISTRIBUTIONS OF OXYGEN SITE TYPES FOR $CaTiSiO_5$, ASSUMING THAT Ti(IV) IS IN AN OCTAHEDRAL COORDINATION GEOMETRY

Si–Ti–Ti	**Ti–Ti–Ti**	**Si–O–Si**	**Si–O–Ti**	**Ti–O–Ti**	**Si nbo**	**Ti(nbo)**	**N**	**R**
+1/3	0	0	−1/3	−2/3	−1	−11/3		
0	0	0	4	1	0	0	4	1
0	0	1	2	2	0	0	3	3
0	0	2	0	3	0	0	2	5
0	1	0	3	0	1	0	4	2
1	0	0	3	0	0	1	3	4
1	0	1	0	2	1	0	1	7
1	1	1	0	0	1	1	2	6
1	1	0	1	0	2	0	0	9
2	0	1	0	0	0	2	1	8
2	0	0	0	1	2	0	0	10

Most columns in the table represent stable oxygen sites, with the formal charge on each site appearing under the site entry. Site designations include (1) μ_3-oxo, where the column entry lists the three cations bonded to that oxygen; (2) μ_2-oxo, where the column entry indicates the two cations in the bridge; and (3) nonbridging oxygens (designated by the cation next to "nbo"). The formula unit for $CaTiSiO_5$ contains five oxygens. Each row in the table indicates a possible distribution of these five oxygens that results in the total anionic network charge of -2 required to neutralize Ca^{2+}. The "N" column indicates the number of oxygens in a given row that are either neutral or slightly anionic and are thus most stable. The "R" column ranks the relative stability of the oxygen distribution in a given row based on maximizing N (with 1 being the most stable).

structures containing Ti(5) [27, 28]. One trick exploited by Ti(5) sites to avoid the creation of nonbridging oxygens involves the placement of two Ti(5) sites adjacent to each other to create an edge-shared titanate dimer. Here, two Ti(5) sites can substitute into a tetrahedral zeolitic structure by forming three $>Si-\mu_2O-Ti<$ bonds, with the fourth site of the "tetrahedron" being occupied by two $>Ti-\mu_2O-Ti<$ bridging oxygens between adjacent Ti(5) centers. (Edge-shared Ti(5) is also seen in alkali metal titanates [28].) These examples illustrate that the anionic charge in silicotitanate ion exchangers is typically distributed between both $>Si-\mu_2O-Ti<$ and $>Ti-\mu_2O-Ti<$ bonds (Table 10.3).

The K_d-versus-pH curve for the most selective silicotitanate for cesium removal (Fig. 10.9) is unusual in that cesium uptake increases and then decreases as the pH decreases [24, 29]. This behavior is in sharp contrast to that observed for strontium uptake by the sodium titanates (Fig. 10.7). The explanation for this behavior involves the competition between Cs^+ and Na^+ for the limited space available within the ion exchanger. The exchangers with the highest K_d values contain a mix of $>Si-\mu_2O-Ti<$ and $>Ti-\mu_2O-Ti<$ oxygen anions. At high pH and sodium ion concentrations, neither site is protonated, and both sites are reactive. However, although both sites prefer Cs^+ to Na^+, at high sodium ion concentrations, Na^+ can outcompete Cs^+ on the $>Ti-\mu_2O-Ti<$ sites, thus crowding out Cs^+ ions attempting to access the $>Si-\mu_2O-Ti<$ sites. At pH values less than 9, the $>Ti-\mu_2O-Ti<$ sites are protonated and rendered inert, which eliminates this competition and allows Cs^+ to occupy all available space within the exchanger. The exceptionally high K_d value for Cs^+ for the $>Si-\mu_2O-Ti<$ sites is retained until the pH is lowered all the way down to around pH 1, where the $>Si-\mu_2O-Ti<$ sites are protonated, eliminating anion-exchange function.

A final consideration of interest regarding silicotitanate ion exchangers involves the conversion of these materials into stable waste forms. It has been shown that Cs-loaded silicotitanates can be converted into ceramic waste forms simply by heating the loaded exchanger above 800°C. X-ray crystallography [30] shows that this heat treatment collapses the eight-membered rings found in the initial ion exchanger into six-membered rings such as those found in the Cs^+-entrapping zeolite pollucite. Although the eight-membered rings are large enough to allow Cs^+ to enter the structure via ion exchange, the openings in the six-membered rings are small relative to the diameter of Cs^+. The heat-treated silicotitanate thus encapsulates the Cs^+. The only way to release the Cs^+ into the environment after heat treatment is to dissolve the Cs-loaded silicotitanate congruently. Because silicotitanates contain titanium rather than aluminum, these zeolites are much more resistant to hydrolysis and dissolution than the more common aluminosilicate zeolites, regardless of pH. The most stable of the

Cs-loaded silicotitanates exhibits a cesium-release rate in water of 10^{-8} g/m^2/day, which is more stable than nuclear-waste glasses containing comparable cesium loadings [31].

10.8 INTERACTIONS BETWEEN Cs^+ AND CLAY MINERALS

The general principles used by engineers to design ion exchangers deliberately for Cs^+ sequestration also explain the behavior of Cs^+ in the environment. Although radioactive $^{137}Cs^+$ is released via reactor accidents such as Chernobyl, and more recently the tsunami-induced accident in Fukushima, Japan, it can also enter the environment through leaks in nuclear-waste storage tanks or from pollution that originated in old nuclear-bomb tests. With a 30-year half-life, $^{137}Cs^+$ can be potentially ingested into plants and animals for several human generations. Therefore, we need to understand how Cs^+ migrates and is dissipated within the soils that surround contaminated sites.

The most prevalent ion exchangers in the environment are clay minerals. The structures, compositions, and ion-exchange behaviors of the clays are discussed in detail in Chapter 18. Ion-exchange reactions involving clays (Fig. 10.11) turn soils and sediments into chromatographic columns that separate and retain cations as fluid flows through the soil. The higher the ion-exchange equilibrium constant (or distribution coefficient) for a given cation, the more strongly it is retained and the more slowly it moves through the environment. For example, measurements have been made regarding the mobile fraction of ^{137}Cs present in lakes near the Chernobyl reactor accident (Table 10.4). The results indicate that

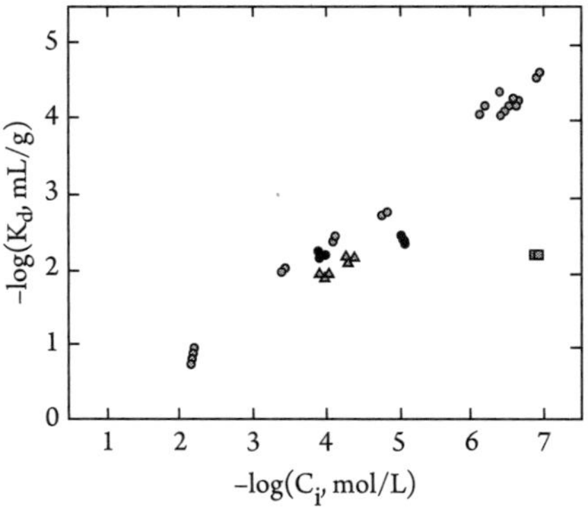

Figure 10.11 Distribution coefficients for Sr and Ba on abyssal smectite-rich clay in 0.68 M NaCl solutions as a function of the log of the initial cation concentration. Circles are for Ba (pH 7.0–8.1); triangles are for Sr (pH 7.1–7.3). Adapted with permission from Erikson, K. Radionuclide Sorption Studies on Abyssal Red Clays, Radioactive Waste in Geologic Storage. *ACS Symp. Ser.* **1979**, *100*, 267–290.

Table 10-4. PERCENT OF THE ^{137}Cs INVENTORY MADE MOBILE TO SOLUTION AS A FUNCTION OF TIME

Time (y)	**Lake 1**	**Lake 2**
1	2.8	2.4
3	1.7	1.4
10	0.4	0.3
30	0.05	0.04

Model estimates of the fraction of ^{137}Cs from the Chernobyl accident that is remobilized from lake-sediment environments each year. The initial ion-exchange reaction has a half-life of minutes to hours, followed by slow uptake by sites in clay that do not exchange with solution. The half-life of the second exchange process is many weeks.
SOURCE: Smith, J. T.; Comans, R. N. J. Modelling the Diffusive Transport and Remobilisation of ^{137}Cs in Sediments: The Effects of Sorption Kinetics and Reversibility. *Geochim. Cosmochim. Acta* **1996**, *60*, 995–1004.

at least two major adsorption sites are present: one site that is quick to immobilize Cs^+ and another site that adsorbs Cs^+ more slowly and irreversibly.

Clues regarding the nature of the two primary adsorption sites for Cs^+ have been provided by experiments conducted on relatively phase-pure clays (Table 10.5). Cesium distribution coefficients have been determined for two extreme clay types: a swelling clay represented by montmorillonite and a nonswelling clay represented by illite [32–34]. In montmorillonite, the predominant anionic sites responsible for ion exchange are $>Al-\mu_2O-Si<$ bridging bonds. As discussed earlier, if swelling is unconstrained, such sites normally exhibit a relative selectivity for cations of $Cs^+ > K^+ > Na^+$, in agreement with experimental results on both pure montmorillonite and abyssal red clay. Because many of these sites are on the basal planes of the clay sheets, access to solution would be largely unimpeded.

One might expect nonswelling illite clay to have less of an affinity for Cs^+ than montmorillonite because the only sites available for ion exchange are edge sites. However, as shown in Table 10.5, the bridging bonds in frayed illite edges exhibit K_d values for Cs^+ that are more than 1000 times greater than those on montmorillonite. The explanation for this behavior involves an unusual steric component to the ion-exchange process. The long-term mechanism of cesium uptake involves the insertion of the ion into the frayed edges of the clay, replacing K^+ or NH_4^+, both of which are ions with a higher solvation energy but similar solvated radius. Because the Cs^+ has a lower solvation energy, it is thought to coordinate well with the distorted oxide rings at the tetrahedral sheets at the edges of the clay. When Cs^+ substitutes for K^+ on plate edges, it forces itself between sheets that

Table 10-5. Values of the Distribution Coefficient (K_d, in milliliters per gram) for ^{137}Cs Adsorbed onto Montmorillonite and Illite Clays Saturated in K^+, Na^+, or Ca^{2+}

Clay	K	Na	Ca
Illite	4350	33,700	24,400
Montmorillonite	713	2636	1450

The experiments were only 2 hours in duration yet already show that Cs^+ has a strong preference for illites. The conditions were nanomolar concentrations of $^{137}Cs^+$, $[Cl^-] =$ 0.01 M, and 1 g clay/L.

From data in Sawney, B. L. Selective Sorption and Fixation of Cations by Clay Minerals: a Review. *Clays and Clay Minerals*, **1972**, *20*, 92–100; and Staunton, S. A. R. Adsorption of ^{137}Cs on Montomorillonite and Illite: Effect of Charge-Compensating Cation, Ionic Strength, Concentration of Cs, K, and Fulvic Acid. *Clays and Clay Minerals*, **1997**, *45*, 251–260.

are held in place by intersheet K^+ ions. In effect, the potassium in the interior of the clay prevents sheets from separating in response to cesium adsorption. Thus, Cs^+ entering at clay edges are adsorbed into cavities that resemble the cage structures in a zeolite. This adsorption results in a slight expansion of the clay near edges and the generation of new peaks in X-ray diffraction patterns. These new peaks can have sufficient intensities to allow scientists to discriminate between sediments that were present during a radioactive leak or bomb test and sediments that were deposited afterward.

10.9 OXIDE ANION EXCHANGERS

Not all hazardous environmental species are cations. In nuclear wastes, $[^{99}TcO_4]^-$ is one of the more prevalent soluble radioactive species requiring sequestration. Such wastes can also contain significant levels of nonradioactive but chemically toxic $[CrO_4]^{2-}$. Removal of such ions from wastes via anion exchange poses two serious challenges relative to cation exchange. First, many of the common anions that compete with $[TcO_4]^-$ for cationic anion exchange sites such as $[HCO_3]^-$, $[H_2PO_4]^-$, and $[HSO_4]^-$ tend to have similar sizes, charges, and z/r values. This means that differences between competing anions can be relatively subtle, resulting in lower selectivity than that seen for simple cations. Second, the solution conditions required to make most oxides function as anion exchangers tend to be relatively acidic. Oxide anion exchangers tend to be unstable under such conditions, and can be susceptible to phase changes and dissolution (see Section 10.6). The key to producing oxide anion exchangers that are both cationic and stable involves creating local bonding configurations in which the oxygen anions in the structure have a positive formal charge without the need for protonation, which tends to promote dissolution of the structure, or phase changes.

The classic oxide anion exchangers are a subset of the clay minerals introduced in Chapter 2 called *layered double hydroxides* (LDHs) (Fig. 10.12) [35–37]. The LDHs are clay-like minerals with compositions given by: $[M^{2+}_{1-x}M^{3+}_{x}(OH)_2]\cdot[A^{n-}]_{x/n}\cdot zH_2O$, where x typically ranges between 0.17 and 0.33. The parent clay consists of dioctahedral sheets based on cations such as Mg(II), Zn(II), or Ni(II). All hydroxyl groups within the parent octahedral sheets are bonded to three divalent cations, and each is neutral (with a formal charge of zero, see Chapters 2 and 18). However, when a trivalent

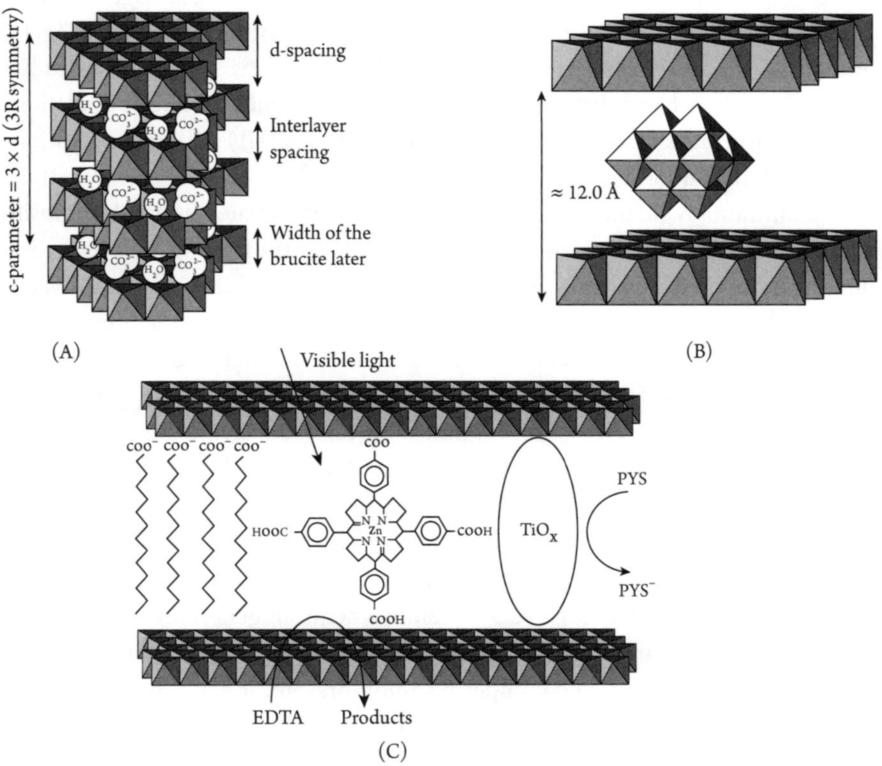

Figure 10.12 Representations of layered double-hydroxide (LDH) anion exchanger with a cationic composition of 2:1 $Mg^{2+} : Al^{3+}$. (A) The octahedral sheets are separated by layers containing water molecules and small anions. (B) LDH material into which an anionic polyoxometalate has been inserted. (C) LDH material in which anionic self-assembled monolayers have been inserted. This inserted layer serves as a host for both light-harvesting porphyrins and TiO_2 nanoparticles for performing photoactivated catalysis. Adapted from structures in Wang, Q.; O'Hare, D. Recent Advances in the Synthesis and Application of Layered Double Hydroxide (LDH) Nanosheets. *Chem. Rev.* **2012**, *112*, 4124–4155, with permission.

cation such as Al(III), Fe(III), Mn(III), or Ga(III) is substituted for one of the divalent cations, the substituted μ_3-OH bridge now has a formal charge of $+1/6$. In other words, the substituted octahedral sheets have a net positive charge that must be compensated for via the intercalation of anions and water between the sheets. Typical intercalated anions have charges ranging from -1 to -3, such as Cl^- and $[NO_3]^-$ ion, $[CO_3]^-$ and $[SO_4]^{2-}$, and $[PO_4]^{3-}$. An example of a naturally occurring LDH is the mineral hydrotalcite $(Mg_6Al_2(OH)_{16}CO_3{\cdot}4H_2O)$.

LDH materials in which x is greater than 0.33 are rare. A simple formal charge explanation for this observation is that $x > 1/3$, some μ_3-OH bridges are forced to bond to two +3 cations. The formal charge on such bridges is $+1/3$, which represents an unstable configuration. The desire to avoid such a configuration has been confirmed by nuclear magnetic resonance studies of the ordering of Mg(II) and Al(III) in LDH materials with an Mg:Al ratio of 2:1 [36]. These studies show that rather than substituting randomly, which would produce populations of μ_3-OH bridges bonded to two and even three Al(III) cations, the cations order such that all bridges are bonded to one Al(III) only. In other words, there is an aluminum-avoidance rule for LDH materials just as there is for aluminosilicate materials (see Chapter 14). The difference is that the driving force for avoidance in LDH materials is the need to disperse a positive charge within a layer as opposed to a negative charge within an extended aluminosilicate framework.

Anion-exchange experiments conducted on the LDH materials show they can exhibit a preference for $[TcO_4]^-$ relative to the $[NO_3]^-$ ion that is prevalent in nuclear waste solutions [37]. The highest value for K_I involving the exchange for these two anions of around 20 (calculated based on a K_d value of 1200 mL/g reported for a nitrate-containing Ni(II)-Al LDH [37]) is one of the highest values yet reported. However, there are two factors that inhibit the deployment of LDH in nuclear waste disposal. First, although a K_I value of 20 is substantial, it is still too low to provide efficient $[^{99}TcO_4]^-$ removal without generating large quantities of Tc-contaminated ion-exchange materials. Second, although the $[NO_3]^-$ ion is the most concentrated competing anion present, it is not necessarily the anion that determines the performance of the exchanger. Other anions can compete for available exchange sites. Extensive adsorption experiments involving a range of anions indicate that both the size and charge of the anion dictate the selectivity sequence on LDH materials. The least competitive ions are halide ions such as Cl^-, followed by monovalent oxoanions such as $[NO_3]^-$ However, LDH exchangers have a strong preference for divalent anions over monovalent anions, which is why naturally occurring hydrotalcite contains $[CO_3]^{2-}$. However, it is important to note that many oxyanions are the conjugate bases of weak acids. The anion charge, as well as the affinity for the exchange site, depends on pH. Nuclear wastes often consist of a highly basic solution with significant carbonate and phosphate concentrations. The conditions are so caustic that these oxyanions are deprotonated, meaning that these, and other highly charged anions, must be removed before $[TcO_4]^-$ can be exchanged.

Recently, a class of highly unusual anion exchangers was discovered based on extended three-dimensional oxide networks [38, 39]. Although anionic networks such as the zeolites are relatively common, cationic networks are quite rare. The baseline material, having the composition $[ThB_5O_6(OH)_6][BO(OH)_2]\cdot 2.5H_2O$, forms a supertetrahedral cationic framework with pores connected by large (0.94×0.78-nm) ring openings (Fig. 10.13). This framework is composed of $B_{10}O_{24}$ clusters (each containing four trigonal and six tetrahedral borate groups) that bridge between Th(IV) centers. Normal borate networks (see Chapter 15) containing tetrahedral borate groups are anionic; each μ_2-oxo-bridge between trigonal and tetrahedral borate groups has a charge of −0.25. However, Th(IV) is so large that it is coordinated to 12 oxygens, and only contributes a charge of +0.33 to

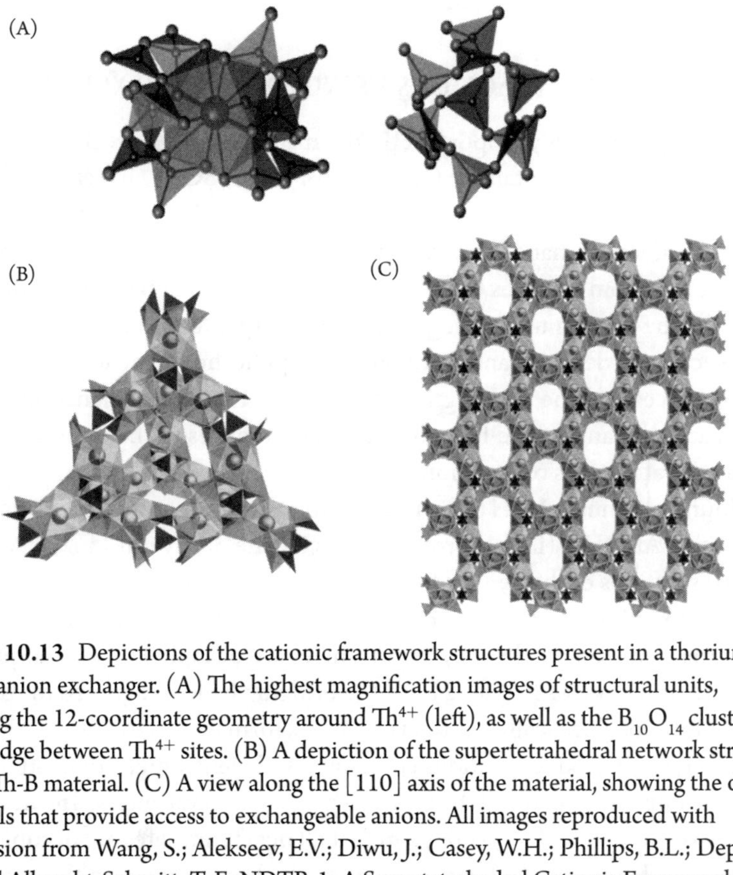

Figure 10.13 Depictions of the cationic framework structures present in a thorium borate anion exchanger. (A) The highest magnification images of structural units, showing the 12-coordinate geometry around Th^{4+} (left), as well as the $B_{10}O_{14}$ clusters that bridge between Th^{4+} sites. (B) A depiction of the supertetrahedral network structure of the Th-B material. (C) A view along the [110] axis of the material, showing the open channels that provide access to exchangeable anions. All images reproduced with permission from Wang, S.; Alekseev, E.V.; Diwu, J.; Casey, W.H.; Phillips, B.L.; Depmeier, W.; and Albrecht-Schmitt, T. E. NDTB-1: A Supertetrahedral Cationic Framework that Removes TcO_4^- from Solution. *Angew. Chem. Int. Ed.*, **2010**, 49, 1057–1060.

Th–O bonds. This local charge means that μ_2-oxo-bridges in the $B_{10}O_{12}$ clusters can bond to Th(IV) to form μ_3-oxo bridges that are only slightly overbonded. Here, the oxygen formal charge is transformed from $-1/4$ to $+1/12$ (i.e., $-0.25++0.33$). Such bonding does not occur for standard +4 cations such as Ti(IV), because these octahedral cations donate too much positive charge (i.e., twice that of 12-coordinate Th(IV)) to produce stable μ_3-oxo bonds in a borate network.

Anion-exchange experiments conducted on the Th–borate framework [39] indicate that the $[H_2BO_3]^-$ group that compensates the positive network charge can undergo anion exchange reactions involving species that are small enough to fit through network ring openings. As expected, the material is capable of exchanging $[H_2BO_3]^-$ for $[TcO_4]^-$ Unfortunately, for sufficiently small anions, the ion exchange equilibrium constants and selectivity sequences seen for the Th–borate material are similar to those of the layered LDH materials. This similarity means that K_d values in tank-waste solutions may be too small to be effective in efficient $[TcO_4]^-$ decontamination processes.

10.10 LITHIUM EXTRACTIONS USING OXIDE ION EXCHANGERS

An unusual, but highly important LDH material is $[LiAl_2(OH)_6]_2(CO_3)\cdot-nH_2O$ and relatives containing OH^-, Cl^-, Br^-, and $[SO_4]^{2-}$ instead of $[CO_3]^{2-}$. One unusual attribute of these Li–Al LDH materials is that, although they are technically anion-exchange materials, they are used to extract a specific cation—namely, Li^+—from complex brine solutions. Such extraction represents a critical source for the lithium used in high-energy-density lithium batteries for electric cars, electronic devices, and load-leveling applications involving renewable energy sources and the power grid (see Chapter 11). The explanation for how LDH materials can be used for cation extraction relates to the unique synthetic and structural features of these ion exchangers.

Lithium–aluminate LDH materials are typically produced via precipitation or by exposing $Al(OH)_3$ (in the form of either gibbsite or bayerite) to concentrated aqueous solutions of the desired lithium salt (e.g., LiOH, LiCl, or Li_2CO_3) [40]. The structure of the parent $Al(OH)_3$ consists of octahedral sheets in which only two-thirds of the octahedral oxygen vacancies are occupied by Al^{3+} to preserve charge neutrality (see Chapter 2). All the bridging hydroxyl groups in $Al(OH)_3$ (all μ_2-OH) are electrically neutral (i.e., have a formal charge of zero). However, if $Al(OH)_3$ is exposed to solutions containing sufficiently high concentrations of Li^+, the lithium ion can fill all the remaining octahedral vacancies. The new structure is isostructural with $Mg(OH)_2$, in that all octahedral sites are occupied. The net positive charge created by incorporating the lithium ion into the octahedral sheets is then compensated by the intercalation of anions between the sheets as in a conventional LDH material.

After the Li–Al LDH has been prepared, the lithium ion and the charge-compensating anion both can be released simultaneously by protonating the material in water under relatively mild pH conditions. The kinetics of the transformation of the protonated phase into gibbsite or bayerite is relatively slow, allowing the protonated form to retain an expanded state. If the protonated phase is then exposed to a brine solution containing any mixture of metal chlorides (most frequently NaCl containing low levels of LiCl), phase-pure $[LiAl_2(OH)_6]Cl.\ nH_2O$ is produced[40]. No other cations are incorporated into this phase. The Li–Al–Cl LDH can be used repeatedly to capture and release LiCl from slightly basic concentrated brine solutions, leading to an efficient lithium-separation processes.

The central question regarding loading and unloading processes involves why Li^+ is the only cation to be captured from the brine. The answer appears to involve both steric and charge effects. Vacant sites within a dense octahedral sheet of oxygen anions have a diameter of 0.12 nm, which is identical to the diameter of Li^+. Although Li^+ fits readily into the vacant sites, Na^+ (diameter = 0.19 nm) and all other monovalent cations are too large to fit within the vacancies. Similarly, all divalent cations with the exception of Mg^{2+} are too large to be captured (see Fig. 10.1). So why isn't Mg^{2+}, with a similar ionic diameter to that of Li^+ (0.13 nm), also retained? The answer lies in simple electrostatics. The Mg^{2+} ion has twice the charge of Li^+. Incorporation of Mg^{2+} into the octahedral vacancies would produce μ_3-OH bridges with a formal charge of $+1/3$ as opposed to the $+1/6$ seen for Li^+. As discussed earlier regarding LDH compositions, bridging oxygen sites with a charge of $+1/3$ or higher are unstable and are not present.

10.11 NANOMATERIALS FORMED VIA ION EXCHANGE: THE PILLARED CLAYS

Although LDH materials may never be deployed in nuclear-waste disposal, research on these materials has exploded in recent years because of the very fact that anion exchange is not highly selective. The lack of selectivity coupled with the fact that LDH materials can swell to accommodate anions with different sizes means that the LDH materials represent a unique platform for creating layered nanomaterials that can intercalate and isolate almost any anion within a controlled nanoscale environment. The range of potential applications for such materials is staggering. Just a few examples of anions that have been intercalated for specific applications [41] include (1) amino acids to form chiral catalysts, (2) thiophenes to form two-dimensional conducting polymers, (3) two-dimensional magnets based on extended Mn(II)-Cr(III) oxalate layers, (4) optically active materials containing dye molecules for dye-sensitized solar cells, and (5) photooxidation catalysts based on a wide range of intercalated anionic polyoxometalates. Simple cation substitutions within the octahedral layers of LDH materials have also proved to be useful for supercapacitors and battery electrodes.

Clays can be pillared in an analogous fashion to their LDH cousins. However, because the network charge in normal clays is anionic, the species that function as pillars for clays are cations. Although the available library of large, stiff cations is less extensive than that of anions, a wide range of interesting clay-based nanomaterials can still be produced via pillaring. For example, large oxide multimers such as $[AlO_4Al_{12}(OH)_{24}(OH_2)_{12}]^{7+}$ can be inserted into the interlayers of swellable smectite clays. This pillaring engenders a porosity that provides the clay with selective catalytic activity based on both the availability of acidic surface sites on the pillars and the ability control site access via layer spacings. Pillared clays can be used to eliminate metal toxicants from polluted ground because contaminants entrapped within the interlayers are then unavailable to biota as poisons [42,43]. Smectite clays can be modified to include amphiphilic molecules such as alkylammonium cations, which are surfactants containing both hydrophobic and hydrophilic parts (see Chapter 9). The alkyl chains impart a hydrophobic character to the clay interlayer, allowing it to store nonpolar pollutants that would normally avoid charged clay interlayer surfaces. For example, hydrophobic contaminants such as dioxanes can be sequestered by clays containing alkylammonium or similar cationic organic molecules.

10.12 OXIDES AS ION AND PROTON CONDUCTORS

Although ion-exchange equilibrium constants determine the extent to which one ion exchanges for another, the kinetics of ion exchange are often determined by the *ionic conductivity* or ion-transport properties of the solid oxide phase. Ionic conductivity can be a rate-limiting process for many chemical phenomena involving oxides, including ion exchange, electrochemical processes (Chapter 11), and even the corrosion of both metals (Chapter 12) and glasses (Chapter 15). In many oxides, this conductivity depends critically on the presence of water and/or hydroxyl groups within the bulk oxide structure.

Ionic conductors can be considered a special class of ion exchangers designed to transport a single ion selectively rather than exchanging one ion for another. Ionic conductors are of critical importance in many energy storage and use technologies that involve electrochemistry, including batteries, ultracapacitors, and fuel cells (see Chapter 11) [44]. In such technologies, ion conductors are often used as either solid electrolytes or as separators between anode and cathode compartments. Because electrons are either consumed or generated at cathodes or anodes, respectively, charge neutrality in both cathode and anode half-cells is conserved via the flow of ions from one compartment to the other through the separator. The ionic conductivity of battery separators needs to be as high as possible to eliminate parasitic resistances that reduce battery performance.

Another critical aspect of separator performance is that it often needs to exclude all ionic species other than the cation that is common to supporting electrolyte solutions in both half-cells. For the example of a Na–S battery, if polysulfide species generated in the cathode compartment migrate to the Na anode, the anode is poisoned and the battery shuts down. Ionic conductivity is also a critical parameter for electroactive oxides used as anodes and cathodes in electrochemical systems. Here, ions must be free to flow within the bulk oxide to compensate for the addition or removal of electrons via redox reactions (see Chapter 11).

Oxides can function as conductors of either anions or cations. The ultimate ionic conductors based on oxides are aqueous solutions [45] in which both cations and anions are free to migrate under the application of an applied electric field. For example, the specific conductivity of a 1-M solution of KCl at 25°C is 0.11 S/cm. To put this number in perspective, the electronic conductivities of metallic Cu, graphite, and pure Si (a semiconductor) are $6 \cdot 10^5$, 730, and 10^{-5} S/cm, respectively [46].

Aqueous solutions illustrate the major factors that control ionic conductivity: (1) mobile ion charges, (2) mobile ion concentrations, (3) the nature of conductive pathways, and (4) the activation barrier to carrier motion. The conductivity per ion is directly proportional to ion charge (or charge equivalents) as well as to ionic concentration. For aqueous solutions, a parameter that includes these factors and allows for direct comparisons between ionic species is the equivalent conductivity (Λ):

$$\Lambda = \frac{\sigma}{[M] \cdot z} \tag{10.6}$$

Here, σ is the conductivity in Siemens per centimeter, [M] is the concentration in moles per cubic centimeter, and z is the ion valence. The current for a material with 1 S of conductance increases by 1 A for every volt of potential placed across it.

An examination of equivalent conductances (Table 10.6) shows that the most mobile ion in water is the proton. Protons can hop from water molecule to water molecule via the Grotthuss mechanism (see Chapter 3). For most other ions, the primary barrier to migration involves the displacement of water molecules that are held together by weak hydrogen bonds. Here, the pathways available for ion motion are almost limitless. However, for all dissolved salts, the equivalent conductance drops with concentration because the presence of ions eventually interferes with and/or blocks the motions of other ions. Equivalent conductivities also show that, although divalent cations carry twice the charge of monovalent ions, they do not produce twice the conductivity because they are less mobile as a result of their more tightly bound solvation shells. However, measurements

Table 10-6. Proton Conductivity (σ) and Associated Activation Energy (E_a) for Some Representative Materials at Room Temperature*

Proton conductor	σ(S cm^{-1})	E_a (eV)
Three-dimensional structure		
$H_2Sb_4O_{11}\cdot 3H_2O$	2×10^{-3}	0.42
$HSbO_3\cdot 2H_2O$	3×10^{-3}	0.20
Tin-mordenite	10^{-2}	
Layered structure		
β''-$NH_4(H_3O)_{0.66}Al_{10.34}Mg_{0.66}O_{17}$	$3\times10^{-5}-1\times10^{-3}$	0.27−0.31
$H_2Ti_4O_9\cdot 1.2H_2O$	10^{-4}	
$HUO_2AsO_4\cdot 4H_2O$	5×10^{-3}	0.27
$HUO_2PO_4\cdot 4H_2O$	5×10^{-3}	0.35
$H_3Sb_3P_2O_{14}\cdot 10H_2O$	4×10^{-3}	0.33
$HSbP_2O_8\cdot 10H_2O$	10^{-2}	
α-$Zr(HPO_4)_2\cdot nH_2O$[a]	10^{-4}	0.3
γ-$Zr(PO_4)(H_2PO_4)\cdot 2H_2O$[a]	3×10^{-4}	0.24
α-Zr sulfophenylphosphonate	1.6×10^{-2}	0.20
γ-Zr sulfophenylphosphonate	1×10^{-2}	0.21
Hydrated oxides		
$SnO_2\cdot 2H_2O$	4×10^{-4}	0.20
$Sb_2O_5\cdot 5.4H_2O$	7.5×10^{-3}	0.16
Heteropolyacids		
$H_4SiW_{12}O_{40}\cdot 28H_2O$	2×10^{-2}	0.4
$H_3PW_{12}O_{40}\cdot 29H_2O$	8×10^{-2}	0.15−0.25

*Adapted from Alberti, G. and Casciola, M. Solid State Protonic Conductors: Present Main Applications and Future Prospects. Solid State Ionics 2001, 145, 3–16. Sources of the data are reported therein.

show that essentially all ions are highly mobile in water, which means aqueous solutions have limited utility as selective separators.

Oxide-anion conductors [47] are being evaluated for alternative energy technologies as separators for solid-oxide fuel cells. In such cells, H_2 or a hydrocarbon fuel is oxidized by the anode on one side whereas gaseous O_2 is reduced by the cathode on the other side. Such materials are already seeing extensive commercial use as oxygen sensors in automotive exhaust systems. Unfortunately, the motion of oxygen anions through a rigid oxide lattice tends to be extremely inhibited. First, anion motion must be preceded by the rupture of M−O bonds within the lattice. Because M−O bond energies tend to be high, such rupture normally requires high temperatures. Second, in a perfect oxide lattice, all lattice sites are occupied, so there is nowhere for the freed O^{2-} ion to go. Both factors lead to high activation energies for O^{2-} motion (e.g., 1 − 1.5 eV for Ca-stabilized zirconia). The best oxygen-ion conductors, such as ZrO_2-based ceramics, are based on highly defective structures containing large concentrations of oxygen vacancies (see Chapter 12). Such vacancies can be created in ZrO_2 by substituting cations with a lower valence (e.g., Ca(II) for Zr(IV)) into the lattice. Although the substituted zirconias are insulators at room temperature, conductivities on the order of 10^{-4} S/cm can be achieved by raising the temperature above 500°C, which is appropriate for some applications [48].

In contrast to the anion conductors, oxide materials can achieve substantial conductivity for both monovalent and divalent cations. In fact, every cation exchanger discussed in this chapter must exhibit appreciable ionic conductivity to function properly. The best cation conductors share many structural aspects with simple cation exchangers. First, mobile cations are not bound within the oxide network but are present as modifiers that compensate the charge of anionic sites within the parent network. If conductivity rather than selectivity is the primary goal, the more diffuse the network charge, the lower the activation barrier to motion. Second, cation conductors contain layered or open-network structures that provide facile pathways for cation motion. Third, the available pathways must not be completely clogged by ions, water, or other species. The drive to maximize conduction and ion-exchange capacity can sometimes create gridlock in an ion conductor just as it can on our freeways.

The principles leading to high cation conduction are illustrated by classic layered sodium β-alumina ceramics [49–51] that resemble the mineral diaoyudaoite [51]' [Fig. 10.14]. These materials are under consideration as separators for high-energy-density Na–S batteries. Batteries such as these are essential because they level the loads for power grids that are supplied by intermittent energy sources, such as wind and solar power. Like many layered ion exchangers (including $Na_2Ti_4O_9$), sodium β-alumina is prepared by adding a modifying oxide (Na_2O) to a network-forming oxide (Al_2O_3). The modifier cations disrupt bridging metal–oxygen bonds and replace them with anionic sites that are compensated by the modifier cation. For modified alumina, the compositions with the highest conductivity have the composition $Na_2O-11Al_2O_3$. This composition creates structures containing spinel blocks that consist of four layers of close-packed oxygen anions, with Al^{3+} filling both octahedral and tetrahedral sites. Each block is propped apart from other blocks by pillars consisting of two $Al(O)_4$ units connected via a μ_2-oxo-bridge, resulting in an interlayer spacing of 0.46 nm. This layer spacing is wide enough to define conduction planes that can contain bare Na^+ (with a diameter of 0.2 nm), but is too narrow to admit hydrated $[Na(OH_2)_6]^+$ (diameter of 0.55 nm). Within the conduction planes there are no nonbridging oxygens or bound hydroxyl groups that might deprotonate and

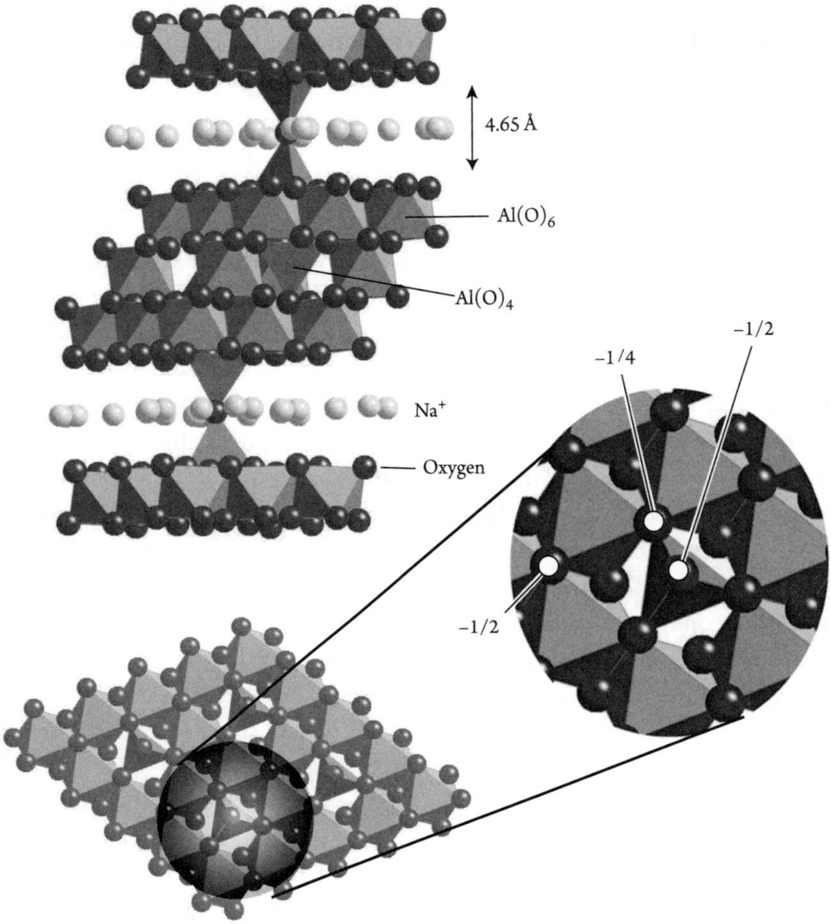

Figure 10.14 Side and top-view depictions of the structure of sodium β–alumina. A side view (top) shows the bridging oxygens that pry adjacent spinel blocks apart and provide open conduction planes that are occupied by sodium ions. The numbers shown in the inset figure correspond to formal charges on the oxygens.

attract migrating cations. Features of this structure that lead to high conduction are as follows: (1) the anionic charge lining the conduction planes is relatively diffuse (with formal charges of $-1/4$ and $-1/2$) and (2) the Na^+ concentration in the conduction planes is sufficiently low that each ion is free to move without being impeded by other ions. The resulting sodium β-alumina crystals exhibit conductivities as high as 0.1 S/cm at temperatures as low as 300°C. Such conductivity is so high as to be comparable to that of an aqueous KCl solution at room temperature.

Oxide materials having among the highest ionic conductivities are the NASICON (for sodium super-ionic conductors) [52,53]. The primary NASICON phases

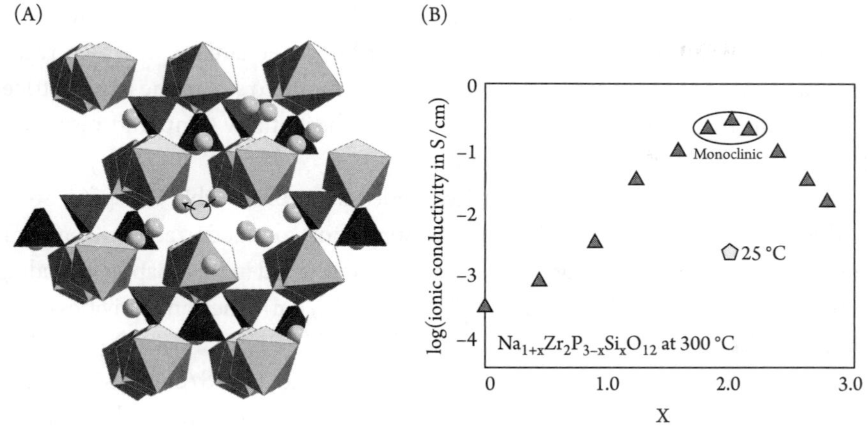

Figure 10.15 (A) A depiction of the three-dimensional network structure of NASICON. (B) The ionic conductivity of Nasicon phases as a function of composition (data from the following sources: Goodenough, J. B.; Hong, H. Y. P.; Kafalas, J. A. Fast Na^+-Ion Transport in Skeleton Structures. *Mat. Res. Bull.* **1976**, *11*, 203–220. Anantharamulu, N.; et al. A Wide-Ranging Review of Nasicon Type Materials. *J. Mater. Sci.* **2011**, *46*, 2821–2837.).

have compositions of $Na_{1+x}Zr_2Si_xP_{3-x}O_{12}$ $(0<x<3)$. For the purposes of direct comparison, $Na_3Zr_3Si_2PO_{12}$ exhibits an ionic conductivity at 447°C of 37.5 S/cm relative to $9.7 \cdot 10^{-2}$ S/cm for sodium-β alumina. One reason for the improvement in conductivity has to do with the fact that the zeolitic NASICON structure (Fig. 10.15) contains a network of channels that allows for facile conductivity in all three dimensions [54], as opposed to the two-dimensional conductivity imposed by the layered structures of β-alumina phases. Because practical oxide materials are polycrystalline, conductive pathways are invariably blocked at grain boundaries in sodium β-alumina to a greater extent than in phase-pure NASICON materials. The second reason for high conductivity has to do with the fact that the network charge in NASICON is even more diffuse than it is in sodium β-alumina. The two primary anionic sites in NASICON are $Zr-\mu_2O-Si$ and $Zr-\mu_2O-P$ having formal charges of $-1/3$ and $-1/12$, respectively.

The final class of ion conductors to be discussed here involves proton conductors based on oxides [55]. Just as protons generate the highest ionic conductivities in water (e.g., $\Lambda = 425$ S-cm^2/mol for HCl vs. 150 S-cm^2/mol for a similar dilute solution of KCl), the best proton conductors exhibit the highest conductivities of any oxide cation conductors. The proton conductors also illustrate how the presence of water within the ion conductor can influence transport properties. The proton is so small that efficient proton transport involves hopping between oxygen anions that are in physical contact with one another (see Chapter 3). This typically means that conduction pathways must be lined with high concentrations of hydroxyl groups. However, the most efficient mechanism for proton transport involves the Grotthuss mechanism, in which rotating water molecules assist in moving protons from site to site in the oxide lattice. This means that proton conductors are at their best at low temperatures where they are able to retain waters of hydration.

Proton conductivities of representative oxides are listed in Table 10.6 [56]. Some of the materials exhibiting the highest values for proton conduction at room temperature are layered structures that have already been discussed in this chapter, including smectite clay ($\sigma = 10^{-4}$ S/cm) and $H_2Ti_4O_9$ (10^{-4} S/cm). The conductivities of hydrated oxides can be higher still, as exemplified by SnO_2-2H_2O and $Sb_2O_5-5.4H_2O$, and zeolitic Sn–mordenite ($4\cdot10^{-4}$ and $7.5\cdot10^{-3}$, and 10^{-2} S/cm , respectively). The highest conductivities that have been reported are for polyoxometalates such as $H_3PW_{12}O_{40}-29H_2O$, which are almost as conductive ($\sigma = 8\cdot10^{-2}$) as aqueous salt solutions. Unfortunately, many of these materials are unstable with respect to temperature and dehydrate, which limits their practical application. Perhaps sufficient stability will be achieved someday in hybrid materials similar to LDH anion exchangers pillared with polyoxometalate ions.

References

1. Press, F.; Siever, R. *Earth, 2nd ed.*; W. H. Freeman: San Francisco, CA, 1978.
2. Griffin, J. J.; Windom, H.; Goldberg, E. E. The Distribution of Clay Minerals in the World Ocean. *Deep-Sea Res.* **1968**, *15*, 433–459.
3. Helferrich, F. *Ion Exchange*; McGraw Hill: New York, 1962.
4. Amphlett, C. B. *Inorganic Ion Exchangers*; Elsevier Publishing: Amsterdam, 1964.
5. Sposito, G. *Chemical Equilibria and Kinetics in Soils*. Oxford University Press: New York, 1994.
6. Crowley, K. D. Nuclear Waste Disposal: The Technical Challenges. *Physics Today* **1997**, *June*, 32–39.
7. Barrer, R. M. *Hydrothermal Chemistry of Zeolites*; Academic Press: London, 1982.
8. Bockris, J. O'M.; Reddy, A. K. N. *Modern Electrochemistry, 2nd ed.*; Plenum Press: New York, 1998; Vol. 1.
9. Shim, S.-H.; Navrotsky, A.; Gaffney, T. R.; and MacDougall, J. E. Chabazite: Energetics of Hydration, Enthalpy of Formation and Effects of Cations on Stability. *Am. Miner.* **1991**, *84*, 1870–1882.
10. Sun, P.; Navrotsky, A. Enthalpy of Formation and Dehydration of Alkaline Earth Cation Exchanged Zeolite Beta. *Micropor. Mesopor. Mater.* **2008**, *109*, 147–155.
11. Sun, P.; Deore, S.; Navrotsky, A. Formation and Dehydration Enthalpy of Ion Exchanged Zeolite Beta. *Micropor. Mesopor. Mater.* **2006**, *91*, 15–22.
12. Yang, D.; Liu, H.; Zheng, Z.; Sarina, S and Zhu, H. Titanate-Based Adsorbents for Radioactive Ions Entrapment from Water. *Nanoscale* **2013**, *5*, 2232–2242.
13. Nyman, M.; Hobbs, D. T. A Family of Peroxo-Titanate Materials Tailored for Optimal Strontium and Actinide Sorption. *Chem. Mater.* **2006**, *18*, 6425–6435.

14. Yang, D.; Sarina, S.; Zhu, H.; Liu, H.; Zheng, Z.; Xie, M.; Smith, S.V.; and Komarneni, S. Capture of Radioactive Cesium and Iodide Ions from Water by Using Titanate Nanofibers and Nanotubes. *Angew. Chem. Int. Ed.* **2011**, *50*, 10594–10598.
15. Dosch, R. G. *The Use of Titanates in Decontamination of Defense Waste*. Sandia National Laboratories Report SAND78-0710; Sandia National Laboratories: Albuquerque, NM, 1978.
16. Tournoux, M.; Marchand, R.; Brohan, L. Layered $K_2Ti_4O_9$ and the Open Metastable TiO_2(B) Structure. *Prog. Solid State Chem.* **1986**, *17*, 33–52.
17. Andersson, S.; Wadsley, A. D. The Crystal Structure of $Na_2Ti_3O_7$. *Acta Cryst.* **1961**, *14*, 1245–1249.
18. Inoue, Y.; Tsuji, M. Studies of the Hydrous Titanium Oxide Ion Exchanger: II. The Equivalence of the Exchange Adsorption of Cations and the Dissociation Constant. *Bull. Chem. Soc. Japan* **1976**, *49*, 111–116.
19. Sasaki, T.; Watanabe, M.; Komatsu, Y.; Fujinki, Y. Na^+/H^+ Ion Exchange Process on Layered Hydrous Titanium Oxide. *Bull. Chem. Soc. Japan* **1985**, *58*, 3500–3505.
20. Lehto, J.; Harjula, R.; Girard, A. The Equilibrium of Strontium Ion Exchange on Sodium Titanate, $Na_4Ti_9O_{20}\cdot xH_2O$. *J. Chem. Soc. Dalton Trans.* **1989**, *1989*, 101–103.
21. Bunker, B. C.; Peden, C. H. F.; Kirkpatrick, R. J.; Turner, G. L. ^{17}O NMR Studies of Titanate Repolymerization in Water. *Mater. Res. Soc. Symp.* **1997**, *432*, 39–43.
22. Anthony, R. G.; Philip, C. V.; Dosch, R. G. Selective Adsorption and Ion Exchange of Metal Cations and Anions with Silico-titanates and Layered Titanates. *Waste Manage.* **1993**, *13*, 503–512.
23. Bray, L. A.; Carson, K. J.; Elovich, R. J. *Initial Evaluation of Sandia National Laboratory-Prepared Crystalline Silicotitanates for Cesium Recovery*. Pacific Northwest Laboratory Report PNL-8847/UC-510; Pacific Northwest National Laboratory: Richland, WA, October 1993.
24. Bunker, B. C. *Evaluation of Inorganic Ion Exchangers for Removal of Cesium from Tank Waste*. Pacific Northwest Laboratory Report TWRSPP-94-085; Pacific Northwest National Laboratory: Richland, WA, September 1994.
25. Hess, N. J.; Su, Y.; Balmer, M. L. Evidence of Edge-Sharing TiO_5 Polyhedra in Ti-Substituted Pollucite, $CsTi_xAl_{1-x}Si_2O_{6+x/2}$. *J. Phys. Chem.* **2001**, *105*, 6805–6811.
26. Speer, J. A.; Gibbs, G. V. The Crystal Structure of Synthetic Titanite, $CaTiOSiO_4$, and Domain Textures of Natural Titanites. *Am. Miner.* **1976**, *61*, 238–247.
27. Su, Y.; Balmer, M. L.; Bunker, B. C. Raman Spectroscopic Studies of Silicotitanates. *J. Phys. Chem. B* **2000**, *104*, 8160–8169.
28. Andersson, S.; Wadsley, A. D. Five Co-ordinated Titanium in $K_2Ti_2O_5$. *Acta Chim. Scand.* **1961**, *15*, 663–669.
29. Bray, L. A.; Carson, K. J.; Elovich, R. J. *Initial Evaluation of Sandia National Laboratory Prepared Crystalline Silico-Titanates for the Recovery of Cesium*. Pacific Northwest National Laboratory Report TWRSPP-93-013; Pacific Northwest National Laboratory: Richland, WA, 1993.
30. Balmer, M. L.; Huang, Q.; Wong-Ng, W.; Roth, R. S.; Santoro, A. Neutron and X-ray Diffraction Study of the Crystal Structure of $CsTiSi_2O_{6.5}$. *J. Solid State Chem.* **1997**, *130*, 97–102.

31. Balmer, M. L.; Bunker, B. C. *Inorganic Ion Exchange Evaluation and Design: Silicotitanate Ion Exchange Waste Conversion*. Pacific Northwest National Laboratory Report PNL-10460; Pacific Northwest National Laboratory: Richland, WA, 1995.
32. Staunton, S.; Roubaud, M. Adsorption of ^{137}Cs on Montmorillionite and Illite: Effect of Charge Compensating Cation, Ionic Strength, Concentration of Cs, K and Fulvic Acid. *Clays Clay Minerals* **1997**, *45*, 251–260.
33. Sawney, B. L. Selective Sorption and Fixation of Cations by Clay Minerals: A Review. *Clays Clay Minerals* **1972**, *20*, 92–100.
34. Smith, J. T.; Comans, R. N. J. Modeling the Diffusive Transport and Remobilisation of ^{137}Cs in Sediments: The Effects of Sorption Kinetics and Reversibility. *Geochim Cosmochim. Acta* **1996**, *60*, 995–1004.
35. Wang, Q.; O'Hare, D. Recent Advances in the Synthesis and Application of Layered Double Hydroxide (LDH) Nanosheets. *Chem. Rev.* **2012**, *112*, 4124–4155.
36. Sideris, P. J.; Nielsen, U. G.; Gan, Z.; Grey, C. P. Mg/Al Ordering in Layered Double Hydroxides Revealed by Multinuclear NMR Spectroscopy. *Science* **2008**, *321*, 113–117.
37. Wang, Y.; Gao, H. Compositional and Structural Control on Anion Sorption Capability of Layered Double Hydroxides (LDHs). *J. Colloid Interface Sci.* **2006**, *301*, 19–26.
38. Wang, S.; Alekseev, E. V.; Diwu, J.; Casey, W. H.; Phillips, B. L.; Depmeier, W.; Albrecht-Schmitt, T. E. NDTB-1: A Supertetrahedral Cationic Framework that Removes TcO_4^- from Solution. *Angew. Chem. Int. Ed.* **2010**, *49*, 1057–1060.
39. Wang, S.; Yu, P.; Purse, B. A.; Orta, M. J.; Diwu, J.; Casey, W. H.; Phillips, B. L.; Alekseev, E. V.; Depmeier, W.; Hobbs, D. T.; Albrecht-Schmitt, T. E. Selectivity, Kinetics, and Efficiency of Reversible Anion Exchange with TcO_4^- in a Supertetrahedral Cationic Framework. *Adv. Funct. Mater.* **2012**, *22*, 1–10.
40. Fogg, A. M.; O'Hare, D. Study of the Intercalation of Lithium Salt in Gibbsite Using Time-Resolved in Situ X-ray Diffraction. *Chem. Mater.* **1999**, *11*, 1771–1775.
41. Rives, V.; Ulibarri, M. A. Layered Double Hydroxides (LDH) Intercalated with Metal Coordination Compounds and Oxometalates. *Coord. Chem. Rev.* **1999**, *181*, 61–120.
42. Sposito, G. *The Surface Chemistry of Natural Particles*; Oxford University Press: Oxford, 2004
43. Bhattacharyya, K. G.; Gupta, S. S. Adsorption of a Few Heavy Metals on Natural and Modified Kaolinite and Montmorillonite: A Review. *Adv. Colloid Interface Sci.* **2008**, *140*, 114–131.
44. Shriver, D. F.; Farrington, G. C. Solid Ionic Conductors. *C&E News* **1985**, *May*, 42–57.
45. Bockris, M. O'M.; Reddy, A. K. N.; *Modern Electrochemistry, 2nd ed.*; Plenum Press: New York, 1998; Vol. 1, Ionics.
46. Weast, R. C., Ed. *CRC Handbook of Chemistry and Physics, 58th ed.*; CRC Press: Cleveland, OH, 1977.
47. Goodenough, J. B. Oxide-Ion Electrolytes. *Annu. Rev. Mater. Res.* **2003**, *33*, 91–128.
48. Yang, Z.; Zhang, J.; Kintner-Meyer, M. C. W.; Lu, X.; Choi, D.; Lemmon, J. P.; Liu, J. Electrochemical Energy Storage for Green Grid. *Chem. Rev.* **2011**, *111*, 3577–3613.

49. Lu, X.; Xia, G.; Lemmon, J. P.; Yang, Z. Advanced Materials for Sodium-Beta Alumina Batteries: Status, Challenges, and Perspectives. *J. Power Sources* **2010**, *195*, 2431–2442.
50. Peters, C. R.; Bettman, M.; Moore, J. W.; and Glick, M. D. Refinement of the Structure of Sodium β-Alumina. *Acta Cryst.* **1971**, *B27*, 1826–1834.
51. Zhu, N.; Guo, F.; Li, Y. Study on Crystal Structure of Diaoyudaoite. *Acta Chim. Sinica* **1992**, *50*, 527–532.
52. Goodenough, J. B.; Hong, H. Y. P.; Kafalas, J. A. Fast Na^+-Ion Transport in Skeleton Structures. *Mater. Res. Bull.* **1976**, *11*, 203–220.
53. Anantharamulu, N.; Koteswara Rao, K.; Rambabu, G.; Vijaya Kumar, B.; Radha, V.; and Vithal, M. A Wide-Ranging Review of Nasicon Type Materials. *J. Mater. Sci.* **2011**, *46*, 2821–2837.
54. Essoumhi, A.; Favotto, C.; Mansori, M.; Satre, P. Synthesis and Characterization of a NASICON Series with General Formula $Na_{2.8}Zr_{2-y}Si_{1.8-4y}P_{1.2+4y}O_{12}$ $(0 < y < 0.45)$. *J. Solid State Chem.* **2004**, *177*, 4475–4481.
55. Kreuer, K. D.Proton Conducting Oxides. *Annu. Rev. Mater. Res.* **2003**, *33*, 333–359.
56. Alberti, G.; Casciola, M. Solid State Protonic Conductors: Present Main Applications and Future Prospects. *Solid State Ionics* **2001**, *145*, 3–16.

11

The Electrochemistry of Oxides

11.1 INTRODUCTION

Many of the critical reactions considered in this book involve the addition or subtraction of protons from oxides. In this chapter, we consider another species that can change oxide charge distributions and reactivity dramatically: the electron. Many oxides contain cations that have access to more than one oxidation state (Fig. 11.1) [1] in water. Cations in these oxides can either donate or accept electrons to change their charge, or *oxidation state. Oxidation* reactions involve the loss of electrons as they are donated to other species, resulting in an increase in the cation charge or valence, whereas *reduction* reactions involve the capture of an electron resulting in a decrease in the cation valence. Below, the basics of electrochemistry are first described in the context of the redox chemistry of water and representative oxide systems. Second, we describe the fundamentals of electron-transfer reactions in oxides and the impact of electron transfer on the acid–base, ion-exchange, and ligand-exchange reactions of the host oxide. Finally, we discuss the behavior of oxides in electrochemical energy storage devices and the role that nanotechnology has in optimizing electrochemical performance.

hydrogen 1 **H** 1.0079																	helium 2 **He** 4.0026
lithium 3 **Li** 6.941	beryllium 4 **Be** 9.0122											boron 5 **B** 10.811	carbon 6 **C** 12.011	nitrogen 7 **N** 14.007	oxygen 8 **O** 15.999	fluorine 9 **F** 18.998	neon 10 **Ne** 20.180
sodium 11 **Na** 22.990	magnesium 12 **Mg** 24.305											aluminium 13 **Al** 26.982	silicon 14 **Si** 28.086	phosphorus 15 **P** 30.974	sulfur 16 **S** 32.065	chlorine 17 **Cl** 35.453	argon 18 **Ar** 39.948
potassium 19 **K** 39.098	calcium 20 **Ca** 40.078	scandium 21 **Sc** 44.956	titanium 22 **Ti** 47.867	vanadium 23 **V** 50.942	chromium 24 **Cr** 51.996	manganese 25 **Mn** 54.938	iron 26 **Fe** 55.845	cobalt 27 **Co** 58.933	nickel 28 **Ni** 58.693	copper 29 **Cu** 63.546	zinc 30 **Zn** 65.38	gallium 31 **Ga** 69.723	germanium 32 **Ge** 72.64	arsenic 33 **As** 74.922	selenium 34 **Se** 78.96	bromine 35 **Br** 79.904	krypton 36 **Kr** 83.798
rubidium 37 **Rb** 85.468	strontium 38 **Sr** 87.62	yttrium 39 **Y** 88.906	zirconium 40 **Zr** 91.224	niobium 41 **Nb** 92.906	molybdenum 42 **Mo** 95.96	technetium 43 **Tc** [98]	ruthenium 44 **Ru** 101.07	rhodium 45 **Rh** 102.91	palladium 46 **Pd** 106.42	silver 47 **Ag** 107.87	cadmium 48 **Cd** 112.41	indium 49 **In** 114.82	tin 50 **Sn** 118.71	antimony 51 **Sb** 121.76	tellurium 52 **Te** 127.60	iodine 53 **I** 126.90	xenon 54 **Xe** 131.29
caesium 55 **Cs** 132.91	barium 56 **Ba** 137.33		hafnium 72 **Hf** 178.49	tantalum 73 **Ta** 180.95	tungsten 74 **W** 183.84	rhenium 75 **Re** 186.21	osmium 76 **Os** 190.23	iridium 77 **Ir** 192.22	platinum 78 **Pt** 195.08	gold 79 **Au** 196.97	mercury 80 **Hg** 200.59	thallium 81 **Tl** 204.38	lead 82 **Pb** 207.2	bismuth 83 **Bi** 208.98	polonium 84 **Po** [209]	astatine 85 **At** [210]	radon 86 **Rn** [222]
francium 87 **Fr** [223]	radium 88 **Ra** [226]		rutherfordium 104 **Rf** [261]	dubnium 105 **Db** [262]	seaborgium 106 **Sg** [266]	bohrium 107 **Bh** [264]	hassium 108 **Hs** [277]	meitnerium 109 **Mt** [268]	darmstadtium 110 **Ds** [271]	roentgenium 111 **Rg** [272]							
			lanthanum 57 **La** 138.91	cerium 58 **Ce** 140.12	praseodymium 59 **Pr** 140.91	neodymium 60 **Nd** 144.24	promethium 61 **Pm** [145]	samarium 62 **Sm** 150.36	europium 63 **Eu** 151.96	gadolinium 64 **Gd** 157.25	terbium 65 **Tb** 158.93	dysprosium 66 **Dy** 162.50	holmium 67 **Ho** 164.93	erbium 68 **Er** 167.26	thulium 69 **Tm** 168.93	ytterbium 70 **Yb** 173.05	lutetium 71 **Lu** 174.97
			actinium 89 **Ac** [227]	thorium 90 **Th** 232.04	protactinium 91 **Pa** 231.04	uranium 92 **U** 238.03	neptunium 93 **Np** [237]	plutonium 94 **Pu** [244]	americium 95 **Am** [243]	curium 96 **Cm** [247]	berkelium 97 **Bk** [247]	californium 98 **Cf** [251]	einsteinium 99 **Es** [252]	fermium 100 **Fm** [257]	mendelevium 101 **Md** [258]	nobelium 102 **No** [259]	lawrencium 103 **Lr** [262]

Figure 11.1 A Periodic Table highlighting the major elements in oxides that are electrochemically active in water.

11.2 ELECTROCHEMICAL REACTIONS

Electrochemical reactions are typically written in the context of the geometry of a battery or a galvanic cell [2] (Fig. 11.2). For the cell shown in Figure 11.2, which contains metal electrodes immersed in aqueous solutions containing solvated cations, the net reaction can be written as

$$Cu^{\circ}(s)+2Fe^{3+}(aq)\rightleftarrows 2Fe^{2+}(aq)+Cu^{2+}(aq) \qquad \mathbf{(11.1)}$$

In this reaction, Cu metal is being oxidized to form Cu(II), whereas Fe(III) is being reduced to form Fe(II). The individual oxidation and reduction reactions leading to Eq. 11.1 are referred to as half-reactions:

$$Cu^{\circ}(s)\rightleftarrows Cu^{2+}(aq)+2e^{-} \qquad E_{o}=-0.34\ V \qquad \mathbf{(11.2)}$$

$$2Fe^{3+}(aq)+2e^{-}\rightleftarrows 2Fe^{2+}(aq) \qquad E_{o}=+0.77\ V \qquad \mathbf{(11.3)}$$

The net result of Eqs. 11.2 and 11.3 in the context of Figure 11.2 is the transfer of electrons from the Cu-containing compartment in which oxidation occurs, which is called the *anode*, to the Fe-containing compartment in which reduction occurs, which is called the *cathode*. This electron transfer generates an electric current, a charge-compensating ion current (in the salt bridge), as well as a voltage.

The expression that relates the voltage generated to the chemical equilibria associated with Eq. 11.1 is called the *Nernst equation* [3] (Eq. 11.4). For a general

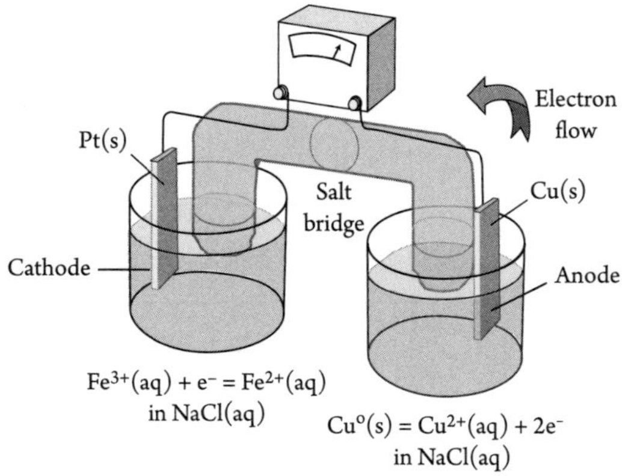

Figure 11.2 A representation of an electrochemical reaction as broken down into half reactions that would occur in an ideal battery. Oxidation (removal of electrons) occurs at the anode; reduction (addition of electrons) occurs at the cathode. Electrons flow in an external circuit from the anode to the cathode, as do charge-compensating cations that flow through an electrolyte bridge between the two electrodes. Adapted from Mahan, B. *University Chemistry, 3rd ed.*; Addison Wesley: Menlo Park, CA, 1975; p 267.

reaction $aA+bB=cC+dD$, the free energy for the reaction is given by $\Delta G=-nFE°$, whereas the cell voltage is given by

$$\Delta E=\Delta E°-\frac{2.303\cdot RT}{nF}log\left[\frac{[C]^c[D]^d}{[A]^a[B]^b}\right] \quad (11.4)$$

In Eq. 11.4, ΔE is the cell voltage, n is the number of electrons transferred, and F is Faraday's constant ($F=96,485$ Coulombs/mole). At room temperature, $\frac{2.303\cdot RT}{F}=0.05912$. The factor of 2.303 arises from the conversion of Napier's logarithms (natural logarithms) to logarithms in base ten. If $n=1$, the electrode responds by 0.059 V for every factor of ten change in the concentrations of products divided by reactants. $\Delta E°$ is the difference between the standard redox potentials for the two half-reactions. The standard redox potential (E°) for each half-reaction corresponds to the voltage that would be generated relative to the $H_2/2H^+$ half-reaction if all concentrations were in their standard states (i.e., 1 M H_3O^+ is equivalent to 1 M H^+; or $H_3O^+=H^+$ in the formalism.) and 1 atm H_2 such that the second term in Eq. 11.4 is equal to zero):

$$H_2(1\,atm)\rightleftarrows 2H^+(1M(aq))+2e^- \quad E_o=0.0\,V \quad (11.5)$$

This E° voltage is not zero by coincidence, but is a reference state that is defined to be zero. Standard reduction, or *redox,* potentials are sometimes referred to by the term E_h, where the *h* refers to the fact that the listed half-reaction is referenced to the hydrogen electrode and half-cell.

Standard redox potentials for all elements and oxidation states [1] are compiled for reactions, written with the electrons are on the left side of the equal sign. The set of such reduction potentials are known as the *electrochemical series* (Table 11.1). Oxidation potentials (i.e., potentials for the reverse reactions) are obtained merely by changing the sign on the reduction potential. For example, the reduction potential that is listed in the electrochemical series for Eq. 11.2 is $+0.34$ V, but because a plate of Cu°(s) metal is shown as being oxidized in Eq. 11.2, $E°=-0.34$ V for the reverse reaction as written. The net reaction in Eq. 11.1 is the sum of Eqs. 11.2 and 11.3, whereas $\Delta E°$ is the sum of the E° values of the two half-reactions (i.e., $\Delta E°=-0.34\,V+0.77\,V=+0.43\,V$).

The sign and magnitude of the redox potential are strong indicators of whether a particular species is prone to being oxidized or reduced. If the net value for $\Delta E°$ is positive, the reaction tends to proceed in the direction shown, whereas if $\Delta E°$ is negative, the reaction proceeds in the reverse direction. As examples, the standard reduction potential for Li(s) is given by

$$Li^+(aq)+e^-\rightleftarrows Li(s) \quad E°=-3.0\,V \quad (11.6)$$

Table 11-1. A Ranking of Pure Metals According to the Energy Available to Oxidize $H_2(g)$ at Standard-State Conditions, Expressed as Volts Relative to the Standard Hydrogen Electrode

Anodic Reaction	**Standard-State Potential (E°/V; for half reaction as written)**
Mg^{2+} (aq) + $2e^- \rightleftarrows$ Mg° (s)	−2.37
Be^{2+} (aq) + $2e^- \rightleftarrows$ Be° (s)	−1.85
Al^{3+} (aq) + $3e^- \rightleftarrows$ Al° (s)	−1.68
Mn^{2+} (aq) + $2e^- \rightleftarrows$ Mn° (s)	−1.19
Zn^{2+} (aq) + $2e^- \rightleftarrows$ Zn° (s)	−0.77
Cr^{3+} (aq) + $3e^- \rightleftarrows$ Cr° (s)	−0.74
Cd^{2+} (aq) + $2e^- \rightleftarrows$ Cd° (s)	−0.40
Fe^{2+} (aq) + $2e^- \rightleftarrows$ Fe° (s)	−0.44
Sn^{2+} (aq) + $2e^- \rightleftarrows$ Sn° (s)	−0.14
Pb^{2+} (aq) + $2e^- \rightleftarrows$ Pb° (s)	−0.13
Ni^{2+} (aq) + $2e^- \rightleftarrows$ Ni° (s)	−0.25
$2H^+$ (aq) + $2e^- \rightleftarrows$ H_2° (s)	0.0
Sb^{3+} (aq) + $3e^- \rightleftarrows$ Sb° (s)	+0.10
Cu^{2+} (aq) + $2e^- \rightleftarrows$ Cu° (s)	+0.35
Ag^{2+} (aq) + $2e^- \rightleftarrows$ Ag° (s)	+0.80
Pt^{2+} (aq) + $2e^- \rightleftarrows$ Pt° (s)	+1.42
Au^{+3} (aq) + $3e^- \rightleftarrows$ Au° (s)	+1.50

Eq. 11.6 indicates that lithium cations are extremely difficult to reduce to lithium metal. In fact, there is an extremely strong driving force for oxidizing lithium metal to form Li^+, which is one reason it is used in batteries. Conversely, the noble metal Pt°(s) has a reduction potential of +1.2 V, which means that Pt°(s) is difficult to oxidize to create either oxides or dissolved cations.

11.3 POURBAIX DIAGRAMS

All the information regarding how the voltages associated with half-reactions change with pH is captured in a graphical plot called a *Pourbaix diagram* [4]. Pourbaix diagrams are the electrochemical equivalent of hydrolysis diagrams and can be used to predict which soluble species, phases, and oxidation states should be stable. In these diagrams, the most abundant aqueous species are shown as a function of pH as well as the applied voltage or reduction potential (commonly referred to as E_h). Pourbaix diagrams were developed initially to predict the corrosion behavior of metals exposed to water (see Chapter 12). These diagrams define those solution conditions leading to either dissolution or the formation of oxide surface films that in some cases can protect or passivate the underlying metal. However, the diagrams are also extremely useful in defining the electrochemical behavior and stability of oxides used in electrochemical processes such as energy storage, water purification, and electrochemical catalysis.

The slopes of lines in a Pourbaix diagram are easy to interpret. A vertical line indicates a pure hydrolysis reaction that does not depend on applied voltage, such as $Fe^{3+}(aq)+2H_2O(l)=FeOH^{2+}(aq)+H_3O^+(aq)$, such as we discussed in Chapter 5 regarding hydrolysis diagrams. Note that the valence of iron remains Fe(III) and the dominant species, Fe^{3+} or $FeOH^{2+}$, depends only on the solution pH. In contrast, a horizontal line is a pure electron-transfer reaction that does not depend on pH, such as $Fe^{3+}(aq)+e^-=Fe^{2+}(aq)$. Lines with intermediate slopes involve both proton and electron-transfer processes. Each line is drawn where the reactants and products have equal activities (i.e., defining the intercepts for each line). However, the most important thing to remember is that each curve is now defined by a new variable, which is the redox potential, E_h, in volts. Next, we discuss the electrochemistry of water and oxides in terms of these diagrams.

11.4 THE ELECTROCHEMISTRY OF WATER

Water itself is electrochemically active and can be oxidized to produce O_2 gas, or reduced to produce H_2 gas via the reactions

$$2H_3O^+(aq)+2e^- \rightleftarrows 2H_2O(l)+H_2(g) \qquad E°=0.0\ V \tag{11.7}$$

$$2H_2O(l)+2e^- \rightleftarrows 2OH^-(aq)+H_2(g) \qquad E°=-0.83\ V \tag{11.8}$$

$$6H_2O \rightleftarrows O_2(g)+4H_3O^+(aq)+4e^- \qquad E°=-1.23\ V \tag{11.9}$$

$$4OH^- \rightleftarrows O_2(g)+2H_2O(aq)+4e^- \qquad E°=-0.4\ V \tag{11.10}$$

Notice that the standard values for the reduction reactions are 0.83 V lower in basic solutions (1 M NaOH) than they are in acidic solutions (1 M HCl), and conversely are 0.83 V higher for oxidation reactions. This difference is a direct consequence of the fact that one proton is consumed for each electron consumed to maintain charge neutrality. Because the number of electrons equals the number of protons transferred, and that both are on the same side of the reaction, the slope of the line is: $\frac{2.303 \cdot RT}{F}=0.0592\ V/pH$. This is why a pH electrode at 298 K, for example, has 0.059 V of signal response for every pH unit change. The net result is a voltage change from pH 0 to pH 14 of $(0.059)(14)=0.83\ V$.

The Pourbaix diagram for water (Fig. 11.3) is particularly important because it defines what reactions can and cannot be expected to occur in an aqueous environment. This diagram is also the key to understanding the conditions under which water can be converted into the environmentally friendly fuels H_2(g) and O_2(g) in fuel cells or via photoelectrochemical processes. The two primary lines shown in the diagram are the line for the oxidation of water (with an intercept value on the ordinate of +1.23 V at pH 0) and the line for the reduction of water (with an intercept value on the ordinate of 0.0 V at pH 0), representing the two most important half-reactions for water. Reduction reactions with voltages that fall between these two curves do not have sufficient energy either to oxidize or reduce water. Therefore, the region between the curves represents the window of thermodynamic stability for water.

Half-reactions that exhibit reduction potentials that fall below the 0.0-V water reduction curve are expected to reduce water electrochemically in reactions that

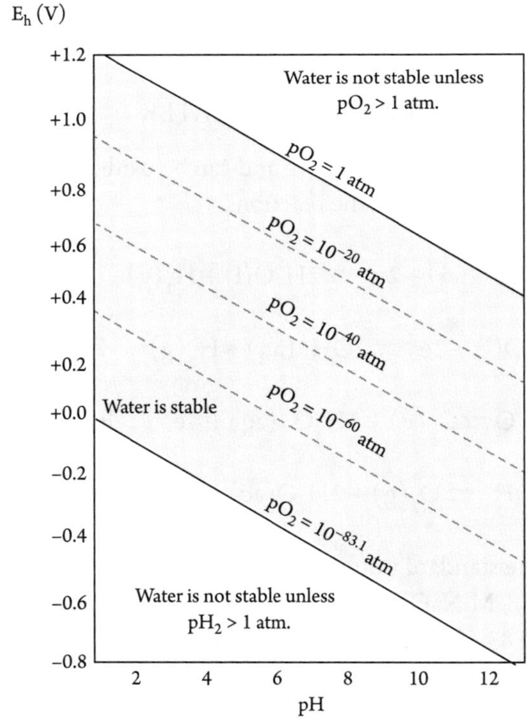

Figure 11.3 The Pourbaix diagram for water, showing the stability limits of water as a function of solution pH. Faint lines show the fugacity, or pressure, or O_2(g) across the diagram from very reducing conditions ($pO_2 = 10^{-83.1}$ atm) to conditions near those in contact with our atmosphere (pO_2 = 1 atm). Adapted with permission from Pourbaix, M. *Atlas of Electrochemical Equilibria in Aqueous Solutions*; National Association of Corrosion Engineers: Houston, TX, 1974.

tend to consume protons and promote alkalization (i.e., increase the solution pH). For example,

$$2Li^\circ(s) \rightleftarrows 2Li^+(aq) + 2e^- \quad E^\circ = +3.0\ V \quad (11.11)$$

$$+ \quad 2H_3O^+(aq) + 2e^- \rightleftarrows 2H_2O(l) + H_2(g) \quad E^\circ = 0.0\ V \quad (11.12)$$

$$= \quad 2Li^\circ(s) + 2H_3O^+(aq) \rightleftarrows 2Li^+(aq) + 2H_2O(l) + H_2(g) \quad E^\circ = +3.0V \quad (11.13)$$

Notice that in Pourbaix diagrams, the curve for Li°(s) intercepts –3.0 V at pH 0. Because 1 V corresponds to 96 kJ/mole, the driving force for Li°(s) to reduce water as it oxidizes is huge (and in fact can be explosive), which is why all lithium-metal batteries require the use of nonaqueous, aprotic solvents. Conversely, half-reactions with E° values greater than +1.23V at pH 0 have the potential to oxidize water to produce O_2 gas. This oxidation often releases protons and promotes acidification. However, with oxides, not all reactions release protons. Consider a counter example where a redox reaction consumes protons. This example is important in lead-acid car batteries:

$$2PbO_2(s) + 8H^+(aq) + 4e^- \rightleftarrows 2Pb^{2+}(aq) + 4H_2O(l) \quad E_o = +1.46\ V \quad (11.14)$$

$$+ \quad 2H_2O(l) \rightleftarrows O_2(g) + 4H^+(aq) + 4e^- \quad E_o = -1.23\ V \quad (11.15)$$

$$= \quad 2PbO_2(s) + 4H^+(aq) \rightleftarrows 2Pb^{2+}(aq) + O_2(g) + 2H_2O(l) \quad E_o = +0.23\ V \quad (11.16)$$

The net reaction indicates that the Pb(IV) cations present in the PbO_2 solid are sufficiently strong oxidizing agents that the +4 oxidation state of lead is not thermodynamically stable in water.

Considering the stability of water relative to O_2 and H_2 gases,

$$O_2(g) + 4H^+(aq) + 4e^- \rightleftarrows 2H_2O(1) \quad E_o = +1.23\ V \quad (11.17)$$

$$+ \quad 2H_2(g) \rightleftarrows 4H^+(aq) + 4e^- \quad E_o = 0.00\ V \quad (11.18)$$

$$= \quad 2H_2(g) + O_2(g) \rightleftarrows 2H_2O(l) \quad E_o = +1.23\ V \quad (11.19)$$

In other words, water is much more stable than mixtures of H_2 and O_2 gases. The release of this energy is why mixtures of the two gases can also be explosive when sparked.

Before leaving the electrochemistry of water, it is important to note two things. First, electrochemical reactions in water are capable of generating species other than $H_2(g)$ and $O_2(g)$. The O^{2-} ion in water can be oxidized to form hydroxyl radical, peroxide (O_2^{2-}), superoxide (O_2^-), and ozone (O_3). Both of these species are thermodynamically metastable with respect to water, and all are either present in low concentrations or undergo gradual decomposition, either by reduction to H_2O or oxidation to O_2. (Although they are not thermodynamically stable, these metastable species have immense importance in biochemistry and reaction kinetics.) For this reason, these species are normally not included in Pourbaix diagrams. Second, the hydroxide and O^{2-} ions in oxide materials can be potentially oxidized and reduced in water, just like water. However, the redox activity of most oxides of electrochemical relevance tends to involve the cations, meaning that cations usually transfer electrons, and not the oxygen anions.

11.5 POURBAIX DIAGRAMS INVOLVING SOLID OXIDES

A representative Pourbaix diagram for a redox-active oxide is shown for the case of nickel in Figure 11.4. First, the diagram captures the fact that nickel can exist in four different oxidation states (0, +2, +3, and +4) as represented by $Ni^\circ(s)$, $Ni(OH)_2(s)$, $Ni_2O_3(s)$, and $NiO_2(s)$. Nickel also forms the stable mixed-valence oxide Ni_3O_4 that

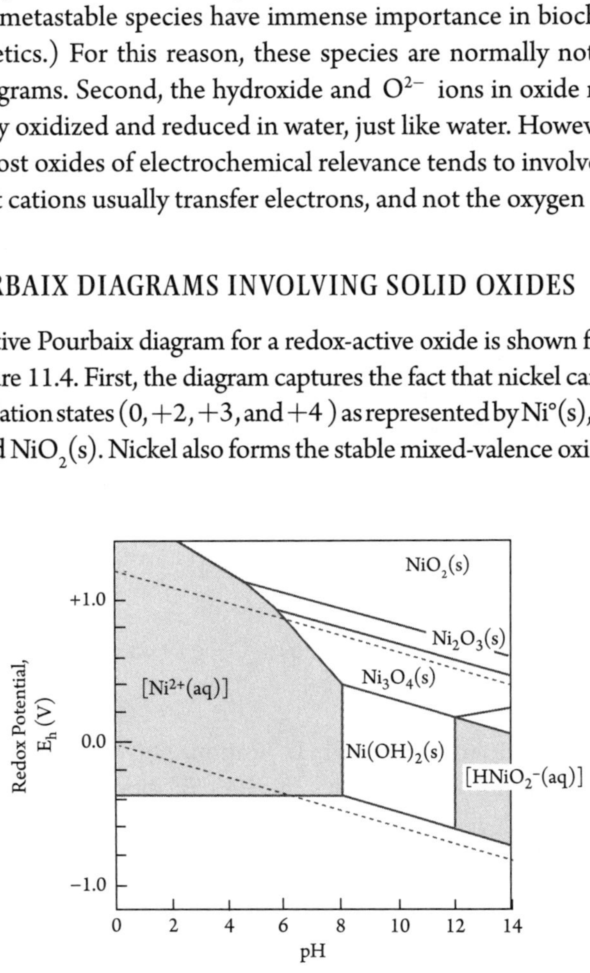

Figure 11.4 A representative Pourbaix diagram for a metal (nickel) that can form oxides with a range of cation oxidation states and total nickel concentration of 0.0001 M. For nickel, these oxidation states range from 0 (for nickel metal) to +4 (for $NiO_2(s)$). Superimposed on the diagram are dashed lines indicating the stability window for water at ambient conditions. Species shown above the top dashed line can oxidize water to form O_2, whereas species below the lower dashed line can reduce water to produce H_2. Adapted with permission from Pourbaix, M. *Atlas of Electrochemical Equilibria in Aqueous Solutions*; National Association of Corrosion Engineers: Houston, TX, 1974.

contains both +2 and +3 cations. Such mixed-valence compounds are of particular interest because of the high electrical conductivity associated with electron hopping from +2 to +3 sites. Second, removal of each successive electron from any cation requires ever-increasing redox potentials. For example, in Figure 11.4 in the basic regime (i.e., at pH 10), increases in redox potentials promote the formation of a succession of phases as each phase reacts with water:

$$\begin{aligned} & Ni^{\circ}(s) \rightleftarrows Ni^{+2}(aq) + 2e^- \\ + \;& 2H_2O(l) + 2e^- \rightleftarrows 2OH^-(aq) + H_2(g) \\ + \;& \underline{Ni^{2+}(aq) + 2OH^-(aq) \rightleftarrows Ni(OH)_2(s)} \\ = \;& Ni^{\circ}(s) + 2H_2O(l) \rightleftarrows Ni(OH)_2(s) + H_2(g) \qquad E^{\circ} = 0.5\ V \end{aligned} \tag{11.20}$$

$$3Ni(OH)_2(s) \rightleftarrows Ni_3O_4(s) + 2OH^-(aq) + 2H_2(g) \qquad E^{\circ} = +0.3\ V \tag{11.21}$$

$$2Ni_3O_4(s) + H_2O(l) \rightleftarrows 3Ni_2O_3(s) + H_2(g) \qquad E^{\circ} = +0.7\ V \tag{11.22}$$

$$Ni_2O_3(s) + H_2O(l) \rightleftarrows 2NiO_2(s) + H_2(g) \qquad E^{\circ} = +0.8\ V \tag{11.23}$$

Third, the successive oxidation of cations results in a loss of protons and, eventually, in an increase in the oxygen content of the solid phase. Fourth, the Pourbaix diagram only indicates which species are expected to be dominant in water. The dashed lines in the diagram indicate the stability window for water, as was shown in Figure 11.3. For nickel (Fig. 11.4), it is clear that at pH 10, water is oxidized at a voltage of +0.64 V, which means that NiO_2 (with $E^{\circ} = +0.8$ V) is not a stable phase in water but will be reduced while oxidizing water to form O_2. $Ni_2O_3(s)$ is expected to have narrow range of stability. Similarly, for soluble species dissolved in water, Ni(II) is predicted to be the most stable oxidation state, Ni(III) is predicted to have limited stability in water, and Ni(IV) is predicted to be unstable.

11.6 THE KINETICS AND MECHANISMS OF ELECTRON TRANSFER REACTIONS

Oxides are used extensively in electrochemical energy-storage systems. Such oxides need to embody several important attributes. First, the cations in the oxide must have access to multiple oxidation states to facilitate oxidation and reduction processes. Second, the redox potential associated with the change in oxidation states needs to be large enough to provide the oxide with a high energy density, whether it is used as an anode or a cathode. However, a favorable redox potential does not guarantee the oxide can be used for energy storage. For practical applications, it must be possible to insert or extract electrons from the oxide at a rate sufficient to generate usable electrical currents. The kinetics of transport and internal redox reactions are just as critical as the thermodynamics of redox processes if all active sites within an oxide electrode are to be used in a timely fashion.

To access electroactive sites within the bulk, the electrons involved in redox processes must be mobile in the solid oxide material. Similarly, protons or other ions must be relatively free to penetrate the solid to compensate for the charge imbalances associated with the addition or removal of these electrons. Table 11.2 lists electrical conductivities and ionic diffusion coefficients associated with a sampling of electroactive oxides to provide a reference point for evaluating the impact of charge transport on battery and ultracapacitor performance. Notice that some oxide materials that might look promising for electrochemical energy storage based on redox potentials alone are expected to exhibit poor performance as a result of limitations imposed by electronic and/or ionic conductivity. The factors that control ionic conductivity are discussed in Chapter 10 (see Section 10.12) and Chapter 12 (Section 12.6). Here, we focus on the basics of electron transport in oxides based on an understanding of the kinetics of electron transport reactions.

Oxides fall into three broad categories based on their electronic conductivity: (1) oxides such as RuO_2, which exhibit near-metallic conductivity with mobile electrons that occupy delocalized orbitals that communicate throughout the

Table 11-2. Room-Temperature Ionic Diffusion Coefficients (D) and Electronic Conductivity (σ) Values for Several Representative Oxides That Could be Used in Ultracapacitors

Oxide	**D (cm^2/s) (Ion)**	**d (Ion)**	**σ/S/cm**	**d (e^-)**	**Limiter**
TiO_2	$4 \cdot 10^{-13}$ (Li^+)	6 nm	<10^{-8}	<2 nm	Both
NiOOH	$5 \cdot 10^{-11}$ (H^+)	70 nm	0.003	50 nm	H^+
$Ni(OH)_2$	$3 \cdot 10^{-10}$ (H^+)	170 nm	$5 \cdot 10^{-5}$	5 nm	e^-
MnO_2	$6 \cdot 10^{-10}$ (H^+)	240 nm	10^{-3}	100 nm	e^-
Nb_2O_5	$6 \cdot 10^{-10}$ (Li^+)	240 nm	<10^{-8}	<2 nm	e^-
V_2O_5	$1 \cdot 10^{-8}$ (Li^+)	1000 nm	10^{-3}	100 nm	e^-
$LiCoO_2$	10^{-9} (Li^+)	300 nm	0.2	20,000 nm	Li^+
RuO_2	10^{-11} (H^+)	36 nm	40	30 nm	H^+

Approximate penetration depths (d) within which electroactive material can be accessed within 1 second are also listed. The species, either ions or electrons, with mobility that limits the speed at which charge can be stored or released appears in the "Limiter" column. Note that all values can be mediated by changing ionic concentrations within the bulk oxide (for D) or by doping the oxide network (for σ). A third strategy for improving performance involves mixing components at nanometer-length scales (see Section 11.12).

solid; (2) insulating oxides such as MnO, which typically contain cations with a single valence and electrons that fill localized orbitals; and (3) semiconducting oxides that contain cations with access to multiple valence states with mobile electrons that occupy relatively localized orbitals. Here, we focus on electron transport and redox reactions in these mixed-valence oxides such as $Na_xMn^{III}_xMn^{IV}_{1-x}O_2(s)$, which represent the bulk of the oxides used in electrochemical systems.

In mixed-valence oxides, electrons hop from a cation with a low valence, such as Fe(II), to an adjacent site with a high valence, such as Fe(III). This electron hop represents an internal redox reaction in which the Fe(II) center is oxidized whereas the Fe(III) is reduced. The products of this *intervalence electron transfer* or *self-exchange reaction* are identical to the reactants except that Fe(II) and Fe(III) have changed places as a result of electron transport. The standard-state Gibbs free energy for this reaction, ΔG^o_r, is zero because the reactants and products are identical–they have the same structures and energies. However, to get from one site to another, the electron has to traverse an activation barrier and this barrier has an energy cost.

Most redox reactions, whether within an oxide, at an electrode surface, or between dissolved species, exhibit activation barriers to electron transfer processes. The nature of these barriers is embodied in the Nobel Prize–winning work of Dr. Rudolf Marcus, as modified by Neal Hush. Although classic Marcus–Hush theory [5, 6] was developed to describe redox reactions involving dissolved metal complexes, it applies equally well to solid oxide materials. The essence of the theory is that the activation barrier corresponds to the energy required to reorganize both the inner and outer coordination spheres around both the oxidized and reduced species in response to an electron-transfer process. Here, we focus on the inner-coordination sphere because it is more applicable to electron transport and redox processes within a bulk oxide.

Oxygen anions in solid iron oxides are more tightly bound to Fe(III) than to Fe(II), with bond lengths of around 1.34 Å and 1.44, respectively. The instant after the electron transfer, all the oxygen anions around both metal centers find themselves in nonequilibrium positions. Metal–oxygen bonds associated with the new Fe(II) must lengthen, whereas those surrounding the new Fe(III) must shorten. The expansion energy for Fe(III) becoming Fe(II) is equal to the contraction energy for Fe(II) becoming Fe(III). The activation barrier corresponds to the energy cost for reorganizing solvent and bond lengths to allow the electron transfer. (Fig. 11.5). These changes in bond lengths are equivalent to symmetric vibrational breathing modes around each metal center. As in a molecular vibration, it is assumed that the M–O bonds are elastic springs with motions described by harmonic potential wells. In a harmonic well, the energy required to expand or contract a bond scales as the square of the displacement of the bond from its

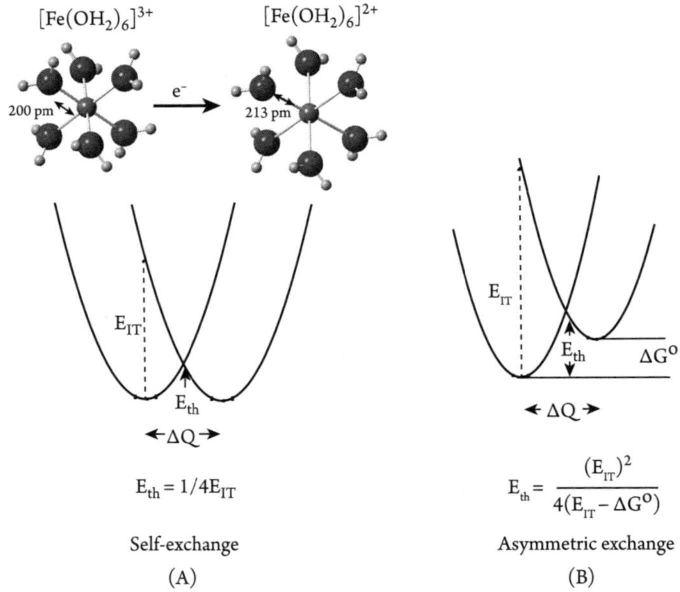

Figure 11.5 (A, B) Marcus-Hush diagrams [5] for an intervalence electron transfer process (A) for self-exchanges, between $[Fe(OH_2)_6]^{3+}$ and $[Fe(OH_2)_6]^{2+}$, and an electron transfer involving two dissimilar metal cations (B). The ions are labeled with bond lengths, and these must adjust during electron transfer. In this diagram, the curves correspond to the energy of vibration of bonds around the stable metal center according to a Hooke's Law approximation. E_{IT} thus represents the energy of the transition for electron transfer between two sites, E_{th} is the thermal energy required for electron transfer, or the barrier energy to electron exchange. For self exchange, $\Delta G = 0$; but, for electron transfer between two dissimilar metal centers, ΔG is not zero. ΔG is the distance between the metal centers. The equations relate the barrier to exchange, E_{th}, to E_{IT}.

equilibrium position, as in Hooke's Law. The displacement distance is defined as the normal coordinate axis shown in Figure 11.5. As a highly localized example, if the μ_2-O bridge between Fe(II) and Fe(III) was provided with sufficient energy to move 0.05 Å away from Fe(III) and toward Fe(II), both bonds would be the same length (1.39 Å). The electron would then sense the same environments around both iron sites, and the electron would be free to hop from Fe(II) to Fe(III) without a further energy penalty. This displacement is equivalent to the distance required to reach the intersection point between the two harmonic wells depicted in Figure 11.5 that defines the activation barrier to electron hopping.

In mixed-valence solids and complexes, light can also stimulate an electron to hop from the Fe(II) to the Fe(III) center by exciting an optical transition from the bottom of one vibrational well to the top of the other (Fig. 11.5). For harmonic potential wells, it can be shown that Hooke's Law dictates that the activation barrier to electron transfer (E_{act}) is related to the energy of this intervalence transfer band (E_{IT}) by

$$E_{act} = \frac{E_{IT}}{4} \tag{11.24}$$

Based on Eq. 11.24, if an intervalence transfer band is observed, it can be used to estimate the magnitude of the thermal energy barrier. The value of E_{act}, in turn, can be used to estimate the rate constant for the electron transfer process using classic activation theory:

$$k_{rate} = \frac{kT}{h} e^{\left(-\frac{E_{act}}{k_B T}\right)} \tag{11.25}$$

For the example of Fe_3O_4 (magnetite), $E_{IT} = 0.6$ eV, which indicates that E_{act} is on the order of 0.15 eV. This low activation barrier is consistent with an electron hop rate of around 10^{10}/second at room temperature.

When the electron transfer occurs between nonequivalent sites, such as between two different soluble complexes or at an electrode–electrolyte interface, one of the harmonic wells is displaced from the other by the free energy difference $\left(\Delta G_r\right)$ between both species before and after the electron transfer. If this free energy difference is known, activation energies for electron transfer rates for the asymmetric case can be estimated. In any case, barriers to electron transfer for both oxides and aqueous complexes span a large range, resulting in electron transfer rates that span many orders of magnitude, reflected by the electrical conductivity numbers tabulated in Table 11.2.

An important conclusion from Hush theory is that the farther apart the participating metal centers are in the lattice, the higher the activation barrier to electron transfer. This phenomenon is important in batteries because, during charging and discharging processes, the relative populations of cations in each oxidation state are changing continuously. For the iron example, when the populations of Fe(II) and Fe(III) are identical, all active sites are immediately adjacent to donor or acceptor sites required to maintain high conduction. However, as one redox state is consumed to form the other state, the minority redox state is eventually diluted to the point where there are no longer continuous pathways between chains of adjacent and alternating Fe(II) and Fe(III) sites. The site concentration at which continuous pathways are broken is called the percolation threshold. This threshold, typically occurs near 13 mole% of the minority species. Below the percolation threshold, electrons are required to hop through more material to reach other active sites, resulting in much higher activation energies and much lower conductivities. Similar percolation thresholds are important to ionic conductivity.

11.7 THE IMPACT OF ELECTROCHEMISTRY ON OXIDE REACTIONS AND STABILITY

Moving electrons into and out of the orbitals of a given metal cation has many more implications than changing local bond lengths or creating currents and voltages in a device such as a battery. Oxidation and reduction reactions change the charge on the cation, which changes the formal charge on all oxygens bound to that cation. These formal-charge changes have a profound influence on both the acid–base and ligand-exchange properties of each oxygen site, as discussed in Chapter 5. Local charge imbalances can also lead to phase transformations. As an example, consider the solubility diagrams corresponding to redox-based conversions between Fe(II) and Fe(III) (Fig. 11.6). First, cation valence dictates the acid–base chemistry of bound water molecules and hydroxyl groups (see Chapters 4 and 5). For an aquated iron cation, increasing the charge from +2 to +3 makes bound oxygens stronger acids and lowers the pH required to deprotonate the first water molecule from around pH 10 to below pH 3. Second, as a consequence of enhanced hydrolysis, condensation reactions to form clusters or solids are also more common for metals having a higher valence in water. For example, Fe(III) forms cluster such as (3,4) below pH 2, whereas Fe(II) exhibits only monomeric solution species. Third, higher valence states exhibit slower ligand exchange reactions, which often results in dramatic decreases in depolymerization and dissolution rates. The net changes in reversible hydrolysis and condensation reactions can have an enormous impact on parameters such as oxide solubility. For example, as shown in Figure 11.6, FeOOH(s) is almost 10 billion times less soluble than $Fe(OH)_2(s)$ in neutral pH solutions.

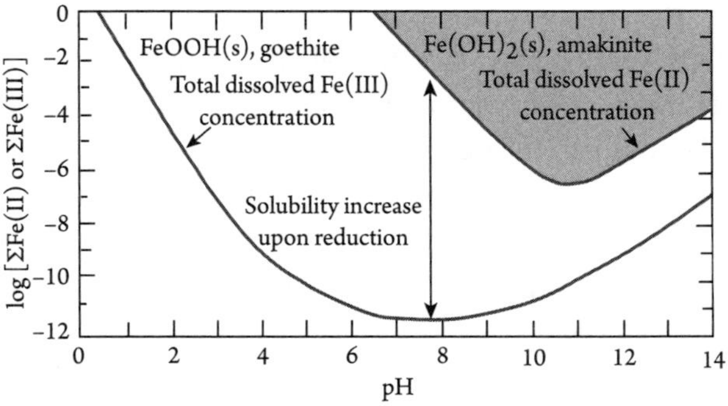

Figure 11.6 The total dissolved concentration of Fe(II) (upper right curve) and Fe(III) (lower curve) show the enormous effect that redox reactions have on the speciation and solubility of metal cations. At pH 7, the oxides of Fe(III) are 12 orders of magnitude less soluble than those of Fe(II). Adapted from Baes, C. F.; Mesmer, R. E. *The Hydrolysis of Cations*; Krieger Publishing: Malabar, FL, 1986, with permission.

For extended solid phases, a fourth consequence of the local charge imbalances that accompany redox reaction involves the promotion of phase changes that either minimize or eliminate the large local charges induced by redox reactions (Fig. 11.7). Formal-charge analyses clearly demonstrate that the oxidation or reduction of metal cations within an oxide lattice will create large local charge imbalances within the material. For example, oxidation of all Ni(II) cations in $Ni(OH)_2$ to Ni(III) would transform the hydroxide from a phase in which all oxygens have a formal charge of zero to a phase where all μ_3-OH groups have a charge of +1/2 (Fig. 11.7). Oxides have several mechanisms at their disposal to rectify such local charge imbalances. First, the reduced or oxidized phase can counteract the changes in charge on redox-active cations by adsorbing or desorbing inert counterions from solution to neutralize the excess charge. This charge neutralization strategy requires that the oxide have a sufficiently open structure to accommodate the entry and exit of modifier cations (see Chapter 10). An example of an oxide that responds to redox reactions via reversible ion intercalation is Li_xCoO_2.

Second, an oxide can adjust local charges via the uptake or release of protons, which involves interconversions between waters of hydration, internal hydroxyl

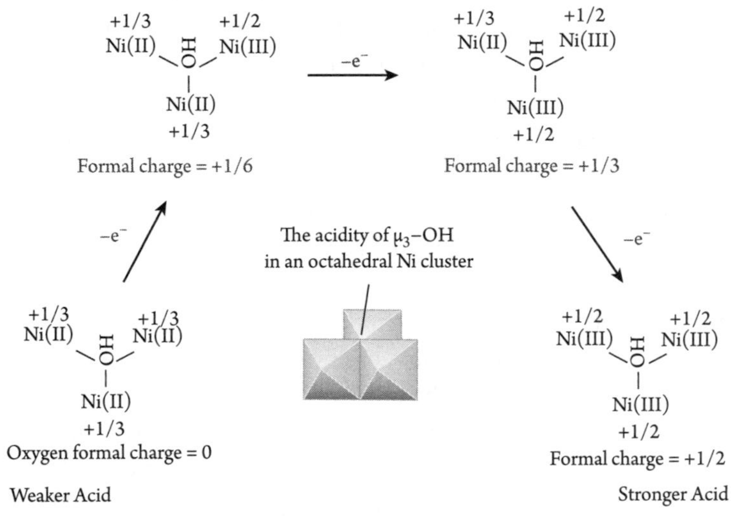

Figure 11.7 The impact of cation oxidation on the local oxygen formal charge in $Ni(OH)_2(s)$. The proton donates a charge of +1. Each octahedral Ni(II) donates a formal charge of +1/3, whereas each oxidized Ni(III) donates a formal charge of +1/2 to the O^{2-} anion, which is bonded to three cations. The net formal charge on the oxygen is listed underneath each species, following a progression in which Ni(II) ions are oxidized sequentially to form Ni(III). The two species on the right are overbonded to the point where they are unstable, and respond via loss of a proton or via conversion into more stable phases such as NiOOH(s).

groups, and/or oxyanions within the structure. However, in contrast to the uptake of species such as alkali metal cations, the protonation reactions and hydrolysis often trigger reversible ligand-exchange reactions, such as condensation or the reverse. A third and related strategy that oxides use to address charge imbalances involves phase changes. With regard to the example of the oxidation of $Ni(OH)_2(s)$, the unstable, local bonding configurations adjust quickly via loss of a proton to form NiOOH; the formal-charge changes are indicated in Figure 11.7. Phase changes can influence both the kinetics of electron transfers and the stability of materials used in devices such as batteries. Sometimes the phase changes are kinetically irreversible. Sometimes there is a large energy difference between the two phases, which adds an additional activation barrier to redox processes. Finally, the volume changes associated with phase transitions can generate stresses within electrode materials that lead to battery failures resulting from physical deterioration.

11.8 ELECTROCHEMICAL ENERGY STORAGE DEVICES CONTAINING OXIDES

Oxides are found in almost every electrochemical energy-storage device, including capacitors, fuel cells, batteries, and ultracapacitors (or electrochemical capacitors). Applications for such devices range from portable electronics to power for transportation systems to the power grids of entire countries. Each of these devices has its own unique energy-storage mechanisms and materials requirements, and each fills a unique niche in terms of how much energy and power can be stored (see [7–10] for general reviews) (Fig. 11.8). Briefly, in aqueous *capacitors*, charge is stored when ions from the electrolyte hop on, or off, the immediate electrode surface in response to an applied voltage. Charge separation is allowed because the adjacent solution has a dielectric constant larger than a vacuum. All charge is stored at the immediate electrode–electrolyte interface in the electrical double layer (see Chapter 8). Double-layer capacitors are characterized by relative low *energy densities*, because the total charge stored is limited to the immediate electrode surface [7]. However, *power densities*, which are determined by how rapidly charge can be stored and/or released, can be extremely high because ions in solution can respond almost instantaneously to changes in applied voltages. *Fuel cells* [8] are galvanic devices that convert the chemical energy of fuels continuously into electrical energy. Instead of having fixed anodes and cathodes, the anode represents one fuel, such as $H_2(g)$, whereas the cathode is another fuel (typically the oxidant $O_2(g)$). Both fuels are consumed in redox reactions that generate an electric current and reaction products (e.g., water). However, oxides are used in fuel cells primarily as separators and catalysts rather than as electroactive materials. Oxide solids do not undergo net electrochemical reactions in

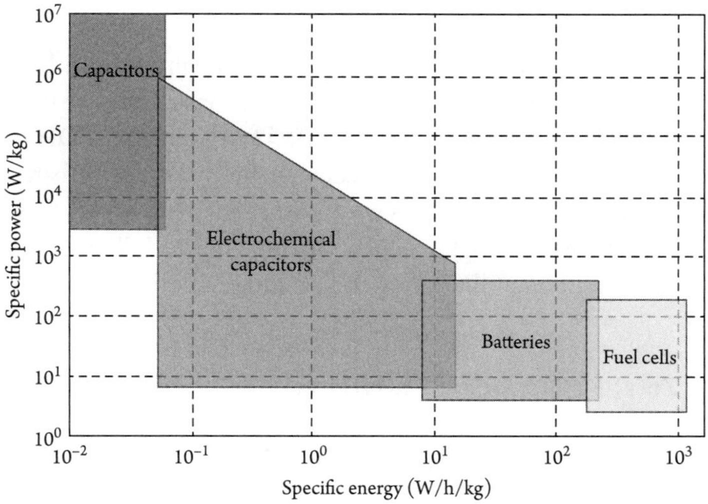

Figure 11.8 A plot [10] showing typical ranges of energy and power densities for capacitors, ultracapacitors, batteries, and fuel cells. These are called 'Ragone' plots.

either capacitors or fuel cells. In the remainder of this chapter we discuss devices, batteries and ultracapacitors, that depend on redox reactions to function.

Most commercial batteries fall into two major categories [9]. *Primary batteries,* such as flashlight batteries, are discharged once and then disposed. *Secondary batteries,* such as lead–acid car batteries, can be recharged repeatedly. In both battery types, anode and cathode materials are oxidized and reduced to liberate or consume electrons. Because almost every metal site in active electrodes can undergo redox reactions, the accessible charge density can be very high, resulting in high energy densities. However, use of all electrochemically active sites in the bulk requires the transport of electrons and counterions within the solid electrode materials, and may also require reversible phase transitions. As described earlier, at least one of these processes is invariably inhibited, limiting the rate at which charging and discharging processes can occur. For this reason, batteries almost always exhibit lower power densities than capacitors.

Ultracapacitors [10] contain materials that are designed to achieve relatively high energy and power densities simultaneously. Ultracapacitors contain the same redox-active materials found in batteries, which allows for high energy densities. However, maximizing capacitance, power density, and discharge rates requires that the electroactive materials be engineered at nanometer-length scales. Primary batteries, secondary batteries, and ultracapacitors each have their own unique requirements for oxide electrodes, as illustrated in the following sections.

11.9 OXIDES IN PRIMARY BATTERIES

The most common primary batteries are alkaline flashlight batteries [9] that contain Zn metal anodes and MnO_2 cathodes. The aqueous electrolyte used in these batteries is typically a concentrated solution (>1 M) of either KOH or NaOH, hence the term *alkaline batteries*. The alkaline–manganese dioxide battery (zinc/KOH electrolyte/MnO_2) battery has become the dominant battery system in the portable battery market (roughly 60% of the total of $25 billion per year). The half-reactions that occur at the anode and cathode, and the net cell reaction are given by

$$Zn(s) + 2OH^-(aq) \rightleftarrows Zn(OH)_2(s) + 2e^- \quad E° = +1.26\ V \qquad \textbf{(11.26)}$$

$$2MnO_2(s) + 2H_2O(l) + 2e^- \rightleftarrows 2MnOOH(s) \quad E° = +0.1\ V \qquad \textbf{(11.27)}$$

$$Zn(s) + 2MnO_2(s) + 2H_2O(l) \rightleftarrows Zn(OH)_2(s) + 2MnOOH(s) \quad E° = +1.36\ V \qquad \textbf{(11.28)}$$

There are several important features about the oxides used in this battery system. The first feature involves the redox potentials shown earlier, which influence both energy density and electrode stability. The good news is that, as a result of the Zn anode, the battery generates a high cell voltage. The bad news is that the oxidation potential for Zn (and alternate anodes such as Mg) is so high that it falls outside of the stability window for water:

$$Zn(s) + 2OH^-(aq) \rightleftarrows Zn(OH)_2(s) + 2e^- \quad E° = +1.26\ V \qquad \textbf{(11.29)}$$

$$2H_2O(l) + 2e^- \rightleftarrows H_2(g) + 2OH^-(aq) \quad E° = -0.83\ V \qquad \textbf{(11.30)}$$

$$Zn(s) + 2H_2O(l) \rightleftarrows Zn(OH)_2(s) + H_2(g) \quad E° = +0.43\ V \qquad \textbf{(11.31)}$$

The situation for Mg metal, with a half-cell oxidation potential of +2.67 V, is worse. The high potential could result in extensive electrode dissolution accompanied by extensive generation of potentially explosive hydrogen gas. How, then, can stable batteries be produced using these anodes? The answer is found in hydrolysis diagrams, which show that $Zn(OH)_2(s)$ and $ZnO(s)$ are relatively insoluble in basic solutions. The zinc oxides and hydroxides (as well as $Mg(OH)_2$) form passivating films that protect the underlying metal and inhibit hydrogen evolution (see Chapter 12). Even so, Zn(s) corrosion and self-discharge during storage can degrade the capacity and lifetime of the battery.

The second feature of oxide electrodes that needs to be accounted for in primary battery performance involves oxide conductivity. In primary alkaline batteries, the cathode is eventually transformed to increasingly reduced solids, from MnO_2 to MnOOH to $Mn(OH)_2$. Use of the total capacity of the cathode requires all phases along the discharge pathway maintain high conductivity. Fortunately, all these phases are mixed-valence compounds that meet conductivity requirements. The starting material, MnO_2, is a semiconductor that can have specific conductivities ranging anywhere from $10^{-6}\,\Omega/cm$ to $10^{-3}\,\Omega/cm$, depending on preparation conditions [11] and Mn(III) content. Small particles containing mixtures of Mn(IV) and Mn(III) are particularly conductive. During discharge, the cathode is first transformed into a solid solution of $MnO_2(s)$ and MnOOH(s), both of which contain both Mn(IV) and Mn(III). When the MnO_2 is consumed, solid solutions of mixed-valence MnOOH(s) and $Mn(OH)_2(s)$ are present in addition to the mixed-valence oxide $Mn_3O_4(s)$. Here, each phase contains Mn(III) and Mn(II). It is only when deep discharge creates relatively pure $Mn(OH)_2$ (with a negligible amount of Mn(III)) that the cathode becomes an insulator, and the battery activity ceases.

The third and perhaps most important feature of oxides in both the anode and cathode involve electrode stability. In fact, stability issues are what prevent primary alkaline batteries from being recharged (which would make them secondary batteries). For the Zn(s) (and Mg(s)) anode, reversible cycling creates stresses in the passivating oxide and hydroxide films that protect the anode metal. These stresses result in mechanical disruption of the passive film, causing electrode failures that can be accompanied by catastrophic hydrogen evolution. $MnO_2(s)$ cathodes are largely constrained for use in primary batteries as a result of a different class of stability issues. As one reduces the Mn(IV) to Mn(III) and finally to Mn(II), the cathode is transformed from $MnO_2(s)$ to MnOOH(s) to $Mn(OH)_2(s)$. The densities of these three phases are 4.9, 4.3, and 3.3 g/cm^3, respectively [1]. Commercial MnO_2 cathodes expand by $\sim$17 vol% when converted into MnOOH(s). With reversible cycling, these stresses lead to cracking and crumbling that promote electrode failures. Hydrolysis and condensation reactions in the MnOOH/MnO_2 solid solution promote irreversible crystallization that drastically lowers electrochemical activity. In addition, Zn^{2+} can migrate to the cathode to form compounds such as $ZnMn_2O_4(s)$, which also degrade battery performance.

11.10 OXIDES IN SECONDARY BATTERIES

Some of the most attractive anode and cathode materials from a purely electrochemical standpoint are constrained for use in primary batteries as a result of limitations associated with oxide stability. In secondary batteries, both cathodes and anodes must be capable of surviving repeated charge–discharge cycles. With the exceptions of the lead–acid battery found in cars and lithium-ion batteries (which rely on nonaqueous solvents), most secondary batteries are aqueous systems containing NiOOH(s) cathodes in combination with Cd(s), Fe(s), or Zn(s) anodes [9]. Such batteries are heavily used by industries in which minimizing weight is important, such as in aeronautics, industrial power, and in commercial and portable electronics. The key to these batteries is the reversibility of the NiOOH(s) cathode [12, 13], involving the half-reaction

$$\mathrm{NiOOH(s)} + \mathrm{H_2O(l)} + \mathrm{e^-} \rightleftarrows \mathrm{Ni(OH)_2(s)} + \mathrm{OH^-(aq)} \qquad \textbf{(11.32)}$$

Nickel hydroxide is just one example of many layered double-hydroxide solids that can be formed via the oxidation or reduction of transition metals. Other examples include the hydroxides of iron, cobalt, and manganese, as well as clay minerals containing these transition metals.

When they undergo a redox reaction, hydroxide and oxide solids compensate for the resulting charge imbalance in three ways: (1) intercalation of charge-compensating ions, (2) adsorption or desorption of protons, and (3) phase changes. How does an oxide pick which pathway to take, and what can be done to maximize reversibility for secondary batteries?

The least reversible systems involve phase changes with substantial structural rearrangement, because reversing such phase changes can be slow or even irreversible. An example of an active hydroxide phase that falls into this category is $Fe(OH)_2(s)$, which is transformed into FeOOH(s) on oxidation. The hydroxide has the planar brucite structure (see Chapter 2) that contains densely packed arrays of μ_3-OH groups. All modifications of the FeOOH(s) structure, including goethite, lepidocrocite, and akageneite, contain corrugated or puckered sheets that are bound covalently to all other sheets, defining tunnels within which protons reside. Conversion from the brucite structure to one of these requires a massive reconfiguration of bonding via hydrolysis and condensation reactions that can be quite slow.

Reversibility can be improved if components are present that interfere with polymerization between adjacent layers. The easiest method for inhibiting polymerization involves populating interlayer spaces with cations or anions that sterically prevent sites on adjacent layers from contacting one another. Such steric interference is illustrated by the sodium titanates discussed in Chapter 10. As long as cations such as Na^+ are present, ions are free to move and exchange with other ions to compensate changes in charge and preserve the initial oxide structure. However, after sites on adjacent layers are protonated, condensation reactions between adjacent layers occur that eventually transform the alkali titanates into denser TiO_2. The key to preserving desired layered structures involves establishing solution conditions under which anionic oxygen sites within the oxide prefer to associate with alkali cations rather than protons. This preference is achieved in solution regimes in which the solution pH is higher than the pK_a of the oxygen anion and/or in regimes in which $[M^+]>>[H^+]$ (see Chapter 10). The need for basic solution conditions is one reason why batteries containing NiOOH normally contain high alkali–metal–hydroxide concentrations (e.g., [KOH] > 1M) [9]. Under such conditions, NiOOH(s) still retains significant proton concentrations, but K^+ concentrations between the layers block polymerization into a denser structure across the layers.

An even simpler method for preserving electrochemical activity involves using oxides that retain isolated layer structures regardless of oxidation state or degree of protonation. For example, cobalt hydroxide can be oxidized from Co(II) to Co(III) to Co(IV) while retaining planar layered structures (Fig. 1.7). Layer stability and closely spaced redox reactions explain why cobalt is used increasingly in commercial batteries [14] and as catalysts. The most reversible oxide systems are those in which structural rearrangements are kept to an absolute minimum. Redox-active clay minerals fall into this category (see Chapter 18). Here, the aqueous zones between individual redox-active octahedral sheets readily admit charge-compensating ions (both cations and anions) to maintain charge neutrality. Although redox stability is improved, disadvantages of clay minerals include the fact that (1) only a fraction of the network-forming cations present can undergo redox reactions, which reduces the available energy density, and (2) although the layered structures are relatively open, counterion diffusion rates can influence the kinetics of reversible charging.

11.11 THE LEAD–ACID BATTERY

By far the most common secondary battery is the lead–acid battery found in almost all our vehicles and backup power systems [9, 15]. Its sales (> $15 billion per year) represent roughly half of all batteries sold in the world. Although the lead–acid battery is perhaps our most familiar battery, the reactions and phenomena that occur within this battery are so complex that most advancements in battery technology have been achieved empirically rather than on the basis of rational analysis [15].

The lead–acid battery consists of a Pb°(s) anode and a lead oxide ($PbO_2(s)$) cathode. Half-reactions and the net cell reaction in this battery are

$$\text{Anode}: Pb^\circ(s) + SO_4^{2-}(aq) \rightleftarrows PbSO_4(s) + 2e^- \quad E^\circ = -0.36\ V \quad \mathbf{(11.33)}$$

$$\text{Cathode}: PbO_2(s) + 4H^+(aq) + 2e^- + SO_4^{2-} \rightleftarrows PbSO_4(s) + 2H_2O(l) \quad E^\circ = +1.7\ V \quad \mathbf{(11.34)}$$

$$\text{Net}: Pb^\circ(s) + PbO_2(s) + 2H_2SO_4(aq) \rightleftarrows 2PbSO_4(s) + 2H_2O(l) \quad \Delta E = 2.06\ V \quad \mathbf{(11.35)}$$

Issues that have been largely surmounted in commercial batteries include the following:

- *Inhibiting the self-discharge reaction and the generation of potentially explosive mixtures of H_2 and O_2 gases.* As indicated in Eqs. 11.33 and 11.34, both Pb°(s) and PbO_2(s) are thermodynamically unstable with respect to water. The instability means that water in contact with the Pb°(s) anode could potentially be reduced to form H_2(g) whereas the electrolyte in contact with the PbO_2(s) cathode could potentially be oxidized to form O_2(g). Through trial and error, battery manufacturers have identified additives that can be used to inhibit these undesired side reactions.
- *Minimizing resistive losses by controlling the conductivity of the PbO_2 cathode.* PbO_2(s) is a defective semiconductor, the conductivity of which is reported to be around 10^{-6} S/cm [16]. However, this conductivity depends critically on the phase (primarily α-PbO_2 and $\beta-PbO_2$), the stoichiometry (ranging from $PbO_{1.85}-PbO_{2.05}$), the presence of impurities (such as Sb from the Pb−Sb anode), and the hydroxyl content, which influences the proton-conduction rates.

However, perhaps the most important, and least understood, issues associated with the lead–acid battery have to do with the fact that the basic charge storage mechanism is not dominated by redox reactions and charge transport within the bulk oxide, but by dissolution and reprecipitation phenomena that occur on electrode surfaces. Optimizing the performance of lead–acid batteries involves not only tuning the conductivity of the oxide, but also mediating nucleation and growth processes. Here, we focus on such phenomena as they impact the PbO_2(s) cathode. The reaction shown in Eq. 11.34 is not a solid-state reaction, but involves the active dissolution and reprecipitation of both PbO_2(s) and $PbSO_4$(s). On battery discharge, insoluble PbO_2(s) is converted to PbO(s), which is highly soluble in concentrated H_2SO_4 battery electrolytes. The Pb^{2+} cations released via PbO(s) dissolution combine with $[SO_4]^{2-}$ to form $PbSO_4$(s), which is sparingly soluble. Conversely, when the battery is recharged, $PbSO_4$(s) is consumed by the production of less-soluble PbO_2(s).

Solution-mediated phase transformations have several critical consequences when it comes to optimizing battery performance. First, high performance requires high surface areas for the active PbO_2(s) material, because electrochemical reactions are confined to oxide–solution interfaces. To achieve these high surface areas, electrodes are prepared via the consolidation of small (<12-μm) colloidal particles (see Chapter 8) onto conductive electrode grids. The particles are applied as viscous pastes consisting of hydrated "leady lead" oxide. These are lead particles that are oxidized to produce hydrated PbO(s) shells surrounding metallic cores. Also in the paste are water, and sulfuric acid. Curing the pastes under just the right conditions (i.e., humid oxidizing atmospheres at moderate temperatures [$T < 90°C$]) leads to the production of high-surface-area mixtures containing α-PbO_2(s), β-PbO_2(s), and $PbSO_4$(s) that adhere strongly to the underlying electrode grid.

A primary factor that controls the cycle life of the battery (range, 50–500 cycles) involves how the porous PbO_2(s) electrode responds to repeated dissolution and reprecipitation processes. As PbO_2(s) is reduced and dissolves to release Pb^{2+} ions into the electrolyte, the solution next to the electrode eventually becomes supersaturated with respect to $PbSO_4$(s). The result is the heterogeneous nucleation and growth (see Chapter 7) of $PbSO_4$(s) crystals on PbO_2(s) surfaces [17]. Lead sulfate is less dense than PbO_2(s) ($\rho = 6.3$ g/cm^3 vs. 9.2 g/cm^3), and occupies almost 50% more volume than the oxide it is replacing. In addition, in contrast to many oxides, $PbSO_4$(s) has a high activation barrier for nucleation, which means that instead of forming many small nuclei, $PbSO_4$(s) forms just a few nuclei that grow to large size. When crystal sizes exceed pore sizes, the lead sulfate begins to pry the host particles apart and causes the electrode to crumble.

Cathodes are synthesized to contain a mixture of α- and β-PbO_2(s) to minimize the damage caused by the reversible appearance and disappearance of $PbSO_4$(s) within the electrode. Although β-PbO_2(s) is more active than α-PbO_2(s) from a purely electrochemical perspective, α-PbO_2(s) has a lower activation barrier for nucleating $PbSO_4$(s). The reasons for this phenomenon are still poorly understood. The presence of α-PbO_2(s) ensures the $PbSO_4$(s) is distributed more uniformly through the cathode in the form of more numerous, but smaller crystals. Practical automotive batteries based on the lead–acid chemistry would not have been possible without the fortuitous discovery of the importance of α-PbO_2(s) to enhanced cycling.

11.12 ULTRACAPACITORS: NANOSTRUCTURED OXIDES FOR ENERGY STORAGE

Ultracapacitors are under development to maintain the high energy densities found in batteries while achieving power densities that approach those found in capacitors. The achievement of supercapacitive behavior requires the design of oxide materials at nanometer-length scales (Fig. 11.9) (1) to maximize the electrode–electrolyte contact area, (2) to maximize solution access to electroactive materials, and (3) to minimize diffusion distances for the transport of both electrons and charge-compensating cations. Before exploring the role of nanotechnology in achieving these goals, it is instructive to examine the behavior of an oxide that represents an ideal ultracapacitor material: RuO_2(s).

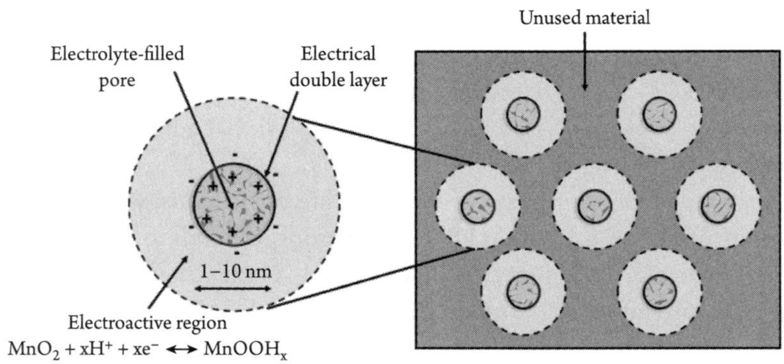

Figure 11.9 A simplified depiction of the role of nanostructures in ultracapacitors. Interconnected pores containing electrolyte solution are shown in stippled gray. Charge is stored on pore surfaces as in a capacitor. However, if pore walls are electrochemically active, much more charge can be stored in electroactive regions adjacent to the pores (light gray). However, if the transport of either electrons or charge-compensating ions is slow, the region that can respond rapidly to redox reactions (yielding high power densities) is a region that is only a few nanometers thick that is adjacent to the pores. Unused material corresponds to the dark-gray areas. If an electroactive oxide can be produced in which all active material is within a few nanometers of the pore, all material is used (i.e., the dark-gray areas are minimized), resulting in high power and energy densities.

$RuO_2(s)$ meets all the basic requirements for an electrochemically active oxide for use in aqueous media. First, it has a redox potential that falls within the stability window for water, as indicated by the idealized half-reaction

$$2RuO_2(s) + 2H^+ + 2e^- \rightleftarrows Ru_2O_3(s) + 2H_2O \quad E_o = +0.94 \text{ V} \quad \textbf{(11.36)}$$

The relevant oxidation states in Eq. 11.36 are +4 and +3, although both higher and lower oxidation states are also accessible. Second, because the dominant oxidation states are +3 and +4, both $RuO_2(s)$ and $Ru_2O_3(s)$ are exceedingly insoluble in water except in strongly acidic solutions. Structures are stable under many potential use conditions.

What makes $RuO_2(s)$ so special is that it can be an exceptional conductor of both electrons and protons within a structure that can accommodate shifts in charge distributions. This exceptional performance is achieved only if the $RuO_2(s)$ is prepared in a hydrated form via precipitation from aqueous solutions. Although hydrated $RuO_2(s)$ is denoted as $RuO_2{\cdot}xH_2O(s)$, all the waters of hydration are really tied up in the form of hydroxyl groups within an ill-defined $RuO_2(s)$ lattice. The capacitance of $RuO_2(s)$ has been investigated as a function of the degree of hydration via controlled thermal dehydroxylations

[18, 19] (Fig. 11.10). The capacitance and power density go through a maximum at a water content of approximately $RuO_2 \cdot 0.5H_2O$, which corresponds to one proton per ruthenium atom, as in RuOOH. Completely hydrated RuO_2 (i.e., $RuO_2 \cdot 2H_2O$, or $Ru(OH)_4(s)$) is an excellent proton conductor, with a proton diffusion coefficient of $10^{-8} - 10^{-10} \, cm^2/second$, but a relatively poor electron conductor. In contrast, pure $RuO_2(s)$ can be an exceptional, near-metallic, electron conductor ($\sigma = 40 \, S/cm$), but is a poor proton conductor ($D = 10^{-11} - 10^{-12} \, cm^2/second$). Optimum electrochemical performance is achieved at intermediate levels of hydration that provide a sufficient hydroxyl concentration to promote facile proton hopping while also maintaining a sufficient population of oxo-bridges to support electron conduction.

At intermediate hydration levels, $RuO_2(s)$ electrodes come close to achieving the theoretical limits for capacitance and power densities, and hold records in terms of the ultimate properties desired for ultracapacitors. Why, then, are

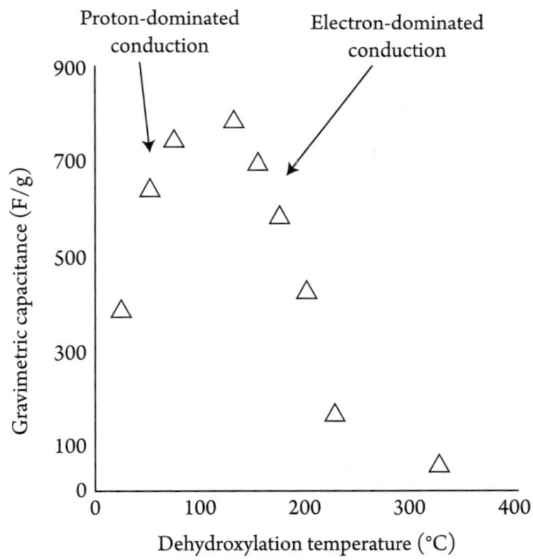

Figure 11.10 Capacitance values measured at room temperature for bulk RuO_2 ultracapacitor materials as a function of dehydroxylation temperature. At low dehydroxylation temperatures, hydroxyl populations are high, leading to high proton conduction. However, hydrous RuO_2 is a poor conductor of electrons. At high dehydroxylation temperatures, RuO_2 forms a dense rutile structure that is an excellent electronic conductor, but a poor proton conductor. Only at intermediate dehydroxylation temperatures are both electronic and ionic conduction sufficient to produce high levels of ultracapacitor performance. Adapted from McKeown, D.A.; Hagans, P.L.; Carette, L.P.L.; Russell, A.E.; Swider, K.E.; and Rolison, D.R. Structure of Hydrous Ruthenium Oxides: Implications for Charge Storage. *J. Phys. Chem. B* **1999**, 103, 4825–4832, with permission.

RuO_2(s) electrodes not seeing widespread commercial use? Ruthenium is exceedingly expensive and is 25 times more rare than platinum (with an abundance in Earth's crust of 0.0004 ppm). Some of its compounds (including oxides) can also be highly toxic. The current goal of ultracapacitor research is to identify strategies that can achieve similar performance characteristics with materials that are cheaper and more abundant. Most of these strategies involve taking more mundane materials and shrinking structural elements down to nanoscale dimensions. Next, we highlight two examples of the impact of nanotechnology on oxides for energy storage, one of which involves the electrode–electrolyte interface, while the other involves the development of nanostructures within the bulk oxide.

If electron or counterion transport is rate limiting, one strategy involves deploying the oxide in structures that are designed to minimize transport distances. If the performance goal is to access all the electroactive material in an electrode within 1 second, a structure must be created in which the rate-limiting mobile species in the bulk can reach the electrode–electrolyte interface in less than a second. Table 11.2 shows the counterion penetration depth for several oxides after one second, calculated using the simple Einstein diffusion equation $X = (Dt)^{1/2}$, where 'X' is a displacement distance by random motions and 't' is time in seconds. If the diffusion coefficient D is comparable with that found for ions in bulk water ($D = 10^{-6}\ cm^2/second$), the penetration depth into the oxide is significant ($X > 10\ \mu m$). However, as diffusion coefficients drop, penetration depths also decrease nonlinearly. For $D = 10^{-14}\ cm^2/second$, which is still relatively high for ionic diffusion through a bulk oxide, the penetration depth is only $X = 1\ nm$. However, even for this low value, all active material can be accessed provided the distance between pores or other electrolyte-containing features is 2 nm or less.

Features with nanometer dimensions can be engineered into many oxides using solution-processing techniques. For the example of RuO_2(s), nanoscale architectures have been produced in layered ruthenates [20], template growth in the pores of anodized aluminum [21], and in Ru–graphite composites [22]. An extreme architecture has been produced via molecular templating in the presence of alkane thiols [23]. In this case, a stable network consisting of interconnected molecular clusters has been produced in a material exhibiting hierarchical porosity (Fig. 11.11). These various materials approach the theoretical limits for power densities and capacitance values of up to 1200 F/g. A wide range of solution-derived nanostructures has been produced in other electroactive materials such as NiOOH(s) [24, 25], representing research on the frontier of electrochemical energy storage.

The second strategy for improving power densities in electroactive materials involves mixing active materials on nanometer-length scales. As shown in Table 11.2, some electroactive materials such as $Nb_2O_5{\cdot}H_2O$(s) are exceptional ionic conductors, but abysmal electronic conductors. Other materials, such as

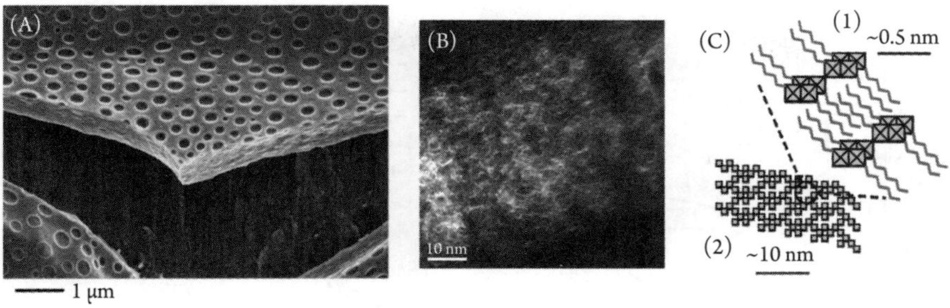

Figure 11.11 Structural features in solution-derived RuO_2 films containing alkane thiols as a function of length scale. (A) A topographical scanning electron microscopic image of a thin RuO_2 film deposited on a titanium substrate showing the large-scale (0.25 μm) pores generated via gas evolution during alkane thiol pyrolysis. (B, C) A transmission electron micrograph (B) of the fine-scale (4-nm) porosity that develops as a result of the interconnection of the molecular clusters (C) (before pyrolysis). Proton nuclear magnetic resonance measurements confirm that the clusters shown (less the hydrocarbon chains) are preserved after pyrolysis.
SOURCE: Reproduced with permission from Brumbach, M.T.; Alam, T.M.; Kotula, P.G.; McKenzie, B.B.; and Bunker, B.C. Nanostructured Ruthenium Oxide Electrodes via High-Temperature Molecular Templating for Use in Electrochemical Capacitors. *ACS Appl. Materials & Interfaces*, **2010**, *2*, 778.

crystalline $RuO_2(s)$, are exceptional electronic conductors, but poor ionic conductors. It has been demonstrated that materials exhibiting desired properties for ultracapacitors can be produced using solution-processing techniques to mix exceptional proton and electron conductors at nanometer-length scales. In the case of $Nb_2O_5{\cdot}H_2O{:}RuO_2$ mixtures [26] (Fig. 11.12), small additions of Ru to Nb_2O_5 can produce nano-networks of extremely small RuO_2 particles within a continuous Nb_2O_5 matrix. If the particle concentrations are above the percolation threshold, insulating Nb_2O_5 is transformed into a material with useful properties for ultracapacitors. Ruthenium-rich compositions separate into larger (5-nm) $RuO_2(s)$ particles separated by grain boundaries consisting of a network of Nb_2O_5. These materials exhibit performance that compares well with that of pure RuO_2. Intermediate compositions have interpenetrating nanoscale networks of Nb_2O_5 and RuO_2. The composition in which Ru:Nb = 1 has a higher capacitance than RuO_2. In fact, the capacitance per ruthenium atom is five time greater than pure $RuO_2(s)$ at rapid charge and discharge rates.

11.13 ENVIRONMENTAL ELECTROCHEMISTRY

Electrochemical systems in Nature are rarely so far out of equilibrium they can accumulate anywhere near the energy levels stored in batteries. However, redox

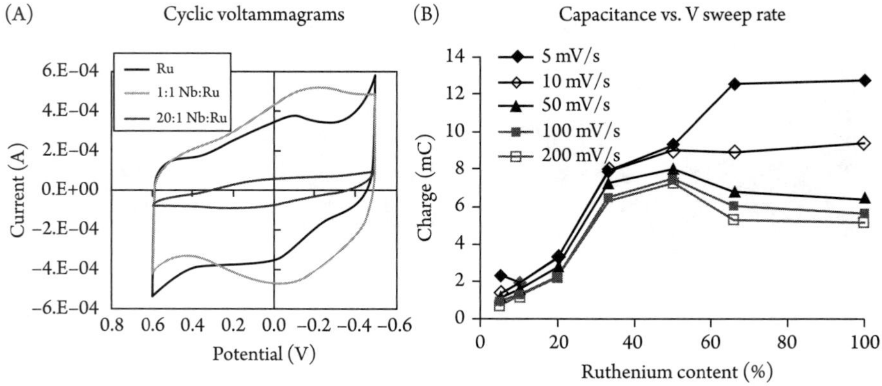

Figure 11.12 Electrochemical characterization of chemically prepared nanostructured oxides containing mixtures of hydrous Nb_2O_5 and RuO_2. (A) Cyclic voltammagrams of the current as a function of applied potential for three of the oxide mixtures. The capacitance is related to the area encompassed by each data loop. Although the 1:1 Nb:Ru film contains only half the ruthenium as pure RuO_2, both films exhibit almost identical electrochemical performance at intermediate sweep rates. (B) Although pure RuO_2 exhibits the highest capacitance at low cycling rates, the mixed Nb–Ru materials exhibit equally good performance at high cycling rates as a result of enhanced proton conductivity. Reproduced with permission from Brumbach, M.T.; Alam, T.M.; Nilson, R.H.; Kotula, P.G.; McKenzie, B.B.; Tissot, R.G.; and Bunker, B.C. Ruthenium Oxide-Niobium Hydroxide Composites for Pseudocapacitor Electrodes.; *Materials Chem. & Phys.*, **2010**, *124*, 359–370.

reactions involving oxides are critical in many natural cycles that impact our daily lives. In fact, civilization exists largely because of electron exchange reactions in soil and rock that help purify our water, remove harmful toxins, and even support single-cell organisms at the base of our food chain. Some of these electrochemical reactions occur as a result of direct reactions on oxide surfaces. More commonly, oxides establish electrochemical cycles by operating in conjunction with microbial catalysis. Finally, semiconducting oxides can be stimulated by light to activate electrochemical processes. All these above environmental reactions are discussed in detail in Chapters 13 and 18.

References

1. Weast, R. C. *CRC Handbook of Chemistry and Physics, 58th ed.*; CRC Press: Cleveland, OH, 1977.
2. Mahan, B. H. *University Chemistry, 3rd ed.*; Addison-Wesley Publishing: Reading, MA, 1975.
3. Bockris, H. O'M.; Reddy, A. K. N. *Modern Electrochemistry 1, 2nd ed.*; Plenum Press: New York, 1998.

4. Pourbaix, M. *Atlas of Electrochemical Equilibria*; National Association of Corrosion Engineers: Houston, TX, 1974.
5. Hush, N. Intervalence-Transfer Absorption. Part 2: Theoretical Considerations and Spectroscopic Data. *Prog. Inorg. Chem.* **1967**, *8*, 344–391.
6. Marcus, R. A. On the Theory of Electron Transfer Reactions VI: Unified Treatment for Homogeneous and Electrode Reactions. *J. Chem. Phys.* **1965**, *43*, 679–700.
7. Conway, B. E. Transition from "Supercapacitor" to "Battery" Behavior in Electrochemical Energy Storage. *J. Electrochem. Soc.* **1991**, *138*, 1539–1548.
8. Li, X. *Principles of Fuel Cells*; Taylor & Francis: New York, 2006.
9. Linden, D.; Reddy, T. B. *Handbook of Batteries, 3rd ed.*; McGraw-Hill: New York, 2002.
10. Conway, B. E. *Electrochemical Supercapacitors*; Kluwer Academic/Plenum Publishers: New York, 1999.
11. Desai, B. D.; Fernandes, J. B.; Dalal, V. N. K. Manganese Dioxide: A Review of a Battery Chemical Part II: Solid State and Electrochemical Properties of Manganese Dioxides. *J. Power Sources* **1985**, *16*, 1–43.
12. Weidner, J. W.; Timmerman, P. Effect of Proton Diffusion, Electron Conductivity, and Charge-Transfer Resistance on Nickel Hydroxide Discharge Curves. *J. Electrochem. Soc.* **1994**, *141*, 346–351.
13. Tessier;C.; Haumesser, C. P. H.; Bernard, P.; Delmas, C. The Structure of $Ni(OH)_2$: From the Ideal Material to the Electrochemically Active One. *J. Electrochem. Soc.* **1999**, *146*, 2059–2067.
14. Desilvestro, J.; Haas, O. Metal Oxide Cathode Materials for Electrochemical Energy Storage: A Review. *J. Electrochem. Soc.* **1990**, *137*, 5C–22C.
15. Ruetschi, P. Review on the Lead–Acid Battery Science and Technology. *J. Power Sources* **1977/1978**, *2*, 3–24.
16. Salkind, A. J.; Kannone, A. G.; and Trumbure, F. A; Lead–Acid Batteries. In *Handbook of Batteries and Fuel Cells*, eds. D.Linden and Reddy, T. B., McGraw Hill: New York, 1995, p. 23.1–23.88.
17. Hattori, S.; Yamaura, M.; Kohno, M.; Ohtani, Y.; Yamane, M.; Nakashima, H. In *Power Sources 5*; Collins, D. H., Ed.; Academic Press: London, 1975, p 139.
18. McKeown, D.A.; Hagans, P.L.; Carette, L.P.L.; Russell, A.E.; Swider, K.E.; and Rolison, D.R. Structure of Hydrous Ruthenium Oxides: Implications for Charge Storage. *J. Phys. Chem. B* **1999**, *103*, 4825–4832.
19. Dmowski, W.; Egami, T.; Swider-Lyons, K. E.; Love, C. T.; Rolison, D. R. Local Atomic Structure and Conductivity Mechanism of Nanocrystalline Hydrous RuO_2 from X-ray Scattering. *J. Phys. Chem. B* **2002**, *106*, 12677–12683.
20. Sugimoto, W.; Iwata, H.; Yusunaga, Y.; Murakami, Y.; Takasu, Y. Preparation of Ruthenic Acid Nanosheets and Utilization of Its Interlayer Surface for Electrochemical Energy Storage. *Angew. Chem. Int. Ed.* **2003**, *42*, 4092–4096.
21. Susanti, D.; Tsai, D. S.; Huang, Y. S.; Korotcov, A.; Chung, W. H. Structures and Electrochemical Capacitive Properties of RuO_2 Vertical Nanorods Encased in Hydrous RuO_2. *J. Phys. Chem. C* **2007**, *111*, 9530–9537.
22. Lin, C.; Ritter, J. A.; Popov, B. N. Development of Carbon–Metal Oxide Supercapacitors from Sol-Gel Derived Carbon-Ruthenium Xerogels. *J. Electrochem. Soc.* **1999**, *146*, 3155–3160.

23. Brumbach, M. T.; Alam, T. M.; Kotula, P. G.; McKenzie, B. B.; Bunker, B. C. Nanostructured Ruthenium Oxide Electrodes via High-Temperature Molecular Templating for Use in Electrochemical Capacitors. *ACS Appl. Mater. Interfaces* **2010**, *2*, 778–787.
24. Chang, J.; Park, M.; Ham, D.; Ogale, S. B.; Mane, R. S.; Han, S.- H. Liquid-Phase Synthesized Mesoporous Electrochemical Supercapacitors of Nickel Hydroxide. *Electrochim. Acta* **2008**, *53*, 5016–5021.
25. Prasad, K. R.; Miura, N. Electrochemically Deposited Nanowhiskers of Nickel Oxide as a High-Power Pseudocapacitive Electrode. *Appl. Phys. Lett.* **2004**, *85*, 4199–4201.
26. Brumbach, M. T.; Alam, T. M.; Nilson, R. H.; Kotula, P. G.; McKenzie, B. B.; Tissot, R. G.; Bunker, B. C. Ruthenium Oxide–Niobium Hydroxide Composites for Pseudocapacitor Electrodes. *Materials Chem. Phys.* **2010**, *124*, 359–370.

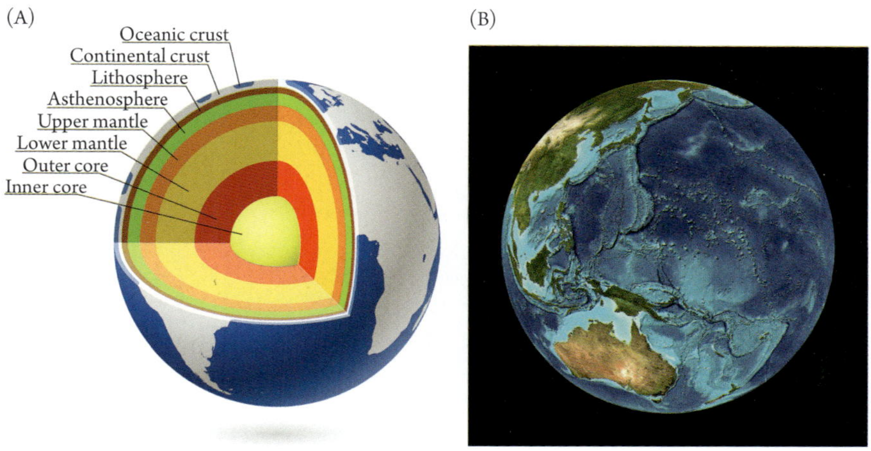

Plate 1 The geochemical importance of oxide–water reactions. (A) A depiction of Earth's interior, showing the different planetary layers. The center of Earth is a solid iron-nickel inner core (r = 1,300 km) that is surrounded by an outer core comprised of liquid iron-nickel (thickness, 950 km). All successive layers (total thickness, 4,150 km), including mantle layers, the asthenosphere, and the lithosphere, are composed primarily of oxides, representing 95% of Earth's total volume (SOURCE: Alhovic/Shutterstock.com). (B) A topographical map of the floor of the western Pacific Ocean. Light blue represents shallow water and dark blue indicates deep regions, such as the Marianas Trench, which has a maximum depth of 11.5 km (SOURCE: Martin Jacobsson/Science Source.com with permission). In the oceans alone, the contact area between the crust and water is more than 361 million km^2 (70% of Earth's surface), indicating the enormous scope over which oxide–water reactions modify our planet. (see Figure 1.1).

(A) (B)

Plate 2 (A) A collection of arrowheads fashioned in North America out of obsidian and chert before the arrival of Europeans (SOURCE: I. Pilon/Shutterstock.com). Humans have used water to assist in the fabrication of stone tools such as these as early as 2.6 million years ago (see Chapter 16 for mechanisms). (B) A view of the prehistoric Stonehenge monument in Wiltshire, England, under falling rain (SOURCE: Pitcha Torranin/Science Source.com with permissions). The oldest slabs in Stonehenge have been exposed to the elements (water in the form of rain, snow, and ice) since they were erected in 3300 BC (see Figure 1.2).

Plate 3 (A) A view of the Long House at Mesa Verde in Colorado. Such cliff dwellings are invariably made of oxide mineral aggregates such as consolidated clay minerals or volcanic tuff (SOURCE: kravka/Shutterstock.com). (B) Clay amphorae uncovered in the excavation of the city of Akrotiri on the Santorini Island in Greece (SOURCE: Will & Demi McIntyre/ Science Source.com with permission). The site was buried by a volcanic eruption around 2000 BC, and thus is quite well preserved. Both the formation and stability of adobe buildings and clay pots are influenced by the colloidal interactions described in Chapter 8 (see Figure 1.3).

Plate 4 (A) A Roman perfume bottle found in Pompeii, Italy, which was destroyed by a volcanic eruption in 79 AD (SOURCE: Patrick Landmann/Science Source.com with permission). Exposure to water for 2,000 years created an iridescent layer of corrosion products on this glass artifact (see Chapter 15 for corrosion mechanisms). (B) Stained glass in the Rose Window of the Notre Dame Cathedral in Paris, France (SOURCE: Vlad G/Shutterstock.com), illustrating the importance of glass in human culture (see Figure 1.4).

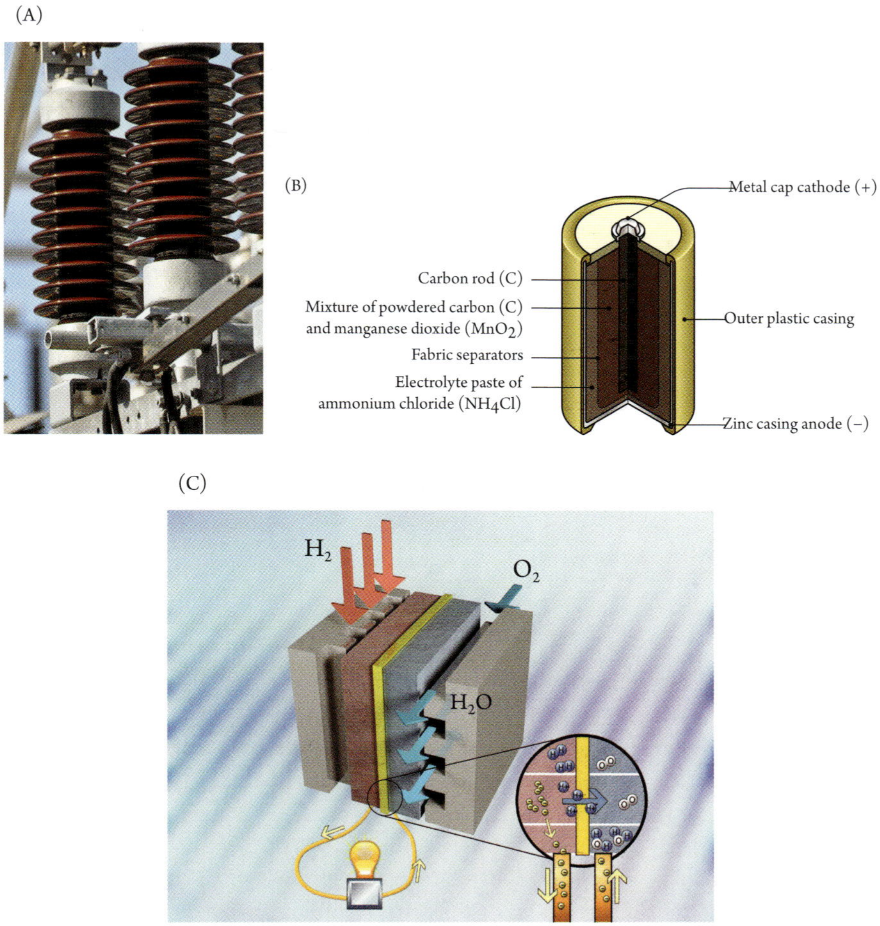

Plate 5 Oxide materials used for the generation, storage, and use of electrical energy. (B) A cross section of a manganese-dioxide dry-cell battery (SOURCE: Charles D. Winters/Science Source.com with permission). (A) A stack of ceramic insulator bodies incorporated into a high voltage power line (SOURCE: D. Russell 78/Shutterstock.com). (C) A schematic representation of a hydrogen-oxygen fuel cell that extracts electrical energy by reacting hydrogen and oxygen fuels with each other to form water. Oxides are often used as catalysts and separators in such devices. (SOURCE: Equinox Graphics/ Science Source.com with permission,) (see Chapter 10) (see Figure 1.5).

(A)

(B)

(C)

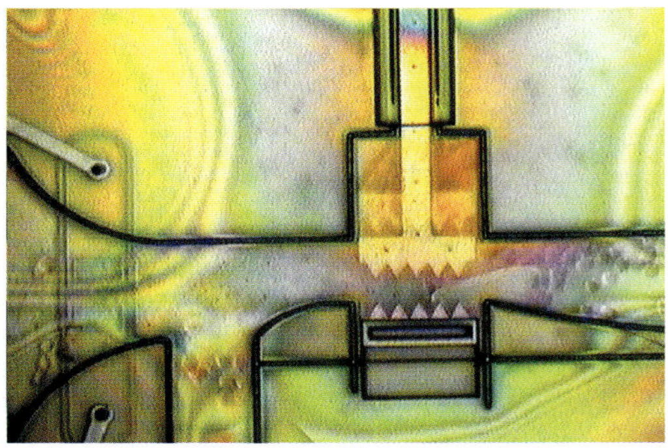

Plate 6 Oxides used in microelectronics applications. (A) A range of computer components (SOURCE: Deserg/Shutterstock.com). Oxide ceramics are used in the packaging of many of these components, as well as in insulating materials. (B)A hard disk memory device (SOURCE: Christian Darkin/Science Source.com with permission). The magnetic materials used in both the storage disk and in the head used to read and write information into the disk are often ferromagnetic ceramic materials such as ferrites. (C) A microjaw device for lysing cells on a microfluidic chip developed at Sandia National Laboratories (SOURCE: DOE/SNL/Science Source.com with permission). Aqueous solutions containing cells enter the region between the microjaws, where an oscillating piston drives the jaws together to smash the cells and release their contents for subsequent on-chip analyses. Moving parts in microfluidic systems can be actuated by piezoelectric ceramics in which applied electric fields induce volume changes and motion. Microfluidic systems based on silicon chips (such as the one shown) rely on the integrity of protective thermal silicon dioxide coatings to protect both static and moving components from attack by water (see Chapters 15 and 16 for mechanisms regarding both corrosion and stress corrosion of silicates in water) (see Figure 1.6).

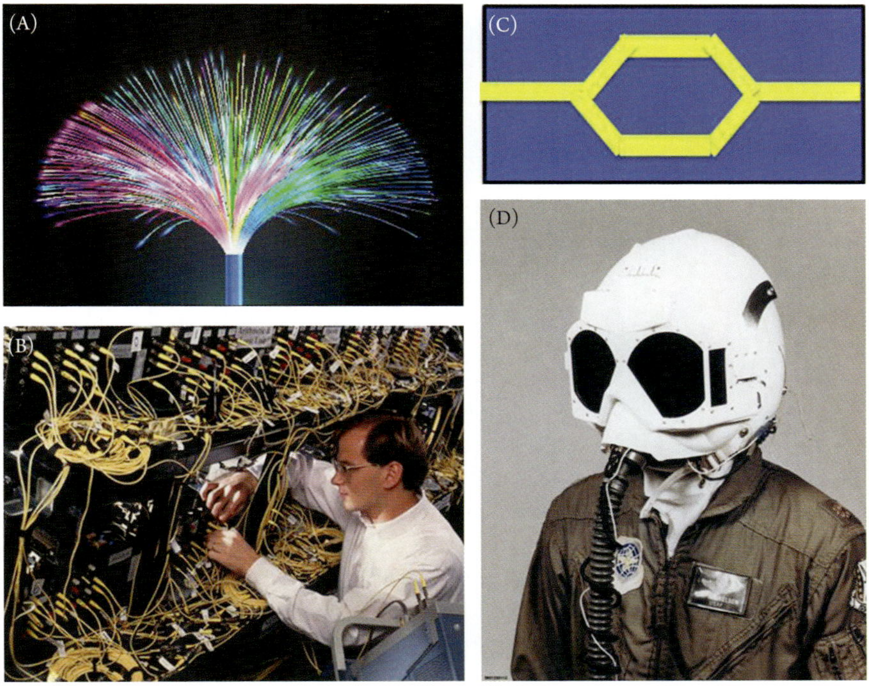

Plate 7 Applications of oxides in optical systems. (A) Passive optical fibers based on silica glass transmit most optical wavelengths with minimal optical losses (SOURCE: D Russell 78/Shutterstock.com). The primary failure mode for such fibers in optical communication systems involves water-assisted stress corrosion cracking (Chapter 16). (B) A technician working on the first optical computer that stores and manipulates data as pulses of light (SOURCE: David Parker/Science Source with permission). (C) The optical switches that guide the light and perform data processing functions are often based on electro-optic ceramics such as lithium niobate ($LiNbO_3$). Shown is a simple schematic of one such optical switch. Light enters the waveguide from the left and is split into two branches. In the absence of an applied voltage, the left and right branches are in phase, so when they recombine on the right, the light is unchanged. However, application of a voltage changes the refractive index and speed of light in the upper branch. With the proper voltage, the two branches can be made to be out of phase, so they cancel each other out on recombination, and no light leaves the device. The on and off states can be used to define 1 and 0 logic states in an optical circuit. (D) Flash-blindness goggles developed for the U.S. Air Force using the electro-optic ceramic PLZT (lead–lanthanum–zirconate–titanate) (Photo courtesy of Sandia National Laboratories). The optical–birefringence properties of PLZT can be switched with an applied voltage. When placed between two polarizer plates, this switching can reversibly change the optical transmission of the goggles from 20% to less than 0.0015% (the state shown) in less than 150 μs (see Figure 1.7).

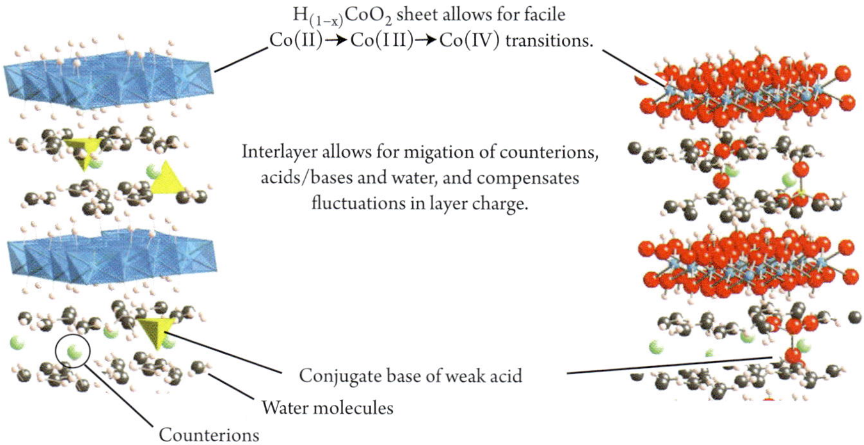

Plate 8 A schematic drawing of the critical elements of cobalt oxyhydroxide CoOOH materials under investigation as electrocatalysts. The cobalt ions within the sheets have access to three oxidation states within the stability window of water—Co(II), Co(III), and Co(IV)—allowing the sheets to be oxidized and reduced under an applied voltage. Thin layers of aqueous solutions exist between the sheets of this clay-like material. When the cobalt ions within the oxyhydroxide layers are oxidized or reduced, changes in layer charge are compensated by the intercalation of counterions that migrate between the sheets (see Figure 1.8).

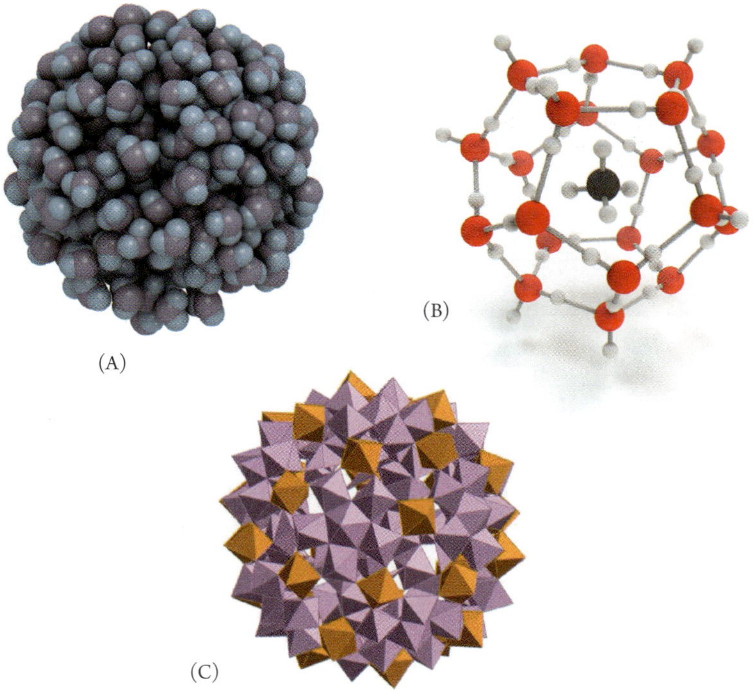

Plate 9 Fundamental oxide reactions (Part Two), proceeding from pure water (Chapter 3) to soluble hydrolysis products (Chapter 5) to extended oxide phases (Chapter 6). (A) A molecular model of a cluster of pure water, showing the orientations of hydrogen bonds between individual water molecules (SOURCE: molecuul.be/Shutterstock.com). (B) A molecular model of methane hydrate, emphasizing the distribution of ring structures between water molecules that encapsulate the methane solute in a zeolite-like cage (SOURCE: Mikkel Juul Jensen/Science Source.com with permission). This model illustrates how the extended structure of water adapts to the presence of ions and other foreign species (Chapter 4). (C) A representation of a water-soluble polyoxometalate cluster that contains 72 molybdenum and 30 iron cations (SOURCE: Reproduced with permission from Balogh, E.; Todea, A. M.; Muller, A.; Casey, W. H. Rates of Ligand Exchange Between >Fe^{III}-OH_2 Functional Groups on a Nanometer-Size Aqueous Cluster and Bulk Solution *Inorg. Chem.* **2007**, *46*, 7087–7092; p. 7088, Fig. 1). This cluster contains many of the oxygen and cation bonding motifs found in extended crystalline oxides (see Figure 1.9).

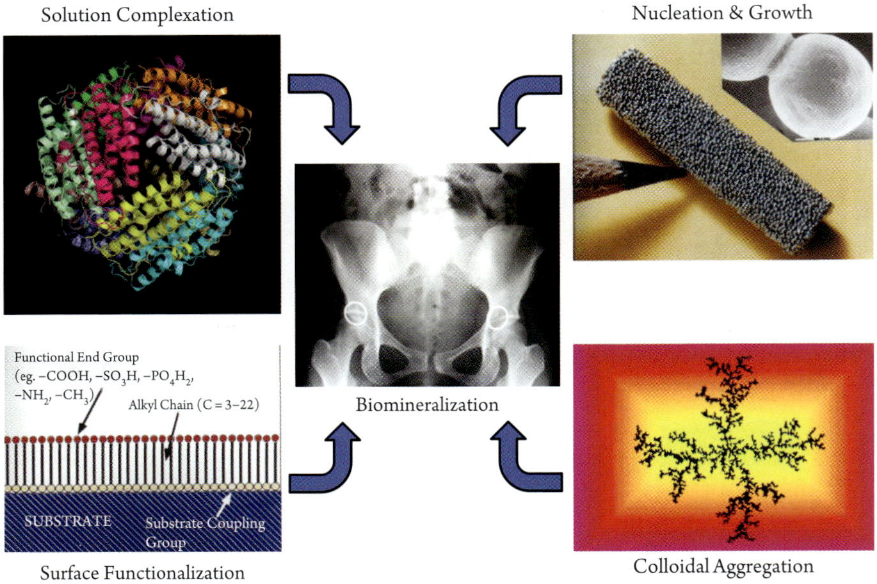

Plate 10 The aqueous synthesis of oxides (Part Three) is illustrated by key attributes of biomineralization processes (Chapter 9), such as those involved in bone formation. (Center) An X-ray image of a broken pelvis (SOURCE: fotografos/Shutterstock.com). (Top left) The aqueous synthesis of oxides typically involves stimulating reactions between soluble precursor complexes by inducing a condition of supersaturation (Chapter 7). Shown is the biological complex ferritin (SOURCE: Laguna Design/Science Source.com with permission). Here, a protein shell stimulates the nucleation of small iron-oxide clusters within a protein cage. (Bottom left) Whether in a protein or on an extended surface, the placement of functional groups can lower the barrier to the nucleation and growth processes leading to oxide formation (SOURCE: Reproduced with permission from Bunker, B. C.; Rieke, P. C.; Tarasevich, B. J.; Campbell, A. A.;Fryxell, G. E.; Graff, G. L.; Song, L.; Liu, J.; Virden, J. W.; McVay, G. L et al. *Science* **1994**, *264*, 48–55; p. 50, Fig. 3). (Top right) The stimulation of nucleation and growth is illustrated by the formation of the bone mineral apatite within a porous titanium bone implant material (Reproduced with permission from Bunker, B. C.; Rieke, P. C.; Tarasevich, B. J.; Campbell, A. A.; Fryxell, G. E.; Graff, G. L.; Song, L.; Liu, J.; Virden, J. W.; McVay, G. L. *Science* **1994**, *264*, 48–55; p. 53, Fig. 12). (Bottom right) Aqueous synthesis can also involve the consolidation of colloidal oxide suspensions via controlled aggregation processes. Shown is the result of a computer simulation of diffusion-limited aggregation (SOURCE: Scott Camazine/Science Source.com with permission). Surface functionalization plays a key role in such aggregation processes (see Figure 1.10).

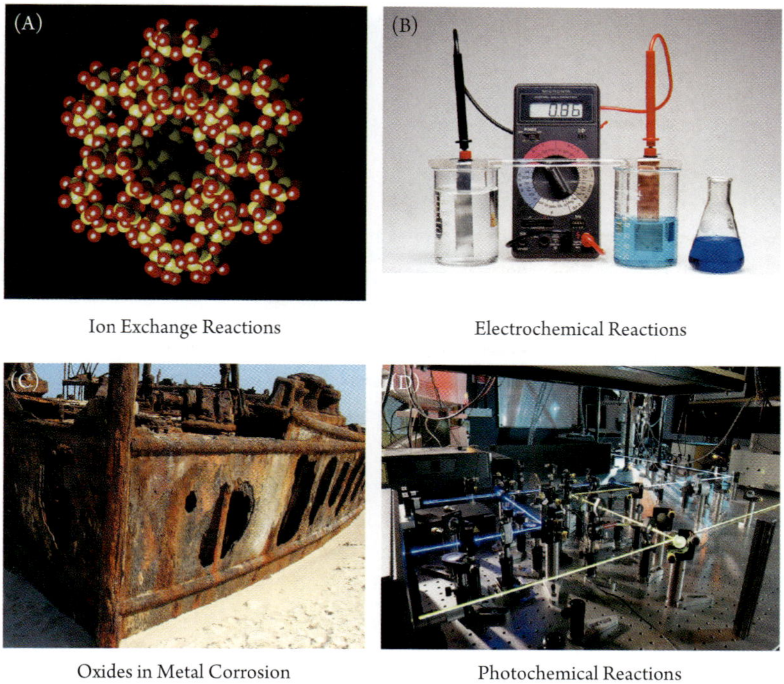

Plate 11 Technologically important oxide reactions (Part Four). (A) A molecular model of the oxide exchange material zeolite Y (SOURCE: Clive Freeman/Science Source with permission). Such structures exemplify the charge distribution and steric attributes that make oxides such as zeolites among the most important materials used in ion exchange and water treatment technologies (Chapter 10). (B) Electrodes immersed in aqueous solutions play a critical role in electrochemical energy storage systems such as voltaic electrochemical cells (SOURCE: Charles D. Winters/Science Source.com). Oxides can be both oxidized and reduced. The coupling of such reactions forms the basis for energy storage systems such as most commercial batteries involving aqueous electrolytes (Chapter 11). (C) The electrochemical conversion of metals into oxides can also be highly destructive, as evidenced by the corrosion seen on the rusted *SS Maheno* shipwreck in Australia (SOURCE: Walter Quirtmire/Shutterstock.com) (Chapter 12). (D) Oxide materials are active in photoelectrochemical reactions used for the harvesting of solar energy (Chapter 13). Shown is flash photolysis equipment (SOURCE: Colin Cuthbert/Science Source.com with permission) used to study such reactions. Here, lasers are used to stimulate optical transitions within the material, and the products of the resulting reactions are identified using techniques such as optical absorption spectroscopy (see Figure 1.11).

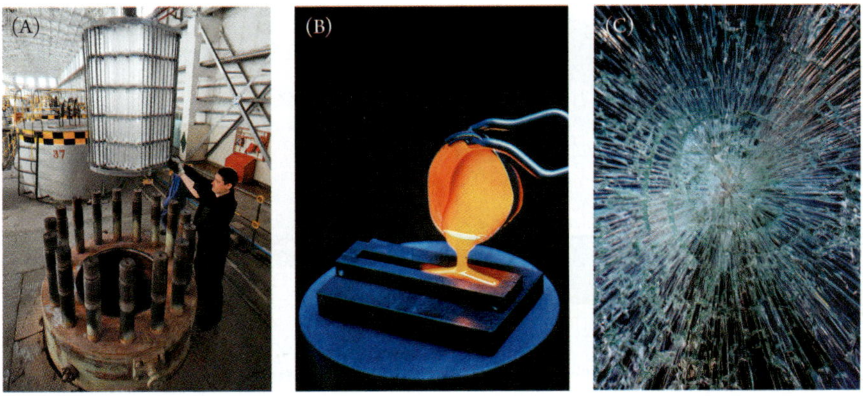

Hydrothermal Synthesis | Nuclear Waste Glass | Glass Fracture (Stress Corrosion)

Plate 12 Reactions involving tetrahedral oxides (Part Five). (A) Autoclaves are key to the aqueous synthesis of tetrahedral oxides at high temperatures and pressures (Chapter 14). Shown is a hydrothermal system used in the synthesis of large quartz crystals (SOURCE: RIA NOVOSTI/Science Source.com with permission). (B) Suppressing the corrosion and dissolution of glass (Chapter 15) is of paramount importance in nuclear waste disposal. Shown is a sample of molten nuclear waste glass being poured into a mold (SOURCE: U.S. Department of Energy/Science Source.com with permission). The resulting glass ingot is used for testing the chemical durability of the waste in a range of hostile aqueous environments to help optimize compositions used in waste repositories. (C) The fracture of glass (SOURCE: bookpc/Shutterstock.com) is often a process that is mediated by water. The role of water is most pronounced in slow crack growth, or stress corrosion cracking, when stress-induced chemical reactions at the crack tip determine failure rates for components such as optical fibers (Chapter 16) (see Figure 1.12).

Plate 13 The environmental geochemistry of oxides (Part Six). Most of the minerals and other materials that cover Earth's surface are produced as a result of weathering reactions involving oxides and water (Chapter 17). When produced, weathered oxides play important roles in the chemistry of all of Earth's major environments (Chapter 18). On land, weathering products such as clay minerals play a major role in mediating environmental contaminants in soil and groundwater. A view of the Chernobyl power plant in Pripyat, Ukraine, 25 years after the nuclear disaster (June 27, 2010) (A) illustrates such environmental recovery (SOURCE: Helen Sergeyeva/Shutterstock.com). Clays and other soil minerals are carried into the atmosphere by the wind, (B) where they play key roles in both water condensation in clouds and photochemical reactions. Shown is a tornado in the process of sucking soil into the air from a Kansas farm field (SOURCE: Jim Reed/Science Source.com with permission). Hot lava and water exhibit extensive reactions on Earth's surface (SOURCE: Claudio Rossol/Shutterstock.com) (C). The reverse reactions occur as weathered and hydrated minerals are pulled beneath the surface in subduction zones that regenerate igneous rocks and release entrapped water in violent volcanic eruptions. All of these processes affect the chemistry of the oceans (D) (SOURCE: Jim Agronick/Shutterstock.com). (see Figure 1.13).

Plate 14 A Landsat 7 satellite view of the mouth of the Mississippi River showing the massive aggregation and sedimentation of suspended clay particles that occurs when colloids suspended in rivers with a low-salt content solutions encounter the ocean, which is a solution with a high salt content. The salt concentrations cause the double layer to shrink sufficiently for attractive van der Waals forces to flocculate the clays, causing them to settle (see Figure 8.5).
SOURCE: NASA/Science Source with permission.

Plate 15 Photographs taken inside two different tanks of radioactive waste at the Hanford site in Richland, Washington. Radioactivity levels are so high in many tanks that images such as these are usually taken robotically (see Figure 8.9).

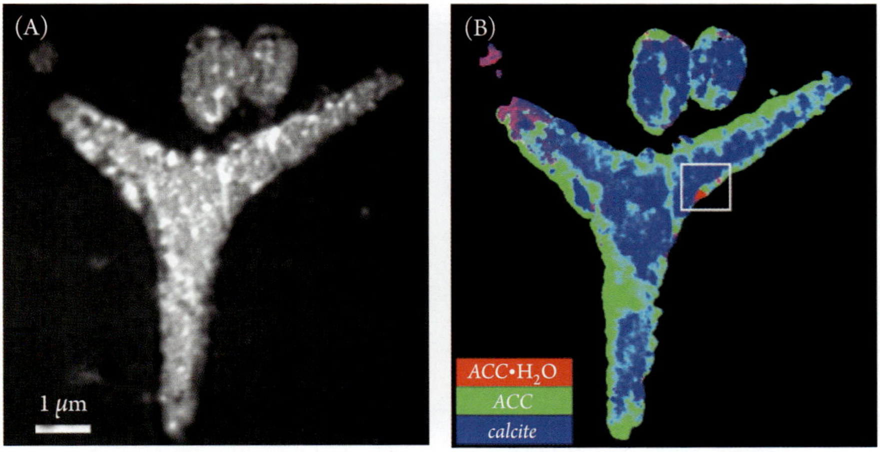

Plate 16 The evolution of calcium carbonate phases in sea urchin (*Strongylocentrotus pupuratus*) spicules as determined by X-ray structural mapping. Sea urchin larvae make a set of spicules that begin as a trigonal form. This trigonal template is first filled with amorphous, hydrated calcium carbonate (ACC) that eventually converts into a single crystal of calcite. For the intermediate stage shown, the gray scales in the inset correspond to different phases (hydrated ACC, dehydrated ACC, and calcite). SOURCE: Images reproduced with permission from Gong, Y. U. T.; Killian, C. E.; Olson, I. C.; Appathurai, N. P.; Amasino, A. L.; Martin, M. C.; Holt, L. J.; Wilt, F. H.; Gilbert, P. U. P. A. Phase Transitions in Biogenic Amorphous Calcium Carbonate. *Proc. Nat. Acad. Sci.*, **2012**, *109*, 6088–6093 (see Figure 9.16).

Plate 17 Corroded metal objects coated by oxide corrosion products. (A) A rusted iron chain (SOURCE: Alistair Scott/Shutterstock.com). (B) A tarnished silver urn (SOURCE: Karin Hindebrand Lao/Shutterstock.com). (C) A corroded bronze statue (SOURCE: Mr Doomits/Shutterstock.com) (see Figure 12.1).

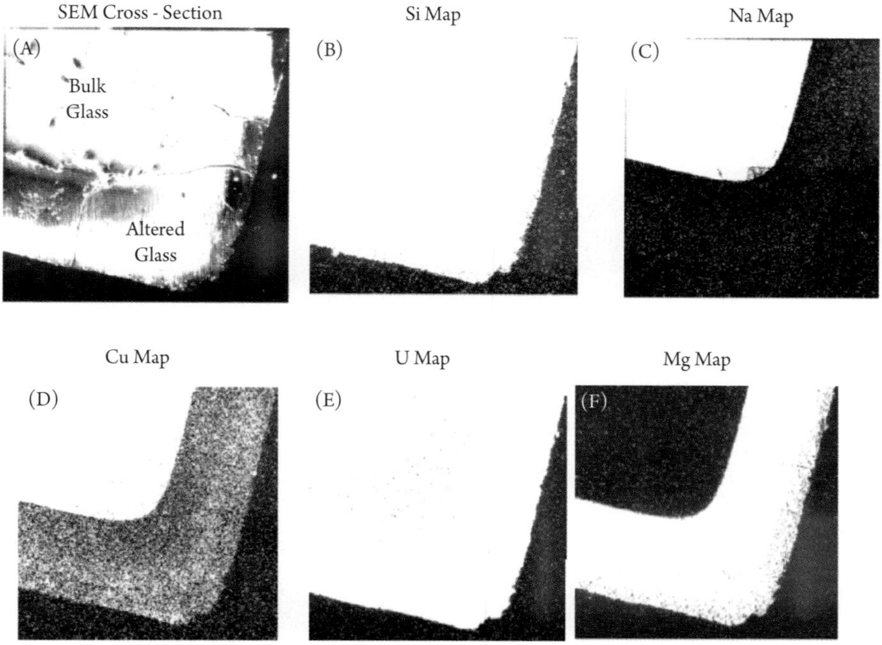

Plate 18 Elemental maps obtained using electron microprobe analyses on a cross-section of a simulated nuclear-waste glass leached in simulated seawater under hydrothermal conditions to accelerate reaction rates (T = 250°C, pH = 5.5, t = 14 days). (A) A scanning electron micrograph (SEM) shows the presence of the leached surface layer, where the thickness of the layer is 170 μm. (B–F) Elemental maps for the same cross-section.
SOURCE: Reprinted with permission from Bunker, B. C., Arnold, G. W., Beauchamp, E. K. Mechanisms for Alkali Leaching in Mixed Na-K Silicate Glasses *J. Non-Cryst. Solids,* **1983**, *58*, 295–322 (see Figure 15.3).

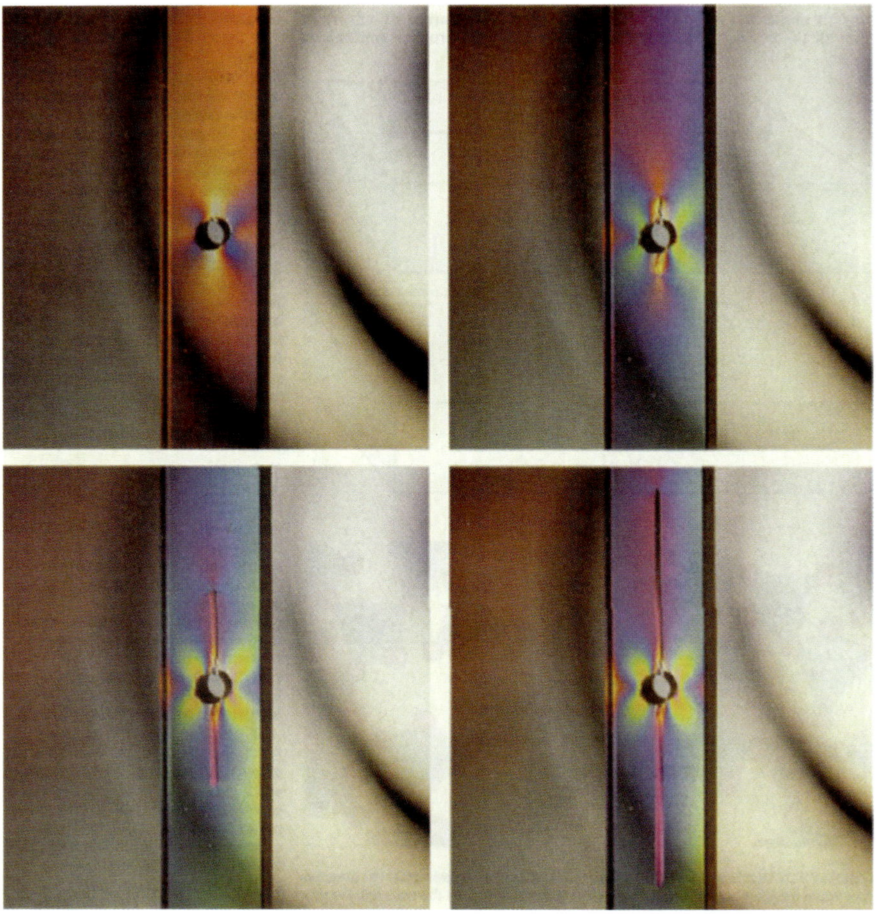

Plate 19 Cracks growing in a laboratory fracture mechanics specimen from a controlled flaw consisting of a hole drilled in the middle of the sample. Compressive loads applied along the long axis propagate two cracks from this flaw in the presence of a controlled vapor pressure of water. Stress patterns (made visible by exposing the sample to polarized light) show stresses are concentrated at the tips of each advancing crack. SOURCE: Reproduced with permission from Michalske, T. A., Bunker, B. C.; The Fracturing of Glass. *Scientific American*, December, 1987, **255**, 122–129 (see Figure 16.2).

Plate 20 A photograph of the Paine Towers in Chilean Patagonia at the southern tip of South America that are exposed to extreme weathering conditions (see Figure 17.1). SOURCE: kavram/Shutterstock.com.

Plate 21 Granite and its components. (Upper left) A granite pluton found in Enchanted Rock Park in Texas (SOURCE: NeonLight/Shutterstock.com.). (Upper right) A closer look at the fine-grained structure of granite (SOURCE: Bragin Alexey/Shutterstock.com). The major mineral that comprise granite are shown in the bottom row: quartz (left) (SOURCE: Epitavi/Shutterstock.com), plates of mica surrounding a central crystal of aquamarine (center) (SOURCE: MarcelClemens/Shutterstock.com), and feldspar (right) (SOURCE: Zbynek Burival/Shutterstock.com) (see Figure 17.3).

Plate 22 The fate of the resistate mineral quartz (top left) (source: MarcelClemens/ Shutterstock.com). First, sand grains are released from granite, where they accumulate to form unconsolidated sand (top right) (source: Fekete Tiber/Shutterstock.com). These sand grains are later cemented together to form sandstone, as illustrated by dramatic formations found in Arches National Park in Utah (source: Tupungato/Shutterstock. com) (see Figure 17.5).

Plate 23 A major weathering pathway involving the hydrolysis and chemical alteration of minerals is illustrated by the conversion of feldspar grains (top left) (SOURCE: J. Palys/Shutterstock.com) into the clay minerals that form mud (bottom left) (SOURCE: Alistair Scott/Shutterstock.com). Muds and other hydrosylate sediments are later compressed and consolidated into stable rock formations such as coastal shale platforms shown on the right (SOURCE: Martyn F. Chillmaid/Science Source.com with permission). Oil trapped by impermeable shale layers can be released using techniques such as fracking (see Figure 17.6).

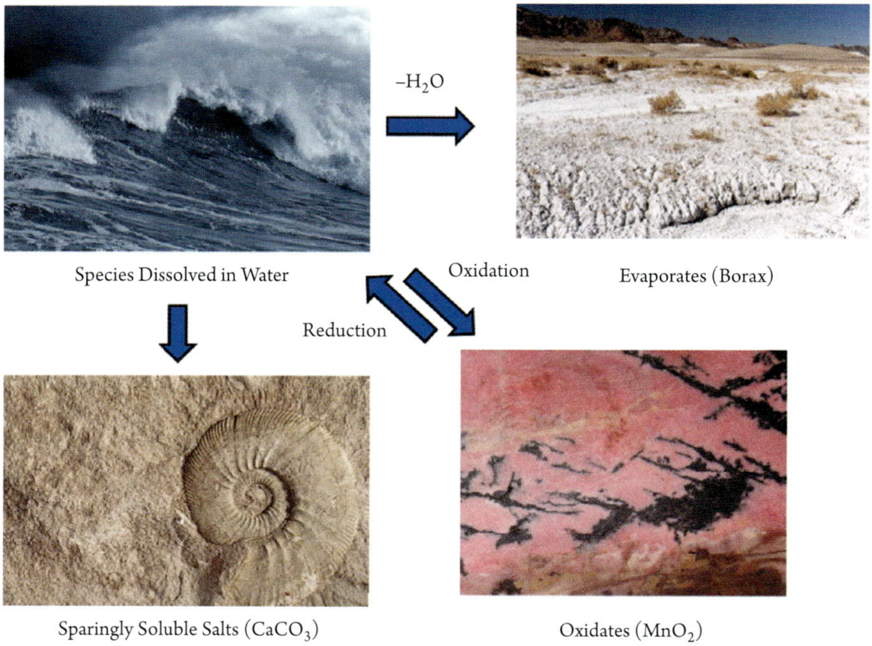

Plate 24 Soluble species released into all bodies of water (top left) (SOURCE: Zacarias Pereira da Mata/Shutterstock.com) by oxide weathering are involved in three major weathering pathways, including (1) precipitation, as illustrated by a fossil snail imbedded in a calcite formation (bottom left) (SOURCE: Doart/Shutterstock.com); (2) reversible reduction and oxidation cycles to form oxidates, as illustrated by the black manganese oxide deposits left on the surface of a pink rhodonite (manganese inosilicate) crystal (SOURCE: Sombra/Shutterstock.com); and (3) evaporation to form evaporates, as illustrated by borax-rich soil found in the Mojave Desert in California (SOURCE: BOB GIBBONS/Science Source.com with permission). (see Figure 17.9)

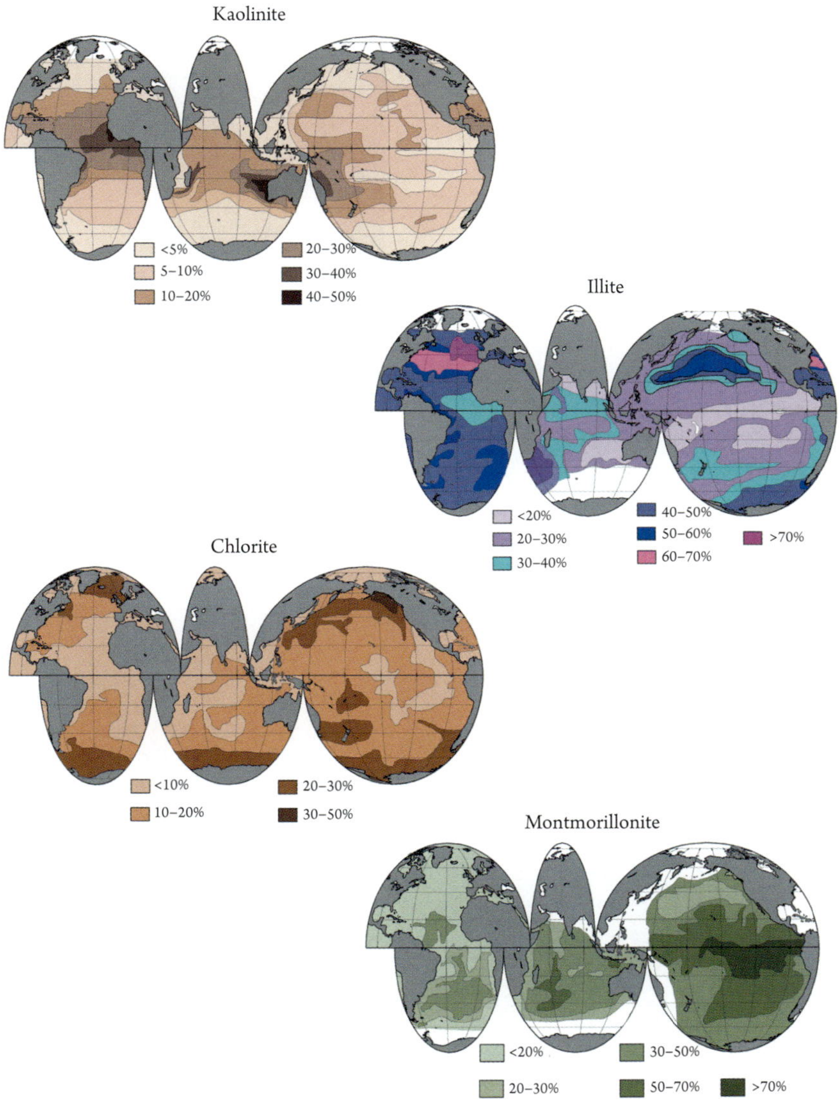

Plate 25 The distributions of each of the major clay families in the major oceanic basins of the world. Adapted with permission from Griffin, J. J.; Windom, H.; Goldberg, E. D. The Distribution of Clay Minerals in the World Ocean. *Deep-Sea Res.* **1968**, *15*, 433–459 (see Figure 18.11).

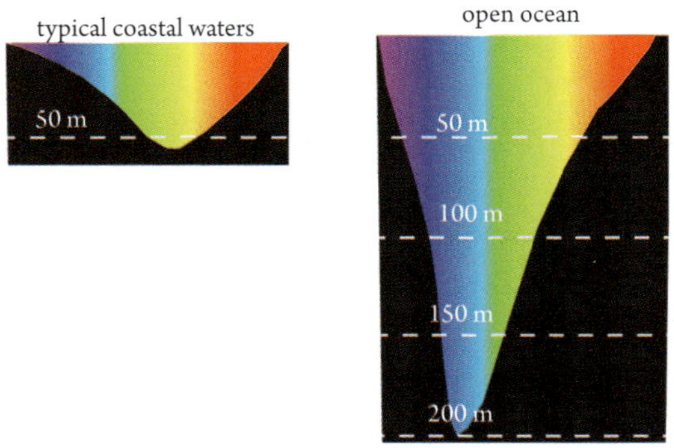

Plate 26 Key features of the photic zones of Earth's oceans. Light penetration depths are depicted for both coastal and oceanic waters as a function of wavelength. (Adapted from an image by the National Oceanographic Atmospheric Administration Ocean Explorer Program, http://oceanexplorer.noaa.gov/explorations/o4deepscope/background/deeplight/media/diagram3.html.) (see Figure 18.12).

Plate 27 Lava entering the ocean near the Kilauea volcano in Hawaii. (Source: CJ Kale/lavalightgalleries.com, with permission) This volcano forms from one of the two dozen hot spots that affect the Earth's crust. (see Figure 18.19).

12

Oxide Films in Metal Corrosion: Oxide Defect Chemistry

12.1 INTRODUCTION

Most metals used by our society corrode, from the mild tarnish on silver to the green patina that coats our copper statues and electrical wiring to the red rust on our cars and iron bridges (Fig. 12.1, Plate 17). Metal corrosion often involves the conversion of metals into either oxides or their soluble hydrolysis products. The metals we use in our structures, as well as in electronic and magnetic devices, are destroyed as a result of this conversion. It has been estimated that the annual cost of metallic corrosion to the U.S. economy is hundreds of billions of dollars (5% of the gross national product). Therefore, enormous efforts have been made to understand why metals corrode, and what can be done to inhibit corrosion processes. Stainless steel is just one example of humankind's attempts to limit corrosion processes. Other examples include the use of inert paints on ships to prevent saline corrosion, coating iron with zinc to galvanize it, and exploiting electrochemical strategies, such as using sacrificial anodes that corrode instead of iron, as a means of protecting more important materials' components.

The number of comprehensive texts and reviews regarding metal corrosion scales with its economic impact, with more than 1000 articles being published on the topic per year. Those of you interested in more comprehensive discussions regarding how specific metals corrode in specific environments such as seawater and acid rain should see other works [1–3]. Our focus in this chapter is

Figure 12.1 Corroded metal objects coated by oxide corrosion products. (A) A rusted iron chain (SOURCE: Alistair Scott/Shutterstock.com). (B) A tarnished silver urn (SOURCE: Karin Hindebrand Lao/Shutterstock.com). (C) A corroded bronze statue (SOURCE: Mr Doomits/Shutterstock.com). (see Plate 17)

to highlight how metal corrosion is mediated by the presence of oxides, with an emphasis on reactions that occur in water. This overview highlights the basic properties of oxide films that give rise to a wide range of complicated metal corrosion phenomena. The discussion draws on many concepts highlighted in other chapters of this book: surface chemistry (Chapter 6), electrochemistry (Chapter 11), and oxide dissolution (Chapter 16).

12.2 THE ELECTROCHEMISTRY OF METAL CORROSION

Metal corrosion often involves electrochemical reactions in which the metal is oxidized by either water or O_2. Those of you seeking an overview of the thermodynamics of aqueous electrochemistry, half-reactions, redox potentials, and electrochemical equilibria involving metals and oxides should refer to Chapter 11. As discussed in Chapter 11, the relevant equilibria used to predict whether a particular metal is stable as a function of both pH and electrochemical potential (E_h) are tabulated as reduction potentials (see Table 11.1). Diagrams of electrochemical stability are depicted graphically in Pourbaix diagrams (see Fig. 12.2 for the case of iron). These diagrams represent the starting point for predicting the pH and E_h regimes under which a given metal exhibits *immunity, corrosion*, or *passivation*. Many metals can exhibit all three of these behaviors. Transitions from one behavior to another are dictated by solution concentrations, pH, and applied voltages and/or redox conditions such as the partial pressure of O_2 gas.

Immunity represents that set of conditions under which no oxidation or corrosion of the metal occurs. Some metals are thermodynamically stable in water. For example, the noble metal gold is unaffected by corrosion over geologic time as long as no exotic complexing ligands or extreme solution compositions are present, as indicated by the reaction (Table 11.1)

$$Au^{3+}(aq) + 3/2H_2(g) \rightleftarrows Au^\circ(s) + 3H^+ \qquad \Delta G^\circ = -433.5 \text{ kJ/mol} \quad \textbf{(12.1)}$$

The standard-state Gibbs energy is so large that this reaction goes strongly to the right under any reasonable pH conditions and partial pressures of H_2. Because this complete equation has the hydrogen electrode included $(2H^+ + 2e^- \rightleftarrows H_2(g))$, the half-cell voltage given by the Nernst relation $(\Delta G^\circ = -nFE^\circ)$ is

$$Au^{3+}(aq) + 3e^- \rightleftarrows Au^\circ(s) \qquad E^\circ = +1.5 \text{ V} \quad \textbf{(12.2)}$$

If one examines a similar reaction for iron metal in water at standard-state conditions:

$$Fe^{2+}(aq) + H_2(g) \rightleftarrows Fe^\circ(s) + 2H^+(aq) \qquad \Delta G^\circ = +78.9 \text{ kJ/mol.} \quad \textbf{(12.3)}$$

Separating out the hydrogen half-reaction, the remaining expression for iron oxidation is

$$Fe^{2+}(aq) + 2e^- \rightleftarrows Fe^\circ(s) \qquad E^\circ = +0.41 \text{ V} \quad \textbf{(12.4)}$$

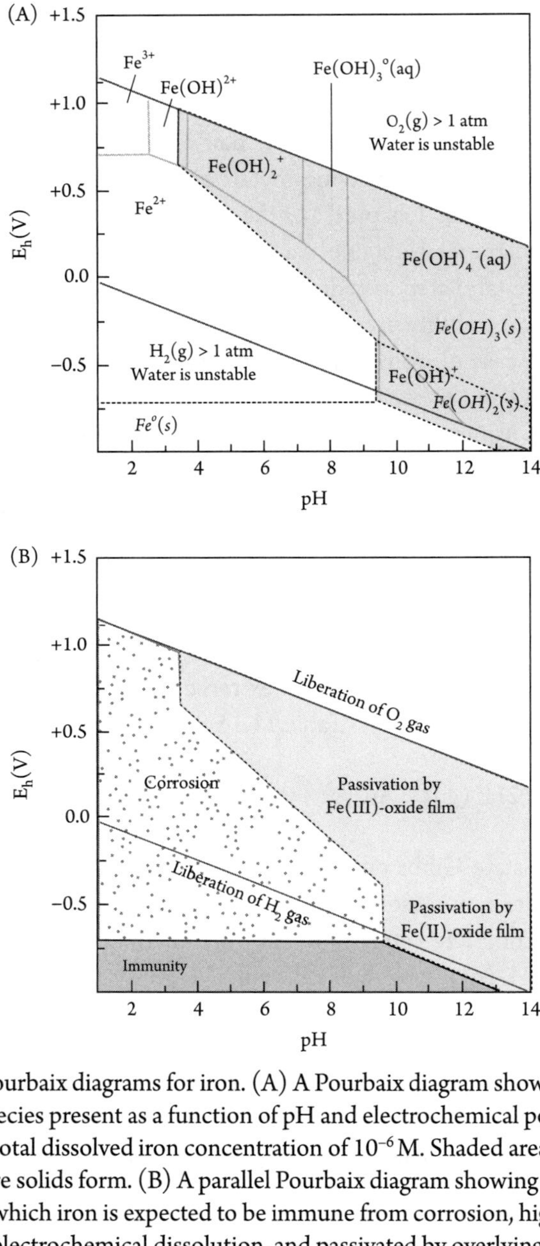

Figure 12.2 Pourbaix diagrams for iron. (A) A Pourbaix diagram showing the predominant species present as a function of pH and electrochemical potential (E_h) for a solution with a total dissolved iron concentration of 10^{-6} M. Shaded areas indicate conditions where solids form. (B) A parallel Pourbaix diagram showing the pH and E_h regimes within which iron is expected to be immune from corrosion, highly susceptible to corrosion or electrochemical dissolution, and passivated by overlying films of insoluble oxides. Diagrams adapted from Pourbaix, M. *Atlas of Electrochemical Equilibria*; National Association of Corrosion Engineers: Houston, TX, 1976, with permission.

The reaction has a positive redox potential. Iron metal is thermodynamically unstable at standard-state conditions and will corrode (see Fig. 12.2) spontaneously in water and generate H_2 gas in the absence of strong intervention. Such intervention might be in the form of an external voltage applied to the iron that exceeds its corrosion potential. Under standard-state conditions (i.e., $P(H_2(g)) = 1$ atm, $[H^+] = 1$ M, $[Fe^{2+}(aq)] = 1$ M), this applied voltage, E_h, needs to be more negative than -0.41V to prevent Fe° oxidation. For other conditions, the necessary applied voltage varies according to the activities of the reactants and products. The general point remains valid that Fe°(s) in contact with water generates $H_2(g)$ and forms oxidized products unless voltages are applied to reverse the process. With these voltages, iron metal can be retained in an immune, pristine state in water (Fig. 12.2).

The *corrosion zone* in the Pourbaix diagram should, more properly, be called the *oxidative dissolution zone*. Here, the metal reacts with water to form soluble ions. For example, in the normal stability field of water (Fig. 12.2) and in acidic conditions, iron metal oxidizes spontaneously to the fully aquated $[Fe(OH_2)_6]^{2+}(aq)$ ion, whereas water is reduced to evolve hydrogen gas. At higher E_h values, the metal oxidizes to produce the more oxidized ferric ion ($[Fe(OH_2)_6]^{3+}(aq)$), whereas at even higher E_h values, the solvent water is also oxidized to produce $O_2(g)$. The solution pH is key. As long as the solution pH is kept sufficiently low so that no insoluble hydrolysis products form, corrosion of bare iron metal undergoes oxidative dissolution and releases solutes. In cases of extensive corrosion in acidic conditions, concentrations of the ions can become high enough to precipitate ionic salts, but solid oxide materials are not usually produced that protect the metal from the ravages of corrosion.

Consider what happens in the *passive* regime. As in the active corrosion regime, iron is electrochemically unstable in water and is oxidized whereas the contacting water is reduced. However, under neutral to mildly basic conditions, the oxides and hydroxides of iron are relatively insoluble and form deposits on the surface of the metal. An example of an oxide-forming corrosion reaction is given by

$$2Fe^{\circ}(s) \rightleftarrows 2Fe^{3+}(aq) + 6e^- \quad \textbf{(12.5)}$$

$$6H_2O(l) + 6e^- \rightleftarrows 3H_2(g) + 6OH^-(aq) \quad \textbf{(12.6)}$$

$$2Fe^{3+}(aq) + 6OH^-(aq) \rightleftarrows 2FeOOH(s) + 2H_2O(l) \quad \textbf{(12.7)}$$

$$2Fe^{\circ}(s) + 4H_2O(l) \rightleftarrows 2FeOOH(s) + 3H_2(g) \quad \textbf{(12.8)}$$

Oxyhydroxide minerals such as goethite (α-FeOOH(s)) are common components of rust, but other solids such as carbonates can also be present. From a purely thermodynamic perspective, even if insoluble oxides are present, the metal continues to corrode until all of it is converted into an oxide. However, from a kinetic point of view, the reaction can be slowed to a negligible rate if the oxide or hydroxide reaction products form a protective barrier between the aqueous solution and the underlying metal. This protective barrier slows the reaction by affecting the rates of solute and electron transport.

The technical term for this protection is *passivation*. In the Pourbaix diagram, passivation is equated with oxide solid formation, but not all oxides form effective passivation layers. Aluminum metal is typically protected from further oxidation by a dense oxide film. Soft iron nails, however, turn to rust inexorably. To achieve an ideal passive state, the oxide must be uniform, nonporous, adherent, and impenetrable to the electrons, ions, and/or water molecules required to sustain corrosion reactions. In water, such conditions are rarely achieved, leading to a wide range of behaviors in addition to passivation, including scaling and pitting. Next, we discuss the kinetics of metal corrosion, followed by an examination of the nature of the oxide films generated during corrosion in air and water, and some of the factors that control reactive transport through such films. As you will see, the oxide films on corroding metals are anything but truly passive. However, before turning our attention to passive films, there is one final topic involving the electrochemistry of corrosion that needs to be described: a phenomenon called *galvanic corrosion*.

12.3 GALVANIC CORROSION

Corrosion reactions that are mediated by an external voltage fall under the heading of *galvanic corrosion*. Pourbaix diagrams clearly show that all redox reactions in water are sensitive to applied voltages, or E_h values. In Pourbaix diagrams, reactions are referenced to the hydrogen electrode, with an E_h value of zero. However, if the metal is in direct electrochemical contact or communication with other metals, these metals can provide a fixed voltage as determined by their own redox potentials. Here, the net redox reaction is represented by the sum of the half-reactions of each of the two contacting metals. The net voltages or potentials experienced by each metal determine whether oxidation or corrosion processes are enhanced or suppressed.

One of the earliest discoveries of galvanic corrosion involved British ships such as the *HMS Alarm*. This wooden ship was sheathed with copper sheets pinned to the underlying wood with iron nails as a means of discouraging burrowing worms and hull rot. However, it was discovered that the iron nails quickly rusted away causing the copper sheets fall off the ship. What the shipwrights failed to realize is that their ship hulls represented gigantic batteries (see Chapter 11) consisting of large copper cathodes and small, sacrificial iron anodes immersed in an electrolyte consisting of seawater, as represented by the coupled redox reactions

$$Fe^{\circ}(s) \rightleftarrows Fe^{2+}(aq) + 2e^{-} \qquad E^{\circ} = +0.41\ V \tag{12.9}$$

$$Cu^{2+}(aq) + 2e^{-} \rightleftarrows Cu^{\circ}(s) \qquad E^{\circ} = +0.35\ V \tag{12.10}$$

The net electrochemical driving force for the reaction between this anode and cathode is +0.79 V, which is almost twice the driving force for corroding the iron in the absence of the copper.

The *HMS Alarm* example highlights several key features associated with the galvanic corrosion of any two metals either in contact or in electrochemical communication in water. First, the electrochemical driving force for interactions between the two metals can be predicted by standard redox potentials, as tabulated in the electrochemical series (Table 11.1). Second, in such interactions, one metal always functions as an anode and the other metal serves as a cathode. The anode is always the metal with the more negative standard reduction potential. The more noble metal (with the more positive redox potential) is always the cathode.

Oxidation occurs at the anode (see Chapter 11), which means that corrosion processes are stimulated or enhanced on the metal in the redox couple having the more negative redox potential. Conversely, the electrons released by corrosion flow to the cathode. These electrons help protect the cathode from corrosion, as well as stimulate reduction reactions involving solution species. This is true even if both metals are thermodynamically unstable in the electrolyte solution. However, it is important to note that the reduction reaction that occurs on the cathode during galvanic corrosion is often the reduction of water to form $H_2(g)$, because high concentrations of redox-active cations such as the $Cu^{2+}(aq)$ depicted in Eq. 12.10 are usually absent.

For engineering materials, which include complex alloys, predictions and prevention strategies regarding galvanic corrosion are often arrived at by consulting the semiempirical galvanic series shown in Table 12.1 [4]. As in the electrochemical series, the driving force for promoting corrosion on the anode and inhibiting corrosion on the cathode depends on the difference in the effective galvanic potentials between the contacting metals. Once again, the metal with the more negative reduction potential functions as the anode and corrodes, whereas the cathode is protected. For example, Type 316 stainless steel (iron with 16%–18% chromium, 10%–14% nickel, and 2%–3% molybdenum) has a galvanic potential of +0.06 V relative to −0.37 V for cast iron. If the two materials are in electrical contact in seawater, the cast iron corrodes rapidly whereas corrosion of the stainless steel is inhibited.

Table 12-1. THE GALVANIC SERIES FOR VARIOUS ENGINEERING MATERIALS IN FLOWING SEAWATER

Engineering Material	**E°/V (relative to standard hydrogen electrode, SHE)**
Magnesium	−1.16
Zinc	−0.79
3003-H Aluminum	−0.55
Cast iron	−0.37
Carbon steel	−0.37
Type 430 stainless steel, active	−0.33
Type 304 stainless steel, active	−0.29
Naval rolled brass, allow 464	−0.08
Copper sheet	−0.12
Copper alloy 443 (Admiralty brass)	−0.05
Type 316 stainless steel (active)	+0.06
Silver	+0.11
Titanium	+0.14
Hastelloy C steel	+0.16
Type 316 stainless steel (passive)	+0.19
Zirconium	+0.20
Platinum	+0.39

The voltages are recalculated to the standard hydrogen electrode.
Adapted from Galvanic Series of Metals and Alloys in Seawater. In *Corrosion: Materials*; Cramer, S. D.; Covino, B. S., Jr., Eds.; ASM International: Materials Park, OH, 2005; Vol. 13B, ASM Handbook; p 672. See also Atlas Steel Technical Note No. 7, Galvanic Corrosion, www.atlassteels.com.au.

Although the galvanic series provides warnings regarding metal corrosion, it also provides engineers with strategies for avoiding corrosion of metal components exposed to water. The example provided by the *HMS Alarm* shows us that an inexpensive and expendable metal such as iron can be used to protect a more valuable metal such as copper. This strategy is clearly evident in the corrosion of the Statue of Liberty. Unfortunately, in the case of the Statue of Liberty, the expendable iron is the support structure, leading to serious mechanical stability problems for the statue. Ships constructed out of expensive stainless steel are often protected by being connected to a post of a cheaper but less noble metal such as zinc. In fact, one of the mechanisms behind the corrosion resistance of galvanized steel has to do with the protection afforded by a sacrificial zinc coating. The other reason why galvanized steel corrodes more slowly than iron has to do with the formation of protective zinc-oxide and zinc-carbonate films. The role of such oxides in the passivation of metals to corrosion is described next.

12.4 THE KINETICS OF METAL CORROSION

Although both the electrochemical and galvanic series (Table 11.1 and Table 12.1) provide information regarding the thermodynamic driving force for corrosion, they do not provide sufficient information to predict corrosion rates. Galvanic corrosion occurs only if electrons are able to pass from the anode to the cathode. This electron flow must be accompanied by ion transport to maintain a neutral charge. The corrosion rate is directly proportional to the net current that passes through this anode-to-cathode circuit. Corrosion rates can be increased by such factors as enlarging the electrode surface areas (consider the relative areas of the copper sheeting on the *HMS Alarm* vs. the exposed area of the iron rivets) or increasing the voltage difference between the anode and cathode. However, in the passive regime, the most important factor controlling corrosion rates is the net resistance to the flow of charge through the system. This resistance is embodied in Ohm's Law: $E = iR$. Here, E is the net galvanic voltage, i is the corrosion current in amperes, and R is the circuit resistance in ohms. In the passive regime, the largest resistances are invariably provided by oxide films on metal surfaces. These oxides can form barriers to the flow of both electrons and ions.

The first detail to note about metal corrosion is that corrosion rates are substantially different in dry air and in water. In dry air, the primary oxidizing agent is O_2 gas. Initially, the O_2 has direct access to the surface and can react directly with the metals via strong bonds. Sometimes the O_2 dissociates and takes electrons from the near-surface metals (Fig. 12.3). Those of you interested in the extensive surface science studies performed to study such reactions are referred to journals such as *Surface Science* and works by others [5–7]. The products of such reactions are pure oxides. When a continuous and impermeable oxide monolayer forms, further oxidation can only occur if electrons, metal cations, and oxygen anions diffuse through the passivating layer. Reaction kinetics are typically followed by analytically profiling elemental compositions as a function of depth and time, via techniques such as ellipsometry or elemental depth profiling [8] via methods such as Rutherford backscattering spectroscopy or secondary-ion mass spectroscopy (SIMS).

Experimentally, two limiting rate laws are typically observed for oxidation in air (Fig. 12.3). At higher temperatures, classic diffusion-limited rates are seen, where the film thickness increases as a function of the square root of time via the Einstein relation

$$x \sim (Dt)^{1/2} \quad \textbf{(12.11)}$$

Here, x is the film thickness (measured in centimeters), D is the diffusion coefficient (measured in square centimeters per second), and t is time (measured in seconds). For the oxidation of iron at an oxygen partial pressure of $1.3 \cdot 10^{-2}$ torr

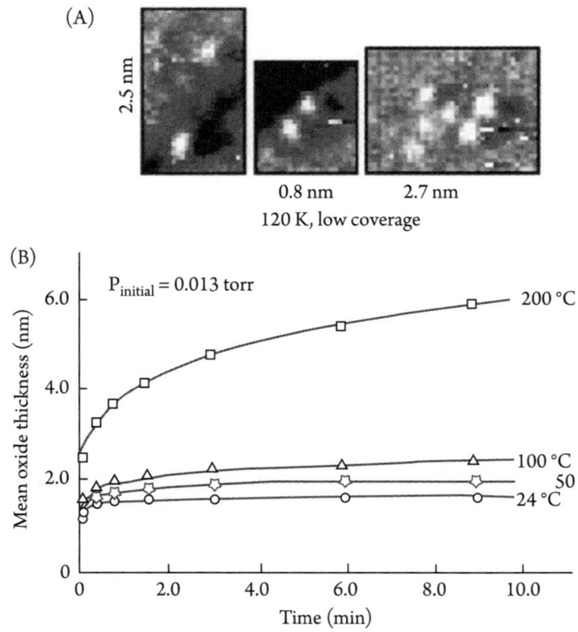

Figure 12.3 The chemisorption of dry O_2 gas on metals. (A) A scanning tunneling microscope image of a partial layer of oxygen on the Cu(110) surface at 120 K. (Adapted from Carley, A. F.; Davies, P. R.; Roberts, M. W. Activation of Oxygen at Metal Surfaces. *Phil. Trans. R. Soc.* **2005**, *363*, 829.) (B) The kinetics of oxide formation on iron metal as a function of temperature at an oxygen partial pressure of $1.3 \cdot 10^{-2}$ torr. (Adapted with permission from Graham, M. J.; Ali, S. I.; Cohen, M. Low Temperature Oxidation (24° to 200°C) and Krypton Adsorption Studies on Polycrystalline and Single Crystal Iron Surfaces *J. Electrochem. Soc.* **1970**, *117*, 513.)

and $T = 300°C$, magnetite (Fe_3O_4) films form at a rate that is consistent with an effective diffusion coefficient of $2 \cdot 10^{-16}$ cm^2/second. As films get thicker, corrosion rates slow until corrosion effectively ceases at film thicknesses that are less than 10 nm.

In low-temperature oxidation, the film thickness increases as a function of the logarithm of time, as rationalized by the Cabrera–Mott model [9, 10]. This model postulates that electrons tunnel through the oxide to create anionic surface sites and leave behind cationic holes at the metal–oxide interface. The net result of this charge separation is the creation of an electric field across the oxide film. Protonation and deprotonation of the oxide generates a surface charge (see Chapter 10) that can also create a gradient in voltage across the film. As the migrating species in metal corrosion are charged, the electric field created across the film can drive diffusional processes. Such *field-assisted diffusion* is also important for corrosion assisted by an externally applied voltage. The Cabrera–Mott expression for field-assisted diffusion is given by [9]

$$\frac{dx}{dt} = N\Omega v \cdot e^{W-qaE/kT} \tag{12.12}$$

where N is the number of potentially mobile ions, Ω is the oxide volume per mobile ion, ν is the vibrational frequency, W is the energy barrier to diffusion, q is the ionic charge, a is half the ion jump distance, and E is the electric field in volts per centimeter (V/x). The important point is that large barriers to ionic diffusion can be mitigated if a sufficiently high voltage is applied to the oxide film. Under cathodic polarization (when the metallic side of the oxide film has a negative charge), anionic species such as O^{2-} are driven toward the solution whereas cationic species such as Fe^{3+} are driven toward the metal. Under anodic polarization, the reverse is true; cations move outward whereas anions move toward the metal through the oxide film.

Corrosion kinetics in aqueous solutions are much more complicated than dry gas-phase corrosion and can vary by many orders of magnitude as a result of variations in pH and E_h. In aqueous solutions, the most common method used to measure the kinetics of metal corrosion involves monitoring corrosion currents as a function of applied voltage using electrochemical cells [1]. A corrosion current of 1 A/cm^2 indicates that a given corrosion reaction is generating approximately 10^{-5} moles of electrons/second/cm^2 of electrode surface. In the case of iron corrosion involving the generation of two electrons per iron atom dissolved, 1 A/cm^2 is equivalent to a catastrophic corrosion rate of $0.37\,\mu m/\text{second}$ or 1.3 mm/hour.

In examining corrosion rates in a pH 8.4 borate buffer solution (Fig. 12.4), several of the regimes of corrosion predicted by Pourbaix diagrams are apparent [11]. Under cathodic (i.e., reducing) polarization conditions, no corrosion occurs, and corrosion currents are negligible. When the protection afforded by cathodic polarization is removed, an active region is observed in which soluble hydrolysis products are formed and removed from the surface. Here, corrosion currents can exceed 1 A/cm^2. This corrosion is so rapid that it is almost controlled by the rate at which reaction products diffuse away from the surface into the adjacent liquid. However, as the applied potential is increased further to reach the anodic regime, the corrosion current drops by orders of magnitude and then levels out. The plateau value of the corrosion current of around $1\,\mu A/cm^2$ indicates the metal has been passivated with a thin oxide film. Now the corrosion rate is only 1 nm/hour.

Additional increases in applied voltage lead to increases in the corrosion current. There are several explanations for this increase, including O_2 evolution, voltage-assisted transport of species through the passive film, and a phenomenon

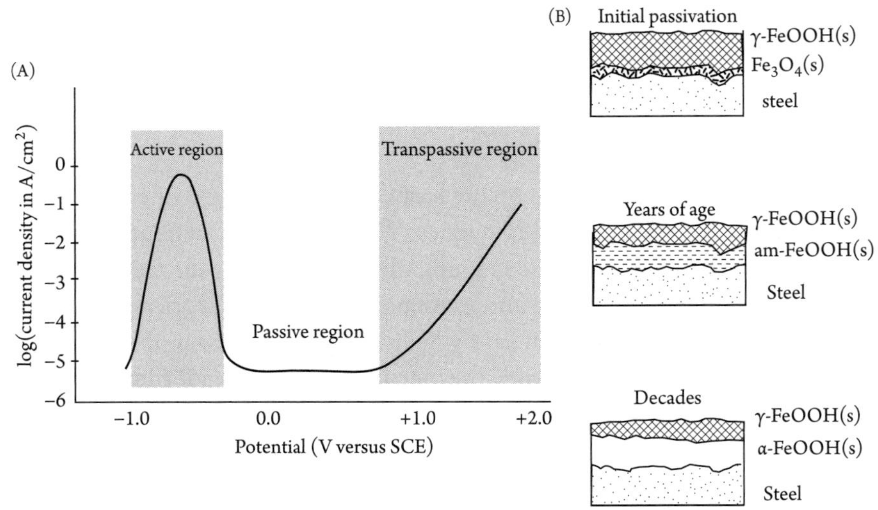

Figure 12.4 (A) The electrochemical current generated by the corrosion of metallic iron as a function of applied voltage in pH-buffered solution. The rate of corrosion is directly proportional to the current density (see text). Under cathodic polarization (far left), iron is a noble metal that does not corrode. However, when the protective voltage is removed, rapid corrosion and iron dissolution take place. However, after the voltage exceeds that required to form insoluble ferric oxides, the iron is passivated by an oxide overlayer. Finally, at the highest voltages, ionic species are driven through the oxide film, which once again promotes rapid corrosion rates. SCE = saturated calomel reference electrode, (Adapted from Yamashita, M.; Miyuki, H.; Matsuda, Y.; Nagano, H.; Misawa, T. The Long Term Growth of the Protective Rust Layer Formed on Weathering Steel by Atmospheric Corrosion During a Quarter of a Century *Corrosion Sci.* **1994**, *26*, 283–299, with permission). (B) Depictions of the corrosion layers that form on iron as a function of time in aqueous environments. (Adapted from Marcus, P.; Oudar, J., Eds. *Corrosion Mechanisms in Theory and Practice*; Marcel Dekker: New York, 1995.)

known as *pitting*. Pitting represents the primary form of corrosion for metals that are protected by conformal passive films. Pitting is clearly indicated by electron currents that deviate considerably from the standard corrosion condition [3]. First, above a voltage called the *pitting potential*, corrosion currents exhibit large, transient current spikes (Fig. 12.5A, B) and an irregular profile of current with time. Second, examination of the metal surface indicates these current spikes are associated with the formation and repassivation of physical pits that penetrate the passive oxide film [12] (Fig. 12.5C). Mechanisms for both uniform corrosion and pitting are described later in the chapter, but the results are the familiar pitted surfaces of metals, such as the turbine blade shown in Fig. 12.5C.

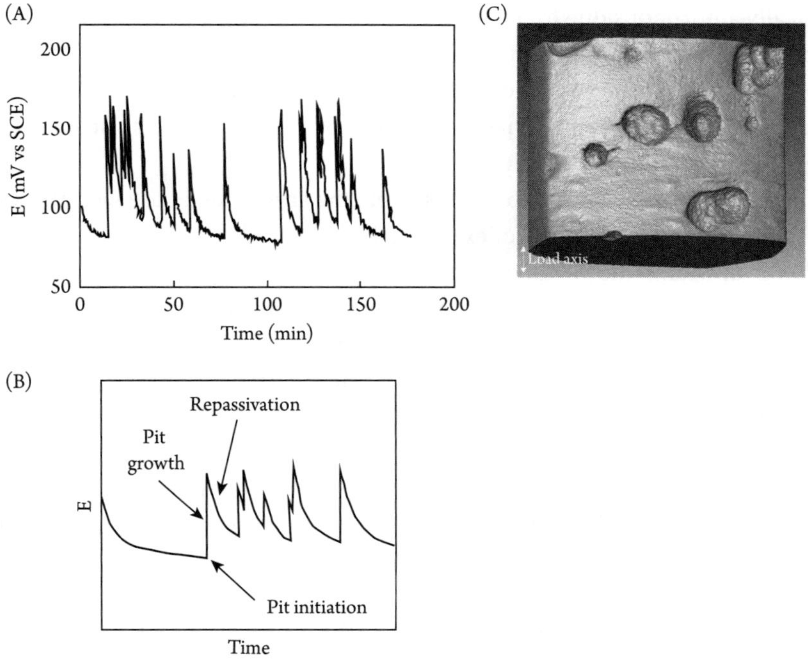

Figure 12.5 Characteristic features of metals that are undergoing corrosion via pitting. (A, B) Corrosion currents exhibit erratic spikes as localized regions of the passive oxide film are destroyed to expose fresh metal underneath. The freshly exposed metal is soon repassivated by a new oxide film, resulting in a drop in corrosion current. SCE = saturated calomel reference electrode, (Adapted with permission from Hashimoto, M.; Miyajima, S.; Murata, T. A Stochastic Analysis of Potential Fluctuation During Passive Film Breakdown and Repair on Iron *Corrosion Sci.* **1992**, 33, 885–904.) (C) A X-ray tomographic image of pits forming on a turbine blade corroded at 90°C in a chloride-containing solution (Adapted with permission from Horner, D. A.; Connolly, B. J.; Zhou, S.; Crocker, L.; Turnbull, A. Novel images of the evolution of stress corrosion cracks from corrosion pits. *Corrosion Sci.* **2011**, 53, 3466–3485.)

12.5 OXIDE PHASES IN METAL CORROSION

The first challenge associated with developing quantitative mechanisms for the various forms of metal corrosion in the passive regime involves identifying which oxide phases are present on the metal surface. Metals that are oxidized by O_2 in ambient dry air are typically covered with a thin conformal native-oxide film. For films on metals that oxidize to a single oxidation state, these films have a nominal composition identical to the standard oxide for that particular metal (e.g., Al_2O_3 covers Al°(s) whereas TiO_2 covers Ti°(s)). However, the films are typically amorphous due to the lattice mismatch between the oxide and the underlying metal. Techniques such as high-resolution transmission electron microscopy and X-ray-absorption fine structure (EXAFS) have been used to establish which crystalline phase comes closest to matching the structures of these amorphous films [13].

Typically, film structures do not resemble the crystalline phase with the highest thermodynamic stability. For example, the film on $Al^{\circ}(s)$ resembles γ-alumina rather than α-alumina [14]. For metals that can be oxidized to more than one oxidation state, the native oxide can contain multiple layers. The more reduced oxides are adjacent to the metal whereas more oxidized compounds are at the immediate oxide–air interface. For example, the oxide that forms on iron can have the composition of magnetite, Fe_3O_4, adjacent to the metal, and γ-Fe_2O_3 at the oxide–air interface [15]. Often there is not a sharp phase boundary separating these different oxide compositions, but a gradient between the reduced and oxidized end members. The nature of the corrosion products can be strongly influenced by trace metals. Finally, native oxides tend to be only a few nanometers thick. Thicknesses are limited because ion and electron transport through the native oxides tends to be extremely slow. In other words, in dry air, the thin oxide films on many metals are indeed passivating.

In contrast to the dry-air case, the phase assemblages associated with corrosion in water tend to be much more complex. For example, extensive studies have shown the rust that forms on iron and steel can contain almost all the major phases of the iron oxides, oxyhydroxides, and hydroxides depending on environmental conditions [16]. These phases are often crystalline and nonconformal. Sometimes the phases assume specific locations with respect to the metal interface. As in air oxidation, reduced oxides are found in direct contact with the metal, whereas more oxidized corrosion products are present at the oxide–water interface. In terms of hydration state, pure oxides are present at the immediate metal interface whereas more hydrated oxides (the oxyhydroxides and hydroxides) are expressed on the exterior of the assemblies of corrosion products. The latter trend is not only driven by the relative access of a given region to water, but also by the fact that hydroxyl groups in contact with the metal are subject to reduction by the metal to form H_2, just as is water. However, corrosion products can also contain random distributions of phases, as exemplified by the thick scale deposits that form on iron pipes exposed to hot water [17] (Fig. 12.6). As this example shows, corrosion products formed in water can be much more voluminous (microns rather than nanometers thick) than those produced in air, especially when they contain random distributions of hydrated phases. The corrosion films can even include bacteria, biofilms, and hosts of mineral solids.

Figure 12.6 A scanning electron micrograph of the mix of iron–oxide corrosion products that formed on a corroded water pipe. The large hexagonal crystals are "green rust," which is a mixed-valence, layered double-hydroxide mineral containing Fe(II) and Fe(III). The smaller particles are FeOOH phases such as lepidocrocite and goethite formed from the oxidation of green rust.
SOURCE: Reproduced with permission from Bingham, J. M.; Tuovinen, O. H. Mineralogical, Morphological, and Microbiological Characteristics of Tubercles in Cast-Iron Water Mains as Related to Their Chemical Activity. In *Planetary Ecology, 6th Int. Symp. Environ. Biogeochem.*; Brierley, J. A.; Brierly, C. L.; Eds.; Van Nostrand-Reinhold: New York, 1983, pp. 239–250.)

12.6 THE DEFECT CHEMISTRY OF OXIDES

What determines whether corrosion products form thin passive films or thick scale deposits? To truly passivate a metal, the overlying oxide film needs to prevent the transport of several key species, including electrons, protons, and metal cations and/or anions. When passive films are present, the rate-limiting step in the corrosion process involves the transport of one of these key ingredients. The metal cations and oxygen anions tend to be the hardest species to move, as both cations and anions are large and firmly attached to the oxide lattice by multiple chemical bonds. For many oxides, ionic transport through a perfect crystal is negligible even at elevated temperatures, which means that corrosion rates and mechanisms are often controlled by the mobility of *defects*. A perfect crystal is akin to a stalled freeway packed with cars during rush hour. These cars can only move after spaces (or defects) open up between cars to allow traffic to flow. Properties controlled by the motion of ionic defects include transport, sintering, and creep, whereas electronic defects determine properties such as electronic conductivity, thermoelectric power, and optoelectronics. As such, comprehensive books have been written on the topic of *defect chemistry* [17–21]. Here, we provide a brief introduction to defects in oxides, with an emphasis on how defects influence corrosion mechanisms.

In contrast to the open oxide structures that are effective materials for ion exchange and/or conduction (see Chapter 10), ionic conduction in dense oxide structures is often dictated by point defects involving discrete lattice sites. The primary point defects in oxides are cation vacancies (V_M), oxygen vacancies (V_O), cation interstitials (M_i), and oxygen interstitials (O_i), where the standard Kroger–Vink notations [21] are used to designate each defect type shown in parentheses. In this notation, V stands for a vacancy, M represents a cation, O is an oxygen atom, and i stands for an interstitial species. For example, O_O represents an oxygen anion on its normal lattice position. *Vacancies* represent ions that are missing from the lattice whereas *interstitials* involve ions that are forced into normally unoccupied voids (e.g., octahedral and tetrahedral spaces between oxygen anions) (Fig. 12.7). Most defects are charged. Interstitial defects tend to have charges corresponding to the parent ion charge (e.g., -2 for O_i and $+3$ for an Fe(III) interstitial). Vacancies have charges that are the opposite of the parent ion charge (i.e., the charge that neutralized the ion before its removal). In Fe_2O_3, oxygen vacancies have a charge of $+2$ whereas iron vacancies have a charge of -3 .

Point defects are not created or transported in a random fashion; they obey a set of rules that follow the general principles of chemical equilibria. First, defect creation must preserve charge neutrality. In stoichiometric oxides, achieving neutrality often involves the pairing of point defects. For bivalent metal oxides, the most common defect pairs involve either Schottky disorder (when the concentrations of cation and anion vacancies are equal) or Frenkel disorder (when the concentrations of interstitial ions and their corresponding vacancies are equal) (Fig. 12.7). Second, defect populations are normally described by equilibrium constants that indicate how the defects were created from the bulk crystal. For example, in a stoichiometric oxide, the creation of a Frenkel defect implies an oxygen ion is moved from its normal lattice position into an interstitial position:

$$O_O \rightleftarrows V_O^{2+} + O_i^{2-} \quad \textbf{(12.13)}$$

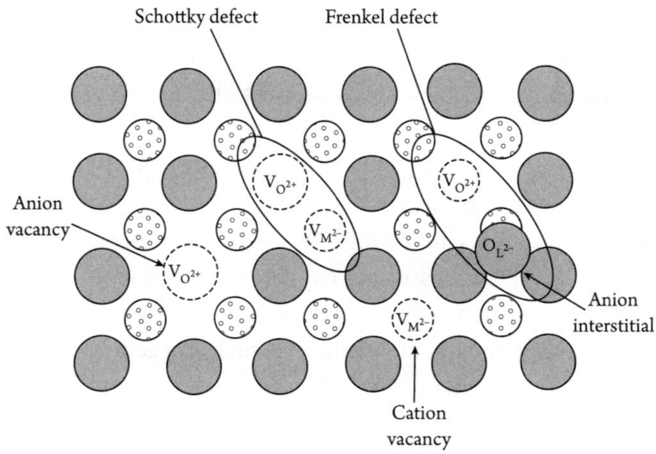

Figure 12.7 Representations of common defects in ideal metal oxide structures, where O_O and M_m are oxygen anions and metal cations on their normal lattice sites, respectively; and V_O and V_M are oxygen and metal vacancies, respectively. Frenkel and Schottkey defects are encompassed by dashed ovals.

$$K_F = \frac{\left[V_O^{2+}\right]\left[O_i^{2-}\right]}{\left[O_O\right]} \approx \left[V_O^{2+}\right]\left[O_i^{2-}\right] \quad (\text{at low defect concentrations } \left[O_O\right] \rightarrow 1) \tag{12.14}$$

$$\left[V_O^{2+}\right] = \left[O_i^{2-}\right] = K_F^{1/2} \tag{12.15}$$

Oxides are not always stoichiometric, but can contain either a deficit or an excess of oxygen, which is equivalent to an excess or deficit of metal cations, respectively. Deviations from stoichiometry are typically controlled by the partial pressure of oxygen. In reducing atmospheres (i.e., low oxygen pressures, or P_{O_2}), oxygen-deficient oxides such as TiO_{2-x} can be produced, whereas at high oxygen partial pressures, metal-deficient oxides such as $Ni_{1-x}O$ can form. An example of the equilibrium constant for creating oxygen vacancies in an oxygen-deficient oxide is

$$O_O \rightleftarrows V_O^{2+} + 2e^- + 1/2O_2 \tag{12.16}$$

$$K_{VO} = \frac{\left[V_O^{2+}\right]\left[e^-\right]^2 P_{O2}^{1/2}}{\left[O_O\right]} \tag{12.17}$$

In extreme cases, losses or gains of oxygen from the lattice can trigger phase transformations:

$$Nb_2O_5(s) \rightleftarrows 2NbO_2(s) + 1/2O_2(g) \tag{12.18}$$

If Nb_2O_5 is considered to be an oxygen-deficient oxide, Nb_2O_{5-x}, the "new phase" NbO_2 corresponds to x = 1, or 20% of the oxygen anions present in the fully oxidized parent Nb_2O_5. In many instances, the electrons generated via vacancy creation reside on metal cations, resulting in cation reduction (see Chapter 11). In this example, all Nb(V) is converted to Nb(IV) via the removal of oxygen anions from the lattice.

Comprehensive diagrams indicating how the concentrations of all defects vary as a function of the partial pressure of oxygen are available for many oxide systems [20]. These diagrams are the defect chemistry analogues of hydrolysis diagrams. A generic diagram for an oxide in which oxygen Frenkel defects are predominant is shown in Figure 12.8. Although it is beyond the scope of this book to derive such diagrams, they serve to make a few key points regarding oxide defect chemistry. First, multiple defect types often coexist in oxides. Second, the concentration of each defect is governed by multiple defect reactions and equilibrium expressions. Third, the predominant defects in a given oxide depend strongly on whether the environment experienced by the oxide is oxidizing or reducing (i.e., has a high or low P_{O_2}, respectively). For the example shown, oxygen vacancies and electrons are the dominant defects present at low P_{O_2}, whereas oxygen interstitials and holes are predominant at high P_{O_2}. This means that a given oxide can be one of three things: an oxygen-poor n-type semiconductor, a stoichiometric insulator, or an oxygen-rich p-type semiconductor, depending on the oxygen partial pressure and the values of the equilibrium constants for various defect formation reactions. The activation energies required to form defects are typically large. There is a strong correlation between the enthalpy of defect formation and the enthalpy for atomization of the host oxide [22] (Fig. 12.9). Defect concentrations tend to be low at room temperature but can increase to high values at elevated temperatures.

In metal corrosion, the two most important parameters associated with point defects are (1) how many defects are present and (2) how mobile the defects are in the absence and presence of an applied field. Defect concentrations and mobilities have been measured for single crystals of most oxides as a function of both temperature and oxygen partial pressure [20]. Generic depictions of defect motion in response to either inherent or applied fields across the oxide are shown in Figure 12.10. Under anodic polarization, the oxide–solution interface is positive and the oxide–metal interface is negative. The resulting electric field drives cationic species such as h^+, H^+, Fe^{3+}, and V_O^{++} through the oxide from the metal toward the solution. Anionic species such as e^-, OH^-, O^{2-}, and $V_{Fe}{}^{3-}$ are driven from the solution toward the metal. Note that, although vacancies can facilitate transport, the vacancies move in the opposite direction of the cations or anions to which the vacancies correspond. Provided that vacancy transport limits the rate, cation vacancies are eliminated at the metal–oxide interface by the metal

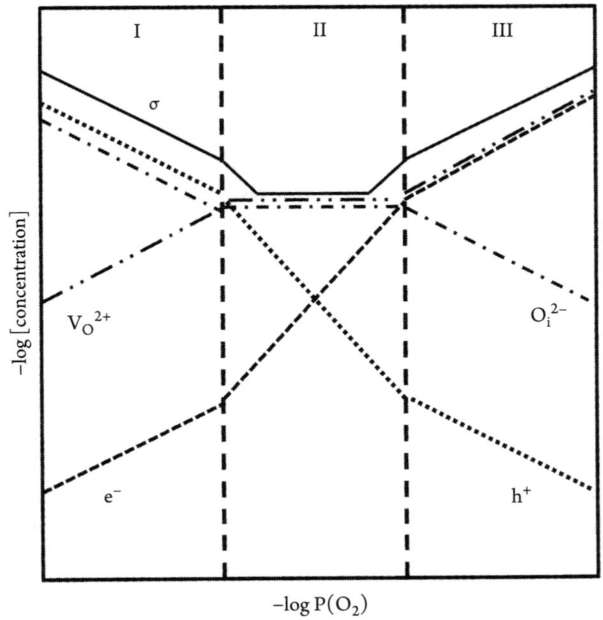

Figure 12.8 An idealized representation of concentration of the dominant defect types found in a pure oxide characterized by the Frenkel disorder as a function of the partial pressure of oxygen (pO_2). The logarithms of concentrations are plotted on the ordinate. Each line in the diagram represents a different species that contributes to the net conductivity of the oxide (solid line). The vertical dashed lines indicate those regimes in which conductivity is dominated by different species. Under oxidizing conditions (i.e., at high E_h or pO_2 in Region I), extra oxygen is stuffed into the lattice, resulting in the creation of oxygen interstitials that are charge-compensated by holes. These holes dominate the conduction in Region I. Under reducing conditions (Region III), oxygen is removed from the lattice to create oxygen vacancies with a charge that is compensated by electrons. In Region III, conductivity is dominated by electrons. Under intermediate conditions (Region II), Frenkel defects are dominant, and there are equal concentrations of oxygen vacancies and oxygen interstitials. In the center of this regime, oxygen ion conductivity (i.e., oxygen hopping involving vacancies and/or interstitials) dominates the net conductivity. Adapted from Tuller, H. L. Mixed Conduction in Nonstoichiometric Oxides. In *Nonstoichiometric Oxides*; Sorensen, O. T., Ed.; Academic Press: New York, 1981; pp 271–335, with permission.

cations created via the oxidation of the metal. Similarly, oxygen vacancies are eliminated at the oxide–solution interface when they react with and are filled by water molecules or hydroxide ions.

Metal corrosion and other electrochemical phenomena are driven by the relative rates at which all charged species move. For example, if ion motion is much

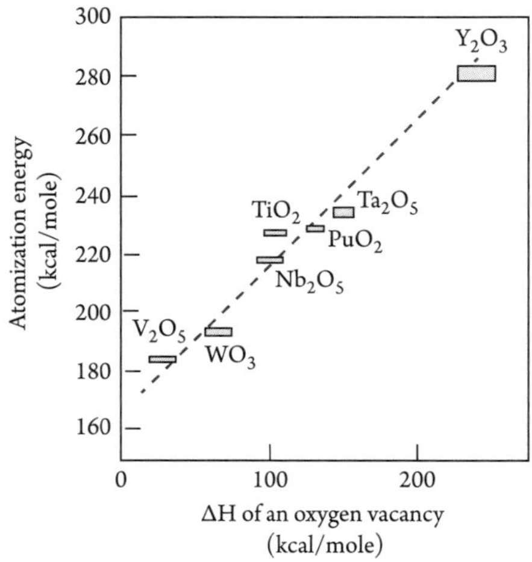

Figure 12.9 A plot of the correlation between the free energy of atomization of an oxide and the ease of oxygen vacancy formation. Note that the activation enthalpies for defect formation are large. Adapted with permission from Kofsted, P. Oxidation of Metals. **1995**, *44*, 3–27.

slower than electron (or hole) motion, voltages applied to oxide-passivated metals can result in the oxidation or reduction of water (generating $O_2(g)$ or $H_2(g)$, respectively) without inducing any apparent changes in the thickness of the oxide film. If the metal cations in the oxide diffuse faster than the oxygen anions, corrosion products will grow at the oxide–solution interface, although if oxygen-anion diffusion is faster, oxide films will grow at the oxide–metal interface. As a final example, it has been postulated that pitting occurs under anodic polarization when cation vacancies are driven to the oxide–metal interface at a more rapid rate than the vacancies can be filled. In this scenario, the vacancies accumulate to form voids that eventually undermine the overlying oxide and lead to the formation of pits. However, as shown in the next sections, this mechanism does not always account for pit formation.

12.7 CORROSION KINETICS VERSUS DEFECT MIGRATION RATES

Now let us examine quantitative data regarding defect concentrations and mobility in single crystals. As one might expect based on the high enthalpies for defect formation, defect concentrations and mobilities both tend to be too low to measure except at elevated temperatures ($T > 1000°C$). Estimates for these parameters below 100°C (i.e., when liquid water is present) are typically arrived at via extrapolations from high temperature results. At one extreme are bivalent metal oxides such as MgO, where oxygen vacancy concentrations are too small to

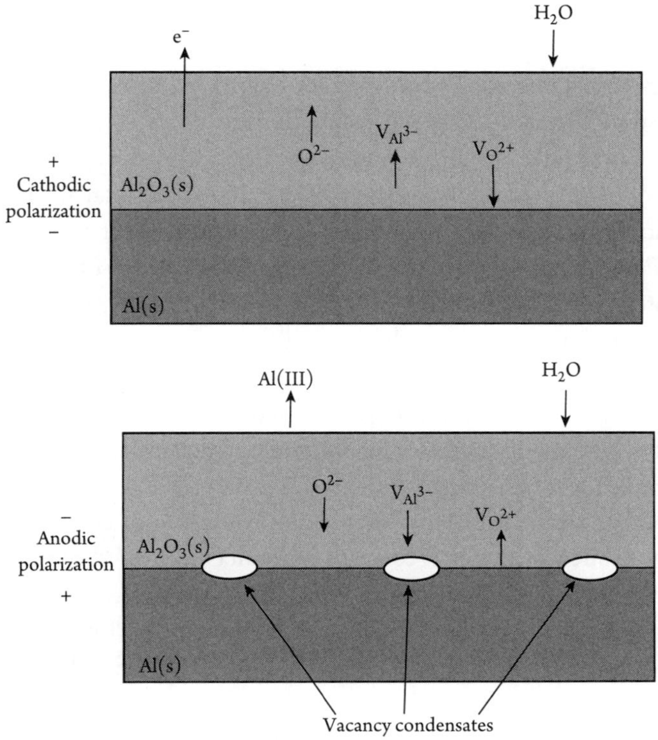

Figure 12.10 A schematic diagram for defect migration and the mechanism for pitting in metals such as Al. In each diagram, the passivating oxide film is shown on top of the underlying metal. Pitting occurs under anodic polarization. Here, positive species such as Al(III) and oxygen vacancies are driven to the surface, where they can react with water. Negative species such as cation vacancies and O^{2-} are driven to the metal–oxide interface. It has been postulated that pitting occurs when cation vacancies at the metal–oxide interface coalesce into voids that undermine the oxide and lead to localized spallation. Adapted with permission from Bunker, B.C.; Nelson, G.C.; Zavadil, K.R.; Barbour, J.C.; Wall, F.D.; Sullivan, J.P.; Windisch, C.F.; Engelhardt, M.H.; and Baer, D.R. Hydration of Passive Oxide Films on Aluminum. *J. Phys. Chem. B* **2002**, *106*, 4705–4713.

measure even at 1500°C. The self-diffusion coefficient for oxygen anions in MgO is $10^{-12}\ cm^2/second$ at 1800°C [23]. Extrapolation of these results to room temperature yields a ridiculously low value of $10^{-54}\ cm^2\ /\ second$. For TiO_{2-x}, it is estimated that x is on the order of 10^{-5} for $T = 1100°C$ in air. These higher vacancy concentrations result in room-temperature oxygen diffusion coefficients for pure, single-crystal rutile on the order of $10^{-35}\ cm^2\ /\ second$. Finally, the iron diffusion coefficient in single crystals of magnetite is estimated to be on the order of $10^{-40}\ cm^2\ /\ second$ at room temperature, which is many orders of magnitude slower than the electron hopping that gives rise to the high electrical conductivity of magnetite.

We are now in a position to compare corrosion rates directly with the rates of ionic transport through oxides corresponding to those of a given metal. For Fe_3O_4, which is known to form continuous films adjacent to Fe°(s) metal, the fastest moving cation in single-crystal Fe_3O_4 at 300°C has a calculated self-diffusion coefficient of around $10^{-22}\,cm^2\,/\,second$ [24]. However, for the Fe_3O_4 that forms on Fe°(s) in air, we have already seen that the kinetics of oxide formation are consistent with a diffusion coefficient of $2\cdot10^{-16}\,cm^2\,/\,second$, which is more than 1 million times faster than what is measured for single-crystal Fe_3O_4. This large discrepancy is not an anomaly for Fe_3O_4, but is a universal observation for all passive films on metals. Whether in water or in air, metal-corrosion rates are almost always many orders of magnitude faster than rates predicted on the basis of simple ionic transport.

Several factors can contribute to enhanced diffusion in corrosion products. First, we have already seen that corrosion products are not ideal single crystals. Conformal passive films are invariably strained and are often amorphous. Such factors can generate higher defect concentrations and thus lower the activation energies for ion transport. Similar factors explain why ionic diffusion rates in single crystals are often lower than rates observed in polycrystalline materials (Fig. 12.11). However, because defect concentrations in thin (1–10-nm-thick) passive films are almost impossible to measure, it is difficult to quantify the extent to which defect populations and transport are enhanced in passive films relative to single crystals.

Secondly, enhancing corrosion rates involves impurities that modify ionic transport in the oxide host. The role of impurities has been studied extensively, particularly in regard to the deliberate addition of dopants to enhance both the ionic and electronic conductivity of oxides. Dopants are used routinely to transform stoichiometric insulating oxides into either n- or p-doped semiconductors. Dopants can have a dramatic effect on ionic conductivity by greatly enhancing defect concentrations, as illustrated by the addition of Y_2O_3 to ZrO_2:

$$Y_2O_3(s) \rightleftarrows 2Y_{Zr}^{-} + V_O^{2+} + 3O_O \qquad \textbf{(12.19)}$$

Eq. 12.19 illustrates that the dissolution of Y_2O_3 into ZrO_2 results in the creation of anionic sites (resulting from the replacement of Zr(IV) by Y(III)) compensated by cationic oxygen vacancies. In this specific case, one oxygen vacancy is created for every two Y(III) cations substituted into the ZrO_2 lattice. Here, vacancy concentrations are no longer controlled by oxygen partial pressure, but by dopant concentrations, which can be quite large. In this example,

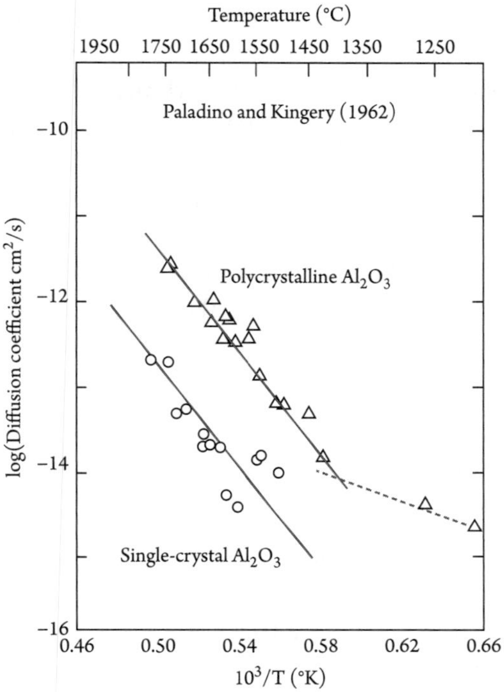

Figure 12.11 Measured rates of oxygen diffusion for in both single-crystal and polycrystalline α-Al_2O_3 (corundum) as a function of temperature. Diffusion is barely measurable at high temperatures and becomes exceedingly slow at room temperature. Adapted with permission from Paladina, A. E.; Kingery, W. D. Aluminum-Ion Diffusion in Aluminum Oxide. *J. Chem. Phys.* **1962**, *37*, 957.

the addition of 1 wt% Y_2O_3 to ZrO_2 enhances the ionic conductivity at 900°C by more than a factor of 1000 [25].

A third factor that can enhance corrosion rates is applicable to oxides containing redox-active cations that can access more than one oxidation state. The classic examples of such oxides are the iron oxide solids. Mixed-valence oxides can be exceptional electron conductors (see Chapter 11). Such electron conduction provides a facile mechanism for interconverting Fe(III) and Fe(II). Ligand exchange rates around Fe(II) centers are invariably many orders of magnitude faster than those around Fe(III). Within the bulk oxide, electron transfers between Fe(II) and Fe(III) have been shown to activate both cation and oxygen anion transport. Thus, metal ion transport is surprisingly rapid in a mineral such as magnetite (Fe_3O_4), in which the structure can accommodate both reduced and oxidized metal cations [26]. Magnetite equilibrates isotopically with aqueous iron species at extremely rapid rates [27]. At the oxide–solution interface, any species capable of reducing Fe(III) to Fe(II) can create surface defects. Facile electron hopping allows these surface defects to migrate rapidly into the bulk via a conveyer-belt mechanism that can fill the oxide quickly with high concentrations of bulk defects. This behavior is best documented in ferric oxide solids in which surface adsorption of soluble Fe(II) species creates Fe(II) centers that migrate through the bulk lattice [28]. Internal redox reactions coupled with long-range electron transport have consequences for corrosion that go beyond the facilitation of defect transport. Depending on local solution conditions, iron oxide crystals can be transformed by dissolving surfaces rich in Fe(II) while depositing on other surfaces as insoluble Fe(III) oxides [29, 30]. A consequence of this long-range redox behavior can cause pitting on one surface while smoothing another surface on the same crystal.

12.8 THE ROLE OF HYDRATION IN CORROSION RATES AND MECHANISMS

Although the role of dopants on the ionic conductivity of corrosion products can be critical for alloys, pure metals do not contain large impurity concentrations. In pure metals, some other factor is needed to account for the observed high corrosion rates. That factor involves a cationic impurity that is common to all oxides and that moves quickly: the proton. Protons or, more correctly, hydroxide ions, are the dominant impurities that lower the stability of most metals in water relative to those in air. Although few hydroxyl groups may be present in the bulk of a pure single crystal, we have already seen (Chapter 6) that oxide surfaces are always susceptible to some degree of hydrolysis in water and humid air. Because the native oxides on most metals are only a few nanometers thick, even limited hydrolysis can create hydroxyl group concentrations that represent the dominant defects present. Among other things, each hydroxyl group represents a broken metal–oxygen bond, facilitating the transport of both the cations and the anions in the lattice. To illustrate the effects of hydroxyl groups on corrosion rates, consider the example of aluminum, which is much less complex than that of iron in terms of both redox chemistry and corrosion products. Here, we examine several facets of $Al^\circ(s)$ degradation in water, including corrosion, pitting, and the anodization of $Al^\circ(s)$ to create ordered nanoporous materials.

Most people have direct experience with aluminum foil. In air, aluminum foil retains its metallic luster even after being put into a hot oven because of the presence of a protective alumina film. However, if that same foil is placed in hot water (particularly saltwater), the aluminum soon becomes brittle and tarnished (Fig. 12.12). Classic studies [31–32] indicate that aluminum corrosion in water is not a continuous process, but proceeds in three distinct stages (Fig. 12.13). First, there is an induction period during which there is no apparent change in the native thermal oxide on the aluminum surface. After this induction period,

Figure 12.12 Photographs of commercial aluminum foil after 1 hour in air at 100°C compared with the same foil after 1 hour in boiling salt water ($[NaCl] = 10^{-2}M$). Note that although the sample heated in air still looks pristine, the sample heated in water is extensively tarnished.
SOURCE: Photographs by Colla, C. A.; UC Davis, with permission.

there is a period of rapid film growth during which the film thickness increases from around 5 nm to more than 100 nm. Finally, the stage 2 film becomes so thick that corrosion rates begin to decrease with time (stage 3). These results have been rationalized by postulating that stage 1 corresponds to hydration of the native oxide, with the induction time corresponding to the time required for a

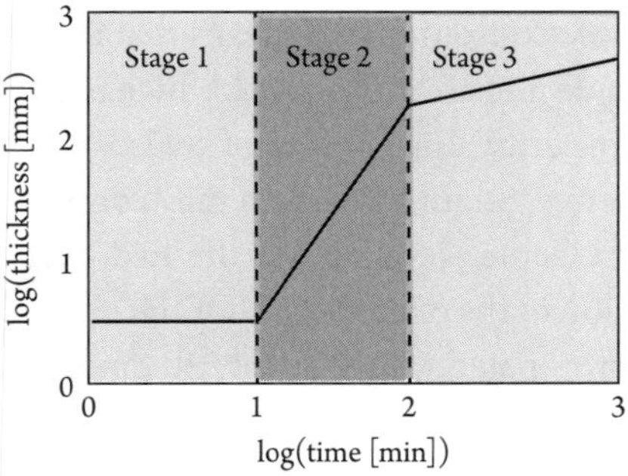

Figure 12.13 The thicknesses of corrosion products (in nanometers) on aluminum metal exposed to 50°C water as a function of time showing three distinct stages to the corrosion process. During an initial induction period (stage 1), the native oxide does not appear to grow at all. However, after the native oxide is hydrated, rapid corrosion can occur (stage 2). Corrosion rates eventually slow during stage 3, because the hydrated oxide film on the surface represents a diffusion barrier, although it is a much poorer diffusion barrier than the initial native oxide. Adapted with permission from Alwitt, R. S. The Growth of Hydrous Oxide Films on Aluminum. *J. Electrochem. Soc.* **1974**, *121*, 1322–1328.

hydrolysis front to diffuse from the oxide–solution boundary to the oxide–metal interface. Because the hydrated phase promotes much more rapid diffusion, rapid corrosion can occur after the native oxide has been breached (stage 2). However, even the hydrated oxide can function as a diffusion barrier after it becomes sufficiently thick, leading to stage 3 behavior.

This sequence of events is observed only if the oxide films remain as intact diffusion barriers. The normal sequence of weathering can be disrupted spectacularly by rubbing a small amount of liquid mercury onto a freshly polished piece of aluminum. The mercury disrupts the oxide film, allowing both water and oxygen from the atmosphere to convert Al°(s) metal to the mineral boehmite, AlOOH(s), at a rapid rate. Over a period of minutes, a pellet of Al°(s) can be converted completely into a long, worm-like structure consisting of boehmite powder.

Details regarding the hydration state and ionic mobility in the oxide films on aluminum resulting from aqueous corrosion have been investigated in more detail using SIMS in conjunction with isotopic labeling [33]. SIMS is one of the few techniques with sufficient depth resolution to probe the composition of oxide films that are only a few nanometers thick. SIMS can detect species such as hydroxide ions at concentrations less than 100 ppm. Finally, with isotopic labeling (i.e., using D and ^{18}O), it is possible to conduct isotopic exchange studies that can help identify whether mobile species are H^+, OH^-, O^{2-}, or H_2O. Here, we highlight some of the key finding that have come out of such SIMS investigations.

First, SIMS results identify several key features regarding the native passive film. The passive oxide on aluminum ranges from around 4 nm thick on single crystals to 6 nm on polycrystalline samples (Fig. 12.14). More important, the SIMS results reveal the hydroxide content is high, ranging from 5 to $10 \cdot 10^{20}\,\text{OH}/\text{cm}^3$. These concentrations means that 0.8% to 1.6% of the oxygen anions in the native oxide film are protonated, which is a much higher concentration than any other impurity. The significant hydroxyl concentration indicates the film is highly defective. Strain in the film may contribute to this high degree of hydroxylation (see Chapter 16). To put the hydroxyl concentration in perspective, total conversion of the oxide into AlOOH(s) would result in 50% of the oxygen anions being protonated, which is only 30 times greater than the concentration seen in the native oxide on Al°(s).

The SIMS results indicate how film thickness and hydroxyl content change when the Al°(s) metal is placed in water in the absence of an applied potential (Fig. 12.14). In the first stage of the reaction, the hydroxyl content of the film increases but the film thickness is largely unchanged. The highest hydroxyl concentration that has been seen as a result of uniform corrosion is around 15%. This value is close to the 12.5% expected for totally hydrated γ–alumina (HAl_5O_8) [34]. Eventually, the hydroxyl concentration levels off and the film begins to thicken. Changes in film thickness with time have been used to

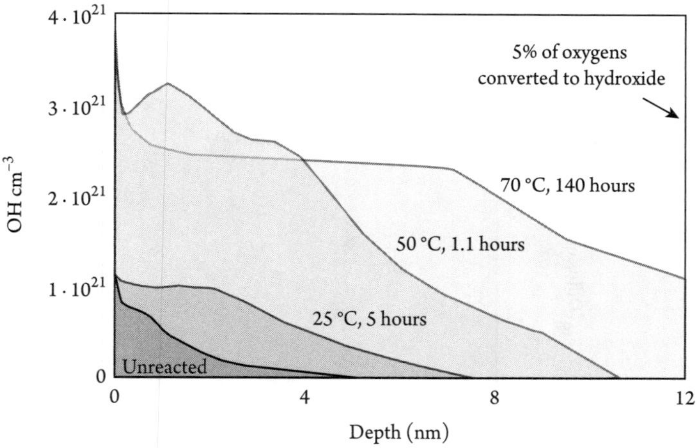

Figure 12.14 Hydroxide profiles (mass 17 + mass 18) for an aluminum film deposited on Si immersed in D_2O for the times and temperatures indicated. The native oxide is 2 to 3 nm thick. The arrow denotes the hydroxide concentration at which 5% of the oxygen ions in the film are protonated. Note that, at first, the hydroxide content increases without resulting in film growth. Next, both the hydroxide content and film thickness increase with time. Eventually, the hydroxide content levels off, although the film continues to thicken.

SOURCE: Reproduced with permission from Bunker, B.C.; Nelson, G.C.; Zavadil, K.R.; Barbour, J.C.; Wall, F.D.; Sullivan, J.P.; Windisch, C.F.; Engelhardt, M.H.; and Baer, D.R. Hydration of Passive Oxide Films on Aluminum. *J. Phys. Chem. B*, **2002**, *106*, 4705–4713.

estimate effective diffusion coefficients for whatever species control the rate of film growth. Different samples (single crystals, polycrystalline Al°(s), and defective films grown in $^{18}O_2$) yield diffusion coefficients ranging from 10^{-19} to $10^{-17}\,cm^2/second$. The magnitude of the diffusion coefficient appears to be dictated by the hydroxyl content of the film (Fig. 12.15). Higher hydroxyl concentrations yield faster diffusion and film growth. This diffusion is many orders of magnitude than diffusion measured as a result of defect migration in single-crystal sapphire (α-Al_2O_3) samples (see Fig. 12.11).

SIMS profiles obtained in conjunction with isotopic labeling reveal that film activity is tied to the mobility of the oxygen sublattice. The oxide film on aluminum is neither passive nor inert. More than 50% of the oxygen in an ^{18}O-labeled film is replaced by ^{16}O after only 2 weeks of storage in ambient air. Immersion of this sample in D_2O for 25.5 hours results in a decrease in the ^{18}O-to-^{16}O ratio in the film surface from 1.1 down to 0.1 (Fig. 12.16A). The results obtained for various samples immersed in both D_2O and H_2O are consistent with a hydrolytic exchange process with rate constants ranging from $2\cdot10^{-4}$ to $3\cdot10^{-5}$ / second. The more rapid exchange rates are only $\sim 10^4$ times slower than the rate of water

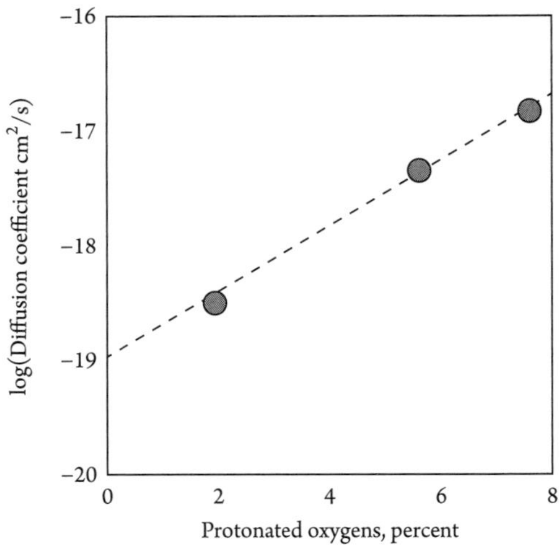

Figure 12.15 The correlation between the effective diffusion coefficient for film growth (based on film thickness measured via secondary-ion mass spectroscopy) and the measured hydroxyl content of the oxide film. Note that corrosion rates increase with the hydroxyl content.
SOURCE: Reproduced with permission from Bunker, B.C.; Nelson, G.C.; Zavadil, K.R.; Barbour, J.C.; Wall, F.D.; Sullivan, J.P.; Windisch, C.F.; Engelhardt, M.H.; and Baer, D.R. Hydration of Passive Oxide Films on Aluminum. *J. Phys. Chem. B*, **2002**, *106*, 4705–4713.

exchange around the $\left[Al(OH_2)_6\right]^{3+}$ (aq) cation (see Chapter 4) of 1.2/second at ambient conditions.

It is an oversimplification to view the thermal oxide as an unchanging structure through which low concentrations of aluminum and oxygen vacancies migrate. Instead, it is clear that reversible hydrolysis and condensation reactions are constantly breaking and reforming bonds within the surface film, enhanced by protons. Almost all bonds present have been broken within less than 1 day. These reversible hydrolysis reactions facilitate the transport of hydroxyl groups through the film. Further evidence for facilitated oxygen transport is obtained by examining films during the film-growth stage of corrosion (Fig. 12.16B). After normalizing the signals for ^{18}O and ^{16}O to be equal at the surface, it is clear that the original ^{18}O-labeled material retains its original thickness, whereas an ^{16}O layer appears and grows underneath the native oxide at the metal–oxide interface. This result clearly shows the oxygen anions are more mobile than the aluminum cations (see Section 10.5).

Finally, the SIMS results reveal what happens to the passive film under both cathodic polarization and anodic polarization above the pitting potential. For

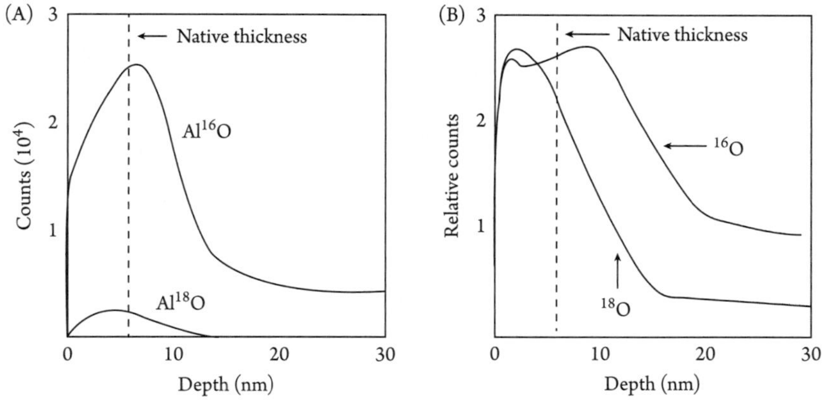

Figure 12.16 Oxygen-isotope exchange experiments between a ^{18}O-labeled native oxide on aluminum and D_2O as determined via secondary-ion mass spectroscopy. (A) The relative amounts of ^{18}O and ^{16}O present after the native oxide is immersed in water for 170 minutes at room temperature shows that, although the oxide has not thickened appreciably, more than 90% of the original oxygen present in the oxide has been replaced by oxygen from the water. (B) Profiles for ^{18}O and ^{16}O normalized to yield the same surface concentration in an aluminum sample immersed in water for 25.5 hours. (Once again, the absolute concentration of ^{18}O is actually around 10% of the ^{16}O concentration at the immediate oxide–solution interface.) These results show that the original ^{18}O-labeled oxide is essentially unchanged in thickness and is immediately adjacent to the solution. However, the net film has almost doubled in thickness. New growth is represented by an ^{16}O-rich layer that forms underneath the ^{18}O-rich surface at the oxide–metal interface. This result shows that oxygen-ion diffusion (most probably involving hydroxyl groups) dominates transport, leading to film growth (see Fig. 12.10).
SOURCE: Reproduced with permission from Bunker, B.C.; Nelson, G.C.; Zavadil, K.R.; Barbour, J.C.; Wall, F.D.; Sullivan, J.P.; Windisch, C.F.; Engelhardt, M.H.; and Baer, D.R. Hydration of Passive Oxide Films on Aluminum. *J. Phys. Chem. B*, **2002**, *106*, 4705–4713.

cathodic polarization, the native oxide shrinks rather than grows. By comparing the rates of film shrinkage with the electrical currents passed through the film, it appears that, under the particular voltage applied (−1.4 V), the electron-rich environment on the metal results in the electrochemical reduction of both the film (converting Al(III) to Al°(s)) and the adjacent water (where $H_2(g)$ evolution is observed).

For anodic voltages above the pitting potential (Fig. 12.17), corrosion currents (note the spikes in current associated with pitting) and SIMS profiles reveal that rapid corrosion occurs ($0.1-0.2\ \mu m$ in only 50 minutes). The fraction of oxygen anions converted into hydroxyl groups in these corrosion products is the highest seen, ranging from 50% to 75%. Although amorphous to X-ray diffraction, these

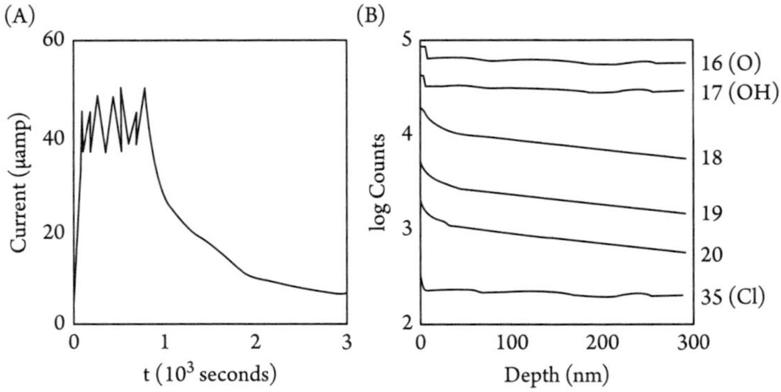

Figure 12.17 Passive film behavior within the anodic-pitting regime. (A) Measured currents and current spikes associated with pitting at a potential of –0.75 V versus a Pt pseudo-reference electrode. (B) Secondary-ion mass spectroscopy depth profiles for different ion masses (shown as numbers on the right axis) after such anodic polarization in water for 3000 seconds. The thickness of the corrosion product layer exceeds the maximum depth profiled (300 nm). Measured hydroxyl concentrations range from 50% to 75% of the oxygen ions present, with lower anodic potentials resulting in higher hydroxyl concentrations. Note that boehmite (AlOOH) has a hydroxyl content of 50%.
SOURCE: Reproduced with permission from Bunker, B.C.; Nelson, G.C.; Zavadil, K.R.; Barbour, J.C.; Wall, F.D.; Sullivan, J.P.; Windisch, C.F.; Engelhardt, M.H.; and Baer, D.R. Hydration of Passive Oxide Films on Aluminum. *J. Phys. Chem. B*, **2002**, *106*, 4705–4713.

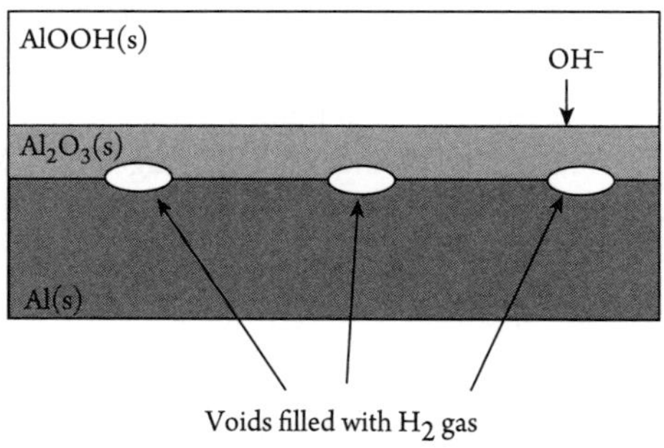

Figure 12.18 A simplified depiction of the mechanism for the pitting of Al(s) as indicated from secondary-ion mass spectroscopy depth profiles. It appears the mobile species leading to corrosion is the hydroxide ion, which is driven from the oxide–solution interface down to the oxide–metal interface. Hydroxyl groups that reach the oxide–metal interface are reduced to generate H_2 gas. Pressures generated by the resulting hydrogen bubbles at the interface eventually lead to spallation and pitting of the overlying oxide film.
SOURCE: Reproduced with permission from Bunker, B.C.; Nelson, G.C.; Zavadil, K.R.; Barbour, J.C.; Wall, F.D.; Sullivan, J.P.; Windisch, C.F.; Engelhardt, M.H.; and Baer, D.R. Hydration of Passive Oxide Films on Aluminum. *J. Phys. Chem. B*, **2002**, *106*, 4705–4713.

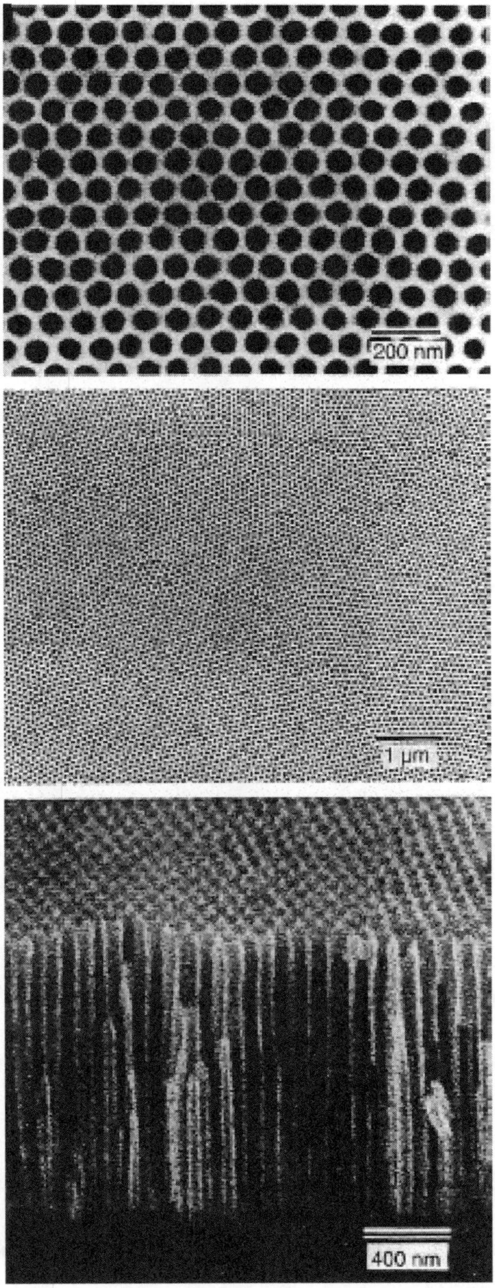

Figure 12.19 Scanning electron micrographs of structures involving anodized aluminum. Top views and a cross-section of an anodized aluminum film that was replicated by Pt metal, showing the close-packed array of pores that penetrate the oxide. (SOURCE: Reproduced with permission from Masuda, H.; Fukuda, K. Ordered Metal Nanohole Arrays Made by a Two-Step Replication of Honeycomb Structures of Anodic Alumina; Science, 1995, 268, 1466–1468.)

products probably represent a mixture of AlOOH(s) and $Al(OH)_3$(s) that is spewed forth as a result of pitting. However, SIMS profiles obtained in the absence of an applied voltage suggest this pitting is not the result of the voltage-driven aggregation of aluminum vacancies to create voids at the metal–oxide interface. The SIMS results show that OH^- is much more mobile than $V_{Al^{3-}}$ in hydrated passive oxide films. Anodic polarization will drive hydroxide ions to the oxide–metal interface. However, AlOOH(s) in contact with Al°(s) is reduced via the reaction

$$2Al(s) + 6AlOOH(s) \rightleftarrows 4Al_2O_3(s) + 3H_2(g) \qquad \textbf{(12.20)}$$

Instead of having vacancies condense to form voids, we now have bubbles of H_2 gas forming at the oxide–metal interface (Fig. 12.18). It is thought that these bubbles eventually generate sufficient pressure to trigger H_2(g) eruptions that burst through the film to produce pits and expose the underlying metal to aqueous attack.

In closing this chapter, we should point out that metal corrosion is not always a destructive phenomenon. In fact, metal corrosion can generate oxide materials that cannot be produced in any other way. A case in point involves the mesoporous materials that fall under the heading of anodized aluminum [35–36]. If aluminum metal is oxidized using just the right buffer solutions and just the right anodic potentials, pitting can be controlled to create specific oxide nanostructures. In fact, it is possible to generate oxide films that contain hexagonal close-packed arrays of nanopores (Fig. 12.19). By controlling the potential and the oxidation time, it is possible to control the film thickness, pore diameter, pore-wall thickness, and pore density in such films. Anodized aluminum has been a critical material for many technologies based on nanotechnology, including membranes for water treatment and purification, and substrates for the electroactive oxides used in devices such as ultracapacitors (see Chapter 11).

References

1. Jones, D. A. *Principles and Prevention of Corrosion*; Prentice Hall: New York, 1995.
2. McCafferty, E. *Introduction to Corrosion Science*; Springer Science + Business Media; 2010.
3. Marcus, P.; Oudar, J., Eds. *Corrosion Mechanisms in Theory and Practice*; Marcel Dekker: New York, 1995.
4. Cramer, S. D.; Covino, B. S., Jr., Eds. Galvanic Series of Metals and Alloys in Seawater; ASM International: Materials Park, OH, 2005; Vol. 13B, ASM Handbook.
5. Henrich, V. E.; Cox, P. A. *The Surface Science of Metal Oxides*; Cambridge University Press: New York, 1996.
6. Henderson, M. A. The Interaction of Water with Solid Surfaces: Fundamental Aspects Revisited. *Surf. Sci. Rep.* **2002**, *46*, 1–308.

7. Thompson, T. L.; Yates, J. T., Jr. Surface Science Studies of the Photoactivation of TiO_2(s): New Photochemical Processes. *Chem. Rev.* **2006**, *106*, 4428–4453.
8. Abou-Ras, D.; Caballero, R.; Fischer, C.-H.; Kaufmann, C.A.; Lauermann, I.; Mainz, R.; Mönig, H.; Schöpke, A.; Stephan, C.; Streeck, C.; Schorr, S.; Eicke, A.; Döbeli, M. Gade, B.; Hinrichs, J.; Nunney, T.; Dijkstra, H.; Hoffmann, V.; Klemm, D.; Efimova, V.; Bergmaier, A.; Dollinger, G.; Wirth, T.; Unger, W.; Rockett, A.A.; Perez-Rodriguez, A.; Alvarez-Garcia, J.; Izquierdo-Roca, V.; Schmid, T.; Choi, P.-P.; Müller, M.; Bertram, F.; Christen, J.; Khatri, H.; Collins, R. W.; Marsillac S.; and Kötschau, I. Comprehensive Comparisons of Various Techniques for the Analysis of Elemental Distributions in Thin Films. *Microsc. Microanal.* **2011**, *17*, 728–751.
9. Cabrera, N.; Mott, N. F. Theory of the Oxidation of Metals. *Rep. Prog. Phys.* **1948–1949**, *12*, 163–184.
10. Chao, C. Y.; Lin, L. F.; Macdonald, D. D. A Point Defect Model for Anodic Passive Films: I. Film Growth Kinetics. *J. Electrochem. Soc.* **1981**, *128*, 1187–1194.
11. (a) Nagayama, M.; Cohen, M. The Anodic Oxidation of Iron in a Neutral Solution: I. The Nature and Composition of the Passive Film. *J. Electrochem. Soc.* **1962**, *109*, 781–790; (b) Nagayama, M.; Cohen, M. The Anodic Oxidation of Iron in a Neutral Solution: II. Effect of Ferrous Ion and pH on the Behavior of Passive Iron. *J. Electrochem. Soc.* **1963**, *110*, 670–680.
12. Hashimoto, M.; Miyajima, S.; Murata, T. A Stochastic Analysis of Potential Fluctuations during Passive Film Breakdown and Repair on Iron. *Corrosion Sci.* **1992**, *33*, 885–904.
13. Floetotto, D.; Wang, Z. M.; Jeurgens, P. H.; Mittemeijer, E. J. Intrinsic Stress Evolution during Amorphous Oxide Film Growth on Al Surfaces. *Appl. Phys. Lett.* **2014**, *104*, 091901–091902.
14. Oka, Y.; Takahashi, T.; Okada, K.; Iwai, S. Structural Analysis of Anodic Alumina Films. *J. Non-Cryst. Solids* **1979**, *30*, 349–357.
15. MacDougall, B.; Graham, M. J. Growth and Stability of Passive Films. In *Corrosion Mechanisms in Theory and Practice*; Marcus, P.; Oudar, J., Eds.; Marcel Dekker: New York, 1995; pp.143–167.
16. Cornell, R. M.; Schwertmann, U. *The Iron Oxides*, Wiley-VCH: Weinheim, Germany, 2003.
17. Bingham, J. M.; Tuovinen, O. H. Mineralogical, Morphological, and Microbiological Characteristics of Tubercles in Cast-Iron Water Mains as Related to Their Chemical Activity. In *Planetary Ecology, Sixth International Symposium on Environmental Biogeochemistry*; Brierley, J. A. and C. L. Brierly, Eds.; Van Nostrand-Reinhold: New York, 1983; pp 239–250.
18. Smyth, D. M. *The Defect Chemistry of Metal Oxides*; Oxford University Press: New York, 2000.
19. Tilley, R. J. D. *Defects in Solids*; Wiley: New York, 2008.
20. Tuller, H. L. Mixed Conduction in Nonstoichiometric Oxides. In *Nonstoichiometric Oxides*; Sorensen, O. T., Ed.; Academic Press: New York, 1981; pp 271–335
21. Kofstad, P. *Nonstoichiometry, Diffusion, and Electrical Conductivity in Binary Metal Oxides*; Wiley-Interscience: New York, 1972.
22. Kofstad, P. Defects and Transport Properties of Metal Oxides. *Oxidation Metals* **1995**, *44*, 3–27.

23. Yoo, H.- I.; Wuensch, B. J.; Petuskey, W. T. Oxygen Self-Diffusion in Single-Crystal MgO: Secondary-Ion Mass Spectrometric Analysis with Comparison of Results from Gas–Solid and Solid–Solid Exchange. *Solid State Ionics* **2002**, *150*, 207–221.
24. Dieckmann, R. Point Defects and Transport in Non-Stoichiometric Oxides: Solved and Unsolved Problems. *J. Phys. Chem. Solids* **1998**, *59*, 507–525.
25. Nasrallah, M. M.; Douglass, D. L. Ionic and Electronic Conductivity in Y_2O_3-Doped Monoclinic ZrO_2. *J. Electrochem. Soc.* **1974**, *121*, 255–262.
26. Gorski, C. A.; Nurmi, J. T.; Tratnyek, P. G.; Hofstetter, T. B.; Scherer, M. M. Redox Behavior of Magnetite: Implications for Contaminant Reduction. *Environ. Sci. Technol.* **2009**, *44*, 55–60.
27. Gorski, C. A.; Handler, R. M.; Beard, B. L.; Pasakarnis, T.; Johnson, C. M.; Scherer, M. M. Fe Atom Exchange between Aqueous Fe^{2+} and Magnetite. *Environ. Sci. Technol.* **2012**, 46, 12399–12407.
28. Larese-Casanova, P.; Scherer, M. M. Fe(II) Sorption on Hematite: New Insights Based on Spectroscopic Measurements. *Environ. Sci. Technol.* **2006**, *41*, 471–477.
29. Rosso, K. M.; Yanina, S.; Gorski, C. A.; Larese-Casanova, P.; Scherer, M. M. Connecting Observations of Hematite (α-Fe_2O_3) Growth Catalyzed by Fe(II). *Environ. Sci. Technol.* **2010**, *44*, 61–67.
30. Yanina, S. V.; Rosso, K. M. Linked Reactivity at Mineral–Water Interfaces through Bulk Crystal Conduction. *Science* **2008**, *320*, 218–222.
31. Alwitt, R. S. The Aluminum–Water System. In *Oxides and Oxide Films*; Diggle, J. W.; Vijh, A. K., Eds.; Marcel Dekker: New York, 1976; 169–254;
32. Alwitt, R. S. The Growth of Hydrous Oxide Films on Aluminum. *J. Electrochem. Soc.* **1974**, *121*, 1322–1328.
33. Bunker, B. C.; Nelson, G. C.; Zavadil, K. R.; Barbour, J. C.; Wall, F. D.; Sullivan, J. P.; Windisch, C. F.; Engelhardt, M. H.; and Baer, D. R. Hydration of Passive Oxide Films on Aluminum. *J. Phys. Chem. B* **2002**, *106*, 4705–4713.
34. Sohlberg, K.; Pennycook, S. J.; Pantelides, S. T. Hydration and the Structure of the Transition Aluminas. *J. Am. Chem. Soc.* **1999**, *121*, 7493–7499.
35. Li, F.; Zhang, L.; Metzger, R. M. On the Growth of Highly Ordered Pores in Anodized Aluminum Oxide. *Chem. Mater.* **1998**, *10*, 2470–2480.
36. Masuda, H.; Fukuda, K. Ordered Metal Nanohole Arrays Made by a Two-Step Replication of Honeycomb Structures of Anodic Alumina. *Science* **1995**, *268*, 1466–1468.

13

Photochemistry and Excited-State Reactions of Oxides

13.1 INTRODUCTION

The applied voltages that drive electrochemical processes (see Chapter 11) are only one of many energy sources that can be used to activate reactions in oxide molecules and materials. Another common energy source that drives many environmental and technological oxide reactions is light from the sun. Water plays a key role in many of these reactions. Imagine that you are on vacation floating in a warm ocean bathed by the sun. Many of the phenomena you experience, from your painful sunburn to the photosynthetic growth of the seaweed you see beneath you, are photoactivated processes. In this chapter, we highlight the roles that oxides play in photon-activated solar energy technologies. Also included are reactions stimulated by other nonthermal energy sources, including electrons in high-energy plasmas. Titanium oxide, found in common white paint, is the basis for much of the discussion, because this oxide is used in many photoelectrochemical energy storage technologies. The photochemistry of colloidal manganese- and iron-oxide particles suspended either in atmospheric droplets or in the upper photic zone of the ocean where the sunlight penetrates are discussed in Chapter 18. Such oxide reactions are important globally in the elimination of pollutants. Both industrial and environmental examples illustrate how oxides participate in a wide range of photoactivated chemical reactions, including the catalytic decomposition of water, photoelectrochemistry, and photoactivated dissolution and precipitation reactions.

13.2 LIGHT AND OTHER SOURCES OF OXIDE EXCITATION

Before exploring excited-state reactions, we need to introduce the energy sources that provide such excitation. In most of this chapter, the excitation source of interest is light. Most of us are familiar with the electromagnetic spectrum [1], in which the energy of a photon is given by

$$E = h\nu = hc/\lambda = hc\omega \tag{13.1}$$

Here, h is Planck's constant ($h = 6.6 \cdot 10^{-34}$ J/second), c is the speed of light ($3 \cdot 10^{10}$cm/second), ν is the frequency of light (measured in Hertz or per second), λ is the wavelength of light (in centimeters), and ω is the wavelength expressed as wave number (measured per centimeter in infrared spectroscopy). The energy of the light itself is normally expressed in terms of electron volts (eV), although kilo-Joules per mole and kilocalories per mole are used to put these energies into a chemical context (Table 13.1). Figure 13.1 delineates different regions of the electromagnetic spectrum and provides representative photon energies. As shown in Figure 13.1, the microwaves that excite the rotations of water molecules in your microwave oven have relatively low energies (around 10^{-4} kJ/mole). The infrared radiation that stimulates O−H stretching vibrations in water molecules (at 3500/cm) has energy comparable with that of hydrogen bonds (around 40 kJ/mole). It is only when we reach visible light (e.g., green light with a wavelength of around 532 nm) that we encounter photon energies that are comparable with the energies of chemical bonds. Visible photons are responsible for driving many of the photochemical reactions of oxides. At even higher energies (e.g., x rays, gamma rays, and other common forms of radiation), we encounter photons that are energetic enough to eject electrons from an oxide or even displace ions from their positions in an oxide lattice. High-energy photons create the radiation damage and the high-energy reactions that are discussed at the end of the chapter.

Table 13-1. Energy Conversion Units

Energy	**eV**	**kJ/mole**	**kcal/mole**	**1000/cm**	**1/100 nm**
1 eV =	1	96	23	11.5	12.7
1 kJ/mole =	0.0104	1	0.24	0.12	0.13
1 kcal/mole =	0.043	4.2	1	0.5	0.55
1000/cm(IR) =	0.087	8.3	2	1	1.1
1/100 nm*(UV/vis) =	0.079	7.6	1.8	0.91	1

IR, (infrared); UV/vis, (ultraviolet/visible). *Energy is proportional to 1/wavelength (e.g., 500 nm = (12.7)(100)/500) = 2.5 eV).

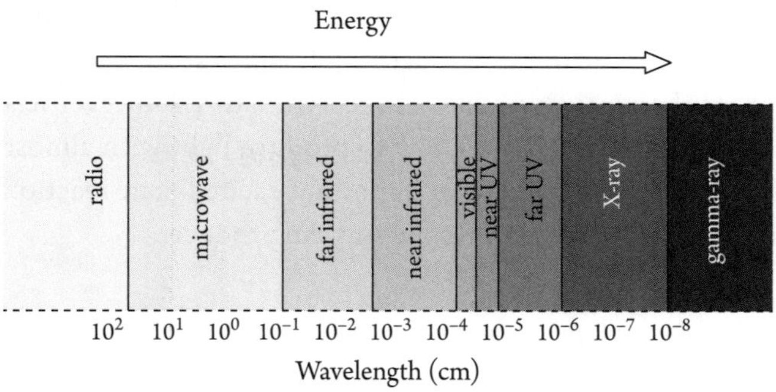

Figure 13.1 The electromagnetic spectrum, indicating frequency ranges (in centimeters) for all common forms of radiation. To convert the abscissa into other wavelength units, please consult Table 13.1.

Particles at these wavelengths, which are energetic enough to damage our DNA, are called *ionizing radiation*.

A second important factor that influences photochemical reactions involves the *availability* of energetic photons. Fortunately, molecules in our atmosphere protect us from the most energetic photons and ionized particles called *cosmic rays*, which emanate from the sun. The bulk of the light reaching our planet corresponds to photons generated by heat in the sun's upper atmosphere. The solar spectrum is equivalent to that of a black-body heat source with a temperature of around 5300°C (Fig. 13.2) [2]. The peak in spectral irradiance is in the visible

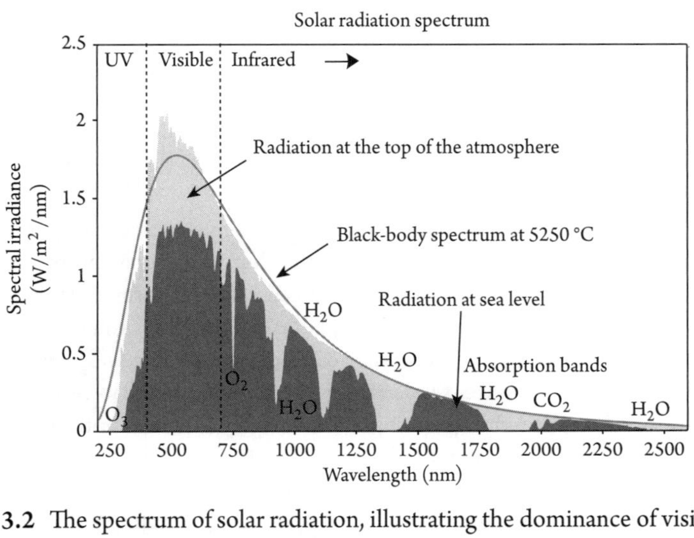

Figure 13.2 The spectrum of solar radiation, illustrating the dominance of visible light and the black-body nature of the light intensity curve. UV, ultraviolet.

SOURCE: Figure courtesy of Robert A. Rohde of Global Warming Art, and Wikipedia, with permission (http://www.globalwarmingart.com/wiki/File:Solar_Spectrum_png).

spectrum between roughly 400 nm and 700 nm, with a significant tail that extends across much of the infrared spectrum. It is this solar spectrum that determines the photochemical reactions that occur in Nature. However, it is important to note that humans have learned how to produce light with almost any wavelength and energy, which means that important excited-state reactions of oxides are by no means limited to the light the sun can produce.

13.3 LIGHT ABSORPTION BY SOLUBLE OXIDE COMPLEXES

Now that we know what energetic photons are available, we need to understand how those photons interact with an oxide material. The simplest method for understanding such interactions involves the use of energy-level diagrams based on both atomic and molecular orbitals [3] (Fig. 13.3). Each atom (or ion) has a set of atomic orbitals (i.e., the s, p, d, and f states that form the basis for the Periodic

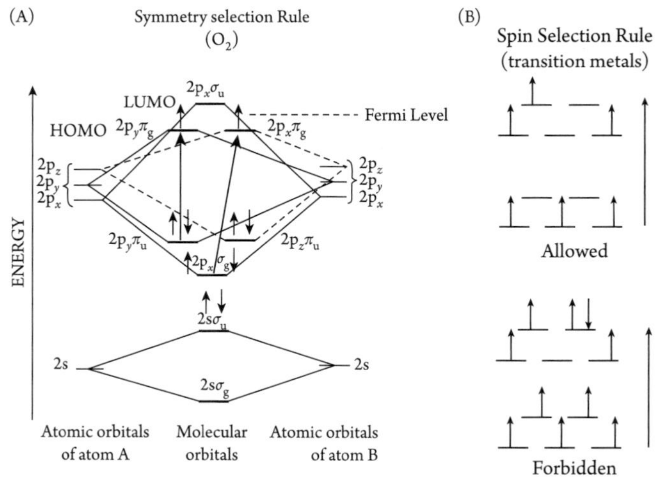

Figure 13.3 (A) Electronic transitions in the O_2 molecule. Molecular orbitals for linear diatomic molecules are here shown with the electron occupancy corresponding to the O_2 molecule. Shown are the highest occupied molecular orbitals (the HOMOs), the lowest unoccupied molecular orbital (the LUMO), and the Fermi level (which is midway between the two). The ground state of O_2 is a triplet state ($^3\Sigma_g$) containing two unpaired electrons, which makes O_2 paramagnetic. Notice that the lowest energy electronic excitation (from the $2p\pi_g$ orbitals to the $2p\pi_u$ orbital) is both symmetry and spin allowed. The two electrons in the $2p\pi_u$ state can be forced to pair up via a spin-forbidden transition ($\Delta E = 1$ eV) to create the singlet oxygen ($^1\Delta_g$), which is important in the photooxidation of organic compounds. (B) Examples of the spin selection rules that control extinction coefficients for the absorption of light by molecular species. Transitions involving no change in spin state are allowed, whereas those involving changes in spin state (such as the pairing of electrons) are forbidden.

Table). Each atom also brings with it a set of electrons to fill those orbitals. The lowest energy orbitals are filled first, leaving many of the higher energy orbitals unoccupied. When atoms interact to form molecules, molecular orbitals form that are linear combinations of the orbitals on each atom. Such combinations typically result in the creation of lower energy states, with peaks in electron density between atoms called *bonding orbitals* and higher energy states with nodes between the atoms called *antibonding orbitals.*

When a photon encounters an atom or molecule, it can be consumed by exciting an electron from one of the low-energy occupied orbitals into a higher energy orbital that is unoccupied. For this to happen, several criteria must be met. First, the photon energy must at least match the difference in energy between the two orbitals involved in the excitation. As indicated in Figure 13.3, these energy differences can be tens or even hundreds of electron volts. Driving such excitations requires light that falls within the X-ray and even gamma-ray regimes, which means that most photochemistry involves transitions between the *highest occupied molecular orbital* (HOMO) and the *lowest unoccupied molecular orbital* (LUMO). Another energy to remember is the midpoint energy between the HOMO and LUMO, which is referred to as the *Fermi level* or *electronegativity of the complex.* The Fermi level can be equated to the redox potential of a 1-electron electrochemical half-reaction involving the reactive species (e.g., $Fe^{3+} + e^- \rightleftarrows Fe^{2+}(aq)$) (see Chapter 11). This equivalency is critical for predicting when photoelectrochemistry will be observed.

The visible light reaching us from the sun (with a high-energy cutoff of around 250 nm or 5 eV, see Fig. 13.2) does not have sufficient energy to excite even the lowest energy electronic transitions for water or the aqueous complexes of most metal cations. The lowest energy transition allowed in water occurs at 167 nm or 7.5 eV [4]. The high transition energy explains why water, most aqueous salt solutions, and for that matter, many oxides and other inorganic compounds are clear, colorless, or white, and are not active in normal photochemical reactions. For the most part, photochemical activity is limited to transitions involving d electrons within transition-metal cations, although heavy elements in groups IIIb and IVb such as Sn and Pb, as well as mixed-valence compounds, are also active (Fig. 13.4).

The second criterion for light adsorption by molecular species involves selection rules based on quantum mechanics (see classic texts such as [5–7]). The two primary selections rules highlighted in Figure 13.3 are that the transition must result in a change in the polarity of the molecule, and the transition must not involve a change in spin state. Application of the first selection rule involves examining the symmetries of the ground-state and excited-state orbitals. Orbitals such as d orbitals, which are centrosymmetric, are given the designation of "g" (based on the German word *gerade,* meaning capable of direct inversion) in contrast to "u" (for *ungerade*) orbitals, such as p orbitals, that do not have a center of

hydrogen 1 **H** 1.0079																	helium 2 **He** 4.0026
lithium 3 **Li** 6.941	beryllium 4 **Be** 9.0122											boron 5 **B** 10.811	carbon 6 **C** 12.011	nitrogen 7 **N** 14.007	oxygen 8 **O** 15.999	fluorine 9 **F** 18.998	neon 10 **Ne** 20.180
sodium 11 **Na** 22.990	magnesium 12 **Mg** 24.305											aluminium 13 **Al** 26.982	silicon 14 **Si** 28.086	phosphorus 15 **P** 30.974	sulfur 16 **S** 32.065	chlorine 17 **Cl** 35.453	argon 18 **Ar** 39.948
potassium 19 **K** 39.098	calcium 20 **Ca** 40.078	scandium 21 **Sc** 44.956	titanium 22 **Ti** 47.867	vanadium 23 **V** 50.942	chromium 24 **Cr** 51.996	manganese 25 **Mn** 54.938	iron 26 **Fe** 55.845	cobalt 27 **Co** 58.933	nickel 28 **Ni** 58.693	copper 29 **Cu** 63.546	zinc 30 **Zn** 65.38	gallium 31 **Ga** 69.723	germanium 32 **Ge** 72.64	arsenic 33 **As** 74.922	selenium 34 **Se** 78.96	bromine 35 **Br** 79.904	krypton 36 **Kr** 83.798
rubidium 37 **Rb** 85.468	strontium 38 **Sr** 87.62	yttrium 39 **Y** 88.906	zirconium 40 **Zr** 91.224	niobium 41 **Nb** 92.906	molybdenum 42 **Mo** 95.96	technetium 43 **Tc** [98]	ruthenium 44 **Ru** 101.07	rhodium 45 **Rh** 102.91	palladium 46 **Pd** 106.42	silver 47 **Ag** 107.87	cadmium 48 **Cd** 112.41	indium 49 **In** 114.82	tin 50 **Sn** 118.71	antimony 51 **Sb** 121.76	tellurium 52 **Te** 127.60	iodine 53 **I** 126.90	xenon 54 **Xe** 131.29
caesium 55 **Cs** 132.91	barium 56 **Ba** 137.33		hafnium 72 **Hf** 178.49	tantalum 73 **Ta** 180.95	tungsten 74 **W** 183.84	rhenium 75 **Re** 186.21	osmium 76 **Os** 190.23	iridium 77 **Ir** 192.22	platinum 78 **Pt** 195.08	gold 79 **Au** 196.97	mercury 80 **Hg** 200.59	thallium 81 **Tl** 204.38	lead 82 **Pb** 207.2	bismuth 83 **Bi** 208.98	polonium 84 **Po** [209]	astatine 85 **At** [210]	radon 86 **Rn** [222]
francium 87 **Fr** [223]	radium 88 **Ra** [226]		rutherfordium 104 **Rf** [261]	dubnium 105 **Db** [262]	seaborgium 106 **Sg** [266]	bohrium 107 **Bh** [264]	hassium 108 **Hs** [277]	meitnerium 109 **Mt** [268]	darmstadtium 110 **Ds** [271]	roentgenium 111 **Rg** [272]							
			lanthanum 57 **La** 138.91	cerium 58 **Ce** 140.12	praseodymium 59 **Pr** 140.91	neodymium 60 **Nd** 144.24	promethium 61 **Pm** [145]	samarium 62 **Sm** 150.36	europium 63 **Eu** 151.96	gadolinium 64 **Gd** 157.25	terbium 65 **Tb** 158.93	dysprosium 66 **Dy** 162.50	holmium 67 **Ho** 164.93	erbium 68 **Er** 167.26	thulium 69 **Tm** 168.93	ytterbium 70 **Yb** 173.05	lutetium 71 **Lu** 174.97
			actinium 89 **Ac** [227]	thorium 90 **Th** 232.04	protactinium 91 **Pa** 231.04	uranium 92 **U** 238.03	neptunium 93 **Np** [237]	plutonium 94 **Pu** [244]	americium 95 **Am** [243]	curium 96 **Cm** [247]	berkelium 97 **Bk** [247]	californium 98 **Cf** [251]	einsteinium 99 **Es** [252]	fermium 100 **Fm** [257]	mendelevium 101 **Md** [258]	nobelium 102 **No** [259]	lawrencium 103 **Lr** [262]

Figure 13.4 A subset of the Periodic Table identifying the most common elements that exhibit electronic transitions within the range of visible light.

symmetry. Allowed transitions involve a change in symmetry (i.e., g→u or u→g), whereas forbidden transitions do not (i.e., g→g or u→u). For atomic orbitals, this means that p→p and d→d transitions are forbidden, whereas p→s transitions are allowed. Regarding the spin-selection rule, any transition that either breaks up or creates electron pairing within the orbitals involved in the transition is forbidden. This selection rule is illustrated in Figure 13.3 for transitions involving d orbitals.

For complexes dissolved in solution, the relationship between the amount of light absorbed in a given transition and the complex's concentration is calculated using the Beer–Lambert Law:

$$A = \log(I/I_o) = \varepsilon bc \qquad \textbf{(13.2)}$$

Here, A is the absorbance, I/I_o is the intensity of the transmitted light relative to the incident light, b is the path length (measured in centimeters), c is the concentration of the dissolved species (in moles per liter), and ε is the molar absorptivity or *extinction coefficient*. Extinction coefficients quantify how effectively a given material absorbs light of a given wavelength. For example, although water is clear to visible light, it is an efficient absorber of ultraviolet (UV) radiation. The UV transition for water at 165 nm has $\varepsilon = 7000$ [4], which means that 90% of UV light having this wavelength is absorbed by the time the light penetrates to a depth of only 23 nm (i.e., $b = A/\varepsilon c = (0.9)/[(7000)(55) = 2.3\cdot10^{-6}\ \text{cm}])$. The UV absorption power of water is what protected early life on Earth from the harmful effects of UV radiation before the development of the ozone layer in the atmosphere.

For fully allowed transitions, values of ε can exceed 10^5. Symmetry-forbidden transitions such as the d→d transitions in fully symmetric octahedral complexes typically exhibit ε values ranging from 10 to 100. However, light absorption by transition-metal complexes can be enhanced by introducing asymmetry into the molecule (see Section 13.7). Finally, transitions that are both symmetry and spin forbidden, such as all d→d transitions in d^5 complexes can exhibit extinction coefficients of less than 0.1.

A final important factor that impacts the efficiencies of photochemical processes involves excited-state lifetimes. Because the excited states are so much higher in energy than the ground state, excited-state lifetimes tend to be short, ranging from nanoseconds to picoseconds. Keeping these states active for longer times is often what makes a good catalyst. Electrons in photoactivated excited states can relax back to the ground state via both radiative (light-emitting) and nonradiative (heat-generating) processes. Charge separation for microseconds, as is shown in the vanadium oxide catalyst shown in Figure 13.5, is a long time. The net result is that the majority of excited-state complexes relax back to the

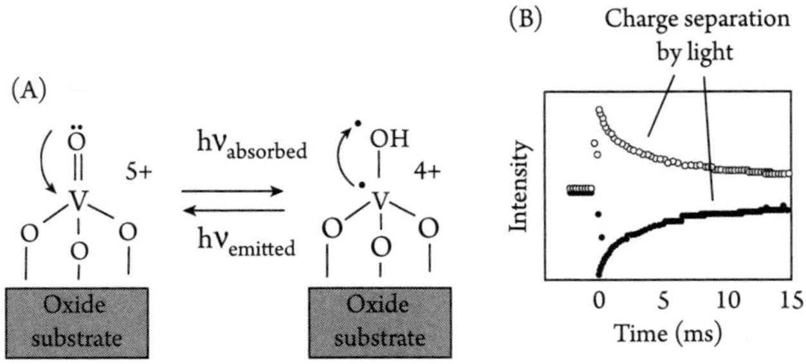

Figure 13.5 Ligand-to-metal charge transfer to the vanadyl oxygen is evident after laser irradiation at 308 nm in vanadium oxide supported on an MgO oxide catalyst. Changes in electron distributions associated with light adsorption are indicated on the left. The transient spectra (right) correspond to channels recording signals at 400 nm and 610 nm wavelengths. Adapted from Garcia, H.; Lopez-Nieto, J.M.; Palomares, E.; Roth, H.D.; and Solsona, B., 2006. Laser Flash Photolysis of Metal Oxide Supported Vanadyl Catalysts. Spectroscopic Evidence for the Ligand-to-Metal Charge-Transfer State *J. Mater. Chem.* **2006**, *16*, 216–220, with permission.

ground state before any meaningful reactions can occur. These short excited-state lifetimes mean the chemical use of photon energy by most soluble oxide complexes is highly inefficient.

Although molecular photoreactions are relatively inefficient, photoactivated states can be very reactive. The states, generate reaction cascades [8] that are important to the elimination of nitrogen, sulfur, and carbon pollutants from air. The simplest reaction to consider is bond dissociation; electrons in the HOMO tend to be in bonding orbitals whereas the LUMO represents nonbonding or antibonding orbital configurations. At the next level of complexity, both the excited electron and the hole that it leaves behind in the parent ground-state orbital are highly reactive. The electron represents a strong reducing agent that can be transferred to adjacent species. Similarly, the hole can abstract an electron from a neighboring species, making it a strong oxidizing agent. Such electron transfers are critical to the light-activated splitting of water and other photoelectrochemical and photocatalytic processes. Even if no immediate reaction occurs, the photoactivated complex can transfer some or all of its energy to another species, placing the other species in a more reactive excited state. However, before discussing oxides for light harvesting, we need to explain why solid oxides are more efficient than most metal complexes for performing photoactivated chemical reactions.

13.4 THE ELECTRONIC STRUCTURES OF OXIDES

Any oxide particle with a radius of more than a few nanometers contains so many atoms and energy states that molecular–orbital descriptions must be replaced by broad, yet discrete, packets of closely spaced energy levels called *bands* [9]. In extended semiconducting oxides, the ground-state band corresponding to the HOMO in a molecular complex is referred to as the *valence band*. As in molecular complexes, electrons in the valence band represent the bonding electrons that hold the solid together. The manifold of higher energy unoccupied orbitals (corresponding to the LUMO in a single molecule) is often referred to as the *conduction band* for the simple reason that these nonbonding orbitals are often delocalized. Electrons in the conduction band are not trapped in bonds and can be much more mobile than electrons in the valence band. The energy gap between the valence and conduction bands is called the *band gap*. In insulating oxides, the band gap is large. For example, MgO has a band gap of 7.7 eV corresponding to far UV light with a wavelength of 166 nm. Semiconducting oxides tend to have smaller band gaps, which allow electrons to be promoted across the gaps by visible light. Green light, with a wavelength near 500 nm, corresponds to a gap of 2.5 eV. For metallic oxides, the gap is so small as to be almost nonexistent. Band gaps for representative oxides are shown in Figure 13.6 [10, 11], and can be referenced with the electromagnetic spectrum shown in Figure 13.1.

The reference point for evaluating the photoelectrochemistry of an oxide is the Fermi level. The Fermi level in an undoped semiconductor corresponds to

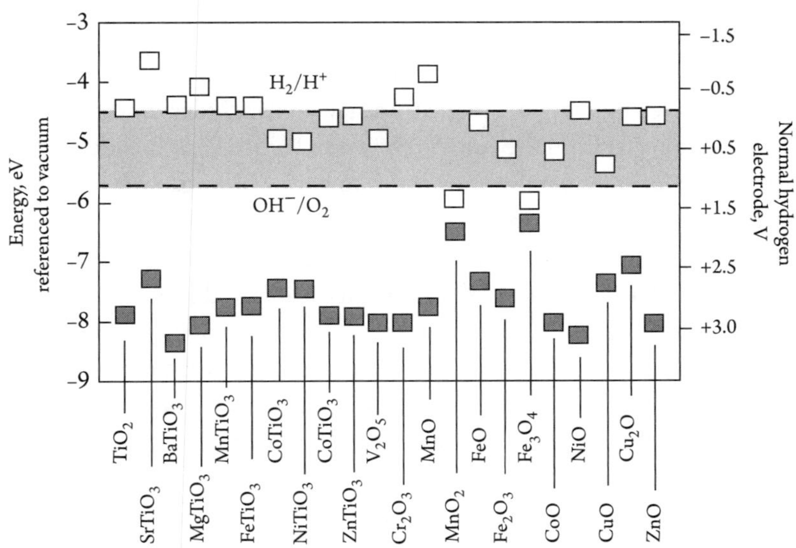

Figure 13.6 Calculated energy positions of the conduction band edges (open squares) and valence band edges (filled squares) at pH 0 for a range of metal oxide semiconductors. The solid lines correspond to the stability limits for water. Adapted from Xu, Y.; Schoonen, M. A. A. The Absolute Energy Positions of Conduction and Valence Bands of Selected Semiconducting Materials. *Am. Miner.* **2000**, *85*, 543–556, with permission.

the midpoint energy between the valence and conduction bands by analogy to the Fermi level in a metal complex. However, the position of the Fermi level can be shifted via doping with an excess of electrons or holes to make n- and p-type semiconductors, respectively. Such doping can be achieved by substituting a cation with a lower or higher valence than the cations in the host oxide, respectively. One could add phosphorus to silicon for an n-type semiconductor, or aluminum to silicon for a p-type semiconductor. In the case of TiO_2, doping is commonly achieved by reducing the oxide in a H_2 atmosphere and thus creating an n-doped semiconductor containing oxygen defects and controlled concentrations of Ti^{3+}. Doping moves the position of the Fermi level. The Fermi level approaches the top of the valence-band energy for p-type semiconductors and the bottom of the conduction-band energy for n-type semiconductors. Because the Fermi level can be related to the electrochemical potential, this parameter is often used to predict whether photoinduced redox reactions should occur.

Two important distinctions need to be made between the bands in oxides and the energy levels in soluble complexes. First, the bands in an oxide are so broad (encompassing a range that can exceed 10 eV, see Table 13.2) that, for all practical purposes, any photon with an energy exceeding the band gap energy is sufficient for promoting electrons from the valence band into the conduction band. Our model semiconducting oxide TiO_2 has a band gap of 3 eV, which means that it is a strong absorber of all photons with a wavelength of 400 nm or less (i.e., blue to UV light). This also means that TiO_2 fails to capture all photons with a wavelength greater than 400 nm, which includes most of the visible photons reaching us from the sun. As shown in Figure 13.2, the ideal semiconducting oxide for solar applications needs to have a band gap of around 1.5 eV, corresponding to the absorption of light with wavelengths less than 800 nm. Pyrolusite, a semiconducting MnO_2 mineral, has a band gap that can be as low as 0.25 eV, allowing it to adsorb all light from the infrared regime (5000 nm) up through the UV regime [12]. However, as shown later, photoactivity is only one of the critical parameters that an oxide must meet to be useful as a photocatalyst. For example, stability to photodissolution makes TiO_2 superior to MnO_2 as a photoactive material when it comes to many technological applications involving water.

A second key difference between a solid oxide and an aqueous solution is that the concentrations of light-absorbing species in the solid are invariably much higher (around 30 M) than concentrations that can be achieved in solution. To make a direct comparison, if the extinction coefficient for each metal center in an oxide is similar to that of a typical transition metal complex ($\varepsilon = 100$), Eq. 13.2 suggests that 90% of the excitation photons impinging on an oxide surface are absorbed within a very thin (100 nm to a few micron) surface layer.

Table 13-2. Absolute Electronegativity (x), Band Gap (E_G), Energy Levels of Calculated Conduction Band Edge (E_{CB}), and Flat-Band Potential at pH_{PZC} (U_{FT}) with Respect to the Absolute Vacuum energy Scale, and Measured or Estimated pH_{PZC} for Selected Semiconducting Metal Oxide Minerals

Mineral	**x** eV	**E_g** eV	**E_{CB}** eV	**U_{ft}** eV	**pH_{pzc}**
$BaTiO_3$	5.12	3.30	−4.58	−4.21	9.00
Bi_2O_3	6.23	2.80	−4.83	−4.82	6.20
CuO	5.81	1.70	−4.96	−4.89	9.50
Fe_2O_3	5.88	2.20	−4.78	−4.69	8.60
$FeTiO_3$	5.69	2.80	−4.29	−4.56	6.30
$Hg_2Nb_2O_7$	6.21	1.80	−5.31	−5.05	6.25
$KTaO_3$	5.32	3.50	−3.57	−3.70	8.55
Nb_2O_5	6.29	3.40	−4.59	−4.16	6.06
PbO	5.42	2.80	−4.02	−4.46	8.29
$PbFe_{12}O_{19}$	5.85	2.30	−4.70	−5.20	7.17
SnO_2	6.25	3.50	−4.50	−4.55	4.30
$SrTiO_3$	4.94	3.40	−3.24	−3.61	8.60
Ta_2O_5	6.33	4.00	−4.33	−3.89	2.90
TiO_2	5.81	3.20	−4.21	−4.16	5.80
V_2O_5	6.10	2.80	−4.70	−4.84	6.54
WO_3	6.59	2.70	−5.24	−5.29	0.43
ZrO_2	5.91	5.00	−3.41	−3.08	6.70

Adapted from Xu, Y.; Schoonen, M. A. A. The Absolute Energy Positions of Conduction and Valence Bands of Selected Semiconducting Materials. *Am. Miner.*, **2000**, *85*, 543–556, with permission.

13.5 BAND BENDING: THE ROLE OF SURFACE CHARGE

The most important attribute that distinguishes oxide surfaces from metal complexes has to do with the oxide's ability to separate photoexcited electron–hole pairs. In a vacuum environment, band energies do not vary much as a function of distance from the surface. In other words, the bands are flat. Under flat-band conditions, oxides share one unfortunate attribute which we saw earlier for water-soluble complexes: excited-state lifetimes are short, and electron–hole recombination leads to poor photoefficiencies. However, when the oxide is placed in an aqueous solution, interfacial acid–base reactions almost always generate a surface charge (see Chapter 10). This surface charge influences the positions of all energy levels in the vicinity of the oxide–water interface. This phenomenon is referred to as *band bending* (Fig. 13.7) [13]. At the

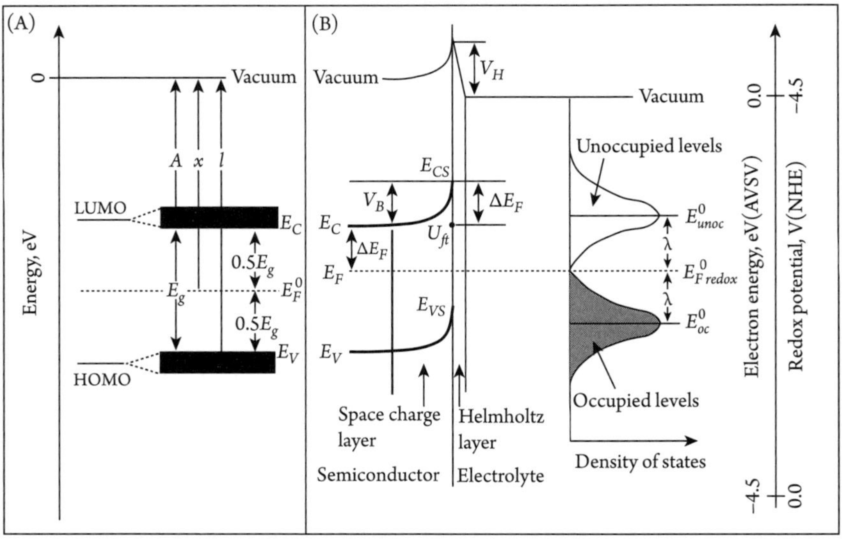

Figure 13.7 (A) A schematic energy-level diagram for a semiconducting oxide showing the positions of the valence and conduction bands (E_C and E_V), the band gap (E_g), and the Fermi level (E_F). Other important parameters include the electron affinity (A), the electronegativity (X), and the ionization energy (I). (B) Positions of the top of the valence band and the bottom of the conduction band for an n-type semiconductor immersed in an aqueous solution showing the band bending (VB) that occurs near the surface. The region within which the bands are bent corresponds to the space charge region. The scale on the far right depicts energies with respect to both the absolute vacuum energy scale and the normal hydrogen electrode. The center depiction is meant to place the same information in the context of electrochemical processes. Adapted from Xu, Y.; Schoonen, M. A. A. The Absolute Energy Positions of Conduction and Valence Bands of Selected Semiconducting Materials. *Am. Miner.* **2000**, *85*, 543–556, with permission.

isoelectric point (or the point of zero charge [PZC], see Chapter 10), flat-band conditions tend to be present. However, for pH values greater than the isoelectric point, the surface is negatively charged. Because electrons are repelled by a negative charge, band energies increase as one moves from the bulk to the oxide–solution interface. In other words, the bands are bent upward (i.e., to higher energies). Conversely, if the surface is positively charged (i.e., for pH values less than the isoelectric point), the bands are bent downward. In both cases, at high values of surface charge (i.e., when all surface oxygen sites are ionized), the difference in band energy between the surface and the bulk can be as much as 0.5 V. This energy difference is sufficient to move electrons and holes within the interfacial zone, as well as to promote surface redox reactions.

The region in which the bands are bent is referred to as the *space charge region*. This region is analogous to the diffuse portion of the electrical double layer encountered on the solution side of the oxide–water interface (see Chapter 10). As in the electrical double layer, the thickness of the space-charge region depends on the magnitude of the surface charge, the dielectric constant of the material, as well as the concentrations and conductivities of charge carriers within the oxide surface. Typical space-charge regions range in thickness from 100 nm up to 1 μm. Note that the space charge region is often comparable in thickness with the near-surface region within which most light is absorbed. The relative thicknesses of the space-charge and light-absorption regions impact the net efficiency of light absorption processes.

Now consider what happens when light excites electrons from the valence band into the conduction band in the presence of a surface charge. Above the PZC, electrons in the conduction band are repelled from the negative charge on the surface, whereas holes in the valence band experience a surface attraction. Conversely, below the PZC, when the surface has a net positive charge, holes follow the bent bands to access lower energies in the interior, whereas electrons are driven to the surface. In either case, charge-induced band bending provides a mechanism for separating electron–hole pairs before they can recombine, allowing these energetic species to drive electrochemical reactions.

13.6 PHOTOINDUCED ELECTROCHEMISTRY

To illustrate how photoinduced electrons and holes perform electrochemistry, let us consider the specific example of using light for the photoelectrolysis of water. The ultimate goal of photoelectrolysis is to harvest light from the sun to generate H_2 and O_2 as fuels that can store solar energy for use in a fuel cell when the sun is not shining. In an analogy to the case of a galvanic cell (see Chapter 11, Fig. 11.2), photoelectrolysis involves two half-cells, one of which contains a photoactive oxide and the other of which contains a platinum electrode.

Consider the case in which the oxide is an n-type semiconductor in which electrons in the conduction band are the primary charge carriers (see Fig. 13.8). For n-type semiconductors, it is desirable for the solution pH to be basic (i.e., greater than the isoelectric point of the oxide). If this is the case, energetic holes are driven to the surface as a result of positive band bending. The holes are then available for the oxidation of water (or hydroxide ions) via reactions such as

$$4OH^- + 4h^+ \rightleftarrows O_2 + 2H_2O \qquad \textbf{(13.3)}$$

In other words, the oxide functions as a photoanode for converting water into oxygen. The electrons in the conduction band are driven to the platinum

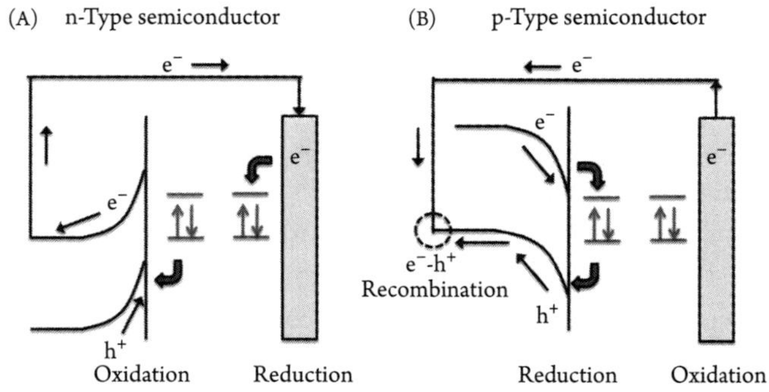

Figure 13.8 (A, B) Energy levels associated with photoelectrochemical reactions using both n-type (A) and p-type (B) semiconducting oxides. In both cases, light excites an electron from the valence band into the conduction band, leaving behind a hole in the valence band. (A) For n-type oxides, a negative surface charge is desired, which creates upward band bending. Here, holes are driven to the surface where they can oxidize solution species with the highest occupied molecular orbitals (the HOMOs) that lie above the valence band of the oxide. Band bending drives electrons in the conduction band into the interior and, eventually, to a counterelectrode. If the electron has a higher energy than the lowest unoccupied molecular orbital (the LUMO) of the contacting solution (resulting from the position of the conduction band and/or band bending), this electron can reduce solution species at the counterelectrode. For the case of pure water (which is both oxidized and reduced) in contact with an oxide such as $SrTiO_3$ (with a band gap of 3.3 eV), the maximum energy efficiency corresponds to the minimum required photon energy divided by the energy of the redox couple (in this case, 1.23 V), or, in this case, 37%. Note that for both oxidation and reduction to occur, both the HOMO and the LUMO of relevant solution species must lie within the band gap of the oxide. (B) For p-type oxides, all the same ground rules and energetics apply, except that now a positive surface charge and negative band bending are desired. Here, electrons driven to the surface of the oxide reduce solution species, and holes are driven into the interior where they react with electrons from the counterelectrode.

electrode via the same positive band bending. The platinum electrode then functions as a cathode for the reduction of water via

$$2H_2O + 2e^- \rightleftarrows H_2 + 2OH^- \tag{13.4}$$

Conversely, if the oxide is a p-type semiconductor (Fig. 13.8), acidic conditions are desired to generate negative band bending. Here, the electrons are driven to the surface to reduce the water to H_2 (making the oxide a photocathode instead of an anode), while the holes travel through the external circuit to oxidize the water to O_2 on the Pt electrode (which is now the anode). The key result is that photoactive oxides can function as either anodes or cathodes, depending on which majority carriers are present and which band-bending configurations are exploited to move photogenerated electrons or holes to oxide electrode–solution interfaces.

We can now evaluate whether an oxide participates in photoelectrochemical processes and how efficient such processes will be. For solar applications, the first factor to be considered in evaluating light-harvesting efficiency is the fraction of solar light that can be captured. For TiO_2, with a band gap of 3 eV, the fraction of the total solar spectrum that can be used is only a few percent (blue and UV light having wavelengths shorter than 400 nm, see Fig. 13.2). The second factor to consider is the number of light-activated electrons and holes that can be exploited before electron–hole recombination occurs. If the light absorption region is thin relative to the space-charge region, most electron–hole pairs can, in principle, be separated before such recombination occurs. Conversely, if the space-charge region is thinner than the light-absorption region, most photons that penetrate the solid below the space-charge region generate electrons and holes that instantly recombine to generate heat without performing any useful chemical work. As far as photoelectrochemical applications are concerned, there is no need for the oxide to be any thicker than the space-charge region, regardless of the thickness of the light-absorption zone. This is why photochemical devices often deploy oxides as thin films, or architectures with numerous stacked rods and/or nanoparticles deposited onto conductive substrates. However, supports must be selected with care, because substrate interactions can also have a major impact on band bending.

The final major factor influencing photo-efficiency involves determining the fraction of the photon energy actually used to perform electrochemistry. For example, although many oxides exhibit band gaps that exceed the energy required to electrolyze water, a large band gap and the mere presence of energetic electrons and holes does not guarantee that solution species are either oxidized or reduced. Just as in the standard electrochemical reactions described in Chapter 11, the holes and electrons must have electrochemical potentials that are appropriate for either oxidizing or reducing these species. These potentials are not equal to the total energy of the absorbed photon (Fig. 13.8).

In the example of water electrolysis, for an electron in the conduction band to reduce water (Eq. 13.4), the LUMO in the water molecule must have an energy lower than that of the bottom of the oxide's conduction band. If this is the case, the photoexcited electron can drop from the conduction band into the lower-energy state of the water molecule (or a hydronium ion), resulting in the desired reduction reaction. The effective electrochemical voltage for such a reaction corresponds to the energy difference between the bottom of the band gap and the LUMO of the water molecule. Similarly, for a hole in the oxide valence band to oxidize water (Eq. 13.3), the energy at the top of the oxide's valence band must be below that of the HOMO of the water molecule.

Again, the voltage difference between the two levels corresponds to the redox potential. In both cases, the effective voltage associated with the photoactivation of the oxide is almost always substantially less than the oxide's band gap energy.

The key to predicting whether photoredox reactions will occur involves establishing exactly where the energy levels in the oxide are relative to those in the species of interest (i.e., a water molecule, in our example). This energy-level matching involves the use of standard states to which all energy levels can be referred. For many semiconductors and selected molecules, a common reference point against which all energy levels can be compared is called the *absolute vacuum scale* [12]. Here, energy levels are measured relative to the resting energy of an electron in a vacuum, which is defined to be zero. The positions of the levels are commonly established via photoemission experiments such as ultraviolet photoelectron spectroscopy (Fig. 13.9) [14].

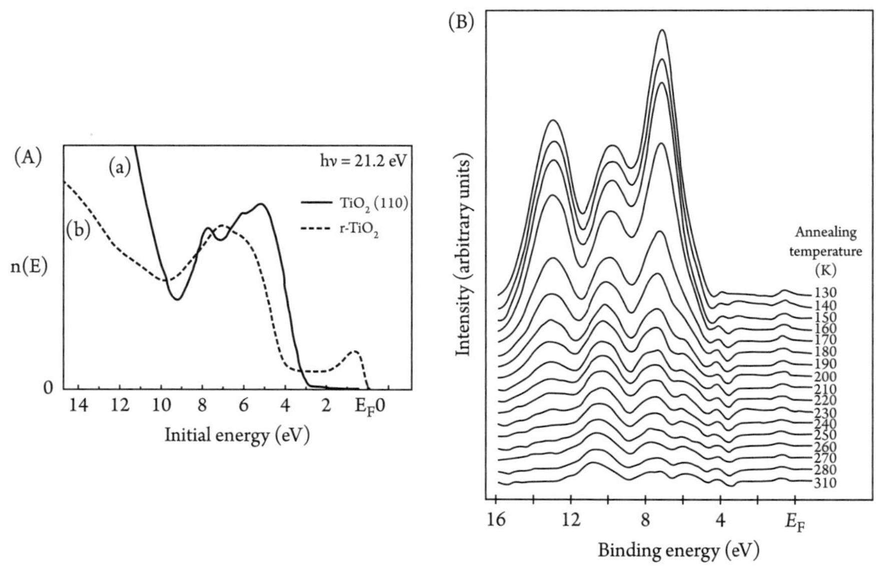

Figure 13.9 Ultraviolet photoelectron spectroscopy (UPS) spectra of TiO_2 surfaces. The UPS spectra of the (110) surface of TiO_2 both before and after reduction. (SOURCE: Reproduced with permission from Sadeghi, H.R. and Henrich, V.E. Electronic Interactions in the Rhodium/TiO_2 system. *J. Catal.*, **1988**, *109*, 1–11.) (Inset) Difference UPS spectra of water adsorbed on TiO_2 (110) at 130 K. Initially, the spectra are dominated by the UPS spectrum for water. However, as the sample is annealed at higher temperatures, the underlying TiO_2 becomes visible as the water desorbs, revealing the presence of species such as surface hydroxyl groups. (SOURCE: Reproduced with permission from Muryn, C.A.; Tirvengadum, G.; Crouch, J.J.; Warburton, D.R.; Raiker, G.N.; Thornton, G.; and Law, D.S.L. TiO_2(100) Structure-Reactivity Relationship. *J. Phys.: Condensed Matter*, **1989**, *1*, SB127–SB132.)

For ions dissolved in water, the effective redox potentials associated with all oxidizing and reducing agents are referenced with respect to the potential required for an electron in solution to react with protons to produce hydrogen (i.e., the potential for the normal hydrogen electrode (NHE), see Chapter 11):

$$2H^+(aq)+2e^- \rightleftarrows H_2(g) \qquad E_h^\circ = 0.0\ V \tag{13.5}$$

It has been established that the Fermi level of the NHE is −4.5 V relative to the electron rest energy used in the absolute vacuum scale [12], which allows one to estimate the positions of energy levels within semiconducting oxides to those in electrochemically active solution species. For example, the redox potentials associated with the electrolysis of water are compared directly with the band energies of selected semiconducting oxides in Figure 13.6. The horizontal lines in Figure 13.6 indicate the potentials required either to oxidize or reduce water. If the valence band edge is less than the Fermi level of the OH^-/O_2 redox couple, holes created in the valence band have sufficient energy to oxidize water to form $O_2(g)$. If the conduction band edge is more than the energy of the H_2/H^+ redox couple, photoexcited electrons are energetic enough to reduce the protons in water to form $H_2(g)$. Because all other redox couples are referenced to the NHE electrode, any other redox reaction that occurs in water (such as $Fe^{3+}(aq)+e^- \rightleftarrows Fe^{2+}(aq)$) can also be mapped onto a similar diagram to determine whether a wide range of other photoredox reactions should occur.

Regarding the photoelectrolysis of water, the first thing to note is that it is rare for electrons in the conduction band to have sufficient energy to reduce water. Among the oxides depicted in Figure 13.6 and Table 13.2, $SrTiO_3$ is predicted to be the most active, followed by MnO and $MgTiO_3$. Experimental results show that $SrTiO_3$ is indeed quite effective for stimulating the photoelectrolysis of water [15]. Conversely, almost every oxide can function as a photoanode for the photooxidation of water. For water oxidation, TiO_2 is predicted to be one of the oxides that makes the most efficient use of absorbed photons. For TiO_2, the energy used for oxidizing water (i.e., the energy difference between the OH^-/O_2 redox potential and the valence band) divided by the minimum required photon energy (i.e., the band gap energy) is 2.1 eV/3.0 eV = 70%. In fact, as a result of efficient photon use and the ability to generate extensive band bending, TiO_2 would be one of the most efficient materials known for the photoelectrolysis of water if only its band gap energy was a better match for capturing light from the solar spectrum. In terms of light harvesting, oxides with a more narrow band gap, such as MnO_2, would at first glance appear to be ideal as oxide photoelectrodes. However, photoredox processes can convert relatively stable and insoluble oxides into soluble species and lead to photoinduced dissolution and oxide decomposition. Such processes are discussed in Chapter 18, where MnO_2 cycling in the ocean by photoreduction is explored.

13.7 THE ROLE OF PHOTOSENSITIZERS: THE GRÄTZEL CELL

So far, we have documented the behavior of two classes of materials that each have a major flaw when it comes to the photoelectrolysis of water. On the one hand, we have soluble metal complexes with orbital energies that can be tuned to match most of the solar spectrum. Unfortunately, these complexes are extremely inefficient at light harvesting because of rapid electron–hole recombination. On the other hand, we have oxide semiconductors such as TiO_2 that are extremely effective in separating electron–hole pairs to perform useful chemical work, but are often inefficient as a result of poor matching between band-gap energies and the solar spectrum. However, these two materials can be coupled to one another for enhanced light-harvesting applications.

The most interesting photoelectrochemical cells created to date rely on *photosensitization*, in which light-absorbing molecules are integrated into oxide electrodes. Here, molecules tethered to oxide surfaces provide efficient light harvesting while the oxide surface separates electron–hole pairs in the photoexcited molecules before they can recombine. Although dye sensitization of oxides has been known since the 1960s, truly efficient cells for photoelectrolysis have been in existence for less than 10 years. Many of these cells are called *Grätzel cells* (Fig. 13.10), after Prof. Michael Grätzel of Switzerland, who helped pioneer their development [16]. Grätzel cells generate electrical currents and voltages when illuminated with light. However, even here the energy generated by these cells is not commonly associated with the electrolysis of water.

The first key to the production of a Grätzel cell involves identifying dyes that can absorb light efficiently across the entire solar spectrum and that can be attached covalently to oxide surfaces. One such Grätzel cell contains a mixture of asymmetric ruthenium complexes with extinction coefficients exceeding 10^4/M/cm for visible light over the entire wavelength range of $390 < \lambda < 750$ nm (Figs. 13.11) [17]. The exceptional light-absorbing capacity of these dyes across the entire solar spectrum has been unmatched since the dyes were first prepared in 1993. These dyes are attached to oxides via covalent oxide anchors such as phosphonate or carboxylate linkages that replace surface hydroxyl groups on the TiO_2.

Typically, the oxides used in Grätzel cells are porous, sintered arrays of nanoparticles deposited as thin films on conductive glass. Nanoporous films are used to ensure the contact area between the solution and the electrode is maximized, all oxide material that can absorb light falls within the space charge region, that holes and electrons can be separated before they recombine, and the

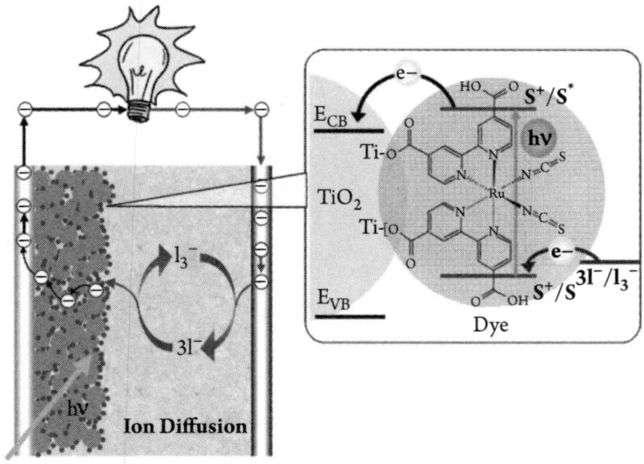

Figure 13.10 A schematic of the energetics of a Grätzel cell showing the role of photosensitization in improving photon efficiencies. Here, visible photons are harvested using dye molecules that are attached to a semiconducting oxide (in this case, TiO_2) surface. Provided that the lowest unoccupied molecular orbital (the LUMO) of the dye lies above the energy of the oxide's conduction band, the oxide can harvest the excited electron from the dye's LUMO. Here, the role of the oxide is to separate electron–hole pairs in the dye rather than to harvest the light. After separation has taken place, the hole left behind in the highest occupied molecular orbital (HOMO) of the dye can oxidize solution species, provided the energy of the HOMO lies below that of the species to be oxidized. The role of SnO_2 is that of a transparent semiconductor to support TiO_2 films or particles.

SOURCE: Reproduced from Grätzel, M.;. Solar Energy Conversion by Dye-Sensitized Photovoltaic Cells. *Inorg. Chem.* **2005**, *44*, 6841–6851, with permission.

effective concentration of photosensitizer molecules is maximized within the light-absorbing regime. For the ruthenium dye shown in Figure 13.11, if $\varepsilon = 10^4$, roughly 60 molecular layers would be required to absorb 90% of the incoming light. For a TiO_2 film consisting of 10-nm particles coated with a monolayer of dye, such absorption requires the film to be on the order of 0.5 to 1.0 μm thick.

The second key attribute of the photosensitizers involves optimizing the match between the HOMO and the LUMO of the dyes, and the valence and conduction bands of the oxide, as well as the redox potentials required to drive the desired reactions (Figs. 13.10, 13.11). In a Grätzel cell for performing oxidations, the LUMO of the dye must have an energy that is greater than the bottom of the oxide's conduction band so that an excited electron can descend from the excited dye into the semiconducting solid. If this criterion is satisfied, and if the anchor between the dye and the oxide does not constitute a resistor that inhibits electron hopping from the dye to the oxide, electrons excited from the HOMO to the LUMO of the dye can be captured by the oxide and swept away via band

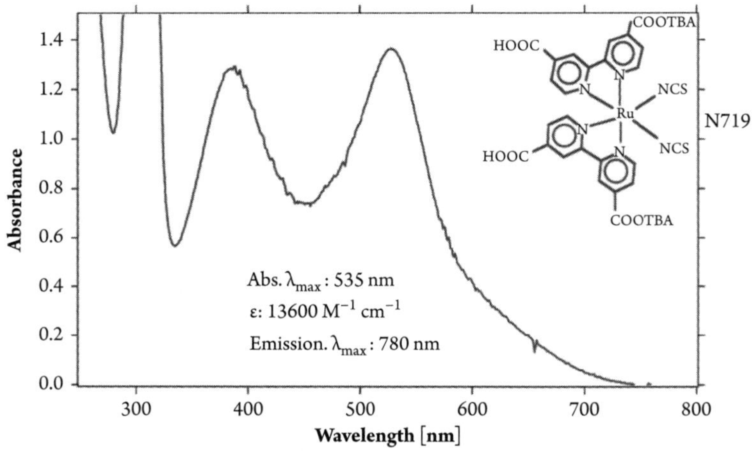

Figure 13.11 The structures and ultraviolet–visible spectra of one of the photosensitizing dyes used to harvest light in Grätzel cells. The solvent was ethanol and the two peaks in the ultraviolet–visible spectra indicate two metal-to-ligand charge transfer bands.
SOURCE: Reproduced from Grätzel, M. Solar Energy Conversion by Dye-Sensitized Photovoltaic Cells. *Inorg. Chem.*, **2005**, *44*, 6841–6851, with permission.

bending. What is left behind is the hole that was created in the HOMO of the dye. This hole must then have a photoredox potential that is sufficiently high to oxidize desired solution species.

The ground rules for using photosensitizers to drive photoelectrolysis of water are illustrated in Figure 13.10, assuming the energy difference between the HOMO and the LUMO of the dye is 1.5 eV (corresponding to 800-nm light). The ideal position of the dye's LUMO needs to be around 0.5 eV more than the bottom of the oxide conduction band to accommodate upward band bending. Given the dye's HOMO–LUMO gap, this means the dye's HOMO needs to be located around 1 eV less than the bottom of the oxide's conduction band. According to Figure 13.10, this also means that holes in the dye's HOMO should have an effective redox potential of around +1 V. However, the potential required to oxidize water is +1.23 V. In other words, the holes created in this particular dye within this particular Grätzel cell do not have sufficient energy to oxidize water. This cell cannot be used for water electrolysis.

Fortunately, Grätzel cells can be used for the photoelectrolysis of other species [17]. The iodide/triiodide redox couple is typically exploited:

$$I_3^- + 2e^- \rightleftarrows 3I^- \qquad E_o = +0.53\ \text{V} \tag{13.6}$$

Because the redox potential for this half-reaction is less than +1 V, the photoactivated holes have enough energy to oxidize I^- to I_3^-. The reverse reaction occurs at the cathode counterelectrode, so neither the oxide nor the redox-active solution species are consumed. Based on measured photocurrents, the photoefficiency of such a cell is more than 10%. This simple system has been optimized such that it can now be plated directly onto glass in buildings, onto plastic sheets, and even onto steel used in roofing. As efficiencies are increasing steadily, solar energy production using oxides in devices such as Grätzel cells could become an integral part of our homes.

13.8 HIGH-ENERGY REACTIONS: ELECTRON- AND PHOTON-STIMULATED PROCESSES

So far, the most energetic activating radiation that we have discussed is a UV photon (240 nm or 5 eV). However, oxides can encounter much more energetic photons, electrons, and ions that constitute high-energy *radiation*. In Nature, oxides are often exposed to radiation. These exposures can modify the structure and chemistry of the oxides via *radiation damage*. For example, the decay of radioactive elements often releases alpha particles, equivalent to an ejected helium nucleus. The damage tracks created by these alpha particles as they pass through a lattice are so persistent they have been used as a dating tool for rocks. One simply counts the number of such tracks and back-calculates the age from a guess at the initial concentration of radioactive element in the mineral. Sites where radiation damage has occurred tend to be so energetic that they react immediately when exposed to acidic aqueous solutions to form an etch pit.

In gaseous environments of technological interest, including catalytic reactors, plasma reactors, and even the engine of a car, these radiation-activated surface sites can be critical to both the reactivity and performance of the oxide. Here, we highlight a few of these radiation-activated oxide reactions, starting with energies that are tens to hundreds of electron volts. Those of you seeking more detailed discussions of radiation damage and radiation-induced chemistry should investigate references [18–20].

The valence and conduction bands that dominate an oxide's interactions with visible light tend to lie from 3 to 8 eV below the energy of an electron in a vacuum (see Fig. 13.6). We now consider interactions with photons that have energies of tens of electron volts corresponding to higher energy x rays. When an oxide is bombarded by such energetic photons (or electrons), it is no longer sufficient to consider just the valence and conduction bands. One must also bear in mind more localized atomic states, called *core states*, that lie at much lower energies (see Fig. 13.3). For example, low-energy electron–energy-loss spectroscopy shows the highest energy core states in TiO_2 (which are the O(2s) state and the Ti(3p) state) have energies that are 21 eV and 32 eV less than the vacuum energy, respectively (Fig. 13.12). To put these numbers into

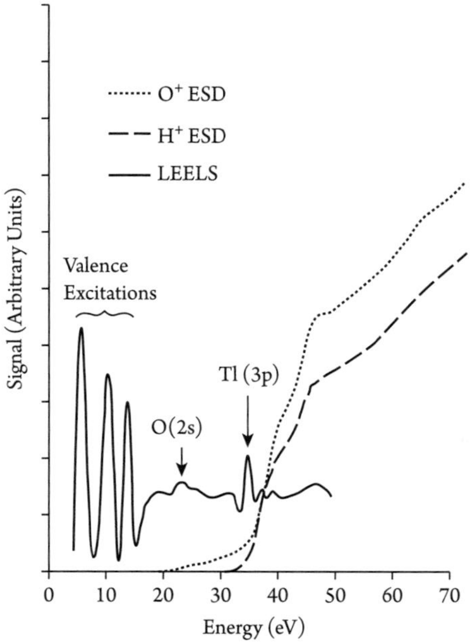

Figure 13.12 The electron-stimulated desorption of both H^+ and O^+ from TiO_2 surfaces as a function of electron energy. (Similar results are observed in photon-stimulated desorption.) The accompanying low-energy electron-loss spectrum (second derivative) shows no desorption occurs via the removal of electrons from either the valence or O(2s) states. It is only when electron energies are sufficient to create core holes in the Ti(3p) state that bond-rupture processes leading to the desorption of cationic species occurs.

SOURCE: Reproduced with permission from Knotek, M. L.; Characterization of Hydrogen Species on Metal-Oxide Surfaces by Electron-Stimulated Desorption: TiO_2 and $SrTiO_3$. *Surface Science*, **1980**, *101*, 336.

chemical perspective, 32 eV = 3079 kJ/mole. What happens when these core states are excited by X-ray photons?

The first processes to consider involve photon-stimulated desorption (PSD) and electron-stimulated desorption (ESD) (Fig. 13.12) [21]. Here, the initial interaction with incoming radiation does not involve the promotion of electrons from the valence band into the conduction band, but rather the ejection of electrons from core atomic levels completely out of the solid. Photons or electrons with energies greater or equal to the energy difference between a core state and the vacuum level can provide such ejection. The extremely energetic core hole left behind is filled immediately via an interatomic Auger process by electrons that reside at the bottom of the higher energy O(2p) valence band. For TiO_2, the minimum consequence of this process is the creation of a reactive hole in the valence band. It is important to note that, in contrast to excitations via visible light, no electrons are excited into the valence band by X-ray photons because the excited electrons have left the solid completely. In other words, X-ray-stimulated electron losses always result in the oxidation of the oxide. For TiO_2, because the valence band is dominated by O(2p) states, the species that is oxidized corresponds to the O^{2-} ion, which is transformed into reactive intermediates such as the surface peroxides and hydroxyl radicals.

After an electron drops from the valence band into a core hole, much of the energy associated with the core hole still remains and must be dissipated. In some cases, this extra energy is released via the ejection of secondary Auger electrons from the solid. For the example of TiO_2 [22], the Ti(3p) core hole has an energy of 34 eV whereas the bottom of the valence band has an energy of around 15 eV. The excess remaining energy then corresponds to 34−15 =19 eV. Because the top of the valence band is 8 eV less than the vacuum level, this remaining energy is sufficient to eject an additional two electrons from the valence band completely out of the solid. This very fast process transforms the native O^{2-} ion in the oxide into the reactive intermediates O^- and O°, and finally to the O^+ ion. The total transformation not only disrupts Ti−O bonds, but results in the photon-stimulated desorption of O^+, which is repelled electrostatically by the neighboring Ti^{4+} ion and expelled. In contrast, lower-energy X rays that create core holes within O(2s) (but not Ti(3p) states) do not stimulate O^+ desorption because the O(2s) hole has an energy of 21 eV. The energy remaining after filling that hole (21−15 = 6 eV) is less than the 8 eV required to eject electrons from the top of the valence band out of the solid.

The stimulated desorption of oxygen from oxide surfaces can have further consequences in terms of surface reactions. For example, low-energy electron–energy-loss spectroscopy measurements [23] show that a Ti^{4+} site that has had its oxygen removed is reduced to Ti^{3+}. This open Ti^{3+} site can coordinate directly to solution species, and functions as a strong reducing agent. If the back-filling ligand is a water molecule, the resulting reaction is

$$2{>}\mathrm{Ti(III)(s)} + \mathrm{H_2O(1)} \rightleftarrows 2{>}\mathrm{Ti(IV)\text{-}OH(s)} + \mathrm{H_2(g)} \qquad \textbf{(13.7)}$$

In other words, oxide surfaces exposed to X rays can be stimulated to promote both oxidation and reduction reactions.

Electrons with appropriate energies can create core holes just like photons can, leading to ESD [21, 22] and reaction cycles similar to those described earlier. In fact, the cross-sections for ESD are much greater than they are for photon-stimulated desorption. Electron-activation processes such as ESD are important in some of the technological application of oxides involving catalytic activation. Traditionally, oxide catalysts have been used to activate gas-phase reactions by heating to temperatures that are sufficiently high to promote desorption of species such as surface hydroxyl groups (see Chapter 6). Such desorption leads to the creation of open, active Lewis-acid surface sites. Recent research indicates that similar levels of activation can be achieved by inserting oxide packing materials into plasma reactors that are operated at room temperature. A major role for plasma reactors involves the breakdown of environmental contaminants. For example, plasma reactors are under consideration for the reduction of nitrous oxides to N_2 as well as the decomposition of hydrocarbons associated with automotive emissions [24, 25].

In a plasma reactor, when an applied electric field exceeds the dielectric breakdown voltage for the gas, electron avalanches occur that create a sustainable plasma discharge in the reactor. The radicals and ions created in the gas phase have sufficient lifetimes to promote desired gas-phase reactions. It has been shown that such plasmas can be created at lower voltages in the presence of oxide packing materials. Experimental results show the packing material acts as a catalyst, with the catalytic effect of a given material being completely different than that observed for the same oxide when used as a thermal catalyst [26] (Fig. 13.13). For example, proprietary zeolite packing materials that have been used for the thermally activated reduction of NO_x to N_2 have been inserted into plasma reactors. In a plasma, NO_x conversions can be induced even at room temperature. The most efficient conversions (60%) are seen at 180°C relative to the 350 to 400°C required in a thermal reactor. In-depth experiments have not yet been conducted to determine reaction mechanisms. However, it appears that catalytic pathways involve both the activation of the oxide via electron-stimulated desorption processes, and the creation of surface defects via the interactions between activated radicals and ions in the plasma and the oxide. The ESD reaction pathway is supported by experiments that have been performed

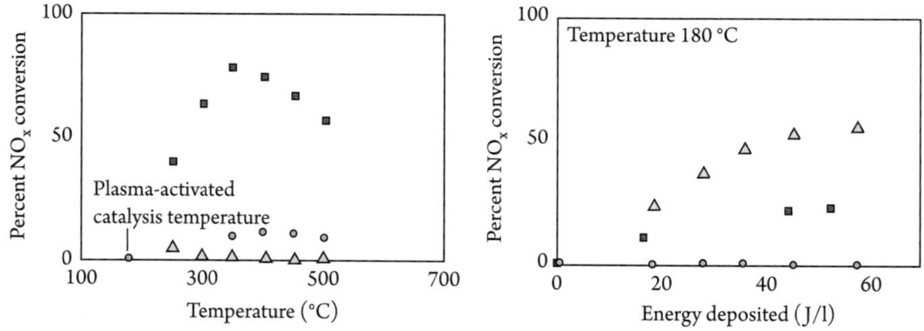

Figure 13.13 A comparison of the efficiency of the conversion of NO_x species to N_2 in the presence of proprietary zeolite packing materials used as catalysts in both thermal (left) and plasma (right) reactors. Adapted from Hsiao, M. C.; Merritt, B. T.; Penetrante, B. M.; Vogtlin, G. E.; Wallman, P. H.; Tonkyn, R. G.; Shah, R.; and T. M. Orlando. Plasma Assisted Oxidation of Propene. *J. Adv. Oxid. Technol.* **1996**, *1*, 79–84.

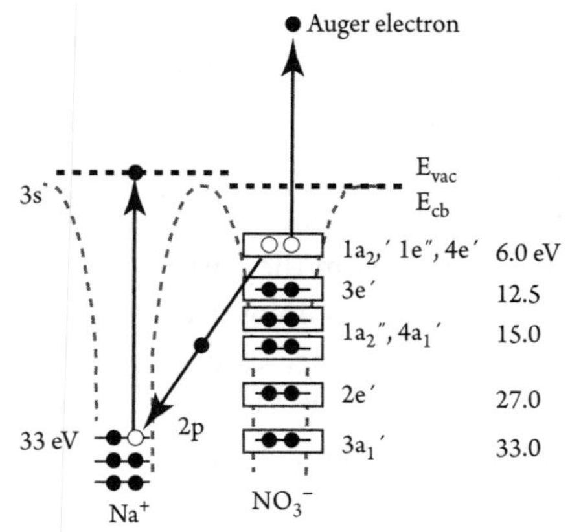

Figure 13.14 An energy-level diagram that shows how electron- or photon-stimulated processes can lead to the decomposition or conversion of contaminants in a plasma environment. The experimental example shown involves the electron-stimulated decomposition of $NaNO_3$. Here, the first step involves the creation of a core hole on Na^+. This core hole is then filled by valence electrons from the nitrate ion. In a plasma reactor, the oxide packing material provides a function similar to that of Na^+, whereas NO_x (or an organic contaminant) is activated in a manner similar to that exhibited by the nitrate ion. Adapted with permission from Knutsen, K.; Orlando, T. O. Stimulated Desorption of $H^+(D^+)$, OH^+ (OD^+), O^+ and NO^+ from Solution-Grown $NaNO_3$ Crystals. *Surf. Sci.* **1996**, *348*, 143–152.

on the electron-stimulated destruction of $NaNO_3$ (Fig. 13.14) [27, 28]. Results show that electrons exceeding the 33 eV required to create Na(2p) core holes promote an intermolecular Auger decay process that leads to the desorption of O^+ and NO^+.

13.9 DAMAGE IN OXIDES RESULTING FROM HIGH-ENERGY RADIATION

This chapter has progressed from reactions that result from relatively low-energy (1–5-eV) molecular excitations involving visible light to high-energy (10–100-eV) core-level atomic excitations involving x rays. However, the energies of these electronic transitions pale in comparison with those associated with the *nuclear transitions* leading to the transformation of one element to another via radioactive decay. The photons (gamma rays), electrons (beta rays), neutrons, and alpha particles (helium nuclei) emitted during decay processes can have extremely high energies—on the order of 1,000,000 eV [29]. When radiation sources with these extreme energies encounter an oxide, they produce a wide range of damage, including heating, the ejections of electrons, the breaking of bonds, the creation of point defects (see Chapter 12), the physical ejection of both cations and anions from the oxide lattice, and even the transmutation of one element to another via nuclear reactions. In fact, ion beams with sufficient energies to induce nuclear reactions have been exploited in a technique for elemental depth profiling called *nuclear-reaction profiling* [30, 31].

Radiation damage is of concern in many applications involving oxides, and is especially true for silicates that form the basis for optical fibers, insulators in microelectronics, and nuclear wastes. In optical fibers, the primary concern involves the creation of radiation-induced color centers that absorb light and interfere with optical transmission [32, 33]. In microelectronics, radiation can create charged ions, electrons, and holes that degrade dielectric properties. Radiation damage interferes with the performance of metal–oxide semiconductor devices such as field-effect transistors. The need to prevent such damage has led to extensive research in creating radiation-hardened microelectronics for space and military applications [34]. In nuclear-waste treatment, glass is a primary candidate as a stable host for long-term disposal (see Chapter 15). Here, the concern is that radiation damage might make the glass more susceptible to dissolution and the release of hazardous species. One reason for this concern is that many of the sites created by radiation damage are highly reactive. However, at the oxide–water interface, most of this damage is repaired almost immediately (see Chapter 6). In fact, reactions between radiation-induced defects and water can also help to mitigate the damage created in optical and microelectronic devices.

Point defects associated with the displacement of cations and anions from their ideal lattice sites include cation and oxygen-ion vacancies, cation and anion interstitials, and defect pairs such as the vacancy–interstitial pair that constitutes a Frenkel defect (see Chapter 12). The need to achieve overall charge neutrality dictates that the formation of such defects is often accompanied by the creation of electrons and/or holes in the oxide lattice. However, radiation damage represents a much more violent and nonequilibrium path than normal defect formation, and results in the creation of defects we have yet to discuss. As in the fracture of an oxide in a high-vacuum environment (see Chapters 6 and 16), bonds are literally ripped apart as a result of radiation damage. When this happens, the two electrons that were in the bond can either be captured by cations (resulting in cation reduction) or by oxygen anions, which leave behind open coordination sites on cations.

Bond rupture can also result in a state in which one bonding electron is retained by each bond fragment, resulting in the formation of charged free radicals on both the cation and the anion. The structures of these radicals and other species containing unpaired electrons have been inferred using electron paramagnetic resonance spectroscopy [35]. Here, we highlight the structures and chemical properties of radical species formed in silica. These species give rise to problematic color centers and charged defect sites that compromise both optical and microelectronic devices. However, it is important to note that similar species can be created by radiation in almost any oxide.

Some of the more important radiation-induced defects in SiO_2 and simple silicates are shown in Figure 13.15 [36]. The basic structure of silica consists of silicate tetrahedra that are held together by $Si-\mu_2O-Si$ bridging bonds (see Chapter 2 and Part Six). One of the more common defects induced by radiation is a Frenkel defect in which one bridging oxygen is displaced from its position between two Si(IV) cations to create an oxygen vacancy next to an oxygen interstitial. The Si(IV) sites adjacent to the oxygen vacancy can react with each other to form a Si–Si bond whereas the free oxygen interstitial can either insert itself into a $Si-\mu_2O-Si$ bond to form a $Si-\mu_2(OO)-Si$ peroxy linkage or can dimerize to form O_2. However, when the broken bonds fail to line up properly, holes can be trapped in what was a bonding orbital on the silicon center. The resulting paramagnetic defect is called an *E′ center*. Similarly, when hydroxyl groups are bombarded by high-energy radiation, the bond rupture process can create hydrogen atoms plus free radicals on oxygen anions called *oxygen hole centers*. The concentrations of E′ and oxygen hole centers can be quite high. For example, E′ center concentrations are around 10^{-5}/Si(IV) for silica that has been exposed to a gamma-ray dose of 10^8 rad [36]. Such concentrations are sufficient to wreak havoc on both the optical and electrical properties of silica and silicate glasses.

As discussed in Chapter 6, broken bonds are highly reactive species. Such sites react with any available species within the oxide or in the environment to

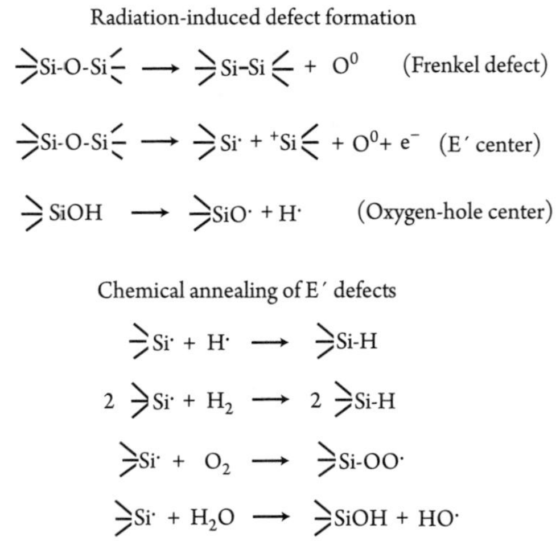

Figure 13.15 Examples of defect formation in oxides (in this case, SiO_2) that are promoted by high-energy radiation. As indicated, many of the bond rupture reactions stimulated by radiation result in the creation of free-radical species within the oxide.

establish more stable bonding configurations. One set of species that is available for reforming chemical bonds includes molecules that are small enough to diffuse through the oxide. Isochronal annealing experiments [37] have shown that the temperature required to eliminate E′ and oxygen hole centers depends on the size of the migrating species (Fig. 13.16). Below 130 K, hydrogen atoms (created via the radiolysis of internal hydroxyl groups) react with E′ and oxygen hole centers to form Si—H and hydroxyl groups, respectively. The H_2 molecule reacts with the same defects to form the same products, plus an additional hydrogen atom that reacts to eliminate a second defect. Molecular oxygen (O_2) reacts with the E′ center to produce a peroxy radical that is also detected via electron paramagnetic resonance spectroscopy. Water reacts with the E′ center to form a hydroxyl group plus a liberated electron. The net effect of all these reactions is the elimination of reactive defects. The good news regarding these reactions is that the unpaired electrons that produce color centers and local charges can be eliminated via

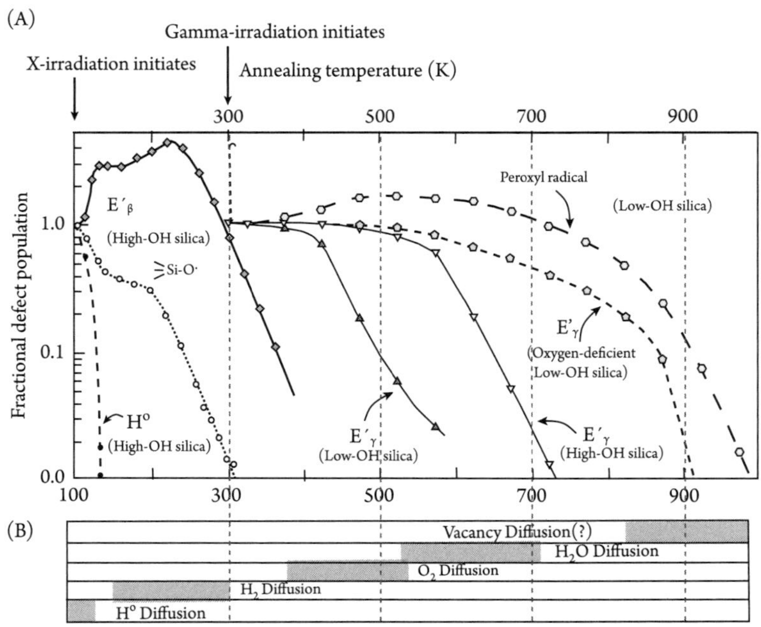

Figure 13.16 Experimental EPR results on the rate of chemical annealing of radiation-induced defects within silica. In all cases, the initial defect concentration is roughly one defect per 10^5 silicon atoms. The bars below the annealing curves indicate the temperature regimes over which a given species is implicated in defect annealing based on diffusion rates. Some of the species involved in annealing (such as hydrogen atoms and H_2) are actually created as a result of the initial radiation damage. Some species (such as the peroxy radical) appear in EPR spectra as the parent radical species (such as the E' center) are removed. (Adapted with permission from Griscom, D. L.; Silica Glasses. *MRS Bulletin*, June 16/Aug15, 1987, 20–23.)

chemically assisted thermal annealing. Such annealing can repair radiation damage relevant to both optics and microelectronics.

The bad news regarding the environmental elimination of defects is that the new bonds that form are often not the Si$-\mu_2$O$-$Si bridging bonds initially disrupted by the radiation. For this reason, researchers have performed extensive studies on glasses that have been damaged heavily by radiation to determine whether such damage compromises the chemical durability of glasses under consideration as hosts for the disposal of nuclear wastes (see Chapter 15). Although we defer a discussion of glass dissolution until Chapter 15, it is clear that dissolution rates of waste-form material could potentially be increased based on three radiation-induced phenomena. First, at high radiation levels, crystalline materials can be disrupted to the point where they become amorphous. Amorphous oxides sometimes exhibit enhanced dissolution rates. For example, vitreous silica dissolves 5 to 10 times faster than quartz and cristobalite, respectively [38]. Second, the replacement of bridging oxygens by nonbridging oxygens (often in the form of hydroxyl groups) can lead to higher dissolution rates after the nonbridging oxygen content exceeds around 10% of the total oxygen sites present [39] (see Chapter 15). At very high hydroxyl group concentrations (and low concentrations of bridging oxygen groups), silica dissolution rates can increase by as much as a factor of 1000. Third, for the case of silicates, it has been shown that radiation damage can lead to the formation of strained silicate rings [40] (see Chapter 16), which can lead to as much as a 10-fold increase in network hydrolysis rates.

All these phenomena require extremely high radiation doses [41]. The dose required to exceed the percolation threshold for ion-track overlap is around $2\cdot10^{20}$keV/cm^3. The doses required for making crystalline oxides amorphous and/or for creating high nonbridging oxygen and/or hydroxyl group concentrations are on the order of 1 to $5\cdot10^{23}$ keV/cm^3. To put these numbers in perspective, ^{137}Cs decays via beta emission with a total decay energy of 1 MeV [42]. A fully loaded cesium ion-exchange material such as a silicotitanate (see Chapter 10) contains on the order of 10^{20} ^{137}Cs/cm^3, which means that by the time all the cesium has decayed (with a half-life of 30 years), the total accumulated radiation dose is on the order of $1\cdot10^{23}$ keV/cm^3. Assuming that none of the damage is annealed away, such a dose might be sufficient, in the worst case, to increase waste-form dissolution rates by a factor of 1000. Such increases in dissolution rates might sound catastrophic. However, as the baseline dissolution rate of a Cs-loaded silicotitanate waste form is barely measurable (around 10^{-8} g/cm^2/day in room-temperature water [43]), a 1000-fold increase still means that radiation-enhanced dissolution rates correspond to etch rates of only tens of nanometers per year. Experiments conducted on synthetic monazite waste forms damaged above the crystalline-to-amorphous threshold show that dissolution rates are still negligible [43]. The incredible durability of oxide waste forms in a radiation environment is in sharp contrast to the irreversible and complete destruction of any organic material or polymer under similar dose conditions.

References

1. Halliday, D.; Resnick, R. *Physics, Part II*; Wiley: New York, 1962.
2. Enge, H. A.; Wehr, M. R.; Richards, J. A. *Introduction to Atomic Physics*; Addison-Wesley Publishing: Menlo Park, CA, 1972.
3. Pauling, L. *The Nature of the Chemical Bond*; Cornell University Press: Ithaca, NY, 1993.
4. Gordon, A. J.; Ford, R. A. *The Chemist's Companion*; Wiley: New York, 1972.
5. Cotton, F. A. *Chemical Applications of Group Theory*; Wiley-Interscience: New York, 1971.
6. Drago, R. S. *Physical Methods in Inorganic Chemistry*; Reinhold Publishing: New York, 1965.
7. Turro, N. J.; Ramamurthy, V.; Scaiano, J. C. *Principles of Molecular Photochemistry: An Introduction*; University Science Books: Mill Valley, CA, 2009.
8. Ronconi, L.; Sadler, P. J. Reaction Pathways for the Anticancer Complex Trans,Trans,Trans-$[Pt(N_3)_2(OH)_2(NH_3)_2]$. *Dalton Trans.* **2011**, *40*, 262–268.
9. Ashcroft, N. W.; Mermin, N. D. *Solid State Physics*; Holt, Rinehart, and Wilson: New York, 1976.
10. Xu, Y.; Schoonen, M. A. A. The Absolute Energy Positions of Conduction and Valence Bands of Selected Semiconducting Materials. *Am. Miner.* **2000**, *85*, 543–556.
11. Osterloh, F. E. Inorganic Materials as Catalysts for Photochemical Splitting of Water. *Chem Mater.* **2009**, *20*, 35–54.
12. Sherman, D. M. Electronic Structures of Iron(III) and Manganese(IV) (Hydr)oxide Minerals: Thermodynamics of Photochemical Reductive Dissolution in Aquatic Environments. *Geochim. Cosmochim. Acta* **2005**, *69*, 3249–3255.
13. Bard, A. J. Photoelectrochemistry. *Science* **1980**, *207*, 139–144.
14. Henrich, V. E.; Cox, P. A. *The Surface Science of Metal Oxides*; Cambridge University Press: New York, 1996.
15. Wrighton, M. S.; Ellis, A. B.; Wolczanski, P. T.; Morse, D. L.; Abrahamson, H. B.; Ginley, D. S. Strontium Titanate Photoelectrodes: Photoassisted Electrolysis of Water at Zero Applied Potential. *J. Am. Chem. Soc.* **1976**, *93*, 2774–2779.
16. O'Regan, B.; Grätzel, M. A Low-Cost, High-Efficiency Solar Cell Based on Dye-Sensitized Colloidal Titanium Dioxide Films. *Nature* **1991**, *353*, 737–740.
17. Grätzel, M. Dye-Sensitized Solar Cells. *J. Photochem. Photobiol. C: Photochem. Rev.* **2003**, *4*, 145–153.
18. Picraux, S. T. Ion Implantation in Metals. *Annu. Rev. Mater. Sci.* **1984**, *14*, 335–372.
19. Ewing, R. C. Plutonium and Minor Actinides: Safe Sequestration. *Earth Planet. Sci. Lett.* **2005**, *229*, 165– 181.
20. Farnan, I.; Cho H.; Weber, W. J. Quantification of Actinide α-Radiation Damage in Minerals and Ceramics. *Nature* **2007**, *445*, 190–193.

21. Knotek, M. L.; Feibelman, P. J. Ion Desorption by Core-Hole Auger Decay. *Phys. Rev. Lett.* **1978**, *40*, 964–967.
22. Knotek, M. L. Characterization of Hydrogen Species on Metal–Oxide Surfaces by Electron-Stimulated Desorption: TiO_2 and $SrTiO_3$. *Surf. Sci.* **1980**, *101*, 334–340.
23. Wang, L. Q.; Shultz, A. N.; Baer, D. R.; Engelhard, M. H. Interactions of Small Molecules with $TiO_2(110)$ Surfaces: The Role of Defects. *J. Vac. Sci. Technol.* **1996**, *A14*, 1532–1538.
24. Penetrante, B. M.; Schultheis, S. E., Eds. *Non-Thermal Plasma Techniques for Pollution Control*; Springer-Verlag: Berlin, 1993.
25. Hsaio, M. C.; Merritt, B. T.; Penetrante, B. M.; Vogtlin, G. E.; Wallman, P. H.; Tonkyn, R. G.; Shaw, R. R.; Orlando, T. M. Plasma-Assisted Oxidation of Propene. *J. Adv. Oxid. Technol.* **1996**, *1*, 79–84.
26. Tonkyn, R. G.; Barlow, S. E.; Orlando, T. M. Destruction of Carbon Tetrachloride in a Dielectric Barrier/Packed Bed Corona Reactor. *J. Appl. Phys.* **1996**, *80*, 4877–4886.
27. Knotek, M. L.; Feibelman, P. J. Stability of Ionically Bonded Surfaces in Ionizing Environments. *Surf. Sci.* **1979**, *90*, 78–90.
28. Knutsen, K.; Orlando, T. O. Stimulated Desorption of $H^+(D^+)$, $OH^+(OD^+)$, O^+, and NO^+ from Solution-Grown $NaNO_3$ Crystals. *Surf. Sci.* **1996**, *348*, 143–152.
29. Weast, R. C., Ed. *CRC Handbook of Chemistry and Physics, 58th ed.*; CRC Press: Cleveland, OH, 1977.
30. Lanford, W. A.; Trautvetter, H. P.; Ziegler, J. F.; Keller, J. New Precision Technique for Measuring the Concentration Versus Depth of Hydrogen in Solids. *Appl. Phys. Lett.* **1978**, *28*, 566–571.
31. Ziegler, J. F.; Wu, C. P.; Williams, P.; White, C. W.; Terreault, B.; Scherzer, B. M. U.; Schulte, R. L.; Schneid, E. J.; Magee, C. W.; Ligeon, E.; Ecuyer, J. L.; Lanford, W. A.; Kuehne, F. J.; Kamykowski, E. A.; Hofer, W. O.; Guivarc'h, A.; Filleux, C. H.; Deline, V. R.; Evans, C. A.; Cohen, B. L.; Clark, G. J.; Chu, W. K.; Brassard, C.; Blewer, R. S.; Behrisch, R.; Appleton, B. R.; and Allred, D. D. Profiling Hydrogen in Materials Using Ion Beams. *Nucl. Inst. Methods* **1978**, *149*, 19–39.
32. Paul, M. C.; Sen, R.; Bhadra, S. K.; Dasgupta, K. Radiation Response Behaviour of Al Co-Doped Germano-Silicate SM Fiber at High Radiation Dose. *Optics Commun.* **2009**, *282*, 872–878.
33. Levy, P. Color Centers and Radiation-Induced Defects in Al_2O_3. *Phys. Rev.* **1961**, *123*, 1226–1233.
34. Hughes, H. L.; Benedetto, J. M. Radiation Effects and Hardening of MOS Technology: Devices and Circuits. *IEEE Trans. Nucl. Sci.* **2003**, *50*, 500–521.
35. Griscom, D. L. Defects in Amorphous Insulators. *J. Non-Cryst. Solids* **1978**, *31*, 241–266.
36. Griscom, D. L. Defects and impurities in Noncrystalline Solids. In *Defects and Their Structure in Nonmetallic Solids*; Henderson, B.; Hughes, A. E., Eds.; Plenum Press: New York, 1976, pp. 323–353.
37. Griscom, D. L. Silica Glasses. *MRS Bull.* **1987**, **7**, *June 16/Aug 15*, 20–23.
38. Iler, R. K. *The Chemistry of Silica*; Wiley: New York, 1979.

39. El-Shamy, T. M.; Lewins, J.; Douglas, R. W. The Dependence on the pH of the Decomposition of Glasses by Aqueous Solutions. *Glass Technol.* **1972**, *13*, 81–87.
40. Devine, R. A. B. Macroscopic and Microscopic Effects of Radiation in Amorphous Silica. *Nucl. Instr. Methods* **1994**, *B91*, 378–390.
41. Arnold, G. W. Ambient Hydration of Near-Surface Region in H/D Implanted Fused Silica. *Nucl. Instr. Methods* **1988**, *B32*, 268–271.
42. Su, Y. L.; Balmer, M. L.; Bunker, B. C. Evaluation of Cesium Silicotitanate as an Alternative Waste Form. *Mater. Res. Soc. Symp. Proc.* **1997**, *465*, 457–464.
43. Sales, B. C.; White, C. W.; Boatner, L. A. A Comparison of the Corrosion Characteristics of Synthetic Monazite and Borosilicate Glass Containing Simulated Nuclear Defense Waste. *Nucl. Chem. Waste Manage.* **1983**, *4*, 281–289.

PART FIVE

Reactions Involving Tetrahedral Oxides

14

Aqueous Polymerization of Silicates and Aluminosilicates

14.1 INTRODUCTION

Part Five of this book is devoted to silicates for several important reasons. First, silicates represent critical components of our planet and our lives. Silicon is the second most abundant element in Earth's crust after oxygen, representing about 28% of the atoms present [1]. As such, transformations of silicate minerals dominate much of the aqueous geochemistry of Earth. Every day, each of us encounters materials and objects the primary constituents of which are silicon oxides and related phases such as aluminosilicates. Granite facings on buildings, bricks, glass, pottery, ceramics, engineered materials used in water purification, catalysis, electronics, and even the optical fibers used in our most advanced communication systems are all silica based. Aluminosilicate minerals are even used as food additives.

A key attribute of silicates that distinguishes them from most of the oxides highlighted in Parts One through Four of this book is that the Si(IV) cation is almost always present in a tetrahedral rather than in an octahedral coordination geometry. Exceptions include a few high-pressure phases such as stishovite [2] (see Chapter 2) and a limited number of chelated Si(IV) complexes (see Section 14.3). The authors know of no stable compounds where Si(IV) is coordinated to only three oxygen atoms. The pathways for both forming and destroying silicate bonds are substantially different than for octahedral metal ions. Ligand-exchange pathways for silicate ions are via nucleophilic attack, where the coordination number increases in a transition state from four to five or even six (see Section 14.3 and Chapters 4 and 5). These contrast with pathways for octahedral metal ions, such as Al(III), where it is easier to decrease the coordination number from six to five or four in dissociative ligand exchange reactions.

Of course, Si(IV) is not the only common element capable of forming tetrahedral oxide species. As outlined in Chapters 2 and 4, any cation with an ionic radius between roughly 0.03 nm and 0.055 nm can fit within the tetrahedral void between four close-packed oxygen anions, as expressed by Linus Pauling's First Rule of coordination chemistry [3] (see Chapter 2). This list includes metals with valences of +7 (in $[MnO_4]^-$), +6 (in H_2SO_4), +5 (in H_3PO_4), +4 (in $[Si(OH)_4]^{\circ}$), +3 (in $[Al(OH)_4]^-$), +2 (in $[Be(OH_2)_4]^{2+}$), and even +1 (in $[Li(OH_2)_4]^+$). In fact, more than 20 elements have cations that can assume a tetrahedral coordination geometry involving oxygen anions, some of which are listed in Figure 14.1.

Tetrahedral Si(IV) reacts in a unique way. Dissolved tetrahedral species of lower- and higher-charged metals, including simple aquo cations such as $[Li(OH_2)_4]^+$ and monomeric oxo-anions such as $[MnO_4]^-$, form solids that are ionic salts. Tetrahedral species of metals of intermediate valence, such as Si(IV) or P(V), polymerize to form larger oligomers or precipitates. Silicates represent one of the few systems in which the tetrahedral geometry of the soluble monomer ion, $[Si(OH)_4]^{\circ}$, is preserved on polymerization. Phosphates are another. However, for many other metals, monomers may be tetrahedral but they form structures of linked octahedra when they polymerize.

The fact that Si(IV) remains tetrahedral under most conditions leads to the formation of a wide range of soluble silicate structures. Soluble silicate species can polymerize to incorporate foreign cations, some of which assume the tetrahedral geometry of the host lattice. The most important of the cations that polymerize with Si(IV) is Al(III). The copolymerization of Al(III) and Si(IV) species leads to a wide range of important aluminosilicate phases. Some of these phases, such as clay minerals, contain Al(III) in octahedral coordination; some phases, such as zeolites, contain tetrahedral Al(III); and some clays can contain Al(III) in both coordination geometries. In this chapter, we explore the formation of all these phases from aqueous solutions, starting with simple silicates and progressing to aluminosilicates. At ambient temperatures, polymerization processes involving silicates and aluminosilicates often yield amorphous phases. Therefore, Section 14.6 describes hydrothermal synthesis conditions used to transform these precipitates into high-quality crystalline materials.

14.2 SOLUBLE SILICATE POLYMERS

At ambient conditions, the hydrolysis diagram for silica in water indicates that the dominant species present in dilute acidic to neutral solutions is the tetrahe-

hydrogen 1 **H** 1.0079																	helium 2 **He** 4.0026
lithium 3 **Li** 6.941	beryllium 4 **Be** 9.0122											boron 5 **B** 10.811	carbon 6 **C** 12.011	nitrogen 7 **N** 14.007	oxygen 8 **O** 15.999	fluorine 9 **F** 18.998	neon 10 **Ne** 20.180
sodium 11 **Na** 22.990	magnesium 12 **Mg** 24.305											aluminium 13 **Al** 26.982	silicon 14 **Si** 28.086	phosphorus 15 **P** 30.974	sulfur 16 **S** 32.065	chlorine 17 **Cl** 35.453	argon 18 **Ar** 39.948
potassium 19 **K** 39.098	calcium 20 **Ca** 40.078	scandium 21 **Sc** 44.956	titanium 22 **Ti** 47.867	vanadium 23 **V** 50.942	chromium 24 **Cr** 51.996	manganese 25 **Mn** 54.938	iron 26 **Fe** 55.845	cobalt 27 **Co** 58.933	nickel 28 **Ni** 58.693	copper 29 **Cu** 63.546	zinc 30 **Zn** 65.38	gallium 31 **Ga** 69.723	germanium 32 **Ge** 72.64	arsenic 33 **As** 74.922	selenium 34 **Se** 78.96	bromine 35 **Br** 79.904	krypton 36 **Kr** 83.798
rubidium 37 **Rb** 85.468	strontium 38 **Sr** 87.62	yttrium 39 **Y** 88.906	zirconium 40 **Zr** 91.224	niobium 41 **Nb** 92.906	molybdenum 42 **Mo** 95.96	technetium 43 **Tc** [98]	ruthenium 44 **Ru** 101.07	rhodium 45 **Rh** 102.91	palladium 46 **Pd** 106.42	silver 47 **Ag** 107.87	cadmium 48 **Cd** 112.41	indium 49 **In** 114.82	tin 50 **Sn** 118.71	antimony 51 **Sb** 121.76	tellurium 52 **Te** 127.60	iodine 53 **I** 126.90	xenon 54 **Xe** 131.29
caesium 55 **Cs** 132.91	barium 56 **Ba** 137.33		hafnium 72 **Hf** 178.49	tantalum 73 **Ta** 180.95	tungsten 74 **W** 183.84	rhenium 75 **Re** 186.21	osmium 76 **Os** 190.23	iridium 77 **Ir** 192.22	platinum 78 **Pt** 195.08	gold 79 **Au** 196.97	mercury 80 **Hg** 200.59	thallium 81 **Tl** 204.38	lead 82 **Pb** 207.2	bismuth 83 **Bi** 208.98	polonium 84 **Po** [209]	astatine 85 **At** [210]	radon 86 **Rn** [222]
francium 87 **Fr** [223]	radium 88 **Ra** [226]		rutherfordium 104 **Rf** [261]	dubnium 105 **Db** [262]	seaborgium 106 **Sg** [266]	bohrium 107 **Bh** [264]	hassium 108 **Hs** [277]	meitnerium 109 **Mt** [268]	darmstadtium 110 **Ds** [271]	roentgenium 111 **Rg** [272]							

lanthanum 57 **La** 138.91	cerium 58 **Ce** 140.12	praseodymium 59 **Pr** 140.91	neodymium 60 **Nd** 144.24	promethium 61 **Pm** [145]	samarium 62 **Sm** 150.36	europium 63 **Eu** 151.96	gadolinium 64 **Gd** 157.25	terbium 65 **Tb** 158.93	dysprosium 66 **Dy** 162.50	holmium 67 **Ho** 164.93	erbium 68 **Er** 167.26	thulium 69 **Tm** 168.93	ytterbium 70 **Yb** 173.05	lutetium 71 **Lu** 174.97
actinium 89 **Ac** [227]	thorium 90 **Th** 232.04	protactinium 91 **Pa** 231.04	uranium 92 **U** 238.03	neptunium 93 **Np** [237]	plutonium 94 **Pu** [244]	americium 95 **Am** [243]	curium 96 **Cm** [247]	berkelium 97 **Bk** [247]	californium 98 **Cf** [251]	einsteinium 99 **Es** [252]	fermium 100 **Fm** [257]	mendelevium 101 **Md** [258]	nobelium 102 **No** [259]	lawrencium 103 **Lr** [262]

Figure 14.1 A Periodic Table of the elements highlighting those elements capable of forming hydrolysis products and/or oxide solids containing tetrahedral cations.

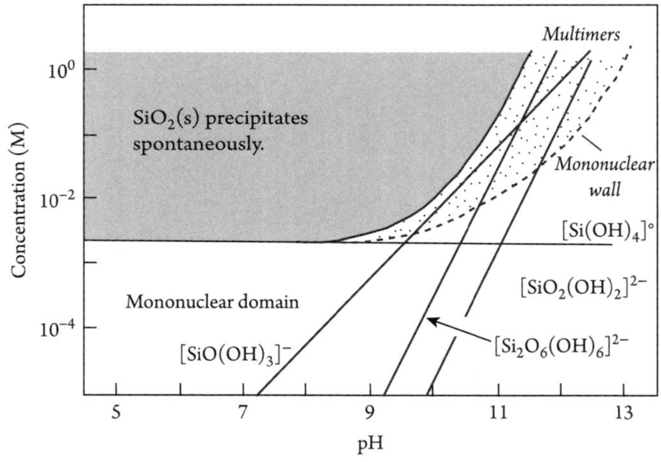

Figure 14.2 A solubility and speciation diagram for silica as a function of pH. Below the lower curve, only monomeric $[Si(OH)_4]^\circ$ and its ionized forms are present. Above the upper curve, the silica is present as insoluble colloidal silica particles. The region between the two curves represents the multimeric domain within which polymeric species are present. Adapted with permission from Stumm, W.; Huper, H.; Champlin, R. L. Formulation of Polysilicates as Determined by Coagulation Effects. *Environ. Sci. Technol.* **1967**, *1*, 221–227.

dral monomer silicic acid ($[Si(OH)_4]^\circ$) [2, 4] (Fig. 14.2). In basic solutions, silicic acid can lose two protons via the reactions [5, 6]

$$Si(OH)_4^\circ \rightleftarrows Si(OH)_3O^- + H^+ \quad pK_1 = 9.46\ (I = 0.5\ M) \qquad \textbf{(14.1)}$$

$$Si(OH)_3O^- \rightleftarrows Si(OH)_2O_2^{2-} + H^+ \quad pK_2 = 12.56\ (I = 0.5\ M) \qquad \textbf{(14.2)}$$

The deprotonation of silicic acid has two major consequences. First, the creation of silicate anions by deprotonation increases the solubility of silica from millimolar concentrations at near-neutral pH to well more than 1 M in strongly basic solutions. Second, the presence of silicate anions stimulates condensation reactions (see Section 14.3) because they can act as effective nucleophiles.

The first step in silicate polymerization involves dimerization and oxo-bridge formation:

$$Si(OH)_4^\circ + Si(OH)_4^\circ \rightleftarrows (HO)_3Si-\mu_2O-Si(OH)_3 + H_2O \qquad \textbf{(14.3)}$$

$$(HO)_3SiO^- + Si(OH)_4^\circ \rightleftarrows (HO)_3Si-\mu_2O-Si(OH)_3 + OH^- \qquad \textbf{(14.4)}$$

Eqs. 14.3 and 14.4 indicate that polymerization should be more pronounced as the concentration of dissolved monomers increases. For example, in Eq. 14.3, dimerization should increase as the concentration of $[Si(OH)_4]^\circ$ squared. This nonlinear dependence on concentration means that dramatic changes in oligomer speciation can occur with small changes in total metal concentration and pH. The onset of dimerization defines a boundary called the *mononuclear wall*. In basic solutions, a region called the *multimeric zone* exists between the mononuclear wall and the true solubility limit above which high-molecular-weight oligomers and amorphous colloidal particles precipitate spontaneously from solution. The multimeric zone contains an extensive range of polymeric species that depends on both pH and the net silica concentration.

The multimeric zone forms the basis for commercial formulations called *soluble silicates* that are widely used as coatings for paints, adhesives, cements, and inhibitors of metal corrosion [7]. Soluble silicates are often prepared by dissolving SiO_2 and alkali metal oxides (M_2O) in water. In such formulations, M_2O additions are used to control the solution pH via

$$Na_2O(s) + H_2O \rightleftarrows 2Na^+ + 2OH^- \quad \textbf{(14.5)}$$

The average molecular weights of species found in representative soluble-silicate solutions are shown in Figure 14.3. Large silicate oligomers can be present in the multimeric zone, but only a few of which have been isolated for structural characterization. For a solution in which $SiO_2/K_2O = 4$, the average molecular weight of 848 indicates that the average polymer contains around 14 SiO_2 units. Figure 14.3 also illustrates that higher pH values (i.e., higher M_2O

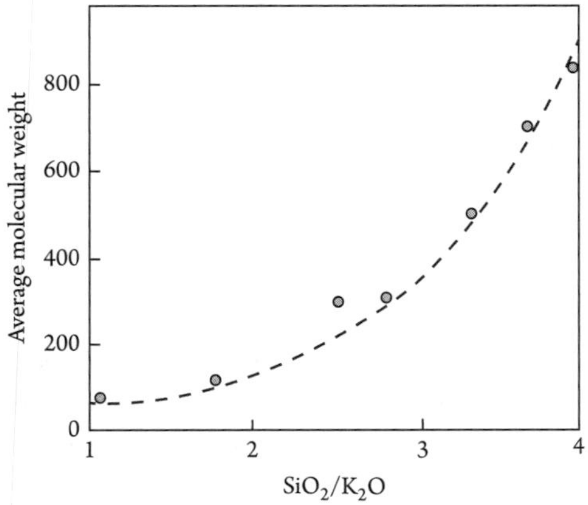

Figure 14.3 The weight–average molecular weights of the species present in soluble silicate solutions as a function of the SiO_2-to-K_2O ratio. The smaller this ratio, the more basic the solution and the more anionic the soluble silicates. Adapted from data in Weldes, H. H.; Lange, K. R. Properties of Soluble Silicates. *Ind. Eng. Chem.* **1969**, *61*, 29–45.

levels) result in smaller polymer sizes, not only because higher $[OH^-]$ concentrations shift Eq. 14.4 to the left, but also for mechanistic reasons to be discussed later.

Molecular spectroscopies such as ^{29}Si and ^{1}H nuclear magnetic resonance (NMR) and Raman vibrational spectroscopy have been deployed to identify some of the structures present in soluble-silicate solutions. A representative ^{29}Si NMR spectrum of a soluble-silicate solution is shown in Figure 14.4. (Complete analyses of these complex spectra are beyond the scope of this book. Those of you interested in more details are referred to [8–14].) In ^{29}Si spectra, the chemical shifts of the peaks provide information regarding the number of bridging bonds around a given Si(IV) center. Relaxation times and coupling between peaks help identify whether multiple silicon sites are present in the same species. Peak intensities are proportional to the species concentrations, so that NMR spectra can yield information directly about equilibria. Finally, isotope-exchange methods involving ^{17}O NMR exist that can provide the rates of ligand-exchange reactions involving these oligomers.

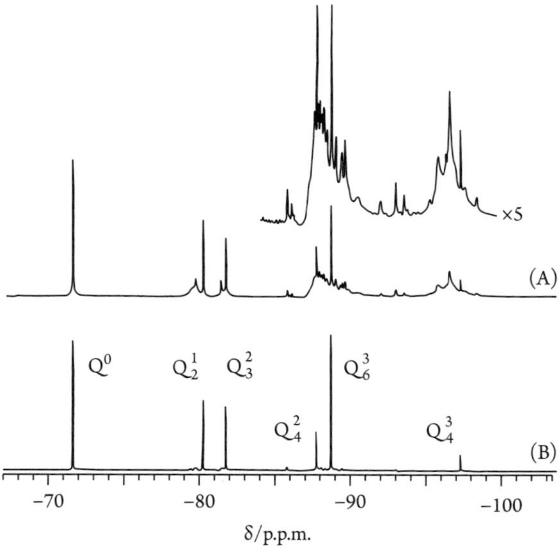

Figure 14.4 ^{29}Si nuclear magnetic resonance spectra of basic solutions containing soluble silicates. The Q notation identifies the number of bridging oxygens for the given Si(IV) species. Note that as the number of bridging oxygens increases, the spectra become more complex, indicating the spectral coupling between sites as well as increased oligomer complexity.

SOURCE: Reproduced with permission from Knight, C. T. G.; Kirkpatrick, J.; Oldfield, E. Silicon-29 Multiple Quantum Filtered N.M.R. Spectroscopic Evidence for the Presence of Only Six Single Site Silicate Anions in a Concentrated Potassium Silicate Solution. *J. Chem. Soc. Chem. Comm.*, **1989**, *1989*, 919–921.

The different chemical shift regimes shown in Figure 14.4 indicate different local coordination geometries as denoted by "Q" notation. Q refers to the number of bridging bonds around a given site: $Q^0 = [Si(OH)_4]^\circ$; the dimer $[(OH)_3Si{-}\mu_2\text{-}O{-}Si(OH)_3]^\circ$ contains two Q^1 sites; Q^2 is a site with two bridging oxygens, as in chains and isolated rings; whereas Q^3 has three bridging and one nonbridging oxygen, as in interconnected rings and cages.

The array of soluble silicate structures identified to date is magnificent, and new species are being discovered almost daily. Figure 14.5 shows only a handful of the small oligomers identified in soluble-silicate solutions at high pH. The structures in Figure 14.5 indicate that silicate oligomerization does not occur in a random fashion, but favors the formation of a few discrete species containing a fixed number (n) of silicate tetrahedra (as is the case for octahedral complexes see Chapter 5). At high pH, species include the n = 1 monomer $[Si(OH)_4]^\circ$, the n = 2 oxo-bridge dimer $[(OH)_3\,Si{-}\mu_2\text{-}O{-}Si(OH)_3]^\circ$, and two trimeric species:

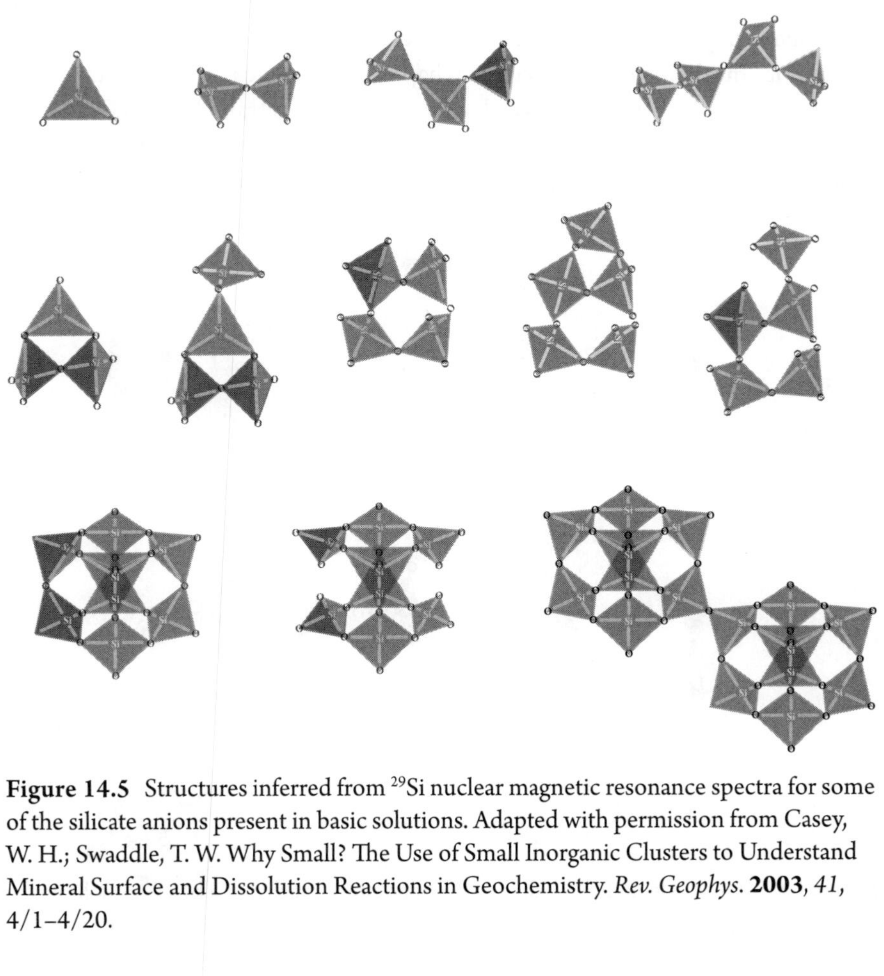

Figure 14.5 Structures inferred from ^{29}Si nuclear magnetic resonance spectra for some of the silicate anions present in basic solutions. Adapted with permission from Casey, W. H.; Swaddle, T. W. Why Small? The Use of Small Inorganic Clusters to Understand Mineral Surface and Dissolution Reactions in Geochemistry. *Rev. Geophys.* **2003**, *41*, 4/1–4/20.

a linear chain and a closed ring. Above n = 3, rings containing three and four tetrahedra predominate. Larger species contain multiple rings, until the number of rings is sufficient to form closed-cage structures. For example, the n = 8 cubane species contains eight Q^3 units in interconnected four-membered rings that form a silicate cube. Linked silicate cubes have also been reported. As species increase in size, Q^1 sites eventually disappear in favor of Q^2 and eventually Q^3 sites. However, until recently, only one soluble-solution species was known to contain Q^4 sites as in bulk silica. This species contains two corner-shared cubane units (Fig. 14.5) [9]. Now, a handful of other soluble Q^4-containing species have been identified in extremely concentrated (ΣSi = 2 M) solutions via NMR methods.

14.3 SILICATE POLYMERIZATION MECHANISMS AND REACTION KINETICS

Understanding the rate and extent of silicate polymerization in water involves understanding mechanistic details that are unique to tetrahedral oxide species. The polymerization of soluble-silicate species occurs via the *associative nucleophilic attack* mechanism introduced in Chapter 4 (Fig. 14.6). The first step in this mechanism involves the attack by a basic molecule on tetrahedral $[Si(OH)_4]^{\circ}$ to produce a reactive five-coordinate intermediate that then expels one of its five ligands to reestablish the original four-coordinate geometry. If the attacking

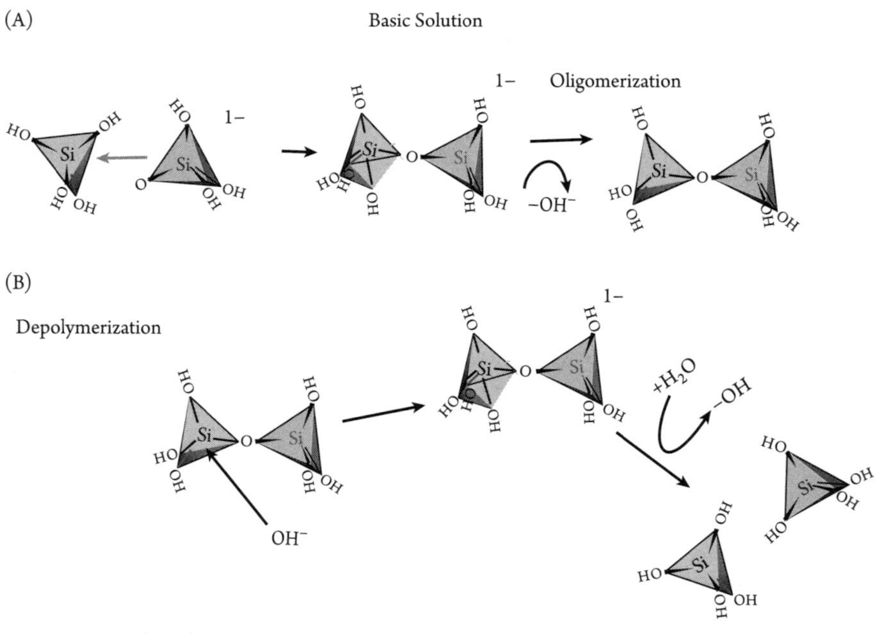

Figure 14.6 (A, B) Depictions of the nucleophilic attack mechanism by which soluble silicate species polymerize (A) and depolymerize (B) in water. The transition state is a five-coordinated Si(IV) which collapses back to a tetrahedral geometry when the ligand leaves.

species is another silicate species and the leaving group is a hydroxyl group, the net result is polymerization (e.g., Eq. 14.4). Conversely, attack by a free hydroxide ion onto the metal center in an oligomer can result in depolymerization back to $[Si(OH)_4]^\circ$. Next, we highlight factors that bias the formation or disruption of bridging bonds from the perspective of both the nucleophile and the metal center that is subject to this nucleophilic attack.

14.3.1 Effects of Nucleophile Basicity and Concentration

From the perspective of the nucleophiles, the two key parameters that control silicate reactivity in water are (1) ligand basicity and (2) relative species concentrations. Ligand basicity can be conceptualized on the basis of formal charge. The oxygen anions in both H_2O and $[Si(OH)_4]^\circ$ have a formal charge of zero, which means that both species are weak bases that are not particularly effective for attacking any tetrahedral Si(IV) centers to form a five-coordinate intermediate. However, the deprotonation of either species to form OH^- or $[Si(OH)_3O]^-$ creates oxygen sites with a formal charge of −1. Both these species are strong bases that can readily participate in nucleophilic attack. Both species compete with each other in attacking tetrahedral silicate sites. If $[Si(OH)_3O]^-$ outcompetes OH^-, polymerization is promoted, whereas if OH^- outcompetes $[Si(OH)_3O]^-$, the net result is depolymerization (Fig. 14.6).

More quantitative estimates for the basicity of the various species present can be arrived at from pK_a values. The pK_a values associated with the removal of the first and second protons from $[Si(OH)_4]^\circ$ to form $[Si(OH)_3O]^-$ and $[Si(OH)_2O_2]^{2-}$ are approximately 9.5 and 12.6, respectively (Eqs. 14.1 and 14.2). These values are to be compared with the dissociation constant for water ($-\log K_w = 14$) and the pK_a for the ammonium cation of 9.2. Brønsted basicity values for the anions are simply arrived at using the relationships $K_b = K_w/K_a$ and $pK_b = pK_w - pK_a$. For example:

$$NH_3 + H_2O \rightleftarrows NH_4^+ + OH^-,\ K_b = \frac{10^{-14}}{6.3\times10^{-10}} = 1.6\times10^{-5},\ pK_b = 4.8 \quad \textbf{(14.6)}$$

$$Si(OH)_3O^- + H_2O \rightleftarrows Si(OH)_4 + OH^-,\ K_b = 3.1\cdot10^{-5},\ pK_b = 4.5 \quad \textbf{(14.7)}$$

$$Si(OH)_2O_2^{2-} + H_2O \rightleftarrows Si(OH)_3O^- + OH^-,\ K_b = 4\cdot10^{-2},\ pK_b = 1.4 \quad \textbf{(14.8)}$$

A smaller pK_b indicates a stronger base. As a common point of reference, $[Si(OH)_3O]^-$ has a basicity that is similar to that of the neutral ammonia molecule dissolved in water. These values lead to a ranking of species basicity:

$$H_2O < NH_3 \sim [Si(OH)_3O]^- < [Si(OH)_2O_2]^{2-} < OH^-$$

This ranking indicates that if the concentrations of OH^- and the silicate monomers are identical, depolymerization should be favored, helping to define the mononuclear wall shown in Figure 14.2. Added OH^- leads to more depolymerization.

The importance of nucleophile concentrations on polymerization is clearly apparent in Figure 14.2. For a solution at equilibrium with $SiO_2(s)$, the concentrations of OH^- and the various silicate monomers are never identical. In fact, silicate concentrations are invariably higher than OH^-. For example, at pH 13, $[OH^-] = 0.1$ M, whereas the position of the mononuclear wall is at 1 M. At pH 7, $[OH^-] = 10^{-7}$ M compared with the $[Si(OH)_4]^\circ$ concentration of $2 \cdot 10^{-3}$ M. Clearly, solution pH is the primary factor that determines the distribution of nucleophilic species, including OH^- and all soluble silicate species. Below pH 8, the weak nucleophile $[Si(OH)_4]^\circ$ predominates. For $8 < pH < 12$, the stronger nucleophile $[Si(OH)_3O]^-$ prevails whereas the strongest nucleophilic $[Si(OH)_2O_2]^{2-}$ dominates in strongly basic solutions ($pH > 12.6$).

The interplay of the basicity and concentrations of the various silicate species is complicated. However, in the most basic solutions, OH^- is so concentrated that it can destroy polymers faster than they can reform. As silicate concentrations increase and/or the pH decreases between the mononuclear wall and the true solubility limit, larger and more complex silicate species form.

Finally, it is important to note that although the concentrations of nucleophilic species are greatly reduced in solutions with a pH less than 9, these low concentrations of silicate nucleophiles and OH^- cannot be ignored. Even at pH 7, the $[Si(OH)_3O]^-$ concentration of $6 \cdot 10^{-6}$ M still represents 0.3% of the total dissolved silicon. This small concentration remains kinetically important to the solution chemistry even if the species is not the dominant form. This small nucleophile is sufficient to promote the formation of small oligomers such as dimers and trimers that have been detected at $5 < pH < 7$ [15].

It is difficult to measure polymerization rates in neutral or acidic solutions because species concentrations are too low for conventional spectroscopies and because the kinetics of reaction are slow. However, reaction rates have been determined in concentrated basic solutions via ^{29}Si and ^{17}O NMR [12, 16, 17]. For example, in a 2.8-M silicate solution with a K_2O-to-Si ratio of 2.3, the rate at which two $[Si(OH)_2O_2]^{2-}$ anions react with each other to form a dimer, has a rate constant $k_1(90°C) = 0.13$ kg/mole-s and $\Delta H^\ddagger = 67$ kJ/mole [18]. The rate constant for dissociation of the dimer (the reverse reaction) is $k_{-1} = 1.4\ s^{-1}$ and $\Delta H^\ddagger = 65$ kJ/mole. Some ^{17}O NMR measurements in small silicate oligomers indicate that the lifetimes of oxygens in bridging and terminal sites are about the same [19], indicating that the five-coordinate intermediate exhibits little selectivity regarding the expulsion of either OH^- or $[Si(OH)_3O]^-$ to form a tetrahedral product. Unfortunately, results for larger molecules cannot be determined via NMR because oxygens in larger silicate clusters are inert to substitution on the NMR timescale of milliseconds to minutes.

A secondary factor that can inhibit the formation of polymeric species in highly basic solutions is simple electrostatic repulsion. The $[Si(OH)_2O_2]^{2-}$ species carries a charge of -2 and is strongly repelled by identical anionic monomers. Such repulsion is similar to that experienced by cationic species that attempt to polymerize with one another across bound hydroxide groups (see Chapter 5). However, this electrostatic repulsion can be minimized if counterions associate with the dissolved species, forming ionically bonded bridges. Here, cation size is important. The rates of silicate condensation in basic solutions have been shown to be faster in solutions containing K^+ rather than Na^+. NMR investigations [20] have shown that alkali cations do indeed influence species distributions in basic soluble silicate solutions (see Fig. 14.5). The degree of polymerization (i.e., Q) increases in the order $Li^+ < Na^+ < K^+ < Rb^+ < Cs^+$. This trend is established because larger alkali-metal cations can associate with multiple silicate anions. The larger the cation, the weaker the interactions, which frees up active sites on the anions to polymerize with one another. Counterions are important in these polymerization reactions because they allow the reactive moieties to approach one another. Similar counterion effects impact polyoxometalate ions, such as electron exchanges between aluminotungstate polyoxometalates (see Chapter 5).

14.3.2 The Role of Metal Center Attributes on Polymerization

From the perspective of the tetrahedral metal center, the nucleophilic attack mechanism is controlled by the ease at which the metal center can be deformed to accommodate the expansion from a four- to a five-coordinate intermediate. For Si(IV) centers, this deformation depends critically on how many of the oxygen anions in the inner-coordination sphere are bridging. The nonbridging hydroxyl groups in $[Si(OH)_4]^{\circ}$ are attached only to the single central Si(IV) site and are relatively easy to displace. The ease of motion makes Q^0 monomers the most reactive silicate species. In contrast, each bridging oxygen in more highly polymerized species is anchored to an additional Si(IV) site, which makes the site harder to rearrange into the five-coordinate intermediate. Thus, high-Q-value sites are less reactive to ligand exchange, and the trend presented here is general:

$$Q^4 < Q^3 < Q^2 < Q^1 < Q^0$$

This trend is followed whether the ligand exchange leads to polymerization or depolymerization, which means that isolated chains and rings (Q^2 units) tend to evolve into interconnected rings and cages (Q^3 units). However, at room temperature, Q^3 units exhibit such slow ligand-exchange kinetics that Q^4 units are rarely seen among the soluble silicates in aqueous solutions.

Another factor leading to the formation of dense silicate structures is the chelate effect described in Chapter 5 in relation to trivalent metals in octahedral complexes. The chelate effect states that the more points of attachment a ligand has with a metal cation, the more difficult that ligand is to displace. For example, it is much harder to disrupt the three bridging bonds connecting a Q^3 unit to a silica surface (resulting in dissolution) than it is to break the single point of attachment between two Q^1 units.

The chelate effect has another profound impact on aqueous silicate chemistry that has only recently been appreciated. Five- and six-coordinate silicate

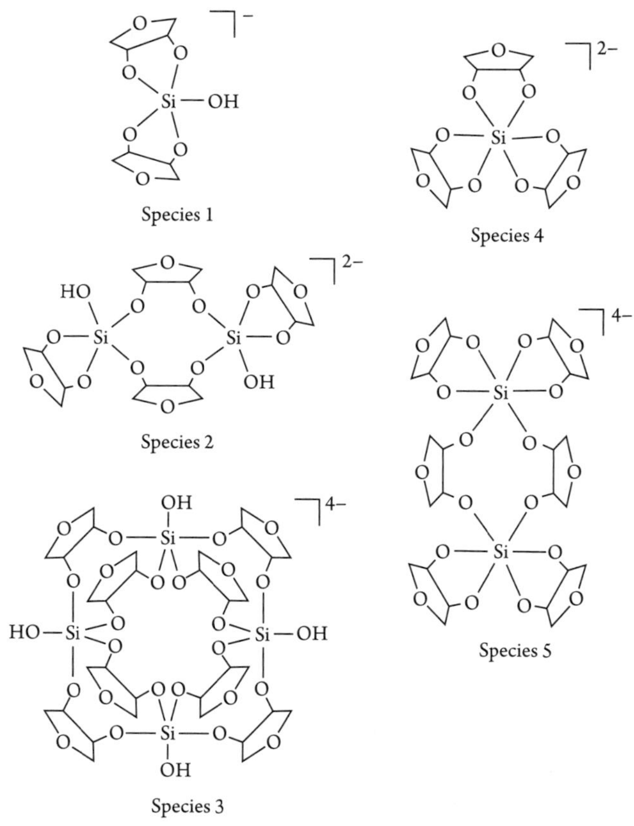

Figure 14.7 Examples of soluble silicate species containing central Si ions that are stabilized in five- or sixfold coordination geometries.

SOURCE: Reproduced with permission from Kinrade, S. D.; Del Nin, J. W.; Schach, A. S.; Sloan, T. A.; Wilson, K. L.; and Knight, C.T.G. Stable Five- and Six-Coordinated Silicate Anions in Aqueous Solution *Science*, **1999**, *285*, 1542–1545.

complexes can form easily at room temperature when stabilized by hydroxyl functional groups contained within chelating ligands such as catechol, polyols, and sugars [21–23]. For example, three catechols coordinate to a Si(IV), each with two hydroxyl groups, to make a six-coordinate silicate complex. Ligand-stabilized five- and six-coordinate Si(IV) centers can form water-soluble dimers and even tetramers (Fig. 14.7). In most instances, ligands that are effective in stabilizing higher coordination numbers around Si(IV) contain multiple hydroxyl groups that must be arrayed in certain positions to act as effective chelating moieties. Similar complexes may play a role in the activity of *silicatein* and other proteins that diatoms and other marine organisms use to form their skeletons in biomineralization (see Chapter 9).

Although a major focus of this chapter is on tetrahedral Si(IV), many other tetrahedral cations form soluble species in water (Table 14.1). Many of these species are also susceptible to polymerization via the nucleophilic-attack mechanism. The relative ease of attack on tetrahedral cations other than silicon depends on the size and field strength of the central cation. For example, Al(III) has a larger ionic radius ($r = 0.051$ nm) than Si(IV) ($r = 0.042$ nm). This means that oxygens in the tetrahedral coordination sphere are farther apart around Al(III) than they are around Si(IV), which provides more space for any incoming ligand. Due to the lower charge on Al(III), Al-O bonds are also weaker than Si-O bonds, making it easier for the incoming ligand to shove them out of the way. Both factors make it easier for a tetrahedral Al(III) site to expand its coordination sphere to accommodate five and six ligands. The ease with which a given tetrahedral cation can increase its coordination number is highly significant for polymerization mechanisms, especially in mixed-cation systems such as the aluminosilicates (see Section 14.5).

14.4 SOL–GEL PROCESSING OF SILICATE MATERIALS

Although the emphasis in this book is on aqueous solutions, it is important to mention briefly hydrolysis and condensation reactions that take place in alcohol solutions containing silicon alkoxides (an alkoxide is the conjugate base of an alcohol). The so-called "sol–gel" processing of chemically prepared silicate thin films and nanomaterials is based on such solutions. Although nonaqueous polymerization is beyond the scope of this book, work done to elucidate reaction mechanisms reinforces the general concepts discussed earlier involving the polymerization of silica in water. Those of you interested in more in-depth discussions of sol-gel chemistry in general and the kinetics of hydrolysis and condensation reactions in sol–gel solutions in particular should consult [24].

As in aqueous polymerization, the most definitive experiments for probing hydrolysis and condensation kinetics have been provided by ^{29}Si and ^{1}H NMR

[25–27]. The three types of reactions required to promote the polymerization of silicon alkoxides in alcohol include

$$\mathrm{SiOR + H_2O \rightleftarrows SiOH + ROH} \qquad \mathrm{k_H} \tag{14.9}$$

$$\mathrm{2SiOH \rightleftarrows Si - \mu_2O\text{–}Si + H_2O} \qquad \mathrm{k_{CW}} \tag{14.10}$$

$$\mathrm{SiOH + SiOR \rightleftarrows Si - \mu_2O\text{–}Si + ROH} \qquad \mathrm{K_{CA}} \tag{14.11}$$

Here, k_H refers to the rate constant for alcohol hydrolysis, k_{CW} is the rate constant for a condensation reaction involving two silanol groups, and k_{CA} defines the rate of condensation between a hydroxyl and alkoxy group. Because of their bulk and lower basicity, alkoxide groups function as protective groups that must first be removed to allow condensation reactions involving more active silanols to occur. The stability of alkoxides in water-free alcohol solutions is the reason why sol–gel processing is often preferred rather than aqueous synthesis, because controlled polymerization occurs only when water is added to the solution.

The hydrolysis and removal of the protective alkoxide groups to form silanols is given by Eqn. 14.9. Hydrolysis requires the presence of water and is sensitive to the effective pH of the alcohol solution. In acidic solutions, protonation of the alkoxy oxygens promotes hydrolysis via alcohol formation. In basic solutions, the hydroxide ion can promote hydrolysis via the standard nucleophilic-attack mechanism discussed earlier. However, because of steric blockage by alkoxy groups, this mechanism is less effective than alcohol formation by protonation. NMR studies show that, in acidic solutions, hydrolysis occurs at a much faster rate than the subsequent condensation reactions. In contrast, hydrolysis is often the rate-limiting step in basic solutions [25–27].

When hydrolysis occurs, condensation reactions lead to oxide polymerization. In sol–gel solutions, such condensation reactions can either involve two silanol groups condensing to produce a water molecule (Eq. 14.10), as in aqueous solutions, or a reaction involving one silanol and one alkoxy group to release an alcohol molecule (Eq. 14.11). Once again, the steric bulk of the alkoxy group dictates that Eq. 14.11 occurs at a substantially slower rate than Eq. 14.10. For example, in a 2.24 M solution of $[\mathrm{Si(OCH_3)_4}]^\circ$ containing $1.64\cdot10^{-3}$ M HCl, $k_H = 0.033$ L/mole-s, $k_{CW} = 1\cdot10^{-4}$ L/mole-s, and $k_{CA} = 1.66\cdot10^{-5}$ L/(mole-s).

Obtaining an accurate description for the evolution of all species in this polymerization process requires rate coefficients for each reaction, involving each independent species. Possible species and reactions at the nearest-neighbor level are depicted in the matrix shown Figure 14.8, where each (X, Y, Z) label represents the number of –OR, –OH, and –OSi functional groups attached to the central Si(IV) atom, respectively. A statistical analysis [26] shows that at the

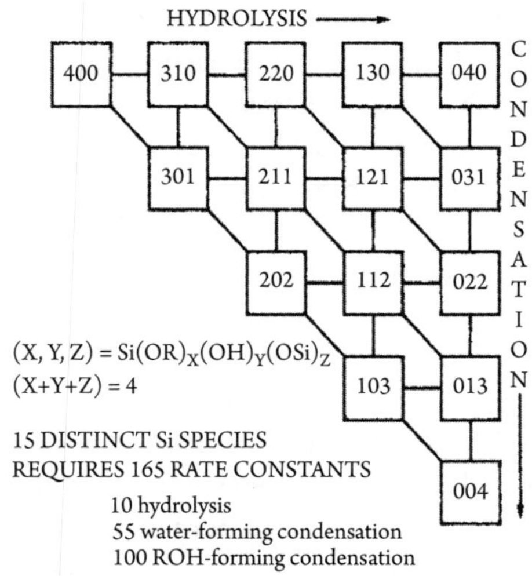

Figure 14.8 A matrix showing the possible chemical species that can be produced during the hydrolysis and condensation of a sol–gel reaction system consisting of $Si(OR)_4$, ROH, and H_2O, where R is the alkoxy group in the alcohol. The notation used in the matrix is shown in the lower left, as are the number of distinct species (15), unique chemical reactions, and their associated rate constants (165). At the next-nearest functional group level, 1365 distinct species are possible, involving 199,290 distinct reactions and rate constants.

SOURCE: Reproduced with permission from Kay, B. D.; Assink, R. A.; Sol-Gel Kinetics. *J. Non-Cryst. Solids*, **1988**, *104*, 112–122.

next-nearest-neighbor functional-group level, there are 1365 chemically distinct Si(IV) species. Predicting the temporal evolution of all species requires knowing 199,290 rate constants, which is clearly an unimaginable number to determine experimentally. However, if one assumes that reaction rates are independent of speciation, one can try to fit experimental speciation data using Eqs. 14.9 to 14.11 [27] only. The first conclusion to emerge from such fits is that the assumption apparently succeeds for acid-catalyzed hydrolysis of alkoxy groups. The second conclusion is that condensation rates depend sensitively on the degree of polymerization around a given silicon atom. Just as in aqueous solutions, monomers (Q^0) react faster than dimers (Q^1), which react faster than Q^2 sites. These results are qualitatively consistent with NMR studies in aqueous solutions on oligomers.

14.5 THE IMPORTANCE OF ALUMINOSILICATE POLYMERIZATION

After oxygen and silicon, the next most abundant elements in Earth's crust are aluminum (8%) and iron (5%) [1]. Oxide structures containing Si(IV) and

Al(III), including clays, micas, zeolites, and feldspars, are the most abundant minerals on the planet. Many, but not all, of the structures found in Nature can be produced experimentally by aqueous reaction under either ambient or hydrothermal conditions. The primary variable that experimentalists cannot reproduce is the long time required for the formation of some of these minerals in a natural setting.

The remainder of this chapter focuses on those structures containing Al(III). However, keep in mind that the behavior of Fe(III) is sometimes similar to that of Al(III) (see Ch. 5). A key feature of aluminosilicates is that Al(III) can be coordinated to four, five, or six oxygens. In some cases, as in the zeolites and the feldspar albite ($NaAlSi_3O_8$), tetrahedral Al(III) sites substitute into a tetrahedral silicate network (see Chapter 2). In some clays, such as kaolinite ($Al_4Si_4O_{10}(OH)_8$), all the aluminum is in octahedral coordination within discrete dioctahedral sheets, whereas all silicon is present in adjacent tetrahedral sheets (Fig. 2.11). In smectite clays (such as beidellite ($Na_x[Al_4][Si_{8-x}Al_x]O_{20}(OH)_4$)) and micas [28], aluminum is present in both the octahedral aluminate and tetrahedral silicate sheets. There even exists a single indicator set of minerals, with the Al_2SiO_5 stoichiometry, that includes three phases of different structure: sillimanite, andalusite, and kyanite. In these Al_2SiO_5 minerals, Al(III) is found in four, five-, and six-coordination to oxygens. The Al(III) coordination, and the mineral species, changes with the temperature and pressure experienced by the rock.

This section provides a framework for understanding how the coordination and distribution of Al(III) in aluminosilicates are affected by aqueous polymerization conditions. Unfortunately, the kinetics of aluminosilicate mineral formation in water can range from hours to thousands of years, and often involve structural rearrangements that take place long after the original solid phase precipitates. For this reason, it is difficult to observe many of the relevant polymerization processes directly. However, it has been possible to observe the formation of small aluminosilicate (and aluminogermanate) clusters under a restricted set of experimental conditions within which monomers and/or reaction products are present at sufficient concentrations to be observed. The Ge(IV) metal center is sometimes used in experimental studies as a surrogate for Si(IV) because it is slightly larger and thus reacts more quickly.

Here, we present fundamental mechanisms for aluminosilicate polymerization, starting with pure silicates and progressing to aluminosilicates. We then provide specific experimental results for the formation of small oligomers and clusters under both strongly basic and slightly acidic conditions. However, it is important to note that, although reactions involving small soluble species provide important clues regarding reaction mechanisms, the extrapolation of these results to every specific condition under which aluminosilicate minerals form is dangerous. This is especially true for reaction pathways involving initial precipitation followed by restructuring within an amorphous, precipitated phase. Our goal here is to describe reasonable pathways, while recognizing that some of the conclusions are speculative.

14.6 THE CHEMISTRY OF MONOMERIC AL(III) COMPLEXES

The starting point for understanding polymerization between silicate and aluminate species is the hydrolysis diagram for Al(III) (Fig. 14.9). This diagram shows that Al(III) is predominantly in octahedral coordination in acidic and neutral solutions. In more basic solutions, Al(III) forms the tetrahedral anion $[Al(OH)_4]^-$, which becomes important even at pH $\sim$7 to 9. The decrease in coordination number is probably stepwise across pH. The first hydrolysis complex has Al(III) coordinated to five oxygens—$[Al(H_2O)_4OH]^{2+}$ [29], Nuclear magnetic resonance data are unavailable for other species until the four-coordinate $[Al(OH)_4]^-$ ion dominates at basic pH. The transitions between six-, five-, and fourfold coordination are triggered by deprotonation-induced changes in the net complex charge. The tetrahedral Al(III) can increase its coordination number much more readily than the tetrahedral Si(IV) species (see Section 14.3).

The mechanisms and kinetics of ligand-exchange processes involving Al(III) reflect the coordination number around the central cation [13] and the ease of coordination number changes. In acidic solutions, the fully protonated $[Al(H_2O)_6]^{3+}$ species undergoes water exchange at a relatively slow rate of $\sim$1.4/second at room temperature [14]. For the first hydrolysis species,

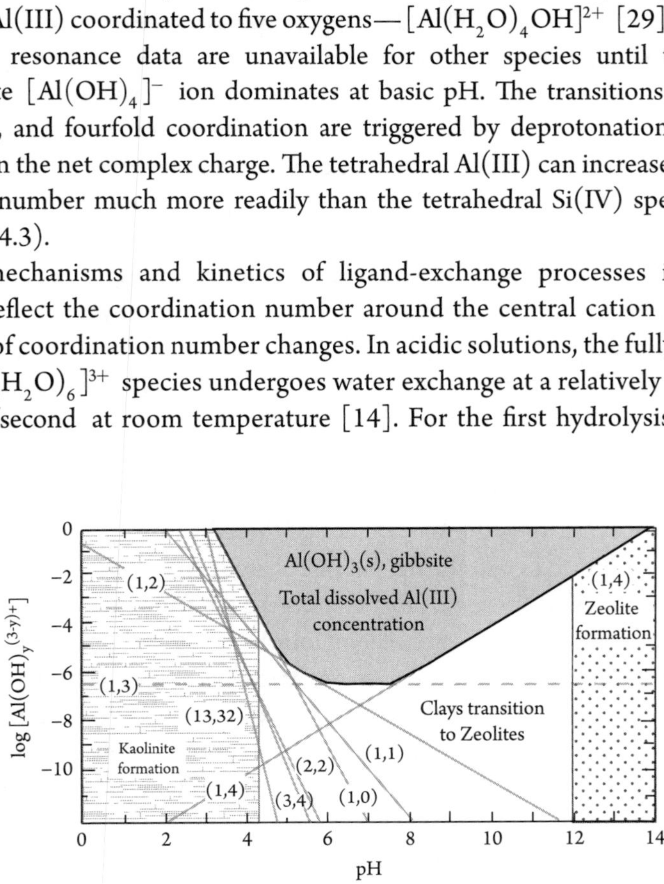

Figure 14.9 The hydrolysis diagram for Al(III), showing the major species present as a function of pH and aluminum concentration. The dashed lines indicate qualitative boundaries between the pH regions within which different aluminosilicate phases are typically formed. Adapted with permission from Baes, C. F.; Mesmer, R. E. *The Hydrolysis of Cations*; Krieger Publishing: Malabar, FL, 1986.

waters bound to $[Al(H_2O)_4OH]^{2+}$, exchange with bulk waters about ~30,000 times faster. In basic solutions, where tetrahedral $[Al(OH)_4]^-$ dominates, oxygen-isotope-exchange rates are so fast they cannot be measured by NMR and are probably at the diffusion-controlled Smoluchowski limit ($\sim 10^9 - 10^{10}$/second). Such rates are more than 1 billion times faster than exchange rates involving tetrahedral silicate species under similar conditions. This difference in rates means that if both Al(III) and Si(IV) are present in a basic solution, Al(III) species are most active for ligand exchange. The relatively nonreactive and anionic silicate species are virtually always the attacking nucleophiles.

Now consider how the various aluminate species polymerize with one another in the absence of soluble silicates. In acidic solutions where octahedral $[Al(H_2O)_6]^{3+}$ dominates (Fig. 14.9), polymerization usually produces solids that retain an octahedral coordination (see Chapter 5), including $Al(OH)_3$ phases such as gibbsite, where all oxygen sites in the structure have a neutral formal charge of zero. Polymerization can also result in the formation of AlOOH phases such as boehmite that contain neutral networks of paired oxygens (i.e., μ_3-oxo with a formal charge of -0.5 paired next to a μ_3-OH with a formal charge of $+0.5$).

Strongly basic solutions contain high concentrations of anionic species such as $[Al(OH)_4]^-$ and, here, polymerization produces hydrated sodium aluminate ($NaAlO_4 \cdot 5/4H_2O$) [30]. The tetrahedral coordination geometry of the $[Al(OH)_4]^-$ monomer is also present in the solid phase. The structure [31] consists of aluminate sheets containing fourfold and eightfold rings connected by $Al-\mu_2O-Al$, with each bridging oxygen having a formal charge of -0.5. The structure also contains one nonbridging hydroxyl group per Al(III) that extends into the spaces between the layers. Each of these hydroxyl groups has a formal charge of -0.25. Although the net anionic charge is neutralized by the presence of one Na^+ per Al(III), the fact that all oxygens are strongly anionic means that local electrostatic repulsion between aluminate structural elements is high. In other words, hydrated sodium aluminate represents a much less stable structure than $Al(OH)_3$. This difference in stability is one reason why extreme solution conditions (11–14 M NaOH and 0.2 M Al(III)) and extensive ion pairing involving Na^+ (see Section 14.3) are required for sodium-aluminate formation. The network charge per oxygen anion is even higher in the tetrahedral network of anhydrous $NaAlO_4$ (where all oxygens in the structure are μ_2-O bridges with a formal charge of -0.5). Hence, $NaAlO_4$ is not produced via aqueous polymerization, but requires subsequent dehydroxylation of the $NaAlO_4 \cdot 5/4H_2O$ phase.

A significant observation is that, although $[Al(OH)_4]^-$ is the predominant solution species above pH 8, $NaAlO_4 \cdot 5/4H_2O$ is not the phase that precipitates from solutions in the window from molar concentrations of OH^- down to pH > 8. Within this window, the primary precipitates are octahedral $Al(OH)_3$ and AlOOH, because these phases have more favorable network charge distributions than sodium aluminate. The main reason why tetrahedral aluminate species often polymerize to form solids with octahedral aluminum, even when the aqueous solution is dominated by tetrahedral $[Al(OH)_4]^-$, has to do with the polymerization mechanism. In contrast to tetrahedral Si(IV) species, tetrahedral Al(III) species form stable five- and six-coordinate hydrolysis complexes and transition easily between these higher coordination numbers. Open coordination sites on both four- and fivefold Al(III) complexes can be filled with waters or hydroxide ions to form octahedra. The prevalent octahedral geometry asserts itself even during the aging of precipitates and gels. For example, solid-state ^{27}Al NMR spectra obtained on precipitates formed via polymerization of the $[AlO_4Al_{12}(OH)_{24}(H_2O)_{12}]^{7+}$ species show that the tetrahedral Al(III) sites at the center of each cluster eventually disappear and are replaced by octahedral sites [32, 33]. It is thus a mistake to think that ions are simply trapped as inert structures when a solid forms. Transitions between coordination states in Al(III) ions in solution, and presumably at the oxide–solid surface, are much faster than the formation rate of solids. The interfacial species are not exotic and have similar acid-base chemistries as the familiar solutes.

14.7 MECHANISTIC CONSIDERATIONS REGARDING ALUMINOSILICATE CLUSTERS

Because they add ligands more easily, soluble Al(III) species are expected to be the host complexes for the addition of the silicate ligands $[Si(OH)_4]^\circ$ and/or $[Si(OH)_3O]^-$ as aluminosilicate species form. Therefore, polymerization chemistry is normally more sensitive to changes in Al(III) speciation than to that of Si(IV). Based on Al(III) speciation, three pH regimes can be defined, each of which is expected to produce different solid phases. With a pH less than 7, where octahedral Al(III) species (and $[Si(OH)_4]^\circ$) predominate, polymerization is expected to yield clays containing relatively pure octahedral Al(III) and tetrahedral Si(IV) sheets. With a pH more than 12, where $[Al(OH)_4]^-$ and $[Si(OH)_2O_2]^{2-}$ dominate, tetrahedral networks containing both Al(III) and Si(IV) are expected. At intermediate pH values (Fig. 14.9), clays can be produced in which the tetrahedral silicate sheets are doped with Al(III) to create anionic layers. Aluminosilicates produced in aqueous media are commonly consistent with these broad observations [34] with exceptions mediated by specific counterion effects.

Now consider what ligand-exchange mechanisms might lead to formation of these phases (Fig. 14.10). With a pH less than 5, the Al(III) complexes are octahedral species but lose protons and bound oxygens as pH increases. The resulting undercoordinated Al(III) can accept an incoming silicate ligand easily and

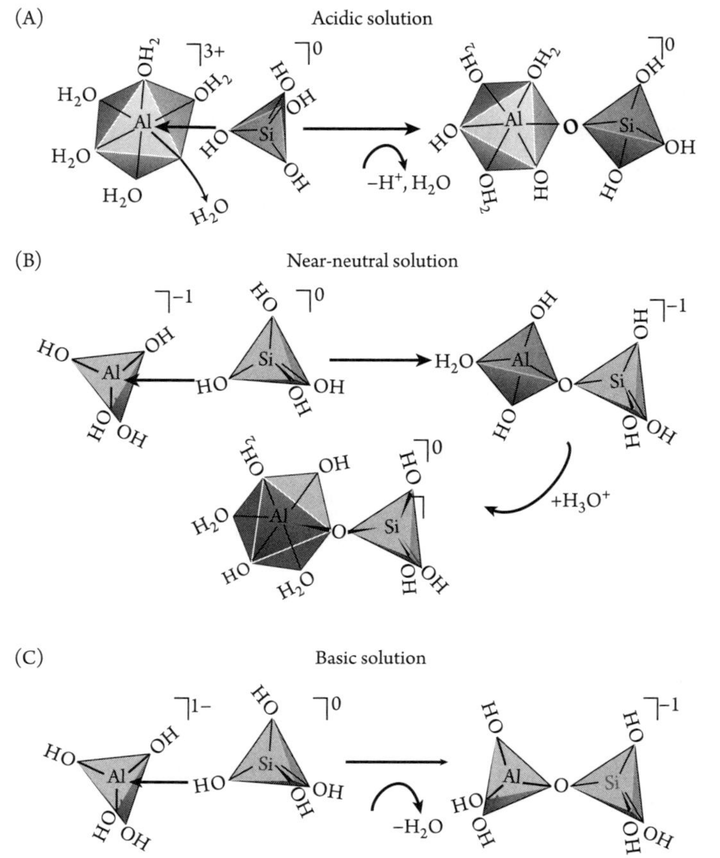

Figure 14.10 Molecular mechanisms proposed to rationalize the initial polymerization to form dimers from aqueous solutions containing Si(IV) and Al(III) monomers as a function of pH. (A) At low pH, polymerization is dominated by the hydrolysis of $[Al(H_2O)_6]^{3+}$, leading to the formation of five-coordinate intermediates that can react with $[Si(OH)_4]^\circ$ to form the dimer via the dissociative ligand exchange processes described in Chapter 5. (B) At high pH, polymerization involves a nucleophilic-attack mechanism in which a silicate anion attacks a tetrahedral $[Al(OH)_4]^-$ ion, which can eject a bound OH^- to reestablish the tetrahedral geometry. (C) However, between pH 8 and pH 12, the Al(III) center does not necessarily eject a ligand to return to tetrahedral coordination, but instead can gain an additional ligand to form an octahedral Al(III) center.

increase its coordination number. This phenomenon provides two potential pathways for clay formation, depending on whether silicate or aluminate monomers serve as entering ligands.

If silicate addition dominates, dimers and other small clusters containing tetrahedral metals and $Al-\mu_2O-Si$ represent the most probable initial reaction products. Further condensation and changes in coordination affecting Al(III) results in stable, larger oligomers. A condensation reaction to form a trimeric cluster containing two octahedral Al(III) and one tetrahedral Si(IV) held together by a μ_3-oxo-bridge creates a stable, neutral oxygen site with a formal charge of zero $(-2+3/6+3/6+4/4=0)$. Such clusters have yet to be isolated from solution. However, reactions between such clusters are expected to produce amorphous materials or clays in which both the octahedral and tetrahedral sheets grow simultaneously.

If aluminate addition dominates, the first extended polymer to form is expected to be dioctahedral Al(III) sheets containing neutral μ_2-OH bridges, such as those found in minerals like gibbsite ($Al(OH)_3$). When formed, these dioctahedral sheets are available to participate in slower ligand exchange reactions in which bridging μ_2-OH groups are gradually bonded to $[Si(OH)_3O]^-$ groups. Such replacement can create the neutral μ_3-oxo bridge connecting two Al(III) and a Si(IV) that are found between the octahedral and tetrahedral sheets in clays. Subsequent condensation reactions would complete the formation of fully polymerized tetrahedral sheets in stacked clay layers by sequential steps.

In strongly basic solutions, the primary monomers present before polymerization are $[Al(OH)_4]^-$ and $[Si(OH)_2O_2]^{2-}$. In this pH regime, $[Al(OH)_4]^-$ is susceptible to nucleophilic attack to form a tetrahedral network, as seen earlier for the formation of hydrated $NaAlO_4$. However, there are several key differences between the aluminosilicate and pure aluminate cases in this basic regime. First, $[Si(OH)_2O_2]^{2-}$ is a stronger nucleophile than $[Al(OH)_4]^-$, as indicated by the oxygen formal charges of -1 and -0.25, respectively; their relative complex charges (-2 and -1, respectively); and their pK_a values. If both species are present in comparable concentrations, this basicity difference suggests that $Al-\mu_2O-Si$ should form in preference to $Al-\mu_2O-Si$ bridges. Second, in solutions saturated with likely solids, higher concentrations of silicate species can be attained (e.g., at pH 13, $[Al]=0.1$ M whereas $[Si]=1.5$ M) than aluminate species.

However, the most important factor driving aluminosilicate polymerization in basic solutions is probably local charge distributions. As discussed for aluminates, the μ_2-O between two tetrahedral Al(III) sites has a formal charge of -0.5. The comparable $Al-\mu_2O-Si$ bridge has a formal charge of -0.25 and is much more stable $(-2+3/4+4/4=-0.25)$. One consequence of this enhanced stability is that tetrahedral aluminosilicate networks form over a much broader range of solution conditions $(pH>12)$ than the highly restrictive conditions (supramolar concentrations of hydroxide ion) required for precipitation of $NaAlO_4$. This difference in bridging-bond stabilities is captured in the well-known Lowenstein aluminum-avoidance rule [35], which states that in tetrahedral aluminosilicate networks, aluminum sites are never adjacent because of electrostatic repulsion. This rule is almost always observed in tetrahedral aluminosilicates. One consequence of this rule is that it is difficult to prepare purely tetrahedral aluminosilicate networks containing $Al:Si>1$, regardless of how the aluminosilicate is formed.

Very little is known about aluminum solute coordination and aluminosilicate solid precipitation at $7<pH<12$, in which tetrahedral $[Al(OH)_4]^-$ predominates. In this window, the availability of Si(IV) species is probably key. As discussed earlier, interconversion among coordination numbers for Al(III) hydrolysis species is undoubtedly very fast. The kinetics of polymerization will also be influenced by reactive Si(IV) species that are present only in minor amounts. High pH favors tetrahedral and deprotonated Al(III) whereas, as the pH drops, higher coordination geometries of Al(III) are more stable. From a formal charge perspective, any octahedral Al(III) sites created would much rather form octahedral sheets than substitute within a zeolitic framework. Tetrahedral Al(III) in oxides should exhibit a strong preference for being surrounded by Si(IV) sites, in keeping with the aluminum-avoidance rule (and minimization of anionic charge on network oxygen sites). The net result is the formation of clays based on octahedral sheets in which there is some substitution of Al(III) into the tetrahedral sheets, consistent with what is typically found in clays.

14.8 ALUMINOSILICATE POLYMERIZATION IN BASIC SOLUTIONS: EXPERIMENTAL RESULTS

In strongly basic solutions, Al(III) and Si(IV) concentrations are sufficiently high that individual species are observable via ^{27}Al-, ^{29}Si- and ^{17}O-NMR spectroscopies. Numerous mixed species exist [18,36,37]. Mixed aluminosilicate species resemble the species seen for the polymeric soluble silicates, with Al(III) substituting easily for some of the Si(IV) sites (Fig. 14.11) in soluble ions. Species identified include the dimer, the linear and ring trimer, and two tetramers (a closed tetrameric ring plus a trimeric ring with a pendant Q^1 silicate unit). An important feature of all mixed species identified is that there are no species containing adjacent tetrahedral aluminum sites, in accordance with Lowenstein's aluminum avoidance rule [35].

Rates of Si(IV) exchange have been measured within the mixed aluminosilicate cluster ions that have been identified in basic solutions [17,36,37]. Results show that, for small acyclic species, the rates of ^{29}Si(IV) exchange around aluminate sites $[Al(OH)_4]^-$ resulting from the reversible formation of $Al-\mu_2O-Si$ bonds are at least $2\cdot10^4$ times faster than exchange rates involving the $Si-\mu_2O-Si$ bonds in pure silicate species. Part of this rate difference is a result of the fact that there is less electrostatic repulsion (as mediated by counterions) between $[Al(OH)_4]^-$ and $[(HO)_2SiO_2]^{2-}$ than between two $[(HO)_2SiO_2]^{2-}$

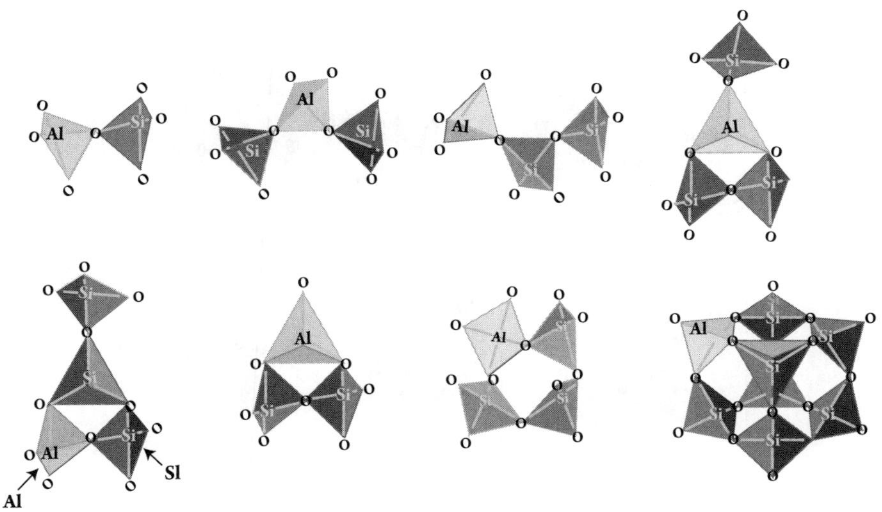

Figure 14.11 Structures proposed for soluble aluminosilicate species based on ^{29}Si and ^{27}Al nuclear magnetic resonance spectra. Note that all mixed species are consistent with the Lowenstein aluminum avoidance rule [34]. Adapted with permission from Swaddle, T. W.; Salerno, J.; Tregloan, P. A. Aqueous Aluminates, Silicates, and Aluminosilicates. *Chem. Soc. Rev.* **1994**, *23*, 319–325.

species. However, the main reason for the enhanced rate is that it is easier to insert a ligand into $[Al(OH)_4]^-$ to form a five-coordinate intermediate than it is on any tetrahedral Si(IV) center. As with pure silicates, reaction rates on aluminate sites decrease as the number of bridging oxygens to a given site (i.e., Q) increases. Rates are rapid for small oligomers yet become immeasurably slow for octamers and larger species.

14.9 ALUMINOSILICATE POLYMERIZATION IN WEAK ACIDS: THE FORMATION OF IMOGOLITE CLAYS

Clay minerals are common in the neutral to mildly acidic conditions where many soils form. In particularly reactive volcanic soils, the initial precipitates have short-range structure even though the X-ray patterns indicate that they are amorphous materials over long distances. One solid found in such soils is an odd, tubular clay-like mineral called *imogolite* with the ideal stoichiometry $(Al_2SiO_3(OH)_4)$. Silicate imogolites form in volcanic soils where rapid weathering of volcanic glasses leads to high local concentrations of dissolved silicon [38]. Rather than forming planar sheets as in normal clays, imogolite consists of hollow tubes with inner and outer diameters of 1 nm and 2 nm, respectively (Fig. 14.12), and lengths as long as several hundred nanometers.

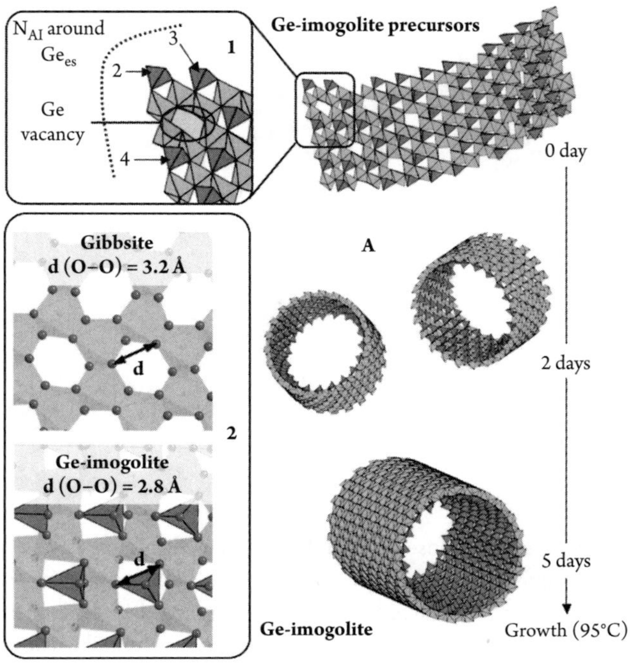

Figure 14.12 A representation of the initial germanium imogolite precursor sheet at high magnification (left). After several days of aging at 90°, this precursor sheet evolves into the final imogolite structure (bottom).
SOURCE: Reproduced with permission from Levard, C.; Masion, A.; Rose, J.; Doelsch, E.; Borschneck, D.; Olivi, L.; Chaurand, P.; Dominici, C.; Ziarelli, F.; Thill, A.; Maillet, P.; and Bottero, J.Y. Synthesis of Ge-Imogolite: Influence of the Hydrolysis Ratio on the Structure of the Nanotubes. *Phys. Chem. Chem. Phys.*, **2011**, *13*, 14516–14522.

The outside of each tube is a curved octahedral sheet with a structure and composition similar to gibbsite ($Al(OH)_3$). The inside of the tube is decorated with isolated silicate tetrahedra. In contrast to a normal clay, in which each Si(IV) associates with one Al(III) in the octahedral sheet and three adjacent Si(IV) centers in a tetrahedral sheet, each Si(IV) center in imogolite is bonded to three Al(III) centers via μ_3-oxo-bridges. These tetrahedra cover the octahedral holes in the $Al(OH)_3$ sheet (Fig. 14.12). Such inorganic nanotubes are of technological interest in applications such as gas storage and catalysis [39, 40].

Mechanistic studies have been performed on imogolite to understand how to control tube dimensions [41]. The most in-depth studies have been performed on aluminogermanate imogolite because Ge(IV) provides much greater electronic contrast relative to Al(III) than Si(IV) [39,40,42,43]. A further benefit is that oxo-bonds to Ge(IV) tend to react more rapidly than Si(IV). This substitution of Ge(IV) for Si(IV) facilitates structural determinations involving X-ray absorption spectroscopy and small-angle X-ray scattering that complement NMR and Raman spectra. The combination of all these analytical methods provides a complete structural picture [41] of these interesting materials. Unfortunately, as a result of low concentrations and/or interference from the large soluble clusters that form initially, definitive determinations of the small species, that are present in neutral to acidic aluminosilicate solutions, are not yet possible. However, some of the attributes of these species can be inferred from the initial large clusters and precipitates. Most of the remaining discussion is based on results obtained for the synthesis of Ge(IV) imogolites from precursor solutions containing Al(III)/Ge(IV) of 2:1 [40]. Although specific details may differ, many of the mechanistic insights apply to the formation of aluminosilicate clays.

Experiments on Ge(IV) imogolite suggest that clays form by a three-step process involving attributes of both simultaneous and sequential polymerization. The first step involves the relatively rapid establishment of a distribution of soluble oligomers that react with each other to form the initial precipitate. This distribution is clearly influenced by both the absolute and relative concentrations of Al(III) and Si(IV) species. However, the results suggest that the primary factor controlling initial speciation is the extent of hydrolysis of $[Al(H_2O)_6]^{3+}$ as mediated by hydroxide additions or solution pH. The corresponding silicate and germanate species are not as significant because they don't change much with pH in this range. Addition of 1 mole OH^-/Al(III) corresponds to a hydrolysis ratio of 1:1 that, on average, creates one active hydroxyl site per Al(III). Similarly, a hydrolysis ratio of 2:1 activates two sites. As OH^- is added to a solution containing $[Al(H_2O)_6]^{3+}$ and $[Ge(OH)_4]^{\circ}$, the first thing that happens is the hydrolysis of aquated aluminum ions to stimulate dimer formation. If ligand addition in dimer formation did not distinguish the incoming nucleophiles, the formation of $Al-\mu_2O-Al$ and $Al-\mu_2O-Ge$ (or $Al-\mu_2O-Si$) dimers would be equally probable. If so, a solution containing Al(III)/Ge(IV) of 2:1 would produce precipitates exhibiting the same 2:1 ratio. However, for a hydrolysis ratio (OH^- / Al(III)) of both 0.5:1 and 1:1, the resulting precipitates actually exhibit Al(III)/Ge(IV) values of around 1.3:1. In other words, $Al-\mu_2O-Ge$ (and by inference $Al-\mu_2O-Si$) dimers form in preference to $Al-\mu_2O-Al$ dimers. The condensation reaction leading to dimer formation has a mechanism in which $[Ge(OH)_4]^{\circ}$ is more active as an entering ligand than the various Al(III) hydrolysis complexes (see Chapter 4, Section 4.7). Electrostatic repulsion between the Al(III) complexes undoubtedly contributes to this preference. These results are consistent with a wide range of experimental results [18], suggesting that the $Al-\mu_2O-Si$ dimer (and by extension the $Al-\mu_2O-Ge$ dimer) is present in aluminosilicate solutions at near-neutral pH.

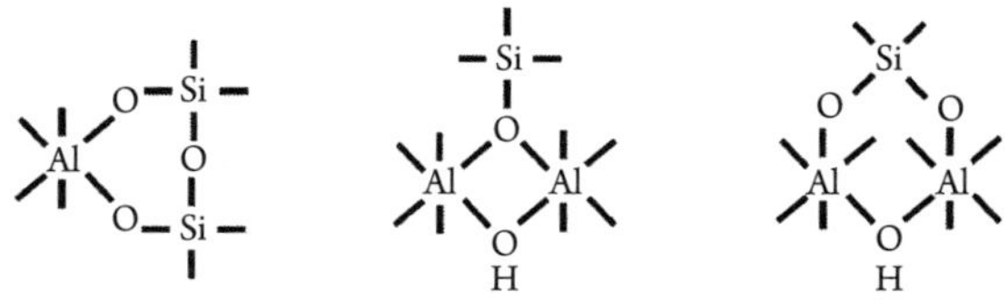

Figure 14.13 Three hypothetical condensed clusters that could form in neutral to acidic solutions containing aluminosilicate dimers and silicate and aluminate monomers. If silicate and aluminate monomers are present in comparable concentrations, the cluster on the left should be less abundant because of the lower reactivity of $[Si(OH)_4]^{\circ}$ relative to the aluminum monomers. A formal-charge analysis (see Chapter 5) suggests that the central cluster should be the most stable trimer, as both bridging oxygens within this cluster are neutral. In contrast the Al-μ_2O-Si bridges in the other two isomers have a formal charge of -0.5. The sheet structures in clay minerals are based upon the central bonding motif.

Now consider what happens when enough OH^- is initially present in monomeric precursor solutions to form oligomers larger than dimers. A few of the aluminosilicate trimeric species that might form as a result of this more extensive hydrolysis are depicted in Figure 14.13.

We invite you to evaluate the relative formation probabilities for other small aluminogermanate and aluminosilicate clusters containing octahedral Al(III). However, based on composition, standard ligand-exchange mechanisms, and formal-charge arguments, the results are most consistent with the formation of aluminogermanate trimers organized around a neutral μ_3-oxo. Although this building block has not been identified in precursor solutions, it represents the most common repeated unit in the large soluble clusters that have been characterized (Fig. 14.12) [40]. A μ_4- oxo between Ge(IV) and Al(III) has been identified in molecules such as $[GeAl_{12}(OH)_{24}(OH_2)_{12}]^{8+}$, which has the structure of the ε-isomer of the Keggin ion [33].

Local structures that are the precursors to adjacent octahedral and tetrahedral sheets in clay probably resemble those formed via the polymerization of oligomeric building blocks, although no ensemble of clay–precursor complexes have yet been identified spectroscopically. Experimentally, aging of clusters in solutions that have a clay-like ratio of metals (e.g., Al : Ge = 1.3 : 1) does not form imogolite, whereas aging of the precursor solid with Al : Ge = 2 : 1 forms imogolite rather than the corresponding clay with that 2:1 ratio of metals. Dioctahedral clays such as margerite have the appropriate stoichiometry, $[Ca_2Al_2Al_2Si_2O_{10}(OH)_2]$, but probably form much more slowly than imogolite. These clay structures are more polymerized and more ordered than Ge–imogolite, but are also much more stable thermodynamically.

The initial material that forms in germanium-imogolite–precursor solutions is not yet imogolite, but consists of clusters containing on the order of 160 Al(III) cations. Most (>93%) of the Al(III) centers reside in octahedral sites organized within the $Al(OH)_3$ sheets. However, in contrast to most clays, the $Al(OH)_3$ sheets are curved (Fig. 14.12). This curvature is caused by the isolated tetrahedral Ge(IV) monomers that occupy one side of the octahedral sheet. Initially, only 80% of these isolated Ge(IV) sites are occupied. Sheet curvature results from the fact that the 0.28-nm O–O separation distance in $[Ge(OH)_4]^{\circ}$ is shorter than the 0.32-nm separation distance in the $Al(OH)_3$ sheet that the $[Ge(OH)_4]^{\circ}$ monomer must bridge (Fig. 14.12). Bridge formation results in a strain that bends the octahedral sheet. One consequence of this strain is reflected in X-ray absorption fine structure results showing that bridge formation in the initial clusters is incomplete. Initially, each Ge(IV) has 4.5 Al(III) neighbors in contrast to the 6.0 neighbors corresponding to a cap that spans each octahedral hole completely.

If the curved preimogolite clusters are aged at 90°C for several days, the clusters gradually evolve into fully formed imogolite tubes. During this aging process, $[Ge(OH)_4]^{\circ}$ that remains in solution slowly fills in vacant sites in the tetrahedral sheet until all isolated sites are occupied. In addition, each of the germanate sites in the tetrahedral sheet eventually forms three bridges to the octahedral sheet to form complete caps over each octahedral hole. Both of these processes increase the bridge density in the tetrahedral sheet, which in turn decreases the radius of curvature of each imogolite cluster. The sheets eventually bend back on themselves to form imogolite tubes (Fig. 14.12) that gradually lengthen, either by adding material on the more reactive tube ends or via the fusion of short existing tubes. Therefore, it is possible to control the tube diameter and length by careful control of aging processes.

The key message from imogolite studies is that clay formation invariably involves parallel processes that span multiple timescales, and includes a wide range of hydrolysis and condensation rates. The most rapid processes involve the formation of soluble clusters from hydrolytically activated monomeric Al(III) species. Cluster formation reflects the Al-to-Si ratio, and these clusters influence the arrangement of silicate units within the initial precipitate. Intermediate-rate processes involve the polymerization of these clusters to form solids that are typically amorphous, but serve as a template for the formation of the final clay. The template contains a higher crosslink density than the soluble oligomers, and reacts so slowly that the template structure is largely preserved during subsequent hydrolysis and condensation reactions. The initial template evolves into the final clay structure via slower hydrolysis and condensation reactions that involve tetrahedral species such as $[Ge(OH)_4]^{\circ}$ and $[Si(OH)_4]^{\circ}$. These reactions fill in structural defects (Fig. 14.14). Even slower condensation reactions involving attached silicate units either bridge between adjacent silicate sites (to complete tetrahedral

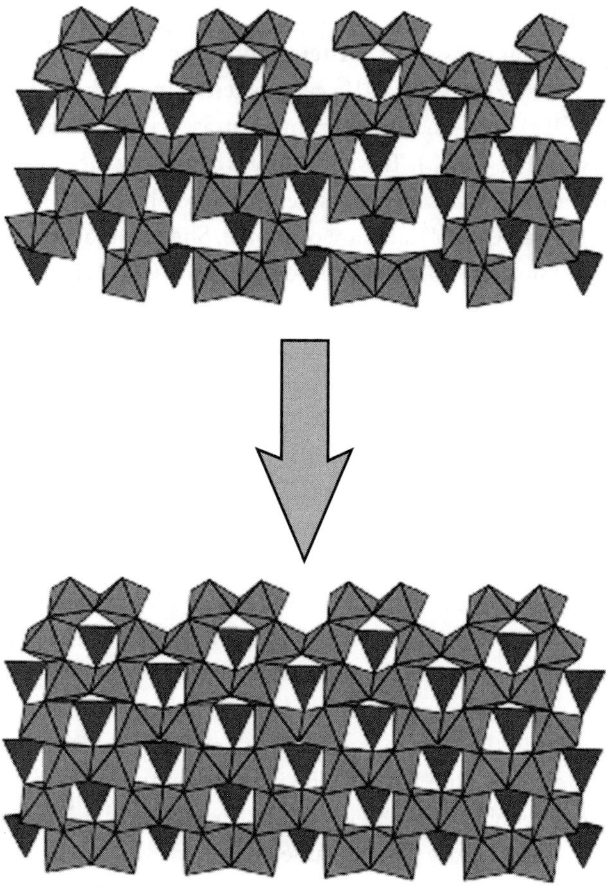

Figure 14.14 As germanium-deficient imogolite clusters fill in, the initial soluble nanoscale sheets curve in on themselves to form imogolite tubes.
SOURCE: Reproduced with permission from Levard, C.; Rose, J.; Thill, A.; Masion, A.; Doelsch, E.; Maillet, P.; Spalla, O.; Olivi, L.; Cognigni, A.; Ziarelli, F.; Bottero, J. Y. Formation and Growth Mechanisms of Imogolite-Like Aluminogermanate Nanotubes.; *Chem. Mater.*, **2010**, 22, 2466–2473.

sheet formation) or cap aluminate sites (as in imogolite) to form the final ordered clay product. In imogolite, these reactions are so slow that the condensed tetrahedral sheets of linked Si(IV) oxides that are found in common clays do not form.

14.10 HYDROTHERMAL SYNTHESIS OF SILICATES AND ALUMINOSILICATES

Oxides in general and silicates in particular tend to precipitate from water under ambient conditions as amorphous or nanocrystalline materials. In Nature, these clusters often aggregate to form gels that are common in polluted waters [32]. These gels then recrystallize slowly into ordered clays and zeolites at ambient conditions. The process can be accelerated by increasing temperature, so formation of crystalline oxides with dimensions greater than 1 μm often requires *hydrothermal* conditions. Hydrothermal conditions do not just refer to water (hydro) that is hot (thermal), but to synthesis conditions during which reactions take place at temperatures greater than the boiling point of water. Maintaining water in a liquid state at temperatures greater than the boiling point requires increased pressure. Hydrothermal pressures are commonly expressed in atmospheres (atm), Pascals (Pa) or kilobars (kbar), where 1 bar = 0.98 atm = 100,000 Pa. To put hydrothermal pressures into perspective, 1 kilobar, or 100 MPa, is roughly the pressure found at the bottom of the Marianas Trench, which is sufficient to crush the hulls of most submarines. Hydrothermal reactions are often conducted under *autogenous* or self-generated pressures, which means the fixed volume of the reactor prevents all the water present from evaporating. Here, pressures correspond to the temperature-dependent liquid–vapor phase boundary [44] (Fig. 14.15). At a sufficiently high temperature, the liquid–vapor phase boundary disappears because the liquid and vapor have exactly the same density. For water, this supercritical fluid forms at a critical temperature of 374°C, corresponding to a critical pressure of 218 atm. The critical volume is 55 cm^3/mole or 18 moles/L, which is roughly one-third the density of liquid water at room temperature.

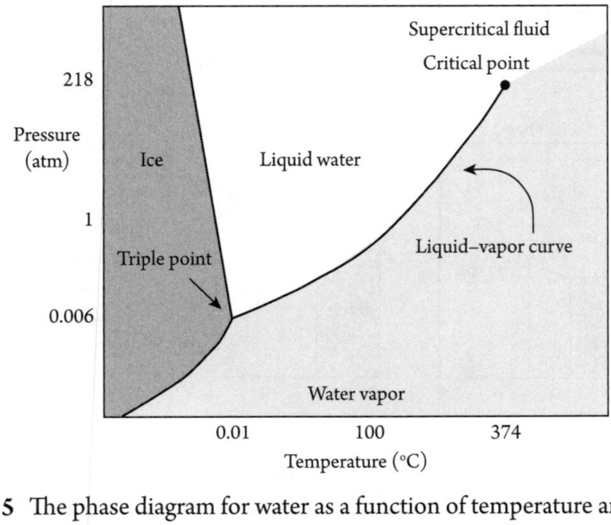

Figure 14.15 The phase diagram for water as a function of temperature and pressure, showing the solid–liquid and liquid–gas phase boundaries, as well as the critical point at which the liquid and gas phases become identical. In a constant-volume reactor, synthesis conditions lie along the liquid–vapor curve until the critical point is reached.

In hydrothermal syntheses, one needs to understand how key variables such as temperature, pressure, pH, and composition influence reaction trajectories and lead to the generation of specific oxide phases. A key advantage associated with elevated temperatures is that heat can overcome the activation barrier to reversible hydrolysis and condensation reactions in water. Rates are much more rapid, and one can control the evolution to thermodynamically preferred phases (Fig. 14.16). In addition, hydrothermal conditions induce dramatic changes in the aqueous solvent environment and acid-base chemistry.

Geochemists interested in understanding the modification of rocks and minerals deep in the Earth have evaluated the fundamental properties of water under extreme hydrothermal conditions (up to 60 kbar pressure and 1200°C) [45–47]. First, the static dielectric constant of pure water, which determines the extent to which water can screen ionic charges, decreases from 78 at room temperature to less than 20 at 300°C (Fig. 14.17). The dielectric constant of supercritical water is highly sensitive to density, which in turn is sensitive to both temperature and pressure. Second, the Brønsted acidity of water can change by orders of

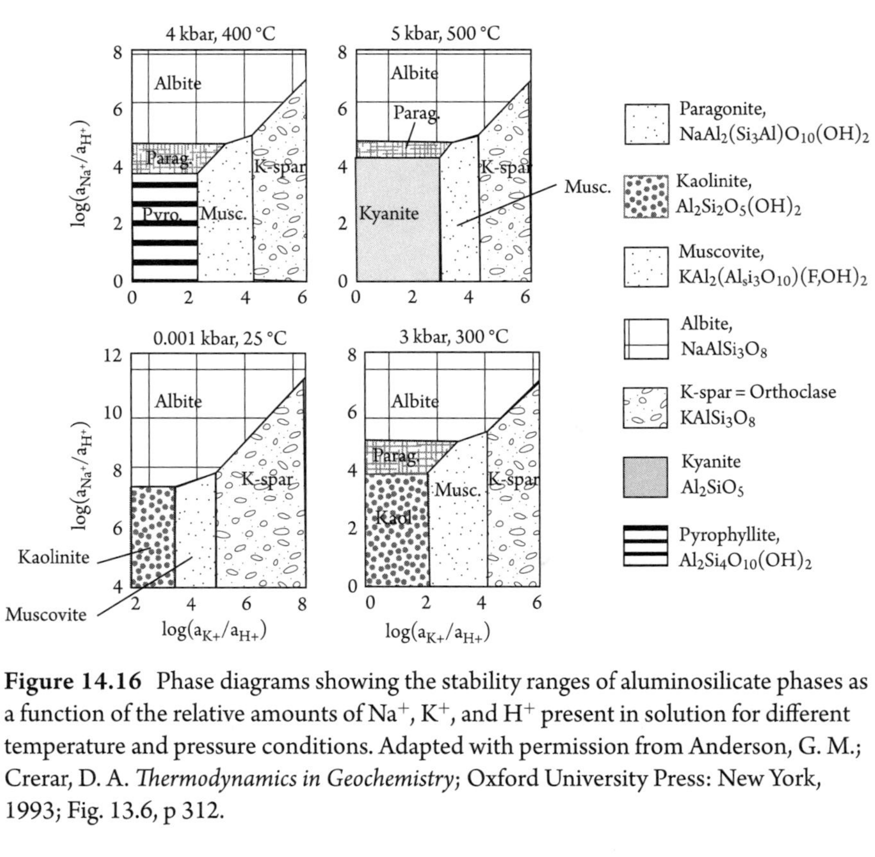

Figure 14.16 Phase diagrams showing the stability ranges of aluminosilicate phases as a function of the relative amounts of Na^+, K^+, and H^+ present in solution for different temperature and pressure conditions. Adapted with permission from Anderson, G. M.; Crerar, D. A. *Thermodynamics in Geochemistry*; Oxford University Press: New York, 1993; Fig. 13.6, p 312.

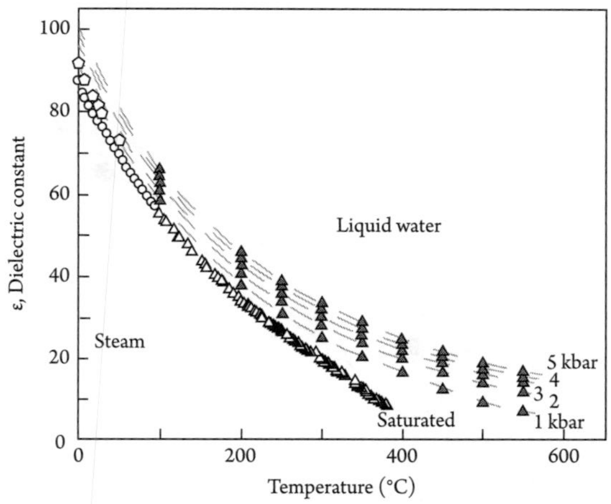

Figure 14.17 Variations in the dielectric constant of water as a function of temperature. Adapted with permission from Helgeson, H. C.; Kirkham, D. H. Theoretical Prediction of the Thermodynamic Behavior of Aqueous Electrolytes at High Pressures and Temperatures I: Summary of the Thermodynamic/Electrostatic Properties of the Solvent. *Am. J. Sci.* **1974**, *274*, 1089–1198.

magnitude with both temperature and pressure (Fig. 14.18). The acidity of waters or hydroxyls bound to hydrolysis products increase as well. Most oxyacids become stronger acids at elevated temperatures and pressures (Fig. 14.18). Mother Nature also favors the dissociation of oligomers into monomers at elevated pressures because monomers pack solvent molecules more efficiently into

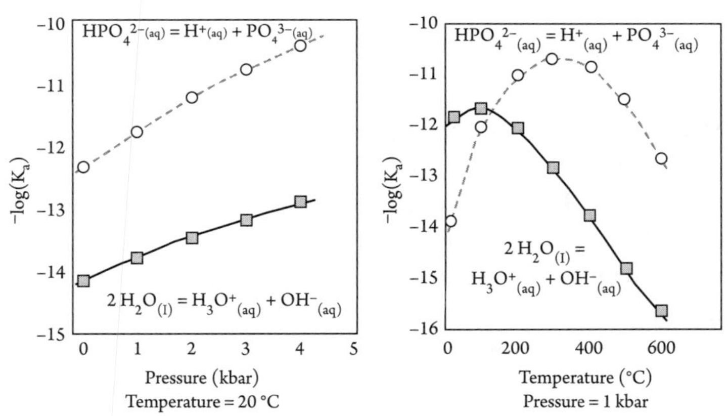

Figure 14.18 Variations in the pK_a of water and phosphoric acid as a function of temperature and pressure [44–46].
SOURCE: Calculations and figure by W. H. Casey using the Deep Earth Water Model of Sverjensky, D. A.; Harrison, B.; Azzolini, D. Water in the Deep Earth: The Dielectric Constant and the Solubilities of Quartz and Corundum to 60 kb and 1200 °C *Geochim. Cosmochim. Acta*, **2014**, *129*, 125.

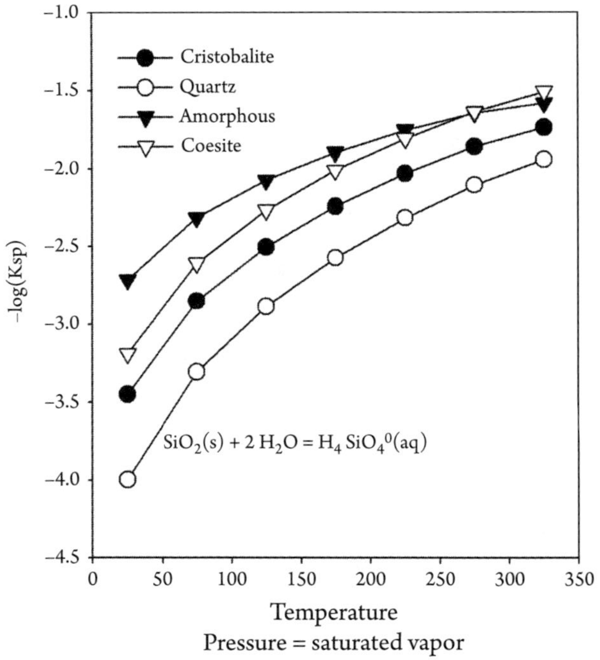

Figure 14.19 The solubilities of different crystalline modifications of silica as a function of temperature.
SOURCE: Calculations and figure by W. H. Casey using the Deep Earth Water Model of Sverjensky, D. A.; Harrison, B.; Azzolini, D. Water in the Deep Earth: The Dielectric Constant and the Solubilities of Quartz and Corundum to 60 kb and 1200 °C *Geochim. Cosmochim. Acta*, **2014**, *129*, 125.

their coordination sphere. Therefore, it should be no surprise that mineral solubility and surface chemistry vary with temperature. For example, the solubility of silica [48] (Fig. 14.19) (and any other material for which the dissolution reaction is endothermic) increases with temperature.

Pressure also modifies the properties of water and reaction conditions, sometimes in strange ways, such as the increased Brønsted acidity just mentioned. Pressure affects the position of the liquid–vapor phase boundary and parameters such as the specific volume and density of water. At low temperatures, pressure can generate up to nine different crystalline modifications of ice (Fig. 14.20) that can persist up to temperatures exceeding 150°C. Pressure has a smaller effect on the aqueous solubility and speciation of silicates than temperature (Fig. 14.21), but both variables have been accounted for in geochemical models that predict solubilities up to 60 kbar in pressure and 1200°C in temperature [45–52].

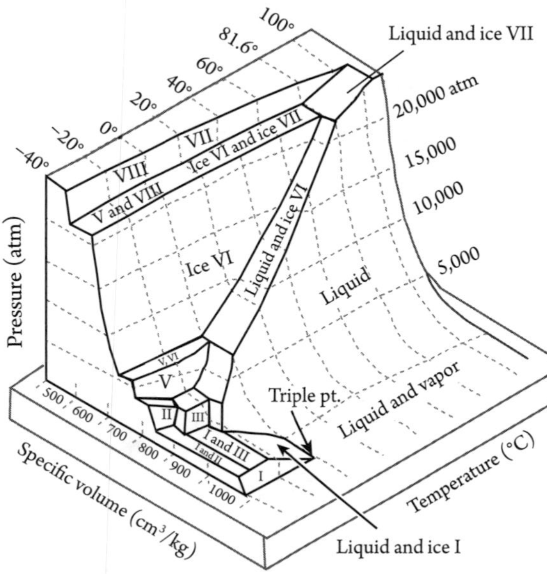

Figure 14.20 The phases of ice as a function of temperature and pressure. SOURCE: Reprinted with permission from Eisenberg, D.; Kauzmann, W. *The Structure and Properties of Water*; Oxford University Press: New York, 1969.

Pressure determines the solubility of volatiles in melts. Here, the most important system involves water in silicate melts (Fig. 14.22). This is important because water represents a common mineralizing solvent used to enhance the crystallization of crystalline silicates in high-temperature-melt processing. In natural geochemical melts, including lava or magma, water can either depolymerize the oxide structure by hydrolyzing bridging oxygen bonds or be stored as an intact molecule that erupts as steam when the pressure is released. This dissolved water represents one of the dominant volatile components that is released when the melt is depressurized, contributing to violent volcanic eruptions. More than 70 mole percent of a silicate melt can consist of water [49], which means that under some hydrothermal conditions, it can be hard to define whether the solvent in the reaction is a silicate melt containing water or an aqueous solution containing large concentrations of dissolved silica.

Because temperature and pressure effects are coupled, phase diagrams for hydrothermal syntheses tend to be exceedingly complex. This complexity is especially evident for the aluminosilicates, for which the number of structures and phases that can be produced is enormous. Recipes developed to generate both thermodynamically stable and metastable phases have generally been arrived at using empirical methods. That being said, trends emerge from an examination of the experimental conditions leading to the formation of a wide

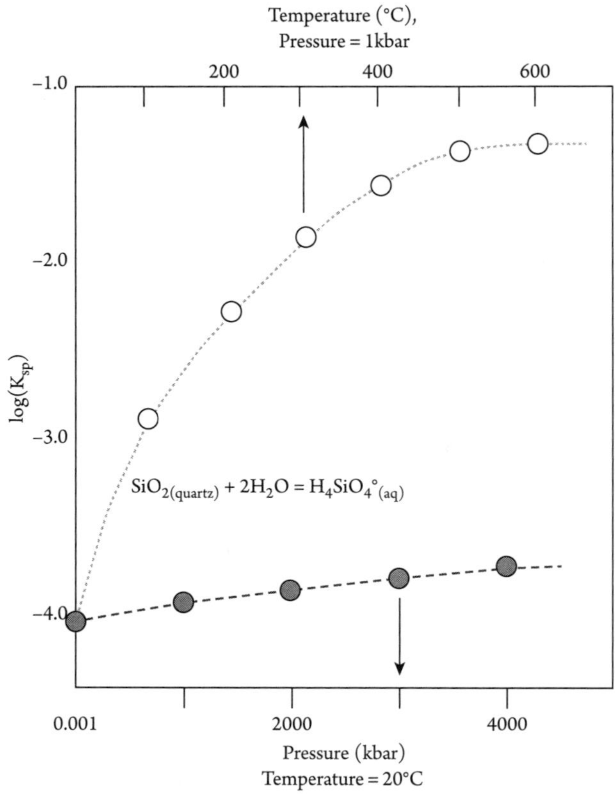

Figure 14.21 The solubility of silica as a function of pressure and temperature. SOURCE: Calculations and figure by W. H. Casey using the Deep Earth Water Model of Sverjensky, D. A.; Harrison, B.; Azzolini, D. Water in the Deep Earth: The Dielectric Constant and the Solubilities of Quartz and Corundum to 60 kb and 1200 °C *Geochim. Cosmochim. Acta*, **2014**, *129*, 125.

range of aluminosilicate materials [18]. In basic solutions, the consequences of increasing the temperature include a shift in aluminum coordination from octahedral to tetrahedral configurations. Temperatures often increase the total dissolved metal concentrations (Fig. 14.19) such that silica solubility under hydrothermal conditions can easily reach molal concentrations [50, 51]. There is a corresponding decrease in the diversity and domain of the soluble silicates, as temperature and pressure destabilize oligomers [52]. In addition, those oligomers and extended structures that are created tend to have smaller rings and higher densities. The latter effects are a consequence of the fact that a solution responds to increases in pressure by transferring bulk waters to ion hydration spheres because they pack more efficiently. Thus, depolymerization decreases the volume of the solution. Finally, the solvation forces that hold oligomers together are weakened at elevated temperatures.

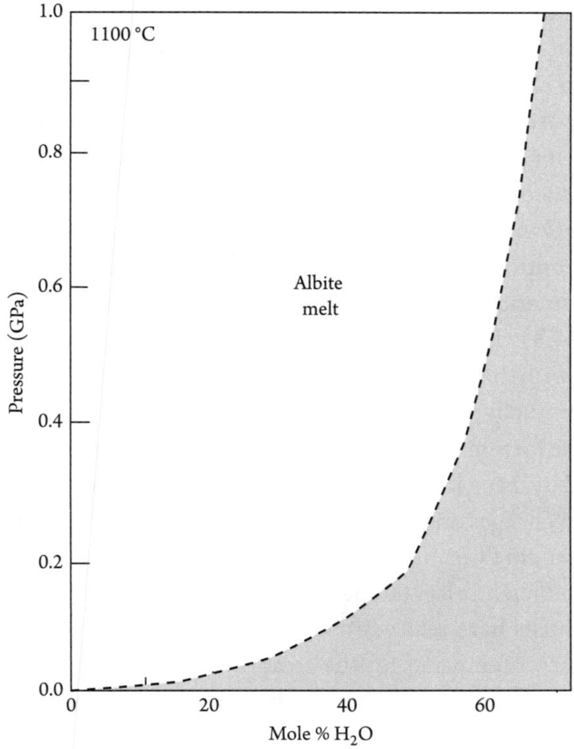

Figure 14.22 The solubility of water in an albite ($NaAlSi_3O_8$) melt as a function of pressure at T = 1100°C. Adapted with permission from Burham, C. W. Water Solutions: Thermodynamic Model for Hydrous Magmas. *Bull. Soc. Fr. Mineral. Cristallogr.* **1974**, 97, 223.

The hydrothermal syntheses of zeolites provide an abundance of examples of how solution composition influences phase formation. The key goal in zeolite synthesis is to produce pure phases in which the following parameters are carefully controlled: (1) the concentration and distribution of cation exchange sites (i.e., the relative concentrations of tetrahedral aluminum and silicon sites), (2) rings size, and (3) the interconnection of rings to form ideal cage and pore topologies that target specific applications such as ion exchange or catalysis. Cation exchange-site concentrations are commonly manipulated by controlling the Al-to-Si ratio in hydrothermal feedstocks, which are typically insoluble gels or glasses. The distribution of Al(III) sites is most difficult to control. NMR results indicate that Al(III) tends to substitute for Si(IV) in a random fashion, with the caveat that Al(III) sites are never directly adjacent, in keeping with the Lowenstein aluminum-avoidance rule (see Section 14.5) [34]. Electrostatic repulsion strongly inhibits the clustering of Al(III) sites in aluminosilicate solids just as it does in aqueous solutions.

Ring, cage, and pore distributions are also strongly influenced by solution composition. In addition to factors of obvious importance, including the Al-to-Si ratio, precursor concentrations, and pH, it is known that nonnetwork cations play a key role in phase formation. Such cations are commonly introduced along with the hydroxide ions used to adjust the solution pH (e.g., Na^+ with NaOH additions). Alkali metal cations are known to influence zeolite network structures strongly [52]. One mechanism proposed to explain the observed effects is that these cations control the initial distribution of polymeric species that are present and these polymeric species then interconnect to form larger silicate structures (see Section 14.3). The other effect associated with the presence of foreign cations has to do with their size. Cations occupy space. As aluminosilicate structures grow, they eventually form cages that must make room to accommodate the volume of these foreign cations. Recognition of this potential template effect has been exploited by using larger organic cations such as the tetramethylammonium ion to form larger cages and rings in zeolites such as ZSM-4. After network structures are formed, the organic cations are typically removed via oxidation by air, conducted at sufficiently low temperatures that the network does not reconstruct. Recently, scientists have realized that surfactants and block copolymers can be used as even larger templates for the creation of much larger porosity in the so-called mesoporous silicate and aluminosilicate materials (see Chapter 9).

References

1. Emsley, J. *Nature's Building Blocks*; Oxford University Press: Oxford, 2011.
2. Iler, R. K. *The Chemistry of Silica*; Wiley: New York, 1979.
3. Pauling, L. *The Nature of the Chemical Bond*; Cornell University Press: Ithaca, NY, 1993.
4. Baes, C. F.; Mesmer, R. E. *The Hydrolysis of Cations*; Krieger Publishing: Malabar, FL, 1986.
5. Bilinski, H.; Ingri, N. A Determination of the Formation Constant of $SiO(OH)_3^-$ *Acta. Chem. Scand.* **1967**, *21*, 2503–2510.
6. Ingri, N. Equilibrium Studies of Polyanions IV: Silicate Ions in NaCl Medium. *Acta Chem. Scand.* **1959**, *13*, 758–775.
7. Falcone, J. S., Ed. *Soluble Silicates*. ACS Symposium Series, Vol. 194; American Chemical Society: Washington, DC, 1982.
8. Harris, R. K.; Knight, C. T. G. Silicon-29 Nuclear Magnetic Resonance Studies of Aqueous Silicate Solutions. Part 6: Second-Order Patterns in Potassium Silicate Solutions Enriched with Silicon-29. *J. Chem. Soc.* **1983**, *79*, 1539–1561.
9. Knight, C. T. G.; Kinrade, S. D. A Primer on the Aqueous Chemistry of Silicon. In *Silicon in Agriculture: Studies in Plant Science*; Elsevier: New York, 2001; Vol. 8, pp 57–84.
10. Knight, C. T. G.; Balec, R. J.; Kinrade, S. D. Aqueous Alkali-Metal Silicate Anions Containing Fully Condensed Four-Coordinate Sites. *Angew. Chem.* **2012**, *51*, 9900–9903.

11. Vallazza, E.; Bain, A. D.; Swaddle, T. W. Dynamics of Silicate Exchange in Highly Alkaline Potassium Silicate Solutions. *Can. J. Chem.* **1998**, *76*, 183–193.
12. Kinrade, S. D.; Knight, C. T. G.; Pole, D. L.; Syvitski R. T. Silicon-29 NMR Studies of Tetraalkylammonium Silicate Solutions: 1. Equilibria, ^{29}Si Chemical Shifts, and ^{29}Si Relaxation. *Inorg. Chem.* **1998**, *37*, 4272–4277.
13. Swaddle, T. W.; Salerno, J.; Tregloan, P. A. Aqueous Aluminates, Silicates, and Aluminosilicates. *Chem. Soc. Rev.* **1994**, *23*, 319–325.
14. Helm, L.; Merbach, A. E. Inorganic and Bioinorganic Solvent Exchange Mechanisms. *Chem. Rev.* **2005**, *105*, 1923–1955.
15. Belton, D. J.; Deschaume, O.; Patwardhan, S. V.; Perry, C. C. A Solution Study of Silica Condensation and Speciation with Relevance to in vitro Investigations of Biosilicification. *J. Phys. Chem.* **2010**, *114*, 9947–9955.
16. Kinrade, S. D.; Knight, C. T. G.; Pole, D. L.; Syvitski, R. T. Silicon-29 NMR Studies of Tetraalkylammonium Silicate Solutions: 2. Polymerization Kinetics. *Inorg. Chem.* **1998**, *37*, 4278–4283.
17. Kinrade, S. D.; Swaddle, T. W. Aqueous Silicate Exchange Dynamics and Silicon-29 Nuclear Magnetic Relaxation: The Importance of Protonation Equilibriums. *Chem. Commun.* **1986**, *1986*, 120–121.
18. Swaddle, T. W. Silicate Complexes of Aluminum(III) in Aqueous Systems. *Coord. Chem. Rev.* **2001**, *219–221*, 665–686.
19. Kinrade, S. D. Oxygen-17 NMR Study of Aqueous Potassium Silicates. *J. Phys. Chem.* **1996**, *100*, 4760–4764.
20. Kinrade, S. D.; Pole, D. L. Effect of Alkali-Metal Cations on the Chemistry of Aqueous Silicate Solutions. *Inorg. Chem.* **1992**, *31*, 4558–4563.
21. Kinrade, S. D.; Del Nin, J. W.; Schach, A. S.; Sloan, T. A.; Wilson, K. L.; Knight, C. T. G. Stable Five- and Six-Coordinated Silicate Anions in Aqueous Solution. *Science* **1999**, *285*, 1542–1545.
22. Kinrade, S. D.; Hamilton, R. J.; Schach, A. S.; Knight, C. T. G. Aqueous Hypervalent Silicon Complexes with Aliphatic Sugar Acids. *J. Chem. Soc. Dalton Trans.* **2001**, *7*, 961–963.
23. Kinrade, S. D.; Balec, R. J.; Schach, A. S.; Wang, J.; Knight, C. T. G. The Structure of Aqueous Pentaoxo Silicon Complexes with cis-1,2-dihydroxycyclopentane and Furanoidic Vicinal Cis-diols. *J. Chem. Soc. Dalton Trans.* **2004**, *6*, 3241–3243.
24. Brinker, C. J.; Scherer, G. W. *Sol–Gel Science: The Physics and Chemistry of Sol–Gel Processing*; Academic Press: San Diego, CA, 1990.
25. Assink, R. A.; Kay, B. D. Sol–Gel Kinetics I: Functional Group Kinetics. *J. Non-Cryst. Solids* **1988**, *99*, 359–370.
26. Kay, B. D.; Assink, R. A. Sol–Gel Kinetics II: Chemical Speciation Modeling. *J. Non-Cryst. Solids* **1988**, *104*, 112–122.
27. Assink, R. A.; Kay, B. D. Sol–Gel Kinetics III: Test of the Statistical Reaction Model. *J. Non-Cryst. Solids* **1988**, *107*, 35–40.
28. Bailey, S. W., Ed. *Micas*; Reviews in Mineralogy, Vol. 13; Mineralogical Society of America, BookCrafters: Chelsea, MI, 1984.
29. Swaddle, T. W.; Rosenqvist, J.; Yu, P.; Bylaska, E.; Phillips, B. L.; Casey, W. H. Kinetic Evidence for Five-Coordination in $AlOH(aq)^{2+}$ Ion. *Science* **2005**, *308*, 1450–1453.

30. Cao, S.; Zhang, Y.; Zhang, Y. Preparation of Sodium Aluminate from the Leach Liquor of Diasporic Bauxite in Concentrated NaOH Solution. *Hydrometallurgy* **2009**, *98*, 298–303.
31. Kaduk, J. A.; Pei, S. The Crystal Structure of Hydrated Sodium Aluminate, $NaAlO_2$–$5/4H_2O$, and Its Dehydration Product. *J. Solid State Chem.* **1995**, *115*, 126–139.
32. Furrer, G.; Phillips, B. L.; Ulrich, K. U.; Pothig, R.; Casey, W. H. The Origin of Aluminum Flocs in Polluted Streams. *Science* **2002**, *297*, 2245–2247.
33. Casey, W. H. Large Aqueous Aluminum-Hydroxide Molecules. *Chem. Rev.*, **2006**, *106*, 1–16.
34. Stringham, B. F. Fields of Formation of Some Common Hydrothermal-Alteration Minerals. *Econ. Geol.* **1952**, *47*, 661–664.
35. Lowenstein, W. The Distribution of Aluminum in the Tetrahedra of Silicates and Aluminates. *Am. Miner.* **1954**, *39*, 92–96.
36. Harris, R. K.; Samadi-Maybodi, A.; Smith, W. The Incorporation of Aluminum into Silicate Ions in Alkaline Aqueous Solutions, Studied by Aluminum-27 N.M.R. *Zeolites* **1997**, *19*, 147–155.
37. Harris, R. K.; Parkinson, J.; Samadi-Maybodi, A.; Smith, W. Exchange Reactions in Aluminosilicate Solutions. *Chem. Commun.* **1996**, *1996*, 593–594.
38. Cradwick, P. D. G.; Farmer, V. C.; Russell, J. D.; Masson, C. R.; Wada, K.; Yoshinaga, N. Imogolite, A Hydrated Aluminium Silicate of Tubular Structure. *Nature* **1972**, *240*, 187–189.
39. Levard, C.; Masion, A.; Rose, J.; Doelsch, E.; Borschneck, D.; Dominici, C.; Ziarelli, F.; Bottero, J.- Y. Synthesis of Imogolite Fibers from Decimolar Concentration at Low Temperature and Ambient Pressure: A Promising Route for Inexpensive Nanotubes. *J. Am. Chem. Soc.* **2009**, *131*, 17080–17081.
40. Levard, C.; Rose, J.; Thill, A.; Masion, A.; Doelsch, E.; Maillet, P.; Spalla, O.; Olivi, L.; Cognigni, A.; Ziarelli, F.; Bottero, J. Y. Formation and Growth Mechanisms of Imogolite-Like Aluminogermanate Nanotubes. *Chem. Mater.* **2010**, *22*, 2466–2473.
41. Wilson, M.A.; McCarthy, S. A. Long-Range Effects of the Aluminum-Avoidance Principle. *Anal. Chem.* **1985**, *57*, 2733–2735.
42. Levard, C.; Masion, A.; Rose, J.; Doelsch, E.; Borschneck, D.; Olivi, L.; Chaurand, P.; Dominici, C.; Ziarelli, F.; Thill, A.; Maillet, P.; and Bottero, J. Y. Synthesis of Ge-Imogolite: Influence of the Hydrolysis Ratio on the Structure of the Nanotubes. *Phys. Chem. Chem. Phys.* **2011**, *13*, 14516–14522.
43. Maillet, P.; Levard, C.; Larquet, E.; Mariet, C.; Spalla, O.; Menguy, N.; Masion, A.; Doelsch, E.; Rose, J.; and Thill, A. Evidence of Double-Walled Al/Ge Imogolite-Like Nanotubes: A Cryo-TEM and SAXS Investigation. *J. Am. Chem. Soc.* **2010**, *132*, 1208–1209.
44. Helgeson, H. C.; Kirkham, D. H. Theoretical Prediction of the Thermodynamic Behavior of Aqueous Electrolytes at High Pressures and Temperatures: 1. Summary of the Thermodynamic/Electrostatic Properties of the Solvent. *Am. J. Sci.* **1974**, *274*, 1089–1198.
45. Shock, E. L.; Helgeson, H. C.; Sverjensky, D. A. Calculation of the Thermodynamic and Transport Properties of Aqueous Species at High Pressures and Temperatures: Standard Partial Molal Properties of Inorganic Neutral Species. *Geochim. Cosmochim. Acta* **1989**, *53*, 2157–2183.

46. Shock, E. L.; Sassani, D. C.; Willis, M.; Sverjensky, D. A. Inorganic Species in Geologic Fluids: Correlations among Standard Molal Thermodynamic Properties of Aqueous Ions and Hydroxide Complexes. *Geochim. Cosmochim. Acta* **1997**, *61*, 907–950.
47. Sverjensky, D. A.; Shock, E. L.; Helgeson, H. C. Prediction of the Thermodynamic Properties of Aqueous Metal Complexes to 1000°C and 5 kb. *Geochim. Cosmochim. Acta* **1997**, *61*, 1359–1412.
48. Newton, R. C.; Manning, C. E. Quartz Solubility in H_2O–NaCl and H_2O–CO_2 Solutions at Deep Crust–Upper Mantle Pressures and Temperatures: 2–15 kbar and 500–900 Degrees C. *Geochim. Cosmochim. Acta* 2000, **64**, 2993–3005.
49. Burham, C. W. Sodium Aluminosilicate ($NaAlSi_3O_8$)–Water Solutions: Thermodynamic Model for Hydrous Magmas. *Bull. Soc. Fr. Mineral. Cristallogr.*, **1974**, *97*, 223–230.
50. Manning, C. E. The Solubility of Quartz in H_2O in the Lower Crust and Upper Mantle. *Geochim. Cosmochim. Acta* **1994**, *58*, 4831–4839.
51. Hunt, J. D.; Manning, C. E. A Thermodynamic Model for the System Near the Upper Critical End Point Based on Quartz Solubility Experiments at 500°C and 5 kbar. *Geochim. Cosmochim. Acta* **2012**, *86*, 196–213.
52. Plyasunov, A. V.; Grenthe, I. The Temperature Dependence of Stability Constants for the Formation of Polynuclear Cationic Complexes. *Geochim. Cosmochim. Acta* **1994**, *58*, 3561–3582.
53. Barrer, R. M. *Hydrothermal Chemistry of Zeolites*; Academic Press: London, 1982.

15

Glass Dissolution and Leaching

15.1 INTRODUCTION

Oxide glasses represent some of the most important and prevalent materials that we encounter in our daily lives. The glass industry in the United States produces more than 75,000 glass products, with annual production estimated to be around 20,000,000 t [1]. Roughly 50% of this production is for glass containers for food, beverages, and other liquids. Everyone relies on transparent glass windows for their homes, cars, and even their cell phones. Fiberglass provides insulation for our homes and businesses. We rely on glass for many optical systems, ranging from eyeglasses to microscope lenses to optical fiber communications. Glass is also an optically pleasing material found in many works of art, including stained glass windows. Glass even plays a role in energy transport and storage, being an important electrical insulator used in devices ranging from transformers to batteries.

Glass compositions need to be optimized for specific applications, with important parameters being melting properties, thermal conductivity, thermal expansion, strength, dielectric properties, and, of course, optical properties. In most of these applications, glass objects encounter water, either to perform their basic functions or as a result of long-term environmental exposure. This means the chemical properties of many glasses also need to be optimized. Fortunately, borosilicate glasses, which represent the most widely used technological glass compositions, tend to exhibit a high level of resistance to aqueous attack (Fig. 15.1).

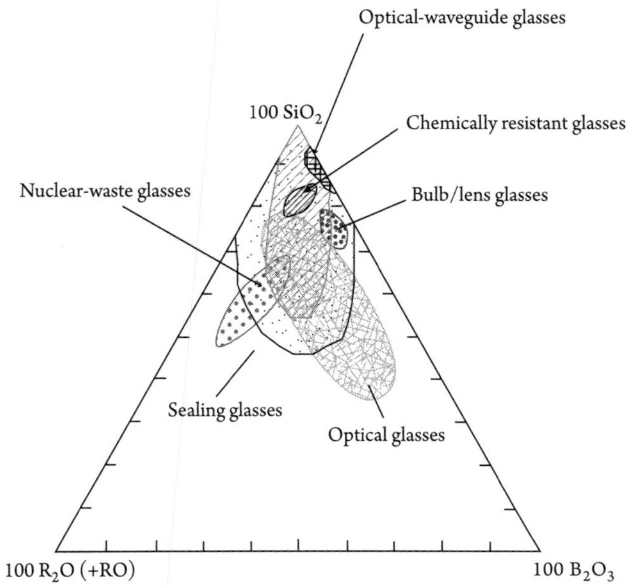

Figure 15.1 Simplified composition regimes for alkali- and alkaline-earth borosilicate glasses showing regimes that are representative of commercial glass types. The teardrop lines that emanate around the SiO_2 corner represent "isodurs" (for equal chemical durability), where each line away from SiO_2 corresponds to an order of magnitude drop in durability. Adapted from Adams, P. B.; Evans, D. L. Chemical Durability of Borate Glasses. In: L.D. Pye, V.D. Fréchette and N.J. Kreidl (Editors) *Borate Glasses: Structure, Properties, Applications*; Plenum Press: New York, 1978, pp. 535–537.

Understanding the kinetics and mechanisms of glass dissolution is critically important to the nuclear power and defense industries, which involves how to dispose of nuclear wastes safely. These wastes can be exceedingly complex, and contain almost every element found in the Periodic Table. The challenge is to incorporate these wastes into solids that encapsulate radionuclides safely for millions of years. Glass is an attractive option as a waste form because glass melts can accommodate almost all the constituents found in nuclear wastes. However, the deployment of glass waste forms requires the ability to predict the stability of the waste out to exceedingly long times based on science-based glass-dissolution models.

Glass dissolution normally follows one of three rate laws [2] (Fig. 15.2), but transitions between the rate laws can be hard to predict. Initially, species such as alkali cations are observed to leach from the glass at a rate that proceeds as a function of the square root of time. This $t^{1/2}$ dependence is consistent with models in which the rate-limiting step is a diffusion-controlled process described by expressions such as:

$$X = \sqrt{Dt} \quad \textbf{(15.1)}$$

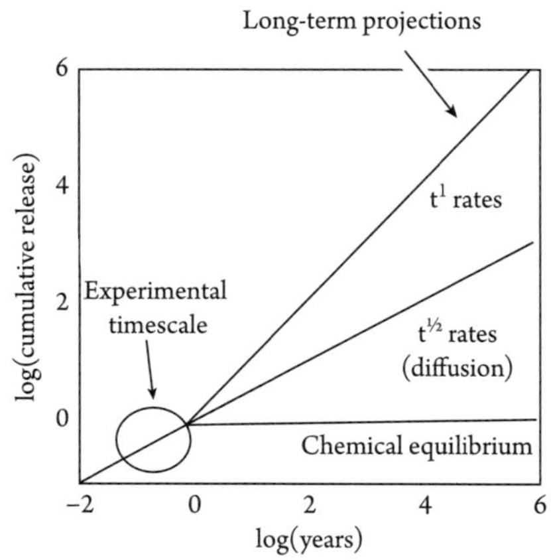

Figure 15.2 The logarithm of the elemental release rate as a function of the log of time corresponding to the limiting kinetic rate laws. Although $t^{1/2}$ kinetics indicative of diffusion control are often observed in short-term leaching experiments, a transition to t^1 kinetics is also common. On reaching a condition of solution saturation, some elements no longer leave the glass. The graph shows that on the timescale of 1 million years, the uncertainty in elemental release rates can be as much as a factor of 1 million in the absence of quantitative predictive models. Adapted with permission from Bunker, B. C. Waste Glass Leaching: Chemistry and Kinetics, *Mater. Res. Soc. Symp. Proc.* **1987**, *84*, 493–507.

Here, X represents the depth into the glass from the glass–solution interface that species such as alkali cations have been depleted, and D is the diffusion coefficient (in units of square centimeters per second), which can vary as the glass structure changes. However, classic glass-dissolution models assume that D is identical to the self-diffusion coefficient for whatever species is being removed from the glass [3], and that the diffusion coefficient is a constant as a function of time.

At longer times, species such as alkali cations enter the solution at a rate that is linear as a function of time. This rate law is consistent with a process dominated by the rate at which the glass network dissolves via reactions that occur at the immediate glass–solution interface. Classic models assume that uniform dissolution occurs when the alkali-depleted surface layer (called the *leached layer*) becomes so thick that the rate at which alkali ions diffuse from the bulk to the glass–solution interface is identical to the rate at which the outer glass–solution interface is dissolving. Under this hypothesis, the leached layer assumes a constant steady-state thickness. Finally, if the glass is being dissolved into a finite solution volume, it is assumed the solution eventually becomes saturated with glass constituents, notably amorphous silica, and that dissolution and alteration processes cease altogether.

Unfortunately, it is rare for laboratory tests of glass dissolution to be carried out for more than a year. Durable waste glasses typically exhibit $t^{1/2}$ leaching kinetics throughout this entire period. In the absence of quantitative models that can predict if, and when, transitions in leaching behavior are expected to occur, extrapolations from 1-year lab tests out to 1 million years can have an uncertainty factor of as much as $\sim 10^6$ with respect to how much radioactive material could be released into the environment (Fig. 15.2). As over 100 million gallons of waste need to be converted to a stable waste form at the Hanford site in Richland, Washington, alone, uncertainties this large are clearly unacceptable.

Experiments on simulated nuclear-waste glasses raise questions regarding the classic glass-dissolution models. Elemental maps for one such glass leached in simulated seawater [5] (Fig. 15.3 and Plate 17) clearly shows that there is a sharp boundary between leached and unleached glass. There is no evidence of diffusional profiles for any species, which would gradually taper off from a high bulk

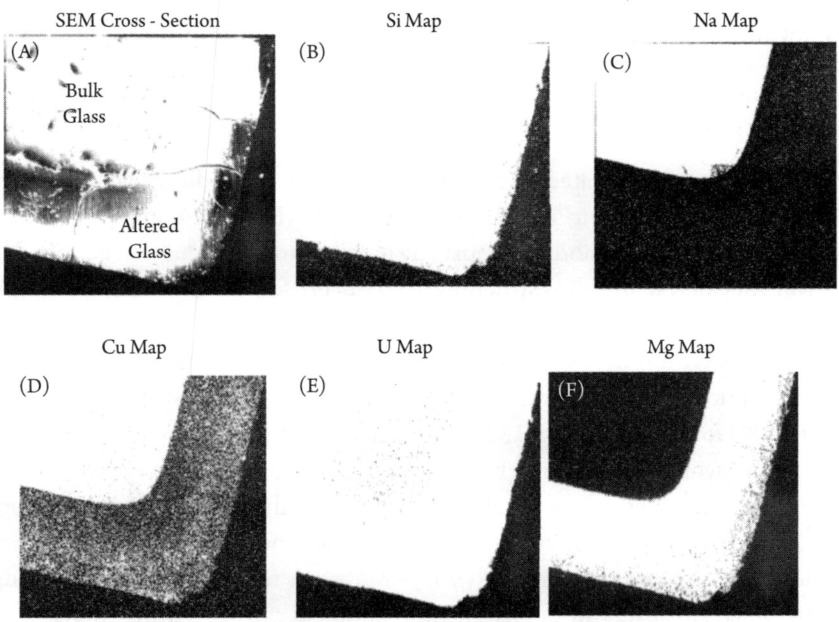

Figure 15.3 Elemental maps obtained using electron microprobe analyses on a cross-section of a simulated nuclear-waste glass leached in simulated seawater under hydrothermal conditions to accelerate reaction rates (T = 250°C, pH = 5.5, t = 14 days). (A) A scanning electron micrograph (SEM) shows the presence of the leached surface layer, where the thickness of the layer is 170 μm. (B–F) Elemental maps for the same cross-section.

SOURCE: Reprinted with permission from Bunker, B. C.; Arnold, G. W.; Beauchamp, E. K. Mechanisms for Alkali Leaching in Mixed Na-K Silicate Glasses *J. NonCryst. Solids*, **1983**, *58*, 295–322. (see Plate 18)

concentration down to a low concentration at the glass–solution interface. Instead, the concentrations of all species are essentially constant within the leached layer. Some species (such as sodium) are completely removed from the leached layer, whereas other species (such as copper) are only partially depleted. Some species such as silicon appear to be unchanged, whereas species such as uranium appear to concentrate in the leached layer. In fact, some elements (e.g., magnesium) appear to enter the leached layer from solution, because they were not present in the initial glass composition. These results clearly show that the "alkali diffusion followed by uniform dissolution" models need to be revisited to maintain safe calculations for waste stability. The purpose of this chapter is to introduce the primary glass-dissolution mechanisms that have been developed based on the molecular structure of glass, and how the initial glass structure evolves during the leaching and dissolution processes. Although the focus of this chapter is on glass, it is important to recognize that the basic chemical mechanisms for dissolution outlined here apply to most oxides, whether they are in a glassy or crystalline state. In many ways, glass dissolution is more useful for understanding dissolution pathways than mineral studies because glass structures can be changed continuously.

15.2 GLASS STRUCTURE AND DISSOLUTION MECHANISMS

Oxide glasses are dominated by cations that can form three-dimensional random networks in preference to crystalline compounds (Figs. 15.4 and 15.5). However, glasses can accommodate almost any metal cation. Cations are identified by the role they play in the glass structure [6]. *Network formers* are tetrahedral (or in some cases trigonal) cations that create random network structures that are typically dominated by μ_2-oxo-bridges to make an anionic framework. By far the most common network-forming cation is Si(IV), although B(III), P(V), and Al(III) also form glasses (Fig. 15.5). *Modifiers* are cations that are not part of the random network, but are present to compensate for anionic oxygen sites. The most common modifiers are the alkali-metal and alkaline-earth cations, although any +1 or +2 cation can play the modifier role. In most glasses, the introduction of modifiers disrupts bridging $M-\mu_2O-M$ bonds to create anionic nonbridging oxygens. However, for borate glasses, modifiers can sometimes increase the crosslink density (i.e., bridging oxygen population) of the glass. *Intermediates* are cations that can be accommodated within an existing glass network, but are normally incapable of forming an amorphous oxide on their own. Octahedral cations fall into this category because octahedral coordination lends itself to the formation of ordered crystalline solids rather than amorphous materials. Some of these elements are identified in various shades of gray in Figure 15.4.

When oxide glasses are placed in water, a number of different phenomena can occur (Fig.15.6). *Hydration* involves the penetration of the glass structure by

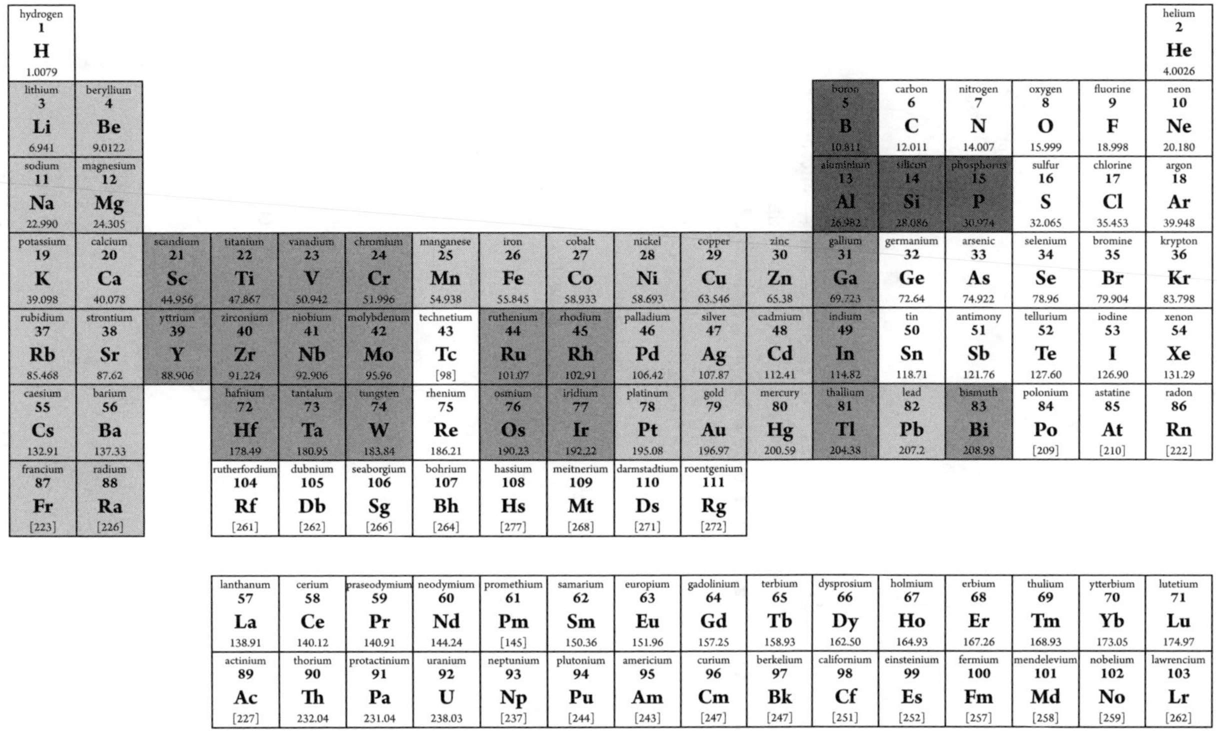

Figure 15.4 Common glass-forming elements. The oxide-forming elements are grouped into three categories based on their role in the glass structure. Network-forming elements are dark gray, and these generally have tetrahedral coordination to oxygens. Intermediate elements, shown in gray, cannot form a glass network on their own, but can substitute into existing glass networks. Intermediate elements typically have a valence of +3 or greater, and often have octahedral coordination geometries. Finally, network-modifying elements, shown in light gray, are present largely for charge compensation and are not part of the network. Some elements play multiple roles.

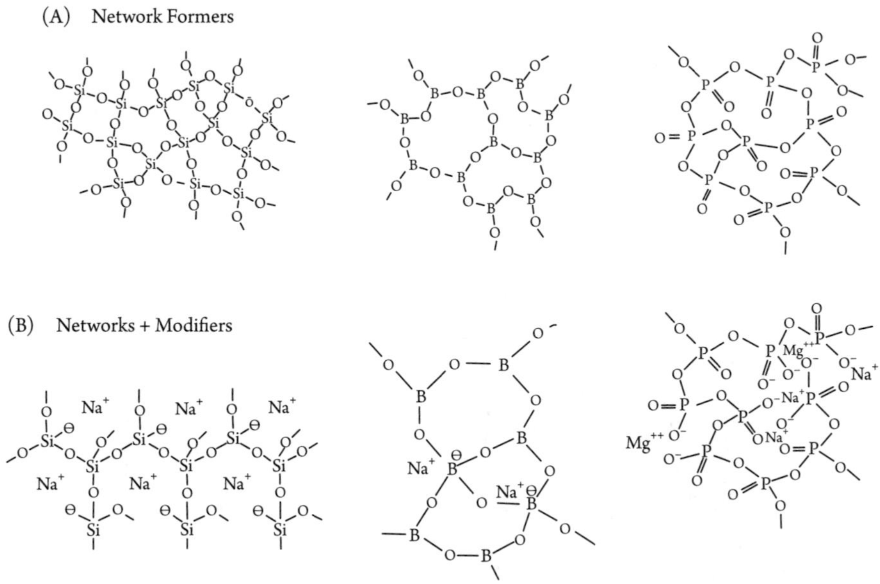

Figure 15.5 (A, B) Representative structures of three glass families—silicate, borate, and phosphate glasses—in the absence (A) and presence (B) of modifier cations. SOURCE: Reproduced from Ray, N. H.; *Inorganic Polymers*; Academic Press: New York, 1978.

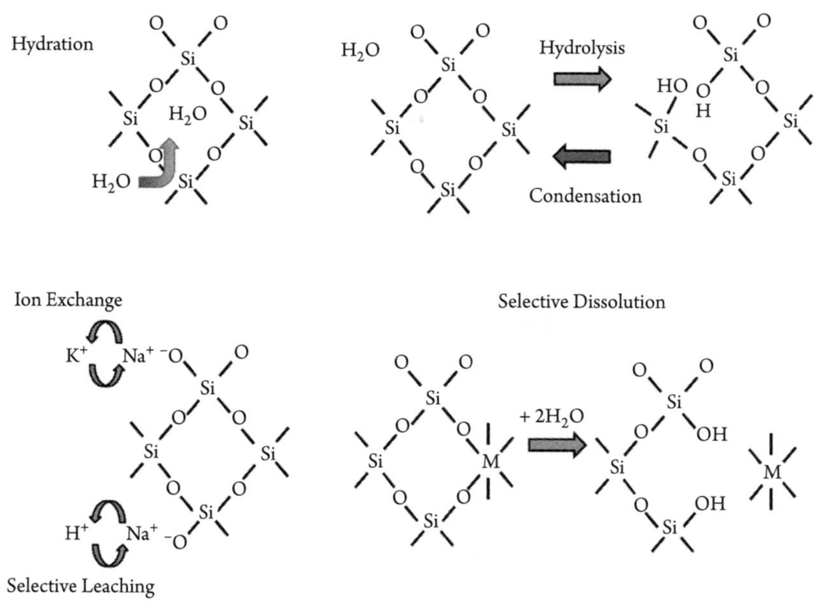

Figure 15.6 Depictions of several classes of fundamental glass-dissolution reactions as illustrated by silica and alkali silicates, including simple hydration (i.e., penetration of the structure by molecular water), reversible hydrolysis and condensation reactions, ion exchange reactions (or selective leaching when the exchanging cation is a proton), and selective dissolution of network-forming cations such as octahedrally coordinated intermediates.

intact water molecules. Hydration not only leads to other glass–water reactions, but also can disrupt interactions between network elements. *Hydrolysis* refers to the direct reactions between water molecules and the covalent M–O bonds that comprise the glass network, such as

$$H_2O + {>}M - \mu_2O - M{<} \rightleftarrows {>}M - OH + HO - M{<} \qquad (15.2)$$

with ">" indicating bonds between the site and the glass network. These reactions may be partly reversible. *Ion exchange* involves a reaction in which one ion from solution replaces a modifier cation in the glass (see Chapter 10). However, when the modifier is replaced with a proton, the reaction is called *selective leaching*. Selective leaching can involve either a water molecule or a hydronium ion. For example,

$$>\text{Si-O}^-\text{Na}^+ + H_3O^+ \rightleftarrows\ >\text{Si-OH} + \text{Na}^+ + H_2O \qquad (15.3)$$

$$>\text{Si-O}^-\text{Na}^+ + H_2O \rightleftarrows\ >\text{Si-OH} + \text{Na}^+ + OH^- \qquad (15.4)$$

Selective-leaching reactions are almost always irreversible. Irreversibility arises because when hydroxyl groups form, condensation reactions (Eq. 15.2 in reverse) or network hydrolysis reactions can occur that obliterate ion exchange sites in the glass. Finally, the term *selective dissolution* is used to describe the selective removal of species (often intermediates) that are initially incorporated into the glass network via $>M - \mu_2O - M<$ bonds. Such removal involves irreversible hydrolysis and condensation reactions. If the most reactive sites in the glass network are the network-forming elements, then the entire glass structure can fall apart and dissolve, leading to *uniform dissolution*. All these reactions can occur either sequentially or simultaneously, which makes glass dissolution a complex process with ensembles of reactions proceeding in parallel. Next, we provide specific examples of the interplay between these reactions as a function of the structure and chemistry of specific families of glass compositions.

15.3 SILICA DISSOLUTION

The structure of amorphous silica glass consists of a random network in which each Si(IV) cation is coordinated tetrahedrally by oxygen anions, and each oxygen anion bridges between two Si(IV) cations as a μ_2-oxo. In the bulk material, the silicate tetrahedra create a random network of interconnected rings. Six-membered rings are most common, but smaller rings containing four and three tetrahedra have been detected via Raman vibrational spectroscopy [7]. All the bridging oxygens in the structure are neutral. On a freshly cleaved surface, undercoordinated Si and O bond fragments are created. On exposure to atmospheric moisture, these fragment react immediately with water vapor to create surfaces that are terminated with silanol groups ($5/nm^2$) in addition to $>Si-\mu_2O-Si<$ bonds. These silanol groups are also electrically neutral. What reactions occur when such a surface is exposed to liquid water? Pure silica contains no modifiers or intermediates, so ion exchange, selective leaching, and selective dissolution do not need to be considered. Relevant reactions include hydration, hydrolysis/condensation, and uniform dissolution.

The first requirement for hydration is the presence of space within a structure that allows for water penetration. In silica and silicates, the rate of molecular diffusion into the glass is imposed by the distribution of silicate ring sizes. The effect of ring size on diffusion has been determined for crystalline aluminosilicate zeolites [8] (Fig. 15.7). When the structure contains voids that are large (>0.7 nm) relative to the kinetic diameter of the water molecule, a water molecule can diffuse through silicates as rapidly as it can through liquid water ($D=3\cdot10^{-5}\,cm^2/s$ at room temperature). However, as ring sizes are reduced to the point where openings are comparable with the molecular diameter of water, diffusion becomes inhibited. For a puckered ring containing six silicate tetrahedra, where the ring opening (0.24 nm) is slightly smaller than the diameter of a water molecule, molecular diffusion can occur only when the ring breathes to provide sufficient room for the water molecule to pass through ($D=2\cdot10^{-13}\,cm^2/s$). If smaller rings are present, molecular water is too large to penetrate the voids.

The zeolite results do not mean that water is prevented from diffusing into silica and other dense oxides. For denser structures, the penetration mechanism shifts from a *hydration* mechanism to a reversible *hydrolysis-and-condensation*

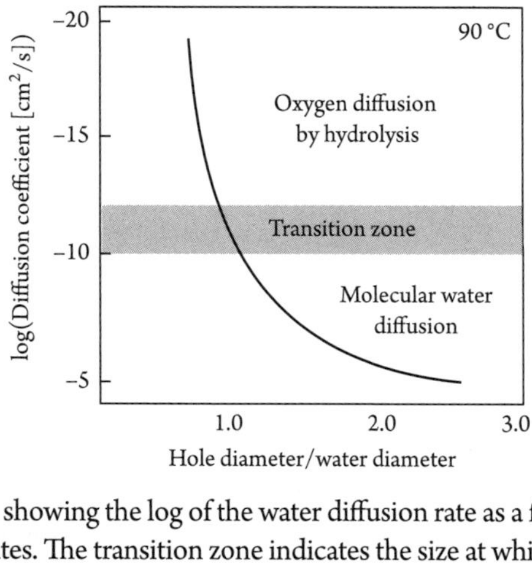

Figure 15.7 A curve showing the log of the water diffusion rate as a function of ring size based on data for zeolites. The transition zone indicates the size at which molecular diffusion is supplanted by diffusion via reversible hydrolysis and condensation. Adapted with permission from Bunker, B. C.; Arnold, G. W.; Day, D. E.; Bray, P. J. The Effect of Molecular Structure on Borosilicate Glass Leaching. *J. Non-Cryst. Solids* **1986**, *87*, 226–253.

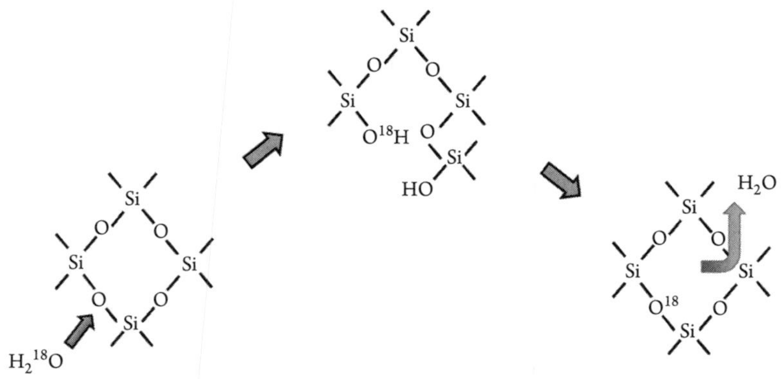

Figure 15.8 A representation of the diffusion of water through silica via the reversible hydrolysis-and-condensation mechanism.

mechanism (Figs. 15.8 and 15.9). Here, a water molecule on one side of a constricting ring reacts with a $>Si-\mu_2O-Si<$ bond, which breaks the bond, opens the ring, and creates two silanol groups. When formed, the silanol groups can react with each other and recondense into a $>Si-\mu_2O-Si<$ bond and reform the silicate ring. However, roughly half the time the water molecule that is released via the condensation reaction finds itself on the other side of the ring from the original direction of approach, resulting in water diffusion. This reaction sequence has been verified in experiments in which silica is exposed to labeled steam ($H_2{}^{18}O$) at high temperatures [9] (Fig. 15.9). Elemental depth profiles clearly show that water is not diffusing as an intact molecule because H penetration greatly outraces ^{18}O penetration. In the experiment, the level of ^{18}O incorporation suggests that, at the immediate surface, roughly 5% of the $>Si-\mu_2O-Si<$ bonds have undergone hydrolysis at some point $([2.4\cdot10^{21}\ ^{18}\text{O-atoms/cm}^3]/[4.4\cdot10^{22}\ \text{O-atom/cm}^3]$ in bulk silica). However, the H profile indicates that more than 90% of the hydrolyzed bonds have recondensed, showing that hydrolyzed silica has the ability to heal itself by repolymerizing.

In hot steam, the reaction products of hydrolysis have nowhere to go. However, in liquid water, $Si(OH)_4$ and larger silicate oligomers can dissolve into the aqueous solution, resulting in uniform dissolution. This is one reason why hydrated layers on silica tend to be exceedingly thin (less than a few nanometers). Silica actually has a relatively high solubility in water [10] $(1.2\cdot10^{-3}\ \text{M}$ at room temperature), which means that when $>Si-\mu_2O-Si<$ bonds are hydrolyzed, large quantities of silica can be dissolved until the aqueous solution becomes saturated. A similar result is observed in leached minerals. However, for tetrahedral silica, the hydrolysis mechanism involves the nucleophilic attack on the silicon cation by water. Because water is not a particularly strong nucleophile, silica

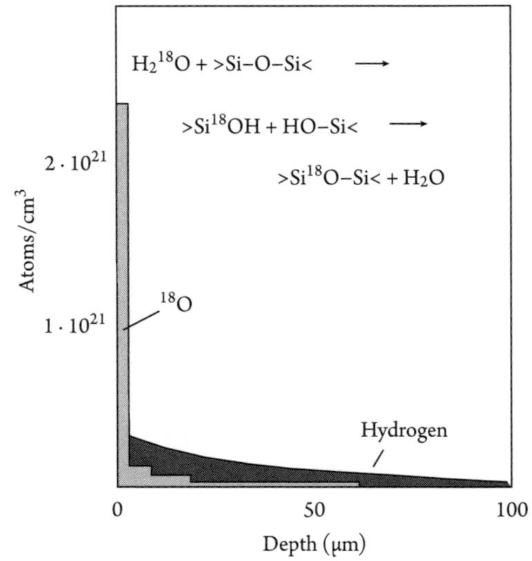

Figure 15.9 Elemental depth profiles for H and ^{18}O obtained using secondary-ion mass spectrometry for a sample of SiO_2 that was exposed to $H_2{}^{18}O$ vapor at 1100°C for 5 hours. If molecular water was the diffusing species, both curves would be coincident. Instead, ^{18}O is confined to the immediate surface, whereas H penetrates to a depth of almost 0.1 mm. Diffusion via reversible hydrolysis and condensation is clearly indicated. Adapted with permission from Roberts, G. J.; Roberts, J. P. An Oxygen Tracer Investigation of the Diffusion of "Water" in Silica Glass. *Phys. Chem. Glasses* **1966**, 7, 82.

dissolution in near-neutral water is exceedingly slow [10]. At room temperature, the dissolution rate is $3.5 \cdot 10^{-14} \dfrac{\text{mole}}{\text{cm}^2\text{s}}$, which translates into an etch rate of less than 1 nm/day. The hydroxide ion is a stronger nucleophile than water. Not surprisingly, silica dissolves more rapidly in basic solutions. The pH dependence that has been reported for basic solutions is that the dissolution rate increases by a factor of 10 as the pH increases by 2 units for pH values of 8 and more. In highly basic solutions such as the KOH–ethanol bath chemists use to clean glassware (which admittedly is nonaqueous), etch rates can exceed $1\ \mu\text{m/day}$, which is sufficient to change the internal volume of a volumetric flask.

15.4 ALKALI SILICATE GLASSES

Now consider the case when alkali-metal or alkaline-earth cations are introduced into silica to make alkali silicate glasses. In glass melts, modifier cations enter the silica structure via reactions such as

$$\text{Na}_2\text{O} + {>}\text{Si} - \mu_2\text{-O} - \text{Si}{<} \rightleftarrows {>}\text{Si} - \text{O} - \text{Na}^+ + \text{Na}^+ - \text{O}^- - \text{Si}{<} \quad \textbf{(15.5)}$$

The reaction of sodium oxide with silica in a glass melt is equivalent to the hydrolysis reaction between water and silica in liquid water. The net result is the replacement of a bridging oxygen site by two nonbridging oxygens, each of which has a charge that is compensated by a sodium ion. For a composition such as $Na_2O-3SiO_2$ or $Na_2Si_3O_7$, the net structure contains two nonbridging oxygens for every five bridging oxygens. The bridging oxygen or crosslink density of $Na_2Si_3O_7$ is 5/7, or 71%, relative to the 100% seen for fused silica. Extensive ^{29}Si nuclear magnetic resonance (NMR) and Raman spectroscopy investigations show that the bridging and nonbridging oxygens are not distributed randomly. This distribution depends strongly on which alkali-metal or alkaline-earth cation is present. Because the dissolution behavior of these glasses is strongly dependent' becomes 'depends strongly on on on the nonbridging oxygen distribution, the nature of this distribution needs to be understood.

Glass scientists use "Q" notation to denote how many bridging oxygens surround a given tetrahedral Si(IV) site in silicate glasses. Possible local sites include Q^4, Q^3, Q^2, Q^1, and Q^0, denoting silicate groups having four, three, two, one, or no bridging oxygens, respectively. Higher alkali contents result in lower crosslink densities and lower average Q values. For glasses with the composition $M_2O-2SiO_2$ or $M_2Si_2O_5$, each silicate tetrahedron has an average of one nonbridging oxygen, which would result in the production of a random network of interconnected Q^3 units. By the time the composition M_2O-SiO_2 is reached, the continuous glass network has been replaced by long chains of Q^2 units. However, the distribution of nonbridging oxygens among the different Q species depends on which modifier is present [11] (Fig. 15.10). Smaller or more highly charged modifiers such as Li^+ or Ca^{2+} can promote a local clustering of nonbridging oxygens via reactions such as

$$2Q^3 \rightleftarrows Q^4 + Q^2 \tag{15.6}$$

The importance of nonbridging oxygen distributions becomes apparent in the discussion of glass-dissolution mechanisms. The other structural difference associated with changes in modifier cations has to do with cation size. Modifier cations occupy space in the glass network, which can bias ring distributions. In addition, when ions are removed via ion exchange or selective leaching, larger cations leave behind larger voids, providing larger openings into which water and other solution species can diffuse.

Alkali silicates can participate in both *ion-exchange* and *selective-leaching* reactions in addition to the reactions described earlier for pure silica. Because ion exchange reactions are discussed in detail in Chapter 10, the emphasis here is on selective leaching leading to the alteration and destruction of the glass. When an alkali-silicate glass is immersed in deionized water, solution analyses clearly show

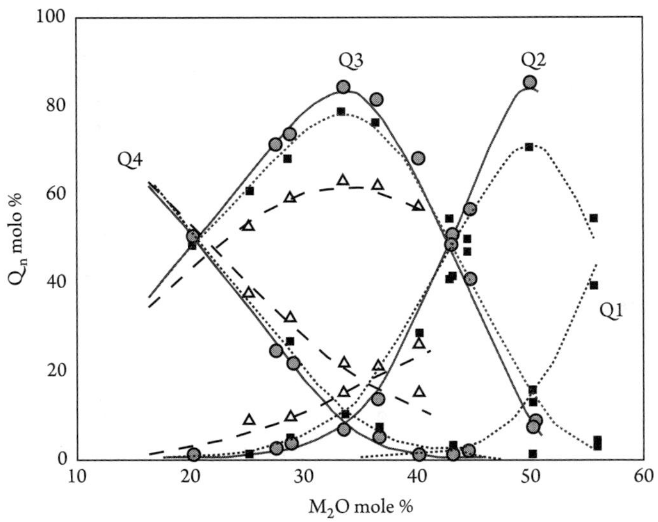

Figure 15.10 The distribution of nonbridging oxygens in alkali-silicate glass as a function of the nature and concentration of the modifier (Li = triangle, Na = square, K = circle). Silicate sites are designated by Q as in the text (e.g., a Q^3 site has three bridging oxygens and one nonbridging oxygen). Adapted with permission from Maekawa, H.; Maekawa, T.; Kawamura, K.; Yokokawa, T. The Structural Groups of Alkali Silicate Glasses Determined From ^{29}Si MAS-NMR. *J. Non-Cryst. Solids* **1991**, *127*, 53–64.

that alkali metal cations are initially released from the glass much more rapidly than any species containing silicon [5, 12]. A visual inspection of the glass (Fig. 15.11) shows that selective alkali leaching produces a distinct leached layer that grows thicker as a function of time. In a fixed volume of deionized water, alkali release is accompanied by the consumption of protons and an increase in the solution pH, as indicated in Eq. 15.3. As the pH increases, the rate of alkali release decreases, as expected, based on ion-exchange equilibria. Alkali release slows and eventually stops when the pH reaches the pK_a of the silanol groups (around pH 10–11), which function as a buffer to resist changes in pH. At pH 12 or more, it is no longer possible to protonate the nonbridging oxygens, so alkali cations are retained instead of being released.

At the high pH values induced by alkali leaching, the rate of silica release increases as a result of nucleophilic attack by the hydroxide ions produced. These two factors led early researchers to conclude that alkali silicates undergo a transition from selective leaching to uniform dissolution. However, to probe dissolution mechanisms thoroughly, it is advantageous to keep the solution pH fixed, because all glass reactions are highly sensitive to changes in pH. The most informative method for studying dissolution at a fixed pH involves using a technique called a *pH-stat titration*. In a pH-stat titration, a pH electrode is inserted into the

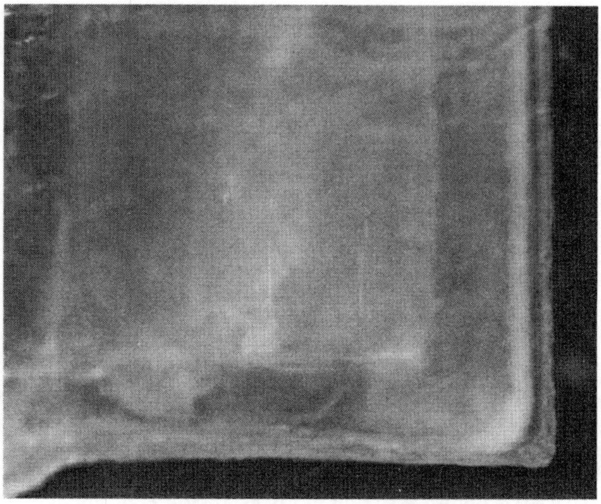

Figure 15.11 A sample of $K_2O-3SiO_2$ glass after several hours of exposure to 0.1 M HCl at 80°C. The thickness of the leached layer is on the order of 1 mm. The photograph was taken through the leaching solution; the leached layer is a gel-like material with poor structural integrity (note the flow of the layer in the bottom left corner).
SOURCE: Reprinted with permission from Bunker, B. C.; Arnold, G. W.; Beauchamp, E. K. Mechanisms for Alkali Leaching in Mixed Na-K Silicate Glasses *J. NonCryst. Solids,* **1983**, *58*, 295–322.

reaction vessel. If this electrode senses any changes in pH associated with alkali leaching, it commands an automatic burette system to add acid to the solution to compensate exactly for the amount of acid consumed by the glass. The kinetics of proton uptake and glass leaching are measured continuously as a function of time at a fixed pH by monitoring how much acid is added to keep the pH constant. When combined with solution analyses and visual inspections of the glass, a comprehensive picture of leaching kinetics versus pH can be obtained.

The first general observation regarding alkali-silicate dissolution at a fixed pH is that silica removal is negligible relative to alkali release except in highly basic solutions (Fig. 15.12). Except in strong bases, alkali-silicate glasses do not dissolve uniformly, which means the explanation for transition from $t^{1/2}$ to t^1 leaching kinetics must be found elsewhere. The exception to this rule involves glasses with an alkali content exceeding that of Na_2O-SiO_2. These do not have a continuous glass network and dissolve as ionic salts. For the mixed-alkali glass shown in Figure 5.11, the silicon dissolution rate is on the order of $\sim 1.7\cdot 10^{-11}\,\dfrac{\text{mole}}{\text{cm}^2\text{s}}$. Although relatively slow, the rate at which the silicate network is dissolving is still almost 1000 times faster than the rate at which fused silica dissolves.

Alkali release is much faster than silica release. With regard to the data in Figure 15.12, the rate of alkali depletion corresponds to the production of an

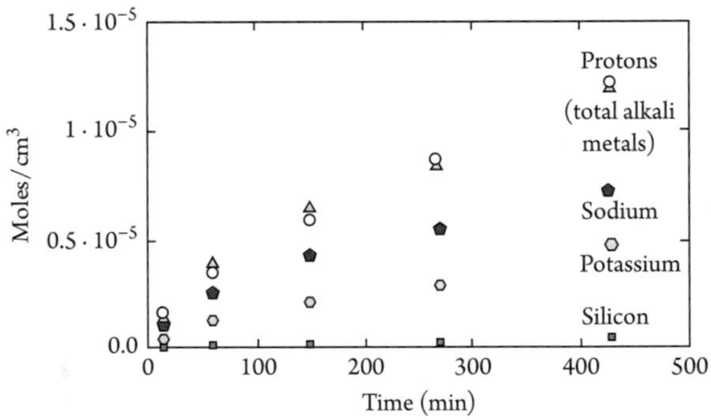

Figure 15.12 Elemental release (and proton consumption) rates for a glass with the composition $0.4K_2O-0.6Na_2O-3SiO_2$ in a pH 5.5 solution at 30°C (pH stat). Note that the rate of proton consumption is identical to the rate of K^+ and Na^+ release, and that the ratio of K^+/Na^+ released is identical to that of the glass composition. Silica release is negligible relative to alkali release. Adapted with permission from Bunker, B. C.; Arnold, G. W.; Beauchamp, E. K. Mechanisms for Alkali Leaching in Mixed Na-K Silicate Glasses *J. Non-Cryst. Solids* **1983**, *58*, 295–322.

alkali-free leached layer at a speed of 2 μm/hour or almost 50 μm/day, compared with the dissolution rate of less than 1 nm/day for fused silica exposed to similar conditions. One of the less durable alkali-silicate glasses, $K_2O-3SiO_2$, forms a leached layer at a rate of 3.6 mm/day in 60°C water at pH 1 [5, 13]. This rapid alkali leaching is a primary reason why simple binary alkali-silicate glasses are almost never used in practical applications. In terms of general trends, alkali leaching is more rapid in acidic than basic solutions, and it increases with the alkali content of the glass. However, at a fixed alkali content, the leach rate depends strongly on which alkali cation is present [5, 14] (Fig. 15.13). Potassium-containing glasses undergo the most rapid leaching, followed by sodium and lithium glasses. As in deionized water, at a fixed pH the kinetics of alkali leaching usually undergo a transition from $t^{1/2}$ to t^1 behavior, as shown for the ideal case of $Na_2O-3SiO_2$ (Fig. 15.14). Next, we describe the factors controlling the rate constants in both the $t^{1/2}$ and t^1 regimes, as well as what causes the transition between these two rate laws.

Because $t^{1/2}$ kinetics are often indicative of a diffusion-controlled process, it is important to establish which species are providing the diffusion control. Early researchers assumed the diffusion coefficient should correspond to the self-diffusion coefficient of the alkali cation in bulk glass [3]. However, studies on the mixed-alkali glasses clearly show that this is not the case. First, diffusion coefficients associated with selective leaching have been compared directly with the self-diffusion coefficients in bulk glass (Fig. 15.15). Depending on glass

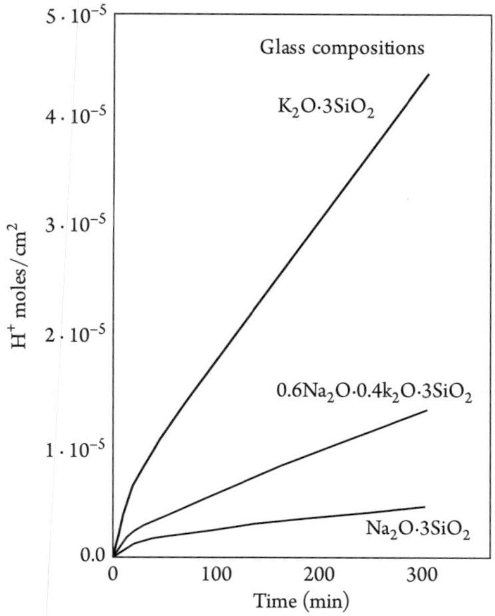

Figure 15.13 Rates of alkali release in pH stat titrations as a function of the nature of the alkali cation (pH 7, T = 22°C). The results show that a glass containing Na^+ is substantially more durable than that containing K^+. Adapted with permission from Bunker, B. C.; Arnold, G. W.; Beauchamp, E. K. Mechanisms for Alkali Leaching in Mixed Na-K Silicate Glasses *J. Non-Cryst. Solids* **1983**, *58*, 295–322.

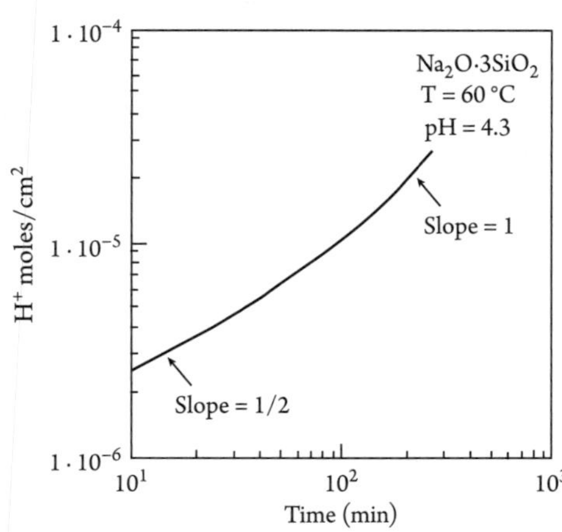

Figure 15.14 A plot of the log of Na^+ release versus the log of time for $Na_2O-3SiO_2$ glass (pH 4.3, T = 60°C) showing the transition from $t^{1/2}$ to t^1 leaching kinetics. Adapted with permission from Bunker, B. C.; Arnold, G. W.; Beauchamp, E. K. Mechanisms for Alkali Leaching in Mixed Na-K Silicate Glasses *J. Non-Cryst. Solids* **1983**, *58*, 295–322.

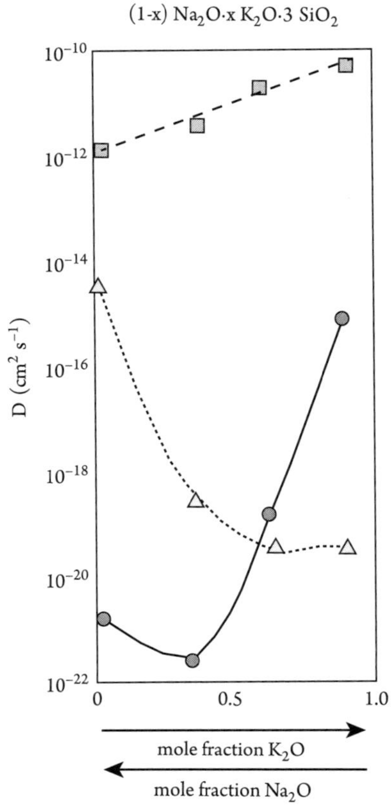

Figure 15.15 A comparison of effective diffusion coefficients for alkali release and alkali diffusion coefficients measured in bulk glass for a range of mixed alkali silicate glasses. The top curve (squares) is for interdiffusion diffusion estimates from leach rates. The curve with triangles is for Na^+ diffusion in bulk glass, whereas the curve with circles is for K^+ in bulk glass. Note that leaching is orders of magnitude faster than bulk diffusion and shows no evidence for the mixed-alkali effect.
SOURCE: Reprinted with permission from Bunker, B. C.; Arnold, G. W.; Beauchamp, E. K. Mechanisms for Alkali Leaching in Mixed Na-K Silicate Glasses *J. NonCryst. Solids,* **1983**, *58*, 295–322.

composition, the diffusion coefficient for leaching is 10^2 to 10^8 times greater than the self-diffusion coefficient. Second, for self-diffusion in bulk glass, it has been determined that the diffusion of one cation can suppress the diffusion coefficient of the sister cation via the so-called mixed-alkali effect [15] by a factor of as much as 1 million. For Na–K glasses, self-diffusion is most rapid for the sodium and potassium end members, and exhibits a pronounced minimum near a 50:50 mix of the two cations. No such minimum is observed in leaching data, where the leach rate shows a steady increase as one moves from the sodium to the potassium end member. Third, Na^+ and K^+ are always removed from the mixed-alkali glasses at a rate proportional to the mole fraction of each ion present (Fig. 15.12). In other words, on a per-mole basis, the leach rates of Na^+ and K^+ are identical. If bulk alkali diffusion were rate limiting, Na^+ and K^+ would exhibit drastically different leach rates for most compositions.

If cation diffusion out of the glass does not limit rates, the next option to consider is the diffusion of water or hydronium ions into the glass. Water can penetrate silicate structures via molecular diffusion if ring openings are greater than or equal to the diameter of the water molecule. For smaller rings, water penetration can also occur via reversible network hydrolysis. For rings containing six silicate tetrahedra, diffusion coefficients for water are near 10^{-12} cm^2/second and are comparable with the values seen for leaching. When the nonbridging oxygen is protonated by either a water molecule or a hydronium ion, the ionic forces holding the alkali ion to the nonbridging oxygen are disrupted, which allows for rapid alkali diffusion. If rates of alkali release are controlled by the kinetics of water diffusion, Na^+ and K^+ release rates should be similar, in agreement with experimental results. Finally, if alkali diffusion were rate limiting, there would be no pH dependence for alkali leaching. Because leaching is clearly enhanced by the presence of hydronium ions, bulk diffusion of alkali cations is not the critical factor controlling leaching kinetics.

Clues regarding what is going on in the t^1 leaching regime are provided by measurements that have been done to determine the thickness of the leached layer as a function of time [5]. When linear t^1 kinetics are in effect, the leached layer grows in thickness as a function of t^1 (Fig. 15.16). This demonstrates that the glass is not dissolving uniformly in this kinetic regime. In fact, it appears the transition from $t^{1/2}$ to t^1 behavior occurs because the structure of the leached layer changes with time to create a porous material that is no longer capable of functioning as a diffusion barrier. In the t^1 regime, the rate of alkali release is controlled by the rate at which the glass restructures itself at the interface between leached and unleached glass. To check this hypothesis, transmission electron microscopy has been used to examine the fine-scale structure of the leached layer [16]. The diffusion coefficient for water transport in silicate materials can be as rapid as that found in bulk water, provided openings are available larger than 0.7 nm. The transmission electron microscopic images (Fig. 15.17) show that, with time, the leached layer starts to develop pores that are even larger than this, and the development of this porosity coincides with the $t^{1/2}$-to-t^1 transition. Eventually, the leached layer assumes a structure that resembles an aggregation of colloidal silica particles, which in fact is just what it is. The interconnected porosity in this layer represents a network of channels through which water and other solution species can diffuse in an unimpeded fashion. Glasses containing K_2O develop such porosity almost immediately,

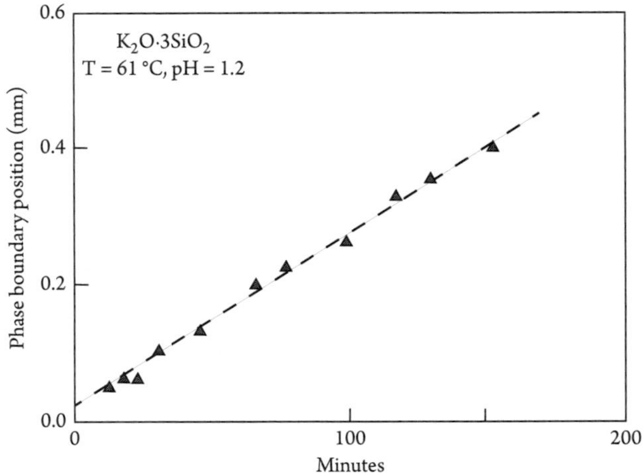

Figure 15.16 The position of the phase boundary between leached and unleached glass (see Fig. 15.10) as a function of time for $K_2O-3SiO_2$ glass (pH 1.2, T = 61°C). The results clearly show that t^1 kinetics are associated with the position of the boundary between leached and unleached glass, and are not a result of uniform glass dissolution. Adapted with permission from Bunker, B. C.; Arnold, G. W.; Beauchamp, E. K. Mechanisms for Alkali Leaching in Mixed Na-K Silicate Glasses *J. Non-Cryst. Solids* **1983**, *58*, 295–322.

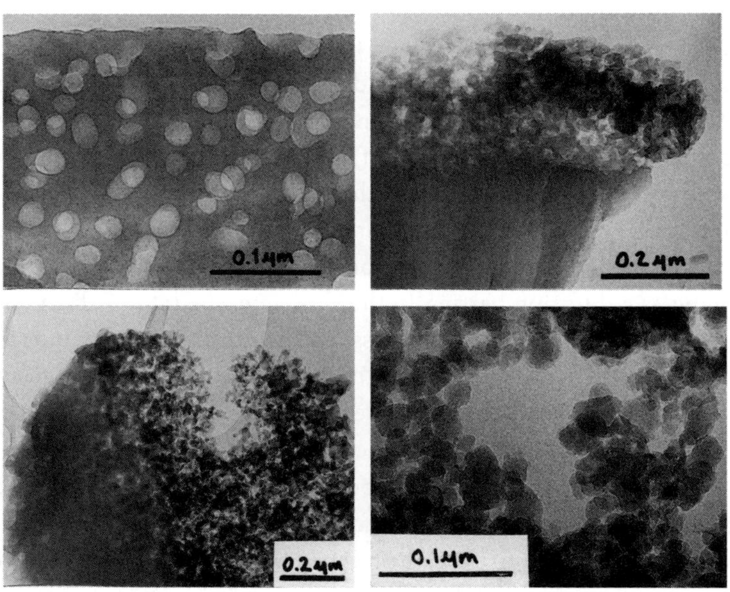

Figure 15.17 Transmission electron micrographs of $K_2O-3SiO_2$ and $Na_2O-3SiO_2$ glass samples after leaching, showing the development of porosity and the ultimate conversion of the leached layer into a gel consisting of colloidal silica particles. The development of porosity corresponds to the transition from $t^{1/2}$ to t^1 leaching kinetics. SOURCE: Reprinted with permission from Bunker, B. C.; Headley, T. J.; Douglas, S. C. Gel Structures in Leached Alkali Silicate Glass. *Mat. Res. Soc. Symp. Proc.*, **1984**, *32*, 41–46.

whereas this porosity is slower to develop in both the Na_2O- and Li_2O-containing glasses.

The model describing the mechanism for the $t^{1/2}$ to t^1 transition is shown in Figure 15.18 [5]. Initially, water and hydronium ions diffuse into the glass structure, resulting in the protonation of nonbridging oxygens and the release of alkali cations. At this stage, the glass network resembles that of bulk glass other than the fact that the alkali cations bonded to nonbridging oxygens have been replaced by protons to form silanol groups. As long as this initial structure is retained, $t^{1/2}$ kinetics are observed. However, the silanol-rich silicate matrix created by alkali leaching is capable of undergoing reversible hydrolysis and condensation reactions. These reactions result in an evolution in the structure of the silicate matrix to form a silica gel consisting of interconnected fused silica particles. This silica matrix is highly porous and permeable. Stresses generated within the layer eventually create a network of cracks called *crazing*. The combination of porosity and crazing leads to the formation of a leached layer with low structural integrity. When the silicate matrix is reorganized, it no longer functions as a diffusion barrier, and t^1 kinetics are observed. In a mature leached glass, the outer leached layer (called *hydrated glass* in Fig. 5.18) represents the porous array of silica particles. An inner layer (called the *transition zone*) represents the area within which the altered glass is still a diffusion barrier, although it is undergoing reorganization. Elemental depth profiles show that this transition zone reaches a constant

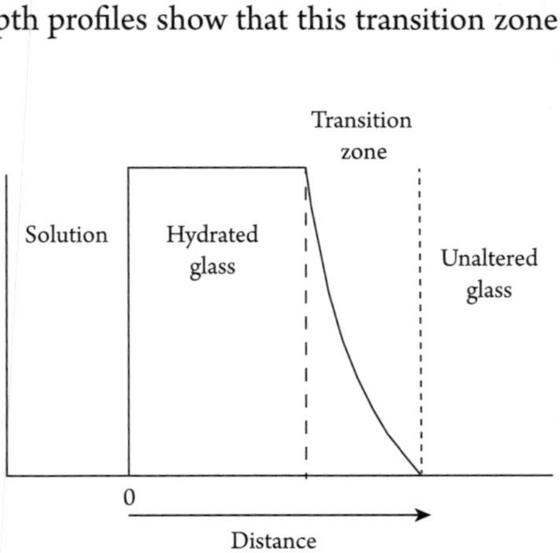

Figure 15.18 A depiction of the water content of a leached glass surface as a means of indicating different zones of relevance to leaching mechanisms. The outer layer of hydrated glass (often colloidal silica) has a relatively constant water content associated with interconnected water-filled porosity. An inner transition zone represents the regime within which selective leaching and network transformations occur. Hydrogen depth profiles of mature leached layers obtained via elastic recoil detection analysis clearly show both zones. Adapted with permission from Bunker, B. C.; Arnold, G. W.; Beauchamp, E. K. Mechanisms for Alkali Leaching in Mixed Na-K Silicate Glasses *J. Non-Cryst. Solids* **1983**, 58, 295–322.

steady-state thickness when the $t^{1/2}$-to-t^1 transition occurs. Thereafter, the transition zone moves into the glass with a t^1 time dependence. In other words, reaction rates are controlled by the constant rate of restructuring and motion of the glass–gel interface rather than uniform dissolution at the glass–solution interface. These processes are obviously affected by the hydrodynamics of the adjacent solution as well as by changes in solution chemistry.

The final factor in alkali leaching that needs to be addressed involves understanding why the silicate network undergoes reconstruction. A related issue is how specific alkali metal cations influence the kinetics of leaching before, during, and after the transition in behavior. The key to understanding reconstruction involves understanding the factors that control reversible hydrolysis and condensation reactions. As outlined in Chapter 14, these reactions proceed via a nucleophilic attack mechanism in which a tetrahedral silicate site reconfigures itself to allow for the formation of a five-coordinate intermediate species with a trigonal pyramidal geometry. For an isolated $[Si(OH)_4]^\circ$ species in solution, rearranging the hydroxyl groups to accommodate the nucleophile is a relatively unconstrained process. However, in moving from $[Si(OH)_4]^\circ$ (a Q^0 species) to Q^1, Q^2, Q^3, and Q^4 geometries, each bridging oxygen adds a local constraint that inhibits the formation of the five-coordinate intermediate. The result is that the kinetics of hydrolysis and condensation reactions are most rapid for the Q^0 species, and systematically decrease as the number of bridging bonds increases (see Chapter 14).

Measurements of depolymerization rates via the silicomolybdate method for detecting monomeric silica [9] (Fig. 15.19) show that, just as was the case in

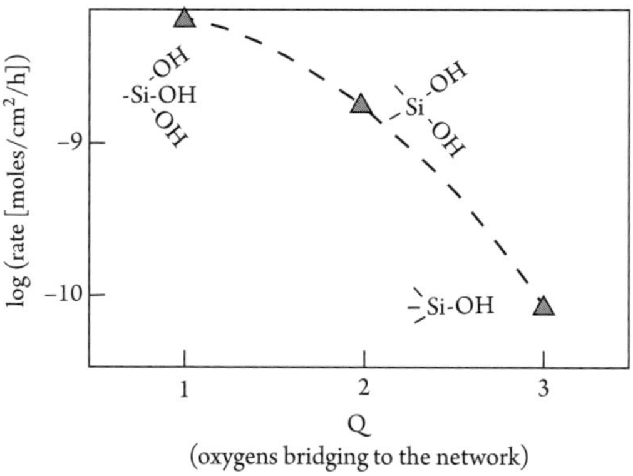

Figure 15.19 The kinetics of monomeric silica release from alkali silicate glasses as a function of the number of nonbridging oxygens per silicate site as measured via silicomolybdate assays. The results clearly show that sites with a low Q value (i.e., more nonbridging oxygens) depolymerize to release silica at a more rapid rate than a site with a high Q value. Graph created from data in Iler, R. K. *The Chemistry of Silica*; Wiley: New York, 1979.

soluble silicate complexes (see Ch. 14), a Q^1 unit is ~10 times more reactive than a Q^2, which in turn is ~100 times more reactive than a Q^3 unit. The net consequence of the relative reaction rates versus Q is that silicate networks in water gradually evolve from a low Q distribution (which is relatively reactive and unstable) to a network that maximizes the population of the most stable Q^4 units (i.e., completely polymerized fused silica particles). When silanol groups are created by selective leaching, they undergo condensation reactions with each other to eliminate nonbridging oxygens and maximize the bridging oxygen population. The process continues until all silanol groups within range of other silanols are consumed. In the final product, most silanols reside at the surfaces of dense silica particles.

One reason why glasses containing Na_2O are more durable than those containing K_2O is that, in the Na_2O glasses, the glass phase separates into regions of Q^2-rich clusters that are encapsulated by regions rich in more durable Q^4 units (see Fig. 15.10). The Q^4 units protecting the Q^2 units represent a more stable distribution than a glass that has a uniform distribution of Q^3 units. The redistribution of Q units is even more pronounced when divalent cations are introduced into the glass. This is one reason why soda–lime glass (which contains CaO as well as Na_2O) is actually quite durable. Soda–lime glass represents almost 90% of all glass produced and is used heavily in drinking glasses, windows, and many other objects we encounter in our daily lives. Another reason for the high durability of soda–lime glass is that the Ca(II) ion has a much higher affinity for anionic nonbridging oxygen sites than Na(I), making it more difficult to leach via replacement by protons (see Chapter 10).

15.5 BORATE GLASSES

Few studies have been published regarding the dissolution behavior of alkali borate glasses, because borate glasses exhibit such poor chemical durability they are impractical for use in most real-world environments. However, the behavior of the borate glasses is relevant to understanding the widely used borosilicate glasses discussed later. Borate glasses develop random networks based on interconnected rings of trigonal planar BO_3 units [17]. In contrast to the silicate glasses, addition of alkali metal oxides to a B_2O_3 melt initially results in the conversion of trigonal borate sites into tetrahedral sites. The introduction of tetrahedral sites increases rather than decreases the crosslink density of the glass. As the mole fraction of the modifier is increased, the mole fraction of tetrahedral sites increases, moving from boroxyl units to pentaborates, triborates, and finally diborates (Fig. 15.20) at a composition of $M_2O-2B_2O_3$. However, electrostatic repulsion between anionic tetrahedral boron sites dictates that the diborates are energetically unstable relative to the triborates and pentaborates. For this reason,

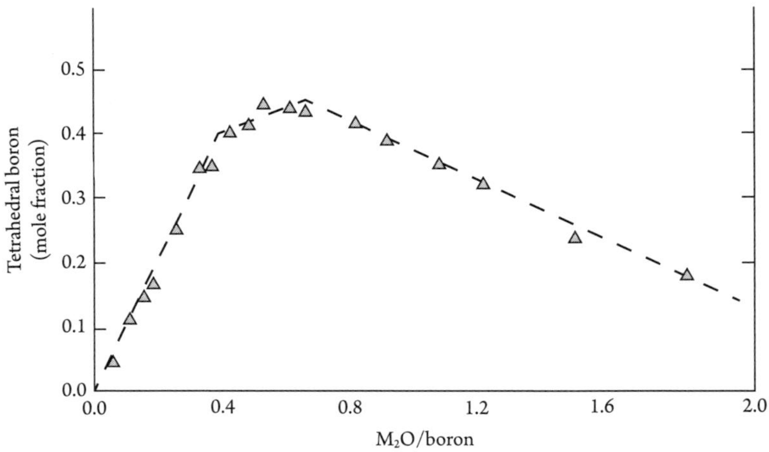

Figure 15.20 The mole fraction of tetrahedral borate sites (N_4) as a function of alkali content as measured via nuclear magnetic resonance spectroscopy. Adapted with permission from Feller, S. A.; Dell, W. J.; Bray, P. J. ^{10}B NMR Studies of Lithium Borate Glasses. *J. Non-Cryst. Solids* **1982**, *51*, 21–30.

as the modifier content is increased further, nonbridging oxygens are created in preference to tetrahedral boron sites. These nonbridging oxygens are arranged progressively into metaborate, pyroborate, orthoborate, and finally loose BO_4 units with increasing modifier content (Fig. 15.20). In the high-modifier-content regime, where nonbridging oxygens are preferred, the tetrahedral borate sites are eventually consumed and the mole fraction of tetrahedral sites relative to trigonal sites (N_4) drops to zero.

The stability of borate glasses in water is in sharp contrast to the stability of silica [18]. Pure B_2O_3 glass dissolves in distilled water (T = 30°C) at a rate of $5\cdot10^{-4}\,\frac{\text{g}}{\text{cm}^2\text{s}}$ or $7\cdot10^{-6}\,\frac{\text{mole}}{\text{cm}^2\text{s}}$. To put this number in perspective, B_2O_3 is etched away at a rate of 17 cm/day or almost 1 cm/hour (i.e., 200 million times faster than silica). This rapid dissolution rate is related to the structure of B_2O_3, which consists of interconnected trigonal planar boroxol rings. This trigonal planar geometry invites nucleophilic attack from both above and below the plane of the ring. Although not a strong nucleophile, even a water molecule finds the open axial sites to be inviting targets for attack. In contrast, tetrahedral borate units must be distorted to open up sites for nucleophilic attack, which makes the tetrahedral units more resistant to hydrolysis than the trigonal units. This mechanism explains the observed trends in dissolution rates for the alkali-borate glasses as a function of alkali content [17] (Fig. 15.21). All the alkali-borate glasses dissolve uniformly, because all these glasses are based on networks containing highly reactive trigonal borate units.

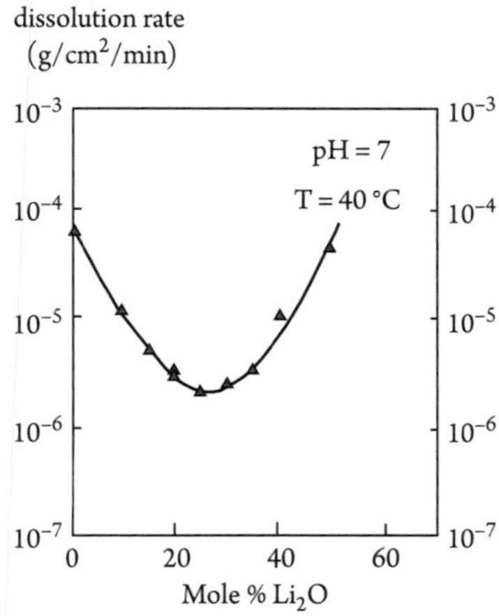

Figure 15.21 The kinetics of uniform dissolution for alkali-borate glasses. Dissolution rates are at a minimum when the concentration of tetrahedral boron in the bulk glass is at a maximum. Adapted with permission from Velez, M. H.; Tuller, H. L.; Uhlmann, D. R. Chemical Durability of Lithium Borate Glasses. *J. Non-Cryst. Solids* **1982**, *49*, 351–362.

However, the addition of alkali improves the durability as long as additions lead to the creation of tetrahedral borate sites. The most durable glasses correspond to the composition $M_2O-3B_2O_3$, which is dominated by triborate units. These units represent the maximum concentration of tetrahedral sites that can be introduced without forcing the sites to share bridging oxygens. For the $Li_2O-B_2O_3$ glasses, this composition dissolves roughly 100 times slower than pure B_2O_3 glass (Fig. 15.21).

15.6 BOROSILICATE AND ALUMINOSILICATE GLASSES

We have seen that alkali-borate glasses dissolve almost as fast as sodium chloride does in water. We have also seen that alkali silicate glasses undergo rapid selective leaching to produce silica gels with negligible structural integrity. Based on these results, one might expect that combining an alkali borate glass with an alkali silicate glass to make an alkali-borosilicate glass would result in disaster. Nothing could be further from the truth. Alkali-borosilicate glasses are among the most durable glasses known, exhibiting dissolution rates almost as slow as that seen for fused silica. Pyrex glasses and other common compositions used for laboratory glassware are borosilicate glasses. How can this be? Once again, the key to understanding the behavior of alkali-borosilicate glasses starts with an appreciation of glass structure.

Boron can enter the structure of a silica-rich glass either by dissolving into the silicate matrix or by forming a separate borate phase (Fig. 15.22). Conversely,

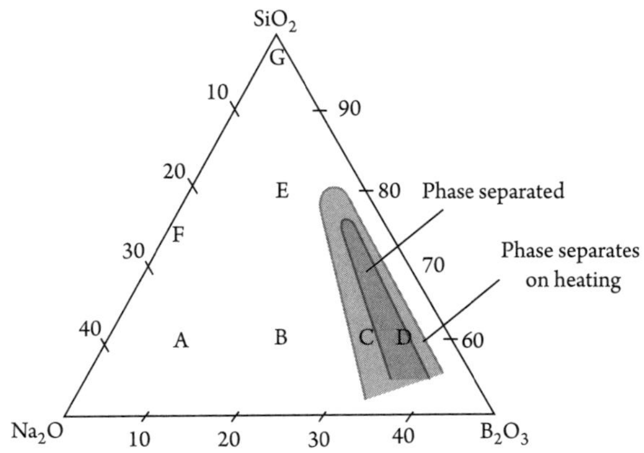

Figure 15.22 A phase diagram for sodium-borosilicate glasses showing the predominant structural units present as a function of composition. The regions within the curved lines correspond to zones within which the glass phase separated into silicate- and borate-rich phases. The letters A-G correspond to specific glass compositions, some of which are discussed in the text. Adapted with permission from Bunker, B. C.; Arnold, G. W.; Day, D. E.; Bray, P. J. The Effect of Molecular Structure on Borosilicate Glass Leaching. *J. Non-Cryst. Solids* **1986**, *87*, 226–253.

silica can have an appreciable solubility in borate or alkali-borate phases. Alkali metal and alkaline-earth cations can dissolve into either the silica-rich phase or the borate phase, and are commonly found in both. As in the alkali-silicate glasses, alkali additions can lead to the formation of anionic nonbridging oxygens within the silica-rich network. Alkali incorporation within the borate phase can lead to the formation of anionic, tetrahedral borate sites. However, when boron dissolves into the silica-rich phase, a new type of anionic site can form that involves tetrahedral boron centers that occupy sites within the covalent silicate network.

A wide range of characterization methods, including Raman vibrational spectroscopy and ^{11}B, ^{29}Si, ^{17}O, and ^{23}Na nuclear magnetic resonance spectroscopy, have been used to obtain local site distributions as a function of glass composition for sodium borosilicate glasses [19]. The trends observed for different structural units as a function of Na_2O/B_2O_3 and SiO_2/B_2O_3 are described next.

In the low-sodium composition regime, the first preference for Na^+ is to enter a borate phase to associate with tetrahedral boron sites. As in alkali-borate glasses, the maximum concentration of tetrahedral boron sites corresponds to the diborate composition $\left(Na_2O-2B_2O_3\right)$ for which $Na_2O/B_2O_3=0.5$. In the low-sodium regime, all borosilicate glasses tend to separate into a sodium borate phase and a silica-rich phase that is relatively alkali free. For example, for the glass $5Na_2O-35B_2O_3-60SiO_2$, (composition D in Figure 15.22) the compositions determined for the two phases are the borate phase $\left(8Na_2O-62B_2O_3-30SiO_2\right)$ and the silicate phase $\left(2Na_2O-8B_2O_3-90SiO_2\right)$.

For intermediate-alkali-content compositions, ion pairs consisting of tetrahedral borate sites and alkali cations insert themselves into the silicate network. The net result is the creation of a network in which all oxygens bridge between two tetrahedral network-forming cations, as in fused silica. However, sodium ions are now present to compensate for the partial negative charges (-0.25 per oxygen) on each $>Si-\mu_2O-B$ bond$<$. Phase separation is typically absent in this composition regime. It was originally thought that the maximum allowable boron content in this pseudo-silica network corresponded to the composition of the mineral reedmergnerite ($NaBSi_4O_{10}$), in which all borate tetrahedra are isolated from one another. However, Raman spectra detect the presence of four-membered rings containing Si_2O_7 and B_2O_7 pairs, as in the mineral danburite $(CaB_2Si_2O_8)$ [19]. Although predicted to be less stable than isolated borate tetrahedra (because each $>B-\mu_2O-B<$ bond has a formal charge of -0.5), danburite units allow for a higher boron content in the silicate network.

At sodium concentrations that exceed the concentrations of the tetrahedral boron sites, nonbridging oxygens are created to charge-compensate the excess Na^+, as in alkali-silicate glasses. In other words, sodium ions are partitioned between both silicate nonbridging oxygens and tetrahedral boron sites. For example, for the composition $30Na_2O-10B_2O_3-60SiO_2$, (composition A in Figure 15.22), roughly two-thirds of the sodium ions are associated with nonbridging oxygens whereas the other one-third is charge-compensated by tetrahedral boron sites in the silicate network.

As expected, the nature and rate of alkali-borosilicate glass dissolution is sensitive to glass composition and structure [20]. The least stable compositions are alkali-poor compositions that undergo phase separation. Here, the primary dissolution mechanism involves the selective dissolution of the alkali borate phase, which is highly susceptible to attack by water. Because the alkali borate phase tends to be continuous and interconnected, the silica-rich phase is normally incapable of providing any protection to the more vulnerable alkali borate phase. For the phase-separated composition: $5Na_2O-35B_2O_3-60SiO_2$, the alkali borate phase dissolves at a rate of around $2\ \mu m/\text{hour}$ at room temperature. Transmission electron microscopy shows that, after leaching, the remaining glass consists of the silica-rich skeleton that has been completely stripped of the alkali borate phase (Fig. 15.23). The kinetics of alkali-borate removal always follow t^1 kinetics, because the release rate corresponds to the rate of hydrolysis of the borate network. The silicate network never forms a diffusion barrier. For the $5-35-60$ glass, the rate of alkali-borate dissolution is roughly 50 times slower than that seen for the most durable lithium borate glass, because the sodium borate phase in 5–35–60 contains appreciable amounts of silica (with a

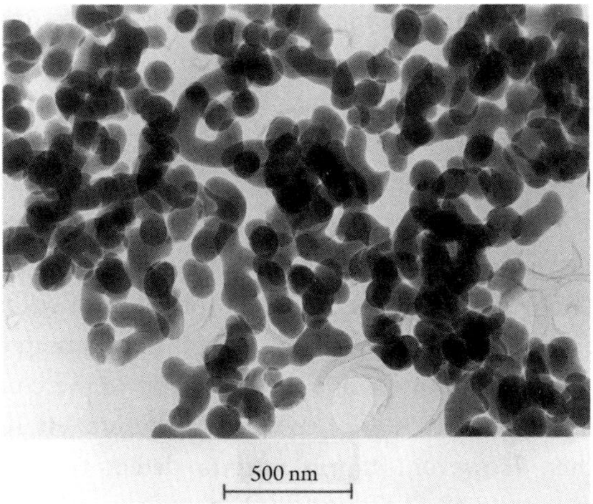

Figure 15.23 A transmission electron micrograph of the phase-separated glass $5Na_2O$–$35B_2O_3$–$60SiO_2$ after leaching. The reactive and soluble sodium borate phase is completely gone, leaving behind a silica-rich phase that is largely untouched.
SOURCE: Reproduced with permission from Bunker, B. C.; Arnold, G. W.; Day, D. E.; Bray, P. J. The Effect of Molecular Structure on Borosilicate Glass Leaching. *J. NonCryst. Solids*, **1986**, *87*, 226–253.

composition of $8Na_2O-62B_2O_3-30SiO_2$), which provides some stabilization to the borate network.

The most durable borosilicate glasses are those in which all boron is present as tetrahedral borate groups that substitute into the silicate network. Although these glasses contain Na^+, the glasses do not experience selective alkali leaching in neutral solutions, but instead undergo uniform dissolution. In other words, all elements leave the glass at the same rate. At short times, a $t^{1/2}$ rate dependence is observed for all species, indicating that a thin hydrated layer forms as a precursor to the onset of linear t^1 kinetics. As the boron content of the network increases, the chemical durability decreases. For example, in neutral pH at 70°C the composition $20Na_2O-20B_2O_3-60SiO_2$ (composition B in Figure 15.22) releases Si at a rate of $4.7\cdot 10^{-11}\,\frac{\text{mole}}{\text{cm}^2\text{s}}$, compared with $8.3\cdot 10^{-12}\,\frac{\text{mole}}{\text{cm}^2\text{s}}$ for $10Na_2O-10B_2O_3-80SiO_2$ (composition E in Figure 15.22) and $1.1\cdot 10^{-14}\,\frac{\text{mole}}{\text{cm}^2\text{s}}$ for fused silica. The reason for the absence of alkali leaching in neutral pH conditions has to do with the large difference in pK_a for $Si-\mu_2O-B$ bonds relative to nonbridging oxygens. Formal-charge arguments (see Chapters 4 and 5) and pH titrations indicate the pK_a for a bridging oxygen between tetrahedral Si and tetrahedral boron is around 3 to 4, consistent with the oxygen formal charge of -0.25. This value is in sharp contrast to the value of around 10 to 11 that is associated with the nonbridging oxygen. In neutral pH solutions, $>Si-\mu_2O-B<$ bonds are not protonated, and the charge-compensating alkali cations do not undergo selective leaching.

As expected based on the pK_a value of the $>Si-\mu_2O-B<$ bond, selective alkali leaching is observed at pH less than 3. All alkali is removed from compositions with a Na:B ratio of 1:1 at low pH, with overall glass-alteration rates that are accelerated relative to those seen under neutral pH conditions. However, the mechanism for alkali release does not involve simple ion exchange, because alkali release is accompanied by boron release. This occurs because proton adsorption on $>Si-\mu_2O-B<$ bonds not only stimulates alkali release, but also results in the immediate hydrolysis of the bridging bond, leading to selective borate dissolution (Fig. 15.24).

The final borosilicate compositions to consider involve glasses in which the total modifier content exceeds the concentration of tetrahedral boron sites. Here, alkali cations are associated with both nonbridging oxygens and tetrahedral boron sites. These particular glasses are the only alkali borosilicate glasses to undergo selective alkali leaching. However, elemental depth profiles (Fig. 15.25) show that, in neutral pH solutions, only a fraction of the alkali is susceptible to removal. The fraction leached corresponds to those alkali cations associated with nonbridging oxygens. For example, for the glass $30Na_2O-10B_2O_3-60SiO_2$, the two-thirds of the Na^+ associated with nonbridging oxygens is removed, whereas the one-third of the Na^+ associated with the tetrahedral boron sites is untouched. Leaching kinetics mirror the behavior seen for the binary alkali-silicate glasses. In other words, the leach rate

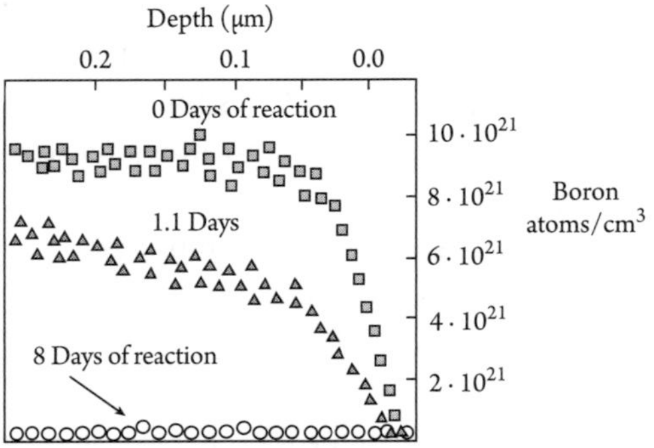

Figure 15.24 Boron depth profiles as a function of leaching time for the glass $20Na_2O-20B_2O_3-60SiO_2$ (pH 12, pH 12, T = 70°C). The results show that after the surface layer is hydrated, $>Si-\mu_2O-B<$ bonds are gradually hydrolyzed, resulting in the selective leaching of both boron and sodium. Although hydrolysis is slow, the only cation remaining after 8 days of leaching is Si(IV). Adapted with permission from Bunker, B. C.; Arnold, G. W.; Day, D. E.; Bray, P. J. The Effect of Molecular Structure on Borosilicate Glass Leaching. *J. Non-Cryst. Solids* **1986**, *87*, 226–253.

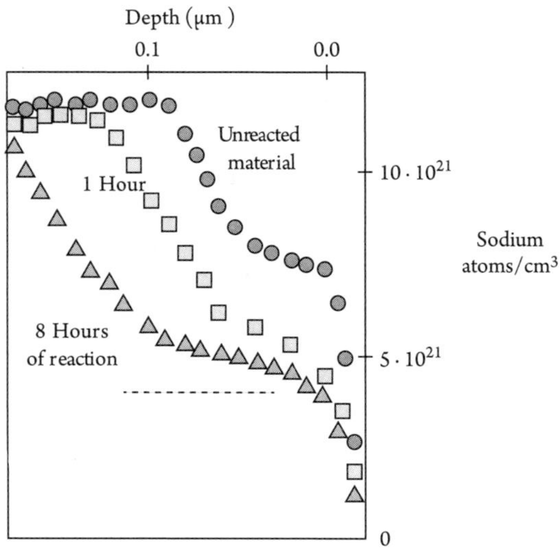

Figure 15.25 Sodium depth profiles for the glass $30Na_2O-10B_2O_3-60SiO_2$ leached at pH 8 and 70°C. In this glass, two-thirds of the Na^+ is associated with silicate nonbridging oxygens whereas the other one-third/3 is associated with tetrahedral boron sites. The profiles clearly show that, at least initially, the only species removed from the glass via leaching at near-neutral pH is the Na^+ associated with the nonbridging oxygens. Adapted with permission from Bunker, B. C.; Arnold, G. W.; Day, D. E.; Bray, P. J. The Effect of Molecular Structure on Borosilicate Glass Leaching. *J. Non-Cryst. Solids* **1986**, *87*, 226–253.

for $30Na_2O-10B_2O_3-60SiO_2$ is similar to that of $20Na_2O-80SiO_2$, which means that when boron is present, much more alkali can be accommodated within a silicate glass without compromising chemical durability. This is important, because the additions of alkali cations and boron to silica impart many favorable properties to the glass, including lower melting temperatures and much lower processing costs.

The aluminosilicate glasses tend to be much more refractory (i.e., have higher melting temperatures) than the borosilicate glasses, and are thus used less frequently. The glass-forming region is also smaller than that of the borosilicates, because compositions containing octahedral aluminum tend to crystallize. However, for those compositions in which the modifier cation content is greater than or equal to the aluminum content, tetrahedral aluminum sites dominate and amorphous glasses are readily formed. Because $>Si-\mu_2O-Al<$ and $>Si-\mu_2O-B<$ bonds in tetrahedral networks have identical structures and formal-charge distributions, glasses containing tetrahedral aluminate sites exhibit dissolution behaviors that mirror those seen for equivalent borosilicate

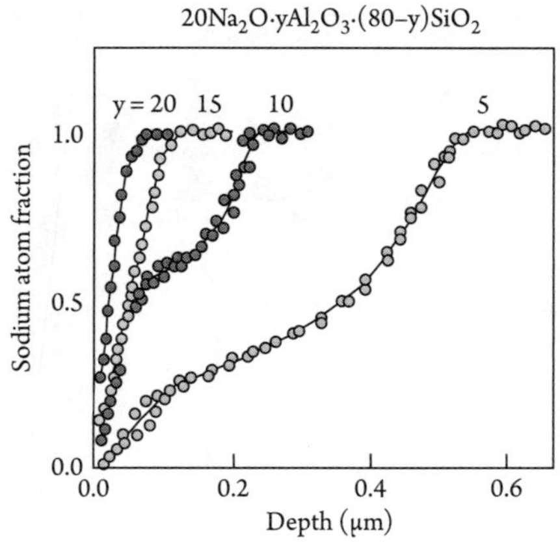

Figure 15.26 Sodium depth profiles for a series of leached sodium-aluminosilicate glasses. As in the sodium borosilicate glasses, the fraction of Na^+ removed via selective leaching in neutral pH solutions corresponds to that associated with silicate nonbridging oxygens. The Na^+ associated with anionic tetrahedral aluminate sites is largely untouched. Adapted with permission from Smets, B. M. J. and Lommen, T. P. Role of Molecular Water in the Leaching of Glass. *Phyics. Chem. Glasses* **1983**, *24*, 35–36.

glass compositions. Elemental depth profiles (Fig. 15.26) show that, in neutral pH solutions, the fraction of alkali leached is that fraction associated with nonbridging oxygens, whereas the fraction associated with tetrahedral aluminate sites is again untouched. The rate at which leached layers form is once again determined by the nonbridging oxygen content. At a fixed alkali content, glasses with a higher aluminum content are more durable that those with a low aluminum content. As with borosilicate glasses, exposures to pH values less than the pK_a of the $>Si-\mu_2O-Al<$ bonds results in protonation of the bridging oxygens, removal of all alkali, and selective dissolution of aluminum from the network via the proton-assisted hydrolysis of $>Si-\mu_2O-Al<$ bonds. However, because these bonds are more resistant to hydrolysis than the corresponding borate bonds, the acid-promoted dissolution of aluminosilicates occurs at a much slower rate.

15.7 PHOSPHATE GLASSES

Phosphate glasses exhibit high thermal expansion coefficients and low melting points, which makes them attractive candidates for glass-to-metal seals to high thermal expansion metals such as aluminum and stainless steel [21, 22]. Unfortunately, the glasses with the best physical properties are highly susceptible to attack by water. In fact, solid centimeter-size glass samples can be transformed

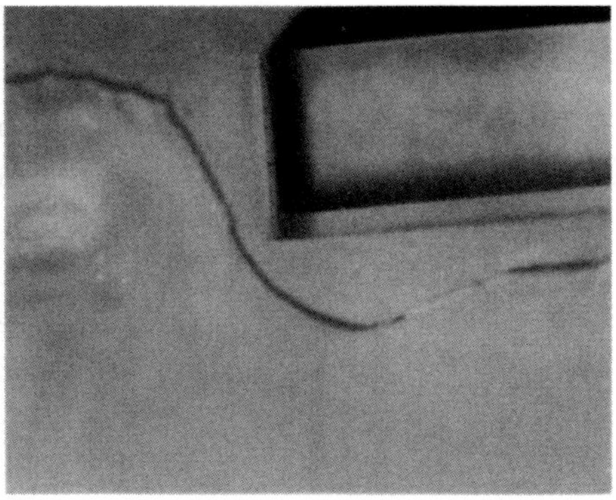

Figure 15.27 A photograph of a 3-mm-wide pristine sample of the phosphate glass $40Na_2O-10BaO-50P_2O_5$ next to a sample exposed to hot, humid air (70°C, 90% relative humidity, t = 16 hours). The photograph clearly shows that humid air is sufficient to convert the glass into a liquid gel.
SOURCE: Reproduced with permission from Bunker, B. C.; Arnold, G. W.; Wilder, J. A.; Phosphate Glass Dissolution in Aqueous Solutions. *J. NonCryst. Solids,* **1984**, *64,* 291–316.

into semiliquid gels after exposure to hot, humid air for less than 1 day (Fig. 15.27) [23]. At the other end of the spectrum, some phosphate glass compositions are so durable in water they are under consideration for use as hosts for nuclear-waste disposal. Next, we describe the molecular structures and reaction mechanisms that give rise to this enormous range in chemical reactivity with respect to water.

The basic structural unit in all phosphate glass networks is a tetrahedral PO_4 group [24, 25]. In P_2O_5, one of the oxygen anions in the inner coordination sphere forms a P = O bond, with the other three oxygen anions forming μ_2-oxo-bridges to neighboring tetrahedra. In other words, although the P_2O_5 network consists of interconnected tetrahedra, the glass has a crosslink density that is only three-quarters that found in SiO_2. As in silicate glasses, the addition of modifiers such as Na_2O or CaO disrupts the bridging $>P-\mu_2O-P<$ bonds and replaces them with nonbridging oxygens. At a P_2O_5 content of 50 mole%, each tetrahedron is, on average, bonded to only two neighboring phosphate groups, resulting in a glass structure consisting of long, linear phosphate chains. Further modifier additions serve to shorten these chains and increase the population of Q^1 groups relative to Q^2 groups.

For simple phosphate glasses, the most critical reactions controlling degradation in water and humid air are hydrolysis and hydration [26]. Starting with hydrolysis reactions, the P = O double bond is a reservoir of electron density that has a peculiar consequence relative to most other tetrahedral cations. The Q^3 units present in the parent oxide P_2O_5 are much more susceptible to hydrolysis than Q^2 units. One of the reasons why P_2O_5 is such an effective drying agent for both gases and liquids is that hydrolysis is nearly instantaneous until all Q^3 units are consumed. The result is a loosely bound network of extremely long linear phosphate chains (i.e., predominantly Q^2 units). However, the $>P-\mu_2O-P<$ linkages within these linear chains are as resistant to hydrolysis as $>Si-\mu_2O-Si<$ bonds in neutral pH solutions. If the kinetics of glass dissolution were controlled by rates of long-chain hydrolysis, dissolution rates at room temperature at pH 7 would be on the order of $1.7 \cdot 10^{-10} \frac{g}{cm^2 s}$. However, P_2O_5 dissolves very rapidly in water because entire chains are hydrated and released into solution without the need for further chain hydrolysis. Here, dissolution rates are determined by chain length, with longer chains resulting in more durable glasses, which is why compositions containing 50 mole% P_2O_5 are the most durable alkali-phosphate glasses. In humid air, the same hydration process occurs; only, instead of dissolving, the reaction products remain as a swollen and hydrated gel (Fig. 15.27).

Highly durable phosphate glasses can be produced by adding elements to the glass that crosslink the phosphate chains together via species that are much more resistant to hydrolysis than the Q^3 units found in native phosphate glass. The simplest form of crosslinking involves the addition of divalent cations to the glass [26]. These cations can interact with nonbridging oxygens on two adjacent chains to form relatively stable ionic chelates. The replacement of monovalent cations by divalent cations can increase glass durability to the point where chain hydrolysis becomes the rate-limiting step for dissolution (Fig. 15.28). Disruption of these divalent cation bridges occurs in strongly acidic solutions only, because the nonbridging oxygens in Q^2 phosphate chains have a low $pK_a (<2)$ and are difficult to protonate.

A second crosslinking strategy involves the addition of trivalent cations such as Fe(III), Al(III), or B(III) to the glass [27]. In such glasses, $P=O$ bonds are replaced by bridging bonds such as $>P-\mu_2O-Al<$ linkages to create tetrahedral networks that resemble fused silica or $AlPO_4$. The Fe(III)-containing glasses are so durable they have been proposed as potential hosts for nuclear wastes. Finally, crosslinks can be generated by replacing divalent oxygen anions in the network with trivalent nitrogen anions [28]. These nitrides bond to three network cations instead of only two. In the phosphorus-oxynitride glasses, nitrogen additions of as little as 10% can improve the resistance of the glass to attack by humid air by a factor of more than 1000 (Fig. 15.29). Phosphorus oxynitride glasses are particularly useful in making glass-to-metal seals to aluminum, because nitrogen incorporation allows the high-thermal expansion coefficient of the parent glass to be retained.

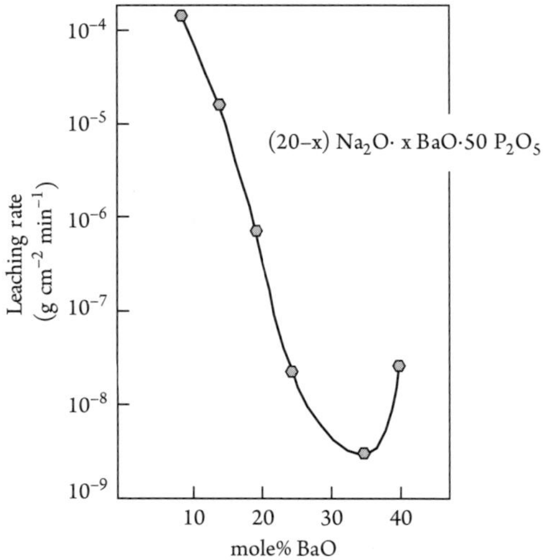

Figure 15.28 Dissolution rates of phosphate glasses as a function of the concentration of Ba(II), which serves as an ionic crosslink between phosphate chains. Above a barium concentration of around 30 mole%, the dissolution mechanism is dominated by chain hydrolysis rather than the release of intact phosphate chains.

SOURCE: Reproduced with permission from Bunker, B. C.; Arnold, G. W.; Wilder, J. A. Phosphate Glass Dissolution in Aqueous Solutions. *J. NonCryst. Solids*, **1984**, *64*, 291–316.

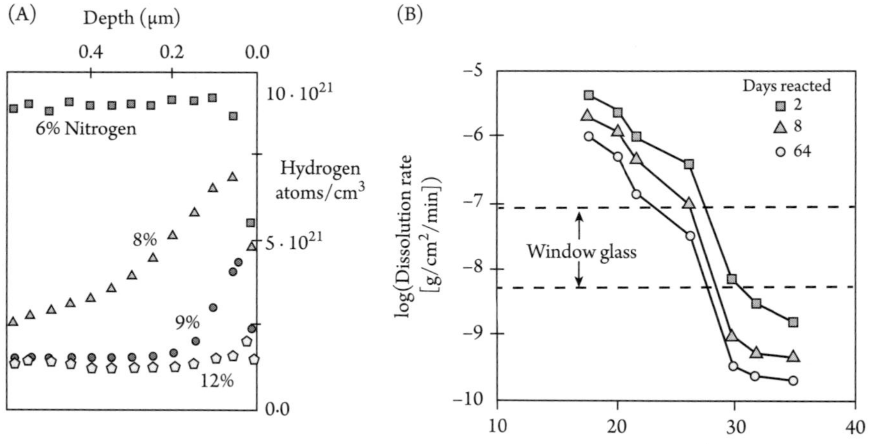

Figure 15.29 The impact of crosslinking on the durability of phosphate glasses. (A) Hydrogen depth profiles of phosphate glasses as a function of the concentration of dissolved nitrogen. (Adapted with permission from Bunker, B. C.; Arnold, G. W.; Rajaram, M.; Day, D. E. Corrosion of Phosphorus Oxynitride Glasses in Water and Humid Air. *J. Am Ceram. Soc.* **2005**, *70*, 425–430.) (B) Measured dissolution rates for phosphate glasses as a function of the concentration of Fe_2O_3 added to the bulk glass composition. (Adapted with permission from Mesko, M. G.; Day, D. E.; Bunker, B. C. Immobilization of CsCl and SrF_2 in Iron Phosphate glass. *Waste Manage.* **2000**, *20*, 271–278.)

15.8 NUCLEAR-WASTE GLASSES

We opened this chapter by introducing some of the unusual aspects of the dissolution of nuclear-waste glasses. Having explored many of the mechanisms for glass dissolution, we are now in a position to revisit these complex glasses to explain the unusual results presented in Figure 15.3. The results in Fig. 15.3 are for a simulated borosilicate-waste glass that was exposed to a seawater simulant under aggressive hydrothermal conditions (i.e., high temperatures and pressures). It is clear there is a sharp boundary between bulk and altered glass, both in visual images and in all depth profiles. We have seen that this boundary represents a reactive front at which the structure of bulk glass is converted into a new matrix consisting of interconnected colloidal silica particles. Most of the silica is retained, which is why little change is detected in the elemental silicon map in crossing the boundary between bulk and altered glass. The almost complete removal of Na^+ from the glass indicates that, although exchange sites such as tetrahedral boron are present in the bulk glass, the only ion exchange sites remaining in altered glass are non-bridging oxygens at the surfaces of the colloidal particles. In seawater, the three primary species competing for these ion-exchange sites are $Na^+(aq)$, $H^+(aq)$, and $Mg^{2+}(aq)$ from the seawater simulant. Under the slightly basic pH conditions found in seawater (pH 8–9), ion-exchange studies on silica gel show that Na^+ should not be able to compete for H^+ on surface silanol groups. However, the $Mg^{2+}(aq)$ ion in seawater has a much higher affinity for silanol groups and is present in sufficiently high concentrations (0.05 M) that $Mg^{2+}(aq)$ can compete with protons. This affinity is why Mg^{2+} is found within the leached layer, although it was absent in the bulk glass composition. Because $Mg^{2+}(aq)$ is so concentrated in the water, it can outcompete ions such as Cu^{2+} for surface sites, although Cu^{2+} still remains in the leached layer to some extent. Finally, uranium is retained and actually concentrated within the leached layer. An examination of the hydrolysis diagram (see Chapter 5) for uranium shows that U(IV) is exceedingly insoluble at pH 8 to 9 ($<10^{-10}$ M), so as the glass hydrolyzes and dissolves, uranium retention and concentration is expected within the leached layer.

Given these results on waste-glass simulants, to what extent should the public be concerned regarding the leaching of radionuclides from waste forms based on glass? The first thing to note is that the hydrothermal conditions under which the glass in Figure 15.3 were tested were extremely aggressive and were intended to accelerate glass alteration processes extensively relative to what glasses would experience under all but the most extreme environmental conditions (e.g., a volcanic eruption). Under more realistic conditions, we need to understand how nuclear-waste glasses are expected to behave.

In most waste glasses, the more prevalent radionuclides are ^{137}Cs and ^{90}Sr. The bad news regarding these elements is that both Cs^+ and Sr^{2+} enter the glass as modifier cations and can be susceptible to selective leaching via proton exchange (or, in seawater, Mg^{2+} exchange). The good news is that ^{137}Cs and ^{90}Sr both have relatively short half-lives (30 years and 28 years, respectively) [29], which means that, every 100 years, roughly 90% of the radioactivity decays away. In other words, nuclear-waste glasses containing Cs^{+} and Sr^{2+} do not need to last forever, only for the several hundred years that it takes the encapsulated ^{137}Cs and ^{90}Sr to decay down to radiation levels corresponding the natural background. Based on what we know about glass dissolution, such expectations are reasonable. If we were to place 1-m^3 of glass in a flowing stream of deionized water (preventing the buildup of reaction products that can slow down dissolution processes), even a poor-performing nuclear-waste glass is expected to dissolve at a rate on the order of $\sim 10^{-13}\,\frac{\text{g}}{\text{cm}^2\text{s}}$ or $3.7 \cdot 10^{-6}\,\frac{\text{g}}{\text{cm}^2\text{y}}$. Given the glass density of around 2 g/cm^3, this means that roughly 0.0003 of 1% of this cube would experience alteration per year. Total alteration of the cube would take around 30 million years. As the volume of water flowing across the cube in that time would be enormous, the concentration of any escaping radionuclide would be less than the natural background.

Not all radionuclides in waste glass have short half-lives. However, it is fortunate that those elements with long half-lives are elements such as uranium, which tend to form insoluble oxides and hydroxides that are even more durable than the glass. However, Murphy's Law states that if something can go wrong, it probably will. For uranium, a potential problem is that glass dissolution and leaching can be highly dependent on complexation, as illustrated by the carbonate complexes. Uranium(VI) has a high affinity for carbonate anions, and forms highly soluble species such as $[UO_2(CO_3)_3]^{4-}$. When these species are present, not only is uranium dissolved, but the anionic carbonate complexes have no affinity for silicate anions, either in glass or in the surrounding environment. For example, a simulated waste glass in contact with illite clay sediment shows extensive uranium leaching from the glass and subsequent rapid uranium migration through the environment when carbonate ions are present [2, 30] (Fig. 15.30). This leaching is in sharp contrast to the situation when Pb(II) is present, which forms insoluble $PbCO_3$ that precipitates within the leached layer, fills in all available porosity, and drastically reduces the glass leaching rates. For this reason, it has been found that encasing waste glass in lead containers can inhibit leaching by orders of magnitude relative to leaching in Teflon containers. Of course, storing nuclear wastes in lead containers in effect replaces a radionuclide contamination problem with a lead contamination problem. A final concern regarding the use of glass as a nuclear-waste form has to do with the effects of radiation damage on dissolution rates. This topic is discussed in Chapter 13.

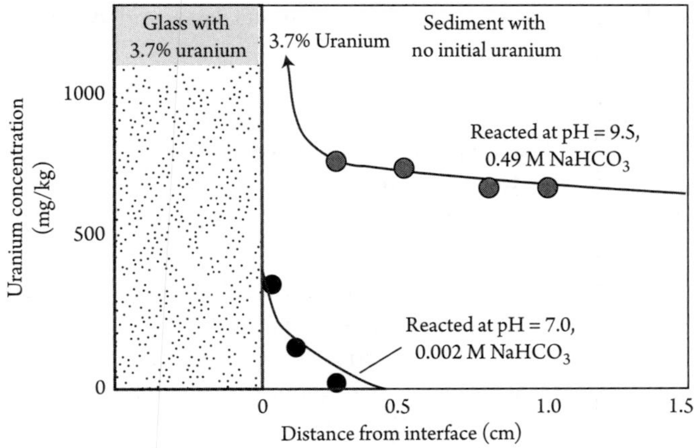

Figure 15.30 A sample of 7668 glass (a waste simulant) in contact with water-containing sea sediment shows that the leaching and environmental transport of uranium from the glass is sensitive to the presence of aqueous carbonate in the environment. This is because uranium forms highly soluble anionic carbonate complexes that have no affinity for anionic sites in the glass or in the clay-rich sediment.

References

1. Adams, P. B.; Evans, D. L. Chemical Durability of Borate Glasses. In: L.D. Pye, V.D. Fréchette and N.J. Kreidl (Editors) *Borate Glasses: Structure, Properties, Applications*; Plenum Press: New York, 1978, pp. 535–5372.
2. Bunker, B. C. Waste Glass Leaching: Chemistry and Kinetics. *Mater. Res. Soc. Symp. Proc.* **1987**, *84*, 493–507.
3. Doremus, R. H. Interdiffusion of Hydrogen and Alkali Ions in a Glass Surface. *J. Non-Cryst. Solids* **1975**, *19*, 137–144.
4. Cohen, B. L. The Disposal of Radioactive Wastes from Fission Reactors. *Sci. Am.* **1977**, *236*, 21–31.
5. Bunker, B. C.; Arnold, G. W.; Beauchamp, E. K. Mechanisms for Alkali Leaching in Mixed Na-K Silicate Glasses. *J. Non-Cryst. Solids* **1983**, *58*, 295–322.
6. Ray, N. H. *Inorganic Polymers*; Academic Press: London, 1978.
7. Galeener, F. L. Planar Rings in Vitreous Silica. *J. Non-Cryst. Solids* **1982**, *49*, 52–63.
8. Barrer, R. M. *Zeolites and Clay Minerals as Sorbents and Molecular Sieves*; Academic Press: London, 1978.
9. Roberts, G. J.; Roberts, J. P. An Oxygen Tracer Investigation of the Diffusion of Water in Silica Glass. *Phys. Chem. Glasses* **1966**, *7*, 82–89.
10. Iler, R. K. *The Chemistry of Silica*; Wiley-Interscience: New York, 1979.
11. Brawer, S. A.; White, W. B. Raman Spectroscopic Investigation of the Structure of Silicate Glasses: I. The Binary Alkali Silicates *J. Chem. Phys.* **1975**, *63*, 2421–2432.
12. Bunker, B. C. Molecular Mechanisms for Corrosion of Silica and Silicate Glasses. *J. Non-Cryst. Solids* **1993**, *179*, 300–308.
13. Das, C. R.; Douglas, R. W. Studies on the Reaction between Water and Glass: Part 3. *Phys. Chem. Glasses* **1967**, *8*, 178–184.

14. Douglas, R. W.; El-Shamy, T. M. M. Reactions of Glasses with Aqueous Solutions. *J. Am. Ceram. Soc.* **1967**, *50*, 1–8.
15. Day, D. E. Mixed-Alkali Glasses: Their Properties and Uses. *J. Non-Cryst. Solids* **1976**, *21*, 343–372.
16. Bunker, B. C.; Headley, T. J.; Douglas, S. C. Gel Structures in Leached Alkali Silicate Glass. *Mater. Res. Soc. Symp. Proc.* **1984**, *32*, 41–46.
17. Feller, S. A.; Dell, W. J.; Bray, P. J. ^{10}B NMR Studies of Lithium Borate Glasses. *J. Non-Cryst. Solids,* **1982**, *51*, 21–30.
18. Velez, M. H.; Tuller, H. L.; Uhlmann, D. R. Chemical Durability of Lithium Borate Glasses. *J. Non-Cryst. Solids* **1982**, *49*, 351–362.
19. Bunker, B. C.; Tallant, D. R.; Kirkpatrick, R. J.; Turner, G. L. Multinuclear Nuclear Magnetic Resonance and Raman Investigation of Sodium Borosilicate Glass Structures. *Phys. Chem. Glasses* **1990**, *31*, 30–41.
20. Bunker, B. C.; Arnold, G. W.; Day, D. E.; Bray, P. J. The Effect of Molecular Structure on Borosilicate Glass Leaching. *J. Non-Cryst. Solids* **1986**, *87*, 226–253.
21. Tindyala, M. A.; Ott, W. R. Lithium-Zinc-Phosphate Glasses. *Am. Ceram. Soc. Bull.* **1978**, *57*, 432–433.
22. Klein, R. M.; Kolbeck, A. G.; Quackenbush, C. L. Glass Formation and Properties in the Aluminum Borophosphate System. *Am. Ceram. Soc. Bull.* **1978**, *57*, 199–202.
23. Bunker, B. C. Unpublished results.
24. Westman, A. E. R.; Gartaganis, P. A. Constitution of Sodium, Potassium, and Lithium Phosphate Glasses. *J. Am. Ceram. Soc.* **1957**, *40*, 293–299.
25. Van Wazer, J. R. Structure and Properties of the Condensed Phosphates: I. Some General Considerations about Phosphoric Acids. *J. Am. Chem. Soc.* **1950**, *72*, 639–644.
26. Bunker, B. C.; Arnold, G. W.; Wilder, J. A. Phosphate Glass Dissolution in Aqueous Solutions. *J. Non-Cryst. Solids* **1984**, *64*, 291–316.
27. Mesko, M. G.; Day, D. E.; Bunker, B. C. Immobilization of CsCl and SrF_2 in Iron Phosphate Glass. *Waste Manage.* **2000**, *20*, 271–278.
28. Bunker, B. C.; Arnold, G. W.; Rajaram, M.; Day, D. E. Corrosion of Phosphorus Oxynitride Glasses in Water and Humid Air. *J. Am. Ceram. Soc.* **1987**, *70*, 425–430.
29. Weast, R. C., Ed. *CRC Handbook of Chemistry and Physics, 58th ed.*; CRC Press: Cleveland, OH, 1977.
30. Bunker, B. C.; Arnold, G. W. The Effect of Solution pH and Ion Concentrations on Leaching of Silicate Glass. *Mater. Res. Soc. Symp. Proc.* **1983**, *15*, 151–158.

16

Stress Corrosion Cracking: Chemically Activated Nanomechanics

16.1 INTRODUCTION

Although dissolution reactions involving water can etch and decompose oxides, truly catastrophic failures of oxide structures usually involve fractures and mechanical failures. Geologists and geochemists have long recognized that water and ice both play key roles in promoting the fracture and crumbling of rock (see Chapter 17). Freezing and thawing create stresses that amplify the rate at which water attacks metal–oxygen bonds at the crack tip. The interplay between water and stressed oxides also leads to common failures in man-made objects, ranging from the growth of cracks from flaws in windshields (Fig. 16.1) to the rupture of optical fibers in communication systems [1]. In this chapter, we outline how mechanical deformations change the reactivity of metal–oxygen bonds with respect to water and other chemicals, and how reactions on strained model compounds have been used to predict time to failure as a function of applied stress.

The basic phenomenon of stress corrosion cracking is illustrated in Figure 16.1. Cracks can propagate through oxide materials at extremely fast rates, as anyone who has dropped a wine glass on the floor can attest. High-speed photography reveals that when glass shatters, cracks can spread at speeds of hundreds of meters per second, or half the speed of sound in the glass. At the other end of the spectrum, cracks in glass can grow from preexisting flaws so slowly that only a few

Figure 16.1 (A) At high-stress intensities, glass and ceramic materials shatter almost instantaneously, as shown by the bullet hole that penetrated this window. However, at low-stress intensities, cracks can grow from existing flaws at much slower rates, as determined by the presence of environmental chemicals such as water. The long cracks radiating from the central bullet hole represent such cracks, which can eventually lead to catastrophic failures of objects such as windshields. (SOURCE: karloss/Shutterstock.com.) (B) A depiction of a signal repeater in an undersea trans-Atlantic fiber optic cable. The rupture of optical fibers resulting from the same slow crack-growth process that degrades windshields can be exceedingly expensive to repair for cables that lie on the ocean floor. (SOURCE: Reproduced with permission from Tuminaro, R. D. Materials Aspects of the SL Lightguide Undersea Cable – 1988 MRS Spring Meeting Plenary Address. *MRS Bulletin*, **1988**, *13* (7), 14–22. Cover artwork.)

chemical bonds are broken at the crack tip per hour. Because mechanical failures are associated with cracking, it is critical for design engineers to understand the factors that control crack growth rates for this enormous range of crack velocities (a factor of 10^{12}). In addition, because it is difficult to measure crack velocities slower than 10^{-8} m/second, it is often necessary to make major extrapolations from measured data to predict the long-term reliability of glass and ceramic objects. Will an optical fiber under stress fail in 1 year or 10 years? Answering this question can require accurate extrapolations down to crack growth rates as low as 10^{-10} m/second.

Surprisingly, glass is intrinsically one of the strongest materials known. Under high-vacuum conditions, flaw-free glass can withstand tensile loads greater than 2 million pounds per square inch (psi), which is more than 10 times the strength of most commercial metal alloys [2]. However, under normal service conditions, glass surfaces contain defects. Tiny surface cracks associated with scratches or pits represent nucleation points for the growth of larger cracks. Under an applied stress, the growth of cracks leads to a weakening of the glass, and ultimately to catastrophic failure. It has long been known that the three factors controlling crack growth rate and time to failure are (1) the flaw size, (2) the magnitude of the applied stress, and (3) the presence of environmental chemicals such as water. By performing experiments under which the crack size, the stress, and the environment are all carefully controlled [3] (Fig. 16.2 and Plate 19), it has been found that (1) at a given stress intensity, cracks grow continuously at a constant rate; (2) crack rates increase as cracks grow, because larger flaws concentrate more stress at the crack tip; and (3) cracks grow until a critical flaw size is reached, resulting in catastrophic failure.

Equations used by engineers to describe the decrease in glass strength with time resulting from slow crack growth or *static fatigue* include [4] (1) the linear elastic–fracture mechanics expression relating the crack-tip stress intensity to the applied stress (Eq. 16.1), (2) empirically derived rate laws for crack extension velocities (Eq.16. 2), and (3) the time to failure (t_f) under constant load conditions, which is the time required for initial flaws to grow to the critical flaw size as obtained by integrating Eq. 16.2.

$$K_I = Y\sigma\sqrt{\pi c} \quad \textbf{(16.1)}$$

$$V = AK_I^n \quad \textbf{(16.2)}$$

$$t_f = \frac{2K_I(2-n)}{AY^2\sigma^2(n-2)} \quad \textbf{(16.3)}$$

In these expressions, K_I is the applied stress intensity, Y is a geometric factor associated with the flaw character, σ is the applied stress, c is the crack length, V is

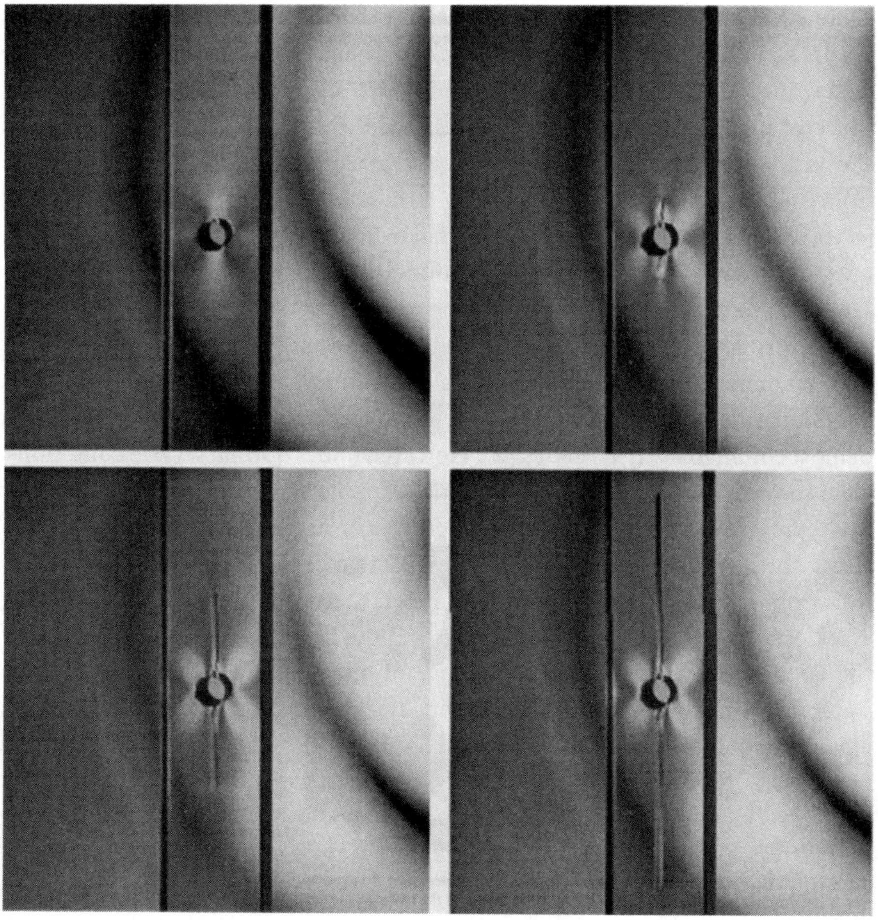

Figure 16.2 Cracks growing in a laboratory fracture mechanics specimen from a controlled flaw consisting of a hole drilled in the middle of the sample. Compressive loads applied along the long axis propagate two cracks from this flaw in the presence of a controlled vapor pressure of water. Stress patterns (made visible by exposing the sample to polarized light) show stresses are concentrated at the tips of each advancing crack (see Plate 19). SOURCE: Reproduced with permission from Michalske, T. A.; Bunker, B. C. The Fracturing of Glass. *Scientific American*, December, 1987, **255**, 122–129.

the crack velocity, and A and n are fitting parameters associated with the empirical crack growth rate law. The parameters A and n both depend on the composition of the oxide; A also depends on the chemical nature and concentration of the environmental agent. Measured crack growth data for silica (Fig. 16.3) show how sensitive the crack velocity is to both the stress intensity and the presence of different concentrations of water. Next, we outline how observed crack growth rates can be rationalized on the basis of chemical reactions between water and strained crack-tip bonds.

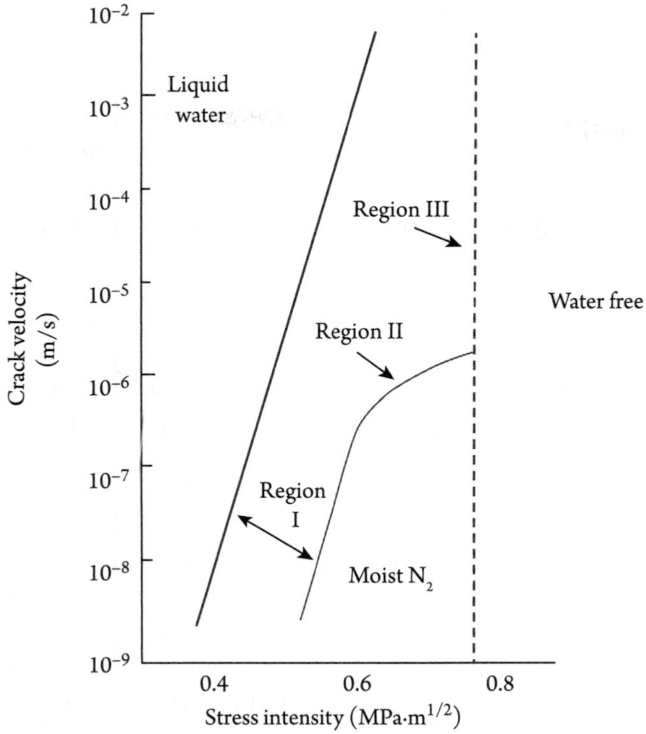

Figure 16.3 Crack velocities measured in fused silica as a function of applied stress intensity for samples exposed to liquid water as well as to air containing trace amounts of water vapor. The dashed line indicates the stress intensity at which samples fail in the absence of water. The data in moist N_2 shows three distinct kinetic regimes: region I, reaction limited; region II, diffusion limited; and region III, catastrophic failure (see discussion in Section 16.5). Adapted with permission from Michalske, T. A.; Bunker, B. C. Steric Effects in Stress Corrosion Fracture of Glass. *J. Am. Ceram. Soc.* **1987**, *70*, 780–784.

16.2 MACROSCOPIC BRITTLE FRACTURE MODELS

Macroscopic models for stress corrosion are based on the pioneering studies of Griffith [5]. In an elastic solid, cracks are predicted to grow when the strain energy released in crack formation is greater than the energy required to create the new surfaces represented by the walls of the crack (see Chapter 6). Alternatively, if the strain energy released is less than the surface energy, crack healing is expected and observed. Reactive chemical species can lower the energy of the fracture surfaces, which in turn lowers the strain energy required to promote crack growth. In the case of silica, the high-energy defect sites associated with ruptured Si—O bond fragments react with water to form a lower energy surface that is passivated with silanol groups. However, although this formalism can be used to rationalize the energetics of slow crack growth, it cannot be used to predict crack growth kinetics.

The next level of modeling complexity involves using the assumption that the crack moves at a rate determined by the rate at which water molecules react with Si–O bonds at the crack tip [6]. In absolute reaction-rate theory (see Chapter 4), the rate of a chemical reaction $(\vec{k})$ is described by the expression

$$\vec{k} = \frac{kT}{h} \cdot e^{\frac{-\Delta G^{\ddagger}}{RT}} \tag{16.4}$$

where k is Boltzmann's constant, h is Planck's constant, R is the ideal gas constant, and $\Delta G^{\ddagger}$ is the free energy of the activated reaction as given by

$$\Delta G^{\ddagger} = \Delta E^{\ddagger} - T\Delta S^{\ddagger} + P\Delta V^{\ddagger} \tag{16.5}$$

where $\Delta E^{\ddagger}$ is the activation energy, P is the applied pressure, $\Delta V^{\ddagger}$ is the activation volume, and $\Delta S^{\ddagger}$ is the activation entropy. Because the activation volume is the difference in partial molar volumes between the reactants and the activation complex, it can be modeled by changes in molecular dimensions along the reaction trajectory. Activation volume models form the basis for all pressure-dependent chemical reaction kinetics, and provide an important formalism to describe reaction rates on strained crack-tip bonds.

Using conventional fracture mechanics formulations, Weiderhorn and colleagues [6] estimated the stress in bonds at the crack tip as a function of applied stress intensity to fit crack growth data to a rate equation analogous to Eq. 16.4:

$$v = v_o e^{\frac{-E^* + bK_I}{kT}} \tag{16.6}$$

Here, v_o, E*, and b are empirical constants. Although empirical, each constant in Eq. 16.6 has a counterpart in absolute-rate theory (Eq. 16.5), where v_o serves the role of $\frac{k \cdot T}{h}$, E* captures the strain-free components of $\Delta G^{\ddagger}\left(\Delta E^{\ddagger} - T\Delta S^{\ddagger}\right)$, and $b \cdot K_I$ encapsulates the strain dependence as captured by the term $P\Delta V^{\ddagger}$. However, although these models provide formalisms for predicting the functional relationships between crack velocities, applied stresses, and chemistry, they do not provide molecular mechanisms that can be used to predict the rates of slow crack growth. Unraveling the meaning of empirical parameters such as E* and b to allow for quantitative predictions of strain-activated reaction rates on crack-tip bonds. Progress has been accomplished using a combination of molecular–orbital calculations and experimental measurements of reaction rates on strained Si–O model compounds.

16.3 MOLECULAR MODELING OF STRAIN-ENHANCED REACTIVITY

Molecular–orbital calculations have been performed to predict how applied stresses activate the reaction between water and $>\mathrm{Si}-\mu_2\mathrm{O}-\mathrm{Si}<$ bonds at the crack tip. Crack-tip bonds can respond to applied stresses either by stretching or by bending. The calculations allow researchers to evaluate how the strain should be partitioned between various stretching and bending modes. In addition, the calculations indicate how different bond deformations should influence local electron density distributions. Finally, calculations have been performed that indicate how $>\mathrm{Si}-\mu_2\mathrm{O}-\mathrm{Si}<$ linkages should react with water based on local bonding configurations.

The three parameters describing local bonding configurations in silicates are the Si–O bond length (d), the O–Si–O bond angle (θ) within a given tetrahedral unit, and the $>\mathrm{Si}-\mu_2\mathrm{O}-\mathrm{Si}<$ bond angle (ϕ) between silicate tetrahedra [7] (Fig. 16.4). An additional reference point is the Si–O bond energy of ~4200 kJ/mole. In unstrained silica, d = 0.163 nm, $\theta = 109.5°$, and $\phi = 144°$. Note that, although d and ϕ vary independently, a contraction of one of the O–Si–O bond angles within a given silicate tetrahedron must be accompanied by an expansion of the two O–Si–O angles adjacent to the contracting oxygen pair

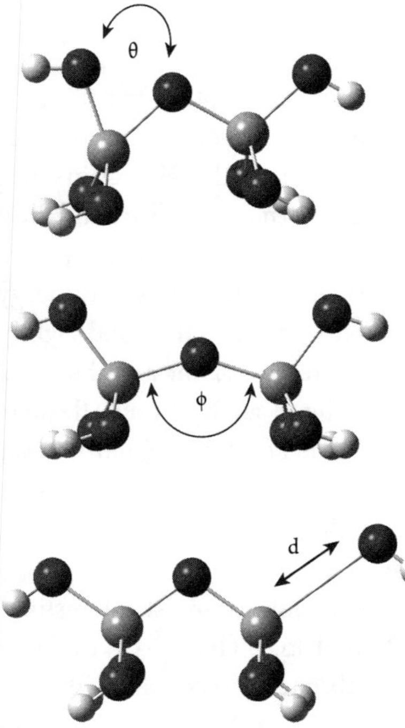

Figure 16.4 A depiction of local bonding configurations in silica, indicating the bond length (d), the O–Si–O bond angle (θ), and the Si–O–Si bond angle (ϕ).

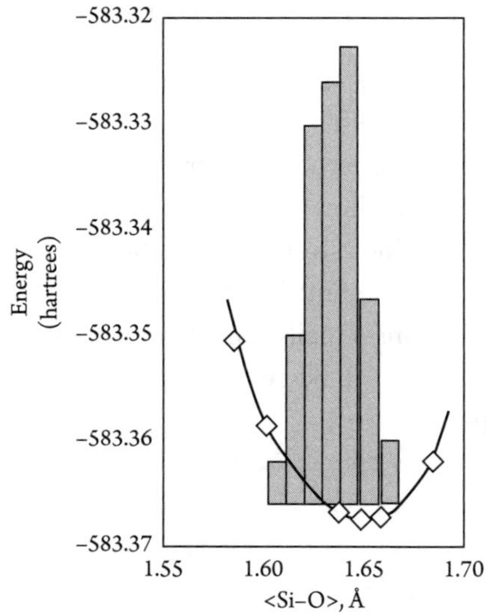

Figure 16.5 The results of molecular-orbital calculations showing how the energy of the Si–O bond changes as the bond is lengthened or shortened. On the energy axis, 1 atomic unit is equivalent to 628 kcal/mole (or 2627.5 kJ/mol). The histogram bars indicate the relative populations of bond lengths measured in a wide range of known silicate minerals. Adapted with permission from Gibbs, G. V. *Am. Miner.* Molecules as Models for Bonding in Silicates. **1982**, *67*, 421–450.

(see Silica-W, discussed later in this chapter). In any case, local strained environments do not necessarily contain single values for d, θ, and ϕ. Starting with the bond length, d, the calculations show that the strain energy-versus-bond length curve is described by a roughly harmonic well around the strain-free bond length [8] (Fig. 16.5). The Si–O bond is stiff, and elongating the bond length by only 0.005 nm is estimated to cost ~21 kJ/mole in strain energy. The calculations also suggest that bond length extension makes the bond slightly more covalent, which would tend to make the Si center less acidic and the O center less basic. In contrast, the $>Si-\mu_2O-Si<$ bond angle is quite flexible. Calculations suggest that this bond can be straightened from its equilibrium value of 144° to its maximum value of 180° at an energy cost of only ~8 kJ/mole. Again, shifts in electron distributions resulting from this distortion make the Si–O bonds slightly more covalent and less reactive.

From a mechanistic standpoint, the most interesting deformation involves changes in the O–Si–O bond angle (Fig. 16.6). Molecular-orbital calculations [9, 10] suggest that, in the absence of water, it is difficult to pull the oxygens in the inner coordination sphere apart. Increasing θ from its equilibrium value of 109.5° to 140° is estimated to require $\sim 100\ \mathrm{kJ/mole}$, whereas straightening the angle out completely to 180° is estimated to cost ~300 kJ/mole. Note that by

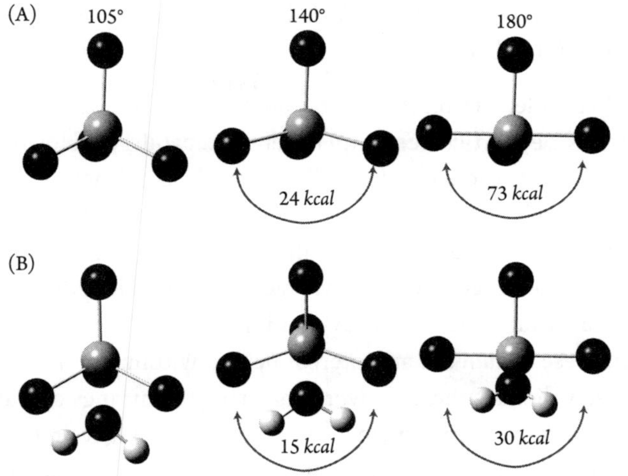

Figure 16.6 (A, B) Depictions of the stress-activated opening of the O—Si—O bond in the absence (A) and presence (B) of water, showing the strain energies required to achieve a given bond angle *Adapted* from Michalske, T. A.; Bunker, B. C. The Fracturing of Glass. *Scientific American*, December, 1987, **255**, 122–129.

increasing one O—Si—O angle to 180°, the adjacent O—Si—O angles are reduced to almost 90°. Calculations indicate that the creation of this highly deformed tetrahedron makes the Si—O bond more ionic, doubling the effective charge on both the Si and the O, and enhancing the acid–base nature of the bond. The same calculations have been performed in the presence of a water molecule. Here, the energy barrier to opening the O—Si—O bond angle is calculated to be only half as high (requiring ~63 kJ/mole and ~215 kJ/mole to open the angle to 140° and 180°, respectively). The strain energies are reduced because a basic lone pair of electrons from the oxygen in water is able to interact with an acidic Lewis-acid site that opens up on the silicon atom. The net result is a set of bonding interaction between the water and the silicon.

The energy gain associated with this bonding partially offsets the energy penalty associated with prying the O—Si—O bond angle apart. Notice that the resulting five-coordinate intermediate resembles the activated complex expected for the nucleophilic attack of water on silicate tetrahedra in the hydrolysis of Si—O bonds (see Chapters 4 and 14). This calculation illustrates that not only can chemicals promote mechanical deformations, but also that the same deformations can promote chemical reactions. However, the calculations still do not address the kinetics at which such reactions should occur. To probe such kinetics, scientists have studied the rates at which water and other chemicals react with model compounds that contain deformed local-bonding configurations.

16.4 REACTIONS BETWEEN WATER AND STRAINED MODEL COMPOUNDS

Watching water molecules react with the single row of bonds at the tip of an advancing crack is beyond current experimental capabilities. However, it is still possible to study the reactivity of O–Si–O bonds as a function of strain using cyclosiloxane compounds that are known to contain distorted silicate bonding configurations. As discussed earlier, in silica, the basic structural units are interconnected silicate tetrahedra with an average Si–O bond length of 0.163 nm, the O–Si–O tetrahedral angle $\theta = 109.5°$, and the Si–O–Si angle $\phi = 144°$ (Fig. 16.4). When these tetrahedra are interconnected within strain-free silicate networks, geometry dictates these networks primarily form rings containing five or six tetrahedra. However, smaller rings can form containing four, three, and even edge-shared rings between two adjacent tetrahedra. Although the ring containing four tetrahedra (n = 4) has almost no strain, the degree of strain for n = 3 is significant, whereas the edge-shared n = 2 ring is highly strained (Fig. 16.7). X-ray diffraction results show that Silica-W [11], which is a silica polymorph containing linear chains of edge-shared tetrahedra ($>Si-(\mu_2O)_2-Si<$), exhibits bonding configurations with d = 0.187nm, an internal $\theta = 87°$, and

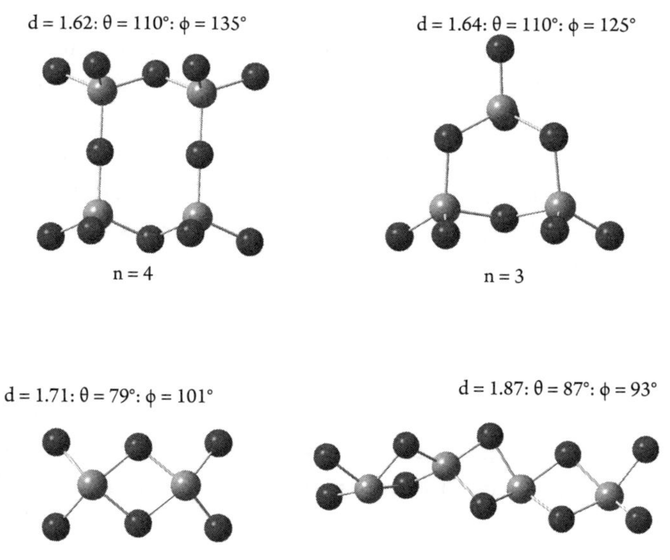

Figure 16.7 Depictions of strained rings in silica as a function of the number of silicate tetrahedra (n) in each ring. Values for d, θ, and ϕ for the n = 4, 3, and 2 rings, respectively, are based on molecular-orbital calculations of isolated H-terminated rings [8], whereas the values for Silica-W are derived from the X-ray crystal structure [10].

$\phi = 93°$. Not only are all bond lengths and angles in Silica-W strained, the oxygen geometry around each silicon atom is splayed out such that the metal cation is more accessible to nucleophilic attack to form a five-coordinate intermediate (i.e., the external O−Si−O angles are more than 140°). As discussed in Chapters 14 and 15, this strained configuration represents a precursor to the classic hydrolysis of $> Si - \mu_2 O - Si <$ bonds via the S_N2 mechanism. Although accurate kinetic measurements have not been made, it is known that Silica-W reacts rapidly with atmospheric moisture to form silica gel [12]. This strained silica polymorph, Silica-W, is much more reactive than vitreous silica, as expected for its unusual bonding configuration.

One consequence of the strain present in smaller silicate rings is that their vibrational frequencies are perturbed relative to those exhibited by unstrained silica. The Raman spectra for vitreous silica [13, 14] exhibit two so-called "defect bands" denoted D_1 and D_2 at 495 cm^{-1} and 606 cm^{-1}, corresponding to rings having n = 4 and n = 3, respectively. Strained n = 3 rings are also found in rock-forming cyclotrisilicates [15], and have been synthesized in individual molecules such as model cyclotrisiloxanes [16]. The n = 2 ring found in Silica-W exhibits infrared bands at 908 cm^{-1} and 888 cm^{-1}. Edge-shared tetrahedra have been synthesized in a series of tetra-substituted cyclodisiloxane molecules [17]. Significant surface concentrations of both the n = 2 and n = 3 rings can be created in native silica surfaces by heating under vacuum to promote dehydroxylation reactions [18] (Fig. 16.8). In high-surface-area silica samples, such as CAB−O−SIL or sol–gel silica, strained rings can be produced readily at

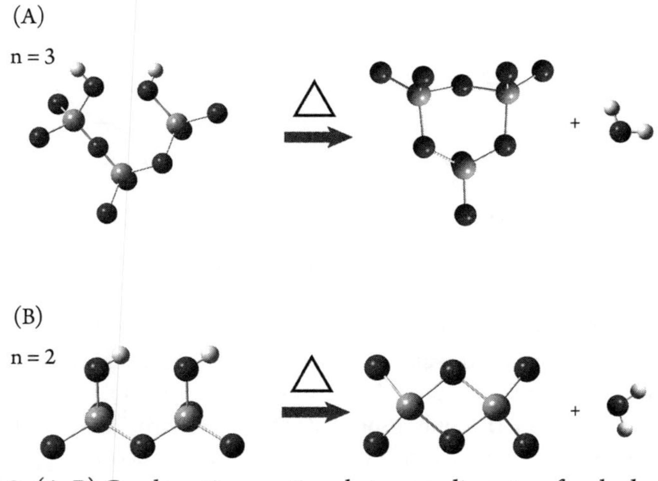

Figure 16.8 (A, B) Condensation reactions between adjacent surface hydroxyl groups can lead to the formation of strained rings containing three (A) and two (B) silicate tetrahedra.

concentrations that are sufficiently high for studying the kinetics of reactions on strained rings in a silica-like environment.

When strained rings are produced in silica surfaces via high vacuum dehydroxylation, the surfaces can be exposed to known concentrations of water or other chemicals to study their chemical reactivity and determine reaction rates. For example, infrared spectra have been used to follow the rate at which edge-shared tetrahedra ($>Si-(\mu_2O)_2-Si<$) disappear when exposed to water, while simultaneously monitoring the rate at which silanol groups (the reaction products) appear [19, 20] (Fig. 16.9). The results show that even at low water concentrations (4 millitorr), the strained edge-shared rings are disrupted in a matter of minutes, with a first-order rate constant: $k_1 = 0.012\ s^{-1}$. As expected,

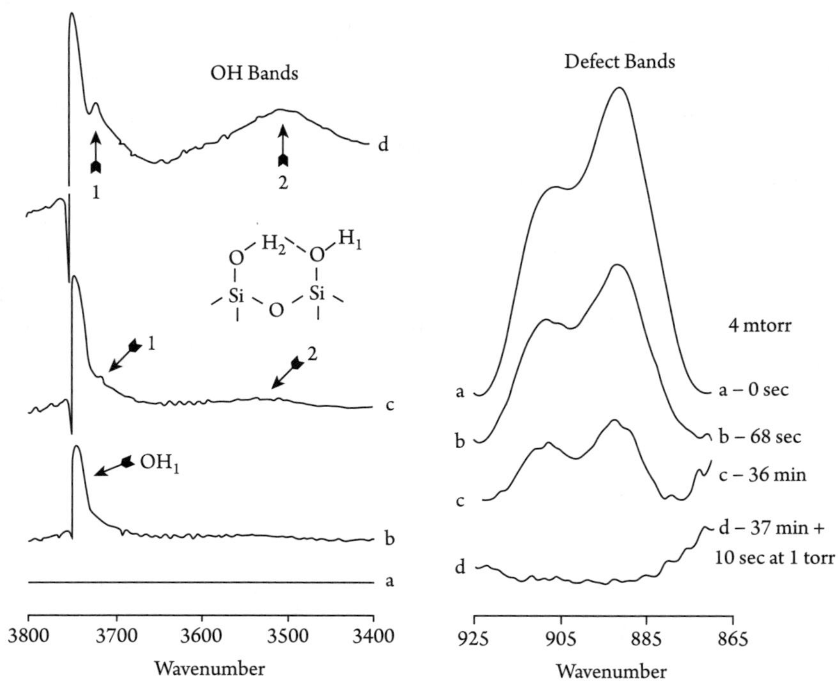

Figure 16.9 Infrared spectra that follow the kinetics of the reaction between water vapor (4 mtorr) and edge-shared tetrahedra in the surface of dehydroxylated silica. On the right are bands associated with the strained n = 2 rings both before (a) and during consumption via reactions with water (b–d). On the left are the infrared (IR) bands associated with surface silanol groups. Initially, only isolated silanol groups are apparent. As the reaction continues, IR bands associated with various hydrogen-bonding configurations between ever-increasing silanol concentrations become apparent.
SOURCE: Reproduced with permission from Bunker, B. C.; Haaland, D. M.; Michalske, T. A.; Smith, W. L. Infrared Spectra of Edge-Shared Silicate Tetrahedra. *Surf. Sci.*, **1989**, *210*, 406–428.

for each ring that is destroyed, two silanol groups are created. Raman experiments conducted on the reaction between water and strained rings containing three tetrahedra exhibit reaction kinetics that are intermediate between the rates seen for the edge-shared tetrahedra and unstrained silica [21].

Extrapolating the observed rates of reactions between humid air and strained rings to rates that would be observed in liquid water or a relative humidity of 100% (assuming the reaction rates are first order with respect to the water concentration) illustrates just how sensitive hydrolysis reaction rates are to strain. At high water concentrations, the rate constant for the edge-shared dimer of 2.9 s^{-1} is 300,000 times faster than the rate constant of $1 \cdot 10^{-5} s^{-1}$ for the hydrolysis of unstrained silica. The strain energies corresponding to this range of hydrolysis rates can be estimated based on molecular–orbital calculations [15]. The calculated strain energies are ~58-kJ/mole Si–O bonds for the edge-shared dimer and 4 kcal/mole for the $n = 3$ ring system. A compilation of hydrolysis rate constants for both strained and unstrained rings (all normalized to 100% relative humidity) show that, as expected, the rate of ring hydrolysis increases exponentially with strain energy (Fig. 16.10). Using the Van't Hoff expression [22] for the dependence of chemical rate constants with pressure:

$$\ln(k) = \ln(k_o) - P\Delta V^{\ddagger}/RT \tag{16.7}$$

where k_o is the zero-stress reaction-rate constant. The slope of the exponential curve yields a value of 2 cm^3/mole for the activation volume for the stress-enhanced

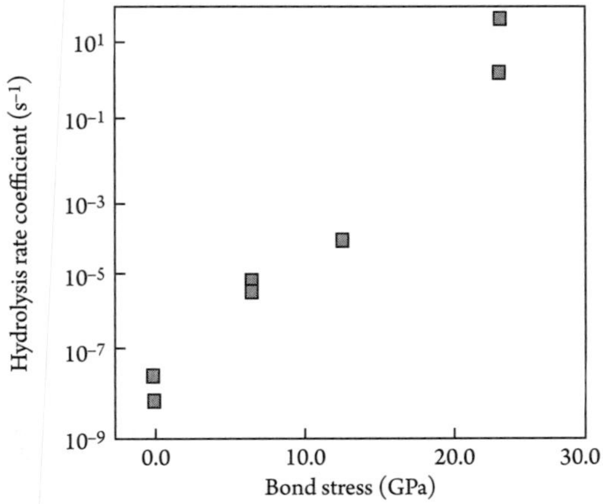

Figure 16.10 A compilation of ring-hydrolysis data versus ring strain. Rate constants are from experimental values reported in the literature, whereas estimates for bond stress are provided by molecular orbital calculations. Adapted with permission from Michalske, T. A.; Bunker, B. C. A Chemical Kinetics Model for Glass Fracture. *J. Am. Ceram. Soc.* **1993**, *76*, 2613–2618.

hydrolysis of Si—O bonds. As a cross-check, fracture mechanics formulations show the activation volume can be described via

$$\Delta V = (b/2)(\pi\rho)^{1/2} \tag{16.8}$$

where ρ is the atomic radius of the crack tip. Assuming that ρ corresponds to the average ring size in silica (containing six silicate tetrahedra), ρ is estimated to be around 0.5 nm, leading to calculated values for $\Delta V^{\ddagger}$ of 2.3 and 4.3 cm^3/mole for soda–lime silicate glass and fused silica, respectively. Using a procedure defined by Hamann [23], these activation volumes can be used to estimate that the transition state required for bond rupture corresponds to an elongation of the Si—O bond by 63%. This bond strain is comparable with that estimated for organic molecules undergoing molecular dissociations [24].

All the reaction-rate expressions for predicting crack growth rates have a functional form in which crack velocities scale with an exponential dependence on the stress intensity. This scaling is in contrast to the empirical engineering power-law expression (Eq. 16.2) in which velocities scale with K_I^n. Although both expressions fit crack growth data at high crack velocities, the expressions diverge strongly at slower crack velocities (Fig. 16.11). This divergence is critical for making lifetime predictions based on extrapolations from measured crack growth rates. Using exponential rate law expressions, the time to failure is given by

$$t_f = -\frac{2}{v_o} e^{-bY\sigma x} \cdot \frac{[bY\sigma x + 1]}{[bY\sigma]^2} \tag{16.9}$$

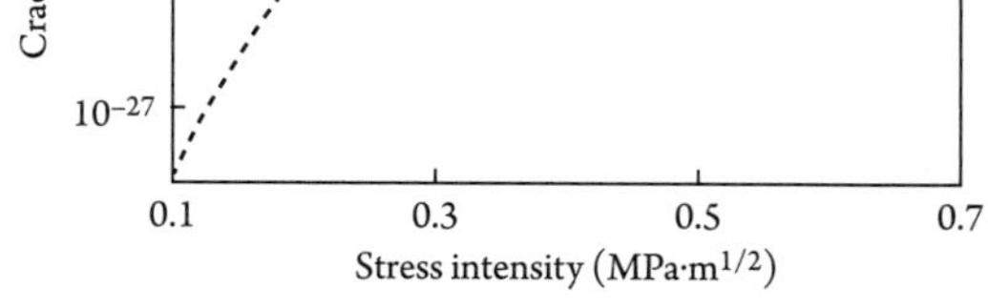

Figure 16.11 Predictions of slow crack growth based on exponential reaction rate law expressions (solid line) versus the empirical power law expression for silica (extrapolating from experimental data shown by open circles). Adapted with permission from Michalske, T. A.; Bunker, B. C. A Chemical Kinetics Model for Glass Fracture. *J. Am. Ceram. Soc.* **1993**, *76*, 2613–2618.

where Eq. 16.9 is integrated between x values corresponding to the initial and critical flaw sizes. It has been shown [25] that the exponential fracture rate model provides much more accurate predictions of the time-dependent strength of silica glass optical fibers, as well as the effects of environmental factors such as relative humidity on measured fatigue behavior. These predictions further validate that the mechanism for slow crack growth involves the rates of reactions on strain crack-tip bonds.

16.5 MOLECULAR REQUIREMENTS FOR STRESS CORROSION

The requirements for molecules that can participate in strain-activated bond rupture reactions (Fig. 16.12) have been elucidated via (1) reactions between a range of molecules with the surfaces of dehydroxylated silica and (2) slow crack growth experiments in controlled chemical environments. Chemically assisted rupture of a metal–oxygen bond requires the attacking molecule to contain species that are capable of satisfying the bonding requirements of each fragment created via bond rupture in a vacuum. For SiO_2, the most effective agents have lone pairs of electrons (Lewis bases) that are capable of nucleophilic attack on the Si site (a Lewis acid), and protons (Brønsted acids) that are capable of bonding to the lone pairs of electrons on the oxygen (Brønsted bases). In other words, active molecules contain acid and base components that can pair readily with the basic and acidic sites associated with the Si–O bond. Hydrolysis and other dissociative chemisorption reactions meet these requirements. Reaction studies on dehydroxylated silica surfaces containing strained rings show that molecules

High Activity	Moderate Activity	No Activity
H_2O Water D = 0.27 nm	CH_3OH Methanol D = 0.36 nm	$CH_3CH_2CH_2{-}NH_2$ n-Propylamine D = 0.46 nm
NH_3 Ammonia D = 0.26 nm	CH_3NH_2 Methyl amine D = 0.36 nm	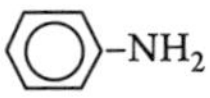 Aniline D = 0.55 nm
	$HCNH_2$ Formamide D = 0.4 nm	

Crack Velocity:	$H_2O = NH_3 > CH_3OH > \text{formamide} >> \text{aniline}$
Reaction Rate:	$CH_3NH_2 > CH_3OH > NH_3 > H_2O$
Size:	$H_2O = NH_3 < CH_3OH < \text{formamide} < \text{aniline}$
Basicity:	$CH_3NH_2 > NH_3 > CH_3OH > H_2O$
Acidity:	$H_2O > CH_3OH >> NH_3$

Figure 16.12 Listings of some of the molecular species that have been tested in slow crack growth experiments, along with their relative sizes, basicity, and acidity [26]. Trends are compared with both surface reaction rates and crack growth velocities for silica.

capable of disrupting strained oxide bonds include H_2O, NH_3 and other primary amines, as well as CH_3OH and other primary alcohols [26].

Three interesting trends emerge in examining crack velocity data for silica in the presence of chemicals other than water. First, the slope (i.e., the value of b) in the exponential crack velocity-versus-stress intensity curves is similar for all active chemicals (Fig. 16.13). Thus similarity implies that, for silica, strain activation is required before active chemicals can interact strongly with crack-tip bonds. The silica result contrasts with results for oxides, such as alumina, in which each active chemical exhibits its own unique stress dependence. It appears that the constant b value is unique to oxides such as silica that contain tetrahedral coordination geometries that react with chemicals such as water via the nucleophilic S_N2 attack mechanism.

A comparison of the acid–base properties of reactive gases with reaction rates on strained surface sites suggests that, to a first approximation, adsorption processes preceding bond rupture are dominated by interactions between basic lone-pairs of electrons on the attacking molecule and the acidic Si centers. For example, on edge-shared tetrahedra, reaction rates follow the

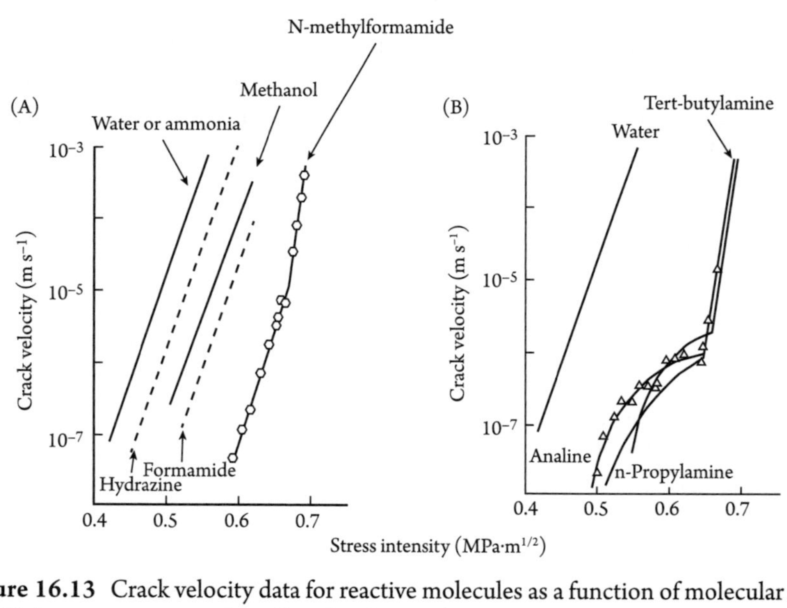

Figure 16.13 Crack velocity data for reactive molecules as a function of molecular size. (A) High concentrations of small molecules result in region I behavior, with an exponential dependence on stress intensity. (B) Conversely, high concentrations of large molecules exhibit region II and region III behavior. Here, region II behavior is an artifact of the presence of trace amounts of water in the solvent rather than being associated with the primary molecule. Adapted with permission from Michalske, T. A.; Bunker, B. C. Steric Effects in Stress Corrosion Fracture of Glass. *J. Am. Ceram. Soc.* **1987**, *70*, 780–784.

trend [26] $CH_3NH_2 > CH_3OH > NH_3 > H_2O$. Basicities follow the trend $CH_3NH_2 > NH_3 > CH_3OH > H_2O$, whereas acidities follow the trend $H_2O > CH_3OH >> NH_3$ [27]. After adsorption, infrared vibrational spectroscopy indicates that dissociation occurs almost immediately for most chemicals. Thus adsorption is the rate-limiting step, leading to the general reaction rate-versus-basicity correlation. However, the infrared studies also show that for the nitrogen donors, subsequent dissociation of the N–H bond can limit the rate in some cases as molecular adsorbates are observed before the formation of dissociation products. In fact, ammonia is such a poor acid that, in the gas phase, there is no evidence it is capable of donating protons at all. The difficulty associated with N–H bond dissociation appears to be the reason why NH_3 falls out of the sequence in the reaction rate-versus-molecular basicity correlation just listed.

Although the intrinsic reactivity of attacking nucleophiles is clearly important, the geometry and surface chemistry of the growing crack both play major roles in determining crack velocities (Fig. 16.14). In contrast to exposed dehydroxylated silica surfaces, the transport of species to the crack tip can be sufficiently slow to limit the rate. Although crack velocities in liquid water follow the exponential relationship described earlier ($V = V_o e^{nK_I}$), more complex curves are observed in dilute vapors of water or any other active chemical [28] (Fig. 16.3). Here, three distinct regions are observed. At low crack velocities (region 1), water molecules are provided to the crack tip faster than they are consumed by crack-growth reactions, and the exponential crack-growth relationship is observed. Eventually

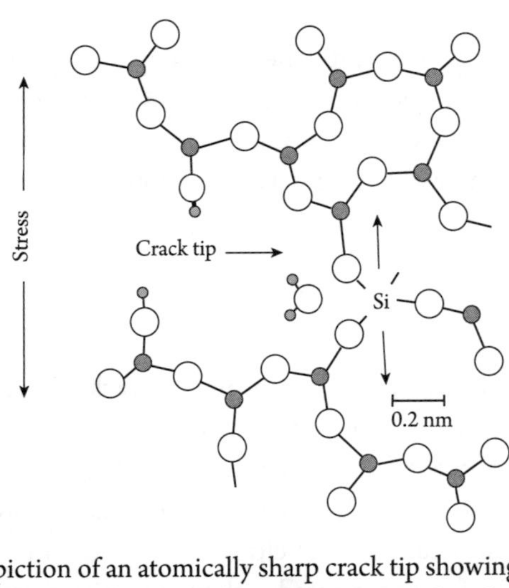

Figure 16.14 A depiction of an atomically sharp crack tip showing a crack wall lined with silanol groups and a small water molecule approaching bonds of the strained crack tip. Adapted with permission from Michalske, T. A.; Bunker, B. C. Steric Effects in Stress Corrosion Fracture of Glass. *J. Am. Ceram. Soc.* **1987**, *70*, 780–784.

cracks grow at a rate comparable with the rate at which water molecules can diffuse to the crack tip (region 2). In this regime, the water concentration at the crack tip decreases as the crack velocity increases, eventually reaching the point where the reaction rate is controlled by the rate of delivery of molecules to the crack tip rather than by the rate of reaction on strained crack-tip bonds. Finally, above a critical stress intensity (region 3), the oxide is ripped apart regardless of whether reactive chemicals are present, resulting in an extremely steep V-versus-K_I relationship that corresponds to the shattering of glass.

Even if reactive species are able to diffuse to the crack tip, crack growth measurements show that the size of the attacking molecule can be just as important as its intrinsic reactivity in controlling crack velocities [29]. Experiments involving molecules with similar reactivities but different sizes show that crack velocities decrease as the molecular size increases. In fact, molecules with a diameter greater than 0.5 nm do not promote slow crack growth in silica at all (Fig. 16.13). It appears there are steric limitations associated with the size of the attacking molecule relative to the radius of the growing crack. The most highly strained bonds are at the immediate crack tip. Molecules that are too large to reach that crack tip cannot participate in a stress corrosion process.

16.6 THE ROLE OF WATER IN COMPLEX CRACK-GROWTH BEHAVIORS

So far, this chapter has dealt with the simplest possible ideal case of chemically assisted crack growth in an oxide, which involves reactions on atomically sharp cracks in fused silica. However, a wide range of crack growth phenomena can occur that deviate from such simple behavior depending on both chemical and kinetic effects within the environment of the growing crack. Here, we highlight a range of more complex crack growth phenomena, including crack healing, crack blunting, crack growth involving stressed alteration layers, and slow crack growth in octahedral oxides.

16.6.1 Crack Healing

We have already seen that water can promote crack growth by reacting with strained bonds at the tip of a growing crack. Paradoxically, water can also participate in the reverse process, which is called *crack healing*. In humid air having relative humidities ranging from 0.01% to 100%, cracks in both soda–lime–silica glass and vitreous silica can actually close and heal even in the presence of an applied load [30]. The force associated with water-assisted crack closure is on the order of 0.15 J/m^2, which is equivalent to the force generated between two silanol-covered silica surfaces (5 OH/nm^2) that interact with each other via hydrogen bonds. It appears that the presence of molecular water at the crack tip assists in forming bridges of hydrogen bonds that close up the crack almost like a zipper. As each hydrogen bond forms at the crack tip, it pulls the walls of the crack closer together and allows the next hydrogen bond to form.

When a crack has closed, it can be reopened by increasing the applied load. For silica glass, the force required for repropagation is identical to the force associated with crack healing ($0.15\ J/m^2$). This observation implies that, when a crack heals, the only thing holding the crack together is hydrogen bonds rather than the original $>Si-\mu_2O-Si<$ linkages. Such cracks can be opened and closed repeatedly with no observed changes in either the healing or repropagation force. The robustness of this reaction indicates that the number of hydrogen bonds across the crack remains essentially constant.

For soda–lime–silica glass, more energy is required to reopen the closed cracks ($1.7\ J/m^2$) than can be accounted for via hydrogen bonding alone. One possible explanation for the increase is that the Na^+ and Ca^{2+} in soda–lime glass form ionic bridges between crack walls. The observed repropagation force compares well with the cleavage energy for layered silicates such as muscovite ($1.2\ J/m^2$), which is held together via similar ionic bridges [31]. It is also possible that the increase is a result of the reformation of a fraction of the $>Si-\mu_2O-Si<$ bonds that bridged the original crack. Such a result might be expected because the baseline soda–lime glass surfaces contain more nonbridging oxygens with lower Q values (see chapter 15) that can condense more readily to form $>Si-\mu_2O-Si<$ bridges across the crack. However, based on the slow kinetics associated with silicate condensation reactions, it is clear that if condensation occurs on the timescale of the crack growth experiments, such condensation is limited at room temperature. Heating cracked soda–lime glass to 600°C does cause an increase in the repropagation force to $2.2\ J/m^2$, suggesting that $>Si-\mu_2O-Si<$ bonds do form across the crack when the hydroxyl groups on either side of the crack sample are dehydroxylated [32]. Complete condensation would restore the glass across the crack completely, resulting in a repropagation force of $8.2\ J/m^2$ (i.e., the energy required to fracture virgin silica glass). To summarize, although visible evidence suggests that cracks can reclose, disappear, and heal, fractured specimens only recover a fraction of their initial strength unless the samples are remelted.

16.6.2 Crack Blunting

Crack propagation can be strongly inhibited by aging silica samples in liquid water for extended periods of time, which causes dramatic increases in fracture strength [33]. The explanation for this behavior involves the dissolution of silica from the walls of the crack followed by reprecipitation in the region of the crack tip. As discussed in Chapter 7 (see Fig. 7.4), the solubility of silica varies as a function of the radius of curvature of the local environment. The crack tip represents a region with a sharp negative radius of curvature. Silica dissolves slowly from the walls and redeposits at the crack tip to create a blunter crack.

According to continuum models such as that of Inglis [34], the stress on the crack tip is given by

$$S = 2 \cdot S_a \sqrt{\frac{C}{P}} \tag{16.10}$$

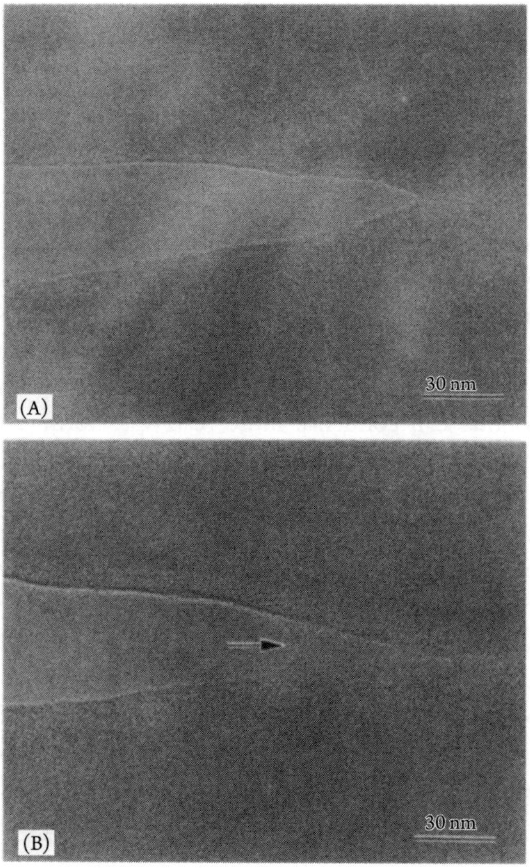

Figure 16.15 (A, B) High-resolution transmission electron micrographs of the crack tip in a pristine silica sample (A) and the same crack after soaking in 90°C water for 5 days (B). The outline of the original crack is still visible, but the crack tip has been filled in with a deposit of silica, resulting in the production of a blunter crack. The blunter tip is more resistant to crack propagation than the original sharp crack.
SOURCE: Reproduced with permission from Bando, Y.; Tomozawa, M. Direct Observation of Crack Tip Geometry of SiO_2 Glass by High-Resolution Electron Microscopy. *J. Am. Ceram. Soc.*, **1982**, *65*, 368–371.

where S is the crack tip stress, S_a is the applied stress, C is the crack length, and P is the radius of curvature of the crack tip. Eq. 16.10 reflects the fact that the applied stress is attenuated over a greater number of crack-tip bonds in a blunter crack, resulting in a lower degree of bond deformation and stress activation.

Experimental evidence for the strengthening of glass samples containing cracks has been obtained using high-resolution transmission electron microscopy (Fig. 16.15) [35]. The results show that by soaking a cracked silica sample in 90°C water for 1 week, silica deposits at the root of the crack and increases the crack-tip radius from 1.5 nm to 5.0 nm. Although the inherent strength of the deposited material is unknown, it is apparent this radius increase is responsible for the high loads required to restart the aged crack.

16.6.3 The Effects of Alteration Layers in Multicomponent Glasses

Multicomponent glasses can dissolve in water in a noncongruent fashion to create alteration or leached surface layers (see Chapter 15). The presence of these leached layers within the environment of the crack tip can influence the nature and kinetics of slow crack growth [36]. Such differences are highlighted by the crack growth-versus-stress intensity data for sodium-silicate and soda–lime glasses relative to the results for fused silica (Fig. 16.16).

At high crack velocities, leachable glasses and silica exhibit a similar linear relationship between the log of the crack velocity and stress intensity that is indicative of stress-activated reactions on strained crack-tip bonds. However, at low crack velocities, strong deviations from this simple relationship are observed for many multicomponent glasses. For some glasses (e.g., soda–lime glass), cracks cease to grow below a specific critical stress intensity. At the other extreme (e.g., $Na_2O-3SiO_2$), cracks grow even in the absence of an applied load. All these results can be rationalized based on the stresses that develop within evolving leached layers. Depending on composition changes, water content, and network restructuring, leached layers can expand or contract relative to the underlying glass to generate either compressive or tensile stresses within the leached layer.

An expression derived to determine the magnitude of the stress intensity (K_{Ir}) resulting from the internal stress generated within the leached layer (S_r) is given by [37]

$$K_{IR} = \pm 0.64\, S_r \cdot \sqrt{T} \tag{16.11}$$

where T is the thickness of the leached layer, and the sign of the stress intensity depends on whether the internal stress in tensile or compressive. When the stress is compressive, as in soda–lime and many borosilicate glasses [38], the stresses in the leached layer work against the applied tensile stress to try to

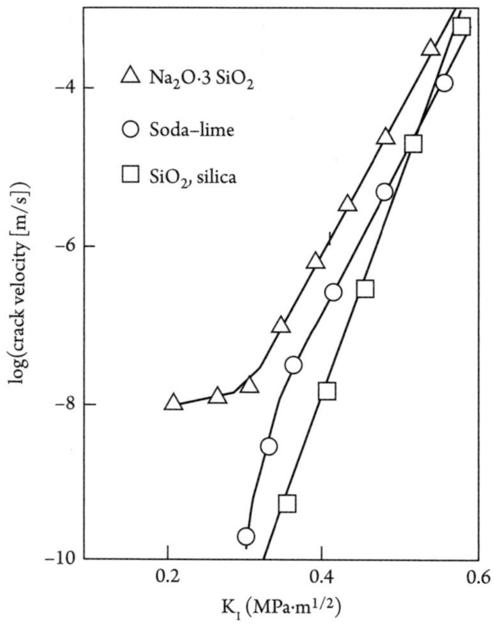

Figure 16.16 Plots of the log of the crack tip velocity as a function of stress intensity for silica, $Na_2O-3SiO_2$ glass, and commercial soda–lime glass immersed in water. At high crack velocities (corresponding to high stress intensities), all three materials exhibit normal strain activation with similar slopes. However, at low crack velocities, when there is sufficient time to produce leached surface layers, the more complex glasses show strong deviations from normal crack growth behavior. For soda–lime glass, the leached layer goes into compression, leading to a stress corrosion limit below which no crack growth occurs (at around 0.3 $MPa \cdot \sqrt{m}$). For $Na_2O-3SiO_2$, the growth of leached layers produces tensile stresses that drive crack velocities at a rate of 10^{-8} m/second in the absence of an applied load. Adapted with permission from Bunker, B. C.; Michalske, T. A. Effect of Surface Corrosion on Glass Fracture. In *Fracture Mechanics of Ceramics*: Bradt, Evans, Hasselman, Lange, Eds.; Springer: New York, 1986; Vol. 8, pp 391–411

squeeze the crack shut. When the leached-layer stress exceeds the applied stress, bonds at the crack tip are no longer pulled apart, and crack growth ceases at the so-called *fatigue limit* (e.g., the K_I value of 0.3 $MPa \cdot \sqrt{m}$ in Fig. 16.16 for soda–lime glass). Conversely, when the stress is tensile, the net stress on crack tip bonds is equal to the leached-layer stress plus the applied stress. Because leached layers tend to be mechanically weaker than unaltered glass (see Chapter 15), the tensile stress can become so large that the leached glass literally tears itself apart even in the absence of an applied load. This fracture mode is clearly seen for materials such as alkali-silicate glasses in which leached layers undergo extensive crazing [39] (Fig. 16.17). However, as indicated in Eq. 16.11, the leached layer

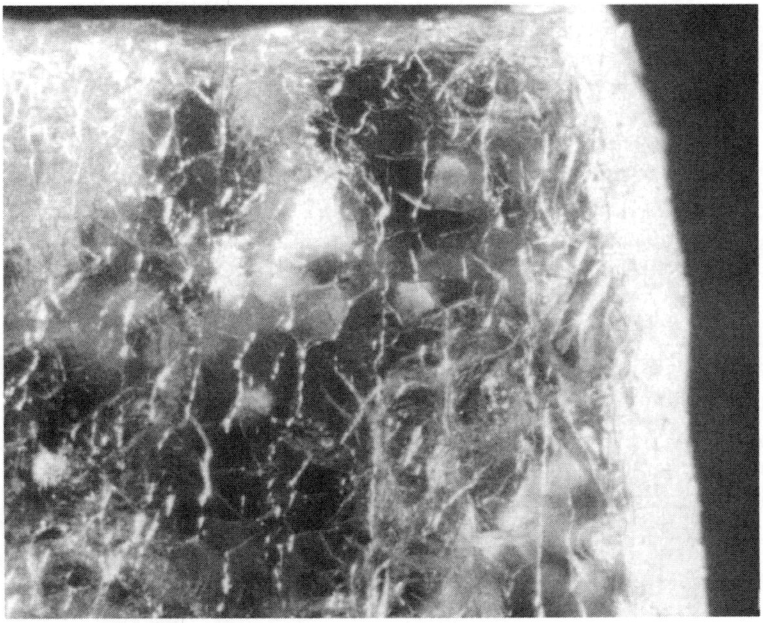

Figure 16.17 A photograph of a crazed sample of $K_2O-3SiO_2$ after several hours of leaching in 0.1 M HCl at T = 50°C. Tensile stresses build up within the leached layer until the altered material releases the strain by crazing.
SOURCE: Reproduced with permission from Bunker, B. C.; Arnold, G. W.; Beauchamp, E. K.; Day, D. E. Mechanisms for Alkali Leaching in Mixed Na–K Silicate Glasses. *J. Non-Cryst. Solids,* **1983**, *58*, 295–322.

must have time to grow to an appreciable thickness for the stress level to build up to the point where this happens. At high crack velocities, cracks grow faster than leached layers develop, which is why leached-layer effects are negligible for rapidly growing cracks.

16.6.4 Slow Crack Growth in Octahedral Oxides

In contrast to the behavior of tetrahedral oxides (Fig. 16.13), octahedral oxides exhibit a different degree of stress intensity activation for each chemical capable of promoting slow crack growth. This difference is illustrated by the crack growth data for single-crystal sapphire [40] (Fig. 16.18), where the slopes (b in Eq. 16.6) for water, ammonia, and hydrazine are 45, 61, and 31, respectively. The sensitivity of the slope to the nature of each distinct molecule suggests that chemicals promoting slow crack growth are adsorbed on crack-tip bonds in the absence of an applied stress.

The differences in crack growth behavior between tetrahedral and octahedral oxides can be rationalized on the basis of the fundamental hydrolysis mechanisms

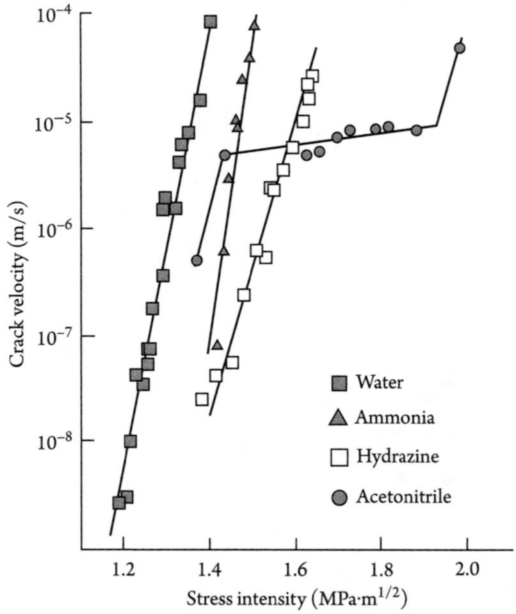

Figure 16.18 Crack growth data for single-crystal sapphire (Al_2O_3) in the presence of different molecular species. Ammonia is strongly activating, giving rise to the steepest slope, followed by hydrazine and water. Acetonitrile does not promote slow crack growth. The activation at low stress intensity is a result of trace amounts of water dissolved in the acetonitrile. Adapted with permission from Michalske, T. A.; Bunker, B. C.; Freiman, S. W. Stress Corrosion of Ionic and Mixed Ionic/Covalent Solids. *J. Am. Ceram. Soc.* **1986**, *69*, 721–724.

introduced in Part One of this book. Aspects of chemically assisted bond rupture resemble ligand-exchange reactions. For tetrahedral oxides, such reactions are most often associative in nature because these metals, such as Si(IV), add a ligand more easily than they lose one and create the five-coordinate intermediate that leads to bond rupture. For octahedral oxides, ligand-exchange reactions tend to involve reduced coordination numbers because cations such as Al(III), when coordinated octahedrally, can reduce their coordination numbers more easily to five than increase them to seven. This trend is reflected in infrared studies that have been performed to study molecular adsorption onto dehydroxylated alumina [41]. Here, dehydroxylation creates penta-coordinate cations at the surface. These metals are coordinately unsaturated and have an open coordination site that can coordinate directly to incoming active molecules. Such dehydroxylated surfaces resemble fracture surfaces near the crack tip. When such surfaces are exposed to active chemicals, infrared vibrational spectroscopy shows the molecules bond directly to open sites without promoting molecular dissociation. However, the adsorption of a Lewis base to the open coordination site removes electron density from all the remaining M—O bonds around a given cation (see Chapter 4), which facilitates the subsequent strain-assisted bond-rupture process. If this model is correct, the more basic the coordinating molecule, the more effective the molecule should be toward activation, and the steeper the slope of the curve in Figure 16.18. A comparison of Figure 16.12 and Figure 16.18 shows that this is indeed the case.

References

1. Tuminaro, R. D. Materials Aspects of the SL Lightguide Undersea Cable: 1988 MRS Spring Meeting Plenary Address. *MRS Bull.* 1988, *13*, 14–22.
2. Weast, R. C., Ed. *CRC Handbook of Chemistry and Physics, 58th ed.*; CRC Press: Cleveland, OH, 1977.
3. Michalske, T. A.; Bunker, B. C. The Fracturing of Glass. *Sci. Am.* 1987, **255** 122–129.
4. Michalske, T. A.; Smith, W. L.; Bunker, B. C. Fatigue Mechanisms in High-Strength Silica-Glass Fibers. *J. Am. Ceram. Soc.* **1991**, *74*, 1993–1996.
5. Griffith, A. A. Phenomena of Rupture and Flow in Solids. *Phil. Trans. R. Soc. London A* **1920**, *221*, 163–198.
6. Weiderhorn, S. M.; Fuller, E. R.; Thompson, R. Micromechanisms of Crack Growth in Ceramics and Glasses in Corrosive Environments. *Met. Sci.* **1980**, *Aug.–Sep.*, **14**, 450–458.
7. Michalske, T. A.; Bunker, B. C. Slow Fracture Model Based on Strained Silica Structures. *J. Appl. Phys.* **1984**, *56*, 2686–2693.
8. Gibbs, G. V. Molecules as Models for Bonding in Silicates. *Am. Miner.* **1982**, *67*, 421–450.
9. Tossell, J. A.; Gibbs, G. V. A Molecular Orbital Study of Shared-Edge Distortions in Linked Polyhedra. *Am. Miner.* **1976**, *61*, 287–294.
10. Revesz, A. G. The Defect Structure of Vitreous SiO_2 Films on Silicon I: Structure of Vitreous SiO_2 and the Nature of the Si–O Bond. *Phys. Stat. Sol. A* **1980**, *57*, 235–243.
11. Weiss, A.; Weiss, A. Uber Siliciumchalkogenide: VI. Zur Kenntnis der Faserigen Siliciumdioxyd-Modification. *Z. Anorg. Chem.* **1954**, *276*, 95–112,
12. Iler, R. K. *The Chemistry of Silica*; Wiley: New York, 1979.
13. Galeener, F. L. Planar Rings in Vitreous Silica. *J. Non-Cryst. Solids* **1982**, *49*, 53–62.
14. Galeener, F. L.; Barrio, R. A.; Martinez, E.; Elliott, R. J. Vibrational Decoupling of Rings in Amorphous Solids. *Phys. Rev. Lett.* **1984**, *53*, 2429–2432.
15. Chakoumakos, B. C.; Hill, R. J.; Gibbs, G. V. A Molecular Orbital Study of Rings in Silicates and Siloxanes. *Am. Miner.* **1981**, *66*, 1237–1249.
16. Balfe, C. A.; Ward, K. J.; Tallant, D. R.; Martinez, S. L. Reactivity of Silicates 1: Kinetic Studies of the Hydrolysis of Linear and Cyclic Siloxanes as Models for Defect Structure in Silicates. *Mater. Res. Soc. Symp. Proc.* **1986**, *73*, 619–625.
17. Michalczyk, M. J.; Fink, M. J.; Haller, K. J.; West, R.; Michl, J. Structural and Chemical Properties of 1,3 Cyclodisiloxanes. *Organometallics* **1986**, *5*, 531–538.

18. Bunker, B. C.; Haaland, D. M.; Ward, K. J.; Michalske, T. A.; Binkley, J. S.; Melius, C. F.; Balfe, C. A. Infrared Spectra of Edge-Shared Silicate Tetrahedra. *Surf. Sci.* **1989**, *210*, 406–428.
19. Bunker, B. C.; Haaland, D. M.; Michalske, T. A.; Smith, W. L. Kinetics of Dissociative Chemisorption on Strained Edge-Shared Surface Defects on Dehydroxylated Silica. *Surf. Sci.* **1989**, *222*, 95–118.
20. Morrow, B. A.; Cody, I. A. Infrared Studies of Reactions on Oxide Surfaces: 5. Lewis Acid Sites on Dehydroxylated Silica. *J. Phys. Chem.* **1976**, *80*, 1995–1998.
21. Morrow, B. A.; Cody, I. A. Infrared Studies of Reactions on Oxide Surfaces: 6. Active Sites on Dehydroxylated Silica for the Chemisorption of Ammonia and Water. *J. Phys. Chem.* **1976**, *80*, 1998–2004.
22. Tallant, D. R.; Bunker, B. C.; Brinker, C. J.; Balfe, C. A. Raman Spectra of Rings in Silicate Materials. *Mater. Res. Soc. Symp. Proc.* **1986**, *73*, 261–267.
23. Laidler, K. J. *Chemical Kinetics*; Harper & Row: New York, 1987.
24. Hamann, S. D. Chemical Equilibria in Condensed Systems. Ch. 2 In *High Pressure Physics and Chemistry*; Bradley, R. S., Ed.; Academic Press: London, 1963; pp 163–207.
25. Boudjouk, P.; Kapfer, C. A.; Cunico, R. F. Synthesis and Reactivity of 1-Silaadamantyl Systems. *Organometallics* **1983**, *2*, 336–343.
26. Michalske, T. A.; Bunker, B. C. A Chemical Kinetics Model for Glass Fracture. *J. Am. Ceram. Soc.* **1993**, *76*, 2613–2618.
27. Bunker, B. C.; Haaland, D. M.; Michalske, T. A.; Smith, W. L. Kinetics of Dissociative Chemisorption on Strained Edge-Shared Surface Defects on Dehydroxylated Silica. *Surf. Sci.*, **1989**, *222*, 95–118.
28. Gordon, A. J.; Ford, R. A. *The Chemists Companion*; Wiley: New York, 1972.
29. Michalske, T. A.; Freiman, S. W. A Molecular Mechanism for Stress Corrosion in Vitreous Silica. *J. Am. Ceram. Soc.* **1983**, *66*, 284–288.
30. Michalske, T. A.; Bunker, B. C. Steric Effects in Stress Corrosion Fracture of Glass. *J. Am. Ceram. Soc.* **1987**, *70*, 780–784.
31. Michalske, T. A.; Fuller, E. R., Jr. Closure and Repropagation of Healed Cracks in Silicate Glass. *J. Am. Ceram. Soc.* **1985**, *68*, 586–590.
32. Deryagin, B. V.; Metzik, M. S. The Role of Electrical Forces in Splitting of Mica Along Cleavage Planes. *Fiz. Tverd. Tela* **1959**, *1*, 1521–1528.
33. Stavrinidis, B.; Holloway, D. G. Crack Healing in Glass. *Phys. Chem. Glasses* **1983**, *24*, 19–25.
34. Ito, S.; Tomozawa, M. Crack Blunting of High-Silica Glass. *J. Am. Ceram. Soc.* **1982**, *65*, 368–371.
35. Inglis, C. E. Stresses in a Plate Due to the Presence of Cracks and Sharp Corners, *Trans. Inst. Naval Architect.* **1913**, *55*, 219–241.
36. Bando, Y.; Ito, S.; Tomozawa, M. Direct Observation of Crack Tip Geometry of SiO_2 Glass by High-Resolution Electron Microscopy. *J. Am. Ceram. Soc.* **1984**, *67*, C-36–C-37.
37. Bunker, B. C.; Michalske, T. A. Effect of Surface Corrosion on Glass Fracture. In *Fracture Mechanics of Ceramics*; Bradt, R.C.; Evans, A. G.; Hasselman, D. P. H.; Lange, F. F. Eds.; Springer: New York, 1986; pp. 391–411 Vol. 8.

38. McMeeking, R. M.; Evans, A. G. Mechanics of Transformation-Toughening in Brittle Materials. *J. Am. Ceram. Soc.* **1982**, *65*, 242–246.
39. Scherer, G. W.; Drexhage, M. G. Stress in Leached Phase-Separated Glass. *J. Am Ceram. Soc.* **1985**, *68*, 419–426.
40. Bunker, B. C.; Arnold, G. W.; Beauchamp, E. K.; Day, D. E. Mechanisms for Alkali Leaching in Mixed Na–K Silicate Glasses. *J. Non-Cryst. Solids* **1983**, *58*, 295–322.
41. Michalske, T. A.; Bunker, B. C.; Freiman, S. W. Stress Corrosion of Ionic and Mixed Ionic/Covalent Solids. *J. Am. Ceram. Soc.* **1986**, *69*, 721–724.
42. Peri, J. B. Infrared Study of Adsorption of Ammonia on Dry Alumina. *J. Phys. Chem.* **1965**, *69*, 231–239.

PART SIX

The Environmental Geochemistry of Oxides

17

The Weathering of Oxides

17.1 INTRODUCTION

Earth's crust is largely composed of oxides, so the biosphere we inhabit is dominated by interactions between oxides, water, and living things. Part Six of this book, on environmental geochemistry, focuses on these interactions and serves as a review of many of the chemical concepts that form the basis for the rest of the book. As such, the final two chapters frequently refer back to previous chapters for more in-depth discussions of specific chemical phenomena. In this chapter, however, we highlight how the diverse environments on the surface of Earth modify the structure, composition, and chemistry of oxide minerals by weathering phenomena. Conversely, in Chapter 18 we explore how oxide minerals and their weathering products modify the structure, composition, and chemistry of the environments they inhabit. These environmental interactions are influenced by life, and are critical to the health and well-being of all living things.

Minerals have a natural life cycle on the surface of Earth. Most oxides emerge from Earth's interior in the form of igneous rocks that form and are stable at the high temperatures and pressures of subsurface environments (see Chapter 18). These minerals usually do not represent phases that are thermodynamically stable in ambient-temperature water. As a result, any pristine rocks exposed to air and water are subject to the physical and chemical degradation processes we call *weathering* (Fig. 17.1 and Plate 20). Weathering processes facilitated by water convert anhydrous oxides formed at high temperatures into hydrous oxides, oxyhydroxides, hydroxides, and dissolved by-products.

Figure 17.1 A photograph of the Paine Towers in Chilean Patagonia at the southern tip of South America that are exposed to extreme weathering conditions. SOURCE: kavram/Shutterstock.com. (see Plate 20)

It has been estimated that volcanic rocks represent only 8% of the rocky outcrops on Earth's surface whereas 26% are more coarsely grained plutonic rocks of igneous origin [1]. The remaining 66% of rocky outcrops represent the decomposition products of these igneous parents, including sandstone (16%), clay-based rocks such as shale (33%), and simple ionic salts such as limestone (16%) and evaporates (1.3%) (Table 17.1). The focus of this chapter is on the physical and chemical processes that form and affect these decomposition products under ambient-temperature conditions. Discussions of the role of the extreme heat from Earth's interior in accelerating or even reversing these weathering reactions are deferred until the closing section of Chapter 18. Such heating ultimately regenerates igneous rocks and other relatively anhydrous minerals, leading to a rebirth of oxides and a completion of the oxide life cycle.

17.2 PHYSICAL WEATHERING

17.2.1 The Physical Transport of Oxides

We begin our journey of rock transformations at mountain peaks, where mechanical fracturing has uncovered igneous rocks and exposed them to water and ice for the first time. Under the force of gravity, virtually all these rocks are

Table 17-1. An Estimate of the Average Composition (Measured in Milligrams per Kilogram) of Elements in Igneous and Three Types of Sedimentary Rocks

Element	Igneous Rock (mg/kg)	Sedimentary Rocks Resistive Rocks Sandstone	Hydrolysate Shale	Precipitate Limestone
Si	285,000	359,000	260,000	34
Al	79,500	32,100	80,100	8970
Fe	42,200	18,600	38,800	8190
Ca	36,200	22,400	22,500	272,000
Na	28,100	3870	4850	393
K	25,700	13,200	24,900	2390
Mg	17,600	8100	16,400	45,300
Ti	4830	1950	4440	377
P	1100	590	733	281
Mn	937	392	575	842
F	715	220	560	112
Ba	595	193	250	30
S	410	945	1850	4550
Sr	368	28	290	617
C	320	13,800	15,300	113,500
Cl	305	15	170	305
Rb	166	197	243	46
Zr	160	204	142	18
V	149	20	101	13
Ce	130	55	45	11
Cu	97	15	45	4.4
Ni	94	2.6	29	13
Zn	80	16	130	16
Nd	56	24	18	8
La	48	19	28	9.4
N	46	—	600	—
Y	41	16	20	15
Li	32	15	45	4.4
Co	23	0.33	8.1	0.12
Nb	20	0.096	20	15
Ga	18	5.9	23	2.7
Pr	17	7.0	5.5	1.3
Pb	16	14	80	16
Sm	16	6.6	5.0	1.1
Sc	15	0.73	10	0.68
Th	11	3.9	13	0.20
Gd	9.9	4.4	4.1	0.77
Dy	9.8	3.1	4.2	0.53
B	7.5	90	194	16

Data from: Hem, J. D. *Study and Interpretation of the Chemical Characteristics of Natural Water*; U.S. Geological Survey, 1989, U. S. Government Printing Office, Washington, D. C. p. 5.

eventually transported back to the ocean to deposit as sediment, but the igneous minerals are rarely unmodified by this journey. Without the grinding effect of ice and the suspending effect of water, the rates of mineral transport from mountains to the ocean basins by wind would be very slow. The most obvious role of ice and water in erosion has to do with the fact that both phases are mobile and can carry solid materials with them. Although ice appears to be a stationary solid to the naked eye, it undergoes viscous flow, with an effective viscosity of around $2 \cdot 10^6$ Pa/s (10^{16} times greater than that of water) [2]. Gravitational flow rates in glaciers range from 10 to 100 m/year, but the glacial masses are enormous. Many glaciers are hundreds of meters thick, miles wide, and tens of miles long. The moving mass of a glacier represents a powerful force that is applied to abrade the underlying rock as *glacial scouring*. The powdered results of this abrasion can be seen from the vast volume of detritus. The material is deposited from glaciers during the last Ice Age and blankets much of the high-latitude world, including the American Midwest and parts of China. Spectacular examples of glacial erosion are provided by the eroded, half-cleaved shapes of massive granite outcrops such as Half-Dome in Yosemite Valley.

Although erosion by ice can be dramatic, water is the dominant medium for transporting solids from the continents to the oceans. The Grand Canyon in Arizona represents just one example of how extensive the erosion of rocks by water can be. A more quantitative measure of the efficiency of the aqueous transport of oxides is given by the *mechanical denudation rate* in tons of rock per square kilometer of Earth's surface per year [3]. Denudation rates for different river systems can vary enormously, ranging from tens to even thousands of tons per square kilometer per year (Fig. 17.2). The efficiency of transport depends on several factors, including the energy of the flowing water and the mechanical properties of the parent oxide source. The denudation rate depends the steepness of river gradients, water volume, and the susceptibility of the rock to erosion. The natural rate of continental denudation by mechanical processes is now estimated to be $1.2 \cdot 10^{16}$ g/year [4]. Unfortunately, human activities involving plows and bulldozers contribute an amount of mechanical denudation to the planet that is roughly equal to the natural value. The total transport resulting from both natural and artificial weathering is now up to $2 \cdot 10^{16}$ g/year.

Although water is the dominant medium responsible for transporting large oxide particles, the impact of air on transport and weathering cannot be ignored. The efficiency of atmospheric transport depends on particle size. Hurricanes and tornados that destroy buildings can move rocks the size of boulders. High-speed winds blowing for several days in the desert can move millions of tons of submillimeter-size sand grains. However, the most important transport mechanism, both in terms of oxide mass and environmental impact, involves micron and submicron oxide particles suspended in the atmosphere in the form of dust. In

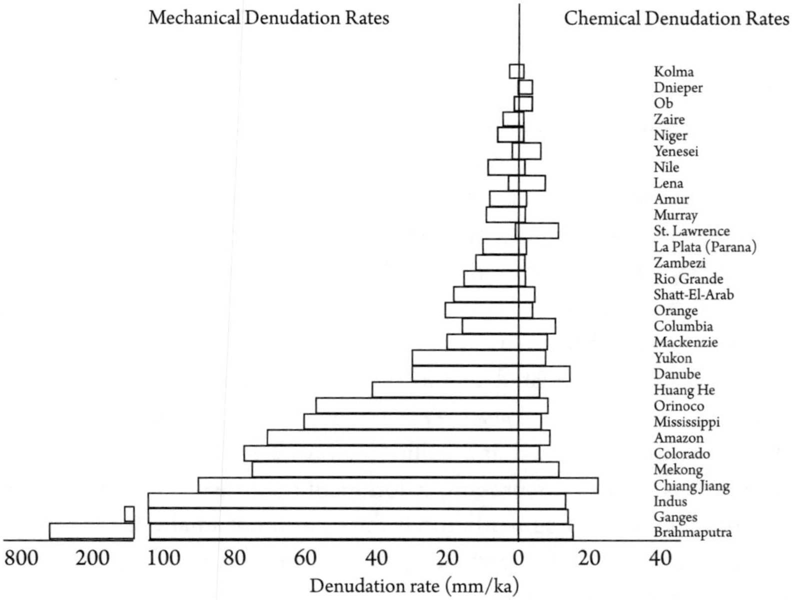

Figure 17.2 A graphical depiction of both mechanical and chemical denudation rates associated with the world's major river basins. Adapted with permission from Summerfield, M. A.; Hulton, N. J. *J. Geophys Res.* Natural Controls of Fluvial Denudation Rates in Major World Drainage Basins. **1994**, *99*, 13874–13883.

dust storms, it is estimated that 1 km^3 of air can hold up to 1000 t of dust particles, corresponding to a solids loading of 0.1 wt%. For large dust storms covering hundreds of square kilometers, the amount of suspended dust in a single storm can exceed 100 million (10^8) t [5]. A second source of oxide dust in the atmosphere is volcanic eruptions. For example, the 1815 eruption of the Tambora volcano ejected 30 to 300 km^3 ($10^{11}-10^{12}$ MT) of rock in the form of ash into the atmosphere [6]. This oxide volume exceeded the amount of material transported by water from the continents into the ocean for that year. The chemical and environmental consequences of this atmospheric dust are discussed in Chapter 18.

17.2.2 Freeze–Thaw Cycles and Stress-Corrosion Cracking

Rocks are brittle materials and can fracture under an applied stress. This is important because the size of an oxide particle determines how easy it is to move, whether transport is facilitated by ice, water, or air. Standard mechanical processes are partly responsible for the breakdown of large boulders into smaller and more mobile materials. Fragmentation during rock slides, avalanches, and glacial scouring are common occurrences. However, water plays a key role in the less dramatic but gradual weakening of rock and in noninstantaneous fracture processes. In fact, the primary processes leading the physical breakdown of rocks involve the stress-induced corrosion cracking phenomena described in Chapter 16.

Water penetrates any cracks or flaws present in rock surfaces, even down to nanometer-size scales. When this water freezes to form ice, the water expands by 9% within confined crack and creates enormous stresses within the crack. Water molecules react with strained crack-tip bonds to promote stress corrosion cracking (see Chapter 16). When subjected to repeated freeze–thaw cycles, even monoliths like the Sierra Nevada and the Andes Mountains (Fig. 17.1) are reduced from massive batholiths or plutons to boulders, pebbles, and millimeter-size particles via water-assisted fracture processes. In rivers and streams, the stress provided by freeze–thaw cycles is replaced by stresses imparted to rocks as they tumble along the streambed and collide with one another. Once again, water-activated stress corrosion represents a dominant fracture mechanism. The consequence of fracture and fragmentation is the creation of smaller particles that are easier to transport. Smaller particles also have more surface area accessible to chemical-weathering processes.

Mechanical fragmentation is not random. Grain boundaries are invariably weaker than pathways through the bulk of adjoining crystalline phases because they tend to contain disordered or amorphous material that is chemically unstable relative to the bulk. The boundaries are also regions of high strain, and are more susceptible to stress-corrosion cracking, pressure-enhanced dissolution, and chemical-weathering reactions such as the conversion of primary minerals into clays. These grain boundaries can be weakened if a chemical reaction has a positive volume, such as the freezing of water into the familiar Ice I structure. For these reasons, mechanical fragmentation often leads to the crude separation of one phase from another. For granite, this means that particles rich in quartz, feldspar, and mica are released from each other and can react with water as independent phases.

17.3 CHEMICAL WEATHERING: THE HYDROLYSIS OF OXIDES

Many minerals at the Earth's surface were formed at high temperatures and pressures and are out of equilibrium with water at ambient conditions. Chemical weathering primarily involves hydrolysis reactions that convert anhydrous oxides such as igneous rocks into oxyhydroxides, hydroxides, and solutes. For metal cations with access to multiple valence states, this hydrolysis can be assisted by electrochemical processes (see Section 17.8).

The place to begin unraveling weathering processes involves an examination of hydrolysis diagrams [7] (Chapter 5). Such diagrams provide the solubility characteristics of the hydrolyzed species of each of the individual metal cations that comprise the oxide. As discussed in Chapter 5, the primary factor influencing hydrolysis product solubility is cation valence. Cations with a valence of +7, +6, or +5 tend to form soluble oxoanions such as $[TcO_4]^-$, $[SO_4]^{2-}$, and $[HPO_4]^{2-}$ However, these anions can pair with other soluble cations to precipitate from solution as insoluble metal salts such as barite, $BaSO_4$. Cations with a valence of +4 or +3 tend to form insoluble oxides such as MnO_2. As examples, the octahedral hydrolysis products of Ti(IV) have a solubility of $3 \cdot 10^{-6}$ M throughout the pH range of natural waters, whereas those of Zr(IV) have a solubility of less than 10^{-10} M. The most abundant cationic metal in rocks, tetrahedral Si(IV), can reach millimolar concentrations as silicic acid in neutral pH solutions if the water equilibrates with amorphous SiO_2 or quartz. The hydrolysis products of the primary +3 cations in igneous rock, Al(III) and Fe(III), are relatively insoluble as hydroxo-bridged minerals such as boehmite (AlOOH) and goethite (FeOOH). Those of Al(III) have a solubility of around 10^{-5} M at pH 5, dropping to 10^{-6} M at the higher pH 8 of seawater, whereas Fe(III) has a solubility of less than 10^{-10} M at near-neutral pH. Most +2 cations are highly soluble in water except in highly basic solutions, whereas most +1 cations form simple aquated monomers that are highly soluble regardless of pH.

Hydrolysis diagrams only contain information regarding the species formed between a single cation and the water molecules and hydroxide ions found in simple aqueous solutions. However, these diagrams do not contain speciation information regarding the soluble complexes that can form involving the myriad other ligands found in natural environments (see Chapter 18). Just because an element forms insoluble hydrolysis products at a given pH does not mean it cannot form highly soluble complexes with other ligands. For example, although uranium oxides are highly insoluble, uranium dissolves rapidly from glass and oxide minerals in the presence of carbonate ions as a result of the formation of soluble species such as $[UO_2(CO_3)_3]^{4-}$ (see Chapter 15, Section 15.8). Ferric iron, Fe(III), has an elevated concentration in seawater as a result of complexation by siderophore ligands that do not appear in hydrolysis diagrams (see Chapter 18). The oxalate anion produced by plants can make soluble $[Al(C_2O_4)]^+$ (aq). Soils can contain dozens of strong ligands produced by the decomposition of organic matter, many of which have ill-defined stoichiometries. Such ligands are the bane of a geochemist's life, because it can be difficult to establish just which ligands are really controlling metal migration in many natural water systems.

There are multiple reasons why hydrolysis diagrams often fail to capture key features regarding the weathering tendencies of mixed-metal oxide minerals. First, mixed hydrolysis products can often exhibit solubility characteristics that are completely different from those of the parent elements. For example, although Al(III) and Si(IV) are relatively soluble in mildly acidic conditions, aluminosilicates such as clay minerals can be highly insoluble. Second, hydrolysis diagrams for single elements ignore the fact that hydrolysis reactions are highly sensitive to local structures, bonding configurations, and formal-charge distributions in mixed-oxide systems. Thus, geochemists construct phase diagrams based on the known solubility characteristics of the dominant mineral products expected to precipitate at a given solution composition. These phase diagrams are constructed from thermodynamic data but are often compared with field measurements of solution chemistry to identify the sequence of precipitation of minerals, or their *paragenesis*.

Fortunately, the solubility characteristics of many common minerals are known, and software now exists that can treat simultaneous equilibria for multicomponent solutions in equilibrium with hundreds of different solids. These programs simulate the evolution of water compositions as solids dissolve or precipitate, as soluble gases enter or leave solution, or as redox conditions adjacent to the oxide change. Such programs solve hundreds of chemical equilibrium equations quickly and simultaneously to provide important guidelines used to help predict chemical-weathering trajectories. A few of these programs are available free of charge through U.S. government agencies and universities [8].

Finally, hydrolysis diagrams carry no information about the kinetics of geochemical reactions. However, a wealth of information has been obtained by geochemists regarding how rapidly many of the most common rocks and minerals dissolve or decompose under a wide range of environmental conditions, including field studies. Because the surface distribution of these rocks and minerals is also known, scientists have been able to estimate the rates at which chemical-weathering phenomena occur (Fig. 17.2). As with physical weathering, chemical denudation rates are defined in terms of the tons of rock that are removed per square kilometer of Earth's surface per year. As with physical weathering, chemical-weathering rates can vary by orders of magnitude, depending on rock type and local environmental conditions such as water availability, pH, and E_h.

Current chemical-denudation rates for highly soluble rocks in ancient salt deposits are relatively rapid (423 t/km^2/year), followed by limestone at 100 t/km^2/year. Silicate rocks, including the igneous rocks and clay minerals, dissolve at a much slower rate of around 20 t/km^2/year. When normalized to the exposures and dissolution rates of all rock types, the net chemical-denudation rate is around 40 t/km^2/year or (given the area of the continents) around $6 \cdot 10^9$ t/year. Although this rate of reaction may seem large, it corresponds to an average retreat of all rock surfaces of only 15 μm/year [9]. To put this rate into geologic perspective, rocks are dissolving at an average rate of 15 m every million years.

Although chemical denudation rates are less than physical denudation rates (Fig. 17.2), the chemical alteration of rock is highly significant in that it defines the oxides that comprise many of the environments in which we live, such as the clay soils in which we grow our crops. Conversely, Earth's environments play a major role in weathering processes. Next, we provide specific examples that illustrate how oxide structure, pH, redox conditions, and solution saturation conditions influence the kinetics of weathering processes for some of the major rock types found in plutonic rocks such as granite.

17.4 A CONCEPTUAL FRAMEWORK FOR OXIDE WEATHERING: THE WEATHERING OF GRANITE

Clearly, it is impractical to try to catalogue how every mineral or rock type (see Table 17.1) is expected to weather in every natural environment, because they vary enormously across the planet. Therefore, in the remaining discussion of chemical weathering, the consequences of igneous rock hydrolysis are discussed in the context of the major minerals (quartz, feldspars, and mica) that comprise granite (Fig. 17.3 and Plate 21). These minerals and their weathering products

Figure 17.3 Granite and its components. (Upper left) A granite pluton found in Enchanted Rock Park in Texas (SOURCE: NeonLight/Shutterstock.com.). (Upper right) A closer look at the fine-grained structure of granite (SOURCE: Bragin Alexey/Shutterstock.com). The major mineral that comprise granite are shown in the bottom row: quartz (left) (SOURCE: Epitavi/Shutterstock.com), plates of mica surrounding a central crystal of aquamarine (center) (SOURCE: MarcelClemens/Shutterstock.com), and feldspar (right) (SOURCE: Zbynek Burival/Shutterstock.com). (see Plate 21)

represent the major broad groupings of both igneous and sedimentary rocks found at Earth's surface [14]. Here, we introduce the rock types represented by the weathering road map discussed in the next paragraph, followed by sections that review aspects of specific reaction pathways as a function of oxide structure, pH, redox chemistry, and reprecipitation processes. The impact of biota on these processes is discussed in Chapter 18.

A simplified road map for the weathering of granite (Fig. 17.4) illustrates how the predominant oxide minerals on Earth's surface are derived from the hydrolysis of igneous rocks. Minerals that are difficult to hydrolyze, such as quartz, rutile, (TiO_2), zircon ($ZrSiO_4$), or sphene ($CaTiSiO_5$), are called *resistates*. Zircon is so resistant to weathering that it is used for dating the oldest rocks on Earth. Some zircons show no evidence of corrosion for the 4 billion years our planet has been in existence. The fate of resistate minerals is represented by the reaction trajectory taken by quartz grains in the weathering road map. Quartz is by far the most

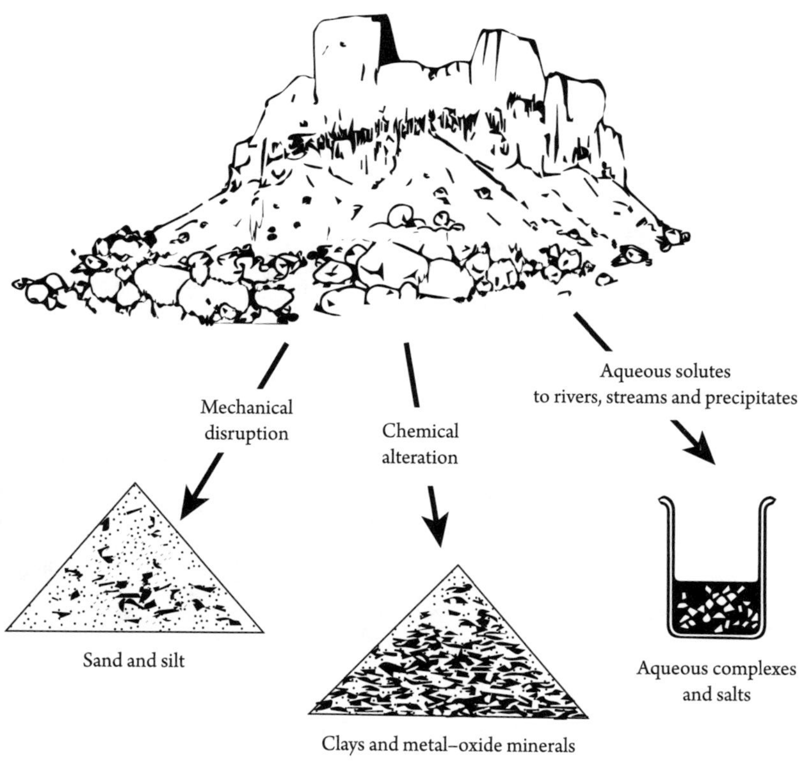

Figure 17.4 A simplified depiction of the weathering of a rocky outcrop (top) showing the release of insoluble grains of quartz and other resistates to form sand (bottom left), mineral hydrolysis to form clays and other hydrosylates (middle), and the release of soluble components to form the oxidates, precipitates, and evaporates (right). Adapted from Press, F.; Siever, R.; *Earth*; W. H. Freeman and Co.: San Francisco, 1978.

prevalent and important member of the resistate family. The road map shows that, although the chemical dissolution rates of resistate minerals can be negligible, physical weathering processes such as stress-corrosion cracking (see Section 17.3) can separate the quartz grains from granite and break them down into finer particles that represent sand. Massive deposits of sand in estuaries and on the seafloor can later be consolidated into the detrital rocks we call *sandstone*. As discussed in Section 17.2, roughly 16% of the exposed rocks on Earth's surface are sandstone and related detrital material.

The most pervasive pathway in the weathering diagram is illustrated by the fate of aluminosilicate minerals such as the feldspars. Silicate minerals containing trivalent and divalent metals tend to be more reactive than the resistates. The feldspars react with water via hydrolysis reactions to form phyllosilicates including the clay minerals, as well as other insoluble oxyhydroxides and hydroxides such as boehmite (AlOOH) and gibbsite ($Al(OH)_3$). These reaction products are classified as *hydrosylates* because they are often formed via hydrolytic transformations involving insoluble minerals. Hydrosylates are the major minerals found in soils, muds, and sediments. As clays evolve, are compressed, and are dehydrated, they eventually form impermeable rocks called *shale* that represent 33% of the rocky outcrops on Earth's surface. When added to terrestrial soils and the vast quantities of sediments in marine basins, hydrosylates represent roughly one-third of the mass of the entire crust, which means the total volume of the hydrosylates exceeds the volume of all our oceans combined.

The third major component of granite is micas. Micas are layered phyllosilicates that are separated from granite during weathering in much the same way as grains of quartz sand. Although micas crystallize from silicate melts, micas have structures and compositions that resemble some of the clay minerals produced via ambient-temperature weathering of aluminosilicate minerals such as feldspars. For example, although illite clay tends to form as much smaller grains than the melt-derived mica muscovite, both minerals represent exactly the same phase. For this reason, we do not distinguish between hydrosylate clays and melt-derived micas in later weathering discussions.

The reduction of redox-active elements, such as the iron and manganese found in both igneous rocks and hydrosylates, promotes the selective leaching of the water-soluble hydrolysis products of Fe(II) and Mn(II). The subsequent oxidation of solutions containing reduced soluble species, primarily by atmospheric O_2, forms colloidal particles of insoluble oxyhydroxides and hydroxides called *oxidates*. Examples of oxidates are minerals such as the FeOOH(s) and ferrihydrite(s) found in iron ore deposits, as well as MnO_2 (pyrolusite). Although not as abundant by mass as the aluminosilicates, the oxidates are extremely important in environmental chemistry because they have high surface areas, are highly reactive, and participate readily in reversible redox and photochemical reaction cycles (see Chapter 18).

The soluble hydrolysis products of +1 and +2 cations leached from rock can combine with soluble anions such as the oxyanions to form ionic salts. If these salts exhibit low solubility in aqueous solutions, such as $CaCO_3(s)$, they are called *sparingly soluble salts*. The most important sparingly soluble salts are carbonates such as calcite ($CaCO_3(s)$) and dolomite ($CaMg(CO_3)_2(s)$) found in the limestone formations that cover 16% of our planet. *Evaporate* rocks consist of water-soluble salts such as NaCl, gypsum ($CaSO_4$), and borates that are left behind when enormous isolated pools of ancient ocean water (such as the Great Salt Lake in Utah) are evaporated completely. Although evaporites are highly soluble in water, they can persist over geologic time as long as local environments remain arid.

17.5 THE GENESIS OF SANDSTONE FROM QUARTZ: THE ROLE OF STRUCTURE IN MINERAL DISSOLUTION

The weathering pathway leading to sandstone formation is illustrated in Figure 17.5 (Plate 22). As discussed earlier, the primary constituent of sand is quartz, which is relatively inert to dissolution or weathering. Grains of quartz and other insoluble minerals represent the fundamental building blocks of sandstone. However, as anyone who has built sand castles on the beach can attest, piles of sand alone have very little structural integrity. The formation of stable sandstone formations requires higher pressures and temperatures to compact the grains and/or cements that bind the sand grains together. In sandstone, these cements are inorganic materials deposited from supersaturated solutions. The most common cement is SiO_2, followed by $CaCO_3$. Cement deposition occurs in regions of high negative curvature at contact points between sand grains. The cements then grow as crystals via heterogeneous nucleation and growth processes (see Chapter 7) in the pores. This deposition process is illustrated for the case of silica in the transmission electron micrograph shown in Figure 16.15 where silica is precipitating in a crack tip

Although the mechanism for forming the silica deposits that consolidate sandstone is clear, the source of the soluble $[Si(OH)_4]^{\circ}$ is not immediately apparent. The primary source of soluble silica is rarely the host sand particles because of the extremely slow dissolution rate of quartz (<0.1 nm/year in ambient-temperature water). Instead, most of the $[Si(OH)_4]^{\circ}$ comes from the weathering of other minerals present in the parent igneous rocks or highly reactive volcanic glasses. It has been known for almost a century that the susceptibility of a silicate mineral to aqueous weathering scales inversely with the temperature at which it was formed [10]. One reason for this observation is that silicate minerals formed at

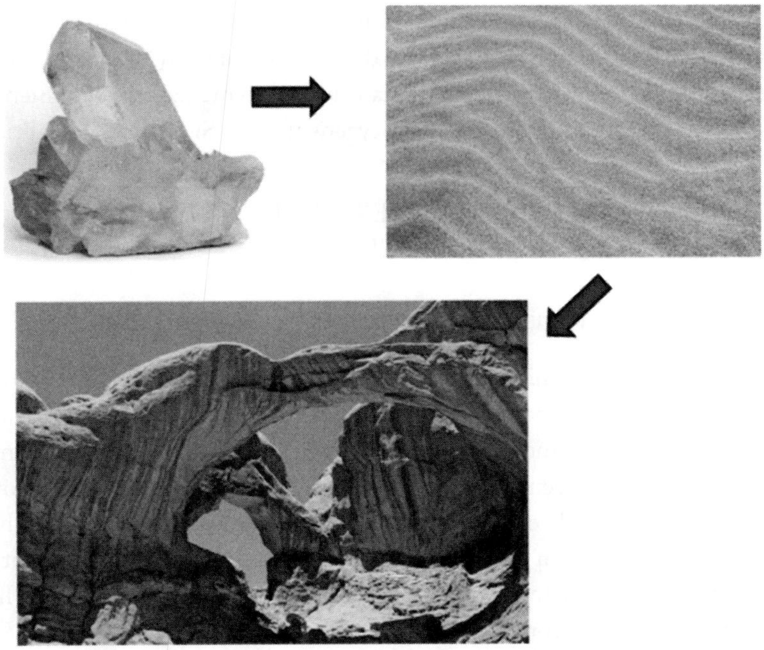

Figure 17.5 The fate of the resistate mineral quartz (top left) (SOURCE: MarcelClemens/Shutterstock.com). First, sand grains are released from granite, where they accumulate to form unconsolidated sand (top right) (SOURCE: Fekete Tiber/Shutterstock.com). These sand grains are later cemented together to form sandstone, as illustrated by dramatic formations found in Arches National Park in Utah (SOURCE: Tupungato/Shutterstock.com). (see Plate 22)

high temperatures tend to contain a higher density of nonbridging oxygen sites than those formed at low temperatures. The local polymeric structures of the silicate minerals largely determine their dissolution rates and the extent to which they add $[Si(OH)_4]^o$ to contacting solution environments, with amorphous volcanic glasses being a particularly reactive source.

The rate at which silicate sites are hydrolyzed and dissolve is related directly to the number of nonbridging oxygens associated with that site (see Part Five). Orthosilicate minerals such as olivine (Mg_2SiO_4, melting point = 1890°C) contain monomeric Q^0 units (i.e., Si(IV) is surrounded by four nonbridging oxygens) and tend to dissolve most rapidly. Olivine dissolution involves the release of both Mg^{2+} and $[Si(OH)_4]^o$ into solution and is often accompanied by the formation of the clay mineral serpentine ($Mg_6Si_4O_{10}(OH)_8$). The next most reactive are the sorosilicates, which contain Q^1 dimers. Isolated rings and the linear chains found in pyroxenes such as $MgSiO_3$ (containing Q^2 groups) are stable at lower temperatures (decomposing to form olivine at 1560°C) and dissolve at even slower rates. Micas and clays, which contain planar sheets of Q^3 units, are even more resistant to aqueous attack. The slowest dissolution rates are exhibited by relatively low-melt tectoaluminosilicates such as the feldspar albite ($NaAlSi_3O_8$, melting point = 1100°C) and quartz that contain fully polymerized three-dimensional frameworks (i.e., Q^4 groups in which all oxygens are in $>Si-\mu_2O-Si<$ bridges).

17.6 THE GENESIS OF CLAY MINERALS FROM FELDSPARS: HYDROLYTIC TRANSFORMATIONS

17.6.1 Structural Transformations Via *in situ* Hydrolysis

Feldspars and other aluminosilicate minerals are hydrolyzed gradually by aqueous solutions to form clay minerals (Fig. 17.6 and Plate 23). However, clay formation involves much more than simple hydrolysis reactions and can involve templating by biomolecules, impinging pore structures and even topotactic (meaning solid-to-solid) conversion. The complexities of hydrolytic weathering are clearly apparent when one compares the structures of a feldspar parent (e.g., albite) and its hydrosylate product (e.g., kaolinite) (Fig. 17.7). Feldspars have a three-dimensional network in which all network-forming cations (Si(IV) and Al(III)) are in tetrahedral coordination. The Al(III) and Si(IV) sites are intimately mixed on an atomic scale, and all are connected to each other via $Si-\mu_2O-Si$ and $Si-\mu_2O-Al$ linkages. In contrast, clay structures contain discrete octahedral and tetrahedral sheets. Here, most of the Al(III) is in octahedral coordination and is segregated into $Al(OH)_3$-like sheets. Most of the Si(IV) is segregated within adjacent tetrahedral sheets. The octahedral and tetrahedral sheets are linked together by neutral μ_3-oxo between two Al(III) octahedra and a Si(IV) tetrahedron. Some clay sheets are terminated with regular structural hydroxyl groups, and large quantities of water and ions often occupy the spaces between the sheets.

Adding to the complexity of the feldspar-to-clay transformation is the fact that clays do not form a single structure, but a bewildering array of layer stacking and compositional motifs (Fig. 17.8) (see Chapter 2) that vary with temperature, pressure, activity of water, and solution composition. Sheet compositions and stacking sequences depend on the composition of the parent rock as well as on local environmental conditions. For example, kaolinite, which consists of 1:1 stacks of adjacent octahedral and tetrahedral sheets, is a common product of feldspar weathering in acidic soils and in tropical climates where weathering is particularly aggressive. However, clays such as pyrophyllite, with 2:1 stacks of octahedral sheets sandwiched between two tetrahedral sheets, can form in silica-rich environments.

The abundance of water is key to the pathways for weathering of minerals. There is a strong thermodynamic driving force for converting most feldspars

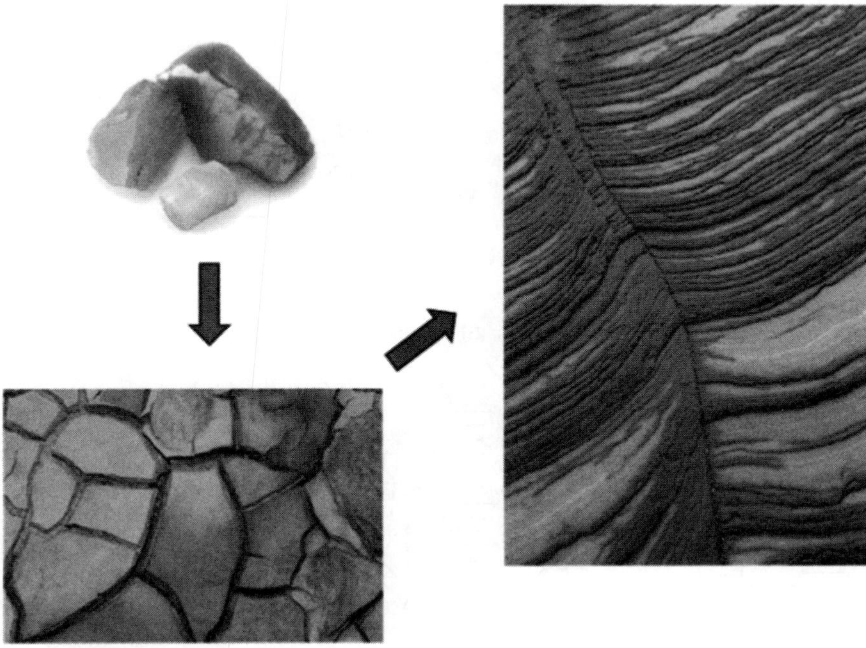

Figure 17.6 A major weathering pathway involving the hydrolysis and chemical alteration of minerals is illustrated by the conversion of feldspar grains (top left) (SOURCE: J. Palys/Shutterstock.com) into the clay minerals that form mud (bottom left) (SOURCE: Alistair Scott/Shutterstock.com). Muds and other hydrosylate sediments are later compressed and consolidated into stable rock formations such as coastal shale platforms shown on the right (SOURCE: Martyn F. Chillmaid/Science Source.com with permission). Oil trapped by impermeable shale layers can be released using techniques such as fracking. (see Plate 23).

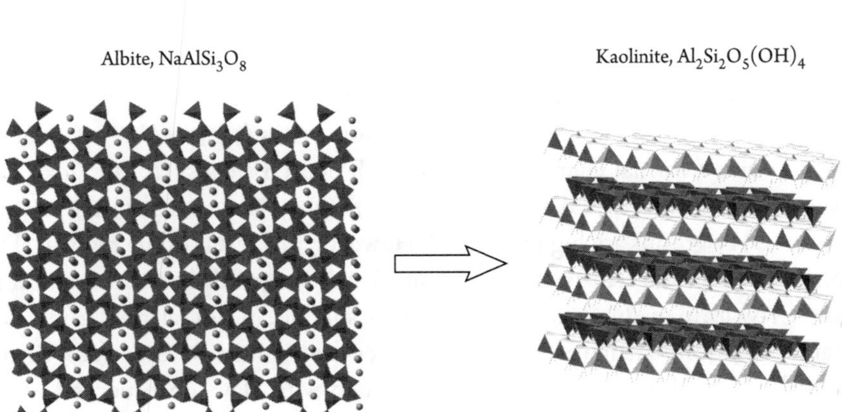

Figure 17.7 Depictions of the structures of the feldspar albite (left) and the clay kaolinite (right) showing the dramatic structural changes that accompany weathering via network hydrolysis.

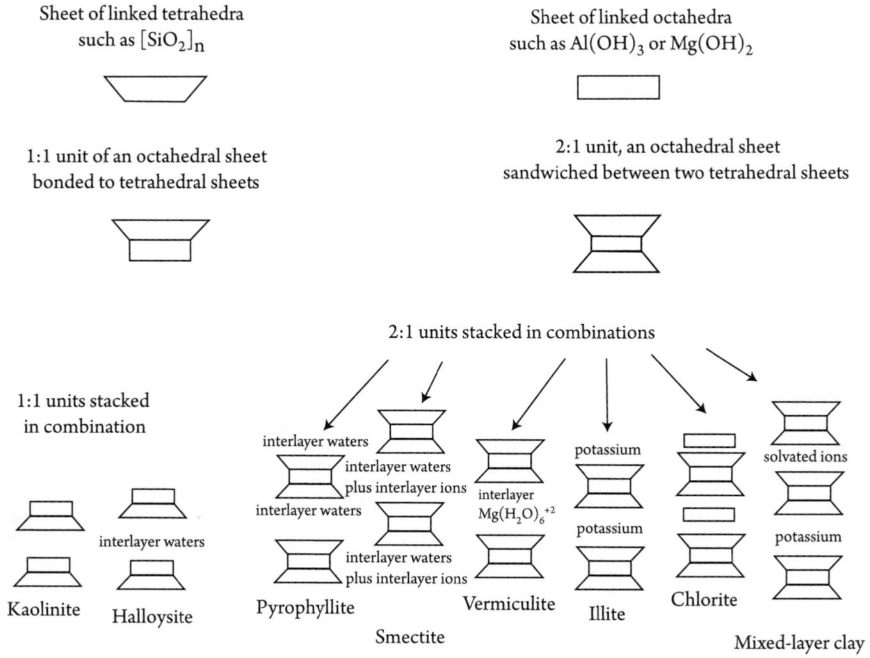

Figure 17.8 Depictions of the stacking sequences for octahedral (rectangles) and tetrahedral (trapezoids) sheets in several common families of clay minerals. Most commonly, the octahedral sheets are dominated by Al(III) or Mg(II), giving rise to gibbsite-like or brucite-like sheet structures, whereas the tetrahedral sheets are made up of linked rings of silicate tetrahedra (see Chapter 2).

into clays via aqueous reactions, but the weathering of many rocks takes place initially in nanopores where water is not abundant. There is much more solid material present than water in most of the early stages of weathering, which is why the textural evidence indicates restructuring *in situ*. From a formal charge perspective, transformations that replace anionic sites (e.g., $>Al-\mu_2O-Si<$ bridges with a formal charge of -0.25) with neutral sites such as μ_3-oxo between the two Al(III) and a Si(IV), $>Al-\mu_2OH-Al<$ bridges, and silanol ($>Si-OH$) groups are highly favored. However, as a result of the high level of connectivity in the feldspars (i.e., Q^4 sites), the rate of this transformation is slow. Clays form readily and directly from the soluble species released from feldspar dissolution, particularly if temperatures are slightly elevated (see Chapter 14). Pore solutions are invariably saturated with clay constituents and, in water-poor settings, there is textural evidence for partial structural breakdown and in situ rearrangements within the solid feldspar structure. Similar transformations are sometimes used in the aqueous synthesis of ceramic materials (see Chapter 7).

As if things aren't complicated enough, the formation of the initial clay mineral does not represent the end of the weathering process. Clays continually adapt their sheet compositions, interlayer counterion compositions, stacking sequences, and water contents to try to come into equilibrium with their local environments. These reactions are often even more sluggish than feldspar-weathering reactions. The sluggish kinetics means the weathering and transformation of clay minerals probably never really stops. Even after the clays are compressed and converted into shale, their compositions and structures are slowly mutable as they equilibrate with the evolving geologic conditions and fluid compositions.

17.6.2 The Role of pH in Hydrolysis Mechanisms

Although it can be difficult to predict the exact composition, layer structure, and extended stacking motifs that will form from specific feldspars or clays in specific environments, it is possible to understand the hydrolysis mechanisms that give rise to such structures (see Chapter 15 for more in-depth discussions). The silicates and aluminosilicates in igneous rocks and clays are amphoteric and susceptible to attack either by protons or hydroxide ions, which is why the rates increase at both low and high pH, with a minimum in near-neutral pH conditions. In basic solutions, bridging $>Si-\mu_2O-Si<$ and $>Si-\mu_2O-Al<$ bonds are disrupted via the nucleophilic attack of hydroxide ions. For example, amorphous silica dissolves 1000 times faster at pH 10 than it does at pH 4. Acidic solutions can promote the selective leaching of modifier (e.g., alkali-metal or alkaline-earth) cations by replacing them with protons on anionic oxygen sites. Such protonation also activates the hydrolysis of bridging-oxygen bonds such as $>Si-\mu_2O-Al<$ sites. The pH at which oxygen sites become protonated (i.e., the pK_a of the oxygen sites) depends on their formal-charge distributions.

Silicate nonbridging oxygens, with a formal charge of -1, are fully protonated at a pH of around 9 or less. For orthosilicate minerals such as Mg_2SiO_4, this protonation results in rapid dissolution and the release of Mg^{2+} and $[Si(OH)_4]^\circ$ into solution. For most other silicates, this protonation creates silanol groups and results in the selective leaching of modifier cations. The silanol groups left behind in the weathered silicate network then tend to condense to form the $>Si-\mu_2O-Si<$ bonds found in amorphous silica. The silicate complexes released by dissolution can subsequently readsorb back onto the mineral surface and aid in the polymerization.

The $>Si-\mu_2O-Al<$ bridging bonds found in tectosilicates such as the feldspars are stronger acids and have a pK_a of around 4 to 5. At pH values less than 4 or 5, the bridging bonds are protonated. Bridge protonation results in the selective leaching of modifier cations, destabilizes the $>Si-\mu_2O-Al<$ bridges, and converts tetrahedral Al(III) into higher coordination numbers at the surface.

Such destabilization can stimulate bond rupture via hydrolysis, leading to structural rearrangements, ligand addition to increase the coordination number of Al(III), and/or the selective leaching of silicate and/or aluminate species.

The protons required to lower the pH and accelerate weathering are provided by other species found in the environment. The most common acid is carbonic acid (H_2CO_3) that forms as a result of the hydrolysis of atmospheric CO_2 gas. The role of carbon dioxide in weathering is exemplified by reactions that decompose a common feldspar (orthoclase, $KAlSi_3O_8(s)$) and a more complex calcium-rich plagioclase feldspar into kaolinitic clay:

$$\begin{aligned}&KAlSi_3O_8(s) + CO_2(g) + 11/2H_2O(l) \rightleftarrows 1/2Al_2Si_2O_5(OH)_4(s)\\&+K^+(aq) + HCO_3^-(aq) + 2H_4SiO_4{}^\circ(aq)\end{aligned} \quad (17.1)$$

$$\begin{aligned}&4Na_{0.5}Ca_{0.5}Al_{1.5}Si_{2.5}O_8(s) + 6CO_2(g) + 5H_2O(l) \rightleftarrows 3Al_2Si_2O_5(OH)_4(s)\\&+2Na^+(aq) + 2Ca^{2+}(aq) + 4H_4SiO_4^\circ(aq) + 6HCO_3^-(aq)\end{aligned} \quad (17.2)$$

These reactions result in the release of both silicic acid and modifier cations (such as K^+, Na^+, and Ca^{2+}) into the environment, and the transformation of feldspars into clay minerals (Fig. 17.5). Other inorganic acids, such as the sulfuric acid released in volcanic eruptions or by oxidation of reduced-sulfur minerals, participate in similar weathering reactions. Organic acids and chelating agents produced by living organisms such as lichens and fungi also accelerate weathering and can influence the types of minerals formed. In addition, organisms affect weathering rates by releasing protons directly to reduce pH or by making specific ligands that attack the oxide structure or ligate the released metals (see Chapter 18). In some cases, the biota cause redox transformations of transition metals in the minerals directly or "template" the formation of new minerals via the distribution of coordinating functional groups on biomolecules, much like the coordination of Si(IV) by sugars (see Chapter 14).

The hydrolysis reactions that occur in basic environments can be illustrated by some of the ligand exchange processes that transform one clay mineral into another. Consider a clay mineral with the idealized composition of ($Al_2Si_4O_{10}(OH)_2$). The tetrahedral sheets in this clay, which commonly forms via the weathering of volcanic ash under fairly neutral conditions, are ideally neutral. However, the sheets are often doped with tetrahedral Al(III), which imparts a slight negative charge to the sheets. This charge is compensated by the structural incorporation of hydrated cations such as Na^+ and Ca^{2+}. The resulting clay can swell readily in water and is highly active in terms of ion exchange and other reactions involving the intercalation of species between the layers. Thus, in nature, the real stoichiometry of our idealized $Al_2Si_4O_{10}(OH)_2$ can approach that of the mineral montmorillonite $((Na,Ca)_{0.33}(Al,Mg)_2(Si_4O_{10})(OH)_2 \cdot nH_2O)$, which is a common smectite clay.

In saline, slightly basic environments, such as in the sediment pores of packed marine sediment, smectites can be converted into illites. Ligand-exchange kinetics among ions in solutions are much faster than solid-state transformations in this environment. The rates of solid-state transformation are slow because there is only a small thermodynamic driving force for recrystallizing one clay into another. Even when assisted by network hydrolysis, the diffusion and reorganization rates for network-forming cations can be exceedingly slow. Nevertheless, burial of sediment presses them down the geothermal gradient to higher temperatures and pressures so that clay recrystallization is evident over geologic time. This extra energy makes it easier to incorporate tetrahedral Al(III) within the Si(IV)-rich tetrahedral sheet, increasing the layer charge. As this occurs, it becomes increasing difficult to keep the interlayer ions hydrated. The common cation that is easiest to dehydrate is K^+ and when ~25% of the Si(IV) in the tetrahedral sheets is replaced by Al(III), the clay collapses to form nonswelling illite clay with the idealized composition $KAl_2(Si_3Al)O_{10}(OH)_2$. The reaction is detectable in the mineralogy of deep sediment in basins.

17.7 THE GENESIS OF THE IRON AND MANGANESE OXIDES: REDOX CYCLING IN WEATHERING

The 3rd and 10th most abundant elements in rock, iron and manganese, have access to multiple oxidation states within the stability window of water, and they alternate among these oxidation states in natural environments. The +2 and +3 oxidation states are most important for iron, whereas accessible oxidation states for manganese include +2, +3, and +4 [11, 12]. This rich redox chemistry means that environments capable of either reducing or oxidizing these elements can have a dramatic impact on weathering and precipitation processes, as well as important environmental reactions involving the solid oxide phases (see Chapter 18) (Fig. 17.9 and Plate 24).

The unique attributes of iron and manganese in environmental chemistry can be understood by examining their Pourbaix diagrams (Fig. 17.10, see Chapter 11). In terms of weathering and dissolution, the key feature to note is that the +2 oxidation states of both iron and manganese generate highly soluble hydrolysis products. This means that reducing environments promote the selective leaching and dissolution of both elements from oxide hosts. For example, for at least half of Earth's history, the atmosphere contained no free O_2, leading to such high concentrations of ferrous iron that Archaen oceans may have been green.

Figure 17.9 Soluble species released into all bodies of water (top left) (SOURCE: Zacarias Pereira da Mata/Shutterstock.com) by oxide weathering are involved in three major weathering pathways, including (1) precipitation, as illustrated by a fossil snail imbedded in a calcite formation (bottom left) (SOURCE: Doart/Shutterstock.com); (2) reversible reduction and oxidation cycles to form oxidates, as illustrated by the black manganese oxide deposits left on the surface of a rhodonite (manganese inosilicate) crystal (SOURCE: Sombra/Shutterstock.com); and (3) evaporation to form evaporates, as illustrated by borax-rich soil found in the Mojave Desert in California (SOURCE: BOB GIBBONS/Science Source.com with permission). (see Plate 24)

During this era, the primary minerals precipitated from aqueous environments were not oxides, but other insoluble phases including $Fe(OH)_2$ (amakinite), $FeCO_3$ (siderite), and sulfide minerals such as FeS (mackinawite or troilite). These minerals continue to be generated in reducing soil environments even today (see Chapter 18).

About 2.4 billion years ago, the planet began to accumulate O_2 in the atmosphere, which is an extraordinarily strong oxidant. Aqueous solutions containing Fe(II) or Mn(II) are oxidized when exposed to atmospheric oxygen. Because the higher oxidation states of both elements form insoluble hydrolysis products (Fig. 17.10), the result is the precipitation of colloidal hydroxide and oxide phases. These oxidates include phases such as goethite (FeOOH), hematite (Fe_2O_3), magnetite (Fe_3O_4), manganite (MnOOH), and pyrolusite (MnO_2). The production of atmospheric O_2 by primitive oceanic plant life eventually precipitated the massive iron-oxide ore deposits that we mine today in places such as

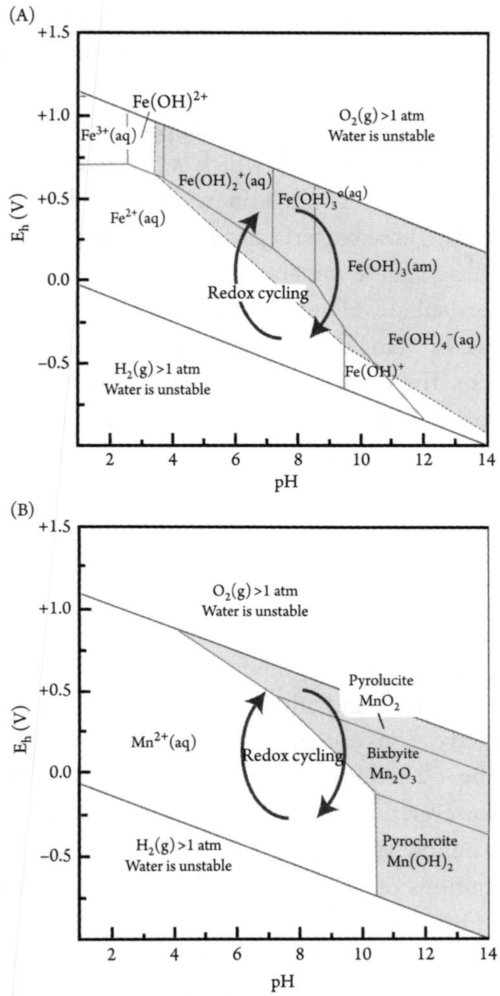

Figure 17.10 (A, B) Modified Pourbaix diagrams showing the major soluble species and phases present for iron and manganese as a function of both pH and E_h for the case when the total metal concentration is 10^{-6} M. Gray fields indicate solutions supersaturated with the indicated mineral. The diagrams are simplified, but show that redox potentials cycle at near-neutral pH in soil. The cycling causes minerals to alternately reductively dissolve and oxidatively precipitate. Adapted from Pourbaix, M. *Atlas of Electrochemical Equilibria*; National Association of Corrosion Engineers: Houston, TX, 1974, with permission.

the Mesabi Iron Range in the United States, where ores of Precambrian iron oxide are shipped as a mixture of minerals called *taconite*.

Colloidal particles of iron and manganese oxides have high surface areas and are small enough to be transported readily through all of Earth's major environments. These oxides participate in many key environmental reactions, including ion exchange (Chapter 10), electrochemistry (Chapter 11), and photochemistry (Chapter 13). In many cases, the solubility of the iron and manganese oxides can be switched back and forth based on changes in the oxidizing power and pH of local conditions. This switching leads to a cycling of the elements between aqueous and solid forms (see Chapter 18), which is characteristic of shallow soils in places such as salt marshes. Here, the iron oxides are reduced to ferrous iron by heterotrophic bacteria. These bacteria also reduce sulfate in seawater ($[SO_4]^{2-}$) to sulfide ($[S]^{2-}$). Iron sulfide minerals like mackinawite precipitate when ferrous iron encounters sulfide. However, if the amount of iron is larger than the available sulfide, the Fe^{2+}(aq) diffuses upward until it encounters dissolved O_2 from the atmosphere. The O_2 reoxidizes the Fe(II) to Fe(III) and precipitates a layer of ferric hydroxide. The rapid cycling of iron leads to a surface of marsh soil that is often orange or rust colored at the immediate surfaces from the FeOOH(s), but is jet black immediately below, from the FeS(s).

17.8 THE GENESIS OF PRECIPITATES: CARBONATE CHEMISTRY AND LIMESTONE FORMATION

Calcium and magnesium represent the fourth and seventh most abundant elements in igneous rock. These and other elements present in oxides as +2 cations form soluble hydrated cations except under highly basic conditions that are rarely found in Nature. The +2 cations are released into aqueous environments via selective leaching. However, the concentrations of the +2 cations in both salt- and freshwater tend to be low (see Chapter 18). A partial explanation for the relatively low concentrations of +2 cations involves the precipitation of sparingly soluble ionic salts from natural solutions. Some of the least soluble salts of +2 cations involve oxyanions leached from rock. For example, barite ($BaSO_4$), the most common barium mineral, has a solubility in deionized water of only 10^{-5} M. The most common phosphate mineral apatite ($Ca_5(PO_4)_3(F,Cl,OH)$) is virtually insoluble—as is its Cu(II)-containing mineral cousin, the beautiful gemstone turquoise. However, because 30 times more Ca^{2+} than $[PO_4]^{3-}$ is present in most igneous rocks, phosphate precipitation removes only a small fraction of the total Ca^{2+} inventory released as a result of weathering.

On a practical note, the effectiveness of aqueous phosphate fertilizers added to soil can be limited by *in situ* precipitation. In many iron-rich soils, the phosphorus added as fertilizer adsorbs to the surface of ferric oxide and ferric hydroxide minerals, rendering them unavailable to plants. The precipitation of insoluble phosphate minerals is an important strategy for mitigating lead contamination. Lead phosphate minerals are extraordinarily insoluble, so a field that is contaminated with old lead paint can be covered with phosphate fertilizer to suppress the dissolved Pb^{2+} concentrations.

Most of the calcium and much of the magnesium on Earth's surface is present as carbonate minerals found in geologic strata. Common minerals include calcite ($CaCO_3(s)$), magnesite ($MgCO_3(s)$), and dolomite ($CaMg(CO_3)_2(s)$). All these phases are precipitated from supersaturated carbonate solutions via heterogeneous nucleation and growth processes (Chapter 7). These precipitation reactions are facilitated by the many organisms capable of producing shells or corals via biomineralization processes (Chapter 9). For example, the famous White Cliffs of Dover were produced during the Cretaceous period by both phytoplankton (coccolithophores) and protozoans (foraminifera). Massive coral reef deposits were created in tropical seas during the earlier Carboniferous and Permian periods. Over geologic time, the volume of limestone that has been precipitated is enormous, amounting to 4% of Earth's entire crust, or $2 \cdot 10^{17}$ MT [9]. This mass is 1000 times greater than the mass of carbonate dissolved in the oceans, and 100,000 times greater than the total mass of CO_2 in the atmosphere. The carbonate solids stored in the crust represent by far the most enormous reservoir of CO_2 in Earth's accessible environments.

Because limestone is so pervasive, it is worth examining the solubility of its major component, $CaCO_{3,}$ in more detail. Calcite solubility is determined by chemical equilibria involving this mineral, the atmosphere, and all natural bodies of water, and thus impacts all the major environments on Earth. Those of you looking for in-depth discussions about the aqueous chemistry of CO_2 and the carbonates should refer to [13–16]. A simplified discussion of carbonate equilibria is presented next.

The soluble Ca^{2+} required for calcite formation is derived from the weathering of rocks. Carbonate-ion concentrations are determined by atmospheric CO_2 levels and the solution pH. The partitioning of CO_2 between the atmosphere and aqueous solutions is given by Henry's Law:

$$CO_2(g) + H_2O(l) \rightleftarrows CO_2(aq) \qquad K_{Henry} = 0.033\ M/atm \qquad \textbf{(17.3)}$$

The dissolved CO_2, which is a linear molecule, is then hydrolyzed into carbonic acid, which is trigonal. Thus, the following reaction

$$CO_2(aq) + H_2O(l) \rightleftarrows H_2CO_3(aq) \qquad K_{hyd} = 1.6 \cdot 10^{-3}/M \qquad \textbf{(17.4)}$$

requires a change in shape of the molecule. Carbonic acid is a diprotic acid. Loss of the first proton involves the following acid–base equilibrium constant, which identifies H_2CO_3 as a moderately strong acid:

$$H_2CO_3(aq) \rightleftarrows H^+(aq) + HCO_3^-(aq) \qquad K_{a1} = 3.2 \cdot 10^{-4} M \qquad \textbf{(17.5)}$$

However, for simplicity, most treatments of carbonate equilibria in aqueous solutions, and particularly treatments of natural water chemistry, combine K_{hyd} and K_{a1} to provide an apparent acid–base equilibrium constant for the first dissociation of dissolved carbonate. This fictitious concentration, $H_2CO_3^*$, is dominated by $CO_2(aq)$, but is actually the sum of both $CO_2(aq)$ and true carbonic acid: H_2CO_3.

$$H_2CO_3^* = CO_2(aq) + H_2CO_3(aq) \qquad K_{a1T} = (K_{hyd})(K_{a1}) = 10^{-6.3} = 5 \cdot 10^{-7}\,M \tag{17.6}$$

Within this formalism, dissolved carbonate is treated as a weak acid, whereas true H_2CO_3 is actually a moderately strong acid that is present in small concentrations (~0.16 percent of the total). Carbonate equilibrium largely controls the pH of natural waters. Although this formalism simplifies equilibrium calculations, it should not be used to predict the kinetics or mechanisms of reactions involving either CO_2 or H_2CO_3 because the reaction rates depend on the shapes of the actual molecules.

The bicarbonate ion can lose a second proton, as described by

$$HCO_3^- \rightleftarrows H^+(aq) + CO_3^{2-}(aq) \qquad K_{a2} = 4.6 \cdot 10^{-11} = 10^{-10.3} \tag{17.7}$$

Finally, dissolved Ca^{2+} and $CO_3{}^{2-}$ are in equilibrium with $CaCO_3$, as described by the solubility product

$$CaCO_3(s) \rightleftarrows Ca^{2+}(aq) + CO_3^{2-}(aq) \qquad K_{sp} = [Ca^{2+}][CO_3^{2-}] = 2 \cdot 10^{-8}/M^2 \tag{17.8}$$

Similar expressions describe the interactions of carbonates with other dissolved cations, which can further influence the concentrations of all these species.

The chemical equilibria just presented has several important consequences. First, if the solution is in contact with the atmosphere and can exchange gases freely, acid–base equilibria dictate that the total concentration of carbonate species in water varies with pH. At high pH values where HCO_3^- and $CO_3{}^{2-}$ form, H_2CO_3 is consumed, allowing more CO_2 to dissolve into solution. High-pH environments such as Mono Lake in California have such large concentrations of dissolved CO_2 that the lake water fizzes when a drop of acid is added. Second, because the equilibrium solution concentration of $CO_3{}^{2-}$ increases with pH, the actual solubility of $CaCO_3$ also depends sensitively on pH and decreases sharply with increasing pH, reflecting the total dissolved $CO_3{}^{2-}$ concentration. This pH dependence is shown assuming an atmospheric CO_2 concentration of 380 ppm and an ionic strength of 1 M in Figure 17.11. Increased CO_2 concentrations cause pH to decrease, not rise.

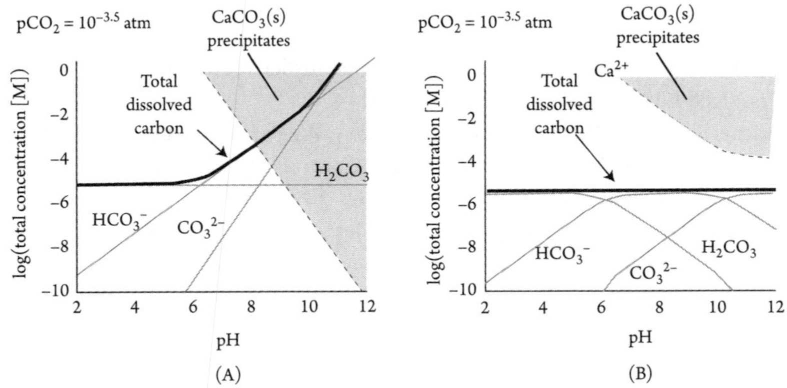

Figure 17.11 Speciation diagrams showing the concentrations of Ca^{2+} and dissolved carbonate species in equilibrium with $CaCO_3$ (calcite) as a function of solution pH. (A) This diagram shows what happens when the solution has unlimited access to atmospheric CO_2 gas. Under this condition, more CO_2 can dissolve as the pH is increased and H_2CO_3 is ionized to form HCO_3^- and CO_3^{2-} ions, as indicated by the solid line for total dissolved carbon. (B) This diagram shows what happens in a closed system in which no communication with air is in evidence, and the total dissolved carbon is constant regardless of pH. In both cases, CO_3^{2-} increases with pH, resulting in the precipitation of $CaCO_3$ at lower calcium concentrations. However, in the "open" system (A), which is representative of ocean surface waters, precipitation of calcium carbonates is sensitive to atmospheric CO_2 levels.

So much CO_2 can dissolve into water that there have been several asphyxiation events of populations near volcanic lakes near areas of crustal rifting. The CO_2 degasses from the lower crust into a lake where it dissolves into the stratified water. When the water is saturated, any small disturbance can cause it to rise and degas. Thus, the degassing of volcanic lakes above areas of crustal rifting, such as Lake Nyos in Cameroon, is catastrophic. The water becomes more buoyant as it forms CO_2 bubbles, leading to even more rapid ascension and even more catastrophic degassing. Villagers near Lake Nyos were asphyxiated in their sleep because the CO_2 displaced the air they needed to breathe.

Both dissolved CO_2 gas and $CaCO_3$ can play major roles in determining the pH of aqueous environments. In pure freshwater, the dissolution of atmospheric CO_2 to form carbonic acid lowers the solution pH to around 5.7. This carbonic acid can accelerate weathering processes (Eqs. 17.1 and 17.2). Conversely, for aqueous environments exposed to significant amounts of $CaCO_3$, dissolution releases carbonate ions into solution. This carbonate ion consumes protons to form bicarbonate, $[HCO_3]^-$, resulting in an increase in solution pH. Proton consumption by dissolved carbonates is one reason why seawater has a higher pH (pH 8.1) than freshwater, and is also why soils rich in limestone have a moderately basic pH (Chapter 18).

References

1. Meybeck, M. Global Chemical Weathering of Surficial Rocks Estimated from River Dissolved Loads. *Am. J. Sci.* **1987**, *287*, 401–428.
2. Horne, R. A., Ed. *Water and Aqueous Solutions*; Wiley-Interscience: New York, 1972.
3. Summerfield, M. A.; Hulton, N. J. Natural Controls of Fluvial Denudation Rates in Major World Drainage Basins. *J. Geophys. Res.* **1994**, *99*, 13874–13883.
4. McLennan, S. M. Weathering and Global Denudation. *J. Geol.* **1993**, *101*, 295–303.
5. Press, F.; Siever, R. *Earth, 2nd ed.*; W. H. Freeman: San Francisco, CA, 1978.
6. Self, S.; Rampino, M. R.; Newton, M. S.; Wolff, J. A. Volcanological Study of the Great Tambora Eruption of 1815. *Geology* **1984**, *12*, 659–663.
7. Baes, C. F.; Mesmer, R. E. *The Hydrolysis of Cations*; Krieger Publishing: Malabar, FL, 1986.
8. Swedish Royal Institute of Technology. http://vminteq.lwr.kth.se// (accessed 10.2.2015) (one useful program among many is Visual Minteq). US Geological Survey's Division of Water Resources. http://water.usgs.gov/software/lists/geochemical (accessed 10.2.2015) (contains other useful programs). The Geochemist's Workbench Home Page. http://www.gwb.com/ (accessed 10.2.2015) (commercial programs available).
9. Berner, E. K.; Berner, R. A. *Global Environment: Water, Air, and Geochemical Cycles*; Prentice Hall: New York, 1996.
10. Goldich, S. S. A Study in Rock Weathering. *J. Geol.* **1938**, *46*, 17–58.
11. Garrels, R. M.; Christ, C. L. *Solutions, Minerals and Equilibria*; Freeman, Cooper: San Francisco, CA, 1965.
12. Schweitzer, G. K.; Pesterfield, L. L. *The Aqueous Chemistry of the Elements*; Oxford University Press: New York, 2010.
13. Palmer, D. A.; Van Eldick, R. The Chemistry of Metal Carbonato and Carbon Dioxide Complexes. *Chem. Rev.* **1983**, *83*, 651–731.
14. Millero, F. J. *The Physical Chemistry of Natural Waters*; Wiley-Interscience Series in Geochemistry; Wiley-Interscience: New York, 2001.
15. Zeebe, R. E.; Wolf-Gladrow, D. *CO_2 in Seawater: Equilibrium, Kinetics, Isotopes, Oceanography*; Elsevier: Amsterdam, 2001.
16. Berner, R.A. Chemical Weathering and Its Effect on Atmospheric CO_2 and Climate. *Rev. Mineral.* **1995**, *31*, 565–583.

18

The Impact of Oxides on Environmental Chemistry

18.1 INTRODUCTION

The ancient Greek philosopher Empedocles defined our environments using the four basic elements of *fire, earth, wind,* and *water*. Although we now know there are at least 118 elements, of which 98 are naturally occurring [1], these ancient descriptions aptly describe the habitats on Earth that are occupied by oxides and living things. Many oxides that comprise Earth's surface are born by the *fire* represented by the massive heat of Earth's interior as mediated by plate tectonics. This heat produces the igneous rocks found in volcanoes and our major mountain chains. *Water* weathers these pristine rocks, which are gradually broken down to form *earth,* which includes the wide diversity of other rock types, soils, and sediments covering the surfaces of our continents and ocean floors. Weathered oxides in the form of dust are blown by *wind* and enter the atmosphere, where they influence the chemistry of the air we breathe and the rainfall that supports continental life. The chemical transformations of oxides are strongly influenced by all the environmental conditions they encounter in their life cycle (see Chapter 17). Conversely, the interactions between oxides, water, and organisms help define many of the environments that allow life on Earth to thrive. These interactions form the basis for this final chapter of our book.

Oxides are present in all our planet's major environments. In this chapter, we explore each of the environments defined by the ancient Greeks in descending order based on their distance from Earth's core. The chapter progresses from the stratosphere (air) to continental surfaces (earth) to our oceans (water) and finally to the subsurface environments of subduction zones such as the Marianas Trench (fire). In each section, we highlight reactions involving the two most important classes of oxides in terms of their environmental impact, both of which are weathering products: (1) the clay minerals and (2) the redox-active colloids of iron and manganese oxides. Clay mineral reactions impact colloidal interactions (Chapter 8), ion exchange (Chapter 10), and the sequestration of environmental nutrients and contaminants. Reactions of the redox-active oxidates of iron and manganese are dominant in terms of reversible and often complex electrochemical (Chapter 11) and photochemical (Chapter 13) processes that take place in natural environments. This chapter is not intended to be comprehensive, but reviews chemical concepts provided in previous chapters in the context of oxides that exist in the natural world. This material is intended to complement the wide-ranging technological roles of oxides discussed in Part Four. Environmental reactions that result in oxide weathering are discussed primarily in Chapter 17.

18.2 WIND: OXIDE CHEMISTRY IN THE ATMOSPHERE

Earth's atmosphere is a critical component of our environment, providing the O_2 we breathe, the fresh water we drink, the climates and weather we experience, and protection from the most intense solar radiation. The three major components of modern air are nitrogen (78%), oxygen (21%), and variable concentrations of water (24 torr or 3% of the air at $T = 25°C$ and a relative humidity of 100%), which means that two of the most abundant species in air (O_2 and H_2O) are oxygen based. Not surprisingly, many of the reaction products generated by atmospheric reactions are other molecular species containing oxygen. However, the primary focus of this chapter is on reactions involving suspended oxide particles in the form of dust. Dust plays a major role in the nucleation of water droplets in clouds, in the fate of atmospheric pollutants, and in human health. The chemistry of the atmosphere is exceedingly complex and is dominated by photoinitiated free-radical reactions. A free radical is a species with a single unpaired electron. Some of the reactions that produce or consume free radicals involve oxide solids. Those of you interested in a comprehensive review of photoinduced free radical chemistry in the atmosphere should see [2].

18.2.1 Interactions of Atmospheric Dust with Water: Clay Minerals

Dust particles injected into the air by pollution, dust storms, or volcanoes (see Chapter 17) can be large or small, but the suspended particles that persist in the atmosphere are the smallest ones. Although visible dust particles have dimensions exceeding 1 μm and can be quite large, most dust particles are only a few hundred nanometers in diameter. Suspended dust has a high surface area (10–100 m^2/g) and can adsorb many of the molecular species present in the atmosphere. The most important of these species is water. In dust-free air, both water droplets and ice crystals form via homogeneous nucleation and growth processes. Parameters that affect these homogeneous nucleation processes include the surface tension (or surface energy) of water (or ice), the temperature, and the degree of supersaturation of the water in the air (see Chapter 7). At the boiling point of water at sea level, the saturation point corresponds to a water vapor pressure of 1 atm (760 torr). This saturation point is referred to as a relative humidity of 100%. At lower temperatures, saturation occurs at lower vapor pressures [3] (Fig. 18.1). Within the troposphere, which is the layer of air closest to Earth, air temperatures drop linearly with distance from the surface by around 5°C/km [3] (Fig. 18.2). As warm, moist air formed at Earth's surface rises, it cools to the point where the air becomes supersaturated in water vapor at the altitude where cloud formation is observed (Fig. 18.1).

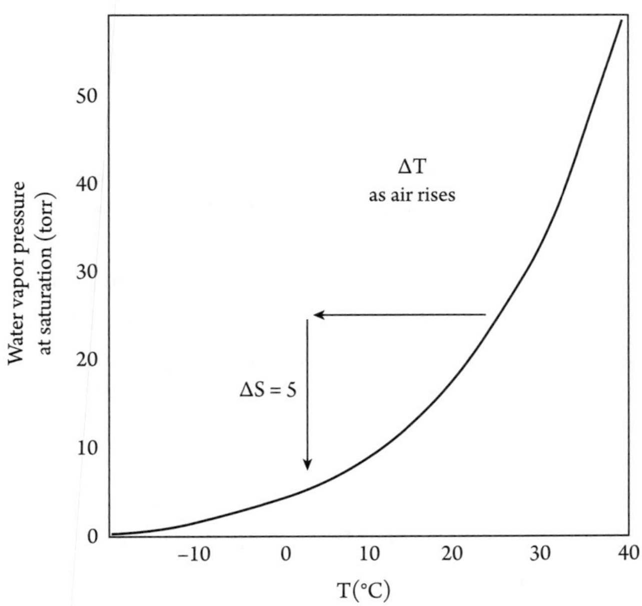

Figure 18.1 The vapor pressure in torr of water vapor is air that is saturated with water as a function of temperature. The arrows indicate that air at 25°C containing 25 torr of water becomes supersaturated as it rises in the air column and cools to 3°C, where a water vapor pressure of only 5 torr corresponds to a relative humidity of 100%, or saturation. This change in conditions leads to air that has a supersaturation value with respect to water of S = 5. Data from Weast, R. C., Ed. *CRC Handbook of Chemistry and Physics, 58th ed.*; CRC Press: Cleveland, OH, 1977.

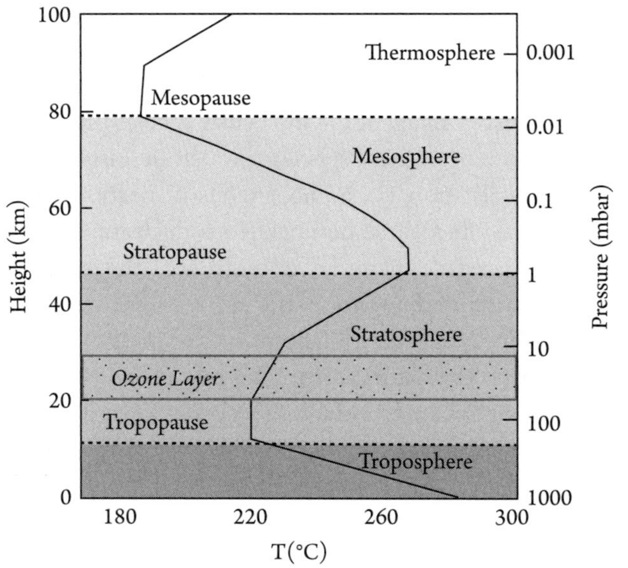

Figure 18.2 An indication of average temperature and pressure conditions encountered in Earth's atmosphere as a function of altitude in kilometers (left y-axis). Pressures in millibars are indicated on the right y-axis, whereas temperatures are indicated on the x-axis in degrees Kelvin.

In the presence of dust, water and ice form via heterogeneous nucleation and growth processes. As discussed in Chapters 6 and 7, oxide surfaces lower the activation barrier to nucleation, thus lowering the energy required to form new water or ice surfaces. Many more water nuclei can be produced on solid surfaces at substantially lower temperatures than in dust-free air. One factor that lowers the interfacial energy for nucleation is the availability of charged or acid–base sites that can accept hydrogen bonds and adsorb water molecules. Oxides and clay particles are covered with such sites. For ice nucleation, a second factor involves the degree of lattice matching between the ice and the particle. As discussed in Chapter 9, the tetrahedral sheets in clay minerals represent an almost perfect lattice match to ice crystals. The close match of lattices means that clay minerals are among the most effective cloud-seeding agents. Cloud-seeding experiments performed using both artificial and Asian dust samples [4] (Fig. 18.3) show that clays and feldspar minerals can stimulate ice nucleation at a relative humidity (or supersaturation level) of only 135% between −45°C and −60°C. Sampling of water droplets and ice crystals in the atmosphere indicates that almost all condensed water phases in air contain a central dust particle. This dust is critical to the formation of rain, snow, sleet, and hail. It also means that water condensation represents a primary mechanism for removing dust from the

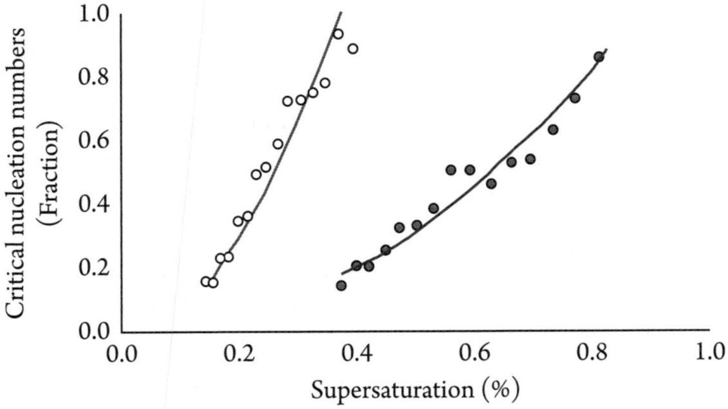

Figure 18.3 The fraction of 200-nm aluminosilicate dust-particle simulants observed to nucleate ice crystals as a function of the degree of supersaturation of water in air for homogeneous nucleation (i.e., ice formation in the total absence of particles). Closed symbols (the curve on the right) represent pristine particles, whereas open symbols (curve on the left) represent particles pretreated with sulfuric acid. Adapted with permission from Archuleta, C. M.; DeMott, P. J.; Kreidenweis, S. M. Ice Nucleation of Surrogates for Atmospheric Mineral Dust and Mineral Dust/Sulfate Particles at Cirrus Temperatures. *Atmos. Chem. Phys.* **2005**, 5, 2617–2634.

atmosphere, because the ice particles or water droplets fall from the sky when they reach a size at which they can no longer remain suspended. Finally, as long as water-encapsulated dust particles are suspended in air, each encapsulated particle represents a microenvironment within which chemistry can occur. Next, we describe a subset of those reactions that involve light from the sun.

18.2.2 Free Radical and Photochemical Reactions Involving Atmospheric Iron Oxides

Chemical reactions in the atmosphere are dominated by free-radical species. Free radicals are created when molecules or particles in the atmosphere are bombarded with energetic photons or electrons (see Chapter 13). These cause bonds to dissociate homolytically to leave one unpaired electron on each atom, in contrast to the more familiar heterolytic dissociation that leaves a cation that is missing electrons and an anion with extra electrons. The dissociation of sodium chloride $(NaCl(s) = Na^+(aq) + Cl^-(aq))$ is an example of a heterolytic reaction. The dissociation of hydrogen gas into atoms $(H_2(g) \rightarrow H\cdot(g) + \cdot H(g))$ is a homolytic reaction.

In the upper atmosphere, cosmic rays and other forms of high-energy radiation from the sun rupture chemical bonds to create free radicals including $\cdot O$, $\cdot N$,

$\cdot$H, and $\cdot$OH. The oxygen atoms react with O_2 to create ozone ($\cdot O_3$) and with nitrogen to form nitrates and NO_x. The ozone layer at an altitude of 25 km shields life on the continents by absorbing harmful ultraviolet (UV) radiation as the ozone dissociates back to $\cdot$O and O_2. Life was only able to emerge onto land in abundance after the ozone layer formed from the energetic dissociation and recombination of O_2. Early Precambrian life was confined to the oceans because UV radiation would have killed any life exposed on Earth's surface.

Studies of gas-phase reactions show that the hydroxyl radical ($\cdot$OH) is a key driver of chemistry throughout the atmosphere [5]. This radical is the most powerful oxidant in natural chemistry, and it reacts with both natural and man-made pollutants. As shown in Figure 18.4, many reactions involving hydroxyl radicals are complex chain reactions that first consume and then regenerate the radical species. Because the concentrations of hydroxyl radicals, oxygen atoms ($\cdot$O), and peroxides are critical for determining the fates of atmospheric pollutants, it is important to understand how such species are produced. Extensive laboratory experiments indicate that primary gas-phase sources for the hydroxyl radical include the reaction between oxygen atoms and water, the photodissociation of HONO, and the reaction of peroxy radicals with NO. These results suggest that hydroxyl radical concentrations should be around $5 \cdot 10^5$ molecule/cm^3 [2]. However, in the troposphere, measured concentrations of $\cdot$OH are closer to $5 \cdot 10^6$ molecule/cm^3, whereas in the stratosphere, concentrations greater than 10^7 molecule/cm^3 have been observed [2]. These results illustrate that atmospheric concentrations of many important radicals are substantially different from those predicted on the basis of models and controlled laboratory experiments. Reactions involving dust particles are responsible for many of these observed discrepancies. Surface reactions on dust particles are now being incorporated into new generations of models describing atmospheric chemistry.

Most dust particles are dense oxides with high specific surface areas. For example, a spherical aluminosilicate particle with a radius of 1 μm contains on the order of $6 \cdot 10^7$ surface sites. In the stratosphere, the concentration of active surface sites on dust particles can exceed the concentration of hydroxyl radicals in the same gas volume. Each site can be activated by high-energy photons.

As discussed in Chapter 13, photochemical activation can create energetic electrons and/or holes in oxide surfaces. At higher photon energies, this activation leads to the photon-stimulated desorption of both neutral and charged species, including oxygen atoms and hydroxyl radicals (Fig. 18.4). The surface sites that are left behind are highly reactive, including free radicals and/or sites with a high degree of electrochemical activity. Dust particles also adsorb many species from the gas phase and can function as heterogeneous catalysts for gas-phase reactions (see Chapter 13). In humid environments, dust particles can adsorb a

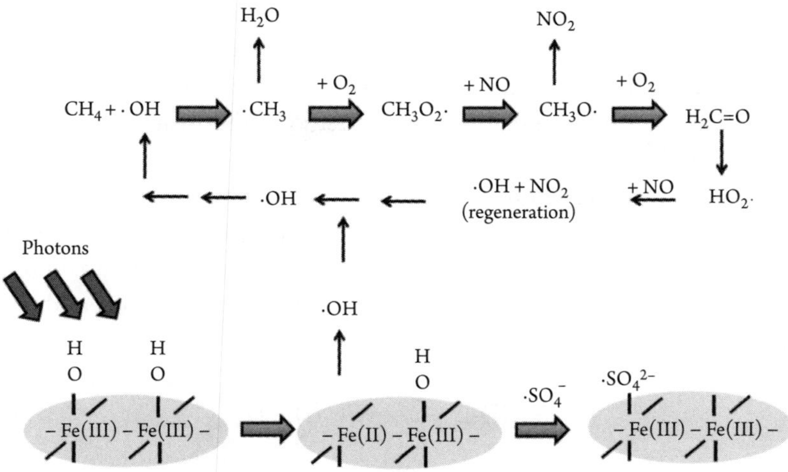

Figure 18.4 Reactions stimulated on Fe(III) centers in dust-particle surfaces by radiation in the upper atmosphere. (Bottom row) High-energy photons impinging on a dust particle (gray oval) participate in bond-rupture reactions. In the example shown, this reaction creates a hydroxyl radical that desorbs into the atmosphere, leaving behind an open coordination position on an iron site that is reduced from Fe(III) to Fe(II). This Fe(II) site can then adsorb and reduce molecular species from the air, resulting in an oxidation of Fe(II) back to Fe(III) (see Eq. 18.8 and accompanying discussion). (Top cycle) The hydroxyl radical ejected from the dust particle is highly reactive and can initiate a wide range of multistep free-radical chain reactions. These reactions often involve organic pollutants and other contaminants in the atmosphere. In the example shown, the primary reaction pathway is a four-step chain reaction that results in the conversion of methane into formaldehyde. (Small arrows point to other side products.) Note that after formaldehyde is formed, free-radical reactions continue that can regenerate the hydroxyl radical and repeat the cycle multiple times.

surface layer of water to create aqueous environments on a micron scale. Photochemical reactions in these microenvironments mirror those that occur in oceans and streams (Section 18.4).

A specific example of a photochemical reaction involving dust is the photo-Fenton reaction for producing free radicals. Fenton reactions can occur for any oxide mineral containing Fe(III), including dusts of both clay minerals and iron oxides. The primary photo-Fenton reaction produces hydroxyl radicals in the presence of light:

$$Fe(OH)^{2+} + h\nu \rightarrow Fe^{2+}(aq) + \cdot OH(aq) \tag{18.1}$$

Here, a photon excites an electron from a surface-bound hydroxyl group or water molecule into an unoccupied orbital of Fe(III), resulting in the reduction of the iron to Fe(II), the oxidation of the bound hydroxide ion (OH^-) to ·OH, and desorption of the oxidized ·OH. This reaction not only creates a reactive hydroxyl radical, but also leaves behind an open coordination site on a reactive Fe(II) center that can undergo further reactions with atmospheric molecules.

The photo-Fenton reaction is not limited to bound hydroxyl groups, but can apply to any molecule with sites of sufficient basicity to form complexes with exposed iron centers on dust particles. The absorbed ligands can be as simple as carboxylic acids or other volatile organic pollutants that can be converted to radicals and ultimately oxidized to form CO_2:

$$\text{Fe(III)-L} + h\nu \rightarrow \text{Fe(II)} + \text{L}\cdot \qquad \textbf{(18.2)}$$

$$\text{L}\cdot + O_2 \rightarrow O_2\cdot + \text{Oxidized ligand} \qquad \textbf{(18.3)}$$

It is important to note that Fenton reactions are not confined to the gas phase, but can occur in aqueous environments ranging from the water droplets in clouds [6] to the photic zone in the oceans, and even to dark environments such as the human lung (discussed next).

18.2.3 Dust Reactions Involving Living Things

The reactions of dust particles encapsulated by water can influence directly the health of both plants and animals. First, consider the case of acid rain and its impact on plant life. One of the primary pollutants in the atmosphere is SO_2, which is emitted along with acidic nitrogen oxides as a result of the burning of coal and other fossil fuels. The SO_2 introduced into the atmosphere by human activities amounts to $1.6\cdot10^8$ MT/year [7]. Sulfur dioxide is also a natural pollutant emitted by volcanic activity. For example, the Mount Pinatubo eruption in 1991 emitted $2\cdot10^7$ MT of SO_2 into the air [8]. In the atmosphere, the SO_2 gas is oxidized within condensed water droplets to form sulfuric acid, whereas the nitrogen oxides are converted into nitric acid. These water droplets then fall to Earth as acid rain. In industrial Germany, this rain resulted in a lowering of the soil pH and the release of soluble aluminum complexes from acid dissociation of clay minerals. The release of these toxic complexes was a direct cause of the "*waldsterben*" (forest death) damage that destroyed parts of the Black Forest in Germany during the late 20th century.

The oxidation of SO_2 by O_2 is catalyzed dramatically by photo-Fenton reactions involving dust particles encapsulated within atmospheric water droplets. It is thus an aqueous reaction. First, SO_2 gas is highly soluble in water, where it is hydrolyzed to form sulfurous acid (H_2SO_3). Deprotonation of sulfurous acid lowers the droplet pH and creates HSO_3^- anions that have a high affinity for coordinating to exposed iron centers. The pH dependence of this affinity is indicated by the speciation diagram for Fe(III) shown in Figure 18.5. The bound SO_3^{2-} is a ligand that can undergo the photo-Fenton reaction, generating $\cdot SO_3^{2-}$ and Fe(II):

$$>Fe(III)\text{-}SO_3 \rightarrow >Fe(II) + \cdot SO_3^- \qquad \textbf{(18.4)}$$

A series of reactions (18.5–18.8), when $\cdot SO_3^-$ is produced, a series of reactions takes place that ultimately create the sulfate ion in sulfuric acid as well as regenerates catalytic Fe(III) species:

$$\cdot SO_3^- + O_2 \rightarrow \cdot SO_5^- \qquad \textbf{(18.5)}$$

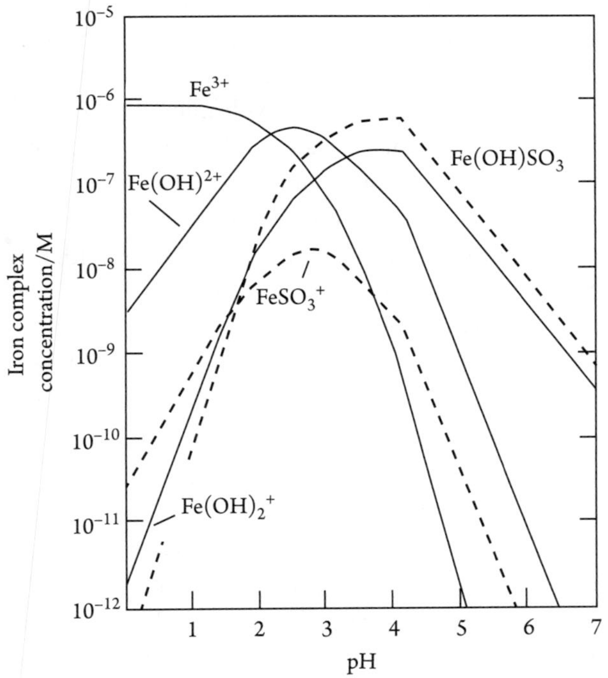

Figure 18.5 Calculated concentrations of the various solution species of Fe(III) in equilibrium with $[SO_3]^{2-}$ in aqueous solutions as a function of pH. The calculations are for solutions in which the total Fe(III) concentration is 10^{-6} M, the total $[SO_3]^{2-}$ concentration is 10^{-5} M, and the ionic strength of the solution is 0.01 M. Adapted from data in Finlayson-Pitts, B. J.; Pitts, J. N. *The Chemistry of the Upper and Lower Atmosphere*; Academic Press: New York, 2000; and Martin, L. R. Aqueous Sulfur(IV) Oxidation Revisited. In *Environmental Oxidants*; Nriagu, J. O.; Simmons, M. S., Eds.; Wiley: New York, 1994; pp 221–268.

$$\cdot SO_5^- + SO_3^{2-} \rightarrow \cdot SO_4^- + SO_4^{2-} \quad (18.6)$$

$$\cdot SO_4^- + SO_3^{2-} \rightarrow \cdot SO_3^- + SO_4^{2-} \quad (18.7)$$

$$\cdot SO_4^{2-} + Fe(II) \rightarrow Fe(III) + SO_4^{2-} (\text{regeneration}) \quad (18.8)$$

This sequence is just one of many reaction channels leading to the observed correlation between Fe(II) and sulfuric acid concentrations in water samples from clouds [2].

Animal life is affected dramatically by dust. For humans, the most-well known consequences are lung diseases such as silicosis. Here, dust particles stimulate nodular pulmonary fibrosis that deteriorates lung function. The greatest risks associated with the inhalation of silica-containing particles are chronic and typically involve workers in an industrial setting (e.g., miners and stone cutters). For shorter term exposures, protective equipment such as dust masks are recommended for workers exposed to quartz, amorphous silica, clay, or mica concentrations exceeding 20 to 30 mg/m^3. The limit for asbestos is only 2 mg/m^3. Active silica in the form of volcanic ash, which contains jagged, unweathered surfaces, is similarly hazardous. To put these concentrations into perspective, the dust level in ambient air is benign (on the order of 0.1 mg/m^3), whereas the 1000 mg/m^3 in a major dust storm is cause for concern. The burden of disease by respirable dusts is enormous. The World Health Organization estimates that fine respirated particles cause several percent of annual mortality by cardiopulmonary disease or from cancers related to the respiratory tract, and about 1% of mortality from respiratory infections in children younger than 5 years [9].

Among the most serious diseases associated with the inhalation of particles is lung cancer. Respiration of asbestos and other iron-containing mineral dusts is particularly pernicious. It is known that many cancers are stimulated by elevated concentrations of free radicals such as ·OH, which is one reason why skin cancers are more pronounced for people living at high elevations, where the atmosphere provides less protection from ionizing radiation. We have already seen that ·OH production is catalyzed by photoactivated Fenton reactions involving Fe(III) centers in materials such as asbestos. However, it is obvious that photoactivation does not play a role within the aqueous environment in a lung. How, then, can the iron centers in inhaled asbestos stimulate ·OH formation?

Ironically, much of the production of ·OH in lungs, as well as in other environments that are shielded from the sun, is catalyzed by Fenton reactions. These "dark" Fenton reactions are not initiated by light, but by environmental chemicals that can function as oxidizing agents. In the atmosphere and natural waters exposed to light, one of the most ubiquitous of these chemicals is hydrogen peroxide (H_2O_2). Hydrogen peroxide is also common in biological fluids. The classic dark Fenton reaction involving H_2O_2 is

$$>Fe^{II}OH + H_2O_2 \rightarrow >Fe^{III}OH^+ + \cdot OH + OH^- \qquad (18.9)$$

Notice that the dark Fenton reaction is almost the opposite of the photo-Fenton reaction. Here, Fe(II) (either on particle surfaces or in dissolved species) is oxidized to form Fe(III). In fact, this reaction represents a primary mechanism for regenerating catalytic Fe(III) sites that were reduced as a result of the photo-Fenton reaction. The coupling of dark and photo-Fenton reactions provides yet another example of how oxides participate in complex catalytic reaction cycles within the environment. In fact, the dark Fenton reaction generates some of the hydroxyl radicals that soil fungi use to degrade organic polymers such as lignin.

Dust ejected from volcanoes can have a large impact on life simply by blocking sunlight. The largest single eruption in recent history, the Tambora volcano in 1815, ejected $\sim 50\ km^3$ of rock in the form of ash into the atmosphere. The initial dense ash cloud covered $1.3 \cdot 10^7\ km^2 mi^2$, and turned night into day as far as 550 km away. This eruption partially blocked sunlight around the globe, resulting in a worldwide drop in temperatures (7°F in New England), resulting in crop failures. Volcanic eruptions transport dust into the upper atmosphere where the dust can take up to 10 years to dissipate. Enormous eruptions of mega volcanoes, such as those associated with the Yellowstone hot spot, can eject up to 5000 km^3 of ash into the atmosphere [10] (see Chapter 17). The drastic temperature drops associated with such eruptions cause massive losses in plant and animal life over large areas, and even in mass extinctions that, in the past, threatened human existence [11].

18.3 EARTH: OXIDE CHEMISTRY IN SOIL

Soils represent some of the most complex environments on our planet. Although all soils result from the weathering of igneous rocks, no two soils are alike (Fig. 18.6). In fact, most soils are strikingly heterogeneous. Most of the physical and chemical properties of soil vary with position. Variable physical properties include soil structure, composition, and mineralogy, as well as water content, flow, and permeability. Variable chemical parameters include solution pH, saturation levels, redox potentials, and organic content. Many of these parameters are controlled by a complex ecosystem of microbial, plant, and animal life. In turn, these parameters influence the weathering and alteration of clays and other

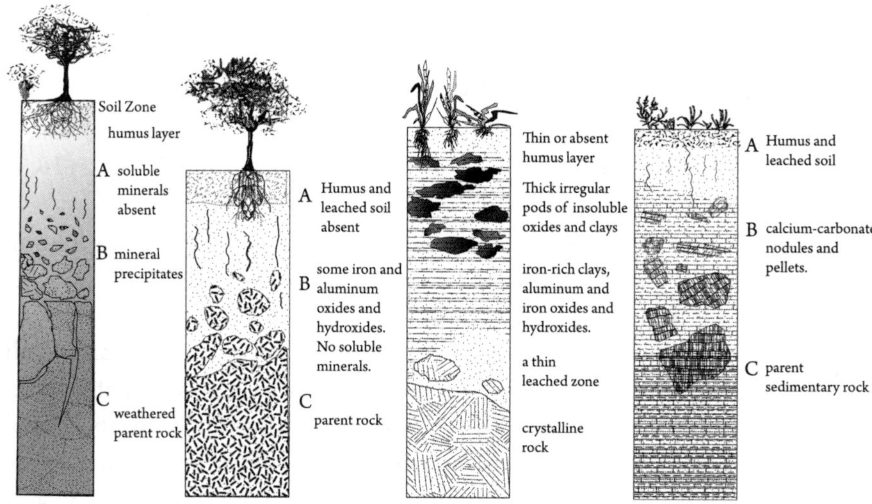

Figure 18.6 Profiles of a range of different soil types showing the vertical differentiation of mineral types and structures with depth. The profiles are strongly influenced by factors such as the parent rock source, the availability of water, soil pH, and the abundance and types of organisms and organic matter present. Representative profiles include soils that form on (from left to right) granite in temperate climates, granite in regions with high rainfall, igneous rocks in tropical climates with high rainfall, and sedimentary bedrock in regions with low rainfall.
SOURCE: Adapted from Press, F.; Siever, R. *Earth*; W. H. Freeman and Co.: San Francisco, 1978.

minerals throughout the soil column. As if things aren't complicated enough, soil environments can vary enormously with time, including climate changes, seasonal variations, diurnal variations, and even sporadic variations such as responses to localized rainstorms. Flooding can change a soil from oxic to anoxic in a matter of hours, resulting in rapid changes in both biota and soil chemistry.

Soil properties vary with depth. In moving from Earth's surface down to bedrock, typical trends include the following. First, particle packing densities are lowest at the surface, resulting in higher water concentrations and flow rates. Second, the redox potential, or E_h, is highest at surfaces exposed to atmospheric oxygen, and decreases with depth, particularly if there is a ready source of organic carbon. Third, biological activity and organic matter tend to be higher near the surface, where access to O_2 and light are maximal. Organic ligands play an important role in oxide chemistry within this near-surface regime. It is important to note that local environmental conditions can also lead to strong horizontal gradients in soil chemistry. Tillage, proximity to rivers, oil spills, and hydrothermal hot spots are just a few examples of phenomena that lead to horizontal gradients in soil chemistry and mineralogy.

Here, we highlight some of the key physical and chemical attributes of soils with an emphasis on the oxides that comprise them.

18.3.1 The Physical Properties of Soils: Colloidal Interactions

The most important components of most soils are clays and their associated organic matter. The fact that clay minerals are usually smaller than 0.4 μm in diameter puts them squarely in the colloidal regime (see Chapter 8). Colloidal interactions dictate a hierarchy of complex soil structures that span multiple length scales. These structures range from the internal structures of individual clay particles to micron-size fractal *aggregates* to larger soil structures called *peds*. (Pedology is the study of soils.) Here, we review the molecular origin of these interactions and the impact that such interactions have at increasing length scales on both soil structures. We also examine interactions between the clay minerals and water.

The primary variable that determines the colloidal interactions of clay particles is its electrostatic charge distribution. This distribution is controlled by the composition of the clay and the charged adsorbates that bond to clay surfaces. The faces of pure octahedral and tetrahedral sheets in clays such as kaolinite are nearly electrically neutral. These layers interact with each other primarily via van der Waals interactions and hydrogen bonding, and do not interact strongly with either water molecules or ions. In extreme instances, the layers can be hydrophobic. However, clays invariably have permanent layer charges resulting from the substitution of cations with a valence that differs from that of the host sheet (e.g., Al(III) for Si(IV)). These substitutions result in charge densities that can range from nearly uncharged 1:1 clays such as kaolinite (typically 0.02 meq/g) to highly charged clays such as smectites (typically 0.8 meq/g). These cationic charges are compensated by the adsorption of +1 or +2 cations between the sheets. The presence of local surface charge not only influences ion adsorption, but also both electrostatic and hydration forces within and between particles (see Chapter 8).

Water adsorption into clay interlayers depends on how much water is available. When dehydrated clays are exposed to atmospheric moisture, water enters the interlayer spaces to solvate the charge-compensating cations. At first, water molecules solvate interlayer cations strongly without changing the clay structure. As more water is added, secondary solvation shells are added, leading to a phenomenon called *swelling*. During the early stages of swelling, the interlayer distance increases in a step-wise fashion. For example, in montmorillonite, stable interlayer distances of 0.95, 1.2, 1.5, 1.9, and 2.3 nm are observed (Fig. 18.7). These distances represent the competition between the driving force for solvation and the forces of interlayer attraction. Forces associated with such

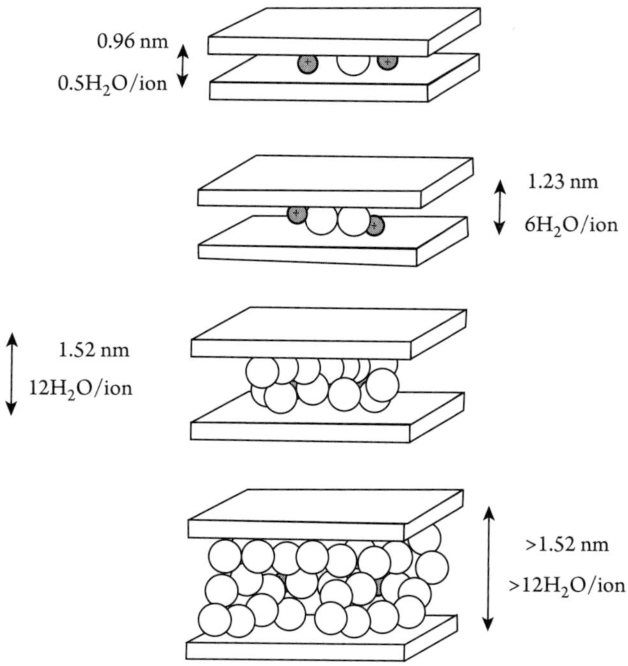

Figure 18.7 Intercrystalline distances within sodium montmorillonite as a function of the amount of water contained in interlayer regions. The distances between layers do not change continuously as more water is added, but increase in the discrete steps shown. The swelling of clay minerals generates stresses within soils that can destabilize slopes or distort overlying buildings. Adapted with permission from Madson, F. T.; Muller-Vonmoos, M.The Swelling Behaviour of Clays. *Appl. Clay Sci.* **1989**, *4*, 143–156.

hydration have been quantified using opposed mica plates in an apparatus that can measure tiny attractive and repulsive forces [12] (see Chapter 8). Above the last swelling transition, interlayer interactions become so weak that swelling continues as a linear function of water activity until an interlayer spacing of 4 nm is reached. At this maximum water content, water in the interlayer regions can represent up to 60% of the total clay mass. Above this water content, the clay exfoliates and can behave like a gel.

The swelling of clays has three important environmental consequences. First, interlayer water helps make soil hospitable for life. As water enters and leaves the clay at a slow rate into adjacent pores, this water represents a stable reservoir on which plant life can draw via capillary tension between intermittent rainstorms. Second, swelling changes the layer spacing, which influences both the extent and selectivity of ion exchange reactions (see Chapter 10). Some clays with a high layer charge, including illite, do not swell at all because interlayer attractions exceed the energies associated with ion hydration. For such clays, ion exchange reactions are limited to plate edges because the interlayer ions are not accessible to ion-exchange processes. Third, swelling changes the physical dimensions of each clay particle. Expandable clays generate enormous stresses within soil as they hydrate and expand. For example, the forces generated within sodium montmorillonite are 400 N/m^2, 110 N/m^2, and 27 N/m^2 for addition of the first, second, and third water layers, respectively [13]. Clay expansion and contraction can generate landslides by reducing the frictional stability of a slope to gravity. Such expansions and contractions can also crack the foundations of our homes and buildings.

At the next higher length scale, interactions *between* individual clay particles become important and lead to the formation of colloidal aggregates (Chapter 8). In dilute suspensions of low ionic strength, anionic clay particles repel each other and are readily dispersed and carried by water. It is estimated that the $3.2 \cdot 10^{16}$ kg/year of river water that flows into the oceans each year carries a clay load of $1.5 \cdot 10^{13}$ kg/year, representing a solids loading of 0.1 wt% [14]. At greater solids loadings and/or solution ionic strength, the electrostatic component of the electrical double layer collapses, allowing van der Waals attractions to overwhelm the electrostatic repulsion. The reversal of force from repulsive to attractive allows the clays to aggregate and settle out of solution. Such reversals occur at the mouths of rivers, where river water encounters the high salinity of the ocean. The result is the formation of the massive river deltas shown in Figure 8.7 and Plate 14.

The primary factors controlling the architectures of the resulting clay aggregates are local charge distributions and particle geometry. Clay edges commonly have higher charge densities than basal-plate surfaces. This assymetric charge distribution means that aggregation tends to form house-of-cards structures in which the edge of one particle is adsorbed to the face of another in a near-perpendicular fashion (Fig. 18.8). The resulting network is extremely porous, with pore diameters that are comparable with the clay plate diameter.

Clay particles can form continuous networks or gel structures at solids loadings as low as 0.1 wt%. However, such structures are exceeding fragile. When aggregates settle out to form soils, the house-of-cards structure starts to collapse. This collapse is mediated by the presence of water as well as natural organic binders that attach mineral grains to each other. Surface tension and capillary forces influence these interparticle interactions as soils swell or collapse, resulting in phenomena such as mudflat cracking (see Fig. 17.6 and Plate 23). As the pressure exerted by overlying soil increases with depth as a result of gravity, the packing density of clay particles increases until the soil structure is dominated by plate-to-plate contacts, as in shale. It has been estimated that recently deposited marine shales contain 70 wt% pore water and 7 wt% interlayer water, with the remainder being swollen clay (13%), nonswelling clay (5%), and nonclay solids (5%) [15].

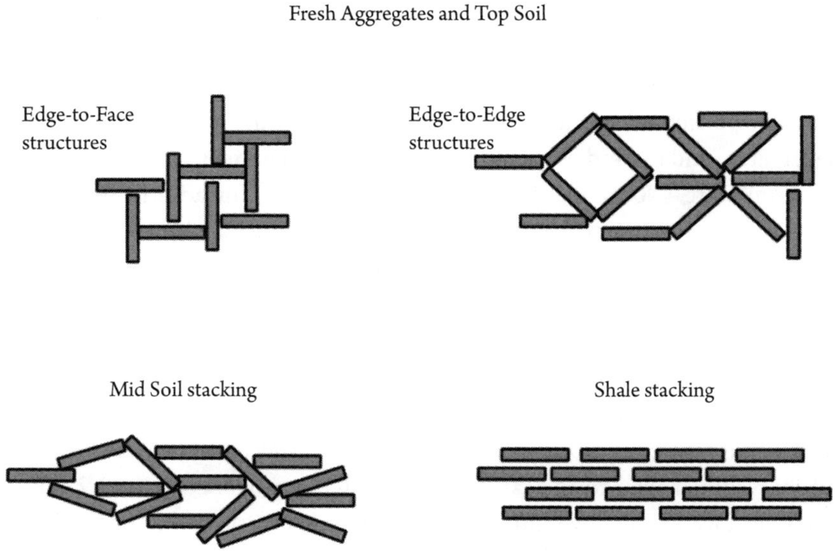

Figure 18.8 Depictions of the structures formed via colloidal interactions between clay particles. Initial aggregates tend to form open edge-to-face and edge-to-edge structures with extensive interparticle porosity. However, as clay aggregates compact in soil, this open structure is compacted until densely packed face-to-face structures form, such as those found in shale.

For most terrestrial soils that sustain life, water content, porosity, and permeability are highest within topsoil. Farmers often engineer these properties to enhance soil fertility.

18.3.2 Acid–Base and Organic Chemistry Involving Soil Minerals

Most soils have a pH between $3.5 < pH < 9$. The optimum pH range for the growth of many plants falls between $5.5 < pH < 6.5$, which is the common range for freshwater in equilibrium with atmospheric CO_2 (see Chapter 17). However, some plants, such as cranberries, prefer a more acidic pH of 4.5, whereas others, such as lilac, can thrive at a basic soil pH of 8. This pH is of primary importance in determining the weathering and biochemistry of soil.

Soil pH is controlled by both external and internal sources. External sources that lower soil pH are atmospheric species such CO_2 (which makes carbonic acid, see Chapter 17) and the sulfuric acid that falls to Earth as acid rain (see Section 18.2). An internal source that increases soil pH is the $CaCO_3$ commonly found in basic soils that overlie limestone bedrock. The release of solutes by biota can also drive up the pH. The dissolution of carbonate minerals releases $[CO_3]^{2-}(aq)$, which takes up protons to form the bicarbonate ion ($[HCO_3]^-$, see Chapter 17). The weathering and chemical alteration of clay minerals also influence the local pH via phenomena such as ion exchange (Chapter 10), protonation and deprotonation of both soluble and insoluble hydrolysis products (Chapter 5), and redox reactions including the oxidation of Fe(II) to Fe(III) (see Fig. 17.6) and of bisulfide ($[HS]^-$) to sulfate ($[SO_4]^{2-}$).

A specific acid-generating redox reaction in clays involves the oxidation of entrapped ammonium ions:

$$NH_4^+(aq) + 2O_2(g) \rightleftarrows NO_3^-(aq) + 2H^+(aq) + H_2O(l) \qquad \textbf{(18.10)}$$

This reaction is significant because (1) the large amounts of ammonium ions that are exchanged as ions into the interlayers of clays and micas represent 20% of the global nitrogen inventory and (2) the nitrate product represents an important soil nutrient, although nitrates can also represent an increasingly troublesome pollutant.

18.3.2.1 The Impact of Life on Soil Chemistry

Organic compounds generated in soils by biological activity can have an enormous impact on factors such as water retention, soil pH, weathering, and colloidal interactions. Decaying plant matter and biological excretions generate an extremely complex mixture of organic compounds called *humus*. Classes of organic compounds in soil include carbohydrates, starches, sugars, proteins, acids, bases, and even waxy hydrocarbons. Humus spans an enormous range of molecular weights that primarily reflect different stages of plant decay. For example, plants make cellulose. Fungi decompose cellulose and lignin to release smaller organic acids and phenolic compounds that heterotrophic bacteria can metabolize. The larger fragments create porous networks and/or serve as surfactants and binders for clay particles (Chapter 8). The smaller fragments, including carboxylic acids, ammonia, and phenols, mediate soil pH and can function as ligands that promote oxide weathering via complexation and the cycling of metals.

The impact of the microbial transformation of plant matter into acids is most apparent in the extensive weathering of soil minerals in moist tropical environments. Here, the release of organic acids is so extensive that most of the $>Al-\mu_2O-Si<$ bonds in clays and other minerals are completely hydrolyzed. At a soil pH of around 5, the solubility of hydrolyzed aluminum is still low ($10^{-5}-10^{-6}$ M), but that of $[Si(OH)_4]^\circ$ is high ($2\cdot10^{-3}$ M). The net result is that silicon and many other hydrolyzed elements are leached away, leaving behind aluminous solids such as boehmite (AlOOH) and gibbsite ($Al(OH)_3$) in a mixture called *bauxite*. Bauxite deposits represent our primary aluminum ores. Ironically, the lush plant life in the tropics has been forced to adapt to living in such mineral-poor, highly leached soils as a result of the abundance of its own acidic decay products.

18.3.2.2 The Impact of Soil Chemistry on Life

Soil weathering has an enormous impact on the release of both vital and toxic minerals into the soil environment. The weathering of fresh volcanic soils provides the richest supply of nutritional monovalent and divalent cations, including K^+, Na^+, Ca^{2+}, and Mg^{2+}. The high flux of these nutrients is why the best wine-growing areas are in Italy, California, and other parts of the world associated with recent volcanic activity. In the case of California vineyards in Napa and Sonoma counties, that volcanism occurred less than 1 million years ago, which is quite recent in geologic time.

Transition metals and other nutrients that form insoluble hydrolysis products at soil pH conditions are extracted actively from the minerals by plants and microbes. Nutrient elements include Fe, Mn, Zn, Cu, Co, and Mo. Extraction strategies include (1) the excretion of protons to make the local pH acidic and promote mineral hydrolysis (see Chapters 5 and 17), (2) the exploitation of simple Lewis bases (e.g., carboxylates or phenols) formed by plant decay that form soluble metal complexes, and (3) the synthesis and excretion of sophisticated chelating agents that organisms have designed to extract specific nutrients selectively from the soil. Classic examples of such ligands are iron-harvesting siderophores (Fig. 18.9).

Many of the larger polymeric or fragmented organic compounds in humus help support plant life as a result of their impact on the colloidal chemistry of clay minerals. The complex chemistry of these fragments is illustrated by just one soluble class of compounds, called *humic acids*, which are generated as intermediate compounds of plant decay (Fig. 18.9). Many humic acids are adsorbed irreversibly by clay particles as a result of hydrophobic forces, extensive acid–base interactions, and the chelate effect. These humic acids are *surfactants* (see Chapter 8) that provide many critical functions within soil environments. First, humic acids and other surfactants define the distribution of acid–base sites (often carboxylate and phenolic residues) on oxide particle surfaces. These acid–base sites determine the surface charges of the underlying oxides, allowing dissimilar oxide components to express the same surface and colloid chemistry. Second, components in humus, or adsorbed humic acids, serve as both network formers and binders that hold aggregates together, increase soil stability, and aid in preserving porosity and water retention.

Humic acids also affect which solution species adsorb onto clay particles. Such adsorption is particularly important for determining the distribution of toxic species in soil environments. For example, although humic acids contain a wide range of anionic sites that can sequester metal cations (Fig. 18.9), they also contain a limited number of cationic sites that can capture and store ionic toxicants

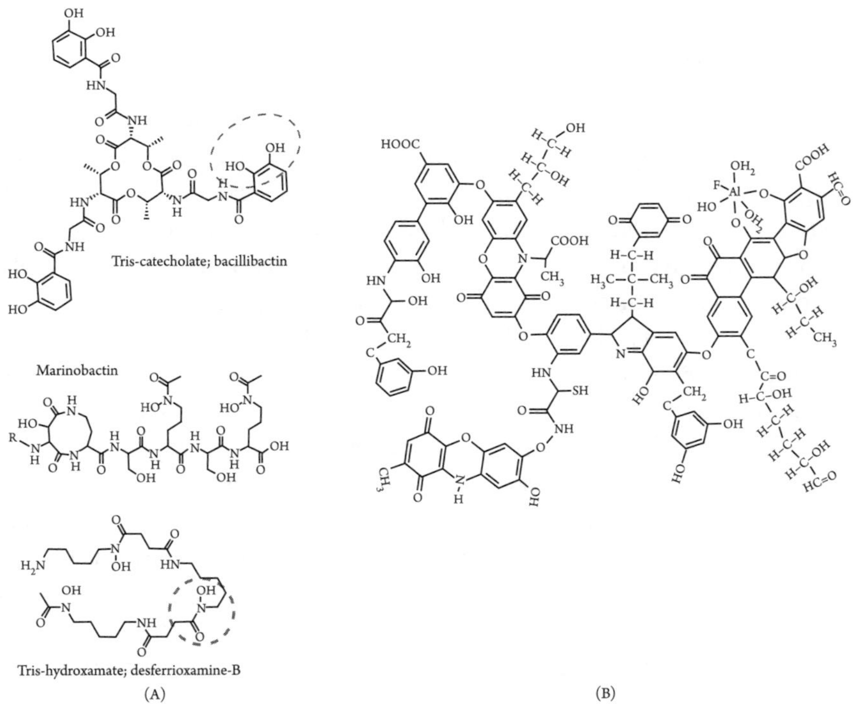

Figure 18.9 Examples of important organic species found in soil environments. (A) The structures of three siderophores created by microbes to chelated and dissolved Fe(III) in aqueous soil environments. The siderophore enterobactin binds to Fe(III) with the stunning stability constant of 10^{52}/M. (Adapted with permission from Sandy, M.; Butler, A. Microbial Iron Acquisition: Marine and Terrestrial Siderophores. *Chem. Rev.* **2009**, *109*, 4580–4596.) (B) An idealized structure of one of many possible natural organic molecules called *humic acids*. Components of a typical humic acid include quinone, phenol, catechol, and sugar moieties.

such as $[TcO_4]^-(aq)$ The hydrophobic zones can dissolve nonionic or nonpolar pollutants such as DDT and carbon tetrachloride (CCl_4). Pesticides such as methyl bromide that are hydrolyzed into harmless molecules in water within a few days can be stabilized and stored in the hydrophobic regions of surfactants on soil minerals for decades. If the soils are stable, these pesticides are persistent. However, colloidal clay particles or polyelectrolyte molecules carrying a cargo of surfactants and pollutants can also be dispersed and suspended in rivers. Clay particles thus provide a vehicle for transporting toxicants over great distances at velocities corresponding to river flow rates. For example, colloidal particles of plutonium oxide or hydroxide that are coordinated to large surfactant molecules can move through a watershed although such particles would normally be immobile.

18.3.3 Electrochemistry in Soils and Mineral Cycling: The Role of Iron and Manganese Oxides

The vigorous redox chemistry that takes place in soils and sediments is essential for terrestrial life. Redox reactions provide nutrients to plants, as well as provide energy to the microbes and animals that inhabit dark soil environments. Redox reactions also represent important pathways for decomposing toxins and pollutants. Oxide and hydroxide minerals, especially those of iron and manganese, play key roles in many of these redox cycles. The oxyanions of sulfur and selenium also exhibit redox activity. Here, we highlight how oxides fit into the complex redox chemistry of soil in terms of energetics, spatial distributions, interactions with life, and cycling. Those of you interested in more in-depth discussions of the redox chemistry of oxides are referred to Chapter 11.

The hierarchy of energies available for soil redox reactions is often manifested in chemically distinct zones of redox activity (Fig. 18.10). The strongest oxidizing agent (i.e., electron acceptor) common to soil and its organisms is O_2. The Mn(IV) found in oxides such as MnO_2 is also a strong oxidizing agent, but not so energetic as O_2. Oxides containing Fe(III) are milder oxidizers, followed by S(VI) in sulfate. The strongest reducing agents (i.e., electron donors) in soil are

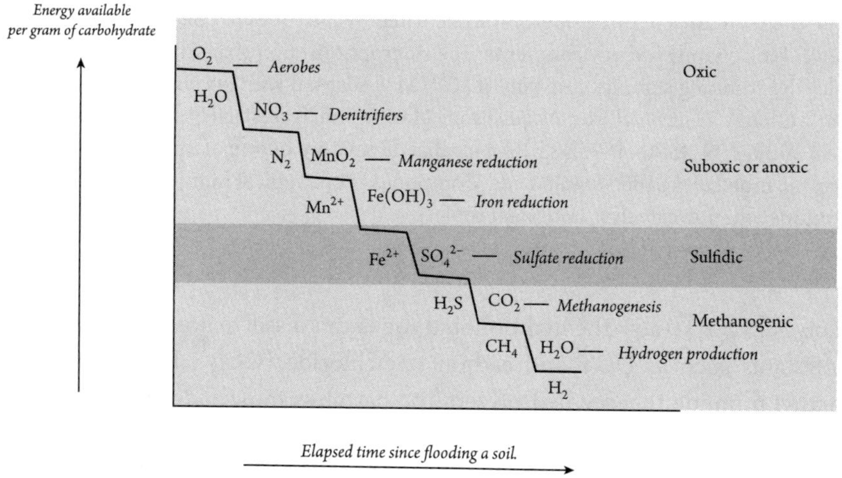

Figure 18.10 Redox couples of importance in soil chemistry in order of decreasing oxidation potentials. The oxidized member of the redox couple is shown above the stair-stepped line, whereas the reduced product is shown below the line. On the right are the typical vertical zones found in soil that correspond to the redox couples shown. The bottom axis also indicates the sequencing in redox reactions that occurs as a function of time after an oxic soil is flooded. Flooding cuts off the soil from a source of atmospheric $O_2(g)$. Adapted with permission from Gao, S.; Tanji, K. K.; Scardaci, S. C. Incorporating Straw May Induce Sulfide Toxicity in Paddy Rice. *Calif. Agricult.* **2003**, 57, 55–59.

usually hydrocarbons and organic matter. Redox reactions in soil are almost always mitigated by microbial enzymes. This is particularly true if multiple electrons are exchanged, such as the eight-electron reduction of sulfate to sulfide. Thus, there is both chemical and microbial stratification in most soils.

The most energetic redox reactions typically occur between the strongest oxidizing and reducing agents. Such reactions continue down the redox hierarchy until the most active reagents are consumed, followed by reactions involving the next most active redox couples. In soils, this means there is definite stratification in terms of which species are produced where within a typical soil column. Bacterial ecosystems influence this stratification. For example, the sulfide ion and oxygen do not usually coexist at the same level. Similarly, if there is an oil spill, aerobic processes involving both O_2 and microbes immediately start to decompose the hydrocarbons until the most energetic oxidants are consumed and the spill is naturally mitigated. Progressively less energy-efficient ecosystems organize themselves to exploit the hydrocarbon source as the electron acceptors are consumed. At the oldest O_2-free parts of the spill, the least energy-efficient methanogic and hydrogen-generating organisms dominate.

In the face of such complexity, it is useful to organize soils and sediments into three broad redox groupings that represent the most common stratification of both redox-active oxides and living things. However, the spatial distribution of these groupings is controlled by chemical gradients regardless of position. The top of the soil and sediment column in contact with the atmosphere is called the *oxic zone* for its access to $O_2(g)$. The oxic zone typically contains elements that are heavily oxidized and form insoluble hydrolysis products such as goethite (FeOOH), hematite (Fe_2O_3), and birnessite or pyrolusite (MnO_2). The electrochemical oxidation of organic matter is most pronounced within the oxic zone, consuming O_2 to yield CO_2 gas.

Below the oxic zone lies the *suboxic zone*, where O_2 concentrations are depleted dramatically relative to the soil surface. Redox processes in this zone are often dominated by redox cycling involving solid metal oxides and bacteria. Here, rather than using O_2 directly in respiration, bacteria exploit the reduction of iron and manganese oxides to oxidize organic matter [16] and induce changes in metal valence. This reductive dissolution of solids occurs because the minerals are used as electron acceptors in bacterial metabolic cycles. Bacteria also use solid MnO_2 as a detoxification agent for removing cell-damaging reactive oxygen species, including superoxide and hydrogen peroxide, and to break up large organic molecules into metabolizable fragments. The extraordinary radiation-resistant microbe *Deinococcus radiodurans* appears to thrive in high-radiation fields by maintaining large concentrations of low-molecular weight manganese complexes to react with harmful free radicals produced by ionizing radiation [17].

Below the suboxic zone lies a region containing no detectable O_2 called the *anoxic zone*. This zone contains species that are highly reduced (i.e., in their lowest oxidation state in water). For example, iron and manganese are in their +2 oxidation states. Water can contain large concentrations of reduced electron acceptors. In extreme cases, water itself can be reduced in this zone to form H_2. Oxyanion species such as sulfate and selenite ions are reduced to sulfide (S^{2-}) and selenide (Se^{2-}) ions. When combined with +2 cations, these anions form insoluble minerals, including siderite ($FeCO_3$), pyrite (FeS_2), and the slightly soluble sulfide mineral mackinawite ($Fe_{1-x}S$).

The sulfate ion is a poor oxidizing agent (Fig. 18.10). However, anaerobic bacteria in this zone can oxidize organic matter using the sulfate ion. The resulting H_2S gas emissions produce the fetid or rotten-egg odors that emanate from O_2-poor environments such as salt marshes. Beyond this sulfidic zone, methane can be produced by reduction of carbonate. This methane is odorless, but combustion of this methane as swamp gas gives rise to will-of-the-wisps that frighten travelers late at night.

Redox stratification can take place as a function of time rather than position, as illustrated by the soil chemistry of flooded fields. During a dry summer, O_2 penetrates soil pores as a gas and makes MnO_2 and Fe(III) solids such as goethite (FeOOH). On inundation, O_2 fluxes are reduced by orders of magnitude, because the diffusion coefficient for O_2 in water is $\sim 10^{-5}$ times smaller than in air. Dissolved O_2 is quickly eliminated by the respiration of soil organisms. Anaerobic processes then take over, such as the reductive dissolution of MnO_2 and Fe(III) solids by soil bacteria coupled to the oxidation of soil organic matter. After inundation, the soil E_h drops considerably, and fluxes of dissolved Mn^{2+} and Fe^{2+} into pore fluid solutions become appreciable. Any toxic metals or pollutants adsorbed to the oxide minerals are released when this reductive dissolution occurs and enter the water as solutes. Abiotic reoxidation of Mn(II) to Mn(IV) is impeded kinetically at the neutral to acidic pH environments in many soils. Here, bacteria and fungi excrete enzymes such as multicopper oxidases to regenerate the black solid MnO_2 at near-neutral pH.

Although the redox cycling of both iron and manganese can help support life in the soil and purify soil environments, these cycles can also have deleterious environmental consequences. For example, the oxidation–reduction cycle of iron has led to the largest mass poisoning of humans in history [18]. The sediment in much of Southeast Asia can be enriched locally in arsenic as a result of the natural weathering of minerals in rock. This arsenic is not toxic to humans unless it is dissolved in drinking water. Decades ago, the British Geological Survey replaced open cisterns with shallow wells in parts of Bangladesh to reduce the propagation of human diseases. However, they were unaware that the shallow wells periodically become anoxic. This anoxia resulted in the dissolution of iron-oxide minerals and the release of their adsorbed arsenic into the drinking water. Between 35 to 77 million people suffer from arsenic poisoning in Bangladesh alone [18] as a result of a poor appreciation for the coupling of toxicant bioavailability and the seasonal electrochemistry of soil and sediment.

18.4 WATER: OXIDES IN THE HYDROSPHERE

The oxide–water reactions that promote weathering have a profound impact on the structure and composition of rock-forming minerals in the environment. Conversely, the weathering of oxide minerals has a profound impact on the composition and chemistry of both freshwater and seawater. Raindrops that fall to Earth normally exhibit a high level of purity that varies with distance from the ocean and from sources of industrial pollution. However, as fresh rainwater accumulates and is transported by rivers and streams, it picks up many elements from the rocks and minerals it encounters. When this water reaches the oceans, it flows into a huge reservoir that collects, concentrates, and maintains elemental concentrations that are substantially different than those found in freshwater. These elements are present as simple ions, dissolved complexes, and/or insoluble colloids that can participate in a vast number of photochemical, electrochemical, acid–base, and ion exchange reactions. Some of these reactions also involve organic matter and living things. Here, we highlight (1) the ion-exchange reactions of clay minerals that influence the composition of both fresh water and seawater, and (2) photochemical reactions involving colloidal oxides that are critical to the oceanic food chain.

18.4.1 The Composition of Natural Waters: The Role of Clay Minerals

The average concentrations of the most prevalent elements in both freshwater and seawater are shown in Table 18.1. Many elements are present only at low concentrations (<1 mg/L or 1 ppm). In some instances, these low concentrations are a reflection of the low natural abundance of the element in the parent igneous rock (see Table 17.1). As discussed in Chapters 5 and 17, another major factor that dictates low concentrations is the low solubility of either hydrolysis products (e.g., for elements present as +3 and +4 cations, such as Fe(III) or Ti(IV)) or ionic salts (e.g., for many metals with a valence of +5 or +6, such as P(V) or S(VI)). Salt precipitation also accounts for the fact that solution inventories of some +2 cations, such as Ba^{2+} (in $BaSO_4$) and Ca^{2+} (in $CaCO_3$), are lower than corresponding concentrations in the parent igneous rocks. Ions that remain are mainly halide ions (which form soluble salts) and elements that form +1 and +2 cations.

Table 18-1. MEAN COMPOSITION OF RIVER WATER AND SEAWATER

Element	**Seawater (mg/L)**	**Mean Rivers (mg/L)**
Cl	19,000	7.8
Na	10,500	6.3
SO_4	2700	11
Mg	1350	4.1
Ca	410	15
K	390	2.3
HCO_3	142	58
Br	67	
Sr	8	
SiO_2	6.4	13
B	4.5	
F	1.3	
N	0.67	1
Li	0.17	
Rb	0.12	
C (organic)	0.10	

SOURCE: Hem, J. D. *Study and Interpretation of the Chemical Characteristics of Natural Water*; U.S. Geological Survey, 1989, U. S. Government Printing Office, Washington, D. C. p. 5.

The concentrations in natural waters of +1 and many +2 cations are controlled by the most prevalent weathering products, which are the clay minerals. The first mechanism by which clays mediate the concentrations of many +1 and +2 cations is ion exchange (Chapter 10). In fact, the oceans of the world can be viewed as an enormous ion-exchange column in which the fluid is seawater and the oceanic clay is equivalent to a column-packing material. The three parameters that control the behavior of this ion-exchange column are (1) the ion exchange capacity of the sediment, (2) the ion-exchange selectivity of the clay minerals present, and (3) the solution concentrations in equilibrium with those minerals.

The ion exchange capacity of the clay minerals is enormous. Clays have a total mass of around $1.5 \cdot 10^{18}$ MT. Most of this clay is in oceanic sediments [19] (Fig. 18.11 and Plate 25). Smectite clays such as montmorillonite are most prevalent, followed by illite, chlorite, and kaolinite. Although clay mineral distributions are complex, to a first approximation montmorillonite is the dominant clay in massive midoceanic plates such as those found in the Pacific Ocean. Illite predominates along major spreading centers such as the Mid-Atlantic Ridge. Chlorite tends to be found in temperate-to-cold climates nearing the poles, whereas kaolinite is generated as a result of the rapid weathering of rocks within tropical climates. Each clay family has a different ion-exchange capacity and selectivity. The

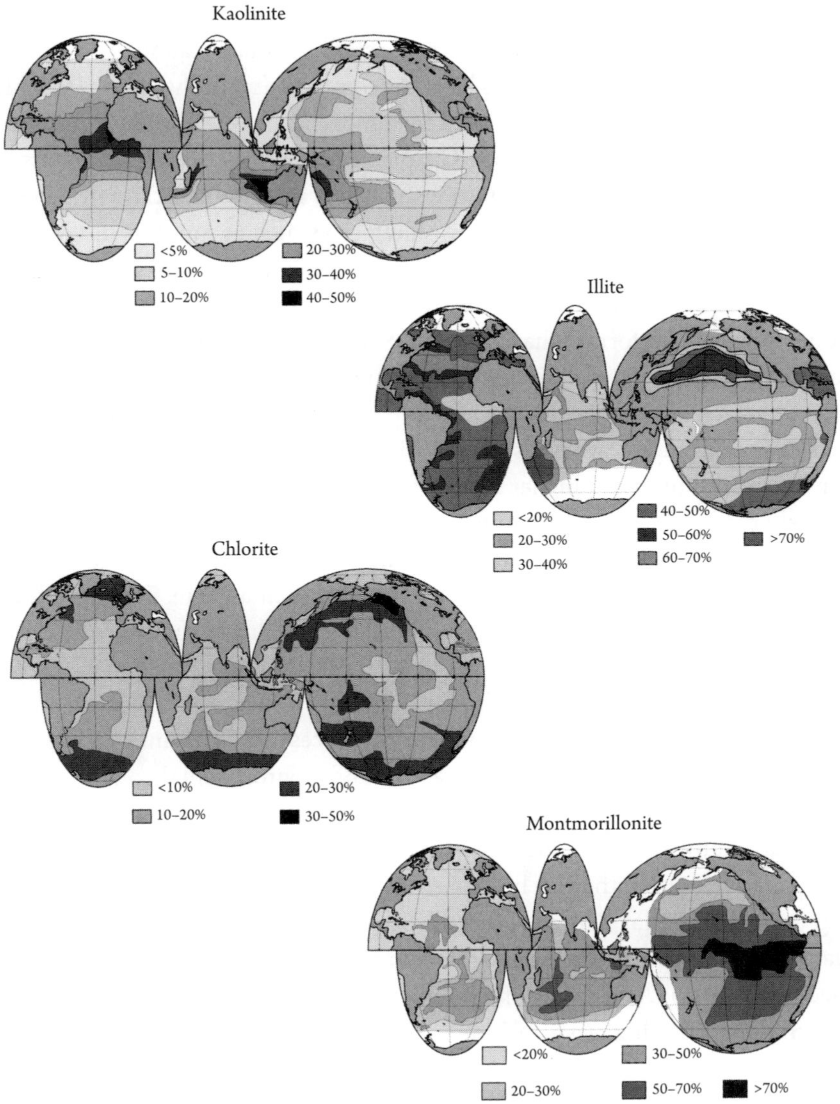

Figure 18.11 The distributions of each of the major clay families in the major oceanic basins of the world. Adapted with permission from Griffin, J. J.; Windom, H.; Goldberg, E. D. The Distribution of Clay Minerals in the World Ocean. *Deep-Sea Res.* **1968**, *15*, 433–459. (see Plate 25)

dominant smectite clays have the highest ion exchange capacity, ranging from around 0.8 to 1.5 meq/g. The total number of cation equivalents in all clays is estimated to be $5 \cdot 10^{21}$ eq. If all these cations were K^+, the mass of exchangeable cations would be $2 \cdot 10^{17}$ MT, which is 12% of the clay mass. This value is much larger than the annual flux of K^+ entering the oceans from rivers $(5 \cdot 10^7 \text{ MT/year})$

[20]. In fact, clays have approximately the same capacity for storing monovalent cations as carbonate minerals have for sequestering divalent cations.

Each of the major clay minerals has a different affinity for the cations to which it is exposed. This affinity is described by ion-exchange equilibrium constants or distribution coefficients (K_d) (see Chapter 10). Both of these parameters are controlled by local charge distributions and (for nonswelling clays or zeolites) the energetics of ion solvation. For swelling clays such as montmorillonite, cations have access to anionic exchange sites with minimal steric (or solvation) constraints. The most common anionic sites in clays are $>Al-\mu_2O-Si<$ bridging bonds, each with a formal charge of -0.25. Because four such sites are required to neutralize a single monovalent cation, these exchange sites prefer to adsorb cations with a low charge density. The typical selectivity sequence for monovalent cations in such clays is $Cs^+ > Rb^+ > K^+ > Na^+ > Li^+$ (see Chapter 10). Such clays also tend to have a higher affinity for monovalent cations than they do for divalent cations. However, it is important to remember that edge sites have different charge distributions than sheet faces, and that steric effects can also influence ionic selectivity, especially for nonswelling clays such as illite. The net effect is that clays remove substantial fractions of the riverine flux of cations released as a result of weathering.

An important aspect of ion exchange in most clays is that the exchange process is reversible, which means clays can serve as sources as well as sinks for adsorbed cations, depending on local concentrations and environmental conditions. Nowhere is the impact of true ion exchange more apparent than when the load of clays carried by a river equilibrates with seawater in estuaries. At this point, ion exchange sites within the clay burden are rich in K^+ and Ca^{2+} relative to Na^+ and Mg^{2+}. When this clay burden enters the ocean, it must now equilibrate with saltwater that is much more saline and richer in Na^+ and Mg^{2+} than the parent stream (see Table 18.1). As a result of this change in environment, some of the ions within the contacting clays, including H^+, K^+, and Ca^{2+}, are released from exchange sites and are replaced by Na^+ and Mg^{2+} until a new equilibrium is established. These ion-exchange reactions are important globally. For example [21], 42% of the Ca^{2+} in riverine smectites ($3.7 \cdot 10^7$ MT/year) is replaced by Na^+ when these clays reach the ocean, which corresponds to about 50% of the global sodium flux and 5% of the global calcium flux [20].

18.4.2 The Photochemistry of Colloidal Iron and Manganese Oxides in the Ocean

Colloidal iron and manganese oxide solids are present in the upper oxic zone of all oceans of the world at concentrations ranging from 0.001 to 0.01 ppm and 0.002 to 0.02 ppm, respectively. Although their concentrations are low, their impact on oceanic chemistry is high. The oxic region has appreciable O_2 and is accessible to light penetration. Therefore, these suspended oxides are active in photoelectrochemical processes (see Chapters 11 and 13), contributing to reactions critical to life on our planet.

The first factor that must be considered to evaluate photoefficiency is the availability of photons. In the ocean, photochemical reactions leading to photosynthesis, the cycling of nutrients, and oxide photochemistry are limited to a near-surface region called the *photic zone* (Fig. 18.12 and Plate 26). The thickness of this zone depends on the fundamental absorption properties of water, as well as the suspended matter present. Optical penetration depths are largest in the open ocean, where solids loadings are minimal. Here, the intensity of light at the surface is diminished to 40% by 1 m and diminishes roughly by a factor of 10 for every 75-m increase in water depth. However, as shown in Figure 18.12 and Plate 26, light intensity does not diminish uniformly as a function of wavelength. Objects in the oceans appear light blue as the longer red wavelengths are completely absorbed in shallow depths, leaving the shorter blue wavelengths to provide illumination. For this reason, photosynthetic organisms tune the absorption spectrum of their light-harvesting molecules according to their habitation depth.

The primary factor that determines how available light is used is the band structure of the oxide (see Chapter 13). The electronic band structure of MnO_2 consist of a valence band dominated by O 2p orbitals and a conduction band dominated by unoccupied manganese 3d orbitals (Fig. 18.13). As depicted in

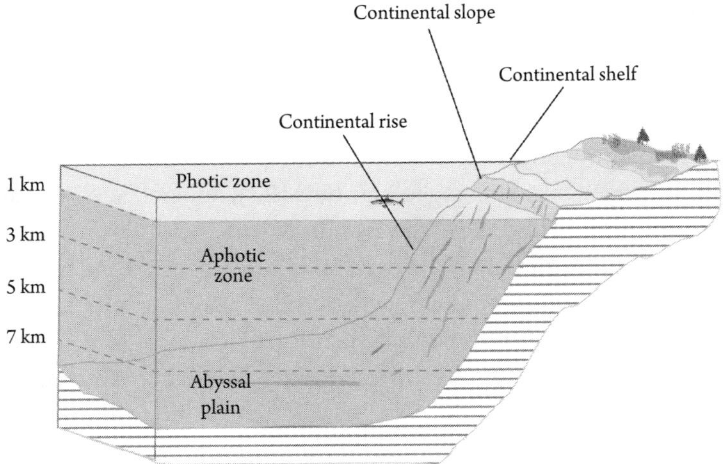

Figure 18.12 A depiction of different zones within the depths of the ocean. Although the photic zone represents an extremely thin layer on the oceanic surface, most of the oceanic food chain is derived from activity within this zone.

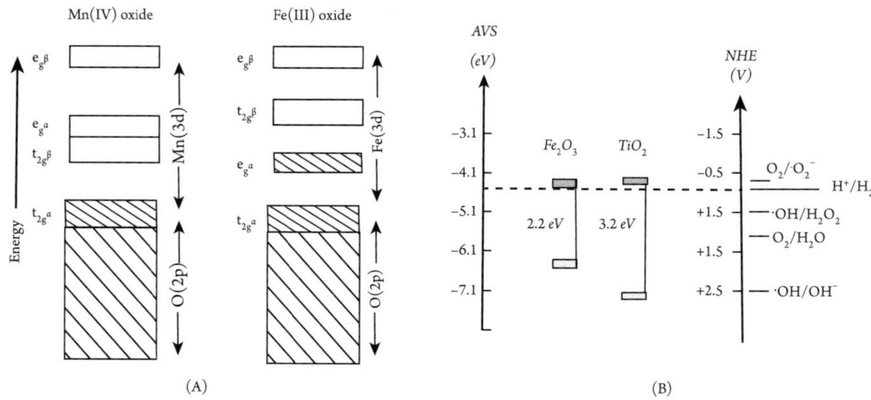

Figure 18.13 (A) The electronic structures of Mn(IV) and Fe(III) oxides. (SOURCE: Reproduced with permission from Sherman, D. M.; Electronic Structures of Iron(III) and Manganese(IV) (Hydr)oxide Minerals: Thermodynamics of Photochemical Reductive Dissolution in Aquatic Environments. *Geochim. Cosmochim Acta,* **2005**, *69,* 3249–3255.) (B) The band structures of hematite (α-Fe_2O_3) and rutile (TiO_2) relative to the energies needed to generate reactive oxygen species. Voltages are reported relative to both the absolute vacuum scale (AVS) and the natural hydrogen electrode (NHE) redox potential scale. (SOURCE: Reproduced with permission from Xu, J.; Sahai, N.; Eggleston, C. M.; and Schoonen, M. A. A., 2013. Reactive Oxygen Species at the Oxide/Water Interface: Formation Mechanisms and Implications for Prebiotic Chemistry and the Origin of Life.; *Earth Planet Sci. Lett.,* **2013**, *363,* 156–167.)

Figure 13.6, the fully dense pyrolusite phase of MnO_2 can have a band gap as small as 0.25 eV (equivalent to a photon wavelength of 5000 nm) [22]. However, the predominant phase found in seawater is the defective birnessite phase of MnO_2 with a layered structure similar to that of a double hydroxide (Chapter 2). Birnessite is reported to have a band gap that ranges from 1.0 to 1.8 eV. Here, we assume that MnO_2 has a band gap of 1.8 eV, which means the oxide should be capable of harvesting much of the visible light spectrum. Because the spectrum of light has changed throughout geologic time [23], minerals such as manganese oxides may have had a key role in the evolution of life.

Manganese oxides participate in a cycle of photoreduction and microbial oxidation that provides energy to help sustain a substantial fraction of primitive life in the ocean. The cycle starts when a suspension of insoluble MnO_2 particles absorbs visible light exceeding the band-gap energy. This light harvesting is assisted by adsorbed photosensitizers (see Chapter 13) that are produced via the decomposition of biota. Electrons in the valence band of MnO_2 are associated with the O^{2-} anions in the lattice, whereas the empty orbitals in the conduction band are associated with the Mn(IV) cations. When MnO_2 absorbs photons that exceed the band-gap energy, the net effect is that the Mn(IV) cations are eventually reduced to Mn(II), whereas the oxygen anions are oxidized to generate surface peroxides, superoxides, and ultimately O_2 gas.

Isotope-tagging studies suggest that the key step in the reductive photodissolution of MnO_2 is the oxidation of surface-associated water molecules to form hydroxyl radicals. Two ·OH combine to form H_2O_2 that then disproportionate (i.e., $2H_2O_2 \rightarrow O_2 + H_2O$) via pathways resembling those used by the catalase enzyme. The net result is the photoactivated dissolution of MnO_2:

$$2Mn(IV) + 4e^- \rightleftarrows 2Mn^{2+} \tag{18.11}$$

$$4O^{2-} + 4H^+ + 2H_2O \rightarrow O_2 + 4OH^- \tag{18.12}$$

$$2MnO_2 + 2H_2O + 4h\nu \rightarrow 2Mn^{2+}(aq) + O_2 + 4OH^- \tag{18.13}$$

The reaction shown in Eq. 18.13 occurs at a slow rate, not only the result of extensive electron–hole recombination, but also because the net reaction is a four-photon (or four-electron and four-hole) process. Layered birnessite-like manganese oxide minerals are well known to produce O_2 slowly in contact with seawater when irradiated [24], especially in the presence of natural photosensitizers [25, 26].

It is important to note that the photoreductive dissolution of MnO_2 is only half a catalytic cycle that takes place in the ocean [25, 27]. The photic zone is an oxidizing environment in which the newly formed soluble Mn(II) species are thermodynamically unstable relative to insoluble Mn(IV) oxides. However, experiments performed in pure water under controlled laboratory conditions show that, even under highly oxidizing conditions, the abiotic rate of Mn(II) oxidation is too slow to counterbalance the normal rate of photoreductive dissolution. Surprisingly, when scientists measured Mn(II) oxidation in oceanic environments they found the rate of oxidative particle formation to be orders of magnitude faster than that measured in seawater simulants in the laboratory.

The rapid oxidation of Mn(II) in the oceans is a result of biotic catalysts such as a bacterial multicopper oxidase enzyme [26, 28] that was recently discovered to be of global importance. In addition, there is evidence that some microbes use the catalytic oxidation of Mn(II) to help them metabolize organic matter. The holes in oxide surfaces (such as superoxide and hydrogen peroxide intermediates) help to break down organic matter to a size that can be metabolized. Thus, MnO_2 solids serve as a "digestive aid" to the microbes. In the eastern Caribbean, the rates of both the oxidation and reduction steps have been measured *in situ*. Dissolution, which only occurs during daylight hours, exhibits rates ranging from $5.6 \cdot 10^{-15}$ mole $L^{-1}s^{-1}$ to $3.75 \cdot 10^{-13}$ mole $L^{-1}s^{-1}$. The rates of manganese oxide particle formation range between $2.8 \cdot 10^{-17}$ mole $L^{-1}s^{-1}$ and $1.25 \cdot 10^{-14}$ mole $L^{-1}s^{-1}$. [27]. Given the fact that particle formation occurs continuously, the net result is a balanced diurnal cycle in which both dissolution and precipitation rates are comparable.

The reductive dissolution of MnO_2 accompanied by the generation of oxidizing agents such as H_2O_2 or O_2 may have profound implications for the origins of life in the oceans. Life originated a full billion years before the oxygenation of the atmosphere. When single cells evolved, the partial pressure of oxygen was much lower than it is today. The lack of oxygen in the atmosphere had two important consequences. First, oxygen was not available as an oxidizing agent in metabolic cycles, which probably involved gases such as CO_2 instead. Second, O_2 was not available for forming a protective ozone layer, so high-energy UV radiation flooded the planet surface. Without this layer, the flux of energetic photons reaching Earth's surface would be comparable to the flux from deep space (10 mW/cm^2) or the flux that bathes the Martian surface today (5 mW/cm^2) [23]. The high flux of energetic photons occurred when the earliest forms of photosynthesis were developing [29]. Although the waters of the ocean provided partial shielding from this radiation, laboratory experiments on iron and titanium oxide materials show there was still sufficient UV light to create discharges of harmful species such as the hydroxyl radical and H_2O_2 even in the absence of O_2 (Figs. 18.4 and 18.13). These would have been important oxidants in the early history of Earth, before the rise of oxygen-generating bacteria about 2.5 billion years ago. The atmosphere at that time would have been rich in CH_4, H_2, H_2O, and CO_2.

One strategy that plant life developed not only to harvest the energetic photons from the sun, but also to exploit the free-radical chemistry of oxides as sources of energy (rather than cell damage) involves photosynthesis. Plants have evolved to incorporate photoactive manganese oxides within light-harvesting cells. Photosynthesis in the thylakoid membranes of all plants is carried out by a construct called *photosystem II* (Fig. 13.16). The metal–oxide core of this photosynthetic engine consists of a mineral-like cluster containing a calcium ion and four manganese ions separated by oxygen ions [30]. This photocenter undergoes four repeated photooxidation reactions to convert H_2O into O_2. It is thought that this sophisticated cluster evolved from the abiotic photoreduction of extracellular colloidal MnO_2 that produces O_2 within the oceans of the world even today [31].

18.5 FIRE: OXIDES AND WATER IN GEOTHERMAL ENVIRONMENTS

Most reactions involving oxides and water are activated by temperature. In Nature, the heat that stimulates these reactions can come from sources that are both external and internal to Earth. The sun warms Earth's surface to provide benign environmental conditions that allow life to flourish. The same cannot be said of Earth's interior. For example, it is estimated that the temperature at the boundary between the inner and outer core is 6000°C, which is as hot as the surface of the sun. The pressure at this boundary is estimated to be 3,300,000 atm. Although much of the interior of our planet is solid, this solid rock is ductile as a result of the enormous temperatures and pressures encountered there, and can flow in response to convection currents and plate-tectonic stresses [32]. Molten rock is also present in the solid mantle and the lower crust. This molten rock, known as *magma,* reaches the surface in specific and, fortunately, limited locations. In this final section of the book, we provide an overview of these geothermal heat sources and how such sources influence chemical reactions involving solid oxides and water.

18.5.1 Plate Tectonics and Geothermal Heat Sources

The conveyor belt that transports oxide materials into and out of Earth's crust is based on the movement of the enormous plates that comprise the lithosphere (Fig. 18.14). The lithosphere consists of Earth's crust and upper mantle, where rocks are relatively brittle. In contrast, the underlying asthenosphere is a viscous fluid that experiences strong temperature and buoyancy gradients associated with internal heat sources. These heat sources include energy remaining from the planet's formation 4.6 billion years ago via gravitational collapse, and heat from the ongoing decay of radioactive elements such as ^{40}K, ^{238,235}U, and ^{232}Th [32]. The resulting temperature gradients create convection cells within the asthenosphere that are in slow but relentless motion throughout geologic time. These convection cells cause the overlying lithospheric plates (which comprise the crust) to move.

New crust is made at the boundaries between moving plates. At *divergent* plate boundaries, such as the Mid-Atlantic Ridge, the East-Pacific Rise, and the East African Rift Valley, plates move away from one another. The space between the moving plates is filled by magma from the upper asthenosphere. Magma originates from the partial melting of hydrated rocks in Earth's mantle. As this magma moves upward to fill the space made available between separating plates, it solidifies into igneous rocks, which near the ocean floor are the silica-poor rocks basalt and gabbro. Old, frozen parts of the upper mantle can be found at the Earth surface as rafted sections of rock that have been thrust onto land, often preserved as hydrated Mg-rich rocks called *serpentinites.*

Moving plates also collide with one another. New crust is also made at these *convergent* plate boundaries. At a convergent margin, one plate is thrust below the other and is *subducted* deep into the earth beneath a wedge of mantle material. As the plate moves deeper into the earth, increased temperatures and

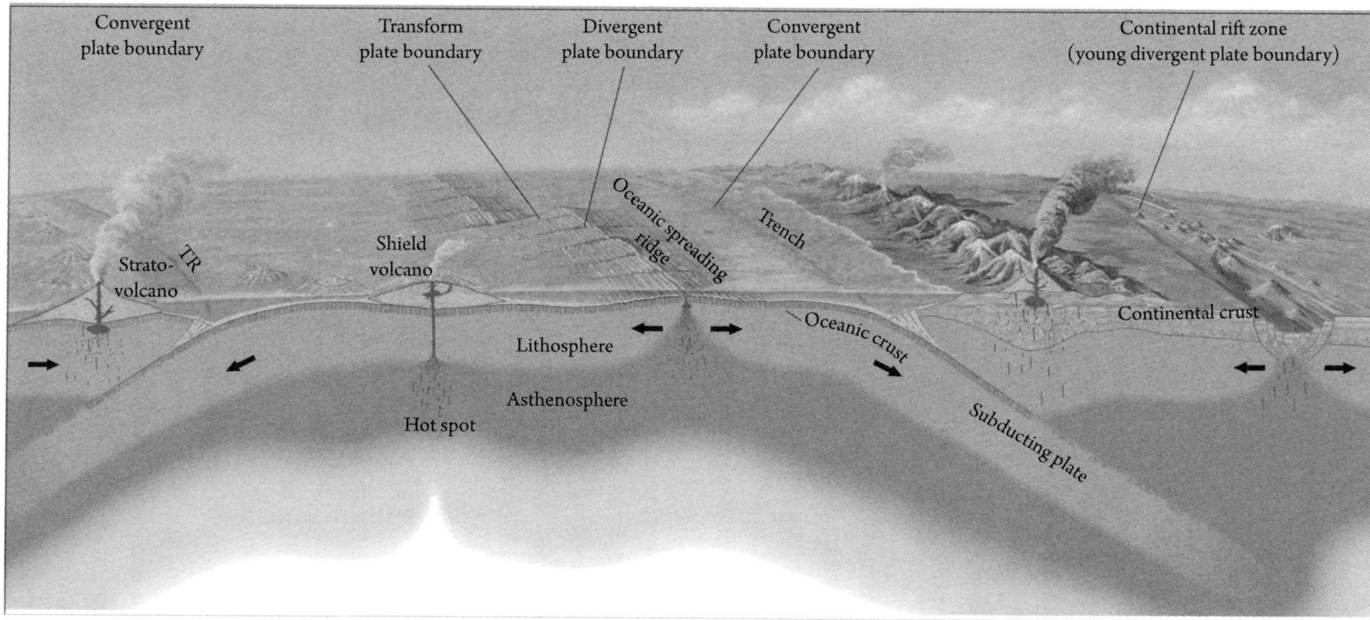

Figure 18.14 A depiction of convective currents in the mantle and the consequences that such currents have on the motion of the crustal plates above. A divergent plate boundary appears in the middle of the diagram whereas a convergent plate boundary and its associated subduction zone is depicted on the right.

SOURCE: Image used with permission from the U.S. Geological Survey and the original artist Jose F. Vigil; http://pubs.usgs.gov/gip/dynamic/Vigil.html.

pressures cause the hydrated minerals to lose their water as the subducting slab becomes dehydrated (subduction zones are discussed later in this section). The water thus released moves upward and lowers the melting point of overlying rocks [33, 34]. Water-induced melting of the mantle above the subducting slab produces magmas with a lower density than the overlying crust. These magmas rise to the surface. If the magma encounters weak crust and rises rapidly, the result can be volcanic eruptions that occur tens to hundreds of kilometers beyond the point of plate collision in what are called *arc volcanoes*. The name arises from the curved distributions of such volcanoes on the globe associated with plate boundaries. If the volcanoes appear in the ocean, the convergent plate boundaries give rise to arcs of islands well away from the subduction zone, such as the Aleutian island chain and the exceptionally violent Indonesian volcanic chain. Arcs are also seen in chains of continental volcanoes such as in the Andes in South America and the Cascade Mountains in the Pacific Northwest of the United States.

If magma rises slowly enough to solidify before it breaks the surface, coarse-grained igneous rocks called *plutonic* rocks are formed. Granite and diorite, consisting of interconnected grains of quartz, feldspars, and mica, are among the most common plutonic rocks. Plutonic rocks form the backbone of most of our major mountain chains, including the Sierra Nevada and the Rocky Mountains in the United States. Anhydrous components of subducting slabs can resist melting. It is thought that these solid, anhydrous slabs accumulate in the colorfully named *slab graveyard* near the boundary between the lower mantle and Earth's core.

Apart from colliding plates, Earth also contains about two dozen midplate hot spots [35] such as those found today in Yellowstone National Park, Hawaii, and Iceland, where heat and magma rise to the surface from deep within the earth near the core–mantle boundary. Hot spots are the most important sources of igneous rocks in terms of their impact on ecosystems and geologic history. Massive lava flows in the middle of a continent are commonly caused by long, narrow hot spots that release enormous amounts of heat from deep within the earth. Hot spots have been implicated in the formation of the Deccan Traps 65 million years ago that are estimated to have erupted $1 \cdot 10^6$ km^3 of basaltic lava that covered up to half the Indian subcontinent. An even larger outpouring of lava formed the Siberian Traps about 250 million years ago that may have once covered as many as $7 \cdot 10^6$ km^2 of what is now Siberia. The formation of these Siberian Traps has been implicated in the largest mass extinction in Earth's history at the end of the Permian Era. It is thought that the Columbia River basalts in the northwest United States originated from movement of Earth's crust over a hot spot that flowed to cover $1.6 \cdot 10^5$ km^2 between 14 to 17 million years ago. Scientists estimate that lava floods from this hot spot exceeded 1000°C and covered enormous areas at speeds of meters per second.

Both arc volcanic chains and hot spots are the points of origin for enormous violent eruptions called megavolcanoes [35] that spew astronomical clouds of volcanic ash into the atmosphere. In recent history, the Yellowstone hot spot has undergone such a massive eruption every 600,000 years or so. The largest megavolcano eruption during the past 2,000,000 years occurred on the Indonesian island of Sumatra 75,000 years ago. The Mount Toba eruption ejected 2800 km^3 of ash into the air compared with the 1 km^3 ejected by the most recent eruption of Mount Saint Helens. The mass of ash from Mount Toba is roughly equivalent to a slab of rock 16 km thick covering 259 km^2. This eruption caused global temperatures to drop by more than 10°C for years, and is thought to have decreased the human population on Earth from more than 100,000 down to less than a few thousand individuals. The Mount Toba eruption represents the closest that humanity has yet come to extinction.

Our primary focus in the remainder of this chapter involves oxide–water reactions that are stimulated by geothermal heat sources. Not all geothermal environments are equivalent. To illustrate this point, our focus is on the two extreme cases represented by divergent and convergent plate boundaries. The divergent plate boundary is represented by the mid-Atlantic spreading center. Here, temperatures and pressures are relatively high compared with Earth's surface. However, a key attribute of the Mid-Atlantic Ridge is that most of it lies beneath the Atlantic Ocean, and thus has access to almost unlimited supplies of water, which means that emerging basalts are exposed to *hydrothermal weathering* conditions. The other geothermal environment considered involves subduction zones that exist at convergent plate boundaries. Here, temperatures and pressures can become truly extreme. Another key attribute of subduction zones is that water supplies are now largely limited to what the hydrated and weathered oxide phases carry with them. Liquid water, at least as it is known on Earth's surface, is not present, which means that hydrated minerals can undergo dehydration reactions that regenerate pristine anhydrous oxide phases. Let us now examine these extreme hydration and dehydration processes in more detail.

18.5.2 Hydrothermal Weathering at Divergent Plate Boundaries

Many of the weathering processes described in Chapter 17 occur under ambient conditions where discernible chemical erosion can require anywhere from hundreds to even billions of years. However, many oxide–water reactions are activated processes that can be accelerated by the application of heat. For example, quartz dissolves 25,000 times faster at 90°C than it does at room temperature. On the continents, the effects of temperature on weathering can be viewed in areas of high geothermal activity such as Yellowstone National Park. Now imagine how rapidly water reacts when it comes into direct contact with hot magma. Such exposures represent weathering events that occur under the ocean at both volcanoes and spreading centers such as the Mid-Atlantic Ridge.

The Mid-Atlantic Ridge is more than 15,000 km long, running from Iceland to Antarctica. As plates move away from each other along this boundary, the newly formed space is filled with oceanic basaltic lavas. Oceanic basalts are the most common volcanic rocks on the planet. Although basaltic melts can be as hot as 1000°C, they typically solidify and cool to around 350°C before coming into direct contact with seawater. Amazingly, because of the high pressures exerted by the overlying water column (around 400 atm, or 40 MPa), the 350°C temperature is usually insufficient to boil seawater, which is maintained in a fluid state. The enormous amount of heat emitted from the hot zone ($9 \pm 2 \cdot 10^{12}$ W) [36–38] drives the hydrothermal convection of seawater through the crust near the spreading center. This convection accounts for about 34% of the total oceanic heat flux and about 20% of the total heat flux from the entire Earth [38]. Water temperatures decrease from the 350°C at the immediate spreading center to ambient temperatures over distances that can be hundreds of kilometers wide. Hydrothermal-weathering processes at spreading centers caused by slow circulation of seawater through hot rock occur over a total area of around 3,000,000 km^2. This hydrothermal activity induces dramatic changes in the composition of both the seawater and the emerging rock [36, 39]. The fluxes of various elements from axial hydrothermal alteration are compared with riverine fluxes in Table 18.2 [36].

Reactions between fresh basalt and circulating seawater strongly influence the global budget of many of the major elements found in seawater [39]. Elemental sources are the parent magmas, whereas elemental sinks consist of weathering products. The products of the hydrothermal alteration of hot igneous basalts are predominantly clay minerals, although zeolites can form as well. An important class of weathering reaction involves the conversion of the orthosilicates in basalt into clay minerals:

$$6(\mathrm{Mg,Fe})_2\mathrm{SiO}_4(\text{olivine}) + 7\mathrm{H_2O}(1) \rightleftarrows 3(\mathrm{Mg,Fe})_3\mathrm{Si_2O_5}\cdot(\mathrm{OH})_4(\text{serpentine}) + \mathrm{Fe_3O_4}(\text{magnetite}) + \mathrm{H_2}(g) \qquad \textbf{(18.14)}$$

This reaction generates hydrogen gas that is important in both the conversion of carbonate ions into methane via Fisher–Tropsch chemistry and in supporting deep microbial life. The most conspicuous manifestation of this hydrothermal activity involves the bizarre ecosystems and biota observed near hydrothermal vents called *smokers*.

Table 18.2 Fluxes of Major Oxide Elements from Submarine Vents and Rivers to, and from, the Oceans

Component	Flanks (g/yr)	Axis Low (g/yr)	Axis High (g/yr)	Riverine Flux (g/yr)
SiO_2	$+1.08 \cdot 10^{12}$	$+1.80 \cdot 10^{13}$	$+6.61 \cdot 10^{13}$	$+3.85 \cdot 10^{14}$
Al_2O_3				
FeO_{tot}	0.0	$+1.65 \cdot 10^{12}$	$+1.37 \cdot 10^{13}$	$+1.65 \cdot 10^{12}$
MnO	$+3.69 \cdot 10^{10}$	$+4.97 \cdot 10^{11}$	$+3.97 \cdot 10^{12}$	$+3.48 \cdot 10^{11}$
MgO	$-2.18 \cdot 10^{14}$	$-4.03 \cdot 10^{14}$	$-1.05 \cdot 10^{14}$	$+2.18 \cdot 10^{14}$
CaO	$+2.64 \cdot 10^{14}$	$+3.03 \cdot 10^{11}$	$+1.23 \cdot 10^{14}$	$+6.73 \cdot 10^{14}$
Na_2O	$-3.84 \cdot 10^{13}$			$+3.53 \cdot 10^{14}$
K_2O	$-3.11 \cdot 10^{13}$	$+6.12 \cdot 10^{12}$	$+3.01 \cdot 10^{13}$	$+1.22 \cdot 10^{14}$
Rb	$-2.22 \cdot 10^{9}$	$+1.37 \cdot 10^{10}$	$+2.39 \cdot 10^{10}$	$+3.16 \cdot 10^{10}$
Cs				$+6.40 \cdot 10^{8}$
CO_2	$-7.04 \cdot 10^{12}$	$-1.76 \cdot 10^{12}$	$-7.04 \cdot 10^{12}$	$+1.41 \cdot 10^{15}$
H_2O				
S	$-3.53 \cdot 10^{13}$	$-1.60 \cdot 10^{13}$	$-4.49 \cdot 10^{13}$	$+2.85 \cdot 10^{13}$
Li	$-1.25 \cdot 10^{10}$	$+4.86 \cdot 10^{10}$	$+4.37 \cdot 10^{11}$	$+9.72 \cdot 10^{10}$
B	$+1.84 \cdot 10^{11}$	$+6.49 \cdot 10^{9}$	$+7.89 \cdot 10^{10}$	$+5.84 \cdot 10^{11}$
Sr	$+2.19 \cdot 10^{11}$	$-9.0 \cdot 10^{8}$	$+4.60 \cdot 10^{9}$	$+2.02 \cdot 10^{11}$
U				$+9.60 \cdot 10^{9}$

Negative signs indicate material is lost to weathering crustal rock.
Data from: Staudigel, H. Hydrothermal Alteration Processes in the Oceanic Crust. In *Treatise on Geochemistry*; Roberta L. Rudnick. R. L.; Holland, H. D.; and Turekian, K. K., v. 3 Eds.; Pergamon Press: Oxford, 2003, p 511–535.

As discussed in Section 18.4, ion-exchange reactions involving clay minerals are important factors that help determine the ionic compositions of the oceans and other major bodies of water. However, the extent to which the hydrothermal formation of clay minerals at spreading centers determines the elemental composition of seawater has only recently been appreciated. Fully 24% of the total flux of silicon to the oceans is generated by hydrothermal basalt-to-clay conversions [20]. The newly formed clays consume enormous quantities of +2 cations, either by the formation of octahedral sheets or by ion exchange processes. Many of these clays contain brucitic sheets of $Mg(OH)_2$. So much clay is formed that these sheets remove up to 87% of the riverine flux of Mg^{2+} into the oceans. In terms of reversible ion exchange, it is estimated that the newly formed clays remove about 20% of the riverine input of sodium. It has not yet been possible to account for all sources and sinks of K^+. However, geochemists now believe that as much as 50% of the global K^+ inventory is trapped irreversibly on exchange sites within nonswelling oceanic clays [20]. Some of these K^+-consuming clays form by weathering processes as spreading centers, including the smectite-to-illite clay transformation discussed in Chapter 17.

Eq. 18.14 is a reflection of the fact that divergent oceanic plate boundaries represent O_2-poor and H_2-rich environments that are highly reducing. Spreading centers release large quantities of Fe(II) and Mn(II), some of which is precipitated immediately as minerals such as the iron sulfides troilite (FeS) and pyrite (FeS_2), which are also found in salt marshes and reduced soils. The remaining soluble Fe(II) and Mn(II) species eventually migrate to the shallower oxic parts of the ocean. There they encounter O_2 and oxidize to form the colloidal iron and manganese oxides found in the oceans (see Section 18.4). Hydrothermal conditions can also stimulate the precipitation of weathering by-products such as solid carbonate and sulfate minerals, including aragonite and calcite ($CaCO_3$), barite ($BaSO_4$), and gypsum ($CaSO_4 \cdot 2H_2O$) (see Chapter 17).

Spreading centers not only exhibit high temperatures, but also strong temperature gradients. Temperatures are highest at the center of the rift and decay to ambient temperatures over distances of kilometers. A conceptual model for the distribution of temperature, rock age, and mass flux is shown for the Iceland geothermal fields in Figure 18.15. Such gradients exert a strong influence on the distribution and time evolution of weathering products that originate from oceanic basalts.

Most of the parameters associated with hydrothermal weathering are influenced by temperature, paralleling trends seen in the hydrothermal synthesis of oxides (see Chapter 14, Section 14.10). One obvious trend is that less water tends to be encapsulated within weathering products formed at high temperatures than low temperatures. For example, zeolites such as chabazite, having a high water content of one water molecule per network-forming cation (ratio = 1) form at temperatures less than 70°C. Heulandite, with ratio = 0.66, forms and is stable up to 150°C, whereas analcime (ratio = 0.33) is a stable product of hydrothermal weathering even up to 230°C.

As discussed in Chapter 14, many other parameters with a strong temperature dependence are critical in determining the trajectories and products of hydrothermal weathering. These parameters include the dielectric constant of water, ionization tendencies, acid–base properties, and hydrolysis-product distributions. Because temperature influences so many weathering variables, large thermodynamic models created by geochemists are essential for understanding mineral origins and evolution, or paragenesis. For example, consider some of the zeolites and other minerals that can form from basaltic volcanic rocks in environments such as those that exist near divergent plate boundaries (P = 10 MPa, T = 50°C and 100°C) (Fig. 18.16). The environments shown in the figure encompass a range of local pH and ion concentrations, including those representative of seawater.

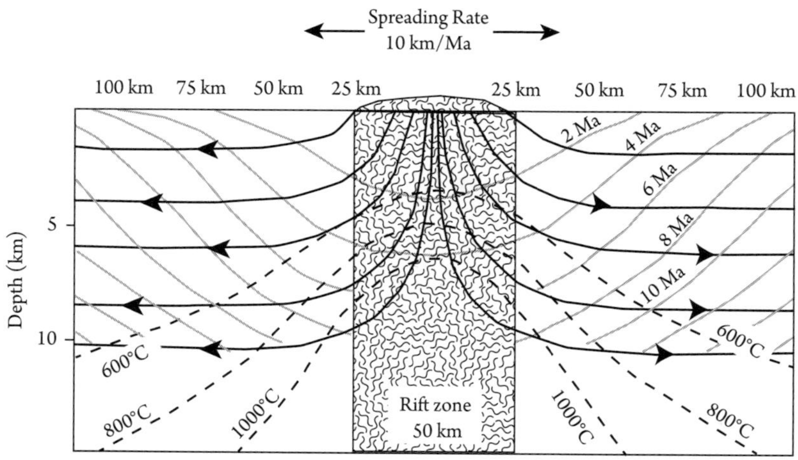

Figure 18.15 A representation of rock motion and temperatures associated with the Icelandic geothermal fields. Hot magma rises and breaks Earth's surface at the center of the rift zone, which can be as wide as 50 km. The resulting pristine igneous rocks then move away from either side of the spreading center at a rate of 10 km/million years (black lines with arrows). Rock ages are indicated by the light gray lines (in million years). Rock temperatures are indicated by the dashed lines. Temperatures at the surface within the rift zone (not shown) can reach 350°C. Adapted with permission from artwork by Prof. Tobias Weisenberger, University of Oulu, Finland; based on Palmason, G. Crustal Rifting, and Related Thermo-Mechanical Processes in the Lithosphere Beneath Iceland *Geologische Rundschau* **1981**, *70*, 244–260.

At low temperature (50°C), the pH (related to the activity of H^+) and ionic concentrations that stabilize the different phases shown in Figure 18.16 are readily understood. At low Ca^{2+} and Na^+ concentrations and/or low pH (i.e., larger values for the activity of H^+), the neutral clay kaolinite $Al_2Si_2O_5(OH)_8$ forms. Here, ion concentrations are too low to stabilize the anionic network charges within zeolitic structures. At higher pH, solutions rich in Ca^{2+} form Ca(II)-containing zeolites such as chabazite ($CaAl_2Si_4O_{12}.6H_2O$), solutions containing substantial concentrations of both Ca^{2+} and Na^+ form mixed-cation zeolites such as thomsonite ($NaCa_2Al_5Si_5O_{20} \cdot 6H_2O$), and Na^+-rich solutions form Na^+-containing zeolites such as natrolite ($Na_2Al_2Si_3O_{10} \cdot 2H_2O$). Increasing the temperature only to 100°C changes all stability boundaries. The lower dielectric constant of water stabilizes neutral networks relative to ionic networks, which expands the stability field of kaolinite to require more basic conditions for zeolite formation. Changes in ion-exchange site selectivity also occur, changing the stability fields of Ca^{2+}-rich and Na^+-rich zeolites relative to one another. It is left to you to imagine how much the phase diagram changes in moving to the 350°C temperatures that represent the middle of the spreading center.

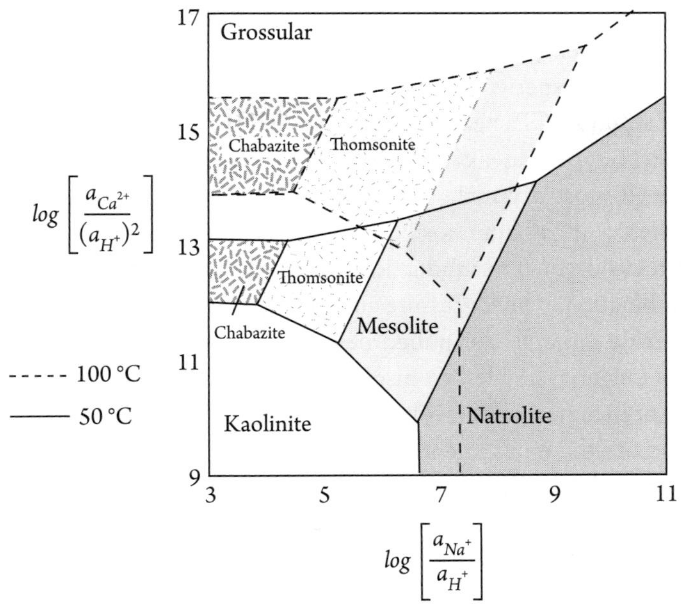

Figure 18.16 A phase diagram showing the stability regions within which zeolites and other aluminosilicate minerals are predominant as a function of temperature (50°C = solid lines, 100°C = dashed lines) and the composition of the contacting aqueous solution. Here, the ions competing for ion-exchange sites leading to phase formation and stability are H^+ (related to the solution pH), Na^+, and Ca^{2+}. The x-axis ($log\left[\frac{a_{Na+}}{a_{H+}}\right]$ reflects the competition between protons and sodium ions, whereas the y-axis $log\left[\frac{a_{Ca+2}}{(a_{H+})^2}\right]$ reflects competition between protons and calcium ions, where "a" refers to the activity of the given cation. The compositions of most phases depicted are provided in the text. Exceptions are the zeolite mesolite ($Na_2Ca_2(Al_2Si_3O_{10})_3 \cdot 8H_2O$) and the garnet grossular ($Ca_3Al_2(SiO_4)_3$). (Adapted with permission from Weisenberger, T.; Spürgin, S.; and Lahaye, Y. Hydrothermal Alteration and Zeolitization of the Fohberg Phonolite, Kaiserstuhl Volcanic Complex, Germany. *International Journal of Earth Sciences,* **2014** 103(8): 2273–2300.

18.5.3 The Thermal Rebirth of Oxides in Subduction Zones

Oxides that are weathered on Earth's surface complete their life cycle when they encounter subduction zones, where enormous slabs of oceanic crust plunge deep within the earth. We have seen that weathering involves the conversion of minerals that were stable at high temperatures and pressures into minerals that are stable in ambient-temperature water. Subduction zones cause weathering reactions to run in reverse. In subduction zones, weathered and hydrated oxide phases that are stable in ambient-temperature water are destabilized in high-temperature and high-pressure environments. Materials that are dragged into the deep crust at subduction zones experience increasingly hostile environments (Fig. 18.17). A pressure of around 1 GPa near the base of an average thickness crust corresponds to 9900 atm. Temperatures can exceed 1000°C. These temperatures are not sufficient to melt most dry rocks at Earth's surface. For example, quartz melts at 1610°C whereas most aluminosilicates must be heated above 1200°C to melt. However, rocks do melt in subduction zones, where melting points are lowered by the combination of high pressure and the presence of water [40].

When oxide minerals are pulled beneath the surface in subduction zones, water is no longer available as a discrete phase such as ice or liquid water. The formation of hydrated minerals by all aqueous reactions that occur on Earth's surface leads to the entrapment of large quantities of water stored within the mineral phases themselves. As these hydrated minerals are subjected to greater pressures and temperatures during subduction, they begin to release their structural water as the activity of water in the adjacent pores is reduced. The sequence of events that occurs in clay minerals with increasing temperature includes (1) pressure expulsion of pore water accompanied by pore collapse ($T<80°C$), (2) dehydration of excess interlayer water ($T<150°C$), (3) loss of structural waters ($80°C<T<300°C$), (4) dehydroxylation reactions between adjacent bound hydroxyl groups (see Chapter 6) ($150°C<T<800°C$), and, finally, (5) melting.

The correlation between water content and temperature in Earth's subduction zones is illustrated in Figure 18.18. Here, the stability regions for some simple minerals are shown as a function of the activity of silicic acid and the activity of water. These reactions are chosen to isolate a key reaction variable—the activity of silicic acid—in the conversion of one mineral to another. For example, the conversion of nepheline to natronite can be written as

$$2NaAlSiO_4(\text{nepheline}) + 4H_4SiO_4(aq) \rightleftarrows Na_2Al_2Si_3O_{10} \cdot 2(H_2O)(\text{natronite}) \quad \textbf{(18.15)}$$

This reaction is very sensitive to temperature and the activity of water, with a major source of water being provided by hydroxyl groups in the silicic acid itself. Notice that at relatively high water activity and low temperature, the stability field for the water-rich zeolite natronite is substantial, and dominates that of the water-poor zeolite analcime ($NaAlSi_2O_6 \cdot H_2O$). However, as the temperature increases, natrolite becomes less and less prevalent until it disappears by 200°C regardless of water activity. Analcime, on the other hand, survives up to 230°C and retains its water of hydration, provided the activity of water is high (see Fig. 18.16).

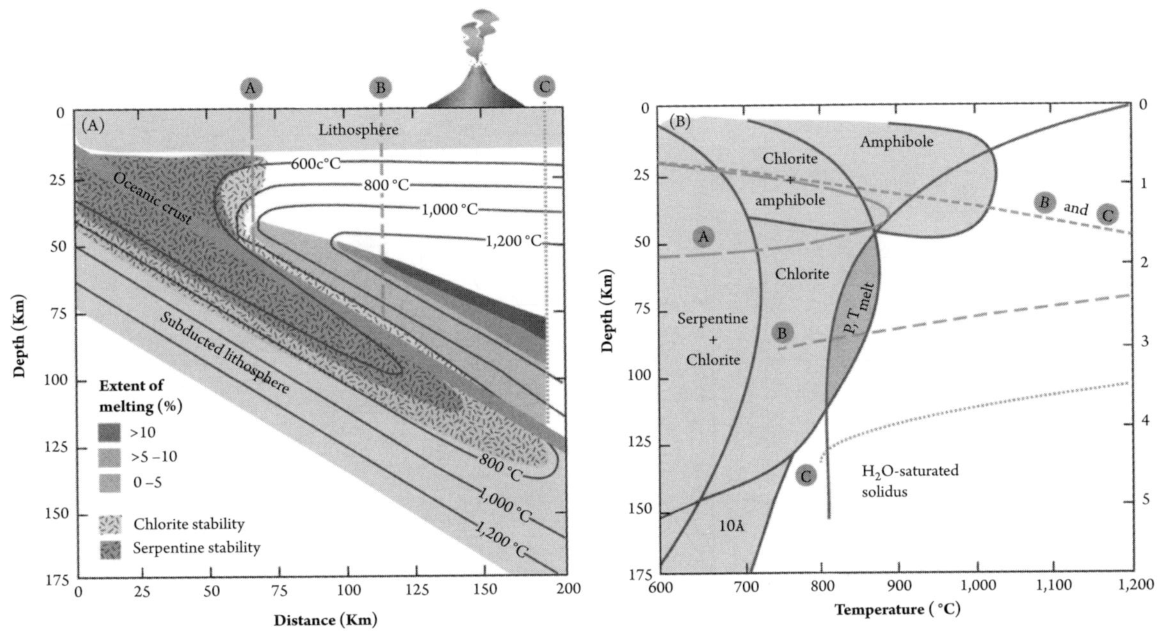

Figure 18.17 (A) A depiction of the temperature and pressure regimes experienced within a descending lithospheric slab at a subduction zone. The primary phases present at different positions are indicated, as is the melting behavior of the rock. (SOURCE: Reproduced with permission from Grove, T. L.; Till, C. B.; Krawczynski, M. J. The Role of H_2O in Subduction Zone Magmatism. *Ann. Rev. Earth Planet. Sci.*, **2012**, *40*, 412–439.) (B) A graphical depiction of the conversion of clay minerals into other phases as a function of temperature and pressure. Mineralogical reactions cause the release of water as clays are converted into anhydrous phases such as amphiboles. This released water acts as a flux to depress the melting temperature of minerals in overlying parts of the slab, creating magmas and volcanic eruptions. Note that at extreme depths chlorite converts into an expandable hydrous material called the 10- Ångstrom phase. (SOURCE: Reproduced with permission from Fumigalli, P.; Stixrude, L.; Poli, S.; Snyder, D. The 10 Angstrom phase—A high Pressure Expandable Sheet Silicate Stable During Subduction of Hydrated Lithosphere, *Earth Planet. Sci. Lett.*, **2001**, *186*(2), 125–141.)

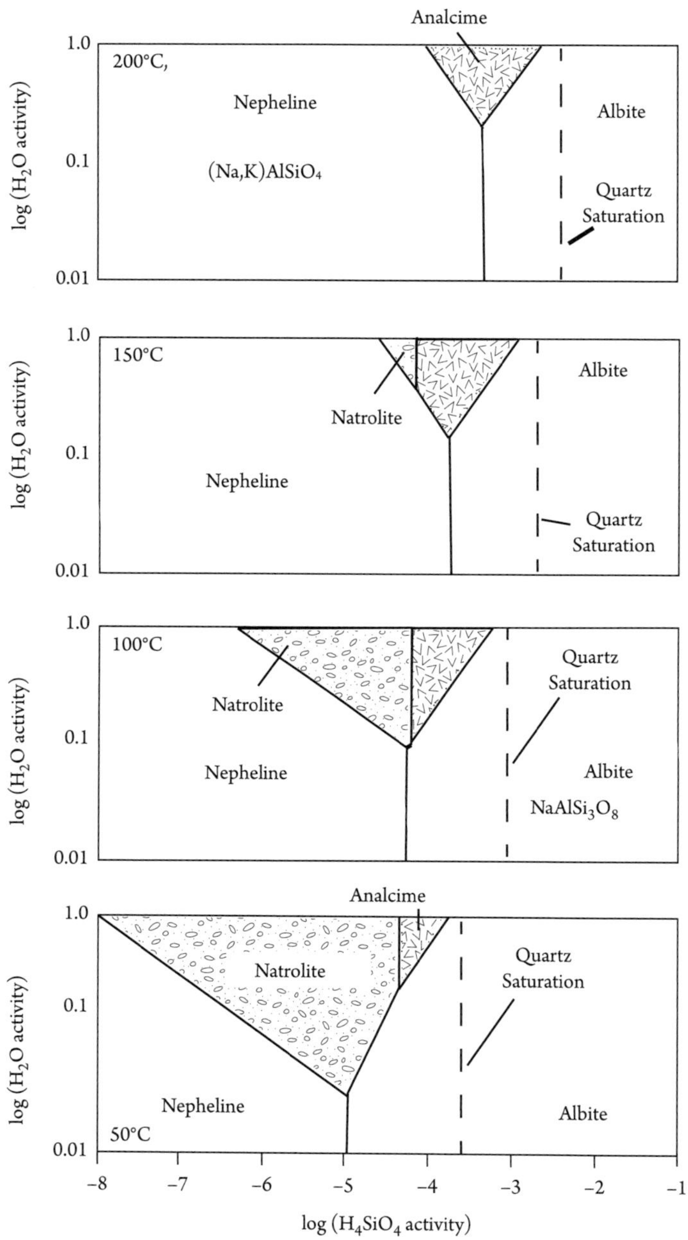

Figure 18.18 Phase diagrams illustrating how silica concentrations, water activity, and temperature influence the stability of several key aluminosilicate phases under the increasingly hostile conditions that can be found in subduction zones. Notice that the stability fields associated with the hydrated zeolite minerals shrink as the temperature increases. Modified with permission of the authors from Weisenberger, T.; Spürgin, S.; and Lahaye, Y. Hydrothermal Alteration and Zeolitization of the Fohberg Phonolite, Kaiserstuhl Volcanic Complex, Germany. *International Journal of Earth Sciences*, **2014** 103(8): 2273–2300

The partial dehydroxylation of serpentine provides another example of how extreme subduction zone temperatures can stimulate reactions. Here, dehydration leads to the release of both water and CO_2, forming an amphibole that is more difficult to dehydrate:

$$\begin{aligned}&6CaCO_3(s)+14SiO_2(s)+5(Mg,Fe)_3Si_2O_5\cdot(OH)_4(\text{serpentine})\rightleftarrows\\&3Ca_2(Mg,Fe)_5Si_8O_{22}(OH)_2(\text{amphibole})+6CO_2(g)+7H_2O(l)\end{aligned} \quad \textbf{(18.16)}$$

Again, dehydration temperatures depend on the compositions of the rock and adjacent fluids, with more tightly bound waters being released at greater temperatures. Reproducing actual subduction conditions is important for estimating realistic dehydration conditions. For example, the hydrous mineral chlorite retains its identity up to several hundreds of degrees Celsius and depths below Earth's surface of tens of kilometers (Fig. 18.17). At even greater depth chlorite converts to another hydrous phase called the 10-Angstrom phase that is mysterious, expandable, and may serve as a large reservoir for water at mantle depths in the Earth.

Water entrapped within mineral phases has important implications for volcanism [40]. The water released via dehydration and dehydroxylation of clays, hydroxides, and hydrated oxides is the major source of water within subduction zones. This high-temperature and high-pressure water has two important consequences for the end of the oxide mineral life cycle and the return of high-temperature oxides to the surface of Earth. First, the hot water molecules are highly reactive and are constantly undergoing reversible hydrolysis reactions with metal–oxygen bonds in the oxide minerals that are present. These reactions lower the effective melting points of the minerals in the subduction zone [40] (see Chapter 14). The water that dissolves into silicates can create a high-temperature flux that serves as a solvent for dissolving other minerals. Water traveling up from the subducting slab of rock lowers the melting temperature of the overlying wedge of trapped mantle materials. This water reduces the melting temperature of these overlying rocks and forms low-density liquid magmas via partial melting. These magmas rise back to the surface to form igneous rocks.

A second consequence of the presence of water in silicate melts is that the water molecules are superheated and are extremely energetic. This water is only retained in the magma because of the extremely high pressures found at the base of Earth's crust. When water-containing magmas reach the surface, this water is released rapidly as superheated steam. Magmas with the highest water content are those formed near subduction zones, resulting in many of the most violent volcanic eruptions on Earth. In a sense, both water and oxides are reborn in such eruptions. The Mount Pinatubo eruption in 1991 released around 10^{10} MT of

Figure 18.19 Lava entering the ocean near the Kilauea volcano in Hawaii. (SOURCE: CJ Kale/lavalightgalleries.com, with permission). This volcano forms from one of the two dozen hot spots that affect the Earth's crust. (see Plate 27)

volcanic ash and $5 \cdot 10^8$ MT of water (5 wt% or 10 vol% of the ash) into the atmosphere [35].

The eruptive reemergence of water and the solid oxides onto Earth's surface (Fig. 18.19 and Plate 27) completes their life cycle. It is fitting that this rebirth constitutes the end of our book on the aqueous chemistry of oxides. We hope you have enjoyed reading it as much as we enjoyed writing it.

— Bruce Bunker and Bill Casey

References

1. Emsley, J. *Nature's Building Blocks*; Oxford University Press: Oxford, 2011; p 699.
2. Finlayson-Pitts, B. J.; Pitts, J. N. *Chemistry of the Upper and Lower Atmosphere*; Academic Press: New York, 2000.
3. Weast, R. C., Ed. *CRC Handbook of Chemistry and Physics, 58th ed.*; CRC Press: Cleveland, OH, 1977.
4. Archuleta, C. M.; DeMott, P. J.; Kreidenweis, S. M. Ice Nucleation of Surrogates for Atmospheric Mineral Dust and Mineral Dust/Sulfate Particles at Cirrus Temperatures. *Atmos. Chem. Phys.* **2005**, *5*, 2617–2634.
5. Atkinson, R. Kinetics and Mechanisms of the Gas-Phase Reactions of the Hydroxyl Radical with Organic Compounds under Atmospheric Conditions. *Chem. Rev.* **1985**, *85*, 69–201.
6. Erel, Y.; Pehkonen, S. O.; Hoffman, M. R. Redox Chemistry of Iron in Fog and Stratus Clouds. *J. Geophys. Res.* **1993**, *98*, 18423–18434.

7. Smith, S. J.; van Aardenne, J.; Klimont, Z.; Andres, R. J.; Volke, A.; Delgado Arias, S. Anthropogenic Sulfur Dioxide Emissions: 1850–2005. *Atmos. Chem. Phys.* **2011**, *11*, 1101–1116.
8. Newhall, C.; Hendley, J. W., II; Stauffer, P. H. *The Cataclysmic 1991 Eruption of Mount Pinatubo, Philippines*; U.S. Geological Survey Fact Sheet; 1989.
9. Cohen AJ[1], Ross Anderson H, Ostro B, Pandey KD, Krzyzanowski M, Künzli N, Gutschmidt K, Pope A, Romieu I, Samet JM, Smith K. The Global Burden of Disease Due to Outdoor Air Pollution. *J. Tox. Environ. Health A* **2005**, *68*, 1301–1307.
10. Mason, B. G.; Pyle, D. M.; Oppenheimer, C. The Size and Frequency of the Largest Explosive Eruptions on Earth. *Bull. Volcanol.*, **2004**, *66*, 735–748.
11. Rampino M. R.; Self, S.; Stothers, R. B. Volcanic Winters. *Annu. Rev. Earth Planet Sci.* **1988**, *16*, 73–99.
12. Israelachvili, J. *Intermolecular and Surface Forces*; Academic Press: London, 1991.
13. Madsen, F. T.; Muller-Vonmoos, M. The Swelling Behaviour of Clays. *Appl. Clay Sci.* **1989**, *4*, 143–156.
14. Lisitsyn, A. P. The Marginal Filter of the Ocean. *Oceanology* **1995**, *34*, 671–682 (English trans.).
15. Burst, J. F. Diagenesis of Gulf Coast Clayey Sediments and Its Possible Relation to Petroleum Migration. *Am. Assoc. Petrol. Geol. Bull.* **1969**, *53*, 73–93.
16. Sunda, W. G.; Kieber, D. J. Oxidation of Humic Substances by Manganese Oxides Yields Low-Molecular-Weight Organic Substrates. *Nature (London)* **1994**, *367*, 62–64.
17. Sharma, A.; Gaidamakova, E. K.; Matrosova, V. Y.; Bennett, B.; Daly, M. J.; and Hoffman, B. M. Responses of Mn^{2+} Speciation in *Deinococcus radiodurans* and *Escherichia coli* to Gamma Radiation by Advanced Paramagnetic Resonance Methods. *Proc. Nat. Acad. Sci. USA* **2013**, *110*, 5945–5950.
18. Argos, M.; Kalra, T.; Rathouz, P. J.; Chen, Y.; Pierce, B.; Parvez, F.; Islam, T.; Ahmed, A.; Rakibuz-Zaman, M.; Hasan, R.; Sarwar, G.; Slavkovich, V.; van Geen, A.; Graziano, J.; and Ahsan, H. Arsenic Exposure from Drinking Water, and All-Cause and Chronic-Disease Mortalities in Bangladesh (HEALS): A Prospective Cohort Study. *Lancet* **2010**, *376*, 252–258.
19. Griffin, J. J.; Windom, H.; Goldberg, E. D. The Distribution of Clay Minerals in the World Ocean. *Deep-Sea Res.* **1968**, *15*, 433–459.
20. Berner, E. K.; Berner, R. A. *Global Environment: Water, Air and Geochemical Cycles*; Prentice Hall: New York, 1996.
21. Sayles, F. L.; Mangelsdorf, P. C., Jr. The Equilibration of Clay Minerals with Sea Water: Exchange Reactions. *Geochim. Cosmochim. Acta* **1977**, *41*, 951–960.
22. Xu, Y.; Schoonen, M. A. A. The Absolute Energy Positions of Conduction and Valence Bands of Selected Semiconducting Minerals. *Am. Miner.* **2000**, *85*, 543–556.
23. Cockell, C. S.; Horneck, G. The History of the UV Radiation Climate of the Earth: Theoretical and Space-Based Observations. *Photochem. Photobiol.* **2001**, *73*, 447–451.
24. Sherman, D. M. Electronic Structures of Iron(III) and Manganese(IV) (Hydr)oxide Minerals: Thermodynamics of Photochemical Reductive Dissolution in Aquatic Environments. *Geochim. Cosmochim. Acta* **2005**, *69*, 3249–3255.

25. Sunda, W. G.; Huntsman, S. A. Photoreduction of Manganese Oxides in Seawater. *Mar. Chem.* **1994**, *46*, 133–152.
26. Tebo, BM, Bargar JR, Clement BG, Dick GJ, Murray KJ, Parker D, Verity R, Webb SM. Biogenic Manganese Oxides: Properties and Mechanisms of Formation. *Annu. Rev. Earth Planet. Sci.* **2004**, 32, 287–328.
27. Waite, T. D.; Szymczak, R. Photoredox Transformation of Iron and Manganese in Marine Systems: Review of Recent Field Investigations. In *Aquatic and Surface Photochemistry*; Helz, G. R.; Zepp, R. G.; Crosby, D. G., Eds.; Lewis Publishers: London, 1994; pp 39–52.
28. Spiro, T. G.; Bargar, J. R.; Sposito, G.; Tebo, B. M. Bacteriogenic Manganese Oxides. *Acc. Chem. Res.* **2010**, *43*, 2–9.
29. Anbar, A. D.; Holland, H. D. The Photochemistry of Manganese and the Origin of Banded Iron Formations. *Geochim. Cosmochim. Acta* **1992**, *56*, 2595–2603.
30. Sauer, K.; Yachandra, V. K. A Possible Evolutionary Origin for the Mn_4 Cluster of the Photosynthetic Water Oxidation Complex from Natural MnO_2 Precipitates in the Early Ocean. *Proc. Nat. Acad. Sci. USA* **2002**, *99*, 8631–8636.
31. Xu, J.; Sahai, N.; Eggleston, C. M.; and Schoonen, M. A. A Reactive Oxygen Species at the Oxide/Water Interface: Formation Mechanisms and Implications for Prebiotic Chemistry and the Origin of Life. *Earth Planet. Sci. Lett.* **2013**, *363*, 156–167.
32. Pollack, H. N.; Hurter, S. J.; Johnson, J. R. Heat Flow from the Earth's Interior: Analysis of the Global Data Set. *Rev. Geophysics*, **1993**, *31*, 267–280.
33. Stern, R. J. Subduction zones. *Rev. Geophysics*, **2002**, *40(4)*, 3-1-3-38.
34. Anderson, R. N.; Uyeda, S.; Miyashiro, A. Geophysical and Geochemical Contraints at Converging Plate Boundaries – Part 1: Dehydration in the Downgoing Slab. *Geophys. J. R. Astr. Soc.*, **1976**, *44*, 333–337.
35. Savino, J.; Jones, M. D.; *Supervolcano*; New Page Books: Franklin Lakes, NJ, 2007.
36. Staudigel, H. Hydrothermal Alteration Processes in the Oceanic Crust. In *Treatise on Geochemistry*; Roberta L. Rudnick. R. L.; Holland, H. D.; and Turekian, K. K., v. 3 Eds.; Pergamon Press: Oxford, 2003, p 511–535.
37. Sclater, J. G.; Jaupart, C.; Galson, D. The Heat Flow through Oceanic and Continental Crust and Heat Loss of the Earth. *Rev. Geophys. Space Phys.* **1980**, *18*, 269–311.
38. Rudnick, R. L.; McDonough, W. F.; O'Connell, R. J. Thermal Structure, Thickness and Composition of the Continental Lithosphere. *Chem. Geol.* **1998**, *145*, 395–411.
39. Berner, R. A. Inclusion of the Weathering of Volcanic Rocks in the GEOCARBSULF Model. *Am. J. Sci*, **2006**, *306*, 295–302.
40. Grove, T. L.; Till, C. B.; Krawczynski, M. J. The Role of H_2O in Subduction Zone Magmatism. *Annu. Rev. Earth Planet. Sci.* **2012**, *40*, 413–439.

Subject Index

Substance Index